ECPPM 2021 – eWORK AND eBUSINESS IN ARCHITECTURE, ENGINEERING AND CONSTRUCTION

PROCEEDINGS OF THE 13TH EUROPEAN CONFERENCE ON PRODUCT & PROCESS MODELLING (ECPPM 2021), 15–17 SEPTEMBER 2021, MOSCOW, RUSSIA

ECPPM 2021 – eWork and eBusiness in Architecture, Engineering and Construction

Editors

Vitaly Semenov
Ivannikov's Institute for System Programming RAS, Moscow, Russia

Raimar J. Scherer
University of Technology, Dresden, Germany

CRC Press
Taylor & Francis Group
Boca Raton London New York Leiden

CRC Press is an imprint of the
Taylor & Francis Group, an **informa** business

A BALKEMA BOOK

CRC Press/Balkema is an imprint of the Taylor & Francis Group, an informa business

© 2021 Taylor & Francis Group, London, UK

Typeset by MPS Limited, Chennai, India

All rights reserved. No part of this publication or the information contained herein may be reproduced, stored in a retrieval system, or transmitted in any form or by any means, electronic, mechanical, by photocopying, recording or otherwise, without written prior permission from the publisher.

Although all care is taken to ensure integrity and the quality of this publication and the information herein, no responsibility is assumed by the publishers nor the author for any damage to the property or persons as a result of operation or use of this publication and/or the information contained herein.

Published by: CRC Press/Balkema
Schipholweg 107C, 2316 XC Leiden, The Netherlands
e-mail: Pub.NL@taylorandfrancis.com
www.routledge.com – www.taylorandfrancis.com

ISBN: 978-1-032-04328-9 (HB)
ISBN: 978-1-032-04329-6 (NIP)
ISBN: 978-1-003-19147-6 (EB)
DOI: 10.1201/9781003191476

Table of contents

Preface	xi
Editor(s) Biography	xv
Organisation	xvii

ICT impacts on BIM standardization, regulation policy and legislative processes

Digital description of the railway telecommunication system for a new data exchange format A. Dsoul, S. Karoui, J.D. Adounvo, P.E. Gautier, J.G. Philibert, C. Carpinteri, L. Lihai, L. Yifan & M. Boutros	3
A critical analysis of linear placement in IFC models Š. Jaud, S. Esser, A. Borrmann, L. Wikström, S. Muhič & J. Mirtschin	12
IFC query language: Leveraging power of EXPRESS and JSON S. Morozov, S. Sazonov & V. Semenov	21
Using uncertainty to link compliance and creativity N. Nisbet	29
BIM model uses through BIM methodology standardization A. Barbero, M. Del Giudice, F.M. Ugliotti & A. Osello	35
Building permit process modeling J. Fauth	42

Key aspects of data integration and management

Towards conceptual interoperability of BIM applications: Transaction management versus data exchange V. Semenov, S. Arishin & G. Semenov	51
A system architecture ensuring consistency among distributed, heterogeneous information models for civil infrastructure projects S. Esser & A. Borrmann	59
A framework for leveraging semantic interoperability between BIM applications M.E. Belsky	67
Towards semantic enrichment of early-design timber models for noise and vibration analysis C. Châteauvieux-Hellwig, J. Abualdenien & A. Borrmann	76
Evaluating SPARQL-based model checking: Potentials and limitations A. Hoffmann, M. Shi, A. Wagner, C.-D. Thiele, T.-J. Huyeng, U. Rüppel & W. Sprenger	83
Interlinking geometric and semantic information for an automated structural analysis of buildings using semantic web T.-J. Huyeng, C.-D. Thiele, A. Wagner, M. Shi, A. Hoffmann, U. Rüppel & W. Sprenger	91
A BIM to BEM approach for data exchange: Advantages and weaknesses for industrial buildings energy assessment M. Del Giudice, M. Dettori, S. Magnano & A. Osello	98
Occupant-centric ontology as a bridge between domain knowledge and computational applications A. Mahdavi, V. Bochukova & C. Berger	106

Application of ontologically streamlined data for building performance analysis 113
D. Wolosiuk & A. Mahdavi

Microservice system architecture for data exchange in the AEC industry 119
G. Šibenik, I. Kovacic, T.-J. Huyeng, C.-D. Thiele & W. Sprenger

Analysis of design phase processes with BIM for blockchain implementation 125
M. Srećković, G. Šibenik, D. Breitfuß, T. Preindl & W. Kastner

Digital traceability for planning processes 132
D. Breitfuss, G. Šibenik & M. Srećković

Models, methods and tools for design, engineering, construction and maintenance

Polyhedral space partitioning as an alternative to component assembly 141
V. Galishnikova & W. Huhnt

A hierarchical kit library to support content reuse for mass customization 147
J. Cao & D.M. Hall

Archi-guide. Architect-friendly visualization assistance tool to compare and evaluate BIM-based design variants in early design phases using template-based methodology 153
K. Jaskula, A. Zahedi & F. Petzold

Navigating the vast landscape of spatially valid renovation scenarios 163
A. Kamari, B. Li & C. Schultz

Approaching the human dimension of building performance via agent-based modeling 171
C. Berger & A. Mahdavi

BIM-based cost estimation in a road project – proof of concept and practice 177
D. Fürstenberg, T. Gulichsen, O. Lædre & E. Hjelseth

Automatic detection of construction risks 184
Q. Cui & A. Erfani

Optimization method for choosing a set of means for probability of failure reduction of critical infrastructures 190
O.S. Burukhina, A.V. Bushinskaya & S.A. Timashev

A new approach for delay analysis process 194
H.A. Dikbas & C. Durmus

Requirements analysis for a project-related quality management system in the construction execution 200
S. Seiß & H.-J. Bargstädt

Applications of artificial intelligence

Metadata based multi-class text classification in engineering project platform 211
M. Shi, A. Hoffmann & U. Rüppel

Using topic modeling to restructure the archive system of the German Waterways and Shipping Administration 216
A. Hoffmann, M. Shi & U. Rüppel

Applying weak supervision to classify scarce labeled technical documents 223
M. Shi, A. Hoffmann & U. Rüppel

Defeasible reasoning for automated building code compliance checking 229
B. Li, C. Schultz, J. Dimyadi & R. Amor

An AI-based approach for automated work progress estimation from construction activities using abductive reasoning 237
K.W. Johansen, R.O. Nielsen, J. Teizer & C. Schultz

Assumption of undetected construction damages by utilizing description logic and fuzzy set theory in a semantic web environment 245
A. Hamdan & R.J. Scherer

An overview of data mining application for structural damage detection in the last
decade (2009–2019)   253
F. Lin, J. Liu & R.J. Scherer

Intelligent structural design in BIM platforms: Optimization of RC wall-slab systems   261
N. Bourahla, S. Tafraout & Y. Bourahla

METIS-GAN: An approach to generate spatial configurations using deep learning and semantic
building models   268
H. Arora, C. Langenhan, F. Petzold, V. Eisenstadt & K.-D. Althoff

Practical experiences from initiating development of machine learning in a consulting
engineering company   274
T. Alstad & E. Hjelseth

Digital twins and cyber-physical systems

A cyber physical system for dynamic production adaptation   283
M. Polter, P. Katranuschkov & R.J. Scherer

A framework for development and integration of digital twins in construction   291
A. Corneli, B. Naticchia, A. Carbonari & M. Vaccarini

A Digital Twin factory for construction   299
C. Boje, S. Kubicki, A. Zarli & Y. Rezgui

Lifecycle oriented digital twin approach for prefabricated concrete modules   305
M. Wolf, O. Vogt, J. Huxoll, D. Gerhard, S. Kosse & M. König

A Digital Twin as a framework for a machine learning based predictive maintenance system   313
C.-D. Thiele, J. Brötzmann, T.-J. Huyeng, U. Rüppel, S.R. Lorenzen, H. Berthold & J. Schneider

Semantic contextualization of BAS data points for scalable HVAC monitoring   320
V. Kukkonen

Integrating sensor- and building data flows: A case study of the IEQ of an office building
in The Netherlands   328
S. van Gool, D. Yang & P. Pauwels

Abstract life-cycle modeling of cyber-physical systems in civil engineering   334
D. Legatiuk, K. Smarsly, K. Lossev & A. Volkov

Photogrammetry, laser scanning and point clouds

Combining point-cloud-to-model-comparison with image recognition to automate progress
monitoring in road construction   343
A. Ellinger, R.J. Scherer, C. Wörner, T. Walther & P. Vala

Research on BIM and virtual pre-assembly technology in construction management   351
D. Han, C.N. Rolfsen, B.D. Engeland, H. Hjelmbrekke, H. Hosamo, K. Hu, T. Guo & C. Ying

The use of the BIM-model and scanning in quality assurance of bridge constructions   357
C.N. Rolfsen, A.K. Lassen, D. Han, H. Hosamo & C. Ying

Application of phase three dimensional laser scanner in high altitude large volume
irregular structure   361
D. Han, C.N. Rolfsen, E. Erduran, E.E. Hempel, H. Hosamo, J. Guo, F. Chen & C. Ying

Automatic detection method for verticality of bridge pier based on BIM and point cloud   367
D. Han, C.N. Rolfsen, N. Bui, H. Hosamo, Y. Dong, Y. Zhou, T. Guo & C. Ying

Application of railway topology for the automated generation of geometric digital
twins of railway masts   373
M.R.M.F. Ariyachandra & I. Brilakis

Project controlling of road construction sites as a comparison of as-built and as-planned models using convex hull algorithm ... 381
T. Walther, M. Mellenthin Filardo, N. Marihal & H.-J. Bargstädt

City and building information modelling

A trend review on BIM applications for smart cities ... 391
A. Pal & S.H. Hsieh

Matching geometry standards for geospatial and product data ... 399
H. Tauscher

City and building information modelling using IFC standard ... 406
V. Shutkin, N. Morozkin, V. Zolotov & V. Semenov

Environmental, social and economic dimensions of sustainability

Evaluating the concept and value of smart buildings for the development of a smarter procurement strategy ... 417
J. Olsen & J. Karlshøj

From linear to circular: Circular Economy in the Danish construction industry ... 423
T.S. Rasmussen, R.J. Esclusa, E. Petrova & K.D. Bohnstedt

A BIM-based tool for the environmental and economic assessment of materials in a building within early design stages ... 431
Q. Han, N. Zhang & C.D. Van Oeveren

Housing energy-efficient renovation adoption and diffusion: A conceptual model for household decision-making process ... 439
H. Du, Q. Han & B. de Vries

Digital technologies as a catalyst to elevating IPD+BIM synergy in sustainable renovation of heritage buildings ... 446
B.F. Brahmi, S. Sassi-Boudemagh, I. Kitouni & A. Kamari

BIM education and training

Analysis of digital education in construction management degree programs in Germany and development of a training model for BIM teaching ... 457
M. Pieper, S. Seiß & A. Shamreeva

Experiences from large scale VDC-education in Norway ... 465
E. Hjelseth & M. Fischer

The efficacy of virtual-reality based safety training in construction: Perceptions versus observation ... 471
M. Poshdar, Y. Zhu, N. Ghodrati, H. Alqudah, J. Tookey, V.A. Gonzáles & S. Talebi

Applying activity theory to get increased understanding of collaboration within the VDC framework ... 479
E. Hjelseth & S.F. Sujan

A conceptual method to compare projects by combining assessment of controllable and non-controllable factors ... 487
S.F. Sujan & E. Hjelseth

The practice of VDC framework as a performance measurement system for projects ... 495
S.B.S. Ahmad & E. Hjelseth

Exploring the degree of automated process metrics in construction management ... 502
K. Rashasingham & E. Hjelseth

Organizational, perceptual and technological issues of BIM adoption

DigiPLACE: Towards a reference architecture framework for digital platforms in the EU construction sector ... 511
A. David, A. Zarli, C. Mirarchi, N. Naville & L. Perissich

"We need better software" – the users' perception of BIM *A. Rekve & E. Hjelseth*	519
Development needs on the way to information-efficient BIM-based supply chain management of prefabricated engineer-to-order structures *P. Lahdenperä, M. Kiviniemi, R. Lavikka & A. Peltokorpi*	527
Analysis of the influencing factors for the practical application of BIM in combination with AI in Germany *A. Shamreeva & A. Doroschkin*	536
A systematic review of project management information systems for heavy civil construction projects *W. Chen, M. Leon & P. Benton*	544
Multi-stakeholder involvement in construction and challenges of BIM implementation *Z. Yazıcıoğlu*	551
Impacts of BIM implementation on construction management processes in Turkey *Y. Beslioglu & İ. Akyaz*	558
Building information modeling warnings towards a deadline *L.V. Damhus, P.N. Gade & R. Qian*	563
The role of trust in the adoption of BIM systems *P.N. Gade, J. de Godoy & K. Otrel-Cass*	569
Author index	577

Preface

As new social, economic, environmental changes, including dramatic epidemic challenges, occur, the role of digital technologies becomes increasingly influential, especially in the development of the Architecture, Engineering, Construction and Facility Management (AEC/FM) industry. They allow a new perspective for sustainable evolution of urban environments and infrastructures that are permanently growing and becoming more complex. Although the digital transformation of the building industry has been an ongoing process over several decades, often hidden under the moniker of Building Information Modelling (BIM), the progress in development and practical adoption of digital technologies does not fully meet the expectations. It certainly has much greater potential that will be realized in the coming years.

The first organizations to introduce digital innovations have good chances to be leaders of the AEC/FM industry. The formula to achieve success and competitive advantage, apparently, is based on the same imperatives that are recommended by Gartner for any organizations adopting digital technologies. These imperatives are dynamism, privacy, augmented intelligence, culture, product management and digital twin; they definitely have need for emerging Information and Communication Technologies (ICT).

Dynamism is crucial for the organization's adaptability and its potential to improve business processes and increase productivity. ICT technologies bring new capabilities and new ways for AEC/FM organizations to succeed. However, rational adoption of the technologies still remains a challenging problem, to a significant degree, due to the multi-disciplinary nature of projects, large volumes of information and its heterogeneity, computationally hard optimization problems, involvement of numerous stakeholders, variety of used software tools and platforms, legacy data to be maintained throughout the entire project lifecycle (design, engineering, construction, operation, demolishment), insufficient levels of standardization, regulation and legislation. These factors prevent the direct use of general-purpose ICT technologies, and the latter must be essentially improved or adapted to meet these requirements. This circumstance opens up a wide field for further research and development.

Modern ICT technologies for management of databases, documents, revisions, workflows, and requirements remain a priority. The need to share information among project participants with own roles, expertise, skills leads to the problem of information consistency and trustworthiness. Involvement of the participants separated geographically, organizationally and technically only exacerbates the problem. Being exchanged, replicated, transformed, and updated in an uncoordinated manner, the information loses consistency and integrity, which makes it useless for further interpretation and processing by BIM tools. The development and deployment of advanced CDE (Common Data Environment) servers providing solid transaction guarantees ACID (Atomicity, Consistency, Isolation and Durability) and BASE (Basically Available, Soft state, Eventually consistent) seems to be a promising approach to management of semantically complex information, combining both BIM data driven by a formally specified schemata and complementary semi-structured or unstructured documents.

BIM interoperability also remains a critically important and still unresolved issue. It relates to the ability of BIM software applications and users to exchange and share project information freely. A rather disappointing experience with openBIM standards by buildingSMART alliance and, particularly, with IFC files has forced the proponents to acknowledge the difference between enabling applications to exchange data (the so-called technical and semantic interoperability), and the desired substantive property of applications to coherently interact with each other while keeping data complete, consistent and meaningful (conceptual interoperability). To move towards this goal and ensure the conceptual interoperability of IFC-compliant applications, not just technical and semantic ones, standardization and certification activities should be thoroughly reviewed.

It is worth acknowledging that the IFC standard seems to be the best solution for achieving the real interoperability between BIM applications produced by different software vendors. It is gratifying to see that the IFC standard continues to evolve, covering more infrastructure applications complementary to traditional disciplines. At the same time, alternative approaches to integrated data management are also being proposed and elaborated. These are federated databases, ontologies for separate BIM disciplines and aspects, linked data technologies, document-oriented information containers. However, they cause extremely hard issues of data consistency, concurrent access, change control, the resolution of which seems still rather nontrivial. All of the above approaches need formal methods to validate and certify software applications, otherwise interoperability can simply be declared, but can never be guaranteed.

The project information privacy is another fundamental problem. Inability to successfully manage privacy puts the entire digital transformation of AEC/FM organizations at high risks. Many employers do not want to give up safety, confidentiality and peace of mind in exchange for convenience and ability to freely share information with partners when collaborating together on multidisciplinary projects. Cryptography, blockchain, and smart contracts are just some of the technologies that need careful validation before being widely adopted into industrial practice.

AEC/FM industry is encouraged to follow the latest ICT technologies like the Internet of Things (IoT), Big Data, Artificial Intelligence (AI), cloud computing, which are often employed together to bring the Digital Twin (DT) paradigm. Being a cyber-physical system (CPS), which integrates computation, control and communication facilities, DT reveals new perspectives on managing the construction, operation and maintenance of the building and its assets. A high level of fidelity is reached by gathering data about physical world objects in real time and automatically converting it into respective digital models. The behavior of the physical objects can be investigated and predicted under changing conditions, proper management decisions are taken to optimize objective functions. DT technologies have a lot of applications for AEC/FM industry. They allow stakeholders to continuously monitor real progress on the construction site against the initial BIM-based planning, to control energy performance of the building under different operation modes and occupation factors, to manage building assets. The methods that can significantly improve the building construction and operation have been well studied, but the services that implement them need more modern deployment platforms and highly automated configuration tools, sometimes called DT factories.

Being scaled up to the city level, digital twins can help in optimizing traffic in urban environments, controlling pedestrian flows in crowded places, managing energy consumption, monitoring air pollution, and so on. Smart cities and smart buildings managed using digital twins are expected to provide better service, improved quality of life and a sustainable future. While BIM is focused on individual buildings, the need of the information integration at the city level and the resulting amalgamation of BIM and City Information Modelling (CIM) become more obvious. However, this aspect is not explored in its full potential yet. Diverse information standards, different computation models and methods should be reviewed and, likely, harmonized to bridge the gap between BIM and CIM technologies.

The information acquisition is necessary for the functioning of CPS services and digital twins. Different sorts of sensors and IoT devices, integrated with BIM and CIM, seem to be the logical choice for monitoring infrastructures and building assets during construction and operation stages. Complementary methods of photogrammetry, laser scanning with point clouds and aerial drone captures have been gaining traction in monitoring urban environments and construction sites. Obtained semi-structured data such as time series, photos, videos and point clouds lacks concise semantics, raising many difficulties when interfacing with BIM and CIM models. Semantic recovery and enrichment of sensed and scanned data is a computationally hard problem often resolved using AI methods. The management of big spatial-temporal data is another serious problem requiring thorough research on special-purpose databases.

As the pressure on the global climate and resources is constantly increasing due to population growth and intensified human activities, more and more attention is paid to sustainable development – a foundational principle that implies ensuring of a civilized quality of life for the future generation. Indeed, being a significant resource consumer, the AEC/FM industry is responsible for significant shares of the total energy use, global greenhouse gas emissions, total waste and raw material. Currently, most of the efforts within the industry are focused on new methods to acquire and use resources within so-called circular economy. It implies studying products, improving processes, increasing energy efficiency, managing supply chains to minimise wasted resources. However, all these efforts are still to be adopted at a larger scale.

Sustainability is a quite complex concept covering both social, economic and environmental pillars and admitting different assessments and criteria such as design quality, functionality, comfort and health, floor area, spatial program, building cost and life-cycle cost, solar insolation, heating/cooling load, energy use intensity and global warming potential. To meet these criteria, different mathematical models, computation, simulation and optimization methods can be effectively employed. At the same time, as some criteria are subjective assessments, the role of social surveys still remains important.

In recent years, AI methods and tools have become an indispensable part of civil engineering, construction management, building operation control and are facing ever new applications. Regardless of what principles the methods are based on: neural networks, swarm intelligence, machine learning, metaheuristics such as evolutionary algorithms, ant colony, simulated annealing; the methods enable solving wide classes of problems of global optimization, reasoning and decision making, clustering and data mining, recognition and classification. Not aiming to provide a comprehensive panorama of AI methods and applications for the AEC/FM industry, we are pleased to present dedicated papers and surveys focusing on AI methods for such actual problems as structural damage detection, optimization of structural design, building code compliance checking, and analysis of semi-structured data. Unfortunately, little attention is paid to the issues of augmented intelligence which is the logical step beyond artificial intelligence in the AEC/FM and, according to Gartner, it will change jobs putting employers side by side with advanced artificial intelligence systems, process and robotics and making new jobs more meaningful and rewarding.

Finally, although mentioned last, culture is identified as the largest barrier to realizing the promise of digital transformations. Culture is not just social behavior and norms, but also the knowledge, beliefs, arts, laws, customs, capabilities, and habits. As culture is acquired through the learning processes, dissemination of the best practices, presentation of emerging ICT technologies at scientific and practical forums, standardization and legislation efforts, new education formats are all of great interest and importance.

This book collects the papers presented at the 13th European Conference on Product and Process Modelling (ECPPM 2021, Moscow, 15–17 September 2021). Covering a wide spectrum of thematic areas overviewed above, the papers are devoted to critically important problems arising on the ongoing way of digital transformations in the AEC/FM industry. High quality contributions hold great promise towards the advancement of research and technological development targeted at the digitalization imperatives.

I want to thank the chairman and founder of the ECPPM conference Prof. Raimar Scherer, the members of the scientific committee whose comments contributed to the quality of the presented papers, my colleague Vasily Shutkin who took over a significant part of the organizational work, and, of course, the authors who found the opportunity, strength and inspiration to complete and present exciting research results in such a difficult and strange time.

Sincerely yours,
Prof. Vitaly Semenov

Editor(s) Biography

Prof. Dr. Vitaly Semenov has headed the Department of System Integration and Multi-Disciplinary Applied Systems at the Ivannikov's Institute for System Programming of the Russian Academy of Sciences since 2015. His main research interests focus on model-driven software engineering methodologies and CASE toolkits for creating advanced digital platforms and integrated systems that meet the requirements of multi-functionality, multi-modality, scalability, interoperability, portability and deployment in heterogeneous environments. He has led more than 25 Russian and international R&D projects and has published 150 articles on fundamental topics of software engineering, data management, computer graphics, and operations research. He is a Professor for computer science at the Moscow Institute of Physics and Technology, and the Higher School of Economics, both leading Russian State Universities. For more than 20 years he has been giving lectures on software and information visualization. Emerging BIM technologies are of particular interest. From 2007 to 2017 he directed the development of 4D modeling and planning system at Synchro Software Ltd. He developed innovative methods for spatial-temporal project planning, near-optimal project scheduling, effective rendering of large pseudo-dynamic scenes, and semantic reconciliation of replicated data, which made it possible to provide competitive advantages of the system. Currently, it is a popular software product that has been successfully used by hundreds of companies around the world to visually plan and manage complex construction projects and large infrastructure programs.

Prof. Dr.-Ing. Raimar J. Scherer is Senior Professor at the Institute of Construction Informatics at the Technical University Dresden responsible for the research. He headed the Institute in 1994–2017. He has more than 35 years of experience in construction IT, including 7 years as full professor for CAD/CAM and structural reliabilty at the University of Karlsruhe and several years of practice in the construction industry including ICT consultancy and a one-year residency with a world leading CAD provider. His research activities include the broad spectrum of construction ICT aspects. He has been involved in 25 BIM related EU and 20 German research projects, of which he co-ordinated 16. He has been a member of the multimodel group of ISO 21597 since 2017 and was member and vice chairperson of the building product model group in ISO 10303 from 1988 to 1999, the forerunner of IFC (ISO 16739). In 1994 he founded the first European BIM conference, the ECPPM. In 1999 he implemented the first BIM Server in Europe, based on the later standardized IFC building core model. He was founder and vice chairperson of the IFC-ST4 structural model group. He developed the simplified MVD/IDM method in 2012 and the multimodel method in 2014. In 2014 he received the ZUSE Medal, the annual highest Award for informatics in Germany. In 2015 he published a book on Information Management Systems in Construction, where the multi model was introduced in a consolidated way.

Organisation

STEERING COMMITTEE ECPPM 2021

Chairperson

Raimar Scherer, Technische Universität Dresden, Germany

Vice Chairperson

Vitaly Semenov, Ivannikov's Institute for System Programming, Russian Academy of Sciences, Russia

Members

Robert Amor, University of Auckland, New Zealand
Ezio Arlati, Politecnico di Milano, Italy
Vladimir Bazjanac, Stanford University, USA
Jakob Beetz, Eindhoven University of Technology, Netherlands
Adam Borkowski, Institute of Fundamental Technological Research, Polish Academy of Sciences, Poland
Michel Böhms, TNO, Netherlands
Jan Cervenka, Cervenka Consulting, Czech Republic
Symeon Christodoulou, University of Cyprus, Cyprus
Attila Dikbas, Instanbul Medipol University, Turkey
Djordje Djordjevic, Institute of Civil Engineering and Architecture, Serbia
Boyan Georgiev, University of Architecture, Civil Engineering and Geodesy, Bulgaria
Ricardo Gonçalves, New University of Lisbon, UNINOVA, Portugal
Gudni Gudnason, Innovation Centre, Iceland
Eilif Hjelseth, Oslo Met, Norway
Noemi Jimenez Redondo, CEMOSA, Spain
Jan Karlshøj, Technical University of Denmark, Denmark
Tuomas Laine, Grandlund, Finland
Ardeshir Mahdavi, Vienna University of Technology, Austria
Karsten Menzel, Technische Universität Dresden, Germany
Sergio Munoz, AIDICO, Instituto Technologia de la Construccion, Spain
Pieter Pauwels, Ghent University, Belgium
Byron Protopsaltis, Sofistik Hellas, Greece
Yacine Rezgui, Cardiff University, UK
Dimitrios Rovas, Technical University of Crete, Greece
Ana-Maria Roxin, National Centre for Scientific Research, France
Vaidotas Sarka, Vilnius Gediminas Technical University, Lithuania
Sven Schapke, Thinkproject, Germany
Ian Smith, EPFL, Ecole Polytechnique Federale de Lausanne, Switzerland
Rasso Steinmann, Institute for Applied Building Informatics, University of Munich, Germany
Väino Tarandi, KTH, Royal Institute of Technology, Sweden
Ziga Turk, University of Ljubljana, Slovenia
Alain Zarli, R2M Solution, France

Founding members

Raimar Scherer, Germany
Jeffry Wix, UK
Patrice Poyet, France
Godfried Augenbroe, The Netherlands

INTERNATIONAL SCIENTIFIC COMMITTEE ECPPM 2021

Robert Amor, University of Auckland, New Zealand
Chimay Anumba, Pennsylvania State University, USA
Ezio Arlati, Politecnico di Milano, Italy
Godfried Augenbroe, Georgia Institute of Technology, USA
Michel Böhms, TNO, Netherlands
André Borrmann, Technische Univesität München, Germany
Tomo Cerovsek, University of Ljubljana, Slovenia
Jan Cervenka, Cervenka Consulting, Czech Republic
Nashwan N. Dawood, Centre for Construction Innovation and Research, University of Teesside, UK
Attila Dikbas, Instanbul Medipol University, Turkey
Robin Drogemuller, Queensland University of Technology UT CSIRO, Australia
Bruno Fies, CSTB, France
Martin Fischer, Center for Integrated Facility Engineering, Stanford University, USA
Thomas Froese, University of British Columbia, Canada
Gudni Gudnason, Innovation Centre, Iceland
Tarek Hassan, Loughborough University, UK
Eilif Hjelseth, Oslo Met, Norway
Wolfgang Huhnt, Technische Universität Berlin, Germany
Ricardo Jardim-Goncalves, Universidade Nova de Lisboa, Portugal
Jan Karlshøj, Technical University of Denmark, Denmark
Peter Katranuschkov, Technische Universität Dresden, Germany
Abdul Samad (Sami) Kazi, VTT Technical Research Centre of Finland, Finland
Arto Kiviniemi, University of Liverpool, UK
Bob Martens, Vienna University of Technology, Austria
Karsten Menzel, Technische Universität Dresden, Germany
Sergey Morozov, Institute for System Programming, Russian Academy of Sciences, Russia
Sergio Munoz, AIDICO, Instituto Technologia de la Construccion, Spain
Svetla Radeva, College of Telecommunications and Post, Sofia, Bulgaria
Yacine Rezgui, Cardiff University, UK
Uwe Rüppel, Technical University of Darmstadt, Germany
Vitaly Semenov, Institute for System Programming, Russian Academy of Sciences, Russia
Miroslaw J. Skibniewski, University of Maryland, USA
Ian Smith, EPFL, Ecole Polytechnique Federale de Lausanne, Switzerland
Rasso Steinmann, Institute for Applied Building Informatics, University of Munich, Germany
Väino Tarandi, KTH, Royal Institute of Technology, Sweden
Helga Tauscher, Dresden University of Applied Sciences, Germany
Walid Tizani, University of Nottingham, UK
Pedro Nuno Mêda Magalhães, Porto University, Portugal

LOCAL ORGANIZING COMMITTEE ECPPM 2021

Ivannikov's Institute for System Programming RAS, Russia
Arutyun Avetisyan
Vitaly Semenov
Vasily Shutkin

Moscow State University of Civil Engineering, Russia
Pavel Akimov
Vera Galishnikova
Makhmud Kharun

ICT impacts on BIM standardization, regulation policy and legislative processes

Digital description of the railway telecommunication system for a new data exchange format

A. Dsoul, S. Karoui, J.D. Adounvo, P.E. Gautier & J.G. Philibert
French National Railway Company, Paris, France

C. Carpinteri
Rete Ferroviaria Italiana, Rome, Italy

L. Lihai & L. Yifan
China Railway SIYUAN Survey and Design Group Co. Ltd, Wuhan, China

M. Boutros
Egis Rail, Paris, France

ABSTRACT: Building Information Modeling (BIM) represents a new approach to manage construction facilities throughout their lifecycle by providing a collaborative platform where building information models can be shared in a 3D digitalized environment. However, the BIM processes are based on fragmented information systems that are disparate and do not allow information exchange among the different software platforms inside the BIM ecosystem. This represents a real challenge for the railway operators and stakeholders who become more and more interested in implementing BIM into their business processes. The scope of this research includes, therefore, a digital description of the telecommunication railway system as part of the IFC Rail project that aims to tackle the interoperability issue in the railway domain by extending Industry Foundation Classes schema, which is an ISO open international standard for data sharing in the building industry, to include the data schema of railway assets as part of the IFC5 release.

1 INTRODUCTION

The French National Railway Company (SNCF) is among the first companies to promote innovation and digital transition of the railway system, which is believed to be mainly achieved by integrating BIM technology into their business process. The first concrete step undertaken along this direction by SNCF and MINnD4Rail, the working group that mobilizes and represents the rail industry in France, was to participate in the set-up of the Railway Room after signing a cooperation agreement with buildingSMART International, the organization that has developed the IFC standard to perform the data exchange between software applications in the construction industry, together with many other national railway companies from Italy (RFI), Austria (ÖBB), Switzerland (SBB), Finland (FTIA), Sweden (Trafikverket) and China (CRBIM). The goal of the Railway Room is creating a comprehensive and applicable digital representation of the entire railway ecosystem that will support all phases of the railway assets lifecycle, which will provide interoperable support systems that can reduce complexity, cost and delay for the railway owners and operators, especially during multi-business projects. The first priority in the Railway Room was, therefore, to tackle the interoperability issue in the railway domain by extending the IFC schema to include the data model of railway entities and concepts, which allows both embracing the benefits of IFC as the main vendor-neutral standard for data sharing in the built asset industry and accelerating the definition and adoption process of the new railway extension schema. This extension is named IFC Rail.

Indeed, the IFC rail project team is working in 4 domains: Track; Energy; Signaling and Telecommunication, in such a way that each domain group can develop its corresponding information delivery manuals (e.g. process map, use cases and exchange requirement) and conceptual models representing the business model of railway domains, independently from IFC. The mapping toward IFC schema will be discussed in the following sections.

This paper presents the results of the work carried out within the telecommunication domain group to develop the conceptual model of the railway telecommunication system, as part of the IFC Rail project. This conceptual model mainly contributes to the definition of the IFC Rail extension schema by defining all new railway entities and concepts that will be mapped as new subtypes or predefined types for an

IFC entity within the IFC schema. The paper also involves the approach adopted to elaborate the conceptual model, which consists of class diagrams that are elaborated according to the Unified Modeling Language (UML™). It is an ISO standardized modeling language that was adopted by the Object Management Group (OMG) since 1997. These diagrams are grouped in packages representing the spatial and the structural models of the railway telecommunication objects that belong to the IFC Rail project scope. In addition, a functional model describing how functions are performed by the telecommunication objects will be also presented in this paper, knowing that currently the functional concepts are not yet part of the IFC Rail candidate standard, but they must be included in the data specification schema during the next phases of the project.

2 CONTEXT OF CONCEPTUAL MODELING IN THE IFC RAIL PROJECT

This paper focuses on the conceptual model of the railway telecommunication system whose structural and spatial aspects were already integrated in the official deliverables of the IFC Rail project phase 1. In order to understand the role of the conceptual model among these deliverables, we need to clarify the overall process of the IFC Rail project, which is given as follows:

Figure 1 illustrates the hierarchy of the conceptual model report inside the overall process and the documentation of the IFC Rail project. The main goal of the project phase 1 is, thus, to deliver the first candidate standard of IFC Rail at the Beijing International Standards Summit in October 2019. To meet this goal, the work plan of the project consists of dividing into two teams: one team works on Information Delivery Manual (IDM), which captures and specifies processes and information flow (Gao & Pishdad-Bozorgi 2019) during the lifecycle of railway facilities; and in parallel, the other team carries out the modeling activities that consist of elaborating the conceptual model of each railway domain (telecommunication, track, energy and signaling), independently from the IFC specification, and then mapping the railway concepts and objects to the IFC schema. The present paper involves, in particular, the conceptual modeling which represents the core activity of the modeling team.

3 METHODOLOGY

The methodology adopted for the modeling activity is based on transforming the data sets that are collected during data mining processes and requirements analysis to a UML model that was specified in the graphical notation of UML. It is named the Railway UML Model and is composed of the conceptual model of railway domains (track; energy; signaling and telecommunication), and IFC Rail model that represents the IFC extension of the conceptual model. The IFC Rail model is realized by two activities:

– IFC extension proposal, in which extensions to the existing IFC schema are proposed, based on business need;
– IFC mapping, in which the business objects and relations are mapped toward the IFC schema, both the existing and the proposed extension;

The IFC Rail model is not part of this paper scope. However, its content should be mentioned in order to understand the link between conceptual modeling and IFC schema.

3.1 Why has the IFC Rail project adopted UML?

The Railway UML model was elaborated in UML, even though the IFC data schema is written in EXPRESS (ISO 10303-11), which is a standardized information modeling language that combines ideas from the entity-attribute-relationship family of modeling language with concepts from object-oriented modeling (Peak et al. 2004). It was formalized in ISO 10303, informally known as the Standard for the Exchange of Product Model Data (STEP). Although EXPRESS is a powerful language, it is relatively unknown to most programmers unlike UML that is widely accepted and supported in the world of general software modeling methods. Beside this, UML, through its class diagrams and its profile extensibility mechanism, is already being used to visually represent Extensible Markup Language (XML) schemas (Arnold & Podehl 1999), which are used to virtually all new work done on developing standard data formats for many domains (Peak et al. 2004). For these reasons,

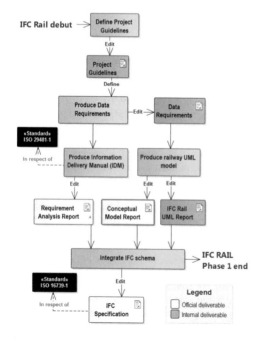

Figure 1. Overall process and relative deliverables of the phase 1 of the IFC Rail project.

the modeling activities have only been done in UML during the phase 1 of the IFC Rail project.

It should be mentioned that the use of UML does not create inconsistency with the existing IFC EXPRESS schema, since a mapping rules standard was already defined in ISO 10303-28, in order to allow generating XML schema definition from the EXPRESS schema.

3.2 Content of the conceptual model and the IFC rail model

The conceptual model contains class diagrams that describe in a logical way:

- Railway objects;
- Objects' attributes, which represent the characteristics and the properties of the railway objects;
- Business relationships and associations between the railway objects.

At the end of the IFC Rail project phase 1, the IFC Rail model will contain solely:

- IFC Rail spatial structure entities, which represent the structure of the whole railway environment and reserved space;
- Mapping of the spatial structure entities toward relative IFC entities;
- All IFC 4.1 and IFC 4.2 entities;
- Mapping of the railway objects toward IFC entities.

4 CONCEPTUAL MODEL OF THE RAILWAY TELECOMMUNICATION SYSTEM

4.1 Modeling views of the conceptual model

The conceptual model contains packages where class diagrams are grouped under a specific modeling point of view. Indeed, each railway object represents a unique entity inside the conceptual model, but it can be defined and represented according to several perspectives coming from how it is seen. For example, consider a router as a whole target system. From a structural viewpoint, it can be decomposed into ports, chassis, routing processor and cooling fans that are characterized by lifespan duration, reliability factor and other properties related to its physical performance under stated conditions. On the other hand, this router can also be represented as a functional entity, regardless its physical structure, where only its roles and intended functional goals are taking into account.

Two major modeling views are adopted by the railway domains (track; energy; signaling and telecommunication) in order to be included, as diagram packages, in the conceptual model:

- Spatial model: describes how railway objects occupy the space. Namely, places, zones and reserved volumes where objects are located;
- Structural (physical) model: is used to express physical breakdown of components and how railway objects are composed;

Functional model was also defined in the IFC Rail modeling guidelines as a representation of functional spaces and groups where railway objects are collected under operational criterion. However this modeling view was not encouraged for the project phase 1 and did not figure in its official deliverables. The main issue with functional modeling is that, despite the numerous researches that have carried out on the functional representations, there is no consensus on the definition of artifacts functionality. Indeed, the functional models have neither standard vocabulary nor constraints on their contents (Kitamura et al. 2002).

The functional model, which will be presented in the following sections, does not respond to this ontological issue, but it allows understanding the functional hierarchies of the modeled systems, which provides pertinent information for problem-solving process such as maintenance. These functional hierarchies describe the decomposition of the required functional goal into sub-functions that can be achieved by agents, which are viewed as the main actor to achieve the function and do not necessarily correspond to the physical components.

4.2 Telecom spatial model

The spatial model, presented in Figure 2, illustrates the UML class diagram that conceptualizes the spatial structure of the railway telecommunication space. The idea here is to define an entity representing the space reserved to house all the telecommunication equipment, which named *Telecom reserved volume*, and then specialize it to as many hierarchical levels as required to reach spatial zones and elements that have independent placement and shape representation. It is important to mention that the entities described in the spatial model are not necessarily tangible, but they all have geometric representation, which defines their bounding elements, so that they can be relatively placed to the railway track alignment and avoid clashes between each other.

The associations among spatial entities of the same hierarchy level, which are represented as a line, describe positioning constraints that should be respected (e.g. the entities *Outdoor laying infrastructure zone* and *Outdoor telecom equipment zone* must touch each other).

There are neither generalization nor aggregation relationships between spatial and physical entities, which means that a physical object cannot represent a component or a specialized form (subtype) of a spatial element and vice versa. However, the spatial model can contain associations to physical entities in order to express their location constraints and where they can be positioned.

4.3 Telecom structural model

The telecom structural model describes the physical layer within the conceptual model of railway telecommunication objects. The entities defined in

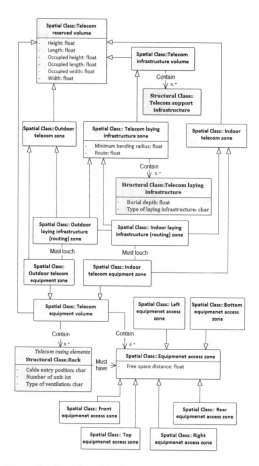

Figure 2. Spatial model of the railway telecommunication zone.

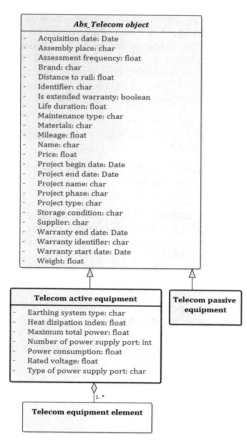

Figure 3. Overview of the root classes of telecommunication physical entities.

this model are decomposed using aggregation relationships that establish a whole-part relation, so that an entity can either be a whole (containing entity), a part (component of another entity) or both.

All entities defined in the physical layer derive from an abstract entity named *Abs_Telecom object*, as shown in Figure 3, which is the root class and the common super type of all railway telecommunication objects. It contains, therefore, all the general properties that are shared by its subclasses.

Abs_Telecom object is specialized by two main entities: *Telecom active equipment*, which is the base class of all telecommunication equipment that requires a power supply to be able to function; and *Telecom passive equipment*, which is the generalization of all passive objects of railway telecommunication domain that do not need to be powered in order to function, such as cables, casing elements, cabling accessories, etc.

4.3.1 *Structural model of Telecom active equipment*
As mentioned before, the entity *Telecom active equipment* is the super type of all active devices that are used in different railway telecommunication systems, such as fixed telephony system, fixed transmission network, mobile network, ticketing system and monitoring system of railway natural disaster and foreign object intrusion. These systems represent functional groups that gather the equipment ensuring the same functional goal. Whereas, from the structural point of view, the active equipment is decomposed into *Telecom equipment element*, which is a physically existent entity that generalizes all the components that make up the telecom active equipment. These components comprise mainly an element named *Connection interface*, which is a set of ports that allow connection to different types of transmission mediums that can be dedicated to power or data communication. Figure 4 bellow illustrates the physical decomposition of *Telecom active equipment*:

4.3.2 *Structural model of Telecom passive equipment*
The passive equipment in telecommunication domain represents the root class of all physical transmission mediums between the telecom active equipment (e.g. cables), the casing elements where the telecom equipment is arranged (e.g. rack, closure, telecom shelter,

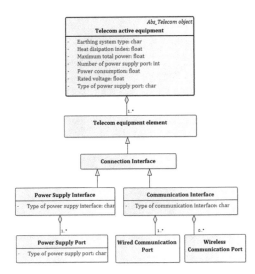

Figure 4. Structural model of *telecom active equipment*.

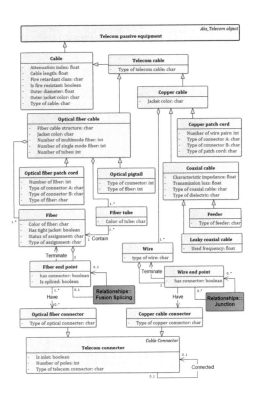

Figure 5. Structural model of physical transmission mediums.

etc.) and the different cabling accessories used for cable management. These elements form the first level of specialization within the hierarchy of Telecom passive equipment and they have relationships with other elements that do not derive from *Telecom passive equipment* but need to be appeared in the structural model; we are talking here about the laying infrastructure where the transmission mediums are deployed and the support infrastructure that holds equipment and cables (e.g. pole and tower).

Figure 5 bellow shows the structural model of *Cable*, which is the fundamental entity type of physical transmission mediums. At the second level of *Cable* specialization, we find two entities named *Optical fibre cable* and *Copper cable*, which are the main types of wired connections used in railway telecommunication domain. These two entities are specified by other subtypes that we need to make explicit in the structural model, since they are widely used and have more specific properties such as *Leaky coaxial cable*, *Optical pigtail*, *Copper patch cord*, etc. Concerning the whole-part hierarchy of the entity *Cable*, it is defined using the aggregation relationships that establish as many levels of physical decomposition as necessary to reach the elementary components of telecommunication cables. The physical constraints over these elementary components are defined using the directed associations as showed in Figure 5; for example, the physical connection between cables that have connectors is represented via a reflexive association named *Connected* that links the instances of the entity *Telecom connector* to each other. The other two reflexive associations named *Fusion splicing* and *Junction* define respectively the physical connection for optical fibre cables and for copper cables whose end points are exposed and do not have connector, but this time, these reflexive associations were converted into association classes (known as reification

in UML), so that they can have their own properties and be viewed as independent entities.

Figure 6 presents the structural model of laying and support infrastructure that are respectively defined by two entities: *Telecom laying infrastructure*; and *Telecom support infrastructure*, which are linked to railway telecommunication objects (*Cable* and *Telecom active equipment*) via directed associations implying positioning constraints.

4.4 Telecom functional model

As mentioned before, the functional modeling is outside the scope of phase 1 of the IFC Rail project and will not be mapped to the IFC model because the functional decomposition cannot be supported by the IFC structure for now, despite the fact that the concept of functional groups does already exist in the IFC model, and it is defined by an entity named *IfcSystem* (subtype of *IfcGroup*), which is an organized collection of related objects that is composed for a common purpose or function. This grouping concept remains incapable of describing the role of each object vis-à-vis another object belonging to the same functional group, and it cannot define the whole-part hierarchy at the functional level, which means that the mapping toward IFC model cannot be done without loss of essential information provided by the current functional model. It remains, therefore, as a platform independent model (PIM).

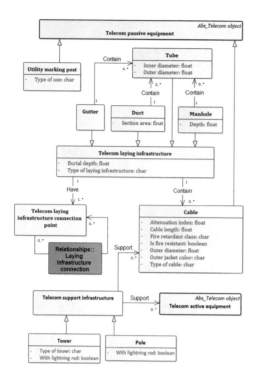

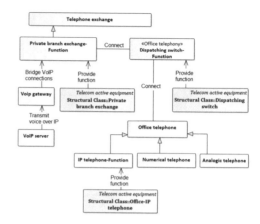

Figure 7. Functional model of office telephony system.

Figure 6. Structural model of laying and support infrastructure.

The approach adopted for the functional modeling is based on three points:

– Each entity within the functional model represents an agent that is viewed as the main actor to achieve a specific function. The agent does not necessarily correspond to a physical element; however, if it turns out that it completely matches the intended goal of a physical entity from the structural diagram, this entity needs to be linked to the agent via a directed association named *Provide function*, since, unlike the structural model, the functional model do not have the notion of entities lifecycle;
– The whole-part (aggregation) hierarchy at the functional level represents that a macro function is achieved by a group of micro functions (Kitamura et al. 2002), which should not be confused with the concept of specialization that consists of specializing a general function into different ways of functional achievement, which means that a specialized function (subtype) is also a general function (super type), but is specific to a particular type of use;
– The named associations among the entities of the same hierarchy level imply dependency and define the role of each entity vis-à-vis another. These relationships play a crucial role in describing how the intended goal of a function group is achieved.

In railway telecommunication domain, many systems are used, which can be either legacy systems

Figure 8. Functional model of railway operational telephony system.

or new generation technologies. There are three basic systems that are widely used almost everywhere in the world and have been approved for the railway telecommunication use; we are talking here about: fixed telephony system, fixed transmission network and mobile network.

4.4.1 *Functional model of fixed telephony system*

The current fixed telephony system provides wired communication services to different railway operators using the circuit-switched telephone networks. It mainly consists of telephone exchanges that insure the routing of telephone calls and communications, and fixed telephones consoles that realize bidirectional voice transmission.

In railway domain, we distinguish between office telephony system, which is a private telephone network dedicated to internal and external telephone calls in offices; and operational railway telephony system that allows communications between railway operators (e.g. switchtender, traffic regulator, railway agents, etc.) in operational centers and also on the railway right-of-way.

Figures 7 and 8 show respectively the functional models of railway office telephony and railway operational telephony systems. It is a functional decomposition, using generalization relationships, of the general functions achieved by the telephone consoles and exchanges. This aims to bring up the entities insuring the particular uses of railway telephony system; for example, in Figure 8, the entity *Operational railway telephone exchange* is specialized into 5 subtypes, each of them is specific to a type of use, such as *Unified telephone exchange*, which allows telephony management in railway regulation control centers due to its high user capacity; *Trackside telephone exchange*, which manages the communications established along the railway trackside; and *Gateway* for new generation railway telephony that bridges traditional telephony connections (i.e. analog data connection) and voice over internet protocol connections (VoIP).

4.4.2 Functional model of fixed transmission network

Fixed transmission network represents all wired networks that provide a data transmission channel using optical fibre cables, copper cables or both. It aggregates many technologies that are based on the multiplexing method; we are talking here about static transmission networks exploiting the time-division multiplexing technique (TDM), such as synchronous digital hierarchy (SDH) system, which is a legacy system for digital data transmission mainly over optical fibre cables; and there is also the wavelength division multiplexing (WDM) technique that is used by most optical fibre networks and allows gaining in speed and performance compared to the TDM network. Beside the multiplexing technologies, we have dynamic transmission network based on packet switching technique, which provides a unified data transport service capable of carrying almost any type of traffic (e.g. voice, internet protocol (IP) packets, etc.). This network is categorised as an IP network, since it encapsulates and routes packets of internet protocol (e.g. internet protocol version 4 and internet protocol version 6), and it is usually deployed to transmit data between the railway traffic regulation centre, the substation control centre and computer based interlocking centre.

Figure 9 shows the functional model of the IP network for railway domain, which consists, indeed, of 3 main functions: packets routing, which is insured by the *Router*; packets filtering that is achieved by the *Firewall*; and IP network termination that is realized using the *Switch*.

4.4.3 Functional model of mobile network

Railway mobile network insures wireless communication by providing a secure platform for voice and data

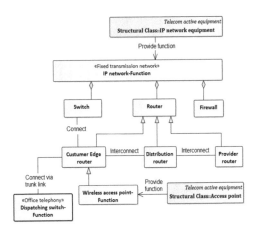

Figure 9. Function model of the IP network (fixed transmission network).

communication between railway operators, including drivers, dispatchers, shunting team members and station controllers. It was developed based on the wireless technologies for digital cellular networks that are used by mobile phones, such as Global System for Mobile Communications (GSM^{TM}), Long Term Evolution (LTE), 5G, etc. However, the main international standard for wireless railway communications that has been selected by various countries across the world, including all member states of the European Union and countries in Asia and northern Africa is the Global System for Mobile Communications for Railway, known as the GSM-R.

The GSM-R architecture is built on the mature GSM technology and also based on the functional specifications of European Integrated Radio Enhanced Network (EIRENE), which is a project lunched by the International Union of Railways (UIC) in order to achieve interoperability across borders while using a single communication platform.

The GSM-R functional model is represented in Figure 10 bellow. It consists of two main functional entities:

– *GSM-R Access Network* that aggregates the global functions of wireless signal receiving/transmitting and radio resource management, which are respectively realized by the *Base transceiver station* and the *Base station controller*;
– *GSM-R Core* Network, which consists of a series of functional entities, including: *Mobile switching center* that provides the switching/routing functions and also allows access to the other communication networks (e.g. fixed telephony system); *Visitor location register*; and *Home location register*, which is a database for mobile user management (Zhong et al. 2018).

The GSM-R has been expanded over time to include also data communications by packet data transport via general packet data service (GPRS).

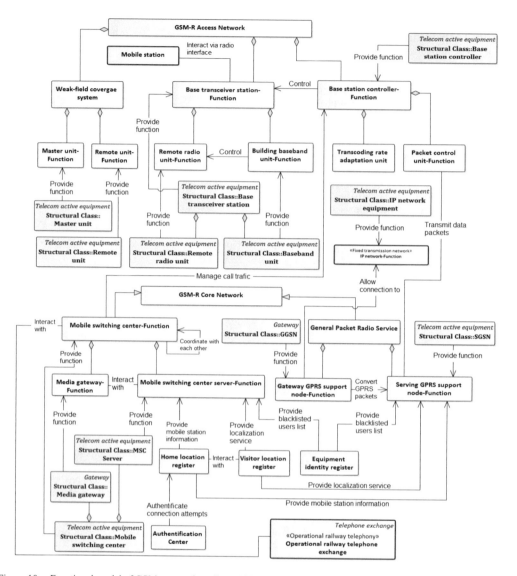

Figure 10. Functional model of GSM system for railway (GSM-R).

5 CONCLUSIONS

The conceptual model presented in this paper represents the first platform independent model (PIM) that provides international consensual description of the railway telecommunication system. Based on this model, new IFC concepts and entities will be added to extend the IFC schema in order to support the railway spatial and structural (physical) objects.

It is necessary to mention that the IFC schema does not define strict hierarchy between entities, which means that the decomposition hierarchy and the logical constraints defined by the association relationships within the structural model and especially the functional model cannot be depicted in the IFC specification. In fact, the IFC mapping only consists of linking the railway objects to both existing and proposed IFC concepts. During the next phase of the IFC Rail project, all the strict relationships between the entities of the structural model will therefore be defined in a Model View Definition (MVD), which is a subset of the IFC schema that describes a data exchange for a specific use (Gao & Pishdad-Bozorgi 2019).

The IFC Rail extension currently reaches candidate standard and it is available on the official website of buildingSMART International (bSI) for review and public consultation. Once approved as bSI final standard, the IFC Rail extension will be ready for software implementation.

REFERENCES

Arnold, F. & Podehl, G. 1999. Best of Both Worlds – A Mapping from EXPRESS-G to UML. In J. Bézivin & P.A. Muller (eds), *The Unified Modeling Language. "UML"'98: Beyond the Notation. UML 1998. Lecture Notes in Computer Science; Proc. International Workshop, Mulhouse, 3–4 June 1998.* Heidelberg: Springer.

Gao, X. & Pishdad-Bozorgi, P. 2019. BIM-enabled facilities operation and maintenance: A review. *Advanced Engineering Informatics* (1474-0346) 39: 227–247.

Kitamaru, Y., Sano, T., Namba, K. & Mizoguchi, R. 2002. A Functional concept ontology and its application to automatic identification of functional structures. *Advanced Engineering Informatics (former Artificial Intelligence in Engineering)* 16 (2): 145–163.

Peak, R., Lubell, J., Srinivasan, V. & Waterbury, S. 2004. STEP, XML, and UML: complementary technologies. *Journal of Computing and Information Science in Engineering – JCISE* (1530-9827) 4: 379–390.

Zhong, Z.-D., Ai, B., Zhu, G., Wu, H., Xiong, L., Wang, F.-G., Lei, L., Ding, J.-W., Guan, K., He, R.-S. (eds) 2018. Dedicated Mobile Communications for High-speed Railway. Heidelberg: Springer.

A critical analysis of linear placement in IFC models

Š. Jaud, S. Esser & A. Borrmann
Chair of Computational Modelling and Simulation, Technical University of Munich, Munich, Germany

L. Wikström
Triona SE, Stockholm, Sweden

S. Muhič
buildingSMART International Ltd, Kings Langley, Hertfordshire, UK

J. Mirtschin
Geometry Gym Pty Ltd., Port Fairy, Australia

ABSTRACT: The open data exchange standard Industry Foundation Classes (IFC) has been recently significantly extended to also cover infrastructure facilities such as roads and railways. The results of these activities form part of version 4.3 of the standard. Linear placement of objects is one of the most important concepts in infrastructure asset modelling. As such, Release Candidate 1 of IFC 4.3 has been critically analysed. In this paper, we address several issues that were identified together with the participants of the IFC Infrastructure Extension Deployment and IFC Rail Phase 2 projects. We present an improved model removing unnecessary doubling of concepts and reusing many already established entities. We showcase the new model on two example scenarios from one of the projects and determine better adherence with the IFC legacy. The proposed improvements have been adopted in Release Candidate 2 of the IFC 4.3.

1 INTRODUCTION

1.1 Background

Building Information Modelling (BIM) is being increasingly implemented in the infrastructure sector within the Architecture, Engineering, and Construction (AEC) domain (Bradley et al. 2016). This calls for the peculiarities common to the infrastructure domain to be introduced to the established workflows, processes, and data models previously only focusing on the building sector (Borrmann et al. 2019).

One of the most important pieces of information about any AEC object is its location and orientation in the three-dimensional (3D) space – the placement of its coordinate system (CS) within some geometric context. This is usually modelled by providing the object's Cartesian coordinates and rotation angles in the geospatial context or relatively to another object's placement (Jaud et al. 2020a).

Infrastructure assets are not residing on small parcels but rather span multiple kilometres connecting cities and industry across the globe. As such, the notion of linear placement has been introduced to specify the position of objects along a linear axis, as opposed to providing the Cartesian orthogonal coordinates. Such an axis is usually defined in the global CS and is declared a (curvilinear) coordinate axis with the stationing coordinate uniquely denoting locations along it. This unique concept has been standardized by the International Standardization Organization (ISO) in ISO 19148 (2012).

For example, consider the two-dimensional (2D) placement of the stations' local CSs $(x,y)_i$ along the railway line as shown in Figure 1. These CSs can be expressed by specifying the Cartesian position of each origin and axes' rotations in the global CS (X,Y) as is usually the case in the building domain.

However, more naturally to the infrastructure domain, first the railway's main curve is defined within the global CS (X,Y). This curve in turn defines the curvilinear axis of the linear CS. The stations' CSs $(x,y)_i$ can then be placed in the curvilinear CS with (i) their stationing coordinate representing the distance along the curvilinear axis, (ii) offsets from the main curve, and (iii) rotation of the axes defined. Note that the example is shown in the 2D plane, but it is similarly valid in the 3D space.

1.2 Problem statement

It is clear that data models need to integrate the concept of linear placement to support infrastructure workflows. In this paper, we take a closer look at one of the most prominent vendor-neutral data

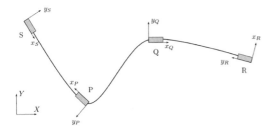

Figure 1. Railway track (black line) defines a curvilinear CS. The four stations (SPQR) have their CSs defined in this CS with their stationing coordinate (distance along the line), their offsets from the line, and rotations of CS's axes (Jaud et al. 2020a).

schemas Industry Foundation Classes (IFC) defined in ISO 16739 (2018). We critically look at how this concept has been introduced in the frame of the recent extensions and explore alternatives.

1.3 Methodology

We follow the Design Science Research methodology (Peffers et al. 2007).

1. Problem identification and motivation (Section 1.1).
2. Definition of objectives for a solution (Section 4.1).
3. Design and development of artefacts solving the problem (Section 4.2).
4. Demonstration of suitability of developed artefacts to solve the problem (Section 5).
5. Evaluation of the solution by comparing the objectives and obtained results (Section 6).
6. Communication of the obtained results (this paper).

1.4 Structure of the paper

The paper is structured as follows. Section 1 presents our motivation, the problem statement, and our methodology. Section 2 briefly summarizes related work. Section 3 presents the necessary theoretical background, describes the IFC standard, and the modelling of placement concepts for the problem at hand. Section 4 lists the issues of the current IFC model and describes a new solution. Section 5 showcases its application to two case studies. Section 6 concludes the paper with a brief discussion.

2 RELATED WORKS

The concept of linear placement has been thoroughly addressed by the Open Geospatial Consortium (OGC) and standardized in ISO 19148 (2012). It specifies a conceptual schema for locations relative to a one-dimensional (1D) object as measurement along (and optionally offset from) that object. The specification is implemented in the Geography Markup Language 3.3 (GML) standard (OGC 2012).

There are three central concepts for specifying a linear position defined: (a) the linear element being measured along, (b) the measure values (captured by a distance expression) specifying the distance along and optionally offset from the linear element, and (c) the method of measurement. Furthermore, in the case where the linear element is a curve in the 2D or 3D Cartesian space, the standard defines the relationship between a linear position and a point in the Cartesian space of the curve with the interface LR_ISpatial. This interface defines two functions that allow the transformation between the linear position and the Cartesian position, which must be implemented by the linear element (ISO 19148 2012).

During recent years, projects conducted by buildingSMART International (bSI) have expanded the well-established IFC data model for infrastructure facilities. The concept of linear placement has been tackled by the IFC Alignment project and a first proposal was included in version IFC4x1 (Liebich et al. 2017). These definitions were reused and built upon by subsequent projects, e.g. IFC-Bridge (Borrmann et al. 2019). The definitions were successfully used by Esser & Borrmann (2019), who converted data from PlanPro and RailML data models into IFC and place railway signals along a railway axis.

The currently running project "IFC Infrastructure Extension Deployment" (Jaud et al. 2020b) has brought inconsistencies in the modelling of linear placement within IFC to light. Additionally, multiple issues have been reported by the software vendors participating in the project. These issues and their resolutions are addressed in this paper.

3 THEORETICAL BACKGROUND

Since the AEC industry operates in the 3D world, we limit our consideration to the 3D Euclidean space. Additionally, we only consider right-handed, orthogonal CSs. We assume that the engineering CS of the project has an already defined relationship with the geospatial context as described by Jaud et al. (2020a). Thus, there exists a global engineering CS (X,Y,Z) within which all other placements reside.

3.1 Placement

The main purpose of placement is to specify the position and orientation of a CS relative to another CS, i.e. to define the relationship between the *origin* and *target* CSs. In this sense, the *origin* CS describes the local CS of the object which is placed within a global CS. For easier notation, we define (x,y,z) as the coordinate axes of the origin CS and (u,v,w) as the coordinate axes of the target CS. The transformation function f connecting these CSs needs a clear definition:

$$\begin{bmatrix} u \\ v \\ w \end{bmatrix} = f\left(\begin{bmatrix} x \\ y \\ z \end{bmatrix}\right). \qquad (1)$$

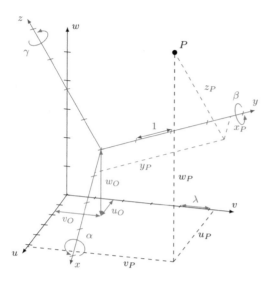
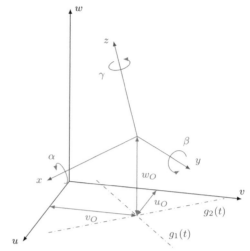

Figure 2. Visualization of the Helmert transformation between two Cartesian CSs from Equation (3) (Jaud et al. 2020a). The 7 parameters are shown in red.

Figure 3. Visualization of the parameters of grid placement as usually defined in AEC domain: the grid axes are defined in $w=0$ plane, i.e. $w_1(t)=w_2(t)=0$, with w_O being defined separately to define 3D location.

Following Equation (1), $[u_O,v_O,w_O]^T = f([0,0,0]^T)$ are the coordinates of origin CS's Point of Origin (PoO) in the target CS.

Additionally, we define $g(t)$ as a C^0 continuous curve in the target CS:

$$g(t) = a + bu(t) + cv(t) + dw(t), \qquad (2)$$

with parameter $t \in \mathbb{R}$ uniquely determining a point on the curve and $\{a,b,c,d\} \in \mathbb{R}$.

We explore three major possibilities of placement as used in the AEC domains in the next subsections.

3.1.1 Orthogonal Placement

The so-called 7-parameter Helmert transformation connects two Cartesian CSs as shown in Figure 2 (Jaud et al. 2020a):

$$\begin{bmatrix} u \\ v \\ w \end{bmatrix} = \begin{bmatrix} u_0 \\ v_0 \\ w_0 \end{bmatrix} + \lambda R(\alpha,\beta,\gamma) \begin{bmatrix} x \\ y \\ z \end{bmatrix}, \qquad (3)$$

where denotes the factor between the Unit of Measurements (UoM) of both CSs, and R defined as:

$R(\alpha,\beta,\gamma)$

$$\begin{bmatrix} c\gamma\ c\beta & c\gamma\ s\beta\ s\alpha + s\gamma\ c\alpha & -c\gamma\ s\beta c\alpha + s\gamma\ s\alpha \\ -s\gamma\ c\beta & -s\gamma\ s\beta\ s\alpha + c\gamma\ c\alpha & s\gamma\ s\beta\ c\alpha + c\gamma\ s\alpha \\ s\beta & -c\beta\ s\alpha & c\beta\ c\alpha \end{bmatrix} \qquad (4)$$

where c and s stand for cos and sin, respectively.

3.1.2 Grid placement

The grid placement is usually used in architecture, where the columns are placed on a grid-like arrangement throughout the building as shown in Figure 3.

The grid is represented with an array of non-colinear curves $g_i(t)$. A chosen intersection of two curves $g_i(t)$ and $g_j(t)$ with ij defines the origin CS's PoO and the transformation follows Equation (3).

3.1.3 Linear Placement

The linear placement is a commonly used method for specifying the location and orientation of objects within the realm of *long* infrastructure assets such as roads and railways. Here, objects are placed relative to the main alignment of the road or railway, which represents a curvilinear coordinate axis as shown on Figure 4. This axis fulfils the requirements for a linear element as defined in ISO 19148 (2012).

The origin CS's PoO is defined in a chosen point along the curve with the transformation following Equations (2) and (3).

3.2 IFC Data model

This section presents the concepts from Section 3.1 as they are incorporated in the *IFC 4.3 Release Candidate 1* (IFC4x3_RC1) published by bSI (bSI 2020a). This version has been superseded by the *Release Candidate 2* (IFC4x3_RC2), which includes the changes as reported in this paper (bSI 2020b).

For easier reading, we adopt *CamelCase* notation for IFC entities. The diagrams are drawn with EXPRESS-G notation as specified in ISO 10303-11, Annex D (ISO 10303-11, 2004).

3.2.1 Geometric representation

Figure 5 shows a diagram of the most important geometric entities used for implementing the concept of placement in IFC. Base entity for any geometry in IFC is the *IfcGeometricRepresentationItem*. From it, entities for points, curves, surfaces, and solids are

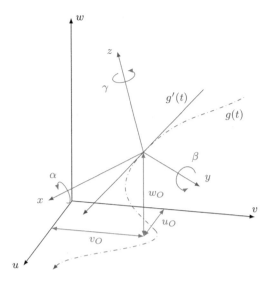

Figure 4. Visualization of the parameters of linear placement: $g(t)$ represents the curvilinear axis in 3D space along which a location can be uniquely defined with parameter t. Additionally, the tangent $g'(t)$ at the chosen t is exemplary shown.

Figure 6. One way of determining a point along a linear element with distance along and an offset vector (ISO, 2012).

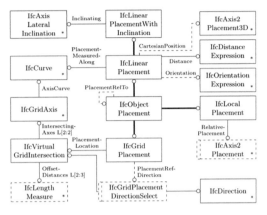

Figure 7. EXPRESS-G diagram of object placement entities from IFC4x3_RC1 that are important to our study.

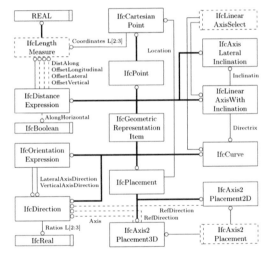

Figure 5. EXPRESS-G diagram of geometric representation entities from IFC4x3_RC1 that are important to our study.

derived as *IfcPoint*, *IfcCurve*, *IfcSurface*, and *IfcSolid*, respectively, with their subtypes.

The abstract super class *IfcPlacement* encapsulates the semantics of placement as stated in Section 3.1. It provides (a) the coordinates of the PoO in its *Location* attribute, and (b) the orientation of the CS axes through its derived classes' attributes *RefDirection* and *Axis* (Figure 5, bottom).

In order to support the notion of specifying a point along a curve, *IfcDistanceExpression* was introduced in IFC4x1 (see requirement (b) from ISO 19148 described above). The entity also enables to optionally offset the point from the curve with longitudinal, lateral and vertical offsets as presented in Figure 6.

Additionally, *IfcOrientationExpression* was introduced to define the orientation of a CS relative to a curve with attributes *LateralAxisDirection* and *VerticalAxisDirection* (Figure 5, left).

As it is usually required for railway engineering, the main axis incorporates lateral inclination in its definition (also called cant). This is modelled in IFC with *IfcLinearAxisWithInclination* that references a curve *IfcCurve* together with the inclination profile *IfcAxisLateralInclination* (see Figure 5, right).

3.2.2 *Object placement*
Each *IfcProduct*'s geometry (e.g. walls, columns, courses, etc.) can be placed in the geometric context by specifying its *ObjectPlacement* attribute. For this, an entity of type *IfcObjectPlacement* must be used (see Figure 7 and Figure 8, top right).

There are three deriving entities, each encapsulating one of the placement possibilities as described in subsections 3.1.1-3.1.3. A placement using Cartesian coordinates as shown in Figure 2 is modelled as *IfcLocalPlacement*, which allows for 2D or 3D local placement (Figure 7, right). Its attribute *RelativePlacement* conveys the coordinates of the origin CS's PoO as well as its orientation in space. The scale factor remains 1 as all *IfcLengthMeasure* instances within one IFC file have the same UoM specified globally.

The grid placement is modelled as *IfcGridPlacement* (Figure 7, bottom) which allows for specifying the 3D location, but only one rotation, i.e. around z-axis (). The PoO is specified by an intersection point (*IfcVirtualGridIntersection*) of two grid axes

(*IfcGridAxis*). The *x*-axis of the origin CS is defined with the direction from the PoO to another intersection point or a direction in the grid's context.

Linear placement is modelled with *IfcLinearPlacement* (Figure 7, top) which combines an *IfcCurve* as the curvilinear axis together with an *IfcDistanceExpression* to obtain a unique point on the axis with optional offsets from the curve resulting in the PoO. The orientation of the CS is determined using an attribute of type *IfcOrientationExpression*. The deriving *IfcLinearPlacementWithInclination* allows to account for the lateral inclination (i.e., cant) of the curvilinear axis supporting the railway domain.

All placements have an optional attribute *PlacementRelTo*, which allows to chain placements and thus establish their intertwined dependencies.

3.2.3 Positioning element

IFC models the geometric context of positioning elements with the entity *IfcPositioningElement* and its derivates (Figure 8, top center). The data model includes a constraint (a WHERE rule) on the inherited optional attribute *ObjectPlacement* that makes it non-optional. Thus, it is ensured that a positioning element always has an assigned placement within the geometric context of the model.

The *IfcGrid* comprises 2 or 3 lists of unique grid axes used for positioning in *IfcGridPlacement*: *UAxes*, *VAxes*, and optional *WAxes*. These are then referenced by the placement entities as described in Section 3.2.2.

The *IfcReferent* can be used a placeholder for additional information along a linear element. For example, a stationing jump can be modelled using the optional *RestartDistance* attribute.

The *IfcLinearPositioningElement* (and derived *IfcAlignment*) points through its attribute *Axis* to the curve along which one can linearly position other objects.

The relationship *IfcRelPositions* models a connection between products and a positioning element (Figure 8, top left). This is useful if the exact placement

of those products (and thus the dependency) cannot yet be expressed with *IfcObjectPlacement*.

4 NEW PROPOSAL

In the course of the deployment activities of IFC Infrastructure Extension Deployment and IFC Rail Phase 2 projects (Jaud et al. 2020b), a multitude of problems and issues were reported to the project teams. Among these were questions and concerns about the linear placement concept, which have been included in our study.

This section describes the problems identified in the model and explores the solution as included in the IFC4x3_RC2 standard (bSI 2020b).

4.1 Problem identification

We identified four major data model issues and additional requirements with the linear placement concept as described in Section 3.2.

4.1.1 Reuse of existing concepts

Borrmann et al. (2017) mandate reuse of existing IFC entities wherever possible when expanding IFC for infrastructure. This prohibits unnecessary duplication of concepts, which would result in redundancy. To put it differently, it omits an increased burden on software vendors when updating their existing IFC interface implementations.

However, clear parallels can be observed in Figure 5 between the left-most and central columns. *IfcDistanceExpression* defines a point similarly to *IfcCartesianPoint*, however it is not inherited from *IfcPoint*. Similarly, *IfcOrientationExpression* defines the orientation of the CS axes but it is not inherited from *IfcPlacement*. The introduction of both entities is violating the recommendations provided by Borrmann et al. (2017).

4.1.2 Method of measurement

ISO 19148 requires a linear placement to define (a) a linear element, (b) a measure value, and (c) a method of measurement (see Section 2). While requirements (a) and (b) are successfully addressed by the attributes of *IfcLinearPlacement*, (c) is implicitly assumed to be an *IfcLengthMeasure*.

The IFC standard foresees that each *IfcCurve* has its parameterization clearly defined and successfully uses this principle in entities such as *IfcPointOnCurve* and *IfcTrimmedCurve*. To uniquely specify the position, *IfcParameterValue* is used which bases the method of measurement on the type of the underlying curve (e.g. $[0,1]$ for straight lines and $[0,2\pi]$ for circles). As such, the implicit assumption mentioned above violates the already established norm in IFC.

4.1.3 Tangent reference

The curve used for linear placement is required to be C^0 continuous (Equation (2)). This means that a tangent

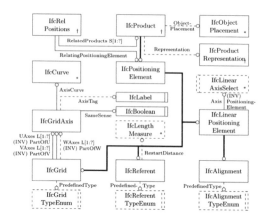

Figure 8. EXPRESS-G diagram of positioning entities from IFC4x3_RC1 that are relevant for our study.

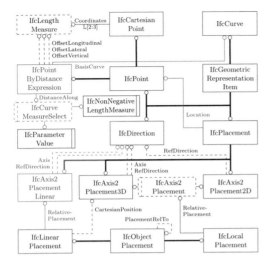

Figure 10. Railway sleepers linearly placed along a straight line with changing inclination values.

Figure 9. EXPRESS-G diagram of changed entities from Figures 5 & 7 marked in red. *IfcGridPlacement* and its attributes not shown for brevity, since nothing changed.

in places without C^1 continuity cannot be uniquely determined and as such the directions of offsets from *IfcDistanceExpression* remain undefined.

Additionally, the CS in which *IfcOrientationExpression* is expressed is unclear. The directions provided do not specify which axes they define, e.g. does the *LateralAxisDirection* attribute define the *x*, *y* or *z* axis direction of the origin's CS?

4.1.4 Chaining of placements

IfcObjectPlacement allows for chaining of placements using the *ObjectPlacement* attribute. The implementation is clear for the case when an instance of *IfcLocalPlacement* is placed relative to another *IfcLocalPlacement* (see Figure 2 and Equation (3)).

However, what does it mean, if an *IfcLinearPlacement* or an *IfcGridPlacement* is positioned relative to another *IfcObjectPlacement*? This is especially critical when considering that both mentioned placements need to account for the geometric context of the *IfcPositioningElement* their curves belong to.

4.2 Solution

This section presents the developed solution to the problems specified in Section 4.1. The solution has been adopted by bSI and is implemented in the Release Candidate 2 of the 4.3 version (IFC4x3_RC2). Figure 9 shows an overview of the changes, where changes are marked with red colour and deprecated entities omitted.

Firstly, the *IfcDistanceExpression* was renamed to *IfcPointByDistanceExpression* and is now inherited from the abstract *IfcPoint* entity. A new attribute *BasisCurve* was introduced to convey along which curve the location is to be measured along. This resulted in the removal of the *PlacementMeasuredAlong* attribute of *IfcLinearPlacement*. The offsets are clearly defined relative to the tangent of the curve at the location specified. This partly addresses the issues from Sections 4.1.1 and 4.1.3.

Additionally, the type of *DistanceAlong* attribute was changed to a new select type *IfcCurveMeasureSelect*. This allows for specifying the measurement method according to (a) the established *IfcParameterValue* with the IFC model or (b) with an absolute length from the beginning of the curve *IfcNonNegativeLengthMeasure* as is usually the case with infrastructure assets. This solves the problem emphasized in Section 4.1.2.

Secondly, a new entity *IfcAxis2PlacementLinear* replaces *IfcOrientationExpression* and derives from *IfcPlacement*. The *Location* attribute of *IfcPlacement* previously pointing to an instance of *IfcCartesianPoint* now allows to reference any *IfcPoint*, thus ensuring that *IfcAxis2PlacementLinear* entities can reference the introduced *IfcPointByDistanceExpression* mentioned above. This is enforced by using a special WHERE rule on *IfcAxis2PlacementLinear*. Consequently, *IfcLinearPlacement* is provided with the necessary semantics to establish a linear placement as explained in Section 2. This partly addresses the issues from Section 4.1.1.

Thirdly, the definition of *Axis* and *RefDirection* attributes of *IfcAxis2PlacementLinear* state that they are defined relative to the tangent of the curve at the specified location. This ensures unambiguity as demanded by Section 4.1.3.

Lastly, a restriction on *IfcObjectPlacement* was introduced, where an instance of *IfcLinearPlacement* cannot be placed relative to another *IfcLinearPlacement*. Rather, only the *IfcObjectPlacement* used by the corresponding *IfcLinearPositioningElement* can be referenced in *PlacementRelTo* attribute of *IfcLinearPlacement*. This addresses the issue described in Section 4.1.4. Consequently, the retrieval of the context of an *IfcGridPlacement* has been significantly simplified. It is not anymore necessary to navigate through *IfcGridAxis* inverse attributes *PartOfU*,

PartOfV and *PartOfW* to obtain the geometric context, as it is specified in *PlacementRelTo* attribute.

In consequence of the changes described, the entities defining inclination were removed as the changes to *IfcPlacement* now allow the specification of inclination angles directly in the context of an *IfcCurve*. The entities *IfcAxisLateralInclination*, *IfcLinearAxisWithInclination* and *IfcLinearPlacementWithInclination* are deprecated in IFC4x3_RC2 with planned removal in the final version of the standard.

Note that other elements not explicitly mentioned in this section from Figures 5, 7 & 8 were left unchanged (like *IfcGridPlacement*). The only exception is the *Axis* attribute of *IfcLinearPositioningElement* whose type got reversed to be *IfcCurve* as it was defined in IFC4x1 by Liebich et al. (2017).

5 CASE STUDIES

We tested our proposal on two unit-test sample files made available by the IFC Infrastructure Extensions Deployment project (Jaud et al. 2020b). The Step Physical Files (SPF) following ISO 10303-21 (2016) with supporting documentation and screen dumps can be obtained from the project's official repository: github.com/buildingSMART/Sample-Test-Files.

5.1 Railway sleepers

The first example consists of ten railway sleepers linearly aligned along and rotating about the axis as seen on Figure 10. Each is slightly more rotated in a clockwise direction than the previous. This example showcases both (i) the location along the alignment as well as (ii) the modelling of different orientation contexts. The SPF file consists of three major blocks: (a) project context with default units, (b) an alignment whose axis consists of a single linear segment, and (c) multiple instances of *IfcBuiltElement* modelling the sleepers with their location and orientation specified w.r.t. the given alignment axis.

We provide a closer look at a linear placement of an individual sleeper in Algorithms 1-3. Algorithm 1 shows an *IfcBuiltElement* instance modelling a sleeper. Each sleeper references the SPF reference *#277* to define its geometrical representation. and is placed in its correct location with the SPF reference *#40* (see also Figure 8, top right).

Algorithm 1. An excerpt from an SPF defining a semantic object for a sleeper together with its geometry from Figure 10. The placement reference *#40* is modelled in Algorithms 2 & 3.

```
#38=IFCBUILTELEMENT('0QLu06Q0LBIfiMID14KKna',
    #1002,'linear positioned: 1',$,$,#46,#40,$,$);
#40=IFCPRODUCTDEFINITIONSHAPE($,$,(#277));
#277= IFCSHAPEREPRESENTATION(#15,'Body','Brep',
    (#651));
#651= IFCFACETEDBREP(#647);
#647= IFCCLOSEDSHELL((...)); //shortened
```

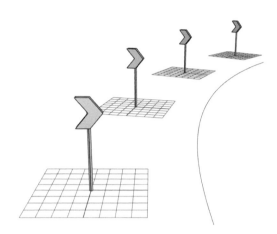

Figure 11. Road signs linearly placed along and perpendicular to the main axis in a curved segment. The horizontal (x,y) planes of their individual local CSs are shown as grids.

The Algorithms 2 & 3 show the most important lines concerning the linear placement of the sleeper. They showcase the proposed changes between the definitions from Sections 3.2 and 4.2 (i.e. IFC4x3_RC1 and IFC4x3_RC2). In both cases, the SPF reference *#20* points to the instance of the alignment axis. The SPF reference *#46* is consumed by an instance of *IfcBuiltElement* as its *ObjectPlacement* (see Algorithm 1).

Algorithm 2. An excerpt from an SPF of a linear placement for a sleeper from Figure 10 following the definitions from Section 3.2 (i.e. following IFC4x3_RC1).

```
#42=IFCDISTANCEEXPRESSION(1.,0.,-0.2,0.,.T.);
#43=IFCDIRECTION((0.,0.1361,0.9907));
#44=IFCDIRECTION((1.,0.,0.));
#45=IFCORIENTATIONEXPRESSION(#44,#43);
#46=IFCLINEARPLACEMENT(#20,#42,#45,$);
```

Algorithm 3. An excerpt from an SPF of a linear placement for a sleeper from Figure 10 following the definitions from Section 4.2 (i.e. following IFC4x3_RC2).

```
#42=IFCPOINTBYDISTANCEEXPRESSION(IFCNON
    NEGATIVELENGTHMEASURE(1.),0.,-0.2,0.,#20);
#43=IFCDIRECTION((0.,0.1361,0.9907));
#44=IFCDIRECTION((1.,0.,0.));
#45=IFCAXIS2PLACEMENTLINEAR(#42,#43,#44);
#46=IFCLINEARPLACEMENT($,$,$,#45,$);
```

5.2 Road signs

The second example models chevron signs positioned on the outsides of horizontal curves of a typical road axis as shown on Figure 11. While the first example showcases rotation of an element about the main axis, the second example models the placement of an object relative to the axis, being arbitrarily rotated and offset from the axis.

The individual signs are modelled as *IfcElementAssembly* instances containing the post and chevron parts positioned locally in the context of the sign (see Algorithm 4). Their geometries are defined once and reused for each sign using *IfcMappedItem*. The position of an individual sign is modelled with *IfcLinearPlacement* with an optional additional *IfcLocalPlacement* instance to rotate the sign to always point towards the curve when positioned on the outside of the axis' curves.

Algorithm 4. A tree view of IFC elements and their attributes representing a sign from Figure 11 (cropped on the right side).

```
#280= IFCELEMENTASSEMBLY('3ZsRKiJhpmC0lFpJHy_xn3',$,$,$,#281,$,
  Placement : #281 = IFCLOCALPLACEMENT(#279,#178);
    PlacementRelTo : #279 = IFCLINEARPLACEMENT(#4,$,#278,$);
      PlacementRelTo : #4 = IFCLOCALPLACEMENT($,#5);
      RelativePlacement : #278 = IFCAXIS2PLACEMENTLINEAR(#277,#1
        Location : #277 = IFCPOINTBYDISTANCEEXPRESSION(IFCNON|
          BasisCurve : #110 = IFCGRADIENTCURVE(#14,(#111,#115,#1;
        Axis : #155 = IFCDIRECTION((0.0,0.0,1.0));
    RelativePlacement : #178 = IFCAXIS2PLACEMENT3D(#164,#155,#17,
      Location : #164 = IFCCARTESIANPOINT((0.0,0.0,0.0));
      Axis : #155 = IFCDIRECTION((0.0,0.0,1.0));
      RefDirection : #174 = IFCDIRECTION((-1.0,0.0,0.0));
  IsDecomposedBy(3)
    #283 = IFCMEMBER('3HDs_AqgUGMyOGXeseK_OG',$,$,$,$,#282,#18
    #286 = IFCSIGN('1EMCzoWtuwgvPd0tNVSEdl',$,$,$,$,#285,#187,$,$)
    #288 = IFCSIGN('0cmjprzXqvhcLurRfHz0Xt',$,$,$,$,#287,#187,$,$);
```

6 CONCLUSION

We have critically evaluated a part of the recent candidate standard extension of IFC (IFC4x3_RC1) as published by bSI (2020a). With the help of participants of the IFC Infrastructure Extension Deployment and IFC Rail Phase 2 projects (Jaud et al. 2020b), several issues about the linear placement concept have been identified as presented in Section 4.1. The proposed simplified model from Section 4.2 addresses the requirements listed while respecting the guidelines provided by Borrmann et al. (2017). The improved model has been adopted as RC2 of IFC4.3.

We showcase the new model on two example scenarios from the projects mentioned above. We argue that the proposed model shown in Figure 9 enables every foreseeable constellation of placement as occurring in the AEC industry – in the building as well as infrastructure sectors. One can observe that the newly introduced inheritance from *IfcPoint* and *IfcPlacement* for *IfcPointByDistanceExpression* and *IfcAxis2PlacementLinear*, respectively, allow for modular software architecture.

The reuse of existing concepts ensures little-to-no effort required by software vendors already supporting these concept templates of the IFC schema. We call for fast adoption of the newly developed IFC4x3_RC2 standard by the industry and active participation of all IFC implementors with the deployment activities.

REFERENCES

Borrmann, A., Amann, J., Chipman, T., Hyvärinen, J., Liebich, T., Muhič, S., Mol, L., Plume, J. & Scarponcini, P. 2017. *IFC Infra Overall Architecture Project: Documentation and Guidelines.* Technical Report. buildingSMART International.

Borrmann, A, Muhič, S., Hyvärinen, J., Chipman, T., Jaud, Š., Castaing, C., Dumoulin, C., Liebich, T., & Mol, L. 2019. *The IFC-Bridge Project – Extending the IFC Standard to Enable High-Quality Exchange of Bridge Information Models.* 2019 European Conference on Computing in Construction, Chania, Crete, Greece.

Bradley, A.; Li, H.; Lark, R.; & Dunn, S. 2016. BIM for infrastructure: An overall review and constructor perspective. *Automation in Construction.* 71, 139–152. DOI: 10.1016/j.autcon.2016.08.019.

bSI 2020a. *Industry Foundation Classes: 4x3 candidate standard 1 documentation.* Online, URL: https://standards.buildingsmart.org/IFC/DEV/IFC4_3/RC1/HTML/, accessed 2020-12-22.

bSI 2020b. *Industry Foundation Classes: 4x3 candidate standard 2 documentation.* Online, URL: https://standards.buildingsmart.org/IFC/DEV/IFC4_3/RC2/HTML/, accessed 2020-12-22.

Esser, S.; Borrmann, A. 2019 *Integrating Railway Subdomain-Specific Data Standards into a common IFC-based Data Model.* In: Proc. of the 26th International Workshop on Intelligent Computing in Engineering, Leuven, Belgium.

ISO 2004. *ISO 10303-11:2004: Industrial automation systems and integration – Product data representation and exchange – Part 11: Description methods: The EXPRESS language reference manual.* Standard International Organization for Standardization, Geneva, Switzerland.

ISO 2012. *ISO 19148:2012: Geographic information – Linear referencing.* Standard International Organization for Standardization, Geneva, Switzerland.

ISO 2016. *ISO 10303-21:2016: Industrial automation systems and integration – Product data representation and exchange – Part 21: Implementation methods: Clear text encoding of the exchange structure.* Standard International Organization for Standardization, Geneva, Switzerland.

ISO 2018. *ISO 16739:2018-1: Industry Foundation Classes (IFC) for data sharing in the construction and facility management industries.* Standard International Organization for Standardization, Geneva, Switzerland.

Jaud, Š., Donaubauer, A., Heunecke, O. & Borrmann, A. 2020a. Georeferencing in the context of building information modelling. *Automation in Construction, 118(103211).* DOI: 10.1016/j.autcon.2020.103211

Jaud, Š., Esser, S., Muhič, S. & Borrmann, A. 2020b. Development of IFC Schema for Infrastructure. In: *Proceedings* of 6*th* international conference siBIM: Structured data is the new gold, pp.27-35. Online.

Liebich, T., Amann, J., Borrmann, A., Chipman, T., Hyvärinen, J., Muhič, S., Mol, L., Plume, J., Scarponcini, P. 2017. *IFC Alignment 1.1 Project: IFC Schema Extension Proposal.* Technical Report. buildingSMART International.

OGC 2012. *Geography Markup Language (GML) — Extended schemas and encoding rules.* OpenGIS Implementation Standard. OGC 10-129r1.

Peffers, K., Tuunanen, T., Rothenberger, M. A., & Chatterjee, S. 2007. A design science research methodology for information systems research. *Journal of management information systems*, 24(3):45–77.

IFC query language: Leveraging power of EXPRESS and JSON

S. Morozov
Ivannikov Institute for System Programming of the Russian Academy of Sciences, Moscow, Russia
M.V. Lomonosov Moscow State University, Russia

S. Sazonov
Ivannikov Institute for System Programming of the Russian Academy of Sciences, Moscow, Russia

V. Semenov
Ivannikov Institute for System Programming of the Russian Academy of Sciences, Moscow, Russia
Moscow Institute of Physics and Technology, Dolgoprudny, Russia
Higher School of Economics, Moscow, Russia

ABSTRACT: Recently, Building Information Modeling (BIM) technologies and, in particular, BIM servers proceeding with IFC-driven product data have become increasingly important in architecture, engineering and construction (AEC). However, a standard query language for IFC data has not yet been established, which prevents the interoperability between BIM applications and wider adoption of the IFC standard. In the paper requirements to an IFC query language are summarized with the focus on the compliance and harmonization with data definition and data manipulation languages which generally are the parts of a common data access interface. As a result of the study, the IFC query language was proposed and formalized. Using EXPRESS constructs and JSON structures, the language allows to specify complex predicative queries on object collections and recursive traversing queries on object networks. Explanatory examples of typical IFC queries are provided to illustrate the proposed language and prove its advantages.

1 INTRODUCTION

Building Information Modeling (BIM) is becoming increasingly important for complex projects and large infrastructure programmes in Architecture, Engineering and Construction (AEC). The main benefits of BIM are improved technological processes and increased productivity due to the replacement of obsolescent practices with 2D drawings and textual documents by advanced software applications proceeding with comprehensive multi-disciplinary information models. As opposed to unstructured or semi-structured documents, high maturity levels of BIM imply the use of structured information similar to that driven by the Industry Foundation Classes (IFC) — a standard vendor-neutral information schema for AEC applications developed by buildingSMART community and certified by ISO (ISO 2018). Providing open specifications for the schema, the IFC standard simplifies the data exchange and sharing between BIM tools dedicated for particular AEC disciplines and produced by different vendors. It is worth to note that the IFC standard borrows the STEP model-driven methodology involving the specification of information schemas using the object-oriented modeling language EXPRESS (ISO 2004), the data storage and transportation using STEP physical files in clear text encoding format (ISO 2016b), and the multi-access to schema-driven product data through Standard Data Access Interface (SDAI) (ISO 1998).

High semantic complexity of multi-disciplinary BIM data and sophisticated techniques to share it among different AEC project stakeholders concurrently lead to the enhanced role of integration platforms and centralized BIM servers (Succar 2009). Among such solutions, both open and commercial IFC servers are of great importance due to the possibilities to integrate a large number of the existing BIM applications that already support open file formats and software interfaces regulated by buildingSMART. In particular, EDMmodelServer™(Jotne IT 2020), GRAPHISOFT®BIMcloud (GRAPHISOFT 2020), Open BIM server (BIMServer.org 2020), IFC Web-Server (IFCWebServer.org 2020), BIMserver.center (Cype 2020), IFChub (IFChub 2015) are available.

However, despite the aforementioned standardization efforts, there is no common understanding and vision of what the IFC query language should be. At the same time, the problem of IFC data requests remains challenging due to the complexity of the data and the need to retrieve and process only those data fragments relevant to the planned project workflows. Model View

Definitions (MVD) (buildingSMART 2020a) help in defining such extraction queries for specific disciplines and applications, but cannot be considered as a functionally complete query language due to mismatch with EXPRESS and lack of functions, parameters, variables, control statements, etc.

It is worth mentioning that query languages are designed in the context of that metamodel on which data models are based. In the case of the IFC standard, its released information schemas are formally specified in the object-oriented modeling language EXPRESS and, therefore, the IFC query language should be compliant or at least harmonized with the EXPRESS language to cover all the variety of IFC datatypes and their semantics.

Much research and development effort has been paid to propose and validate BIM query languages. However, until today, the BIM community has not established or recognized a standard query language, and each IFC server implementation uses its own language for requesting, filtering, and retrieving data driven by IFC schemas. Let us look at well-known query languages, dividing them into a few categories.

The first category languages are based on the EXPRESS metamodel and are designed for querying arbitrary data whose model is specified in the EXPRESS language, including, of course, IFC-driven data. Employing the EXPRESS metamodel does not mean that the languages should use the EXPRESS syntax. As an example, the EQL language constructs (Huang 1999) are similar to SQL, and the PMQL (Adachi 2002b) is based on XML.

The compliance with the EXPRESS language and, in particular, the support of logical expressions for defining object predicates is an obvious advantage of this approach. However, the EXPRESS language is poor of facilities to define the path queries necessary to traverse objects along given associations and extract the requested data similar to the way as it is prescribed by MVD templates. To compensate this shortcoming, EXPRESS-based query languages are usually extended by special path constructs. So, the PMQL defines a 'cascades' operator for propagating recursively along IFC objects' graph.

Graph-based query languages more focus on path queries than EXPRESS, but they are quite limited for predicate definitions. These languages are more user-friendly due to graphic notations, which make the queries more intuitive and understandable compared to strict syntax text (Tauscher et al. 2016). Exploiting a visual programming paradigm and expressive visual metaphors (Preidel et al. 2017; Wülfing et al. 2014), the languages are becoming used for large groups of AEC players (architects, engineers, planners, accountants, managers, etc.) who do not have or begin programming experience. At the same time, the graphic notations do not prevent their translation to machine-interpretable text. For example, the graph-based queries can be translated to the special textual notation (DSL) for further effective processing by BIM servers (Tauscher & Crawford 2018).

The BIMQL language (Mazairac & Beetz 2013) was developed as a general, domain-specific, platform-independent solution for the open source BIMserver. It employs syntax close to natural languages, which simplifies its understanding. The language allows recursive path queries using so-called shortcuts. However, predicate expressions in BIMQL are not EXPRESS-compliant and can lead in misinterpretations for complex datatypes. The expressions are limited to simple comparisons bracketed by logical operations. The right part of each comparison can only be a literal of integer, real or string type. Another drawback is that the query expressions become too cumbersome if the required shortcuts are not predefined.

A special category is formed by high-level domain-specific languages dedicated to perform queries that are of interest to some restricted field or problem. The 3D Spatial Query Language (Borrmann & Rank 2009) is intended to retrieve geometrical and topological information from BIM projects. The BERA language (Lee 2011) is designed to perform rule checking and validation in architecture domain. MMQL (Fuchs & Scherer 2014) focuses on extraction of construction specific information from multi-models including IFC. QL4BIM (Daum et al. 2015) is aimed at spatial analysis using IFC and CityGML models in an integrated context.

Finally, another approach to retrieving IFC data realizes an attracting idea to translate the IFC schema to an alternative data model and to re-express original queries in terms of the new model. This may be rational if a general-purpose database management system with advanced query planning and executing facilities is used to store and retrieve IFC-driven data. For example, the IFC schema and related queries can be translated to SQL, SPARQL or XQuery statements. However, the metamodel impedance is a serious obstacle to any implementation of this approach regardless of what particular information technology is planned to employ.

Converting the IFC schema to SQL statements (ISO 2016a) is a non-trivial task, although object-relational mapping patterns have been carefully studied (Klein et al. 2001; Scott 1999; Semenov et al. 2004). The main reasons are the complexity of IFC data types as well as unavoidable performance and scalability issues due to the need to permanently transform and execute object queries in terms of relational operations (Adachi 2002a; Lee et al. 2014; You et al. 2004).

SPARQL is a declarative graph-based query language adopted by W3C for Semantic Web and ontological systems specified using RDF or OWL (W3C 2013). The IFC-to-RDF transformation has been standardized (buildingSMART 2020b). However, querying IFC objects in terms of RDF triples complicates the use of SPARQL in the construction domain and requires language extensions (Zhang & Beetz 2016). Nevertheless, there are some successful reported adoptions for specific BIM tasks (Liu et al. 2016).

XQuery is a declarative functional language standardized by W3C for querying and filtering XML data (W3C 2017). In spite of XML being one of the official formats for the IFC schema and data representations (buildingSMART 2020b), it is relatively rare used in BIM applications and is inferior to the JSON format due to the slower parsing and processing by modern Web services (Crockford 2006). Nevertheless, XQUERY extended by custom spatial predicates was applied successively to retrieve construction specific spatial information (Nepal et al. 2012).

Thus, the development of a standard IFC query language remains an actual problem. In Section 2 we summarize requirements to an IFC query language. The proposed language is conceptualized and formally specified in Section 3. Some explanatory examples of IFC queries and their JSON bindings are provided in Section 4. In Conclusions, the advantages of the IFC query language are shortly enumerated with the focus on efficient execution of the specified queries.

2 GENERAL REQUIREMENTS TO IFC QUERY LANGUAGE

Let's try to summarize requirements for a new IFC query language taking into account the comparative study of the approaches presented above.

The IFC query language should satisfy the following general requirements:

– *it should be compliant or at least harmonized with the EXPRESS language* to avoid problems of metamodel mismatch and IFC datatype misinterpretation. This requirement is considered as a key principle for further development. In addition, complex predicative queries on collections of objects can be specified naturally using the full power of EXPRESS;
– *it should be easy to invoke and define queries* within statements of data manipulation languages and methods of data access interfaces. Whereas BIM Service interface exchange (BIMSie) de facto defines a standard JSON-based manipulation language for IFC-driven data in the cloud (NIBS 2019), it is desirable that queries could be also presented in accordance with JSON format;
– *it should be easy to define queries of various kinds*, including queries by types (with or without subtypes), by exact values of the object attributes (including unique identifiers), complex predicative queries using logic and arithmetic operations, collection iterating, functions and other constructs provided by advanced object-oriented data modeling languages;
– *it should support path queries* for traversing complex object-oriented data along associations and retrieving the requested data. Path queries can help in defining MVD's, some of which have been standardized by buildingSMART and some are still under development. While EXPRESS lacks such capabilities, the JSON format is well suited for structuring path queries;
– *it should allow the reuse of already defined queries*, for example, by referencing commonly shared information resources and employing query parameterization;
– *it should be programmatically extensible*, for example, to execute spatial-temporal requests and to solve various problems of quantitative and qualitative reasoning;
– *it should use strictly defined syntax and semantics* for unambiguous interpretation and translation of queries into/from alternative representations, including statements of visual programming languages;
– *it should enable efficient parsing, planning and execution of queries* taking into account CPU power and RAM limits;
– *finally, it should be versatile enough* to be used in various BIM applications running under heterogeneous hardware and software platforms.

3 CONCEPTUALIZATION AND FORMALIZATION OF THE LANGUAGE

The IFC query language should be considered and specified as a part of data manipulation languages and software interfaces to access IFC-driven data. It is worth to remind that the STEP SDAI (ISO 1998) is specified by defining a complete set of methods, their semantics and signatures with all input and output parameters. Because the parameters can have complex structures, they are specified in dedicated EXPRESS schemas. Such conceptualization and formalization approach has an obvious advantage: the interface declared in an abstract implementation-neutral form allows alternative bindings in the imperative programming languages C, C++, Java.

In this study we follow the same approach, providing a formal specification of the data types needed to define queries. This allows one to use alternative encodings for IFC queries defined in one language. For example, the encodings may be based on STEP physical file format, XML, or JSON.

Here, an assumption regarding the granularity of the requested data and the retrieval method should be made. We assume that the method accepts a query formatted in JSON as input and returns a set of the requested object handlers as output. Depending on the implementation, object handlers can be unique object identifiers, references, pointers, or any other system structures that provide direct access to attributes of individual objects. Having an object handler, it becomes possible to get the values of the object attributes: all or some of the selected ones. Depending on the implementation, these can be not only explicit attributes, but also attributes declared as DERIVED or INVERSE in the IFC schema. The retrieval method and object handlers are part of the data access interface rather than the query language itself. However, in further conceptualization and formalization we do not distinguish between IFC objects and their handlers. For brevity, we omit the formal specification of IFC objects here, referring to the definitions provided by the SDAI interface for entity instances.

The key concepts of the IFC query language are formalized as EXPRESS entities and are shown in the Listing 1 below. The *QueryDefinition* entity is intended to define simple queries that allow you to specify the object constraints of several predefined types. These are constraints by entity datatypes, specified object predicates and parametric logical functions. To define complex queries that retrieve objects of different entity extents along extensive networks of association routes, multiple *QueryDefinition* instances are usually constructed in combination with auxiliary *Path* instances. Such instances are intended to propagate the retrieving process to collections of associated objects and to recursively invoke the following queries given as *QueryDefinition* instances. The *Main* entity is intended to initialize entries to the entire retrieving process and begin it. To avoid ambiguity in the further interpretation, we consider that such an entry is always unique.

Listing 1. EXPRESS schema of the IFC query language constructs.

```
TYPE ParameterType = ENUMERATION OF
(BooleanType, LogicalType, IntegerType, RealType,
StringType, BinaryType, ListOfBooleanType, ListOf
LogicalType, ListOfIntegerType, ListOfRealType,
ListOfStringType, ListOfBinaryType, DefinedType);
END_TYPE;

TYPE DefinedTypeValue = STRING; END_TYPE;

TYPE ParameterValue = SELECT (BooleanValue,
LogicalValue, IntegerValue, RealValue, StringValue,
BinaryValue, ListOfBooleanValue, ListOfLogical-
Value, ListOfIntegerValue, ListOfRealValue, ListOf-
StringValue, ListOfBinaryValue, DefinedTypeValue);
END_TYPE;

TYPE ParameterName = STRING; END_TYPE;

ENTITY FormalParameter;
    Name: ParameterName;
    Comment: OPTIONAL STRING;
    Kind: ParameterType;
    DefaultValue: OPTIONAL ParameterValue;
UNIQUE
    UR1: Name;
END_ENTITY;

ENTITY ActualParameter;
    Name: ParameterName;
    AssignedValue: OPTIONAL ParameterValue;
    AssignedParameter: OPTIONAL ParameterName;
UNIQUE
    UR1: Name;
WHERE
    WR1: EXISTS(AssignedValue) XOR
EXISTS (AssignedParameter);
END_ENTITY;

TYPE LogicalExpression = STRING; END_TYPE;
TYPE LogicalFunction = STRING; END_TYPE;
ENTITY QueryDefinition;
    Name: STRING;
    Comment: OPTIONAL STRING;
    ObjectType: OPTIONAL STRING;
    IncludeSubtypes: OPTIONAL BOOLEAN;
    PredicateExpression: OPTIONAL
        LogicalExpression;
    Parameters: OPTIONAL SET[1:?] OF
        FormalParameter;
    PredicateFunction: OPTIONAL LogicalFunction;
    Paths: OPTIONAL SET[1:?] OF Path;
UNIQUE
    UR1: Name;
WHERE
    WR1: NOT(EXISTS(ObjectType) XOR
        EXISTS(IncludeSubtypes));
    WR2: NOT( EXISTS(Parameters) XOR
        EXISTS(PredicateFunction));
    WR3: EXISTS(PredicateExpression) XOR
        EXISTS(PredicateFunction);
END_ENTITY;

ENTITY QueryInvocation ABSTRACT SUPER-
TYPE OF (ONEOF(Main, Path));
    URL: OPTIONAL STRING;
    QueryName: STRING;
    Parameters: OPTIONAL SET[1:?] OF
        ActualParameter;
END_ENTITY;

ENTITY Path SUBTYPE OF (QueryInvocation);
    PathExpression: OPTIONAL STRING;
END_ENTITY;

ENTITY Main SUBTYPE OF (QueryInvocation);
    FromObjects: OPTIONAL SET[1:?] OF
        EntityInstance;
END_ENTITY;
```

For each query, the *QueryDefinition* entity defines a unique name, an optional comment explaining its purpose, an object type extent or selection statement produced by an IFC entity datatype, a specifier clarifying whether subtype objects are requested, a predicate imposing additional constraints on the requested objects, and references to *Path* instances that define paths for possible propagation of the retrieval process to collections of associated objects. The predicate can be defined either as a simple expression or as a parametric function.

The *PredicateExpression* attribute is defined as a string containing a logical expression according to the syntax specified in Chapter 12 of the EXPRESS standard (ISO 2004). The expression must be the same in syntax as WHERE clause with the implicit parameter SELF that corresponds to the analyzed entity instance. An entity instance is considered to satisfy the imposed predicative constraint if its logical expression is valid for the instance and takes TRUE on evaluation. The expression may include all the constructs valid for logical expressions in the EXPRESS language, in particular, logical and arithmetic operations, collections iterating, calls of built-in and schema-defined functions and procedures, etc.

The *PredicateFunction* attribute is defined as a string containing a logical function according to the syntax specified in Chapter 15 of the EXPRESS standard (ISO 2004). The first formal parameter of the function should have the type given by the *ObjectType* attribute, other should have names and types declared by the *Parameters* set. Each *FormalParameter* includes a name, a comment explaining its role, an optional default value, and a type that can be either predefined in EXPRESS, or a list of values of such predefined type, or a user-defined type in accordance with the underlying information schema such as IFC. The values for user-defined types are given as strings.

Being a supertype of *Main* and *Path*, the *QueryInvocation* entity defines a common set of attributes necessary to specify a query context, to set its actual parameters and to transfer control to the following *QueryDefinition* instance. This query definition shall be referenced by its name specified in the *QueryName* attribute. If the query definition is located in another information resource, *URL* path to it should be additionally given. If the query definition uses a parametric logical function for specifying object constraints, its invocation should also include the set of actual parameters defined by the attribute *Parameters*. Actual parameter can be given as a literal value of the proper parameter type or as a name of the parameter defined in the preceding query and reassigned for invocation of the following query. Formal and actual parameters are matched by unique names. If any actual parameter cannot be found, its default value is applied. The invocation is canceled every time neither the actual parameter nor its default value is defined.

The *Main* entity additionally defines the optional *FromObjects* attribute. The attribute is used to provide an initial set of objects from which the retrieval process can begin. If the attribute is not set, the retrieval process begins from an object extent of the entity type given by the optional *ObjectType* attribute in the corresponding query definition. If both attributes are unset, the query is applied to the entire data set. If both attributes are set simultaneously, the initial set of objects shall first be checked against the entity type, then — against a predicative constraint if given by the *PredicateExpression* or *PredicateFunction* attribute, and finally be selected for further retrieval. The optional attribute *IncludeSubtypes* prescribes whether the objects of subtypes of the given entity shall be selected or not. Objects that do not satisfy all the constraints imposed shall be excluded from the result set.

The *Path* entity additionally defines the *PathExpression* attribute that is used to specify the association route along which the retrieval process should be propagated. The attribute is given as a string that concatenates the names of the associations on the route. The names are separated by the dot symbol in accordance with EXPRESS syntax. Due to the specialization relationships between IFC datatypes, path expressions may not be valid for some intermediate objects, which prevents further propagation and completion of the request. Such objects shall be ignored, and only the objects reached by the terminal association are selected for further analysis. The selected objects are suffered to checks against the constraints given in the following query definition and the propagation process is recursively repeated.

Finally, notice that the language constructs allow one to define cyclic queries, which can be useful, for example, to request object collections reached by different ways. Cyclic queries are formed every time when a *Path* instance references to a preceding query definition. To prevent endless loops at execution runtime, the objects that have been already traversed are tracked and excluded from the further propagation process.

4 SOME EXAMPLES OF IFC QUERIES

The proposed IFC query language is general enough to specify both simple queries by attributes and entity extents, and complex predicative and recursive queries used to solve some computational problems, for example, spatial-temporal modeling and reasoning problems. The language does not prevent the use of the specified queries as ad-hoc requests or as stored procedures that, when precompiled, provide more efficiency. Finally, being conceptualized and formalized the language admits alternative encodings for queries. For the above reasons, JSON-based encoding seems preferable for BIM services and applications and will be used in the examples presented. For brevity, we omit issues on how to produce an object notation for EXPRESS schema-driven data and just provide explanatory examples of typical IFC queries.

4.1 Sample dataset

Let the entire IFC dataset include the following entity instances:
#179= IFCANNOTATION('07p2XhKoTFRAvuxKPgbU$v', #42,'Model lines:285328',$,$,#148,#173);
#173= IFCPRODUCTDEFINITIONSHAPE($,$, (#166));
#166= IFCSHAPEREPRESENTATION(#109,'Annotation', 'Annotation2D',(#156));
#156=   IFCGEOMETRICCURVESET((#200,#201, #202));
#250= IFCCARTESIANPOINT((1.0,1.0));
#251= IFCCARTESIANPOINT((-1.0,1.0));
#200= IFCPOLYLINE((#250,#251));
#252= IFCCARTESIANPOINT((-1.0,-1.0));
#201= IFCPOLYLINE((#251,#252));
#253= IFCCARTESIANPOINT((1.0,-1.0));
#202= IFCPOLYLINE((#252,#253));

As seen, the annotation #179 is geometrically represented by the curve set #156. The set consists of the polylines #200–#202 connecting the Cartesian points #250–#253.

4.2 Query by GUID

The most typical query is a request to get an instance of the *IfcRoot* entity by a given globally unique identifier (GUID). The abstract entity defines the unique attribute *GlobalId* for representing GUID values in instances of specialized entities such as *IfcProduct, IfcProcess,* or *IfcBuildingElement*. This query can be naturally defined by specifying the parametric function *EqualGUID* as follows:

```
{
"type" : "QueryDefinition",
"name" : "Instance by GUID",
"objectType" : "IfcRoot",
"includeSubtypes" : true,
"parameters" : [ {
    "name" : "GUIDValue",
    "kind" : "StringType"
  } ],
"predicateFunction" : "FUNCTION EqualGUID(root: IfcRoot, GUIDValue: STRING) : LOGICAL; RETURN (root.GlobalId = GUIDValue); END_FUNCTION;"
}
```

To retrieve the annotation #179 that is of *IfcProduct* entity, the defined query should be invoked with the assigned *GUIDValue* parameter:

```
{
"type" : "Main",
"queryName" : "Instance by GUID",
"parameters" : [ {
    "name" : "GUIDValue",
    "assignedValue" : "'07p2XhKoTFRAvuxKPgbU$v'"
  } ]
}
```

4.3 Query by entity extent

Another typical query is a request to get all instances belonging to an entity type extent. The query definition to get all polylines from the entire dataset looks as follows:

```
{
"type" : "QueryDefinition",
"name" : "Polylines",
"objectType" : "IfcPolyline",
"includeSubtypes" : false
}
```

The extent type is given by the attribute *ObjectType*. As the entity *IfcPolyline* has no subtypes the value of the attribute *IncludeSubtypes* has no matter here.

4.4 Query by predicate

Predicative queries can be defined by specifying a logical parametric function or a logical expression. The provided above example of the query by GUID uses the first way. The second way is illustrated by the following query intended to find 2D points inside a unit circle:

```
{
"type" : "QueryDefinition",
"name" : "2D Points inside circle",
"objectType" : "IfcCartesianPoint",
"includeSubtypes" : false,
"predicateExpression" : "(Dim = 2) AND (Coordinates[1]**2 + Coordinates[2]**2 <= 1)"
}
```

The specified logical expression is given by the *PredicateExpression* attribute. Being invoked and evaluated for the sample dataset, the query result will contain the points #250–#253.

4.5 Query by route

Traversing queries are constructed by defining several queries intended to retrieve particular object collections. To extend the retrieval process across all such collections, the query definitions must be supplemented by auxiliary path queries. Let's consider how a typical route query can be specified using the language proposed:

```
{
"type" : "Main",
"queryName" : "Products",
"fromObjects" : [ "#179" ]
}
```
```
{
"type" : "QueryDefinition",
"name" : "Products",
"objectType" : "IfcProduct",
"includeSubtypes" : true,
"paths" : [ {
    "type" : "Path",
    "queryName" : "Polylines with points",
    "pathExpression" : "Representation.Representations.Items.Elements"
  } ]
}
```
```
{
"type" : "QueryDefinition",
"name" : "Polylines with points",
"objectType" : "IfcPolyline",
"includeSubtypes" : false,
"paths" : [ {
    "type" : "Path",
    "queryName" : "Cartesian points",
    "pathExpression" : "Points"
  } ]
}
```
```
{
"type" : "QueryDefinition",
"name" : "Cartesian points",
"objectType" : "IfcCartesianPoint",
"includeSubtypes" : false
}
```

The query is aimed at the extraction of all polylines and Cartesian points that form the shape of the annotation #179 in the sample dataset. It is assumed that the OID of the requested annotation is known in advance, and therefore the annotation can be obtained directly using the specified attribute *FromObjects* of the main query. The query execution begins from the main query invocation, and then it is propagated along the specified route, where the intermediate objects #173, #166, #156 are passed and the terminal polylines #200–#202 are included into the final result. Then the retrieval process extends to the associated Cartesian points, which are also added to the result.

4.6 Query by spatial constraint

Here is an example of a spatial query that extracts those 2D polylines that intersect a given line. To reuse a once-defined query for various purposes, it must be specified as a parametric query. For the example query, the parameters may be a 2D point lying on the line and a 2D direction of the line. In the definition below, we use the built-in EXPRESS functions, such as SIZEOF and QUERY, to define a proper logical function:

```
{
"type" : "Main",
"queryName" : "Polylines intersection",
"parameters" [ {
    "name": "LinePoint",
    "assignedValue" : [1.6, 1.2]
    },
    {
    "name" : "Direction",
    "assignedValue" : [1.0, 0.0]
    } ]
}
{
"type" : "QueryDefinition",
"name" : "Polylines intersection",
"objectType" : "IfcPolyline",
"includeSubtypes" : false,
"parameters" : [ {
    "name" : "LinePoint",
    "kind" : "ListOfRealType" },
    {
    "name" : "Direction",
    "kind" : "ListOfRealType"
    } ],
```
"predicateFunction" : "FUNCTION LineIntersection (Polyline: IfcPolyline, LinePoint: LIST[2:2] OF REAL, Direction: LIST[2:2] OF REAL) : LOGICAL; RETURN (SIZEOF (QUERY (Point<*Polyline.Points |(((Direction[1] * (Point.Coordinates[1] - LinePoint[1])) + (Direction[2] * (Point.Coordinates[2] - LinePoint [2]))) * ((Direction[1] * (Polyline.Points[1].Coordinates[1] - LinePoint[1])) + (Direction[2] * (Polyline.Points[1].Coordinates[2] - LinePoint[2]))) <= 0))) > 0); END_FUNCTION;"
}

Although we presented a relatively simple example, the proposed language allows specifying complex spatial-temporal queries of the geometric and scheduling information available in IFC-driven datasets. It is important that such requests can be defined using the same principles.

5 CONCLUSIONS

Thus, the IFC query language was proposed and developed as a result of the comparative study of known implementations and the efforts undertaken to harmonize it with data definition and data manipulation languages, which is considered as the key principle. The language meets various requirements, including:

– the compliance with the EXPRESS language, in which the IFC schema is formally specified,
– the ability to define complex predicative queries on object collections and recursive traversing queries on object networks,
– the ability to invocate both ad-hoc queries and parameterized stored procedures directly from JSON-based data access interfaces,
– efficient parsing, planning and executing of queries taking into account CPU power and RAM limitations.

The language interpreter was implemented and tested as a part of the IFChub collabolation platform (Semenov & Jones 2015a, b). The interpreter has demonstrated relatively high performance when requesting large IFC-driven data consisting of tens of millions of objects. This was achieved due to various code optimizations, among which the most important are indexing by identifiers and entity datatypes, as well as object caching. The validation carried out and the obtained performance results confirm the feasibility of the proposed IFC query language and its prospects for widespread adoption in BIM software applications and BIM servers.

REFERENCES

Adachi, Y. 2002a. Overview of IFC Model Server Framework. *eWork and eBusiness in Architecture, Engineering and Construction: Proceedings of the 4th European Conference on Product and Process Modelling (ECPPM 2002)*, Portorož, Slovenia: 367–372.

Adachi, Y. 2002b. Overview of Partial Model Query Language. Secom Co., Ltd./VTT Building and Transport, Espoo, VTT-TEC-ADA-12.

BIMServer.org. 2020. GitHub—opensourceBIM/BIMserver: The open source BIMserver platform page. https://github.com/opensourceBIM/BIMserver

Borrmann, A. & Rank, E. 2009. Topological analysis of 3D building models using a spatial query language. *Advanced Engineering Informatics* 23(4): 370–385. DOI: 10.1016/j.aei.2009.06.001.

buildingSMART. 2020a. Model View Definitions (MVD) —buildingSMART International page. https://www.buildingsmart.org/standards/bsi-standards/model-view-definitions-mvd/

buildingSMART. 2020b. IFC Formats — buildingSMART Technical page. https://technical.buildingsmart.org/standards/ifc/ifc-formats/

Crockford, D. 2006. JSON: the fat-free alternative to XML. *Proceedings of XML 2006*, Boston, USA.

Cype. 2020. BIMserver.center. BIM in the cloud page. https://bimserver.center/en

Daum, S., Borrmann, A. & Kolbe, T.H. 2015. A spatio-semantic query language for the integrated analysis of city models and Building Information Models. *Proceedings of the 3D Geoinfo 2015*, Universiti Teknologi, Malaysia.

Fuchs, S. & Scherer, R.J. 2014. MMQL — A language for multi-model linking and filtering. *eWork and eBusiness in Architecture, Engineering and Construction: Proceedings of the 10th European Conference on Product and Process Modelling (ECPPM 2014)*, Vienna, Austria: 273–280.

GRAPHISOFT. 2020. GRAPHISOFT®BIMcloud Overview page. https://www.graphisoft.com/bimcloud/overview/

Huang, L. 1999. EXPRESS Query Language and Templates and Rules: Two Languages for Advanced Software System Integration. PhD Thesis, Ohio University.

IFChub 2015. IFC server for construction and built environment page. http://ifchub.com/

IFCWebServer.org. 2020. IFCWebServer — Open BIM Server and Online BIMViewer page. https://ifcwebserver.org/

ISO. 1998. ISO 10303-22:1998. Industrial automation systems and integration — Product data representation and exchange — Part 22: Implementation methods: Standard data access interface. Geneva: International Organization of Standardization.

ISO. 2004. ISO 10303-11:2004. Industrial automation systems and integration — Product data representation and exchange — Part 11: Description methods: The EXPRESS language reference manual. Geneva: International Organization of Standardization.

ISO. 2016a. ISO/IEC 9075-1:2016. Information Technology — Database languages — SQL — Part 1: Framework (SQL/Framework). Geneva: International Organization of Standardization.

ISO. 2016b. ISO 10303-21:2016. Industrial automation systems and integration — Product data representation and exchange — Part 21: Implementation methods: Clear text encoding of the exchange structure. Geneva: International Organization of Standardization.

ISO. 2018. ISO 16739-1:2018. Industry Foundation Classes (IFC) for data sharing in the construction and facility management industries — Part 1: Data schema. Geneva: International Organization of Standardization.

Jotne IT. 2020. Jotne IT EDMmodelServer™(ifc) page. http://www.jotneit.no/edmmodelserver-ifc

Klein, L., Stonis, A. & Jancauskas, D. 2001. EXPRESS/SQL white paper. LKSoftWare GmbH.

Lee, G., Jeong, J., Won, J., Cho, C., You, S.J., Ham, S. & Kang, H. 2014. Query performance of the IFC Model Server using an object-relational database approach and a traditional relational database approach. *Journal of Computing in Civil Engineering* 28: 210–222. DOI: 10.1061/(ASCE)CP.1943-5487.0000256.

Lee, J.K. 2011. Building environment rule and analysis (BERA) language — and its application for evaluating building circulation and spatial program. PhD Thesis, Georgia Institute of Technology.

Liu, H., Lu, M. & Al-Hussein, M. 2016. Ontology-based semantic approach for construction-oriented quantity take-off from BIM models in the light-frame building industry. *Advanced Engineering Informatics* 30(2): 190–207. DOI: 10.1016/j.aei.2016.03.001.

Mazairac, W. & Beetz, J. 2013. BIMQL — An open query language for building information models. *Advanced Engineering Informatics* 27(4): 444–456. DOI: 10.1016/j.aei.2013.06.001.

Nepal, M.P., Staub-French, S., Pottinger, R. & Webster, A. 2012. Querying a building information model for construction-specific spatial information. *Advanced Engineering Informatics* 26(4): 904–923. DOI: 10.1016/j.aei.2012.08.003.

NIBS. 2019. BIM Service interface exchange (BIMSie) Project — National Institute of Building Sciences. https://www.nibs.org/page/bsa_bimsie

Preidel, C., Daum, S. & Borrmann, A. 2017. Data retrieval from building information models based on visual programming. *Visualization in Engineering* 5, 18. DOI: 10.1186/s40327-017-0055-0.

Scott, W.A. 1999. Mapping objects to relational databases. An AmbySoft Inc. White Paper.

Semenov, V., Morozov, S. & Porokh, S. 2004. Object-relational mapping strategies: pattern-based systematization and analysis. *Proceedings of the Institute for System Programming* 8(2): 53–92.

Semenov, V. & Jones, S. 2015a. IFChub: An Innovative Collaboration Platform for Architecture, Engineering and Construction. *Proceeding of the 15 International Conference on Construction Applications of Virtual Reality*, Banff, Canada: 216–224.

Semenov, V. & Jones, S. 2015b. IFChub: Multimodal Collaboration Platform with Solid Transactional Guarantees. *Digital Proceeding of the 32 CIB W78 Conference 2015*, Eindhoven, The Netherlands: 667–675.

Succar, B. 2009. Building Information Modelling framework: A research and delivery foundation for industry stakeholders. *Automation in Construction* 18(3): 357–375. DOI: 10.1016/j.autcon.2008.10.003.

Tauscher, E., Bargstädt, H.J. & Smarsly, K. 2016. Generic BIM queries based on the IFC object model using graph theory. *Proceedings of the 16th International Conference on Computing in Civil and Building Engineering (ICCCBE 2016)*, Osaka, Japan: 905–912.

Tauscher, H. & Crawford, J. 2018. Graph representations and methods for querying, examination, and analysis of IFC data. *eWork and eBusiness in Architecture, Engineering and Construction: Proceedings of the 12th European Conference on Product and Process Modelling (ECPPM 2018)*, Copenhagen, Denmark: 421–428.

W3C. 2013. SPARQL 1.1 Overview. https://www.w3.org/TR/sparql11-overview/

W3C. 2017. XQuery 3.1: An XML Query Language. https://www.w3.org/TR/xquery-31/

Wülfing, A., Windisch, R. & Scherer, R.J. 2014. A visual BIM query language. *eWork and eBusiness in Architecture, Engineering and Construction: Proceedings of the 10th European Conference on Product and Process Modelling (ECPPM 2014)*, Vienna, Austria: 157–164.

You, S.J., Yang, D. & Eastman, C.M. 2004. Relational DB implementation of STEP based product model. *CIB World Building Congress 2004*, Toronto, Canada.

Zhang, C. & Beetz, J. 2016. Querying linked building data using SPARQL with functional extensions. *eWork and eBusiness in Architecture, Engineering and Construction: Proceedings of the 11th European Conference on Product and Process Modelling (ECPPM 2016)*, Limasson, Cyprus: 27–34.

Using uncertainty to link compliance and creativity

N. Nisbet
AEC3 UK Ltd, UK

ABSTRACT: Compliance to regulations, requirements and recommendations remains a weak point for the design, construction, and operation in the built environment. The RASE mark-up methodology has been demonstrated to significantly raise the accuracy and efficiency of the capture and evaluation of regulations and requirements. This paper extends the relevance of RASE beyond regulations and requirements into two new areas, firstly where uncertainty may impact decision making and secondly where design is supported by constraints. Examples are taken from UK Building Regulations Approved Documents Part M on Accessibility.

1 BACKGROUND

1.1 Impetus

There is a broad consensus that the built environment sector and the design and construction sector in particular has a poor profile both extrinsically and intrinsically. In comparison with other sectors it is noted that it has delivered poor performance in economic, social, and environmental terms. Examples include the poor short-term and long term-valuation of companies and multi-national corporations focused on construction. In social terms, the UK and Italy have seen grim failures to provide even the most basic level of safety. Environmentally the sector which typically accounts for around a tenth of the GDP base produces four tenths of all environmental impacts. Internally it is reported to be poor in terms of accuracy, efficiency, and responsiveness. The sector has proved unable to accurately predict or manage time, cost, or quality. It has stagnant or falling productivity. Responsiveness to changes in need has been slow, even if responsiveness to economic fluctuations has been brutal and abrupt.

1.2 Systematic engineering

The conventional approach to securing the performance of complex systems has been to adopt systems engineering. An engineered system, can be defined as a combination of components that work in synergy to collectively perform a useful function (Sys Eng 2020). Systems engineering takes as a central tenet that design requirements can be stated and disaggregated and so tested against the reaggregation of design solutions. This pattern should be reapplied during manufacturing phase and during the operational phases. However, in the built environment the design requirements, include legislation and regulations, requirements and recommendations are not checked repeatedly. Legislation and regulation have expanded rapidly. In the UK it is estimated that there are 1600 pages of primary regulations (Hackitt 2019). This body of regulations is not static with most documents being issued or revised on a 3–4 year timescale. The primary requirements will come from the client or from bodies of sector knowledge such as the UK school Generic Design Brief (EFA 2013). The industry is also subject to growing bodies of recommendations emanating from advisory schemes such as insurers', and near-Government bodies' such as the UK Health and Safety Executive, and BREEAM.

1.3 Relevant developments

The growth in use of object-orientated building information modelling (BIM) has not been matched by growth in the systematic validation at design, construction, or in-use. The checking processes have remained essentially unchanged. Whilst some electronic submission of PDF documents is occurring, the checking activity remains a manual, intuitive and highly skilled process at a time when the availability of such staff is necessarily falling due to the aging of the current skill base and the growth of rival sectors. This can be seen in some high-profile failures and also anecdotally in everyday experience.

1.4 Intention

AEC3 has attempted to learn lessons from these experiences to specify a new approach to the delivery of systematic engineering for the built environment sector in design construction and use. The normative documents are definitive and should be the focus to establish the primary feed-back loop. This doesn't preclude the possibility of a secondary feedback loop that may revise or refine the client's requirements or even the bodies of knowledge. For example, in the UK there is a small body of Governmental clarifications on the planning and building regulations, but it does require that any such changes be formally documented. Compliance is more of a management issue over the aggregation of individual technical checks and must include processes around exemptions and exceptions. Any solution may need to draw on other descriptive resources, such as product specifications and product data, alongside the BIM models. When considering construction or in-use, it is necessary to consider other resources such as site-diary entries, and live monitoring, sensors and IoT devices.

A long-term solution is needed that is viable in technical, commercial, and political terms (DCOM 2019). All three criteria make it necessary to separate the normative material from the executing engine, and from the target descriptions. Technically this reduces the amount of document specific work and allows for the use of robust general-purpose rule engines. The separation may require the use of 'data dictionaries' with mapping rules to relate document terms with target description terms. Commercially this separation creates a viable international business model for the core rule-engine whilst the reporting may need to be locally configurable. Politically implementation must be able to happen within the communities responsible for design, construction, and use, but equally, it must be able to happen within the communities responsible for the generating and enforcing normative demands.

2 APPLICATION OF RASE

2.1 Obtaining a normative statement

The outcome from RASE (Requirement, Applicability, Selection and Exception) is always a single statement expressible in propositional logic. The variables in the statement represent distinct 'metrics'. Metrics are resolved by examination of the descriptive resources. Intermediate results and the overall outcome are 'objectives'. Objectives are resolved by the recursive resolution of the subsidiary objectives and metrics.

RASE takes as its starting point any written normative document. The topic experts are asked to mark-up the document in terms of the logical roles of the phrases and sections. This mark-up of phrases and sections is relatively simple because there are only four roles (colors) that can appear. Phrases and sections narrow, broaden, or exclude the scope of attention. RASE is defined by the tagging: Any narrowing is tagged Applicability (green). Any broadening is tagged Selection (purple). Any exclusion is tagged Exception (orange). Normative phrases and sections are tagged as Requirements (blue).

The significance of the RASE mark-up is twofold. On the one hand it is easily reviewed and endorsed by the topic experts who can identify any key phrases that have not been marked up and review the mark-up that has been added. On the other hand, it determines a single logical structure that captures the exact intent of the regulation in a form that can be evaluated.

The proof of concept demonstrated for US ICC at Orlando for their 2006 conference was implemented using a commercial HTML authoring tool. This allowed the rapid development of the RASE conventions and the associated presentational stylesheet that highlights the presence of RASE mark-up. These conventions were then implemented in the "ICC SmartCodes Editor", a proprietary tool which operated on documents held in the ICC's own XML based document management format (Nisbet et al. 2008). In work for the US Army Corp of Engineers, RASE was applied to both the written normative document and to the published table (30 x 1000 cells) of design requirements for specific room types (East & Nicholas 2010). Tables represented a specific challenge for RASE as they are non-linear text. Initially the solution was to develop a specific XSLT transformation to systematically represent each tabulated requirement as a separate clause. Later developments have made this extra process redundant.

2.2 Exploitation

RASE has been applied, in many different contexts including regulatory and recommendatory roles, in different languages and through various engines and using various dictionary and descriptive resources. Given a marked-up normative resource, trials were conducted with various XSLT transformations applied to the HTML to prepare the rules in a form that could be delivered to a number of rule-engines.

The first target for automated compliance checking was BIM models shared in the buildingSMART IFC format (ISO 16739). As any application that had the ability to read IFC BIM models would also have the ability to read IFC Constraint models, it was pragmatic to automatically transform the marked-up regulation into the equivalent IFC Constraint model which was accessible to three distinct implementations:

1. A modified version of the Singapore ePlancheck using EPM Jotne Model checker (AEC3 2004);
2. Solibri model checker (Solibri 2006) with a specially developed add-on extension;
3. AEC3 XABIO Compliance1 (AEC3 2010) command line tool.

Java Drools is a standardized library of java functions for rule execution. For the UK InnovateUK RegBIM (Beach et al. 2013) project, a sequence of transformations was applied to capture rules of a regulatory and two advisory schemes and generate Drools coding and visualizations.

DMN rule tables are the preferred rule language to support BPMN syntax diagrams for business process management. A transformation is used to generate a DMN file using the four logical operators applied to multi-variate lists of operands and using three-valued logic (true, unknown, false). The demonstration was

Table 1. Resources marked up using RASE.

Short name	Description
US IECC (ICC 2006)	International Code Council International Energy Conservation Code
US UFC 4-510-01 (WBDG 2010)	Whole Building Design Guide for Unified Facility Criteria for Medical facilities
KR BA 34 1 (KR 2019)	South Korean Building Act factory escape regulation
FR DGUHC 2007 53 6 F1 (DGUHC 2007)	French building regulations on communal stairs
Various UK, US and other classification schemes	Each classification forms a cascade of definitive rules where typically each sub-category is an exception to the category above.
UK BREEAM (BREEAM 2016)	Environmental assessment. (With special adaptions to account for the maximum and actual points awarded.)
UK CfSH (CfSH 2006)	Code for Sustainable Homes, defining levels of sustainability. (Withdrawn)
RH1-TRH-XX-XX-RP-C-0001 (STREAMER 2013)	Descriptive survey report on the fabric of a general hospital building
UK BRAD M 2 13 (BRAD 2012)	Building Regulation Approved Document on Accessibility covering entrance doors
UK RIDDOR (HSE 2019)	Descriptive mandatory reporting of construction accidents, taken as circumstances requiring preventive measures.
UK CDM 2015 4 22 (CDM 2015)	Statutory instrument covering the management of construction excavations

Figure 1. UK BR AD M 2 13 with RASE mark-up.

intended to complement the investigation by the Norwegian Government of manually coded DMN rules on fire prevention.

Procedural code can be generated by capturing individual steps as function calls. The style of this output can be Visual Basic or any other procedural programming language. These types of output have not been extensively tested because they necessarily assume a large number of supporting libraries and functions.

These approaches have been applied to a number of normative resources summarized in Table 1.

2.3 *Summary of experiences*

The RASE methodology has been found to be efficient and accurate independent of the quality of the norm, in terms of structure, complexity and mixing tables with text. It has been used with a number of different languages including English, French, Russian and Korean and been found to be independent of grammar and character-set. Work to date has been focused on exploiting IFC BIM along with formulae and plain language definitions. As part of the consideration of health and safety, and the management of existing facilities, RASE has been extended to render operable other resources. Dictionary content can be plain text with Definitions instead of Requirements and descriptive documents can be plain text with Descriptions instead of Requirements. These extensions suggest that RASE can be exploited to accurately and efficiently capture most written knowledge.

AEC3 has now developed a workbench tool "AEC3 Require1" (AEC3 2020) to bring together the functionality and experience described above. It supports the mark-up of normative documents, the development of the dictionary resource and the effecting of automated code compliance checking of multiple norms against multiple descriptive resources including IFC building models and user provided supplementary information. A key motive was the need to avoid existing rules-engines that variously do not support, or did not support well some pre-requisites such as three valued logic.

2.4 *Example*

The text of 'UK BRAD M 2 13' can be taken as an example and marked-up as in Figure 1. This mark-up automatically implies a single statement that can be expressed in a number of forms including the logical (true/unknown/false) proposition shown in Figure 2.

Alternatively, the original mark-up resource can be used to directly drive the checking of any descriptive resource such as a BIM model seen in Figure 3. Every entity in the BIM can be checked against the norm and the results aggregated to give an overall result as seen in Figure 4.

Other reports can be developed summarizing or tracing the decision process, highlighting any salient unknowns or metrics leading to failure.

3 UNCERTAINTY

3.1 *Sources of uncertainty*

There may be circumstances when the logical result may be better understood if some indication is given of the confidence attaching to it. Both checkers and designers benefit from identifying the robustness of the result.

Skepticism may be applied to a particular value or measurement in the descriptive resource. A specific

Figure 2. UK BR AD M 2 13 in propositional calculus.

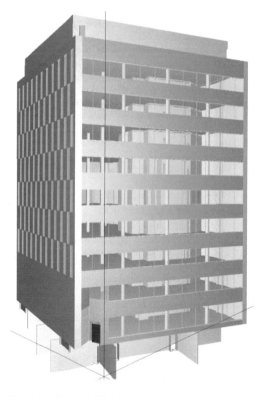

Figure 3. Example BIM (courtesy of MACE Ltd).

Evaluation of UK-BR-AD-M-2-13 on MON-MAK-XX-XX-M3-ARCHT has evaluated to TRUE

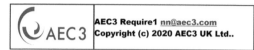

Figure 4. Summary result from AEC3 Require1.

be a significant but not absolute correlation with the metrics. There may be obvious but undocumented exceptions.

A general skepticism may be useful until confidence in automated code checking is developed. It may serve to draw attention towards those results where the outcome, especially a pass, is vulnerable to more specific sources of uncertainty or errors. Compared to logical metrics, a lower level of confidence could be placed on continuous (real) metrics. There may be a need to allow for general uncertainty arising from continuous metrics, especially when the documented value is close to the target value. This section suggests functions that can be used to represent this.

3.2 Other approaches

There are a number of non-binary approaches to confidence, uncertainty and probability, and there is a lot of discussion in the literature on semantic separation of these concepts – even whilst continuing to apply similar mathematics. Some approaches seek to replace the logical variable with a numeric quantity, such as fuzzy-logic. The literature contains many speculations on the appropriate equations to apply when processing this numerical qualification.

3.3 Supplementary approach

For design and checking purposes, there is more value in tracking the uncertainty separately but alongside the logical decision making. This ensures that the application can still give a primary logical answer, even if it is then qualified by a secondary comment such as "87.36% confident". The equations chosen should deliver a continuously decreasing value as the number of determining metrics and objectives increase and the handling of the four logical operators should be consistent. For example, we can use the two self-consistent confidence calculations in Table 2.

3.4 Example

We can apply a degree of skepticism to an example BIM model by assigning a 95% confidence to all metrics measured off the BIM. The same process that evaluates the overall logical outcome can also be used to track the impact of that skepticism. Results obtained by enhancing the reporting tools in AEC3 Require1 are shown in Figure 5.

The resulting confidence in the result on the glazed entrance door is still as high as 54% as the outcome

metric may involve accessing an uncertain value. In BIM authoring tools it has been unusual to document uncertainty. IFC 4x3 does contain proposals to associate shape and positional uncertainty to objects using a specific property set. A metric may be challenged, and the checker may call on supplementary evidence.

A specific objective may be subject to doubt. This may arise from a looseness of the normative requirement, as sometimes found in third-party or non-contractual recommendations, where there may

Table 2. Confidence calculations.

Decision basis	Examples	Confidence calculation
On one metric or objective	false requirement, true exception, true selection, false application	Current confidence
On one or more metrics and objectives	some unknowns, true requirements, false exceptions, false selections, true applications	Product of current confidences

Entity	Name	Outcome	Confidence
IfcWall	A-G252-WallInternal-100mm	PASS	95.00%
IfcDoor	A-G322-Timber-Door-DH-OP-1250x2500	PASS	90.25%
IfcDoor	A-G321-DoorGlazing	PASS	54.04%

Figure 5. Individual results with confidence measures.

is determined by the relatively small number of 12 metrics.

4 CREATIVITY

4.1 *Supporting design*

Logical statements, such as those generated by RASE, can be mathematically differentiated in an analogous way to the differential calculus of real functions. For a given state of the descriptive resource, every variable metric will have a logical value and so the overall objective will have a logical value. In considering any one variable, its connection to the overall result can be detected, and so the relationship of its state to the overall state can be obtained. In different circumstances a change to a single metric may have no impact, a direct correlated impact, a reverse correlation, or it may trigger the propagation or absorbance of unknowns up towards the overall results. This technique has been exploited to supplement results on non-compliant designs, so that when a facility was found to fail, the decisive metrics can be listed as in Figure 6.

Logical differentiation can be applied proactively to allow any decisive metrics, having a non-zero differential, to be highlighted to the user or prioritized by an algorithm. Where the overall outcome is 'fail', metrics that could improve the decisions can be highlighted in italic (red) as in on the left in Figure 7. Where the overall outcome is 'pass', metrics that are critical to maintaining that situation can be highlighted in italic (green) as on the right of Figure 7.

4.2 *Design feedback*

Logical differentiation exposes the Boolean (cliff-edge) behavior of the metrics and the overall outcome, shown as the line with (blue) squares in Figure 8.

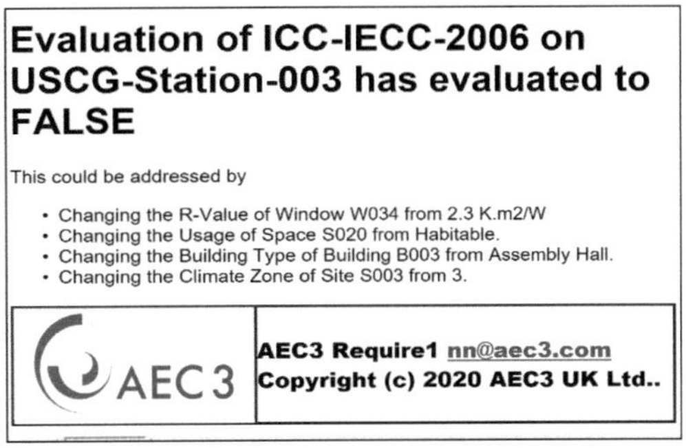

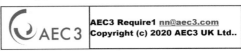

Figure 6. Failure report with decisive metrics listed.

A-G3–A-G321-DoorGlazing (IfcDoor)

Name	Value
GlobalId	3tM7tTRzP23ODneKtdIKZn
Name	A-G321-DoorGlazing
Description	MON-MAK-A-ES-M3-FAC:67501
ObjectType	A-G321-DoorGlazing
OverallHeight	2.6 m
OverallWidth	0.5 m

UK_DoorCapture (IfcPropertySet)

Name	Value
Access Statement	none
effective clear width	*0.45 m*
force required to open door at leading edge	25 N
is external	True
is self closing	True
lower vision panel base height	0.1 m
lower vision panel top height	0.8 m
powered door opening closing system used	True
upper vision panel base height	1 m
upper vision panel top height	2.1 m
part of accessible entrance	True
direction of approach to door	Straight On
used by general public	True
width of door access route	3 m

A-G3–A-G321-DoorGlazing (IfcDoor)

Name	Value
GlobalId	3tM7tTRzP23ODneKtdIKZn
Name	A-G321-DoorGlazing
Description	MON-MAK-A-ES-M3-FAC:67501
ObjectType	A-G321-DoorGlazing
OverallHeight	2.6 m
OverallWidth	2.6 m

UK_DoorCapture (IfcPropertySet)

Name	Value
Access Statement	*none*
effective clear width	*2.5 m*
force required to open door at leading edge	*25 N*
is external	*True*
is self closing	*True*
lower vision panel base height	*0.1 m*
lower vision panel top height	*0.8 m*
powered door opening closing system used	*True*
upper vision panel base height	*1 m*
upper vision panel top height	*2.1 m*
part of accessible entrance	*True*
direction of approach to door	*Straight On*
used by general public	*True*
width of door access route	*3 m*

Figure 7. Presentation of decisive metric.

Logical differentiation is less informative in situations where the metric is actually driven by a continuous (real) measure, such as the width of a door, for example *(clear width - 1000mm >= 0)*. The measure can be transformed by selecting a function that can expresses the pressure or compliance such as *atan (clear width - 1000mm)*. This transformation brings the measure into a range of -1 to 1 but with sensitivity to its proximity to 0 as an S-curve marked with (orange) triangles in Figure 8. An automated design process is then able to detect the direction of change required and its urgency, and a human designer is able to see or perhaps feel that pressure haptically through the mouse/cursor feedback. More information about the pressure is illustrated by the hump-backed derivative curve shown with (grey) dots and the switchback second derivative shown with (yellow) crosses in Figure 8.

These pressures are aggregated across any number of continuous design parameters, even if they

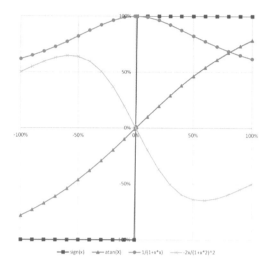

Figure 8. Interpretations of continuous metrics.

are conflicting, to generate a direction vector with the best chance of improvements for the design. This allows a set of normative documents and requirements collectively to influence the design process.

5 CONCLUSIONS

We have demonstrated that the RASE methodology can systematically supplement the logical outcome with some more nuanced numeric values. This allows both design and checking processes to be enhanced. This in turn should encourage the adoption of systematic engineering in AEC practice.

REFERENCES

AEC3 2020. AEC3 Require1 tool. Accessed March 2020. http://www.aec3.eu/require1/AEC3_Require1.html

Beach, T.H., Kasim, T., Li, H., Nisbet, N. & Rezgui, Y. 2013. Towards automated compliance checking in the construction industry, *Lecture Notes in Computer Science*, 8055 (2013) 366–380 ISSN 0302-9743 10.1007/978-3-642-40285-2_32

BRAD 2012. Access to and use of buildings: Approved Document M. Accessed March 2020. https://www.gov.uk/government/publications/access-to-and-use-of-buildings-approved-document-m

BREEAM 2016. BRE Environmental Assessment Method. Accessed March 2020. https://www.bregroup.com/products/breeam

CDM 2015. The Construction (Design and Management) Regulations. Accessed March 2020. https://www.hse.gov.uk/construction/cdm/2015/index.htm

CfSH 2006. Code for sustainable homes: technical guidance. Accessed March 2016. https://www.gov.uk/government/publications/code-for-sustainable-homes-technical-guidance

DCOM 2019. D-COM Research Network. Accessed April 2020. http://www.dcom.org.uk/downloads

DGUHC 2007. Direction Générale de l'Urbanisme, de l'Habitat et de la Construction. Accessed March 2020.

East E.W. & Nicholas, N. 2010. Facility Capacity Analysis. *CIB W78 2011 28th International Conference – Applications of IT in the AEC Industry*.

EFA 2013. Generic Design Brief. Accessed March 2020.

Hackitt, D.J. 2019. Building a Safer Future Independent Review of Building Regulations and Fire Safety: Final Report. Accessed March 2020.

HSE 2019. Reporting of Injuries, Diseases and Dangerous Occurrences Regulations. Accessed March 2020. https://www.hse.gov.uk/riddor
https://assets.publishing.service.gov.uk
https://assets.publishing.service.gov.uk/government/uploads/system/uploads/attachment_data/file/707785/Building_a_Safer_Future_-_web.pdf
http://reglementationsaccessibilite.blogs.apf.asso.fr/archive/2010/09/22/temp-360a91f7f40917bd6213b0a2b2665a0e.html
https://www.wbdg.org/ffc/dod/unified-facilities-criteria-ufc

ICC 2006. *ICC International Energy Conservation Code ISBN-13: 978-1-58001-315-4.*

ISO 16739. *ISO 16739-1:2018 Industry Foundation Classes (IFC) for data sharing in the construction and facility management industries — Part 1: Data schema.*

KR 2019. Building act. Accessed March 2020. https://elaw.klri.re.kr/eng_service/lawView.do?hseq=31602&lang=ENG (now clause 50).

Liebich 2002. Speeding up the building plan approval. *eWork and eBusiness in AEC*.

Nisbet, N. 2010. AEC3 BimServices.BimServices – Command-line and Interface utilities for BIM. Accessed March 2016. http://www.aec3.com/en/6/6_04.htm

Nisbet, N., Wix, J. & Conover, D. 2008. The future of virtual construction and regulation checking, in Brandon, P., Kocaturk, T. (Eds), *Virtual Futures for Design, Construction and Procurement*, Blackwell, Oxfordshire. doi: 10.1002/9781444302349.ch17.

Solibri 2020. Accessed March 2020. www.solibri.com

STREAMER 2013. EU Streamer. Accessed March 2020. http://www.streamer-project.eu/

Sys Eng 2020. System engineering. Accessed February 2020 https://en.wikipedia.org/wiki/Systems_engineering

WBDG 2010. Whole Building Design Guide. Unified Facilities Criteria (UFC). Accessed March 2020.

BIM model uses through BIM methodology standardization

A. Barbero, M. Del Giudice, F.M. Ugliotti & A. Osello
Politecnico di Torino, Turin, Italy

ABSTRACT: Digital transformation is influencing the strategy to develop virtual repository able to collect data from different disciplines and domains in a useful way for the building lifecycle. In this framework, Building Information Modelling (BIM) can be the innovative methodology to optimize the overall workflow including a proper definition and management of geometrical and alphanumerical contents. The article aims to investigate the meaning of standard according to specific model uses that goes through the identification of the owner's objectives and their operational declination by means of a defined protocol of activities and tailor-made solutions. The study presents a progressive increasing of the complexity of the BIM system, which, starting from the definition of an As-is model, is enriched through the integration with other data domains and improves its usability through customized virtual experiences.

1 INTRODUCTION

The current fragmentation of the workflow related to the design and construction and operation and maintenance processes, both in technical and collaboration terms, lead to an inadequacy of the available information of the heritage. Realizing a digital model representative of the As-built state of a new construction, in fact, provides different solutions and approaches compared to the setting up of an As-is model functional to management and maintenance.

The reason lies in the different way the information is produced according to the specific objectives connected with the construction phases. For a new building, the planned information content is designed to fulfil the user's requirements and to achieve a certain performance. This implies the creation of a detailed As-built model aimed at the construction site. For existing buildings, the information production is aimed at developing a model for the Operation and Maintenance (O&M) step. This As-is model is adopted for updating over time, describing the state of the art of the building at a certain time of the survey.

The need to create a different detailed model from the As-built for existing assets is mainly due to the reliability of As-built data compared to the actual state. In this context, the crucial points of BIM model implementation are based on the set of model uses and objectives that characterize the content of parametric BIM objects. Following this perspective, a lot of research is oriented to define transversal standards to facilitate information management by following international and national legislations.

Currently, the state of the art of this area of investigation provide several definitions, interpretations and enrichments of the Level of Geometry (LOG) and Level of Information (LOI) within the O&M step.

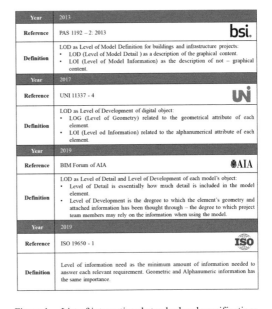

Figure 1. List of international standard and specifications.

Figure 1 summarizes the main references. These references constitute an initial benchmark to guide the implementation choices of the models, however they need more articulation for practical use. For example, it is not exhaustive to establish a Level of Detail (LOD) for the entire structure to be modeled, but it is necessary to identify different standards for each object in order to control the reliability, granularity and use of the information associated with it. It is therefore appropriate to define information specifications for each project, considering the objectives, documentation and professional and economic resources available. This is also expressed in ISO 19650-1:2019 which introduces the concept of the Level of Information Need. This aspect could be strictly related to the concept of heterogeneous LOD applied to the entire model obtained by application of a specific LOG and LOI for each element category (Barbero et al. 2019). It is strictly related to define standards for the production of the model and the objects according to BIM procurement specifications and documents that allows the definition of operational guidelines.

Currently the documents required to produce a BIM model are mainly two:

– Employer Information Requirement (EIR) that clarifies the employer's requirements during services' procurement. It should include levels of modelling detail, training/competence requirements, ordinance systems, exchange formats or other employer-mandated processes, standards or protocols (BIM Dictionary 2020).
– BIM Execution Plan (BEP), based on EIR, that defines how the information modelling aspects of a project will be carried out. It includes other documents that clarify the roles and responsibilities, standard to be applied and procedures to be followed (BIM Dictionary 2020).

Clearly, both of the two documents include BIM standards for modelling and information management for private and public sector. Their definitions change in function of the specific national legislation: for example, the Italian technical report UNI11337 – 6 defines the content of the "Capitolato Informativo" (CI), based on the EIR's structure. These documents represent the basis for the identification of the graphic and alphanumeric content required for a properly structured BIM model during the services' procurement. However, this content, as previously mentioned, needs to be enriched with several operational requirements to achieve the defined practical uses (Ashworth et al. 2016). BIM uses for Facility Management (FM) require indeed both the data content upload and extraction according to the day-to-day needs as the model represents a valuable database for maintenance activities. For this reason, the concept of BIM guidelines was born that contain the dispositions of the BIM procurement, enriched with a series of technical standards on which are based operating protocols that should be followed to achieve defined BIM uses.

2 METHODOLOGY

This paper aims to evaluate the creation of a BIM methodology standardization in the meaning of defining the activities necessary to start a digitization process aimed at using BIM models in management activities. Specifically, the definition of different model uses is preparatory to the correct geometric and alphanumeric definition of the objects. In this contribution the digital twin of the building has been assessed in function of the role within maintenance

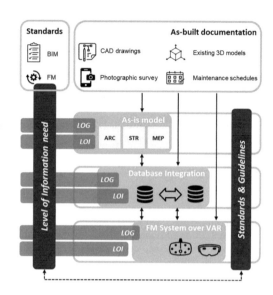

Figure 2. Methodological workflow.

activities, identifying the milestones to be satisfied for the development of proper guidelines. They can be considered as an integration of the contractual documents that allow the definition of the operational requirements for the achievement of the owner's objectives. They consist of a sort of protocol issued by the employer to allow the creation of models. They are developed through continuous updating and implementation loops during the entire BIM process, thanks to the collaboration and the joint analysis with all the actors involved. According to various BIM standard investigated, the proposed workflow starts from the analysis of the objectives and uses of a certain project, starting from the employer's requirements. Figure 2 summarizes in a methodological schema the main contents that have to be followed for the BIM development for FM. One of the most important actions to ensure decision in the BIM model development consists of involving the owner at the beginning of the entire process. In this way, these decisions are "based upon accurate and relevant information and data, and their impact on operational needs has to be understood before they are committed to construction work and/or installation" (BSi 2015). Therefore, the proposed methodology begins from the definition of the level of information need (ISO 19650 2019), highlighting the model purposes in order to define BIM model uses. In this contribution, three model uses were selected (Kreider et al. 2013): i) As-is model; ii) Database integration; iii) FM system over Virtual and Augmented Reality (VAR). Clearly, each of these uses requires different kind of Level of Graphical information (LOG) and Level of Information (LOI) useful to achieve the BIM objectives. The development of the BIM model oriented to each use culminates with the definition of model requirements that have to be included in standards and guidelines, creating BIM specifications for the O&M step.

This kind of output aims to ease the supplier in following the procedures indicated in the EIR. To achieve the goal of information management during the FM step, the three model uses have been analyzed basing on the exploiting of data, according to the end user. Therefore, a set of activities related to the development, management and visualization of the model have been listed in Figure 3.

The results produced by each activity have been collected and summarized for the drafting of an operating standard to support the parties involved in the BIM process.

The above methodology can be generalized to each case study. The model uses mentioned in the following paragraph refer in particular to a case study summarizing the obtained results as an example. The case study concerns an existing venue for outdoor sports, a huge structure characterized by many aspects that require specific building registry definition and maintenance activities. Autodesk Revit software has been used for this study for its diffusion on the international market and for its multidisciplinary nature.

	Activities	As-is model	Database Integration	FM system over VAR
1	Statement of purpose	x	x	x
2	As-built documentation analysis	x		
3	National regulation adoption	x	x	
4	Collaborative BIM Working	x		
5	Graphical and information data delivery	x		x
6	Data sharing	x	x	x
7	Model breakdown structure	x		
8	Modeling rules	x		
9	Model accuracy	x		
10	Folder Structure and Naming Conventions	x		
11	Model checking	x		x
12	Code checking	x		
13	Data validation	x		
14	Import/export data transfer		x	
15	Database integration with other datasets		x	
16	Geometrical data update		x	x
17	Alphanumerical data update		x	x
18	Presentation style	x	x	x
19	Data communication			x
20	Supporting maintenance activities		x	x

Figure 3. List of activities for model uses.

2.1 As-is model

As O&M represents a significant part of the building lifecycle, the creation of an effective BIM model can contribute to streamline processes helping the Facility Manager/Department to control costs and manage data. For this purpose, it is not enough to implement all the objects and their properties, but it is necessary to organize information so that it can be functional in retrieving data for specific activities. The model breakdown, and consequently the database that can be extracted, is a decisive factor to consider carefully at the beginning of the modelling phase, as subsequent variations can be very complex and often result in a loss of information. Different strategies are possible depending on the complexity of the building and its purpose of realization. In the case study analyzed, a federated model (Barbero et. al. 2018) has been set up to articulate the model to the multiple disciplines investigated. Despite the fact that the parametric model is a database in itself, consulting it may not be immediate, especially for people who are not expert in using this kind of software. In these terms, governing the elements in a unique way through coding, classification, decomposition and georeferencing systems that facilitate their precise identification and management is fundamental. At the same time, the use of schedules, themed plans, and three-dimensional views is exploited to promote a structured building registry and ready and user-friendly access to the data. The utilization of schedules facilitates the listing of rooms for spaces management, likewise, building components for refurbishment or energy efficiency evaluation and assets for maintenance activities. By an appropriate use of shared parameters and equations, it is possible to make the most of their format for each different purpose, from analysis to managing or reporting. In the case of existing buildings, it is extremely useful to map the main components subject to periodic maintenance, by adopting a reasonable Level of Detail in function of the complexity of the structure.

Generally, reinforced concrete structures do not require special maintenance. Therefore, they can be modelled with a low Level of Detail giving them the correct function, the type of use (e.g. foundation, elevation) and materials. With regard to steel structures, the identification and coding of the most significant structural elements and connections (e.g. reticular beam, column, and node) is relevant.

Despite the representation of these elements being very complex, it is not useful to achieve a high level of graphic detail for existing buildings, but it is appropriate to link the elements to technical details and maintenance procedures.

According to the architectural part, the model must provide reliable information regarding the areas and surfaces of the environments and materials used. For this reason, maximum attention needs to be paid during the modelling phase to ensure a correct calculation method. Walls and windows are some examples of critical elements. The outer layer of the wall defines the finish, so modelling it as needed becomes important (e.g. tiles to be cleaned, surface to be painted). While for the fixtures, the possibility to estimate the incidence of the glazed part on the frame is useful for cleaning. In this way, it is possible to distinguish between the opaque surface and the transparent surface for a facade.

Through specific plugins like Autodesk Roombook, Areabook, and Buildingbook or computational design platform such as Dynamo for Autodesk Revit, an accurate model take-off can be achieved. The summary of the room-related surface areas and interior finishes of walls, floors and ceiling elements as well as

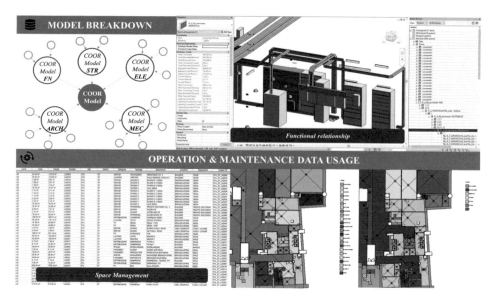

Figure 4. As-is model data usage.

a comprehensive floor area calculation and material-related quantities of constructive building parts can be exported, providing valuable quantitative indications for cleaning and maintenance activities. As BIM can play a big role in space management, the model has to properly map spaces and include all the information needed to ensure a great operational control in terms of use, occupancy, and maintenance of the building. Enriching the database with a punctual and up-to-date knowledge of these entities allows managers to control services in an increasingly thoroughness and to obtain facility management and key performance indicators (KPIs) useful for evaluating cost chargebacks, the utilization rate as well as performance measurement of the maintenance activities (Osello & Ugliotti 2017). Furthermore, information about systems component and equipment are crucial. Thanks to the parametric nature of the objects, it is possible to identify the functional relations between the elements, allowing to map and manage circuits and branches. For example, in the case of a lighting device, it is possible to know to which electrical switch is connected, to which electrical panel and consequently to which electrical substation. Since in the case of an existing building is very difficult to get reliable scheme and technical drawings, the essential aspect lies in establishing the functional and spatial relationships among the elements, not detailing the graphical representation.

2.2 *Database integration*

Starting from the literature definition of a BIM model as a geometric and alphanumeric data repository, another possible BIM use within the O&M field is represented by the Database integration (Kensek 2015). This case can be achieved in the same way using the protocol and specifications indicated in the BIM guidelines. The Database integration is based both on the need to update data from the As-is model and increase information related to the specific aim of the project. This aspect is strictly connected to one of the major strengths of the BIM methodology represented by the uniqueness of data, which must be maintained during the building lifecycle.

The database extrapolation is the first step to manage BIM data in different external or integrated management platforms. For this research, the Open Database Connectivity (ODBC) structure has been investigated as it allows to test both possibilities that have been recalled. Monitoring the effective transition of the information contained in a BIM model is an essential phase as it has an impact on subsequent actions. Obtaining data in a usable format is part of the standard identification of the guidelines. The second step is therefore represented by the identification of the platform that is necessary to use to reach the specific purpose of the BIM model. In this study, MS Access has been used for the visualization and updating of data, while an Integrated Workplace Management System (IWMS) has been identified for the integration with other management information, according to different model uses.

The ODBC export activity of the database can be done, as know from the operational guides of Autodesk Revit, through the Revit DB-Link plug-in with the direct connection with MS Access or by the general ODBC connection format. As the purpose is the employment by different actors, the definition of each Autodesk Revit type of parameter should be done according to the effective daily operational usage of information. For this reason, two other aspects become essential: the bidirectional data update and the possibility to create a custom mask, that enables information consultation in a simplified way. The first one is

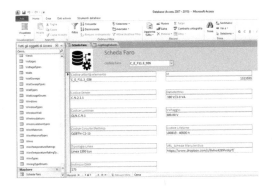

Figure 5. The consulting mask in MS access for data integration (Barbero 2016).

ensured by the Import/Export tool of the ODBC format that allows to bring in the BIM model all the implementation that has been done by users directly in MS Access, overwriting the entire database. The second one is based on the flexibility of the data content and structure in the management activity. The requirements to ensure data usability by different kind of users and their ability to visualize and modify this information constitute an important issue for FM environment. For these reasons, specific MS Access queries have been created to connect, between them, different data spread over separated tables.

The ODBC exchange format respects the table structure of a BIM model database, and every single table could be imported and exported massively through the direct integration with MS Excel. User-friendly masks have been designed to allow data consultation. Figure 5 shows as an example, the consultation interface set for lighting devices. It has been created with specific labels that allow the selection of individual information of an element by its own identification code. This field acts as input for the display of the other object's parameters since it's a primary key. All this information is editable directly from the MS Access query thanks to the insertion of the update data in the relative fields. Managing the mask structure, the user could identify and visualize specific fields that will be visible and used to show the desired information from the data content. In this way, data visualization and its updating could be done with some specific operation without the interaction with the table structure of the database. Initially, these updates will be only saved in the ODBC connection format without real-time effects on the BIM model. For its synchronization, it will be necessary to follow the import procedure as previously mentioned. During this input activity, the data owner could check what information has been updated, preventing data loss, monitoring its change during different O&M activities.

On the other side, an ODBC Database integration can be achieved with IWMS systems, an advanced technology designed to manage more effectively the core functional areas within an enterprise, including FM management, overcoming the simple data visualization. For this purpose, the BIM building registry must be transferred/synchronized according to the final database structure. One of the main issues is represented by the object – level association strictly connected both to 2D visualization performance and maintenance activities. Furthermore, this kind of platform is essential to investigate the correct exportation of the Autodesk Revit parameter type and their mapping in the IWMS structure.

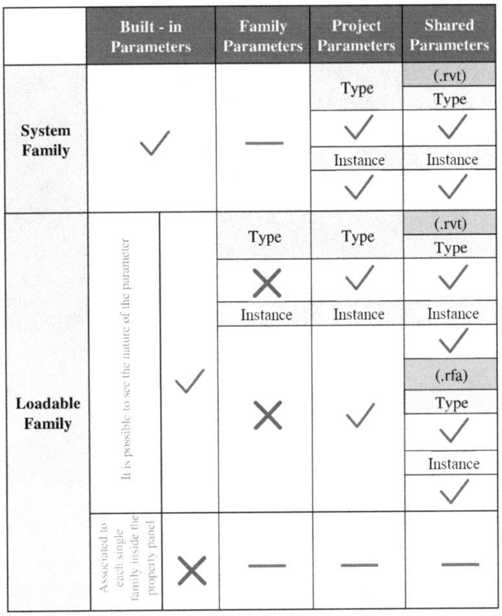

In this context, the knowledge and definition of denomination rules for parameters become essential to avoid special characters that can generate possible export errors. For this research, a series of tests have been done to identify the correct interoperability of different type of attributes: i) Built-in parameters that already exists in a Autodesk Revit file; ii) Family parameters which are created and related to a family (.rfa); iii) Project parameters which are generated directly inside a BIM model file (.rvt); iv) Shared parameters that belong to an independent file and can be shared among different models and families.

In addition, another factor considered is related to the "Type" or "Instance" nature of the parameter. This aspect affects both the table in which a parameter is located in the BIM database and the corresponding table in the IWMS platform. The achieved results are visible in Figure 6. For example, to ensure a correct transfer of shared parameters they need to be created in the .rvt file as a project parameter even if they are contained within a family (.rfa).

Figure 6. ODBC exchange tests.

2.3 *FM system over VAR*

The last case concerns the connection with tools able to communicate more directly and effectively through VAR technologies (Swanström Wyke et al. 2019). This

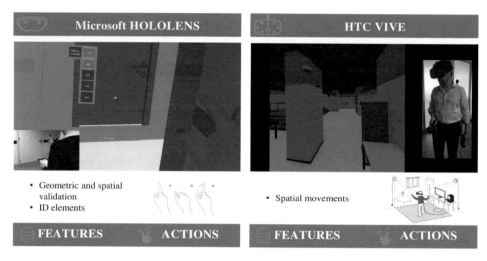

Figure 7. Virtual data usage.

approach aims at highest usability of the graphical representation for visualization purposes but finds its maximum potential in the dynamic interrogation of objects. Currently, the interoperability process is not able to transfer the associated database, therefore a subsequent programming activity is required. For this reason, at the moment, only the requirements related to the graphic component of objects can be identified. The tested model uses are focused, on the one hand, on verifying the correspondence between the digital representation and reality and, on the other hand, on the creation of discovery virtual tours. In the first case, the geometric model check is carried out by simultaneously displaying the real configuration through the employment of Mixed Reality (MR) applications.

Specific instrument such as Microsoft HoloLens allows to overlap the virtual model into real one according to a real scale environment (Viale 2019), as visible on the left side of Figure 7. To achieve this use, great attention must be paid to the multidisciplinary management of the model and the reliability associated with the elements.

The second application conceives the model as a visual cognitive resource that can be explored through Virtual Reality (VR) tools such as HTC Vive viewer. The issue is to set up a navigation mode functional for maintenance and training. For this purpose, it is necessary to pay more attention compare to an As-is model in the realization of the geometric component not only increasing of Level of Detail, but also in terms of graphic performance. The more the model is likely to be realistic, also in terms of appearance, colours and materials, the more usable the customized exploration will be. As shown in the right side of Figure 7, the setting of the visibility of objects is used, for example, to make accessible systems and components that are not visible in reality, such as the Heating, Ventilation and Air Conditioning (HVAC) conduit behind ceilings, and that can therefore be consulted. Furthermore, the representation of the furnishing elements, often overlooked in the digital restitution of buildings, becomes a fundamental element for the management of interior design configurations according to the event scenarios.

3 RESULTS AND CONCLUSION

This paper aims to evaluate the challenge for developing BIM guidelines able to support the definition of a building registry in line with technical due diligence. Different analyses have been done to identify the main issues that characterize the workflow to preserve and to manage geometric and alphanumeric contents. This investigation process has facilitated the fulfillment of the model uses defining the specific requirements of each by the analysis of the protocol and its activities. As shown in the cases discussed above, a progressive increasing of the BIM system complexity is achieved, starting from the As-is model definition to Database integration and virtual experiences setting. The synoptic overview above shows the current strengths and weaknesses characteristic of each purpose and the related modelling efforts to achieve a correct data extrapolation.

The resulting BIM guidelines are tailored to the individual project and the owner's purposes, overcoming the actual lack of standardization among BIM procurement documents. In this way it is possible to customize them, answering to the increasing complexity of BIM systems, analyzing BIM uses at the beginning of the building lifecycle.

The proposed methodology can contribute to the continuous research definition and refinement of the second level of maturity of BIM aimed at a collaborative and interoperable use of data.

Model Uses	Strenghtness	Weaknesses
As-is model	Digital building registry; Data management optimization; Discipline model coordination; KPIs control;	High modelling effort; Employee BIM training; Tailored rulesets; Resistance of change;
Database Integration	Efficient data update; Data source integration; Bidirectionality data sharing; User-friendly interface;	Data exchange issues; Open standard limitation; Lack for employee expertise; Daily database updating;
FM system over VAR	Virtual tour; Virtual/Real checking; Effective data visualization; FM process optimization;	Data loss Not fully automation data update; Increase of model accuracy; Lack of rules for enabling FM with VAR;

Figure 8. Strengths and weaknesses matrix.

4 AUTHORSHIP

The authors have agreed on the contents, the methodological approach and the final considerations presented in this research. Specifically, Matteo Del Giudice introduced the method in paragraph 2, Francesca Maria Ugliotti described the model uses in Paragraph 2.1, Andrea Barbero the Paragraph 2.2 and Anna Osello the Paragraph 2.3.

REFERENCES

Ashworth, S., Tucker, M. & Druhmann, C. 2016. The Role of FM in Preparing a BIM Strategy and Employer's Information Requirements (EIR) to Align with Client Asset Management Strategy. *15th EuroFM Research Symposium.*

Barbero, A. 2016. BIM 4D – pianificazione e gestione della manutenzione: il caso studio dello Juventus Stadium. *Master degree thesis*, Politecnico di Torino.

Barbero, A., Del Giudice, M. & Manzone, F. 2018. BIM model methods for suppliers in the building process. eWork and eBusiness in Architecture, *Engineering and Construction*, proceedings of ECPPM 2018: 291–296.

Barbero, A., Ugliotti, F.M. & Del Giudice, M. 2019. BIM-based collaborative process for Facility Management/Impostazione di un processo collaborativo BIM per il *Facility Management*. Dienne Volume 4: 6–14.

BIM Forum 2019. 2019 LOD Specification.

BSi 2015. BS 8536-1:2015 Briefing for design and construction. Code of practice for facilities management (Buildings infrastructure). Retrieved January 6, 2016, from www.shop.bsigroup.com.

ISO 2019. 19650-1:2019, Organization and digitization of information about buildings and civil engineering works, including building information modelling (BIM) – Information management using building information modelling.

Kensek, K. 2015. BIM guidelines inform facilities management databases: A Case Study over Time. *Building*: 899–916.

Kreider, R.G. & Messner, J.I. 2013. The Uses of BIM Classifying and Selecting BIM Uses Version 0.9 Penn State: Computer Integrated Construction.

Osello, A. & Ugliotti, F.M. (edited by) 2017. BIM: verso il catasto del futuro. Conoscere, digitalizzare, condividere. Il caso studio della Città di Torino. Roma: Gangemi Editore spa.

PAS 1192 – 2: 2013. Specification for information for the capital/delivery phase of construction projects using building information modelling.

Swanström Wyke, S. C. & Svidt, K. 2018. Virtual Reality based Facilities Management planning. *CIB Proceeding*, 36th CIB W78 2019 Conference: 965–973.

Viale, R. 2019. BIM and Mixed Reality for Facility Management. Case study: Allianz Stadium. *Master degree thesis*, Politecnico di Torino.

UNI 2017. UNI 11337-4:2017. Edilizia e opere di ingegneria civile – Gestione digitale dei processi informativi delle costruzioni – Parte 4: Evoluzione e sviluppo informativo di modelli, elaborati e oggetti.

UNI 2017. UNI 11337-6:2017. Edilizia e opere di ingegneria civile – Gestione digitale dei processi informativi delle costruzioni – Parte 6: Linea guida per la redazione del capitolato informativo.

Building permit process modeling

J. Fauth
Bauhaus University, Weimar, Germany

ABSTRACT: To implement BIM models successfully in an established procedural environment it is necessary to understand the current state of processes and their influences. Significant decisions are necessary until building permitability is determined. Automated model checking can only help with a certain number of regulations, which have quantitative content with measurable parameters. Regulations containing qualitative content still require manual reviewing by the authority. Detailed processes regarding decision making are currently not investigated or published. Building permit processes differ even within the authorities of a state or a country and have been mostly developed traditionally. A decision model is proposed as a decision-supporting instrument which shows the authorities' reviewer possibilities and alternatives of actions during a building permit review. This procedure aims at more transparency and objectivity. The proposed approach provides a starting point for implementing and using BIM. It also serves as a basis for further research in an international context of standardization of building permitting.

1 INTRODUCTION

Building Information Modeling (BIM) is increasingly implemented in the Architecture Engineering and Construction sector. Some pioneer countries like Singapore adopted BIM already in their building permit review (Borrmann et al. 2015). It can be seen as an exemption. A lot of building authorities are still working paper based. At the same time, the building permit is decisive for the realization of a construction project. Obtaining a building permit is a crucial aspect of the construction process. (Ponnewitz 2019, Plazza et al. 2019). It needs to be examined why it is so complicated to use BIM models for building permit review and how to support the authorities.

Current research approaches focus on informatics such as overall automation solutions using open BIM. Therefore, model checkers are an innovative tool to check different regulations automatically (Tulke 2015). However, the trend towards performance-based building regulations is in contrast to this method (Pedro et al. 2011). Performance-based building regulations aim at individual and qualitative contents and decisions (Fiedler 2015, Schleich 2018, Meacham 2010, Holte Consulting 2014). In contrary to quantitative, clearly measurable contents, qualitative contents cannot be checked automatically by a model checker (Nawari 2018, Nawari & Alsaffar 2015, Ponnewitz & Bargstädt 2019). Model checkers focus on the examination of content-based regulations and thus also have limitations regarding formal issues. This aspect is also supported by the current legal situation (e. g. in Germany), which does not accept decisions made by third parties, including machines like computational applications (Grüner 2016). The situation especially applies for individual decisions or deviations.

For a prompt full automation in administrative procedures the legal basis as well as the social acceptance is missing. By a step-by-step integration of automation, a still missing confidence in the technology can be developed. Therefore, it is especially important to act with transparency and traceability (Etscheid 2018).

A further challenge is shown in the form of a great diversity in the building permit authorities and their procedures. Results from conducted interviews show building authorities have concerns and scepsis about digitalization and BIM. In addition, there is the problem that every building project with every plot of land has a unique character. Nevertheless, building authorities should profit from BIM models in future. Therefore, a general basis and understanding from a procedural point of view is necessary before the integration of digital methods like BIM is possible. The proposed research focusses on the current processes within building permit authorities.

The organizational and procedural perspective is not yet considered in previous research as detailed as it is needed for decision-making of building permitability. The individual process steps in the building permit review are only marginally prescribed by law. This is one reason for the frequent lack of transparency. A further reason for the lack of transparency in the process is the high number of individual factors which influence decisions. The lack of transparency impacts reviewer in the authorities as well as all other involved stakeholders. This complicates especially the situation of the building applicants, as they have to deal with the circumstances of each individual authority again and again.

This research aims to develop a general decision model that shows the processes within building permit authorities to assess the building permitability. The model represents the actual state of the authorities. It serves as a basis for a step-by-step implementation and gives assistance using a BIM model in the building permit assessment by the authorities. The objective is to optimize the building permit review process and to increase transparency in the building permit process to the advantage of all parties involved.

2 METHODS

2.1 Building permitting seen as a decision

The determination of the building permit ability is based on decisions. Especially in case of non-by-right building applications in form of modifications or variances, a plan reviewer has to consider the individual case before he or she is able to make a decision. A by-right building application means a building project which fits with all regulations perfectly. There is no leeway for any interpretation or question regarding the approval. All other building applications can be defined as non-by-right applications. Among other deviations and exceptions, this also affects leeway in the law.

A decision can be projected onto the basic elements of a decision model to identify which factors are essential for a decision. Figure 1 shows that objectives, action alternatives (in form of processes) and environmental factors are required to make a decision. With the knowledge of these aspects, a decision support for plan reviewer can be provided. A decision model does not dictate the decision, but it makes the decision maker aware that there is a decision to be made. This strategy excludes an implicit or unconscious decision. An unwanted routine in the process is thereby decimated. Thus, a certain objectivity is created (Laux et al. 2012).

2.2 Empirical study

Because of the gap of detailed process description of building permitting and to satisfy the detailed procedural approach, own data was collected. In order to determine the exact current status, the empirical study was conducted in form of qualitative expert interviews. This procedure is used to formulate a theory (Gläser & Laudel 2010). In this study, qualitative expert interviews were necessary because of the high degree of individuality and the need for explanation in various areas. For example, if a process was not described in sufficient detail, direct enquiries could be made. In addition, the topic has a high sensitivity character especially in the authorities. This means that plan reviewers are inhibited from giving detailed answers. In a personal conversation, the interviewer can still obtain satisfactory answers. With a quantitative survey the achieved results could not have been determined. Experts included authorities, planners and project developers as well as other experts such as lawyers. They were chosen because of their deep and detailed understanding and expertise of the as-is procedure in the authorities. Although experts in different countries were interviewed during the data collection, the following results accelerate the investigations in Germany. In total, 100 experts were interviewed for the study. Of these, 69 interviews took place in Germany. There is no publication yet which can be compared which the collected qualitative data material.

The valuable results of the empirical study show processes, action alternatives and influencing parameters which are necessary to comprehend how a decision is made. The plan reviewers' action to decision-making was investigated by means of queried examples concerning deviations from building law. Findings are visualized as individual processes based on each authority. They are illustrated in a uniform form using Business Process Modeling and Notation language. In summary, it was found that building permit ability shows a high degree of complexity, with insufficient structuring and standardization.

2.3 Analysis of building permit processes

Individual processes could be identified from the interpretation of the results of the empirical study. The analysis of these individual processes aims at the formalization of a general process. Therefore, individual sub-processes from the individual processes have been identified, ordered and generalized. The result is a general process which includes all circumstances possible during a building permit review. Hence, it is necessary to identify action alternatives. These give the plan reviewer a choice how to decide. Furthermore, it shows the plan reviewer that there is a selection of measures to be considered at all. In this context, it should be noted that one action alternative can always be a failure of the building permit. An example can be seen in the area of the participation of public authorities. Here the reviewer has the choice between the specific authorities or other experts he or she involves.

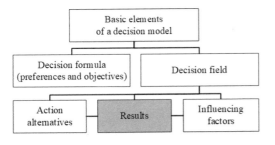

Figure 1. Basic elements of a decision model (according to Laux et al. 2012).

3 BUILDING PERMIT PROCESS MODELING

3.1 Building permitabilty system

A general decision model was developed. The model consists of four subsystems: Product system, objective system, action carrier system and action system. These subsystems are necessary for the determination of the building permitability as a decision. The action system is described in more detail in this paper.

3.2 Action system

The action system contains all processes of building permitting within an authority. It is presented in the form of a process organization. For the sake of clarity, the processes are transferred to a hierarchy of levels which bases on the research by previous research of the author (Ponnewitz & Bargstädt 2019). Therefore, processes are divided into sub-processes.

The action system considers both formal and material information, since both areas are decisive for building permit ability. Formal information concerns the form of the procedure. This means, for example, that certain process steps must be followed. Material information deals with the content of the construction project. They complement each other. Figure 2 shows which processes are assigned to the formal area and which processes are considered materially. The material information is reviewed in the process content review. Formal review (including completeness check), Assignment (which reviewer will get the case), participation (of public interest organization and equals as well as municipality) and finally the issuing of the letter approval (with reasoning the positive or negative decision).

With the material determination it comes to a conformity review. Conformity review means the examination if the submitted content of the building project meets the expectations of the law. Here the model is divided into each individual paragraph to be examined.

The arrangement after the existing paragraphs is to be regarded as user-friendly, because it facilitates their used expirations for the reviewers. These are already known to the reviewers. Each paragraph is treated equally. It does matter if the content of the paragraph is based qualitatively or quantitatively. Former identified action alternatives are directly integrated into the processes.

3.3 Examples

To get a better understanding, examples of subprocesses are provided. Figure 3 shows an example of a formal process. The example shows the review for completeness of the submitted building application documents. The planning documents including the BIM model, the statistical data entry form, the technical evidences and the forms are requested. In case of incompleteness, the missing information will be submitted in a further process. Once this has been done, a further process is continued.

Figure 4 shows a subprocess as part of the major process Assignment. The assignment of building applications to a plan reviewer can be either authorized or automated. Authorized means that a supervisor or an employee upstream in the process assigns the building application documents to one or more plan reviewers. Automated here means that the construction application documents are passed through an internal system.

During a building permit review, various experts are involved and is called as participation, as can be seen in Figure 5. These are, on the one hand, the municipality to which the building project is planned and, on the other hand, public interest organizations and equals. For the participation of public interest organizations and equals, a selection must be made. After that, all received comments are collected and evaluated. If necessary, the contents are transferred in the letter of approval.

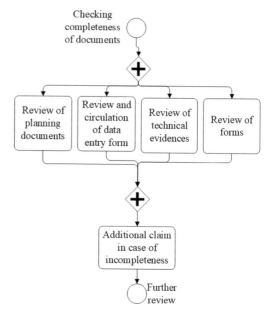

Figure 3. Example of a formal process – Checking completeness of documents submitted.

Figure 2. Major processes of a building permit review (according to Ponnewitz & Bargstädt 2019).

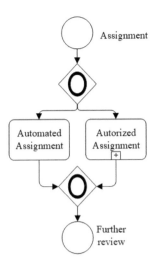

Figure 4. Example of a formal subprocess – Assignment.

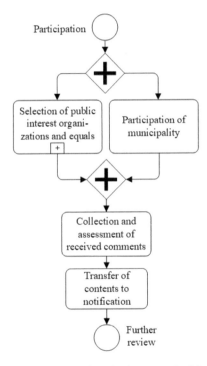

Figure 5. Example of a formal subprocess – Participation.

An example with material content is shown in Figure 6. The example illustrates only an excerpt of a conformity review. Each conformity review begins with a review of the respective paragraph. After that, it is checked whether there is a request for deviation. If there is no request for deviation, the question follows whether a deviation is still present. If so, then the plan reviewer must ask himself how marginally this deviation affects the special situation and the respective objective of the law and whether the plan reviewer can tolerate the deviation. As a result, the plan reviewer can make a direct decision on the building permitability, offer a hearing with the building applicant or call a meeting with colleagues or supervisors. In this example, a meeting is held before a decision is made on building permitability. Subsequently, the reasons are recorded, regardless of whether the decision was positive or negative.

Issuing a letter of approval, shown in Figure 7, provides two options: a positive or a negative permit. The positive permit is followed by the formulation of requirements. With the negative permit a justification of the decision is required. Eventually, in both cases the letter of approval is issued before the building permit assessment is completed.

3.4 Manual BIM-supported review

This approach is a manual, but nevertheless BIM-supported review. To use a BIM model procedural to support a decision by the reviewer the following manual BIM processes for conformity review are defined:

(1) Filtering of relevant and needed objects regarding every certain paragraph in the BIM model
(2) Using tools and functions the software application provides. This step is named possibilities for determination.

The possibilities for determination are summarized and divided in two sections:

a. Selection (of objects to show properties of objects, e. g. fire resistance classes) and
b. Calculation (using tool functions in the software like measuring)

(3) Alignment of calculations and other information or visualizations provided by the BIM model.
(4) Decision and documentation

Related to the example of distance areas the following steps need to pass with the support of a BIM model:

- Filtering the objects plot of land (IfcSite), building (IfcBuilding) and exterior walls (IfcWall) in the BIM model. Thereby the situation in the BIM model is already visualized.
- The distance between property boundary and outer dimension of the building need to be calculated using the measuring tool in the software.
- Comparing the processed calculation with given information by the applicant and law content.

4 DISCUSSION

A process model could be derived from a developed decision model, which includes the processes for the assessment of building permitability. The decision model integrates not only the current processes of a review, but also action alternatives. These alternatives are relevant for both the reviewer in the building permit authorities and the applicants. Architects, engineers

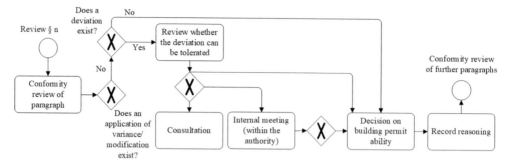

Figure 6. Excerpt of a conformity review of a paragraph.

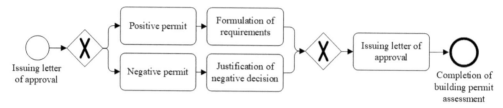

Figure 7. Example of a formal subprocess – Issuing letter of approval.

and building owners can use the knowledge of the action alternatives as an instrument of argumentation of their modification explanations.

The process/decision model results in a standardization of the procedure, while flexibility with regard to the individual building authority organization is ensured. This strengthens the certainty of the decision-making process and contributes transparency in the process. Based on the decision model, a BIM-oriented, manual building permit check can be supported. Thus, qualitative content can also be checked with the help of BIM. The results serve as an interim solution between the paper- or pdf-based building permit review nowadays and a full automated model checking in future. The model can be used immediately and does not contradict any law.

The results provide a starting point for the digital transformation of building permitting. The developed process model offers basic knowledge especially for researchers of their countries to find out their building permit procedures. With the awareness how the building permit procedures in the authorities in cities and municipalities work, an approach of further ambitions of BIM implementation is given. Make a decision and document reasoning.

5 SUMMARY AND CONCLUSION

BIM usage in building authorities is lacking. The reasons for the lack of implementation of BIM in the building permit process so far are the difficulty of automated checking of performance-based regulations, an absence of conviction on the part of authority employees and the legal basis. A further aspect is the complexity associated with the building permit review, where structuring is missing.

The paper presents an action system based on a decision model approach which support the manual building permit review in a BIM-supported way.

With the proposed approach, detailed processes are shown, which are relevant for a decision in building permit determination but have not been found in the literature so far. The research offers a novel approach for the reviewers' support.

The approach considers the procedural perspective, decision-relevant processes and action alternatives. It contributes the BIM standardization and implementation in the field of building permitting research.

Model checkers are not able to check performance-based regulations yet. Moreover, an automated decision cannot be legally applied in building permit authorities currently. Model checkers may not be used by law, so a rapid alternative or additional solution can help to optimize the building permit process and make it more transparent.

For this reason, other starting points for BIM-based building permitting need to be considered, for example, the translation of legal texts and their automated verification. Precisely since the trend is towards performance-based regulations and a large number of deviations occur, a more intensive consideration focuses the process-based connection with manual BIM checking. Even if full automation cannot (yet) be applied, this does not mean that a BIM model cannot be used in building permitting in present.

Some building authorities are skeptical or have fears about digitalization and BIM use. In order to gradually introduce the building authorities to BIM models, it is important to offer an independent solution that is both directly used and legally compliant.

The data from a conducted empirical study concern a certain scope mainly in Germany. The data collection could not include all building authorities and is limited on this point. In a follow-up study, the model could serve as a basis for a quantitative survey in various countries. This could prove the established theory more extensively. Related to the informatics point of view, it is still research for BIM implementation in case of qualitative aspects missing and need to be considered in future research as a complementing project.

REFERENCES

Borrmann, A. et al. 2015. Einführung. In: Borrmann et al. (eds.), *Building Information Modeling*: 1–21. Berlin: Springer Vieweg.

Etscheid, J. 2018: Automatisierungspotenziale in der Verwaltung. In: Mohabbat Kar, R.; Thapa, B.; Parycek, P. (eds): *(Un)berechenbar? Algorithmen und Automatisierung in Staat und Gesellschaft*. Berlin: Fraunhofer Verlag: 126–158.

Fiedler, J. N. 2015. *Modernisierungsszenarien des Baubewilligungsverfahrens unter Berücksichtigung neuer technologischer Hilfsmittel*. Vienna: Technische Universität Wien, dissertation.

Grüner, J.: BIM im Baugenehmigungsverfahren. In: Eschenbruch, K. & Leupertz, S. (eds.): *BIM und Recht*. Köln: Werner Verlag: 216–233.

Holte Consulting 2014: *Status survey of solutions and issues relevant to the development of Byggnett*. Oslo. https://dibk.no/globalassets/byggnett/byggnett_rapporter/byggnett-status-survey.pdf [Accessed 15.09.2017]

Laux, H. et al. 2012: *Entscheidungstheorie*. Berlin Heidelberg: Springer Verlag.

Meacham, B. J. 2010. *Performance-Based Building Regulatory Systems – Principles and Experiences*. Interjurisdiction Regulatory Collaboration Committee (IRCC).

Nawari, N. O. 2018. *Building Information Modeling – Automated code checking and compliance processes*. Boca Raton: CRC Press.

Nawari, O. N. & Alsaffar, A. 2015. Understanding Computable Building Codes. In: *Civil Engineering and Architecture*, 3 (6): 163–171. DOI: doi: 10.13189/cea.2015.030601

Pedro, J. B. et al. 2011: Comparison of building permit procedures in European Union countries. In: *COBRA 2011, RICS International Research Construction and Property Conference, Salford, 12–13 September 2011*: 415–436.

Plazza, D. et al. 2019. BIM for public authorities: Basic research for the standardized implementation of BIM in the building permit process. In: *IOP Conference Series: Earth and Environmental Science*. 11–14 September 2019, Graz. Vol. 323: article nr. 012102. IOP Publishing. DOI: 10.1088/1755-1315/323/1/012102

Ponnewitz, J. 2019. *Die BIM-basierte Baugenehmigungsprüfung – eine Grundlagenbetrachtung*. In: Haghsheno et al. (eds), *30. BBB-Assistententreffen 2019, Karlsruhe, 10–12 July 2019*: 234–246. Karlsruhe: KIT Scientific Publishing.

Ponnewitz, J. & Bargstädt, H.-J. 2019. The building permit – how to standardize traditionally established processes. In: *The Evolving Metropolis. Proceedings of the 20th Congress of International Association for Bridge and Structural Engineering (IABSE)*, New York City, 4–6 September 2019, Vol. 114: 1561–1565.

Schleich, M. 2018. *Kosteneinsparpotenziale einer effizienteren Landesbauordnung – Ökonomische Analyse der Bauordnung für das Land Nordrhein-Westfalen im Vergleich*. Wiesbaden: Springer Vieweg.

Tulke, J. 2015. BIM zur Unterstützung der ingenieurtechnischen Planung. In: Borrmann et al. (eds.): *Building Information Modeling*: 271–282. Berlin: Springer Vieweg.

Key aspects of data integration and management

Towards conceptual interoperability of BIM applications: Transaction management versus data exchange

V. Semenov
Ivannikov Institute for System Programming of the Russian Academy of Sciences, Moscow, Russia
Moscow Institute of Physics and Technology, Dolgoprudny, Russia
Higher School of Economics, Moscow, Russia

S. Arishin
Ivannikov Institute for System Programming of the Russian Academy of Sciences, Moscow, Russia

G. Semenov
Information Technologies, Mechanics and Optics University, Saint-Petersburg, Russia

ABSTRACT: Building information modelling (BIM) originates as an innovative methodological approach to enhance products, processes and to reach greater efficiencies in the architecture, engineering, construction and facility management industry. Although BIM tools have been successfully adopted, their integrated use encounters numerous software interoperability issues. In the paper, these issues are examined by addressing to the conceptual multi-level interoperability model. The presented data exchange scenarios explain why the expectations from the use of open BIM standards and Industry Foundation Classes (IFC) have not yet fully met. As a result, the requirements for an interface to access IFC-driven data are formulated. The interface provides project, revision and transaction management methods that allow the synchronization of data rather than the direct exchange of IFC files. The correct implementation and use of the interface by software vendors could predetermine a commonly accepted way from technically interoperable applications to substantively interoperable IFC-compliant integrated systems.

1 INTRODUCTION

Building information modelling (BIM) originates as an innovative methodological approach to enhance products, processes and to reach greater efficiencies in the architecture, engineering, construction and facility management industry (AEC/FM) through digitalization, extended collaboration between project stakeholders and integrated management of project information throughout the entire building lifecycle from inception, design, evaluation, construction to operation and demolition.

Digital representation of building information allows passing data across multi-disciplinary software applications and bringing the users new opportunities for working together. Since the end of 2000, BIM has been successfully adopted in the AEC/FM industry (Celnik et al. 2014). Initially being applied for 3D design and engineering, nowadays BIM is effectively used for project planning and scheduling, construction management, cost analysis, facility operation and asset management (Leite & Akinci 2012; Liu & Akinci 2009).

BIM implies the circulation of data among users without any data loss, misinterpretation, re-entry, and correction. However, these expectations run into obstacles due to software interoperability issues (Muller et al. 2017). Intensive investigations in the domain of BIM interoperability have been done. Grilo & Jardim (Grilo & Jardim 2010) addressed the need for enhancing interoperability in BIM to achieve higher value levels in the AEC/FM sector. Karan & Irizarry (Karan & Irizarry 2015) presented an extending interoperability framework for preconstruction operations using geospatial analyses and semantic web services. While the number of BIM applications is permanently increasing, the interoperability becomes one of the most crucial factors affecting the adoption of BIM technologies.

For today, Industry Foundation Classes standard (ISO 16739) seems to be the best solution for achieving the real interoperability among BIM tools developed by different software vendors. Indeed, IFC appeared in 1994 as an initiative by International Alliance for Interoperability (recently, buildingSMART) to define common multi-disciplinary information schema that, in combination with the STEP family standards (ISO 10303), could be adopted by wide industry and academy communities and be used as a consensus for achieving BIM interoperability. Remind that STEP is organized as a series of description and implementation parts which specify EXPRESS data modeling language, clear text encoding format (so-called SPF or STEP physical file), XML representations of EXPRESS schema and data, standard data access interface (STEP SDAI, ISO 10303-22) and its bindings in C++, C, Java languages. Obeying to the STEP regulations, IFC standard gives rise to constructive ways to exchange and share BIM data. To a certain degree, IFC can be considered as an adaptation and extension of the STEP parts for the AEC/FM industry.

Despite many years of efforts and shipping several IFC releases, the expectations from its use have not yet been fully met. Complementary standards such as Information Delivery Manual (ISO 29481), International Framework for Dictionaries (ISO 12006), Model View Definitions (MVD), BIM Collaboration Format (BCF), BIM Service interface exchange (NIBS BIMSie) did not improve the situation significantly.

Many researchers and practitioners report numerous occurrences of data loss, data misinterpretations, and roundtrip failures (non-recovery of the original data revision after successive export and import operations) due to imperfect IFC file exchange facilities. At the same time, IFC servers are very rare and long for the expenses to integrate BIM tools. In addition, they do not provide ACID (*Atomicity, Consistency, Isolation* and *Durability*) and BASE (*Basically Available, Soft state, Eventually consistent*) guarantees that are commonly recommended for any information system and database management system (Semenov 2017). Indeed, without ensuring the data consistency, and the transaction isolation and serialization, the widespread use of BIM servers seems doubtful, since the results of collaboration sessions can be ambiguous and unpredictable. Indeed, IFC data must be kept consistent regardless of whether one or more applications commit transactions.

Thus, the practical use of IFC standard is currently rather limited, especially in those cases when it is required to guarantee the completeness and trustworthiness of exchanged and shared BIM data. The goal of the paper is to analyze the reasons for the current situation and to explain the existing misconceptions around IFC standard by addressing to the conceptual interoperability model (Tolk & Muguira 2003). This model identifies seven levels (more exactly, layers) of interoperability: zero, technical, syntactic, semantic, pragmatic, dynamic and conceptual ones.

In order to meet the expectations, IFC-compliance should be understood in a more precise sense and be implemented in such a way as to satisfy the conceptual interoperability requirements. In the paper we argue the BIM tools capable only to exchange IFC files cannot meet the requirements at the pragmatic, dynamic and conceptual levels. Moreover, semantic errors are often detected in existing IFC file exchange facilities. This remains a serious problem since visual reviews of the exchanged data by means of IFC-compliant applications cannot be satisfactory due to the imperfect implementations and high semantic complexity

Table 1. Semantic complexity growth of the IFC schema.

IFC SCHEMA (ADOPTION YEAR)	2.3.0.1 (2007)	4.0 (2013)	4.0.2.1 (2017)	4.1 (2018)
ENTITIES	653	766	776	801
TYPES	327	391	397	400
EXPLICIT	1471	1661	1686	1737
DERIVED	57	59	62	62
INVERSE	115	149	153	158
FUNCTIONS	38	42	47	47
ENTITY RULES	339	639	652	658
TYPE RULES	24	24	25	25
UNIQUE RULES	17	4	4	4
GLOBAL RULES	2	2	2	2

of the IFC information schema. Table 1 demonstrates significant evolution and complexity growth of the IFC schema since 2007 when its version 2.3.0.1 was adopted.

Semantic mismatch of the underlying data models employed in BIM tools is another problem (Muller et al. 2017; Trzeciak & Borrmann 2018). Usually, BIM tools are based on own proprietary data models distinct from the IFC schema. Internal data of a BIM tool is converted to an IFC schema-driven representation every time when its export operation is invoked. A similar reverse process is initiated when the import operation is activated. The necessary processes of bi-directional converting with all risks of data loss and misinterpretation give grounds to many interoperability anomalies caused by direct IFC file exchange.

To avoid these anomalies, IFC standard should be used as a median representation schema so that BIM tools can synchronize their own data. BIM tools should be able to respond to the external changes that have occurred with the median representation and bring their own data into line with it. And vice versa, if any changes have occurred with the internal data, BIM tools should be able to update the external median representation. Such synchronization operations are naturally expressed in terms of transactions with replicated heterogeneous data and lead to a paradigm shift for achieving BIM interoperability in practice.

In the presented paper, we prove that in order to be conceptually interoperable, BIM tools should share common IFC data, managing transactions, rather than exchanging the entire data. Moreover, this should be done under the transaction concurrency control provided by IFC servers with ACID/BASE guarantees.

The paper is structured as follows. In Section 2, typical IFC file exchange scenarios are presented with a focus on the interoperability issues. For this purpose, an interoperation diagram is introduced. The interoperability issues are investigated in Section 3 by addressing to the classical LCIM model. Section 4 provides explanatory scenarios how the issues can be resolved by using transaction management facilities. Finally, in Section 5 we summarize the requirements for BIM servers and, particularly, for data access interfaces that should be implemented by the servers and used by BIM tools. A particular attention is paid to principal distinctions from STEP SDAI and BIMSie interfaces. In Conclusions we outdraw the perspectives for further standardization of the interface and its adoption by software vendors to leverage great IFC's potential.

2 IFC DATA EXCHANGE AND ANOMALIES

In this section, we will look at the typical scenarios of direct exchange of IFC data between BIM software applications that are supposed to be IFC-compliant or certified to be IFC-compliant. Unfortunately, the official certification process remains quite limited. Basically, it confirms the fact that an application is capable to export and/or import IFC files as well as to process and visualize typical IFC data such as beams, columns, slabs, and walls properly. Nobody can guarantee that no data are lost and all data are processed correctly even if the software application has been certified. In other words, the certification process is "more of a test of the ability to exchange information via IFCs rather than the quality of the exchange" (Lipman et al. 2011).

In the scenarios presented, the functionality of the applications and their focus on specific AEC/FM domains are not essential for further consideration. It does not matter whether the applications belong to the same domain or different ones, and solve the same domain-specific problem or different ones. It is only significant that the applications can interact with each other by importing and exporting IFC files, committing transactions and synchronizing their own data.

2.1 Interoperation diagram

To present the scenarios, in which the interacting applications are involved, as well as arising anomalies we use a form of sequence diagrams called hereinafter the interoperation diagram Each diagram depicts the applications involved in one common scenario and the events occurring during the scenario realization. Being arranged in a timeorder sequence the events shows the application operations activated at some points in time as well as their actual parameters expressed in terms of the application states Typical operations are file export/import, synchronize to/from, and transaction start/commit.

An interoperation diagram shows, as parallel vertical lines (*lifelines*), different applications that interact with each other and, as balls, their states that undergo revisions resulting from the interaction and, as horizontal arrows, the operation calls in the order in which they happen. This allows the specification of simple runtime scenarios in a graphical manner.

Rectangles drawn on the top of lifelines represent interacting applications, ovals on top of lifelines — some data, i.e. IFC files stored on local hard disk or in remote repository. Leaving the lifeline name blank can mean anonymous applications or data. If the lifeline is just a data, it shows a sequence of its revisions marked by black and light balls. Each black ball indicates a new major revision on the main branch of the revision tree, each light ball – a minor revision on some side branch. If the lifeline is an application, it demonstrates both a sequence of data revisions and a sequence of operations producing and processing the revisions.

Horizontal arrows denote operation invocations with the names and actual parameters. The operations are assumed synchronous, however the standby intervals during which the applications stand idle waiting for relevant events are explicitly marked. The intervals are marked with hourglass and brick symbols, thus indicating the need to wait or interrupt/resume the application session. While the lifelines are drawn thin, solid segments indicate the effective intervals during which the applications produce own data revisions by running independently.

In the scenarios in Figures 1–3 we denote software applications with capital letters 'X', 'Y', 'Z'. The notation X_{op} means an operation op invoked by the application X. The application data is denoted by corresponding lower case letters so that the notation $x_{1.1}$ means a root revision of the data x and $x_{1.2}$, $x_{1.3}$ and so on — the next succeeding revisions. Thus formed branch (with two digits in the number) is called main branch. Minor branches (with four or more digits in the number) are produced by applications running concurrently rather than sequentially.

2.2 Data loss, misinterpretation, roundtrip and breakup anomalies

Let us identify the main types of interoperability issues or so-called anomalies, by considering a simple scenario in the Figure 1. For definiteness and clarity, we will consider that the application X is an architectural design system that employs constructive solid geometry modeling (CSG) to represent building elements. The application Y is assumed to be a graphic application oriented on boundary representations (BRep) of building elements. It is worth noting that the IFC schema allows alternative and simultaneous use of CSG and BRep models and therefore is the best solution for median integration of the applications X, Y.

Even if all CSG models were correctly exported by the application X to an IFC file, there would be high risks that the application Y will not be able to correctly import all of them into its internal representation. In fact, GSG to BRep transformations are often error-prone due to the computational accuracy (Grimsdale & Kaufman 2012), and therefore the application Y has the dilemma of either skipping some geometric models or not applying unsuccessful CSG operations and fixing approximate intermediate results. In the first case,

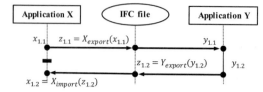

Figure 1. Typical IFC file exchange scenario and interoperability anomalies.

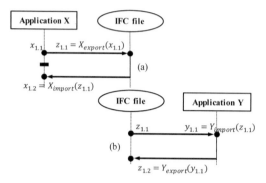

Figure 2. Roundtrip anomalies resulting from IFC export/import (a), and import/export (b) operations.

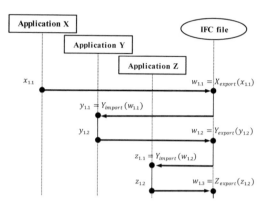

Figure 3. Downstream workflow using IFC file exchange.

an anomaly of data loss is identified, in the second – an anomaly of data misinterpretation or its incorrect processing.

If we try to bring the application data to a consistent state by exporting from the application Y and importing into the application X, the situation only gets worse. Firstly, some of the CSG models were lost or damaged. Secondly, having BRep tessellations, the original CSG models cannot be reliably restored. This means that in the best case, instead of the original models, polyhedral primitives will be obtained, thereby limiting the user in the possibilities of their further modification and parameterization by means of the application X.

Non-recovery of the original data revision after successive export and import operations is identified as a roundtrip anomaly. This phenomenon often occurs even with a single application if its own data model differs from the IFC schema. Data loss and misinterpretation are the main reasons of roundtrip anomalies resulting from successive IFC export/import (see Figure 2, a) and import/export (see Figure 2, b) operations. Sometimes, the divergence of data revisions can be detected visually if the application provides a graphical user interface. Sometimes, it can be done automatically using IFC file comparison utilities (Lee et al. 2011) or appropriate facilities of the application.

Finally, there is another, so-called session breakup anomaly also caused by a model mismatch. This anomaly is marked by a brick symbol, thus indicating a possible breakup in the data revisions and the need to interrupt/resume an application session. Notice that IFC import is usually understood and implemented as an operation of creating a new project or updating an existing project with fully replacing content. User-specific data accumulated during the application session and associated with the existing project is completely lost. Such situation is disappointing for users due to the need for repeated work which can require tremendous effort and resources.

Breakup anomaly can occur even in cases where other interoperability issues are fixed. Let us return to the scenario in Figure 1 and assume that all CSG and BRep model transformations are correct and reversible. At the same time, architectural design systems often use their own libraries of elements and materials, the definitions of which and references to which are not present in the IFC schema. It means that references to elements and materials, if established during the application session, will be irretrievably lost when importing IFC file.

The anomalies mentioned above look surprising but not fatal only if the IFC file transfer between applications happens once, i.e. downstream workflows as shown in a scenario in Figure 3. However, this way of organizing collaborative work is quite burdensome, since it completely eliminates concurrent work of users, and requires repeated work from the initial stages, even if local corrections are necessary.

These limitations are essential for effective adoption of BIM technologies aimed at minimizing construction project makespans, resources and costs. Therefore, we propose another approach based on IFC data synchronization and transaction management rather than direct IFC data exchange.

3 CONCEPTUAL INTEROPERABILITY MODEL

As mentioned, experience with IFC files has been disappointing and forced the proponents to acknowledge the difference between enabling applications to exchange data, so-called technical interoperability, and substantive interoperability – the desired outcome of exchanging meaningful data so that coherent interaction among applications takes place. IFC file

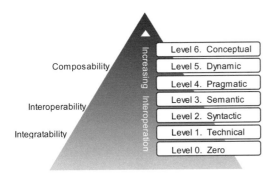

Figure 4. Levels of conceptual interoperability model (LCIM).

exchange does not meet the expectations due to the low levels of the interoperation maturity. Because the concepts of technical interoperability, operational interoperability, integratability and composibility are often confused with each other, the LCIM model will be considered as it was originally introduced and documented (Tolk & Muguira 2003).

This model identifies the following seven levels (or layers) of interoperability among participating systems:

Zero level that excludes interoperability as such and is applicable to stand-alone systems.

Technical level deals with communication infrastructure allowing systems to exchange carriers of information. For this purpose the underlying networks and protocols are unambiguously defined. This is the domain of integratability.

Syntactic level provides a structure to exchange information. On this level a common data format with predefined symbols within protocols is used. This level belongs to the domain of interoperability.

Semantic level implies a commonly shared meaning of the information units like objects, properties, requests, messages, parameters. This level avoids any misinterpretations of data and ensures the common understanding of terms.

Pragmatic level recognizes the context in which the information is exchanged. It is reached when the interoperating systems are aware of the methods and procedures that each system is employing. This level implies a common understanding of the use of the exchanged data.

Dynamic level recognizes various system states while participating systems interact with each other over time. The systems become able to comprehend the state changes and to take advantage of those changes. This level therefore ensures a common understanding of effects of the data exchange.

Conceptual level covers all assumptions and constraints to be taken into account when exchanging and processing the information. This is the highest level of interoperability that represents the harmonized data driven by formally specified models.

It is obvious that BIM applications capable only to import and export IFC files have low LCIM levels. The *technical level* is provided by operating systems under which the applications run, the *syntactic level* — by standard SPF format on which IFC files are based, the *semantic level* — by standard IFC schema whose terms are formally specified and documented. Unfortunately, BIM applications do not meet the conceptual interoperability requirements needed for effective building and deploying collaborative environments. This circumstance may explain the unfulfilled expectations of using the IFC standard.

The *pragmatic level* requirements can not be met due to the lost context in which IFC data is exchanged. It is impossible to recognize the original request and options by which a particular IFC file was exported. It is also not clear whether the file contains the entire IFC data or only partially extracted data obtained by means of object queries.

The same uncertainty arises when importing IFC files. It may seem strange but the operational semantics is not rigorously defined. Sometimes the import operation is implemented in such a way as to create a new project in the application, sometimes to add new data to an existing project, sometimes to update the project data in accordance with the changes that have occurred in the external IFC data.

The last mentioned aspect brings us closer to understanding the importance of the *dynamic level*. As known, popular BIM applications use own data models and proprietary file formats. To meet this level requirements BIM applications must be capable to bring own data in correspondence with the external changes and, reversely, to interpret own changes in terms of external data.

The role of the IFC standard is reduced to the definition of a common multi-disciplinary information schema within which the exchange of semantically equivalent data between BIM applications becomes possible. It would be desirable for the IFC schema to cover as many AEC/FM disciplines and aspects as possible, but it should be recognized that a certain part of data will always be application-specific. IFC schema cannot substitute other representations and the users should keep both IFC files and proprietary BIM files when working in different applications.

Therefore, the IFC schema should be used as a median data integration schema through which the applications can synchronize their own data each with others. That requires the applications be able to identify data changes and support the needed synchronization operations. Hereafter, we suggest that the operation *synchFrom* enables BIM application to respond to external changes that have occurred with the median representation and bring its own data into line with it. And vice versa, the operation *synchTo* enables BIM application to update the median representation if any changes have occurred with its own data. Finally, the operation *merge* enables to consolidate concurrent changes occurred with the median representation and present them as if they were

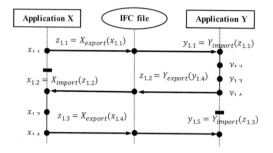
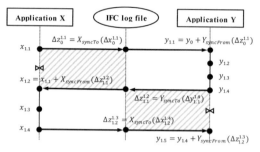

Figure 5. Collaborative scenario using IFC file exchange.

Figure 6. Collaborative scenario using IFC data synchronization and pessimistic concurrency control.

produced by one single application. All the introduced operations are expressed in terms of transaction logs.

Finally, the *conceptual level* requirements are also critically important to facilitate IFC-based interoperability in the AEC/FM industry. Embracing multiple disciplines such as 3D CAD, HVAC design, electrical design, formwork design, structural analysis, energy simulation, quantity take-off, cost estimation, production scheduling, steel and timber frame construction, facility management the IFC schema is quite complex from a semantic point of view (see Table 1). The total number of data types and semantic rules has exceeded a few thousands. The rules when applied to millions of data elements can induce billions of algebraic constraints that must be always satisfied. Otherwise, IFC data loses consistency and becomes useless for further processing. To meet the requirements of the conceptual level, BIM applications must control the consistency of IFC data when reading or updating it.

4 IFC DATA SYNCHRONIZATION SCENARIOS

The above-mentioned anomalies are overcome by managing transactions and synchronizing heterogeneous data governed by different BIM applications. For definiteness, consider that transactions may be either read or write and a new data revision appears whenever BIM application commits a successful write transaction. Such transactions may be triggered by user manipulations, import operations or in some other way. Read-only transactions do not produce revisions. Existing revisions can never be changed. Therefore, the full transaction history is represented as a data revision tree with the root corresponding to the initial dataset and leaves — to its terminal representations. The revisions are represented as complete datasets or as transaction logs that store only the changes committed by successfully completed transactions. In any case, we assume that a transaction log Δx_i^j between any two given revisions x_j, x_i can always be obtained or calculated.

Returning to the scenarios with direct IFC file exchange and considering a scenario in Figure 5, different anomalies can appear if the underlying data models of the applications X and Y differ from IFC schema.

The application X produces the first data revision $x_{1.1}$ and then exports it to IFC format as a file $z_{1.1}$. When importing the file, the application Y generates its own data revision $y_{1.1}$, which undergoes certain changes during the application work and then is exported back to IFC format as a file $z_{1.2}$. Similar actions are repeated in the application X. In addition to the possible data loss and misinterpretation, the roundtrip and breakup anomalies with the revisions $x_{1.1}, x_{1.2}$ and $y_{1.4}, y_{1.5}$ are also inevitable.

In order to ensure the continuity of work with the application data, synchronization operations should be used instead of IFC export and import. Figure 6 presents the same scenario realized using *synchTo* and *synchFrom* operations. Instead of IFC files, the applications X and Y exchange IFC transaction logs and use them to bring their own data in line with each other.

Initial revisions of the applications are synchronized relative to empty datasets and therefore the revision $y_{1.1}$ is the same that could be obtained by equivalent export and import operations. Subsequent changes fixed by the revisions $y_{1.2}, y_{1.3}, y_{1.4}$ are translated into terms of IFC schema and passed to the application as a transaction log $\Delta z_{1.1}^{1.2}$. The application converts the IFC transaction log into its own transaction and produces a new revision $x_{1.2}$ by committing it. It is important that, since the application transaction captures changes relative to the initial revision $x_{1.1}$ the continuity of work is not broken. Application-specific data that cannot be represented by IFC entities and transferred in IFC files remains in the revision $x_{1.2}$ without any additional rework. The same conclusions apply to the application Y, in which successive revisions $y_{1.4}, y_{1.5}$ are synchronized with the revisions $x_{1.2}, x_{1.4}$ of the application X.

As in the previous scenario in Figure 5, concurrent working remains limited due to the need to wait for the results of one application to continue working in another even if a pre-planned workflow allows concurrent access to fragmented data. If one user resumes work without synchronizing with the results of another, he will inevitably encounter an anomaly of consolidation, expressed in the loss of part of the results.

One of the possible approaches to avoid consolidation anomalies is a pessimistic concurrency control with locks on IFC data revisions. In this case, it is

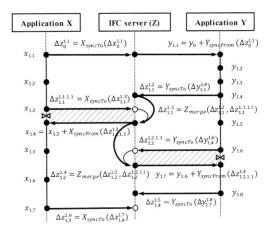

Figure 7. Effective collaboration scenario using IFC data synchronization and optimistic concurrency control.

possible to ensure ACID properties when proceeding with semantically complex IFC data (Semenov & Jones 2015). Unfortunately, locking entire IFC data revisions makes concurrent working impossible.

Another approach to avoid consolidation anomalies or, at least to mitigate their effects, is an optimistic concurrency control. Its implementation implies the availability of the *merge* operation applicable to concurrent changes. A collaboration scenario presented in Figure 7 extends the previous ones, allowing the applications running concurrently and producing new revisions in parallel. At the same time, if one application fixes a new IFC revision and a concurrent revision by another application is detected, the changes need to be merged and the resulting IFC revision to be stored. Optimistic control takes additional CPU resources and risks, but it is more suitable for collaborative environments while providing BASE guarantees (Semenov & Jones 2015).

In the presented scenario, the application Y fixes an external revision $z_{1.2}$ first and it gets to the main branch of the application Z (shown as a dark ball on its lifeline in the interoperation diagram). The application X fixes an external revision second, therefore it gets to a minor branch (shown as a light ball on the lifeline) and takes the number $z_{1.1.1.1}$. The invocated merge operation produces the resulting revision $z_{1.2}$. The application X synchronizes own data with it, while the application Y continues to work. When the application Y fixes an external revision, it gets to a minor branch and takes the number $z_{1.2.1.1}$. The merge operation is used to consolidate the concurrent changes and produce the resulting major revision $z_{1.4}$. The applications work simultaneously, while the needed downtime for synchronization and merging of results is less than in the scenario in Figure 6.

To support effective collaboration scenarios, various services are required to manage revisions, identify objects, execute queries, check consistency, control transaction concurrency, authenticate users and manage privileges. Since the services should be applied in a coordinated manner, their implementation and deployment is appropriate on centralized IFC servers, which allows BIM tools to access IFC data through standardized interfaces. Properly designed and implemented the interfaces allow IFC servers and client applications to interoperate at the conceptual level by avoiding the mentioned-above anomalies.

5 IFC DATA ACCESS INTERFACE

Unfortunately, STEP SDAI (ISO 10303-22) and BIM-Sie interfaces are quite limited for the purposes declared. Here we omit the detailed overview of the standards and emphasize the principal bottlenecks.

First of all, both standards do not support the concept of data revision, but it is the key to building lifecycle management. STEP SDAI excludes it at all. BIMSie replaces it with the concept of versioned objects that prevents the data integrity.

Second, both standards allow transactions, but do not clarify how ACID/BASE transaction guarantees are ensured, under which transaction concurrency control policies and which requirements to server implementations it becomes feasible.

Third, both standard interfaces do not provide commands necessary for synchronization operations. Their implementations would imply explicit manipulation of transaction logs in order to bring the application data to a state concordant with the external changes.

Therefore, a new standard interface for IFC-compliant applications should be developed by specifying the following groups of commands to access IFC-driven project databases:

– commands to get database specifications (e.g. supported schemas, concurrency policies, consistency techniques, etc.);
– commands to manage sessions and hold transactions (assign roles, grant privileges, error logging, start, commit, roll back, etc.);
– commands to manage projects and directories (e.g. create, move, rename, delete, etc.);
– commands to manage revisions (e.g. compare, merge, submit/receive IFC file, etc.);
– commands to get schema specifications (e.g. entities, types, subtyping relations, attributes, rules, etc.);
– commands to execute queries by identifiers, entities, predicates, routes, etc.;
– commands to manipulate objects (e.g. get explicit, derived, inverse attributes as well as to create, delete, modify objects);
– commands to check consistency (e.g. local, global and uniqueness rules, type domains, aggregate cardinalities, referential integrity).

The standard should also specify the requirements to IFC model servers, including revision and change management, ACID/BASE transaction guarantees, concurrency control policies, consistency checking.

6 CONCLUSIONS

Thus, the general requirements for the interface to access IFC-driven data are formulated. The interface provides project, revision and transaction management methods that allow the synchronization of data and its coordinated sharing by BIM tools rather than the direct exchange of IFC files.

The correct implementation and proper use of the interface must ensure the conceptual interoperability of IFC-compliant applications. Further its standardization and adoption by software vendors could improve the current situation significantly and predetermine a commonly accepted way of transition from standalone and technically interoperable applications to substantively interoperable systems and advanced collaborative environments in the AEC/FM industry.

REFERENCES

Celnik, O. & Lebègue, E. 2014. *Building Information Modeling (BIM) & digital model for architecture, building and construction*. Paris: Eyrolles.

Grilo, A. & Jardim-Goncalves, R. 2010. Value proposition on interoperability of BIM and collaborative working environments. *Automation in Construction*. Vol. 19: 522–530.

Grimsdale, Richard L. & Kaufman, Arie, 2012. *Advances in Computer Graphics Hardware V: Rendering, Ray Tracing and Visualization Systems*, Springer Science & Business Media.

ISO 10303. ISO 10303 - Automation systems and integration — Product data representation and exchange.

ISO 10303-22. ISO 10303-22 - Industrial automation systems and integration — Product data representation and exchange — Part 22: Implementation methods: Standard data access interface.

ISO 16739. ISO 16739-1:2018 - Industry Foundation Classes (IFC) for data sharing in the construction and facility management industries — Part 1: Data schema.

Karan, E. & Irizarry, J. 2015. Extending BIM Interoperability to Preconstruction Operations Using Geospatial Analyses and Semantic Web Services. *Automation in Construction*. Vol. 53: 1–12.

Lee, Ghang & Won, Jongsung & Ham, Sungil & Shin, Yuna. 2011. Metrics for Quantifying the Similarities and Differences between IFC Files. *Journal of Computing in Civil Engineering*. 25: 172–181.

Leite, F. & Akinci, B. 2012. Formalized Representation for Supporting Automated Identification of Critical Assets in Facilities during Emergencies Triggered by Failures in Building Systems. *Journal of Computing in Civil Engineering*. Vol. 26: 519–529.

Lipman, R. & Palmer, M. & Palacios, S. 2011. Assessment of conformance and interoperability testing methods used for construction industry product models. *Automation in Construction*, 20(4): 418–428.

Liu, X. & Akinci, B. 2009. Requirements and Evaluation of Standards for Integration of Sensor Data with Building Information Models. *International Workshop on Computing in Civil Engineering*: 95–104.

Muller M. & Garbers A. & Esmanioto F. & Huber N., Rocha Loures E. & Canciglieri Jr O. 2017. Data interoperability assessment though IFC for BIM in structural design – a five-year gap analysis. *Journal of Civil Engineering and Management*. Vol. 23: 943–954.

NIBS, buildingSMART alliance, BIM Service Interface Exchange (BIMSie) project, National Institute of Building Sciences, 2015. Online. https://www.nibs.org/page/bsa_bimsie. Accessed 30.03.2020.

Pazlar, T. & Turk, Z. 2008. Interoperability in practice: geometric data exchange using the IFC standard. *International Journal of Information Management* 13: 362–380.

Semenov, V. 2017 Product Data Management with Solid Transactional Guarantees. *Transdisciplinary Engineering: A Paradigm Shift Series Advances in Transdisciplinary Engineering*: 592–599.

Semenov, V. & Jones, S. 2015. IFChub: Multimodal Collaboration Platform with Solid Transactional Guarantees. *Digital Proceeding of the 32 CIB W78 Conference*. 667–675.

Tolk, A. & Muguira, A. 2003. The Levels of Conceptual Interoperability Model, *Fall Simulation Interoperability Workshop*: pp. 1–11.

Trzeciak, M. & Borrmann, A. 2018. Design-to-design exchange of bridge models using IFC: A case study with Revit and Allplan. *eWork and eBusiness in Architecture, Engineering and Construction*: 212–220.

A system architecture ensuring consistency among distributed, heterogeneous information models for civil infrastructure projects

S. Esser & A. Borrmann
Technical University of Munich, Munich, Germany

ABSTRACT: The application of suitable data structures is an essential aspect for novel digital workflows in engineering and design processes of the Architectural, Engineering, and Construction (AEC) industry. Since model-based data exchange gets increasingly adopted by the industry, feasible and more effective methods must be considered to improve data exchange in the future. While the concept of federated model integration and container-based collaboration as demanded by ISO19650 is well established and widely adopted, it shows a number of deficiencies, in particular when it comes to consistency preservation across the domain models and the handling of design updates. Current practice relies on the exchange of complete domain models which requires the manual identification of design changes by all other stakeholders. Consistency is checked merely by collision detection, which however can cover only geometric aspects. To overcome these limitations, this paper proposes a comprehensive system architecture as well as techniques to identify updates in models and federate such update information by means of update patches. To this end, specific focus is put on possible mechanisms to detect changes and to integrate update patches in the receiving application.

1 INTRODUCTION

The building industry is heading towards more and more digitized processes. Especially for civil infrastructure assets, a very large number of engineers, clients, contractors, and authorities are involved during several planning stages. Each of these disciplines has specific requirements for capturing, storing, and presenting relevant information, ranging from pure semantic information over schematic system design to 3D geometry representations. In this collaboration, all experts contribute to the common goal of a coherent and error-free design of the built asset. Hence, combining heterogenous knowledge representations is a significant challenge that has not been entirely solved so far.

It is well established in today's BIM practice, that engineers work in individual environments and upload their domain models to a common project data platform (also known as "Common Data Environment (CDE)"), thus implementing the concept of federated data models as described by UK BIM Level 2 (NBS 2020) and ISO19650 (CEN 2019). Such an approach enables each engineer to work with applications that suit his needs the best. The actual integration is realized by means of model coordination which is dominantly based on collision detection of 3D geometry. Many other logical dependencies between objects in the domain models remain untracked, especially as an object might have different representations in different domain models. Thus, the concept of collaborative but separated work leads to severe limitation in consistency preservation.

Linked data approaches can help to manage relationships between these various representations (Beetz & Borrmann 2018; Zhang & Beetz 2016). However, it is often useful or even obligatory for modeling and design tasks to fully integrate foreign models into the engineer's design environment. The design of security equipment for railway systems is a vivid example: the security equipment engineers who design necessary signals and sensors along a new railway track must rely upon a given alignment axis and a linear reference systems (LRS). The design of the track alignment, however, is typically a task of another subdomain in railway engineering, considering vehicle dynamics, existing facilities, terrain and land ownership. Thus, consistency preservation mechanisms between the alignment designer and the security engineer are a crucial factor for a successful and faultless design of the complete railway system. To achieve this high level of data coherence among all disciplines, all involved parties must be notified on updates in foreign models which their domain models are based upon.

According to BIM Level 2, project participants propagate such model changes by uploading the complete updated domain model to the CDE again. This approach suits the basic needs of information provision but does not unlock the potential of truly integrated digital workflows. The main deficit lies in the fact that each participant must manually incorporate the model changes into their own domain models. This becomes particularly prone to errors, as such models contain an extensive amount of components, different types of geometry, and complex relationships. Hence, incorporating update information in a domain-specific design environment.

Looking more in-depth to the specifics of civil infrastructure projects, linear reference systems (LRS) is an essential denominator among all civil infrastructure domains. Since these projects typically extend over several kilometers, also data integration over spatially distributed submodels must be considered.

The currently available IT solutions for CDE are mostly based on file management. According to ISO19650, the concept of "containers" allow the bundled management of inter-related files. File-based approaches, however, have severe limitations with respect to consistency preservation as the smallest accessible information unit is the file. Individual updates on object level can thus neither be identified by the CDE, nor be propagated to other domain models. On the other extreme, the concept of a central database fully resolving and managing all objects of all domains has been discarded as unpractical, mostly due to reasons of ownership and liability.

To overcome the discussed limitations, we propose a distributed system architecture that respects the principle of discipline-oriented collaboration with loosely coupled, federated domain models and heterogenous information models, but provides the means for fine-grained, object-level integration and consistency preservation by a patch-based update deployment mechanism. Heading towards an increasing number of federated models, this can significantly reduce the exchanged data and allows automated consistency preservation by integrating updates. Thus, instead of the full model, only the change operations must be re-evaluated. Nevertheless, a patch-based approach requires the correct interpretation of a patch, which must be ensured on the receiver's side (e.g., providing integration rules).

Furthermore, the proposed patch-based mechanism should have a generic form that can be used independently from any specific data exchange format. Contrary to building design processes where three-dimensional geometry is the leading common denominator among domains, significantly differing information representations are in use for civil infrastructure design. These representations range from detailed 3D geometries and simple semantic information up to simple geometry but complex and heavily coupled semantic data.

Our paper summarizes existing approaches for shared collaboration in distributed systems and presents a conceptual approach for distributed CDE systems. The proposed architecture represents a possible approach towards implementing BIM Level 3 principles, realizing a deep integration of domain models. A specific focus is put on the requirements of civil infrastructure projects which have particular requirements regarding spatial placement and varying representations of individual semantic objects.

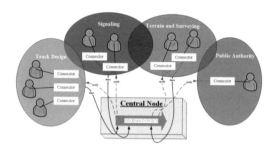

Figure 1. Collaboration system with one central node and distributed domain nodes.

2 OVERALL SYSTEM ARCHITECTURE

The paper at hand presents a possible approach for a BIM Level 3 collaboration architecture. Several stakeholders coming from various domains are working collaboratively together. Each modeler is responsible for the domain-specific models he is creating. Such model is created in authoring applications that suit the domain requirements best.

The proposed CDE is a distributed, loosely coupled system basically following the peer-to-peer architecture (Chen & Hou 2014). Each authoring application forms an individual peer in the system. To this end, existing software systems must be extended by connector modules that provide means for communicating within the distributed system. This module must hold a representation of the domain model, must be capable to identify changes performed by the specialist in order to propagate them into the P2P network. At the same time it must be able to receive notifications from other domain nodes. In deviation from pure P2P architectures, the CDE has a central node that holds all shared domain models and provides an information bus. A possible domain scenario for civil infrastructure projects is illustrated in Figure 1.

The integrity of the distributed, heterogenous model is preserved by a patch-based update mechanism to federate design changes. Such patches can be utilized to ensure consistency between different representations among different domains. Practically spoken, this approach enables each domain to represent the domain-specific knowledge in a data model that suits its specific needs. However, if updates are applied to a domain model, all other domains are informed. This allows the domains which are impacted by the update to integrate the respective changes in their own model. This integration can be performed automatically by defined rules, or with manual interventions and under active control of the respective specialist.

A major component of the proposed architecture is the information bus. The concept is well-known from distributed application architectures and enables clients to communicate via centralized communication medium.

In the proposed architecture, each domain node can connect or disconnect to the information bus. The connection between a domain application and the information bus is managed by a connector module. This module contains methods to track local changes and enables the user to share the updates once the domain model reaches an sufficiently mature state for sharing. Both communication approaches, request and reply as well as publish and subscribe are supported (Oki et al. 1993). Furthermore, domain nodes can be introduced as either master or slave nodes. Master clients can send updates to the information bus at any time whereas slave nodes can only reply to update messages that are relevant for its individual domain.

A basic assumption of the proposed architecture is the availability of heterogenous, domain-specific data models. Common anchor points define denominators that help to achieve consistency between various data models. In the context of civil infrastructure projects, such anchor points are defined using the following concepts:

- **Project breakdown structure**. This concept splits a built asset in multiple containers. Components can be assigned to these containers to ease the navigability of the overall model. Examples are the separation of a building into building storeys or the use of several containers that separate elements along a railway track.
- **Reference systems**: As explained in the introductory section, (linear) referencing systems are of essential importance in the design processes of linear stretched infrastructure. LRS can build a reliable concept among several data representations (Esser & Borrmann 2019).
- **Classfication systems**: Established classification systems like Omni- and UniClass can also help the reach consistency among various data sets Chang et al. (2009) have investigated research in commonly used classification systems in railway systems that assists in maintenance and operation issues.

3 LITERATURE REVIEW AND RELATED CONCEPTS

The management of complex interdependencies has been subject to numerous investigations in the context of BIM.

The UK BIM task group has defined different levels of BIM implementation (Bew & Richards 2008): BIM level 0 describes conventional collaboration approaches to design and collaborate within a project. Data exchange among involved parties is realized by data structures that are either proprietary or their interpretation not automatable. The predominant communication medium is large printed 2D plans. BIM level 1 improves the situation by introducing file-based exchange scenarios. Such data are still not fully automatable for receivers since no or only weak requirements formalize data handover scenarios.

Nevertheless, BIM Level 1 marks the milestone from purely paper-based to file-based information exchange. Additionally, 2D geometry is often extended using the third spatial dimension. Moving on, BIM level 2 adds increasing structure to exchanged design data. Such formalization is based on object-oriented approaches to separate information in logical units. Besides, using any kind of database system also improves the quality of information exchange. All these initiatives result in model-based data exchange, which further enhances the formalization and automatization of sending and receiving data sets. Comparing BIM level 2 and 3, the latter focuses on increasing data exchange consuming web technologies.

In comparison to level 2, exchanged information must be still structured and formulated according to an agreed data model (Counsell 2012, Eadie et al. 2015).

The novel concept in a BIM level 3 implementation is the object-based communication including the identification and federation of updated information instead of re-sending complete instance models. Research publications have already shown that collaboration based on domain-specific models should be preferred over working in a single but centralized model. Furthermore, this principle leads to the situation that each involved party can work in their preferred design environments and only have to ensure suitable data exchange interactions with other engaged experts.

Windisch et al. (2012) have defined a framework to filter domain models for further usage in foreign domain environments. However, their approach does not manage updates on the filtered models, which is one of the critical aspects of a BIM level 3 environment. Adoption of such filtering techniques for BIM models might help to identify relevant updates.

Semenov & Jones (2015) have pointed out that none of the existing BIM CDEs provide functionalities for the continuous integration of committed updates.

Transmitting only model updates rather than full files were already subject to earlier investigations. Researching a more generic field of (weak) coupling of distributed data systems, Crooks et al. (2016) have developed an approach called TARDiS, which abstracts update information as simple as possible and at the same time still interpretable. Their concept ensures context reasoning as well as respecting the core principles of distributed systems like ALPS and the CAP theorem.

In addition to TARDis, an older but still valid approach on branched versions among several versions of XML-based representations was introduced by Vagena et al. (2004). Their paper proposes a compromise of a so-called log approach and a snapshot approach. The log approach starts with an initial state of the XML document and only stores the update. In comparison, the snapshot approach always stores the full document which consumes more storage but provides faster results when querying for a specific intermediate version.

Dawood et al. (2019) are facing the problem of continuous model updates using natural language processing and updating the data model accordingly.

In the context of combining various design options, Mattern & König (2018) have proposed a graph-based data system and a possible extension to an existing data model. Their proposal intends to harmonize content from various design options authored by different domains. Since their proposal is limited to a single data model (namely IFC), the issue of deploying update information among different data models remains unsolved.

Moving from various design options to refined building designs among different project stages, Abualdenien & Borrmann (2019) have proposed a meta-model to face the challenges of emerging levels of developments in building models. This approach is an essential contribution to the presented situation in civil infrastructure projects but cannot take the variety of domain-specific data models into account that are currently in use for civil infrastructure design tasks.

In conclusion, interacting with change requests in complex building information models or other types of structured data is not an entirely new discipline. However, none of the referenced publications is fully capable of ensuring consistency between different data models that represent correlated built assets.

4 CONCEPTUAL APPROACH

The following sections provide further explanations about specific exchange steps and functionalities to reach consistency among distributed data. Upfront, a short overview of computational principles summarizes basic concepts, which are preliminary assumptions. Subsequently, these technical principals are evaluated in the context of shared data environments for railway and civil infrastructure projects.

4.1 Concurrency principles

It is assumed that each domain has a well-defined data model that suits its specific needs. The domain data models are supposed to follow the principles of object-oriented programming. Furthermore, it is supposed that the schema definitions of all data models are known and available (but not necessarily vendor-neutral). Therefore, also applications with a proprietary backend data model can be considered once these tools provide the implementer an accessible application programming interface (API).

In projects, design tasks require local copies of foreign domain models. Thus, updates made to such locally stored replicas can lead to inconsistencies and contradictions. The concept of concurrency control can help to resolve such contradictory model stages distributed over several clients and domains. In general, conconcurreny can be performed in different approaches. With the pessimistic concurrency control conflicts are avoided in advance and only certain changes are allowed. In the optimistic concurrency control, conflicts in the project information are identified and resolved eventually. Transferring these approaches to programming paradigms of database systems, the pessimistic concurrency control is aligned with the ACID principles whereas the optimistic concurrency control can be interpreted in the context of BASE principles.

4.1.1 ACID principles

ACID principles define baseline to interact with a database system. The abbreviation ACID represents:

Atomicity: A database interaction consists of several steps which are performed one after the other. The principle of atomicity states that all actions must be performed correctly during one transaction period. If any problems occur during the transaction process, all executed actions must be rolled back to reach the initial and valid state inside the database system;

Consistency: Each interaction with the database must result in a valid state. This principle can be interpreted in two different approaches: an adequate representation in a technical manner or an accurate description in terms of the engineer's knowledge. This set of fundamentals can be formulated in rules that validate an incoming transaction request;

Isolation: An update on the database is decoupled from any other action that is performed on the database system. Intermediate results cannot be accessed. This principle ensures repeatable queries that result in the same response;

Durability: successfully committed transactions are not reverted in case of any external effects such as power loss.

In summary, no access is granted to the database as long as any other running transaction is executed. The policy ensures consistent data at any query time (since queries are a transaction in itself) but limits the system to a few amounts of operations per timeframe.

4.1.2 BASE principles

Contrary to the ACID principles, the BASE approach does not guarantee permanent and up-to-date consistency of the database. The abbreviation BASE stands for:

Basic Availability: the stored data is not blocked during running transactions. Thus, datasets are available most of the time;

Soft state: store operations do not have to be consistent during the writing operation and do not directly federate to all existing replicas of the data set;

Eventual Consistency: Consistency among replicas is achieved eventually at a later point in time (Robinson et al. 2015).

Hence, users can interact more frequently with the system. The advantage of this approach is better scalability for large data sets and improved performance. However, complex inconsistencies an contradictions (like described in the optimistic concurrency control approach) might occur during synchronization if distributed replicas have changed in themselves since the last synchronization. BASE principles are applied in all modern web systems where multiple users communicate with servers in parallel (e.g., webshops, mailing systems, video platforms, chats).

4.2 Adoption and conclusions

In the context of collaborative platforms for building and civil infrastructure projects, a mixture out of both approaches, optimistic and pessimistic replication, will serve the users' requirements best. In general, pessimistic replication guarantees a valid centralized dataset since federated information is directly incorporated. However, such push operations to the database might happen irregularly due to various reasons. The frequency of commits might depend on domain-specific design tasks and relevant local iterations or the individual behavior of the engineer.

An approach of optimistic replication can lead to unsolvable clashes during delayed synchronizations but make data better available using local replicas. Thus, federated update information must be instantly checked when a new change request is sent to the centralized data system.

As a result of the presented principles, the next section introduces a system architecture that advances existing CDE technologies by extending them by a bus network architecture.

5 IMPLEMENTATION APPROACH

Based on the explained definitions of consistency, the proposed system should extend existing approaches known from BIM Level 2 implementations. The management of models as the smallest unit must be overcome by a finer level of object-based communication. Information stored in a single asset is consistent in itself and represents the result of updates, which were applied to a model. Federated assets update distributed local replicas by extending existing CDE platforms by an bus network architecture. We propose a hybrid system that provides all project members with constant information and update flow. This integration can be performed automatically by defined rules, or with manual interventions and under active control of the respective specialist.

5.1 System architecture

The complete system can be broken down into smaller units of client-server interactions. The following sections explain required operations in detail. Figure 2 visualizes the overall architecture of the proposed system.

The central part of the collaboration system comprise of an project server that contains the information bus. It is initialized once a new project is set up. Domain applications identify updates on their domain models and share the applied changes to the information bus. Various types of notifiers are used to classify update information. Domain nodes can subsribe to specific update messages and can decide whether the provided update information presented on the bus has to be incorporated in their own replicas. Besides, client nodes can behave in various manners (e.g., only as notification bots in communication applications or for decoupled downstream use cases).

Thus, client nodes can either be equipped with a federator and a notifier method or only provide one of these interfaces. Authoring tools subscribe to specific subjects and get notified if events of these subjects happen.

The development of our system architecture focuses on the exchange of updates applied to domain models. However, additional types of data (documents, raw binary data, etc.) will exist in parallel to the actual models. Hence, the common data environment that is based on the information bus, also provides further methods to store and exchange accompanying documents that represent results out of various design stages. Such documents include but not limited to guidelines, codes, tendering documents or billing sheets.

The following steps build the technical base to identify, deploy and integrate model changes between distributed replicas of domain models.

5.2 Update identification

To identify update information that has to be dispatched among the project, it is essential to track (a) relevant assets and changes inside such assets and (b) relationships between related assets.

An asset can be defined as a generic data ressource that represents a logical piece of model information (e.g., products, spatial structure containers, components of (linear) reference systems). Both assets and relationships carry additional attributes. The set of assets and relationships results in a graph representation that is aligned with but not necessarily a 1:1 replica of the tracked domain data model. Various domain data models distinguish their classes in the root layer and additional resource layers. Classes that inherit from root entities and thus have a unique identifier are suitable candidates to be represented as an individual asset node.

Two approaches for identifying updates between an initial and an updated model can be defined:

5.2.1 Comparison of an initial an updated model state using graph representations

Graph representations of instance models extend simple text comparisons by the extended use of relationship information between assets. Subgraph isomorphism can help to detect changes between an initial and an updated graph representation (Ullmann 1976). Hidders (2001) has worked on a generic approach utilizing object-oriented data models with graph representations.

Simple graph comparison between an initial and an updated model might not supply the complete

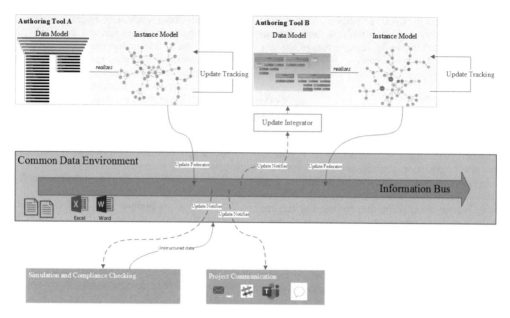

Figure 2. Technical realization of the proposed BIM level 3 system architecture.

procedure of modification steps that are necessary to re-construct the modification steps in a receiving platform.

5.2.2 Update tracking during the design process

Contrary to the approach mentioned above, performed updates can be directly tracked inside the modeling or simulation applications. This approach requires additional plugins in each authoring and modification tool that should either identify or integrate update patches. Koch & Firmenich (2011) have defined a generic description for models and applied model changes.

Further research has to be conducted to verify the applicability of both approaches. Although the latter approach contains the potential for live collaborative work sessions, it might be beneficial to include an explicit share operation that initializes an update federation process to the information bus. Such a gate is crucial since design and modeling tasks in the civil infrastructure projects sometimes require iterative calculations during the design phase. Thus, not every iteration loop should be propagated instantly to all other project participants. Besides, the information aggregation per asset node requires further investigation to ensure compatibility with a broad range of domain data models that are currently used in projects.

5.3 Update federation

Besides the identification of model changes, the concept for information federation is crucial for the proposed system.

The information bus, which was introduced in Section 2, federates the update information among project members. Besides the federation, the central node should also hold the latest instance of all domain models. The provision of all recent models is already implemented in varios CDEs currently available on the software market. If a new actor joins the project, he can directly pull the latest model versions using existing approaches related to BIM Level 2 systems.

Aditionally, each federated update patch can be tested against individual criteria to ensure the integration feasibility in receiving replicas. Besides, the provision of all recent models inside the central node enables further services within the platform (e.g., visualization, clash detection or issue management) (Oki et al. 1993).

Once a local replica on the client system is stored, the interpretation of incoming patches (modifying the fetched model state) can start. Providing a particular model state on a centralized platform is a technological approach, which is already known from BIM Level 2 systems. However, providing the latest state of the available instance model and not only the complete history of federated patches helps in several regards: Incoming patches can be instantly tested and verified before deploying them to clients. Furthermore, new project members can easily fetch the latest model to start their collaboration interactions without re-interpreting a long list of already performed updates.

Since not every single update information might be relevant for each domain or project stakeholder, participants should get the opportunity to decide on specific types of updates that have an impact on their design and modeling tasks.

5.4 Information integration

Due to the enormous number of components that have complex relationships within the considered models, a fully-automated update integration cannot be realized. In addition, civil infrastructure projects often comprise of models that do not only carry domain-specific content but also have a specific spatial context (e.g., a detailed bridge model which should be integrated into a railway track model for visualization and clash detection purposes). Furthermore, different projection and coordinate systems used for georeferencing have to be taken into account as well (Jaud et al. 2019).

Hence, scripting interfaces help to interpret and automate incoming update patches in receiving applications. Recurring events that don't cause any contradictions in the receiving model are automatable. Besides, additional scripts evaluate the relevance of incoming updates for specific design tasks.

6 SUMMARY

The proposed system architecture extends existing BIM Level 2 approaches towards BIM Level 3 features. The smallest unit of data exchange is not a model but a set of updated objects, which reduces data traffic and makes updates easier to interpret in receiving applications. To improve information federation, a bus architecture enhances data availability and serves as the key concept in an extended CDE platform. Each domain can subscribe to update events that are relevant for its own design tasks.

The separation of update identification, federation, and integration enables a agile approach that supports existing applications and enables users to stick with their familiar tools. Besides, established data models remain untouched and are integrated in the communication procedure.

7 OUTLOOK

Due to the legacy of fragmented and cross-disciplinary collaboration, it is essential and challenging at the same time to maintain existing and established approaches as good as possible and formulate extensions based on this heritage. However, tools from giant software vendors like Microsoft, Google and Apple already demonstrate how shared collaboration within complex data works.

REFERENCES

Abualdenien, J. & Borrmann, A. 2019. A meta-model approach for formal specification and consistent management of multi-LOD building models. *Advanced Engineering Informatics*, 40, 135–153. https://doi.org/10.1016/j.aei.2019.04.003

Beetz, J. & Borrmann, A. 2018. Benefits and limitations of linked data approaches for road modeling and data exchange. *Workshop of the European Group for Intelligent Computing in Engineering*, 245–261.

Bew, M. & Richards, M. 2008. BIM maturity model. *Construct IT Autumn 2008 Members' Meeting. Brighton, UK*.

CEN. *DIN EN ISO 19650-2:2019*. 2019.

Chang, T., Lee, S. & Cho, M. S. 2009. A Study of the Information Classification for Railway Industry. *IJR International Journal of Railway*, 2(1), 37–42.

Chen, H. M. & Hou, C. C. 2014. Asynchronous online collaboration in BIM generation using hybrid client-server and P2P network. *Automation in Construction*, 45, 72–85. https://doi.org/10.1016/j.autcon.2014.05.007

Counsell, J. 2012. Beyond level 2 BIM, web portals and collaboration tools. *Proceedings of the International Conference on Information Visualisation*, 510–515. https://doi.org/10.1109/IV.2012.88

Crooks, N., Pu, Y., Estrada, N., Gupta, T., Alvisi, L. & Clement, A. 2016. TARDiS: A branch-and-merge approach to weak consistency. *Proceedings of the ACM SIGMOD International Conference on Management of Data*, 26-June-20, 1615–1628. https://doi.org/10.1145/2882903.2882951

Dawood, H., Siddle, J. & Dawood, N. 2019. Integrating IFC and NLP for automating change request validations. *Journal of Information Technology in Construction*, 24, 540–552. https://doi.org/10.36680/j.itcon.2019.030

Eadie, R., Browne, M., Odeyinka, H., McKeown, C. & McNiff, S. 2015. A survey of current status of and perceived changes required for BIM adoption in the UK. *Built Environment Project and Asset Management*, 5(1), 4–21. https://doi.org/10.1108/BEPAM-07-2013-0023

Esser, S. & Borrmann, A. 2019. Integrating Railway Subdomain-Specific Data Standards into a common IFC-based Data Model. *26th International Workshop on Intelligent Computing in Engineering*. Leuven, Belgium.

Hidders, J. 2001. *A Graph-based Update Language for Object-Oriented Data Models* (University Press Facilities, Eindhoven, the Netherlands). https://doi.org/10.6100/IR551259

Jaud, Š., Donaubauer, A. & Borrmann, A. 2019. Georeferencing within IFC?: A Novel Approach for Infrastructure Objects. In Y. K. Cho, F. Leite, A. Behzadan, & C. Wang (Eds.), *Computing in Civil Engineering 2019: Visualisation, Information Modelling, and Simulation*. Atlanta: American Society of Civil Engineers.

Koch, C. & Firmenich, B. 2011. An approach to distributed building modeling on the basis of versions and changes. *Advanced Engineering Informatics*, 25(2), 297–310. https://doi.org/10.1016/j.aei.2010.12.001

Mattern, H. & König, M. 2018. BIM-based modeling and management of design options at early planning phases. *Advanced Engineering Informatics*, 38(July), 316–329. https://doi.org/10.1016/j.aei.2018.08.007

NBS 2020. BIM Levels explained - Definitions for levels of BIM maturity from Level 0, through Level 1, Level 2 and Level 3 and beyond. Retrieved March 28, 2020, from https://www.thenbs.com/knowledge/bim-levels-explained

Oki, B., Fluegl, M., Siegel, A. & Skeen, D. 1993. Information bus an architecture for extensible distributed systems. *Operating Systems Review (ACM)*, 27(5), 58–68. https://doi.org/10.1145/173668.168624

Robinson, I., Webber, J. & Eifrem, E. 2015. Graph Databases. In M. Beaugureau (Ed.), *Joe Celko's Complete Guide to NoSQL*. https://doi.org/10.1016/b978-0-12-407192-6.00003-0

Semenov, V. & Jones, S. 2015. IFChub?: Multimodal Collaboration Transactional Guarantees Platform with Solid. *Proc. of the 32nd CIB W78 Conference 2015, 27th-29th October 2015, Eindhoven, The Netherlands*, 667–675.

Ullmann, J. R. 1976. An Algorithm for Subgraph Isomorphism. *Journal of the ACM (JACM)*, 23(1), 31–42. https://doi.org/10.1145/321921.321925

Vagena, Z., Moro, M. M. & Tsotras, V. J. 2004. Supporting branched versions on XML documents. *Proceedings of the IEEE International Workshop on Research Issues in Data Engineering*, 14, 137–144. https://doi.org/10.1109/RIDE.2004.1281713

Windisch, R., Katranuschkov, P. & Scherer, R. J. 2012. A generic filter framework for consistent generation of BIM-based model views. *European Group for Intelligent Computing in Engineering, EG-ICE 2012 - International Workshop: Intelligent Computing in Engineering*, 1–12. Munich.

Zhang, C. & Beetz, J. 2016. Querying Linked Building Data Using SPARQL with Functional Extensions. *Proceedings of 11th European Conference on Product and Process Modelling*, (September).

A framework for leveraging semantic interoperability between BIM applications

M.E. Belsky
SmartSeeBIM Ltd, Moscow, Russia

ABSTRACT: AEC/FM industry experiences significant challenges in full adoption of BIM despite its obvious benefits. Deficient semantic interoperability between BIM applications is one of the factors causing the industry dissatisfaction. The IFC schema is the open, international, vendor agnostic standard for data exchanges between BIM applications. It is a generic, rich and flexible standardized data model. However, the IFC specification lacks formal rigidness to unambiguously capture the full semantics needed for seamless and reliable information exchanges. A framework for new approach to overcome that drawback and improve the overall semantic interoperability is outlined in the paper. The novelty of the approach is that it places the onus for interpretation of exchanged data on an importing application which is an opposite to current practices when an exporting application is responsible for a semantic content of IFC exchange file. A pilot technical implementation was developed to test the applicability of the approach.

1 INTRODUCTION

Building Information Modeling (BIM) interoperability issues between different stakeholders in Architecture, Engineering, Construction and Facility Management (AEC/FM) projects may be seen as a modern story of the Tower of Babel. AEC/FM projects today are very complex undertakings, professionally highly specialized and fragmented, and require seamless and reliable collaboration among the project's teams. Individual software applications, which are used by the teams, generate and store information in their native product data model schemata and formats, imposing challenges for interoperability. This requires the ability to interconnect, share, and use the information seamlessly and efficiently, which becomes a major challenge. There is no single BIM software platform or system that can support all functionalities required for the AEC/FM industry needs (Shafiq et al. 2013). Therefore, interoperability among heterogenous software applications used in the industry becomes crucial. The annual cost of deficient interoperability in the AEC/FM industry in United States was estimated at over $15 billion in 2002 (Gallaher et al. 2004). A study conducted by McGraw-Hill in 2009 states that 8 out of 10 users of BIM software tools in the United States consider lack of adequate interoperability between software applications to be the limiting factor in achieving the full potential of BIM (Youngs et al. 2009). Lack of interoperability leads to wasteful activities and impedes value creation due to the loss and distortion of mismatched information, and uncertainty about reliability of data (Poirier et al. 2014). Therefore, there is a major need for improving interoperability among plethora of various BIM software applications and platforms.

There are various barriers to interoperability and different methods and approaches for enabling it. In general, interoperability comprises both technical integration and information integration of software systems (Peltomaa et al. 2008). Most of the current solutions point only at technical mechanism of data integration for achieving system, structural and syntactic interoperability, while information integration, focused on semantic interoperability that involves the interpretation and understanding of exchanged data (i.e. intent), remains a critical issue (Chander et al. 2017; Jim et al. 2017; Pollock 2001). Semantic interoperability accentuates the importance of the information and centers on enabling content, data, and information to interoperate with software applications outside their source (Pollock & Hodgson 2004).

The paper is organized as follows. In the next section the author examines the general concepts of semantics and interoperability with special focus on semantic interoperability. Next, the paper provides an overview of major existing approaches for enabling semantic interoperability among heterogenous BIM applications. Then, the author presents a framework for novel approach for improving this type of interoperability. The approach is illustrated with an example of interpretation of exchanged information in the architecture domain. Finally, conclusions and discussion of challenges are presented to stimulate further research.

2 BACKGROUND

2.1 Semantics

Semantics is the study of meaning. It has a long and not straightforward history. Semantics studies can be traced back to the first studies of language by man. Aristotle's first ruminations on language included questions about meaning in language. Semantics has several definitions. For example, Saeed (1997) opines that *"semantics is the study of meaning communicated through language"*. According to Loebner (2013), *"semantics is the part of linguistics that is concerned with meaning"*. Kreidler's (1998) definition is also useful: *"Linguistic semantics is the study of how language organizes and expresses meaning"*. Anyway, semantics concerns with 'meaning' that is organized, expressed and communicated by language.

In the context of the paper a formal approach to semantics was adopted. Formal semantics describes the meaning of language using the descriptive apparatus of formal logic and follows the principle of compositionality, i.e. the meaning of the whole is a function of the meaning of the parts. Raison d'etre of language is communication. Meaning is at the beginning and at the end of communication process, and mental representation, i.e. language is a way of re-presenting a world in our minds. Formalized representations allow us to conceive hypothetical scenarios, complex reasoning patterns, inferencing, conditions, etc. (Martinez 2002).

2.2 Interoperability

Interoperability can be defined from different perspectives. The European Interoperability Framework (IEF), part of the eEurope Action Plan 2005 adopted by the European heads of state at the Seville summit in June 2002, defines interoperability as *"the ability of information and communication technology (ICT) systems and of the business processes they support to exchange data and to enable the sharing of information and knowledge"* (Framework 2004). The ISO/AEC 33001 defines interoperabilty as *"the ability of two or more systems or components to exchange information and to use the information that has been exchanged"* (ISO 2015). These definitions are readily applicable to the AEC/FM industry. Although there are many other defintions of interoperability, its types and levels, it can be reasonably argued that interopoerability in general is about exchange and interpretation of data and information at various levels by heterogenuos software applications in one or another way. For example, the Knowledge Industry Survival Strategy (KISS) has developed a classification for BIM interoperability. It has five different levels: file and syntax, visualization, semantic, alternative representations, and parametric modeling levels (Steel et al. 2012). Information exchanges at these five levels are to be implemented to achieve an ultimate BIM interoperability. Sheeth (1999) has identified four types of interoperability needed by computer systems that still hold true today: system, schematic, syntactic, and semantic interoperability types. Fully interoperable systems must address these four types of interoperability in order to exchange information without loss of meaning and intent. Interacting systems are considered essentially useless if the receiving system cannot interpret data as it was intended by the sending system although they could successfully connect and transfer data (Costin & Eastman 2019). The primary challenge in determining whether two or more systems are interoperable is in determining levels at which the systems can communicate and if they share the same framework from which they share and interpret information (Diallo 2016). The Level of Conceptual Interoperability Model (LCIM) identifies seven levels of interoperability: Level 0 – no interoperability, Level 1 – technical interoperability, Level 2 – syntactic interoperability, Level 3 – semantic interoperability, Level 4 – pragmatic interoperability, Level 5 – dynamic interoperability, and Level 6 – conceptual interoperability. LCIM provides a metric for determining if two or more systems are interoperable. Dobrev et al. (2007) opines that full semantic interoperability can be achieved only if pragmatic, dynamic and conceptual levels of interoperability are implemented to some extent.

2.3 Semantic interoperability

Generally, semantic interoperability aims at exchanging information among heterogenous software applications in a meaningful way and with minimum human intervention. It provides the ability to attach meaning to exchanged data. This is used to structure and organize domain knowledge about an object or a phenomenon in such a way that software applications can automatically process and integrate large amount of diverse information (Kalfoglou 2009). One of the ways for achieving interoperability at the semantic level is to make sure that the relationship between different domains is maintained during data transfer (Peachavanish et al. 2006). Once data is exchanged, the receiving application must be able to parse the meaning of data as intended by the sending application. Semantic interoperabilty assures that the exchanged content is understood in the same way in communicating systems, including by those humans interacting with the systems in a given context.

At present semantic interoperability among two or more heterogenous applications is always partial because software developers encode lots of hidden knowledge in the implementation of application algorithms (Dobrev et al. 2007). Semantic interoperability is the single most important interoperability challenge in BIM yet to overcome.

2.4 Industry Foundation Classes and interoperability issues

BIM was introduced to the AEC/FM industry based on the premise of creating, storing and managing information throughout the project lifecycle in an integrated way (Eastman et al. 2011). Due the professionally diverse nature of the industry data sharing and exchange among multiple heterogenous software applications become an inevitable need (Ramaji & Memari 2016; Tolman 1999). To address data interoperability between diverse software tools the Industry Foundation Classes (IFC) schema was developed. The schema acts as a medium for bidirectional data sharing and exchange between heterogenous applications. It is an open and neutral format, so that multiple applications can use it. The initial specification for IFC was created in 1995 to address the sharing of information to facilitate and support more efficient workflows and data exchanges (Liebich et al. 2006). IFC schema was developed by buildingSMART - an open, neutral and international not-for-profit organization (buildingSmart n.d.). It is a common data schema intended to hold interdisciplinary information for the project lifecycle in a building information model and exchange it among heterogeneous software applications used by the AEC/FM industry. The IFC schema is written in EXPRESS data modeling conceptual language. IFC entity instances represent geometry, relationships, processes, materials, performance, fabrication, and other information needed for design, production, construction and maintenance of built assets (Leibich & Wix 1998; Liebich et al. 2006). After over two decades of the IFC schema development it is widely recognized as the common data exchange format for interoperability within the industry (Eastman et al. 2011) and is registered in the International Organization for Standardization (ISO 16739-1:2018). It is a rich product modeling schema, but highly redundant and flexible, and lacks formal logic rigidness, offering different ways to define objects, relationships, properties, and attributes in support of different data uses (Venugopal et al. 2012). Thus, data exchanges selecting from the redundant data representations have serious problems of mismatch, resulting in unacceptably low confidence level due to loss, distortion, and misinterpretation between exported and imported data.

Difficulties in interoperability among diverse BIM applications have posed a barrier to the adoption of BIM by the industry (Eastman et al. 2010; Olofsson et al. 2008). This fact leads to a vicious cycle: interoperability between BIM applications does not have a satisfactory level of reliability for full adoption of BIM by the industry, and in turn BIM should be more widely adopted to in order for interoperability to be improved (Muller et al. 2017).

Interoperability on the data level can be described as a process of data export and import between heterogenous software applications. This process includes two mappings: one mapping is from the native schema of software application to IFC model (file) for export, and another mapping is from IFC model to native schema of importing software application. Interoperability issues are likely to arise over these mappings and can be partially related to the lack of semantic uniformity in the way BIM software applications map to and from IFC entities and properties (Venugopalet et al. 2012). The IFC specifications have changed considerably over the years to improve interoperability. Each new release of the IFC schema included not only additional entities and attributes but also changes in data structures. However, these efforts are useful only for data interoperability and do not address interoperability issues fundamentally (Lai & Deng 2018).

3 METHODS FOR INTEROPERABILITY IN BIM

3.1 Model View Definitions

In order to implement effective data exchanges between BIM applications (i.e. data interoperability), it is imperative to define relevant subsets of the IFC schema that are appropriate for a given exchange. A subset of the entire schema which satisfies a particular model exchange is called Model View Definition (MVD). Such subsets allow extraction of 'model views', which are akin to database views. Therefore, MVDs can be defined as virtual, specialized and structured subsets of data compiled dynamically from databases (Venugopal et al. 2010). Studies on data exchanges show that without well-defined MVDs, the exchanges are vulnerable to errors, omissions, contradictions and misrepresentation because they reduce or simplify the information (Bazjanac & Kiviniemi 2007). In 2010, buildingSMART developed the IFC Certification 2.0 procedure for the IFC 2X3 Coordination View Version 2 (CV2.0), which is an MVD intended to promote consistent and reliable implementations of the IFC specification by many software vendors across multiple software platforms. However, its purpose is limited to support of clash detections and other basic building system coordination tasks, which require exchange of geometry, material properties and object identities. Moreover, the MVD approach has some drawbacks. In developing an MVD for a given exchange, a developer must determine which appropriate level of semantics will be embedded in the exchange model. At one end of the spectrum, the exchange model can include only basic solid geometry and material properties. In this case, for any use beyond a simple geometry clash detection, importing software would need to interpret the geometry and material properties to compile the meaning of exchanged objects using internal representations of its own native schema. The export routines at this level are simple and the exchanges are generic. At the other end of the spectrum, an exchange file can carry various semantic constructs such as hierarchies that define design intent, procurement groupings, production methods and phasing, and other pertinent information about a building and its parts. In this case, the importing software can generate native objects in its own schema with minimum effort. The export subroutines at this level, however, must be carefully tailored for each use case, the reason being that the information must be structured in a way suitable for the importing applications. Another drawback of the MVD approach is that software companies are required to provide matching IFC export and import subroutines. This requirement appears to resurrect the original challenge for interoperability between multiple heterogenous software applications that gave impetus to the IFC schema development, which was the need for $n*(n-1)/2$ separate exchanges for different tools.

3.2 Ontologies and semantic web

Ontologies can be seen as means to describe important aspects of a domain of interest and are used to define standardized and machine-readable definitions and concepts of that domain. Ontologies can be used for a variety of reasons, such as capturing, sharing, reusing, and analyzing domain knowledge (Costin 2016). Ontologies are widely used in information retrieval, product model data schema development, etc. (Gómez-Pérez et al. 2004; Gruber 1995). An ontology provides a shared vocabulary, which can be used to model a domain. In general, an ontology consists of five components: terms, definitions, properties types, classes, and a class hierarchy. These components are the basis for all ontologies (Gruber 1995). Ontologies stem from description logics, which are formal knowledge representation languages and assertions, in which the building blocks include terminology, instances, and rules (Baader et al. 2007).

One of the most prominent methods for achieving semantic interoperability between heterogenous BIM applications is the Semantic Web technology or Web Ontology Language (OWL) (OWL 2 Web Ontology Language n.d.). Semantic web technologies are becoming utilized in the AEC/FM industry by using the OWL and resource description framework (RDF). Main constructs in OWL are concepts, individuals and properties. A concept denotes a construct that can have members belonging to that concept and sub-concepts in order to form hierarchies. The IFC schema can even be translated to OWL (i.e. ifcOwl) (Pauwels & Terkaj 2016). IfcOWL is by far the most advanced effort to leverage the IFC schema onto the ontology level (Beetz et al. 2009). However, because most of the generated ontology elements are strictly related to the IFC schema, its flexibility in different application scenarios is somewhat restricted (Costin & Eastman 2019).

One of the major issues with ontologies development is the definition of scope. It is impractical to develop a single, large, one-fits-all ontology that represents the universe of discourse. Practical application involves the usages of multiple, smaller ontologies to fit a certain domain, hence a shared conceptualization. Therefore, ontology matching, and mapping have become crucial as it plays an important role in uniting heterogeneous ontologies to work together (Cheng et al. 2008; Paolucci et al. 2002).

3.3 Translators and mapping

Converting data from the sending format to the receiving format is another popular approach to interoperability among heterogenous software applications. This process is referred to as translators or mappings depending on the type of application.

On the application level (e.g. BIM authoring tools), the translator converts one software product model (source format) into another product model (receiving format). Translators can be either one-way or bidirectional. Direct data exchanges between heterogeneous source (export) and receiving (import) software applications are typically performed through proprietary written translators that have their own data structures. This inevitably leads to serious exchange issues (Eastman et al. 2011). Although well-defined and properly implemented translators can effectively support interoperability between different applications, they, however, have intrinsic limitations. Translators are usually written in ad-hoc manner. For instance, translators depend on source and receiving applications, and as a result, when an application is updated the translator has to be also updated. Moreover, translators are labor intensive and prone to errors.

Mapping is typically performed on the data level or the semantic level. On the semantic level the mapping is referred to as ontology matching or reconciliation. Ontology reconciliation is a relatively new but important method to translate between ontologies in the context of the AEC/FM industry (Pauwels et al. 2017). However, ontology reconciliation constitutes only a fragment of more complicated task concerning the alignment, articulation and merging of ontologies (Kalfoglou & Schorlemmer 2005).

Translating and mapping require a significant human involvement. Natural Language Processing (NLP) techniques coupled with machine learning are possible solutions for automating these processes. NLP based methods rely on the semantic interoperability and can be used for data and ontology mappings in AEC/FM subdomains (Le & Jeong 2017; Zhang & El-Gohary 2017).

3.4 Open application programming interfaces

Automated Programming Interfaces (API) is one of the open-world responses to address information exchanges issues among heterogenous BIM applications. Many popular BIM authoring applications such as Revit or Tekla Structures have open APIs. Interfaces generally specify how different software applications can interact directly with each other. APIs enable application-to-application communication by having the subroutine definitions, variables, protocols and tools accessible to use. Software companies provide support to interoperability of their platforms with external applications through interfaces to help read and write data in their native format. API provides encapsulation mechanism for underlying information and serves as a means to modify underlying information schema and particular implementations without directly affecting third-party developers or end users (Zamanian & Pittman 1998). Significantly, open data from various heterogenous sources can be aggregated using programming interfaces (Costin & Eastman 2019). APIs approach to interoperability, however, abidingly has limitations. The main limitation is the dependence of plug-in or extension on the software it is interfacing with. This includes the need to update the plug-in whenever the software is updated due to issues of backward compatibility (Oti et al. 2016).

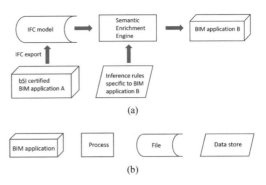

Figure 1. Semantic enrichment information workflows: a) direct native import; b) legend.

4 PROPOSED SOLUTION

4.1 Overview

The proposed framework for leveraging semantic interoperability among heterogenous BIM applications stemmed from the original approach for semantic enrichment of IFC models developed by Belsky et al. (2016). Semantic enrichment generally refers to an automatic or semiautomatic addition of meaningful information to a digital model by software tool, for example rule processing engine that can deduce new information by processing inference rules applied to the model. The basic idea is that importing application, when mapping an IFC model to native schema, can apply its domain knowledge (i.e. ontology), encapsulated in the form of inference rules, to the model and interpret semantics of exchanged data accordingly. Thus, in doing so the onus for semantic content of an IFC exchanged model is moved from the exporting to the importing application.

The novelty of the proposed approach is that the forward reasoning inference engine for semantic enrichment is incorporated directly within the receiving BIM application through its API. A semantically enriched model produced by the engine is output directly in the native schema of the receiving application. Figure 1 describes the configuration of the semantic enrichment for direct import information workflow. A certified BIM application 'A' exports an IFC model that complies with IFC 2X3 Coordination View 2 or IFC4 Reference View (buildingSmart n.d.). Then, missing semantic concepts and information required by the receiving BIM application 'B' are deduced by the inference engine and imported as the application 'B' native objects.

Inference rules are defined as *IF-THEN* rules using a predefined set of object types and operators expressed in a format easily comprehensible by receiving BIM application domain experts who are not programmers and are skilled in describing domain ontology of the application. The operators include functions for reading the existing IFC model, testing for geometrical and spatial topology relationships, and for creating new objects, properties and relationships (Belsky et al. 2016). Inferred semantic concepts conform to the definitions of the receiving application. Rules for deducing missing or misinterpreted information are specific to the importing BIM application rather than just domain-specific. Concepts, properties, and relationships used to describe the importing application ontology are either universal or domain-specific. The universal concepts, properties, and relationships are universal within the universe of discourse, which in this context is the broad AEC/FM domain as defined by the IFC schema. The domain-specific concepts, properties, and relationships are defined for the receiving application.

4.2 Prototype implementation

A prototype forward chaining inference rule processing engine was implemented in C# programing language in the Microsoft Visual Studio 2019 integrated development environment supporting .NET Framework 4.7. The engine can reason using tests of facts that depend on different objects data types. The spatial topology of building elements is defined by their relative location with regards to one another, so that more than one object data type is often needed in IF-THEN rule clauses. The same is true for other object relationships which are used in inference rules. The engine was incorporated in Autodesk Revit 2019 through its API (RevitAPI.dll and RevitAPIUI.dll) as a plug-in. This BIM authoring tool was chosen to 'host' the inference engine because it is widely used in disciplines such as architecture, structure, and MEP. Its native schema supports a wide variety of semantic concepts used in these domains. An input IFC file is assumed to be exported from a buildingSmart certified BIM modeling tool. This means, among other things, that building elements are explicitly represented as IFC instance entities (subtypes of IfcBuildingElement entity) with distinct 3D geometry and material properties. The input IFC file supplied with missing and/or required semantic concepts is imported into Revit.

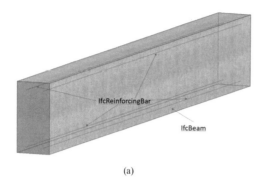
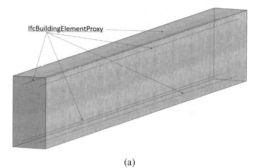

(a)

#244= IFCBEAM('2g_KlIHU97jPVTjpIk4yS4',#42,'M_Concrete-Rectangular Beam:400 x 800mm:411578',$,'400 x 800mm',#242,#233,'411578',,BEAM.);
#310= IFCREINFORCINGBAR('2g_KlIHU97jPVTjpIk4yc4',#42,'Rebar Bar:30M : Shape M_00:411962: 1',$,'Rebar Bar:30M:412582',#292,#307,'411962',$,30.,0.000706858347057703,5920.,,NOTDEFINED.,$);
#346= IFCREINFORCINGBAR('2g_KlIHU97jPVTjpIk4ykK',#42,'Rebar Bar:30M : Shape M_00:412458: 1',$,'Rebar Bar:30M:412582',#329,#343,'412458',$,30.,0.000706858347057703,5920.,,NOTDEFINED.,$);
#367= IFCREINFORCINGBAR('2g_KlIHU97jPVTjpIk4yls',#42,'Rebar Bar:30M : Shape M_00:412488: 1',$,'Rebar Bar:30M:412582',#350,#364,'412488',$,30.,0.000706858347057703,5920.,,NOTDEFINED.,$);
#388= IFCREINFORCINGBAR('2g_KlIHU97jPVTjpIk4yIP',#42,'Rebar Bar:30M : Shape M_00:412519: 1',$,'Rebar Bar:30M:412582',#371,#385,'412519',$,30.,0.000706858347057703,5920.,,NOTDEFINED.,$);

(b)

Figure 2. Reinforced concrete beam: a) 3D view; b) IFC representation.

(a)

#245= IFCBUILDINGELEMENTPROXY('2G_KLIHU97JPVTJPIK4YS4',#43,$,$,$,$,$,$);
#323= IFCBUILDINGELEMENTPROXY('2G_KLIHU97JPVTJPIK4YC4', #43, $,$,$,#300, $,'411962',$,30., $,.NOTDEFINED.,$);
#350= IFCBUILDINGELEMENTPROXY('2G_KLIHU97JPVTJPIK4YC4', #43, $,$,$,#300, $,'411962',$,30., $,.NOTDEFINED.,$);
#371= IFCBUILDINGELEMENTPROXY('2G_KLIHU97JPVTJPIK4YC4', #43, $,$,$,#300, $,'411962',$,30., $,.NOTDEFINED.,$);
#395= IFCBUILDINGELEMENTPROXY('2G_KLIHU97JPVTJPIK4YC4', #43, $,$,$,#300, $,'411962',$,30., $,.NOTDEFINED.,$);

(b)

Figure 3. Re-exported reinforced concrete beam: a) 3D view; b) IFC representation.

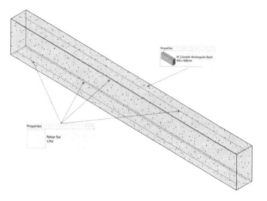

Figure 4. Reinforced concrete beam in Revit.

4.3 Illustration

As a result of multiple bi-directional information exchanges among different BIM applications, interoperability issues such as data loss and misinterpretation commonly arise when a BIM application imports IFC model created by other applications and export it back to IFC format (Lai & Dang 2018). A domain specific application can correctly interpret information from its own domain. However, information from other domains, not closely related to the application's domain, may be lost or misinterpreted due lack of related knowledge in the application native data schema. The applicability of the proposed framework for re-interpretation of exchanged information by the importing BIM application based on its domain ontology is illustrated for reinforced concrete domain.

4.3.1 Classification of proxy elements

Concrete elements with embedded lateral steel reinforcing bars are typical concepts in the reinforced concrete domain. Reinforced concrete elements are usually exported from dedicated BIM applications such as Tekla Structures, Revit, ArchiCAD, etc. as instances of subtypes of abstract entities *IfcBuildingElement* and *IfcReinforcingElement*. For example, a concrete beam reinforced with embedded steel bars, see Figure 2, is usually represented by the instances of *IfcBeam* and *IfcReinforcingBar* entities when exported in the IFC format from a BIM modeling tool.

An AEC/FM project collaboration requires information exchanges among different disciplines. The exported IFC model, then, can be imported into a BIM application used in a different domain, such as MagiCAD in the MEP domain, for further development. The importing application is not familiar with the concept of reinforced concrete. As a result, this concept is lost or misinterpreted when the receiving application maps the content of IFC model to its native schema and exports it back to the IFC format. In that case the re-exported IFC model can contain *IfcBuildingElementProxy* instances instead of *IfcBeam* and *IfcReinforcingBar* instances (see Figure 3). This means that the original semantic representations for reinforced concrete pieces were lost, although their geometry and material properties were preserved.

The use of proxy elements is a common practice in information exchanges among heterogenous BIM applications when exchanged semantic concepts are out of the scope of domain of receiving application (Eastman et al. 2011). The proxy elements, however, can be re-classified by the importing application based on its domain ontology. In this example Autodesk Revit 2019 is used as the importing BIM tool within the reinforced concrete domain. The task for the pilot inference rule processing engine is to re-classify the exchanged proxy elements as a concrete beam reinforced with structural steel rebars and add them to a Revit model. Inference rules need to re-classify proxy elements as the reinforced concrete beam can be written in pseudo-code by the domain's expert as follows:

IF
    (<object1> *is_a* 'proxy element'
OR
    <object1> *is_a* 'concrete beam')
AND
    <object2> *is_a* 'proxy element'
AND
    <object1> *made_of* 'concrete'
AND
    <object2> *made_of* 'steel'
AND
    (<object2> *is_contained_in* <object1>
OR
    <object2> *is_overlapping_with* <object1>)
AND
    <object1> *is_horizontal*
AND
    <object1> *is_1_dimensional*
AND
    <object2> *is_horizontal*
AND
    <object2> *is_1_dimensional*
AND
    | <object1>.Length - <object2>.Length|
    < tolerance>
THEN
    <object1> *is_a* 'concrete beam'
    <object2> *is_a* 'structural bar'

In this example, universal object operators (topological and geometric operators) are denoted by *Italic*; application-specific object concepts are shown as underlined; logical operators are shown in **bold** font; and IFC schema entities are shown in regular font (<object1> and <object2> are instances of subtypes of *IfcBuildingElement* entity). The universal *is_a* operator in 'IF clause' checks if the object type is equal to a given type. In the 'THEN clause' this operator assigns a new type to the object. The universal *made_of* operator checks for material properties of the object. The *is_contained_in*, *is_overlapping_with*, *is_1_dimensional* and *is_horizontal* are universal topological and geometric operators. The *is_contained_in* and *is_overlapping_with* operators check if the 3D shape of object2 is contained in or overlapping with the 3D shape of object1 (i.e. one volume is inside or overlapping with another one); the *is_1_dimensional* operator checks whether the object's one dimension is bigger than its other two dimensions; and the *is_horizontal* operator checks for spatial orientation of the object. The last rule in the 'IF clause' checks whether a difference in the length of two objects is less than a given tolerance.

All operators return Boolean constants: TRUE or FALSE. The engine executes the 'THEN clause' statements if all propositions in 'IF clause' are TRUE. The custom-built forward-chaining rule processing engine applies user defined inference rules to the IFC model. The processing continues until no further inference is possible, i.e. no new facts about a model can be found. Figure 4 summarizes results of executing the inference rules for the reinforced concrete beam. The proxy elements were re-classified as concrete *M_Concrete-Rectangular Beam* and *Rebar Bar* Revit type objects and output into the Revit project model.

5 DISCUSSION AND CONCLUSIONS

Enhancing semantic interoperability among heterogenous BIM applications is an important step towards achieving full potential of BIM and removing barriers to its overall adoption by the AEC/FM industry. There are many issues and costs that are associated with non-interoperable applications. However, the existing solutions to achieve semantic interoperability are also costly, time consuming and their development require substantial amount of effort.

Therefore, it is important to determine the best course of action for achieving acceptable level of semantic interoperability. MVDs, ontologies and open APIs are all valuable and much-needed methods for improving the interoperability in BIM. Regardless of which method is used to improve the semantic interoperability, the use of standards is critical since interoperability requires agreement and consensus among users from different domains. Ontologies and Semantic Web are a great resource to share and exchange domain knowledge. However, the openness of creating an ontology can cause major challenges with its sharing and reuse. Ontologies represent the physical world and people might have different perspectives on how to capture and organize the information.

According to existing bi-directional IFC exchanges it is recommended to propose a new method for achieving effective semantic interoperability. The framework proposed in the paper can help to overcome the lack of rigorous data standards for information exchanges and the drawbacks of the use of stand-alone ontologies developed in ad-hoc manner by placing the onus for semantic interoperability on the receiving application. This is achieved by enabling the receiving application to interpret imported data according to its domain ontology expressed in the form of inference rules compiled by a domain application expert. The proof-of-concept test demonstrated how the goal of placing the onus for information exchanges on receiving applications can be achieved in practice. Inference rules could be easily extended or updated as the host application develops.

The proposed framework is rooted in the semantic enrichment approach for inferencing lost and/or misinterpreted semantic concepts required by the importing application domain from explicit and implicit information contained in the IFC exchange model. The framework is shown to significantly contribute to improving semantic interoperability among heterogenous BIM applications. The approach is based on spatial topology relationships among 3D objects and their geometric properties which are universal across different subdomains of the AEC/FM industry. The proposed framework can also be used as part of code compliance checking platforms and MVD export validation tools. An important advantage the proposed approach has over other existing tools for MVD output validation and code checking (which only check syntax, structure of the concepts, and their attribute values) is in its ability to be embedded in a tool and to check the spatial topology of the building elements in the imported IFC model as well.

6 FUTURE DEVELOPMENT

The use of ontologies appears to be the most promising method to enable semantic interoperability across the multi-disciplinary domains of the AEC/FM industry by introducing a formal and explicit specification of shared conceptualizations. However, the use of stand-alone ontologies for a specific purpose represents a great difficulty in merging and reusing them for other applications. Therefore, development of rigorous data exchange standards and best practices for developing ontologies are crucial for future development and expansion of interoperable software applications.

REFERENCES

Baader, F., Calvanese, D., & Mcguinness, D. 2007. *The Description Logic Handbook: Theory, Implementation, and Applications.* Cambridge: Cambridge University Press, 2nd Edition.

Bazjanac, V., & Kiviniemi, A. 2007. Reduction, simplification, translation and interpretation in the exchange of model data. *CIB W78,* (pp. 163–168).

Beetz, J., van Leeuwen, J., & de Vries, B. 2009. IfcOWL: A case of transforming EXPRESS schemas into ontologies. *Artificial Intelligence for Engineering Design, Analysis and Manufacturing, 23(1),* 89–101.

Belsky, M., Sacks, R., & Brilakis, I. 2016. Semantic enrichment for Building Information Modeling. *Computer-Aided Civil and Infrastructure Engineering, 31(4),* 261–274.

buildingSmart. n.d. *buildingSmart International.* Retrieved March 03, 2020, from https://www.buildingsmart.org/

buildingSmart International. n.d. *buildingSMART International Software Certification Program.* Retrieved March 10, 2020, from https://technical.buildingsmart.org/certification/

Chander, R., Mukherjee, S., & Elias, S. 2017. An applications interoperability model for heterogeneous internet of things environments. *Search Results,* 163–172.

Cheng, J., Lau, G., Pan, J., Law, K., & Jones, A. 2008. Domain-specific ontology mapping by corpus-based semantic similarity. *NSF CMMI Engineering Research and Innovation.* Knoxville, TN.

Costin, A. 2016. *A new methodology for interoperability of heterogeneous bridge information models.* Atlanta: Georgia Institute of Technology.

Costin, A., & Eastman, C. 2019. Need for Interoperability to Enable Seamless Information Exchanges in Smart and Sustainable Urban Systems. *Computing in Civil Engineering, 33 (3).*

Dastbaz, M., Gorse, C., & Moncaster, A. 2017. *Building Information Modelling, Building Performance, Design and Smart Construction.* Springler.

Diallo, S. 2016. On the complexity of interoperability. *Modeling and Simulation of Complexity in Intelligent, Adaptive and Autonomous Systems 2016 (MSCIAAS 2016) and Space Simulation for Planetary Space Exploration (SPACE 2016)* (pp. 1–6). Pasadena, Ca: SprindSim.

Dobrev, P., Kalaydjiev, O., & Angelova, G. 2007. From Conceptual Structures to Semantic Interoperability of Content. *International Conference on Conceptual Structures* (pp. 192–205). Berlin: Springer.

Eastman, C., Jeong, Y., Sacks, R., & Kaner, I. 2010. Exchange model and exchange object concepts for imple-mentation of national BIM standard. *Computingin Civil Engineering, 24(1),* 25–34.

Eastman, C., Teicholz, P., Sacks, R., & Liston, K. 2011. *BIM-handbook: A guide for building information modelingfor owners, managers, designers, engineers and contractors.* New York: Wiley.

Framework, E. I. 2004, November. European Interoperability Framework for pan-European eGovernment services. Belgium: the Europian Communities.

Gallaher, M. P., O'Connor, A. C., Dettbarn, J. L., & Gilday, L. T. 2004. *Cost Analysis of Inadequate Interoperability in the U.S. Capital Facilities Industry.* Gaithersburg, MA: National Institute of Standards and Technology.

Gómez-Pérez, A., Fernandez-Lopez, M., & Corcho, O. 2004. *Ontological Engineering.* London, UK: Springer-Verlag.

Gruber, T. 1995. Toward principles for the design of ontologies used for knowledge sharing. *International Journal of Human-Computer Studies, 43(5–6),* 907–928.

ISO. 2015. *ISO/IEC 33001:2015(en).* Retrieved March 01, 2020, from the International Organization for Standardization: https://www.iso.org/obp/ui/#iso:std:iso-iec:33001:ed-1:v1:en

ISO. 201). *ISO 16739-1:2018.* Retrieved March 4, 2020, from ISO: https://www.iso.org/standard/70303.html

Jim, H.-J., Seo, D., Jung, H., Back, M.-K., Kim, I., & Lee, K.-C. 2017. Description and classification for facilitating interoperability of heterogeneous data/events/services in the Internet of Things. *Neurocomputing,* 13–22.

Kalfoglou, Y., & Schorlemmer, M. 2005. Ontology Mapping: The State of the Art. *Semantic Interoperability and Integration.* Dagstuhl, Germany: Internationales Begegnungs- und Forschungszentrum.

Kalfoglou, Y. 2009. *Cases on Semantic Interoperability for Information Systems Integration: Practices and Applications: Practices and Applications.* New York: Hershey.

Kreidler, C. W. 1998. *Introducing English Semantics.* London: Routledge.

Lai, H., & Deng, X. 2018. Interoperability Analysis of IFC-based Data Exchange between Heterogeneous BIM Software. *Civil Engineering and Management, 24(7),* 537–555.

Le, T., & Jeong, H. 2017. NLP-Based Approach to Semantic Classification of Heterogeneous Transportation Asset

Data Terminology. *Computing in Civil Engineering, 31(6)*.

Leibich, T., & Wix, J. 1998. Highlights of the De-velopment Process of Industry Foundation Classes., (p. European Conference on Product and Process Modeling). Watford, UK.

Liebich, T., Adachi, J., Forester, J., Hyvarinen, J., Karstile, K., & Wix, J. 2006. *Industry Foundation Classes IFC.* International Alliance for Interoperability.

Loebner, S. 2013. *Understanding Semantics.* London: Routledge.

Martinez, M. S. 2002. Semantica Inglesa. *La semantica en el ingles profesional y academico*. Asociacion Espanola de Linguistica Aplicada.

Muller, M., Garbers, A., Esmanioto, F., Huber, N., Loures, E., & Canciglieri, O. 2017. Data interoperability assessment though IFC for BIM in structural design – a five-year gap analysis. *Civil Engineering and Management, 23 (7)*, 943–954.

Olofsson, T., Eastman, C., & Lee, G. 2008. Editorial - Case studies of BIM in use. *Information Technology in Construction, 13*.

Oti, A., Tizani, W., Abanda, F., Jaly-Zaba, A., & Tah, J. 2016. Structural sustainability appraisal in BIM. *Automation in Construction, 69*, 44–58.

Paolucci, M., Kawamura, T., Payne, T., & Sycara, K. 2002. Semantic Matching of Web Services Capabilities. *International Semantic Web Conference*, (pp. 333–347). Sardinia, Italy.

Pauwels, P., & Terkaj, W. 2016. EXPRESS to OWL for constructionindustry: Towards a recommendable and usable ifcOWL ontology. *Automation in Construction, 63*, 100–133.

Pauwels, P., Zhang, S., & Lee, Y.-C. 2017. Semantic web technologies in AEC industry: A literature overview. *Automation in Construction (73)*, 143–165.

Peachavanish, R., Karimi, A., Akinci, B., & Boukamp, F. (2006). An ontological engineering approach for integrating CAD and GIS in support of infrastructure management. *Advanced Engineering Informatics, 20(1)*, 71–88.

Peltomaa, I., Helaakoski, H., & Tuikkanen, J. 2008. Semantic interoperability: Information integration by using ontology mapping in industrial environment. *International Conference on Enterprise Information Systems*, (pp. 465–568). Barcelona, Spain.

Poirier, E., Forgues, D., & Staub-French, S. 2014. Dimensions of interoperability in the AEC industry. *Construction Research Congress*. Atlanta, GA.

Pollock, J. T. 2001. The Big Issue: Interoperability vs. Integration. *eAI Journal*, 48–52.

Pollock, J. T., & Hodgson, R. 2004. *Adaptive Information: Improving Business Through Semantic Interoperability, Grid Computing, and Enterprise Integration.* New Jersey: Wiley.

Ramaji, I., & Memari, A. 2016. Interpreted Information Exchange: Systematic approach for BIMto engineering analysisinformation transformations. *Computing in Civil Engineering, 30(6)*.

Saeed, J. I. 1997. *Semantics*. Oxford: Blackwell.

Shafiq, M. T., Matthews, J., & Lockley, S. R. 2013. A study of BIM collaboration requirements and available features in existing model collaboration systems. *Journal of Information Technology in Construction*, 148–161.

Sheth, A. P. 1999. Changing Focus on Interoperability in Information Systems:From System, Syntax, Structure to Semantics. In *Interoperating Geographic Information Systems* (pp. 5–29). Boston : Springer.

Smith, K. D., & Tardif, M. 2014. *Building Information Modeling: A strategic implementation guide for architects, engineers/.* New Jersey: Wiley.

Steel, J., Drogemuller, R., & Toth, B. (2012). Model interoperability in building information. *Software & Systems Modeling 11*, 99–109.

Tolman, F. 1999. Product modeling standards for the building and construction industry: Past, present, and future. *Automation in Construction, 8(3)*, 227–235.

Venugopal, M., Eastman, C., Sacks, R., & Teizer, J. 2012. Semantics of model views for information exchanges using the industry foundation class schema. *Advanced Engineering Informatics, 26(2)*, 411–428.

Venugopal, M., Eastman, C., Sacks, R., Panushev, I., & Aram, S. 2010. Engineering Semantics of IFC product model views. *CIB W78 – Information Technology for Construction.* Cairo, Egypt.

W3C. n.d. *OWL 2 Web Ontology Language*. Retrieved March 6, 2020, from https://www.w3.org/TR/owl2-overview/

Youngs., Jones, S., Bernstein, H., & Gudgel, J. 2009. *The Business Value of BIM: Getting Building Information Modeling to the Bottom Line.* New York: McGraw-Hill Cos.

Zamanian, M., & Pittman, J. 1998. A software industry perspective on AEC information models for distributed collaboration. *Automation in Construction, 8(3)*, 237–248.

Zhang, J., & El-Gohary, N. 2017. Integrating semantic NLP and logic reasoning into a unified system for fully-automated code checking. *Automation in Construction, (73)*, 45–57.

Towards semantic enrichment of early-design timber models for noise and vibration analysis

C. Châteauvieux-Hellwig
Technical University of Applied Sciences Rosenheim, Germany

J. Abualdenien & A. Borrmann
Technical University of Munich, Germany

ABSTRACT: Low carbon footprint and high sustainability characterize timber construction. Accordingly, architects and engineers are increasingly using it in their designs. However, a major challenge in timber construction is to provide a sufficient sound insulation. In contrast to masonry and concrete construction, timber construction lacks software tools to evaluate sound insulation and usability with regard to vibrations. Using Building Information Modelling (BIM), the design phases already incorporate model-based calculations for sound insulation prognosis. However, performing such evaluations requires specific information that describes the junctions between elements. For example, relevant factors are: the type of elements in the junction, the connecting elements like screws or angle brackets, and the decoupling materials used. Typically the building information model lacks this information; hence, we present a semantic enrichment approach overcoming these limitations.

1 INTRODUCTION

Managing the carbon emissions resulting from the architecture, engineering and construction (AEC) represents a major factor in mitigating these emissions globally (Wong 2013). The sustainability of timber construction encourages architects and engineers to choose this construction method for their projects. However, there is a lack of specialized evaluation tools for the seamless application of this construction method. The innovative approach of Building Information Modelling (BIM) opens up the possibility to integrate model-based evaluations and analysis for a complete noise control prognosis as well as for the compliance with the usability requirements directly into the design phases via BIM-compatible interfaces.

In this paper, we use the architectural discipline as a basis, which contains the necessary positions and dimensions of walls and ceilings. In this regard, the acoustic engineer analyses the architectural model during the design phases to perform the calculation. Such prognosis requires the selection of the separating elements between the different rooms and analyses all junctions between each separating element and its flanking element. For every junction, the acoustic engineer needs to find all transmission paths and assign them a vibration reduction index K_{ij} to calculate the sound transmission index R.

In the future, the goal is to automate this procedure. The determination of junctions out of an IFC data model is part of a thesis project at the University of Applied Sciences Rosenheim in cooperation with the Technical University of Munich.

2 IFC IN TIMBER BUILDING DESIGN

In solid wood constructions, the timber constructor is responsible for the detailed planning of the construction site. Therefore he creates manufacturing plans for the building components, according to the architect's design plans. The manufacturing model is created with programs that are capable of both CAD (computer automated design) and CAM (computer-aided manufacturing). After drawing CNC machines (Computerized Numerical Control) uses the plans directly for manufacturing. However, linking the manufacturing model to the architectural model is not yet a common practice (Huß 2017).

The vendor-neutral IFC data exchange format is available for construction projects to be independent of software manufacturers. In addition to the geometric representations, the IFC data model stores information describing the semantics of the different components. These semantics are defined as topological relationships as well as a set of Property-Sets which can hold multiple properties. Although these building models define the material information they are not provided by a fixed catalogue. Additionally, multiple properties need to be collected from manufacturers or assumed by experience. Currently, there are some properties available in the common Property-Sets. For example, the

Property-1Set `Pset_MaterialWood` defines the properties of timber elements. However, it relates to properties of structural analysis and is not suitable for sound insulation in timber construction. Further, the current version of the standard does not sufficiently represent acoustic properties.

3 SOUND TRANSMISSION IN WOODEN BUILDINGS

The prognosis of sound insulation includes the calculation of airborne sound insulation and impact sound insulation. The calculations are performed in building construction according to (ISO12354-1 2017) and are frequency-dependent from 50 to 5000 Hz:

$$R_{ij} = \frac{R_i + R_j}{2} + \Delta R + \overline{D_{v,ij}} + 10lg\frac{S_S}{\sqrt{S_i \cdot S_j}} \quad [dB] \quad (1)$$

With
$\overline{D_{w,i,j}}$ velocity level difference [dB]
R_i, R_j sound insulation of element i and j [dB]
S_S surface of separating element
S_i, S_l surface of element i/j [m²]
$\Delta R = \Delta R_i + \Delta R_j$ improvement of sound insulation due to wall linings, screeds or suspended ceiling [dB]

For timber construction, the calculation models are still in development (Rabold 2017; Rabold 2018a). The existing models need to be optimized: The unique features of timber construction concern the building components, which have a much lower mass than solid components, and the connection situation of components to each other.

3.1 Flanking transmission

Not only the separating component directly transmits the sound but also the flanking elements. Figure 1 shows the sound transmission paths depending on the direction and excitation.

3.2 Vibration reduction index Kij

The calculation takes into account the joints based on the vibration reduction index K_{ij}, which depends on the direction of the junction, the design of the connection details, and the component types used. All elements in the sending room have the index i and all elements in the receiving room have the name j.

$$K_{ij} = \overline{D_{vlj}} + 10lg\frac{l_{ij}}{\sqrt{a_i \cdot a_j}} \quad [dB] \quad (2)$$

With
$\overline{D_{v,lj}}$ velocity level difference [dB]
l_{ij} junction length between element i and j [m]
a_i, a_j, equivalent absorption length of element i/j [m]

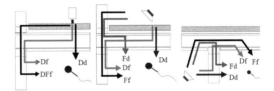

Figure 2. Schematic representation of the transmission paths Ff, Df, Fd and DFf in timber construction: impact sound insulation (left), airborne sound insulation through a slab (middle) or a wall (right).

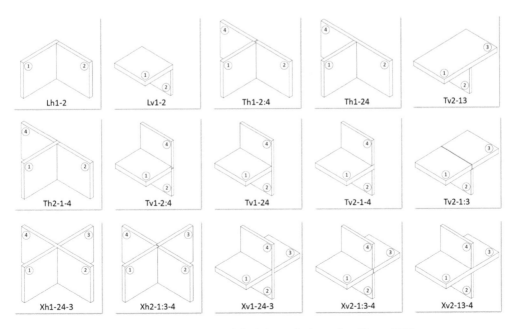

Figure 1. 15 types of junction without consideration of elastic layers for decoupling (Timpte 2016).

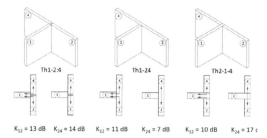

Figure 3. Types of T-junction of wall elements in CLT construction without consideration of elastic layers for decoupling (Timpte 2016).

3.3 Details for Kij

The length of the junction l_{ij} between both elements is a geometrical length. The equivalent absorption length for the element depends i on the area of the element and its structural reverberation time. The velocity level difference $\overline{D_{v,ij}}$ is measured in laboratory.

The vibration reduction index is also investigated from existing measurements and simulations. Factors of influence for K_{ij} are the construction material of the elements, the dimensions of the elements, the mass ratio between the elements, the execution of the junction details, elastic layers and their stiffness, the excitation and the direction of the junction.

3.4 Types of junction for CLT constructions

In the beginning, we only consider constructions of CLT (cross laminated timber). Starting from one to three possible flanking elements in a junction, there are 15 possible combinations of junctions (see Figure 1). The distinction between the types of junction is essential. The flanking transmission depends on how and where the elements met. This is shown in Figure 3 for a T-junction. The 3 different combination of elements leads to very different vibration reduction indices K_{ij}. There exist more combinations if we consider putting elastic layer for decoupling in the junction or if the walls are fixed with angles instead of screws. Each of those details needs to be considered to find the correct value for K_{ij}.

4 PREVIOUS WORK

The University of Applied Sciences in Rosenheim has worked for many years on the acoustics in timber buildings. The aim of the project "vibroacoustic" (Rabold et al. 2018b) was to lay the scientific foundations for a calculation based on FEM and other model approaches of the vibroacoustic properties of a wooden building. One outcome of this project was a spreadsheet tool to simplify the calculation of the sound insulation of elements for a single value. Another result was the creation of the online database (Vabdat 2018), which stores frequency-dependent measurement and calculation results from vibroacoustic properties of construction products and building parts.

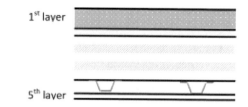

B_sCE60_ipMF6_bCLT162_frM27||iMW20_bGP12

sCE60: screed, cement, 60 mm
ipMF6: impact insulation, mineral fibre, 6 MN/m³
bCLT162: board, cross laminated timber, 162 mm
frM27: framework, metal, 27 mm
|| in the same layer
iMW20: insulation material, mineral wool, 20 mm
bGP12: board, gipsum, 12 mm

Figure 4. Description of a building element using his layers with the abbreviation of layer in (Vabdat 2018).

In the database, every building element ("Bauteil") is formed by different layers of construction products ("Bauprodukt/-stoff"). The single layers get abbreviations depending on their material. The name of the building elements is then put together from the name of the single layers of construction products. Thus, the name of the building element describes its construction (see Figure 4).

Furthermore, the database also stores the vibration reduction index from specific junctions. It also stores the junction type according to Figure 1, information about elastic layers and the fastener. The values for the vibration reduction index K_{ij} are stored frequency-depend and for every transmission path.

5 JUNCTIONS IN IFC

The information provided by the IFC model enormously depends on who built the model and how. If the model comes from an early planning phase, it has fewer details than a model created for manufacturing. Since the planning of the sound insulation should take place at an early stage of the planning, we assume the IFC data model to have only rudiment information about the elements and their junctions.

Additionally, junctions do not explicitly exist in the IFC standard. But there are many different possibilities to form relations between objects with the IfcRelConnects class. Relevant classes are IfcRelConnectsElements to connect elements together and IfcRelContainedInSpatialStructure to position elements in a space or building storey.

IfcRelConnectsElements includes two classes: IfcRelConnectsPathElements where the connection of two elements is described by the attributes AtStart, AtEnd and AtPath. In IfcRelConnectsWithRealizingElements the elements are connected the same way, but it is possible to add elements that realize the connection. In both options, `IfcConnectionGeometry` stores the connection geometry. Depending on the quality of the model, this information may not be available.

The relation does not describe a junction because it can only be a relation between two elements: 3 relations describe a junction with 3 elements and 6 relations describe a junction with 4 elements.

Besides, there is no intended connection between walls and ceilings. Such a connection would require additional attributes to specify if the connection is `AtBottom` or `AtTop` of the wall. A connection can still be forced without specifying the connection type or with AtPath to characterize the connection on the slab.

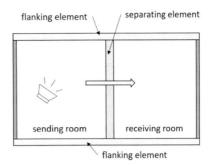

Figure 5. Schematic representation of the separating element and the flanking elements: a standard, rectangular wall has 4 flanking elements: 2 walls and 2 slabs (floor and ceiling).

6 FLANKING ELEMENTS

The identification of the flanking elements plays a central role in the calculation of sound insulation. Flanking elements are all elements, which have a common edge with the separating element. For a wall as a separating element, this is usually 2 walls and 2 slabs. For a slab as a separating element, this is then 4 walls. A flanking element is one continuous element or it consists of 2 different elements that both met at the junction. The approach used to identify the flanking elements depends on the information provided in the IFC-Model.

Furthermore, a sending room and a receiving room need to be identified. The separating element is in between both rooms, for example, for a wall as a separating element, both rooms must be on the same floor. In this case, it does not matter which of the rooms is named sending or receiving room. If the separating element is a slab, then the room existing on the higher floor is automatically the sending room and the room under the slab is the receiving room. In this regard, the separating element and four flanking elements characterize both connected rooms. Additionally, the last elements are parallel to the separating element.

6.1 With IfcRelConnectsPathElements

If a relation between the separating element and another element exists as `IfcRelConnectsPath Elements`, this is used to determine the flanking elements.

6.2 With IfcRelSpaceBoundary

It is possible to find all flanking elements if the adjacent room to the separating element is an `IfcSpace` and `IfcRelSpaceBoundary` describes the adjacent building elements. Then only the elements related to the space need to be analysed for distance to the separating element.

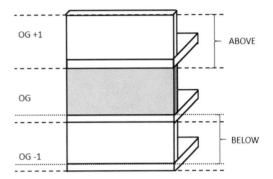

Figure 6. Elements filtered before distance checking: consider all walls and slabs one storey above the separating wall, the slabs and walls in the same storey and walls one storey below.

6.3 Semantic enrichment by geometry analysis

If the IFC data model does not provide any of the above semantical information, a geometric analysis of the model defines the flanking elements.

At first, the model is filtered by building storeys to reduce the number of elements for the analysis. If the separated element is a wall, the filter considers walls and slabs on the same storey and slabs above. For a slab as separating element, walls of the same storey and of the storey below are filtered out.

Then elements that are close to the separating element are considered: The distance between the separating element and each of the other elements is calculated to find any adjacent elements. Afterward, the position of the element, depending on the separating element, is computed and stored for each adjacent element. Algorithm 2 shows the calculation of the distance and the determination of the position in X-Y-plane. Algorithm 1 defines if the element is above, below or in the same height as the separating element.

In combination, both algorithms determine the position of an element in comparison to the separating element as North (higher on the y-axis), South (lower on the y-axis), West (lower on the x-axis), East (higher

```
double string position p
1:   foreach element (E)
2:     if E.type = slab
3:       if E.storey = SE.storey
4:         p = +Under
5:       else if E.storey = SE.storey +1
6:         p = +Below
7:       end if
8:     else if E.type = wall
9:       if E.storey = SE.storey
10:        p = p
11:      else if E.storey = SE.storey +1
12:        p = +Above
13:      else if E.storey = SE.storey -1
14:        p = +Below
16:      end if
17:    end if
```

Algorithm 1. Assignment of position p in Y-Z-plane with help of storeys for the Element E in comparison to the separating element SE.

on the x-axis), Above (higher on the z-axis), Below (lower on the z-axis) and the combinations. The reference coordinate system is the coordinate system of the separating element.

7 DEFINING JUNCTIONS

The currently developed prototype works for straight elements with a rectangular shape. Regardless of whether the separating element is a wall or a slab, it has 4 edges. For each separating element, four junctions are created. Each junction gets information about the flanking elements involved and about other separating elements like elastic layers and fastener elements.

Not only elements, which touches the separating element, are considered as flanking elements, but also adjacent elements with a certain distance (see Figure 7). It can be the case if the modeling of the elements is not entirely correct and the elements do not touch each other, even if they should. Sometimes intermediate layers like elastic layers are in between or even a facing layer if it is modeled separately from the flanking element. If the algorithm finds such an element, other elements in the same direction with a distance $d < d_1 + d_2 + d_3$ with d_1 — distance to the first element d_2 — thickness of the flanking element d_3 — distance to the next flanking element, $d_3 < 0,1\,m$ could be adjacent and relevant for the junction. Here the material helps to distinguish between a flanking element or an intermediate layer.

A new object is defined to describe the 4 junctions of the separating element. The junction elements contain 4 slots each. Intermediate layers in the junction are also saved in the corresponding slot.

A set of rules finds the adjacent element, their position and stores them in the correct slot of the junction

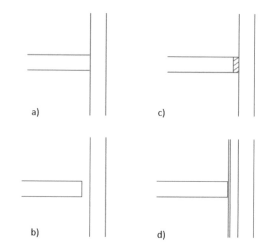

Figure 7. Consideration of junction for an element which is a) touching the separating element (d = 0) or adjacent (0 m < d < 0,1 m) to the separated element with b) air in between, c) an elastic layer or d) a facing layer.

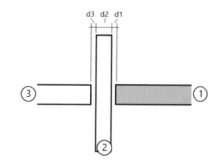

Figure 8. Distances considered to define a X-junctions with the separating element in grey (1) and the added flanking elements in white (2, 3).

object. Algorithm 3 shows an example of defining a T-junction with 3 elements.

8 MATERIAL LAYER

Each element needs a sound transmission index R, which depends on his material and the structure of the different layers. In the IFC data model, IfcRelAssociatesMaterial assigns the material to an element. The IfcMaterialLayerSet assigns different materials to one element consisting of different layers.

The sound transmission index R_i considers the main element, while ΔR_i takes into account the facing layers like a screed or a suspending ceiling. For double-layered walls, R_i considers both parts, but the vibration reduction index K_{ij} is only relevant for the shell on the inside. To find the correct values the online database (Vabdat 2018) can be used.

	double distance d, string position p
1:	foreach element (E)
2:	if angle to separating element (SE) = 0 OR = 90
3:	if E.Max.Y <= SE.Min.Y
4:	if E.Max.X <= SE.Min.X
5:	$d = \sqrt{(E.Max.X - SE.Min.X)^2 + (E.Max.Y - SE.Min.Y)^2}$
6:	$p = SouthWest$
7:	else if E.Min.X >= SE:Max.X
8:	$d = \sqrt{(E.Min.X - SE.Max.X)^2 + (E.Max.Y - SE.Min.Y)^2}$
9:	$p = SouthEast$
10:	else
11:	$d = SE.Min.Y - E.Max.Y$
12:	$p = South$
13:	end if
14:	else if E.Min.Y >= SE.Max.Y
16:	if E.Max.X <= SE.Min.X
17:	$d = \sqrt{(E.Max.X - SE.Min.X)^2 + (E.Min.Y - SE.Max.Y)^2}$
18:	$p = NorthWest$
19:	else if E.Min.X >= SE.Max.X
20:	$d = \sqrt{(E.Min.X - SE.Max.X)^2 + (E.Min.Y - SE.Max.Y)^2}$
21:	$p = NorthEast$
22:	else
23:	$d = E.Min.Y - SE.Max.Y$
24:	$p = North$
25:	end if
26:	else
27:	if E.Max.X <= SE.Min.X
28:	$d = E.Max.X - SE.Min.X$
29:	$p = West$
30:	else if E.Min.X >= SE.Max.X
31:	$d = E.Min.X - SE.Max.X$
32:	$p = East$
33:	else
34:	$d = 0$
35:	$p = 0$
36:	end if
37:	end if

Algorithm 2. Calculation of distance d and assignment of position p in X-Y-plane (not considering above and under) from the element E in comparison to the separating element SE.

	junction j
1:	if E1.type = wall
2:	AND if E1.distance < 0,5
3:	AND angle (SE,E1) = 90
4:	AND if E1.position = South
5:	AND if E1.X.Max = SE.X.Max
6:	then
7:	$\{j.type = Lh1 - 2$
8:	$j.slot1 = SE$
9:	$j.slot2 = E1$
10:	if E2.type = wall
11:	AND if E2.distance < 0,5
12:	AND angle (SE,E2) = 90
13:	AND if E2.position = North
14:	AND if E2.X.Max = SE.X.Max
16:	Then
17:	$\{j.type = Th1 - 2 - 4$
18:	$j.slot3 = E2\}$
19:	}

Algorithm 3. Definition of a T-junction for a separating element (SE) of type wall and a first flanking element E1 and the second flanking element E2.

Original IFC file content:

```
#182= IFCWALL('0T1$T1hUf8 f2ueO1CDddP',#42,
              'Basiswand:CLT80:2394561',
              $,'Basiswand:CLT80:2392169',
              #143,#176,'2394561',.NOTDEFINED.);

#294= IFCSLAB('0sBzAasRvAof649twrHZzP',#42,
              'Geschossdecke:STB 200:2397895',
              $,'Geschossdecke:STB 200',
              #257,#290,'2397895',.FLOOR.);
```

Semantic Enrichment:

```
#391= IFCRELCONNECTSPATHELEMENTS('1Fteoyk6r14
      Ot5BkeHj9bQ',#41,'0T1$T1hUf8 f2ueO1CD
      cD7|0T1$T1hUf8_f2ueO1CDcDw','Structural',
      $,#294,#182,(),(),.ATSTART.,.ATPATH.);
```

Figure 10. Semantic enrichment for the connection between a wall and a slab.

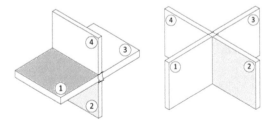

Figure 9. Definition of slots for junctions: a slab as separating element is in slot 1, a wall as separating element is in slot 2.

9 SEMANTIC ENRICHMENT

To supplement the IFC file with semantically useful content, Belsky uses the concept of semantic enrichment (Belsky 2015). In contrast to the MVD approach where the software companies develop an IFC export based on the MVD, the semantic enrichment uses a domain-specific rule-sets to infer semantics from geometry and material properties of the IFC data model. The task for semantic interpretation is then in the importing application and not at the exporting software.

For example, in (Belsky 2015), a rule-set is used to identify the joint of precast concrete walls that were initially declared as `IfcColumn` and rename them to `IfcFastener`.

Figure 10 shows an example using semantic enrichment for the junctions of elements. It illustrates how the connection between a wall as `IfcWall` and a slab as `IfcSlab` can easily be executed by

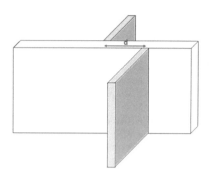

Figure 11. Example of a special junction: offset between the flanking elements (blue and green wall). For d < 0,5 m the green wall is considered being on the exact opposite of the blue wall and the junction is handle as an X-junction (ISO12354-1 2017).

using `IfcRelConnectPathElements` even if the Attribute AtPath is not a perfect description of the junction position.

10 NEXT STEPS

The next steps are to generate the complete rule-sets to identify all junction types and demonstrate the workflow with prototype software. Then the inferred information needs to be added to the IFC data model. The semantic enrichment can either use existing entities like `IfcRelConnectsPathsElements` or create new PropertySets.

Afterward, particular junctions that are not symmetrical, like in Figure 11, need to be considered to extend the use of the developed prototype for more applications. Also, exceptional cases with the separating element being bigger or smaller than the separating area between the sending room and the receiving room will be a major challenge to solve in the future.

ACKNOWLEDGMENT

The authors would like to thank the Bavarian State Ministry of Education and Cultural Affairs, Science and the Arts for the financial support of the program for the promotion of applied research and development at Universities for Applied Sciences (funding reference number: VIII.2-F1116.RO/17/2).

REFERENCES

Belsky, M., Sacks, R., Brikalis, I., 2015. Semantic enrichment for Building Information Modelling, *Computer-Aided Civil and Infrastructure Engineering.*

Huß, W., Stieglmeier, M., 2017. *leanWOOD, Buch 1 – Teil A leanWOOD – Herausforderunen & Motivation.* München.

ISO12354-1, 2017. *Building acoustics – Estimation of acoustic performance of buildings from the performance of elements – Part 1: Airborne sound insulation between rooms.*

Rabold, A., Châteauvieux, C, Schramm, M., 2017. *Vibroakustik im Planungsprozess für Holzbauten – Modellierung, numerische Simulation, Validierung – Teilprojekt 4: Bauteilprüfung, FEM Modellierung und Validierung.* Rosenheim.

Rabold, A., Châteauvieux, C., Mecking S, 2018a. Nachweis von Holzdecken nach DIN 4109 – Möglichkeiten und Grenzen. *DAGA 2018.* Munich.

Rabold, A.; Wohlmuth, B.; Horger, T.; Rank, E; Kollmannsberger, S.; Frischmann, F.; Paolini, A.; Schanda, U.; Mecking, S.; Kruse, T.; Châteauvieux-Hellwig, C.; Schramm, M.; Müller, G.; Buchschmid, M.; Winter, C., 2018b. *Vibroakustik im Planungsprozess für Holzbauten – Modellierung, numerische Simulation, Validierung.* cooperation project TU München, TH Rosenheim, ift Rosenheim.

Sacks, R. Ma, L., Yosef, Raz, Borrmann, A., Daum, S., Kattel, U., 2017. Semantic Enrichment for Building Information Modelling: Procedure for compiling inference rules and operators for complex geometry. *Journal of Computing in Civil Engineering Vol 31 Issue 6.*

Timpte, A. 2016. *Stoßstellen im Massivholzbau – Konstruktionen, akustische Kenngrößen, Schallschutzprognose.* Master Thesis, TH Rosenheim.

Vabdat, 2018. vibro-acoustical database, *https://www.vabdat.de/,* TH Rosenheim.

Wong, J. K.; Li, H.; Wang, H.; Huang, T.; Luo, E.; Li, V. (2013): Toward low-carbon construction pro-cesses: the visualisation of predicted emission via virtual prototyping technology. In: Autom. Constr. (33), S. 72–78.

Evaluating SPARQL-based model checking: Potentials and limitations

A. Hoffmann, M. Shi, A. Wagner, C.-D. Thiele, T.-J. Huyeng & U. Rüppel
Technische Universität Darmstadt, Institute of Numerical Methods and Informatics in Civil Engineering, Germany

W. Sprenger
Zentrale Technik, Digitalisierung & Softwareengineering (DS) · Digitalization & Software Engineering BIM 5D®, Ed. Züblin AG, Germany

ABSTRACT: Model checking is an important task in the BIM collaboration process to prevent expensive planning errors. The submodels of the individual disciplines are transferred into a coordination model. Part of the transfer is a conversion into an exchange format. The exchange format allows the import into the model checking application. In the model checking application routines are performed to check the model against collisions and building regulation violations. During the transfer into the exchange format, information may get lost, especially with parameters that are not yet part of the exchange format supported by the authoring software. In recent years, ontologies have been investigated as a feasible approach to combine the submodels, since they model data in a flexible manner. Hence in the conversion process to an application-specific ontology, the data structure of the submodels can widely persist, which could lead to smaller information loss in comparison to converting the data into a standardized exchange format. The evaluation of the geometric properties of the building is indispensable for detecting and analyzing collisions. The basis for the connection of the different sub models could be the BOT (Building Topology Ontology), which defines the topological structure of a building and can be used to represent further building information by linking it with other ontologies. The relevant geometric relationships for the collision model checks have to be derived with a geometry kernel. For the research in this paper pythonOCC, a wrapper for the geometry kernel Open CASCADE is used with the Semantic Web's own query language SPARQL, queries can be formulated to analyze the collision relationships in combination with other semantic information. These queries can be used to verify model correctness. By connecting the information from different domains, more sophisticated tests are possible than in an exchange format dependent model checking application. The goal is to integrate the developed functionalities into a project platform. This platform is based on an extensive project description in an ontology-based data model and is connected to different authoring tools for the exchange of information.

1 INTRODUCTION

The goal of the research project *SCOPE* is to develop a software platform that can be configured modularly and is designed for the needs of the civil engineering domain ("SCOPE – Semantic Construction Project Engineering" 2020).

Due to the diversity and individuality of the processes in individual construction companies, it is difficult to provide a uniform solution that can be adapted to the needs of each company. Each company uses an individual combination of authoring software applications, which requires data models that differ in level of detail and context.

For the exchange of data, there are unified interfaces, such as open data formats, of which the most mature for the BIM methodology is IFC. Uniform interfaces have the disadvantage that they are designed for certain standardized processes. The interface tries to consider all possible data from the internal data model of the software. Sometimes data is not required for the use case at hand, or a combination of data from other sources is necessary to obtain the required modeling depth. This results in lengthy conversion chains, which means that the documentation of various application interfaces and data formats has to be dealt with. Individual solutions emerge that become obsolete after software updates.

In recent years, there have been repeated ambitions to simplify the conversion chains on the basis of ontologies: The internal data schema of the software application's proprietary format is modeled outside the application using RDF (resource description framework) via graph-forming triples. Subsequently, both data models are linked by so-called ontological matching. In a simple case, triples represent the equivalence of two graph nodes (Euzenat & Shvaiko 2013). Ontologies, thus, offer the possibility to partially map the original conversion process in the data model.

This disclosure of the data model offers the possibility to enrich the graph with further knowledge. For example, rules can be defined to check the overall model integrity. The advantage compared to an application like Solibri is that the rules can be based directly on the authoring tool's proprietary format and, thus, correlations can be checked with less data conversions. An abstraction in IFC is not necessary. Especially in AECOO, rules are based on geometric relationships. The geometric relationships are formed between the individual instances that form the model. To identify these relationships, at least a uniform abstraction into a geometric model is necessary.

In the *SCOPE* project, this uniform abstraction is implemented based on Open CASCADE geometry kernel commands, which are mapped to an ontology. The geometric descriptions are assigned to individual instances that can be uniquely identified by Uniform Resource Identifiers. The alignment towards other ontologies, which are based on proprietary formats, can, thus, be established via these URIs. The uniform geometric abstraction simplifies the identification of geometric relationships between the instances, which can enrich the entire data model, consisting of the different ontologies based on the proprietary export formats. Subsequently, rules can be formulated that are based both on the geometric relationships between the objects and on the individual expression of the objects in the data model of the proprietary authoring software.

This paper discusses the potential and limit of ontology-based coordination models in respect of performing rule checks via Semantic Web technologies. To evaluate both aspects of combining heterogeneous data and rule checking, multiple use cases are considered. Existing practices for Linked Building Data are applied to structure building data and connect objects to their geometry description in a uniform manner. Moreover, the domain of structural engineering is considered within the use cases with the assumption that the internal database of the SOFISTIK software application is translated into an ontology.

Next, the methodology of model checking, including currently implemented approaches, is elaborated, before going into more detail for collision and proximity queries. Thereafter, the assumed data model for this paper's analysis is presented and a concept for a Semantic Web-based rule checking engine is proposed. Based on this concept, the considered use cases are demonstrated and discussed.

2 MODEL CHECKING

In (Eastman et al. 2009), the four essential functionalities of a rule-based model checking system are elaborated: The first functionality is the **rule interpretation**. A distinction between the context of the rule and the checked property is made. In the first step, potential instances of the model to which the rule refers, are identified and, in the second step, the instances' properties checked against the rule. The rule can be checked logically, resulting in either *true* or *false*. It can also be evaluated whether the result refers to an existing instance or all instances. The context of the rule can be hard-coded and/or allow a parametric freedom of choice.

The second functionality is the **preparation of the building model**. This functionality involves providing the model with the required properties to check the rule. Next to those properties that are already present in the building model, implicit properties need to be derived from the model to complete the rule's requirements. For example, the distance between different instances resulting from their explicit positions may be needed for executing the rule. The model preparation can go so far that completely new models have to be derived or separate model views with their individual properties need to be created for rule checks. This second functionality also includes model enrichment with analytical data from, for example, energy or structural modeling.

The third functionality represents the **rule execution**. Before the actual execution of the rule, the model should be checked for syntax and version first. Moreover, the completeness of the applied rules has to be ensured.

The last functionality is the **rule check reporting**. It should be possible to display each instance of the model to which a rule is applied. In the case of spatial conflicts, for example, graphical reporting using a camera position and a coordinate is useful. In general the textual description of the original rule should always be accessible from the reporting result.

Solibri and Navis Works represent two established software solutions for model checking that have proven themselves in the industry for several years. Integrating model checking firmly into the planning process supports collaboration of stakeholders from different domains that are not working on the same model. With Solibri, one unit in the organization takes over the model checking with Solibri ("Solibri Workflow" 2020). The individual stakeholder models are exported from the authoring software tools according to the IFC Coordination View and imported into Solibri via the coordination unit. Within this unit, the validity and completeness of the individual models are examined before the models are checked for rule violations. If rule violations are detected, the user categorizes and summarizes them and forwards them either in the BIM Collaboration Format (BCF) or as a PDF report to the responsible planning offices, which have to coordinate among themselves. Due to the lack of support of the BCF format by authoring software applications ("Supporting Applications" 2019), the PDF version is easier to be used. Solibri offers rule sets that can be composed of parametrically changeable predefined rules. For free creation of custom rules the Solibri API has to be used, which is not open (Sydora & Stroulia 2019).

The nondisclosure of the data model of proprietary applications is contrasted with prototypical work in research projects based on Semantic Web technologies. In these research projects, the disclosure and reuse of data models for rule formulation and storage of building information based on ontologies is proposed.

Pauwels (Pauwels & Zhang 2015) present the essential strategies for ontology-based rule checking. The first strategy, *"Hard-coded rule checking after querying for information"*, involves implementing an application that interprets the rules described in RDF and executes them on its internal data model. In the second strategy, *"rule-checking by querying"*, building information is stored as ontology and the rules are coded as SPARQL-query. In the third strategy, *"Semantic rule checking with dedicated rule languages"*, the rules are recorded in a RDF-based notation and rule engines are used to interpret the rules and extend the model with the results of the rule evaluation. Hence, the rule engine is responsible for the tasks that were executed by the applications in the previous strategies.

All three strategies assume that the data model, to which the rules are applied, is complete. In practical application, the model preparation of the second rule-based model checking systems' functionalities should be performed as well, since rules in civil engineering are based on derived information, such as distance, collision, etc. To keep the application flexible and performant, depending on the rule type, custom code has to be executed to obtain this information.

3 COLLISION AND PROXIMITY QUERIES

Collision and proximity queries aim to identify information about the spatial relationship between two objects (Lin & Manocha 2008). They return either Boolean or numerical values that characterize the spatial relationship. A collision check is usually defined to return a boolean value that specifies whether two objects overlap or not. A distance function returns the length of the shortest connecting line segment between two objects. For both functions there are approximate procedures with varying accuracy and computing time efficiency.

The objects can be represented by geometric primitives depending on the method. For a coarser representation, they can be convex polytopes. In detail, geometric objects can be represented more precisely by representing the surface as a polygon mesh. To avoid having to re-evaluate geometric primitive pairs, sets of geometric structures are combined and a bounding volume is formed around them. The bounding volumes are represented as nodes of a tree structure (for example, k-d trees and octrees, R-trees and their variants, cone trees, binary space partitioning trees), where the root node contains all primitives of an object and the lowest child node the smallest geometric representation (for example, triangles, polygons, or Non-uniform rational B-Splines) (Larsen et al. 2000). Bounding volumes can be represented by various geometric concepts, e.g., axis-aligned bounding boxes, oriented bounding boxes, spherical shells and k-DOPs.

The strategies for querying for collisions and distances are similar: The bounding volume hierarchies of two objects are traversed simultaneously. In a distance function, the smallest distance at the current state is saved in a variable. Iteratively, the bounding volumes are used to check whether the distance between the geometric primitives contained in the node and the object to be checked is smaller than the saved distance variable. If this is the case, the node is further subdivided, otherwise it is discarded. The collision function checks for possible collisions instead of smaller distances.

The calculation of the collision itself depends on the selected bounding volume. For example, if axis-aligned bounding boxes are selected, it is sufficient to check the minimum and maximum values on the axes of the global coordinate system. To check the collision of two triangles it is sufficient to look at the plane that is spanned by the edges of the triangle of one object and check whether it intersects with the edge of another object (Nour 2016). An example for a domain-specific implementation, is the clash detection of the BIM server, which is based on this principle (Helm et al. 2010).

Finding the minimum distance between two geometric primitives is a search problem, for which different algorithms have been developed depending on the geometry, based, for example, on the Minkowski Sum of the geometric primitives (Lin & Manocha 2008).

4 DATA MODEL

The upcoming concepts and evaluation are based on a Semantic Web-based data model that spans across different domains of the AECOO. The Building Topology Ontology (Rasmussen et al. 2020) serves as the central element of this multi-domain data model, see Figure 1.

BOT is designed to link the various disciplines of the Architecture, Engineering, Construction, Owner Operator sector (AECOO) and focuses on mapping only the interdisciplinary topological relationships between the elements and spaces. Geometries and domain-specific descriptions can be linked via ontologies. Thereby, the BOT is kept simple compared to the IFC schema and analytical queries are easier to formulate and understand. Their properties describe spatial, or more precisely, topological relationships (e.g. bot:intersectingElement, bot:containsElement), which can be used in subsequent work to optimize collision queries or to check the correctness of these relationships.

Geometric representations can also be connected with a bot:Element, see Figure 2. To map this relationship in a uniform manner, the relation omg:hasGeometry of the Ontology for Managing Geometry (OMG) can be used (Wagner et al. 2019a). The reference geometries are represented by Open CASCADE kernel commands. When executed, they construct a Brep that can be used for collision checking.

The general building element properties, are described via the Building Product Ontology (BPO,

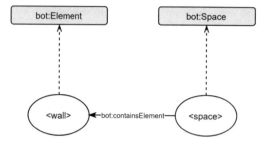

Figure 1. A wall in the BOT- data model.

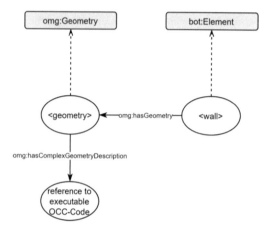

Figure 2. Referencing the geometric description.

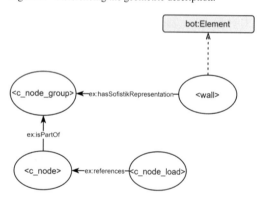

Figure 3. Adding general building properties.

Wagner et al. 2019b). The connection to the BPO is realized via the gr:hasMakeOrModel property of the GoodRelations-Ontology (Hepp 2011), see Figure 3. The data from the software applications of the different AECOO disciplines could be linked to BOT and BPO. To verify the approach, analytical data from the specialist simulation tool for static calculations Sofistik are assumed to be stored in an ontology.

For this purpose the SOFiSTiK Central DataBase (CDB) is assumed to be accessed and the datasets and python objects, that are oriented towards the internal Sofistik data model, are transferred (SOFiSTiK AG, Oberschleissheim, Germany 2018). These objects are then converted one-to-one into an example ontology

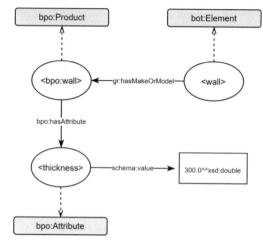

Figure 4. Example for referencing analytical data.

(ex:) and linked to the corresponding bot:Element, see Figure 4.

For the alignment the example property ex:hasSofistikRepresentation is assumed. Within this example, the forces acting on a wall could be accessed in the ontology, as well. In the FEM analysis, the wall is represented by a network of nodes and, subsequently, various classes of the *cnode* type are available for the storage of the nodes, which link the coordinates and the associated forces as attributes.

5 CONCEPT

The described concept considers collaboration processes involving multiple AECOO domains and stakeholders. More specifically, the concept is based on a central project platform, which describes the data in RDF as graph-forming triple using the BOT data schema. Different stakeholders, such as architects, HVAC planners or structural engineers, access the platform and store building elements specific to their specialist role that is structured according the discipline-specific ontologies.

The purely geometric 3D models are serialized as Open CASCADE compliant kernel commands or directly as Breps, serialized either as an ontology or text-based, and referenced via the OMG. In parallel, the structural analysis is performed based on the architectural model. The analysis' results are inserted in the project database in an object-oriented manner, and assigned to the individual bot:Elements and referenced by the geometric representations. The Model Checking unit accesses the enriched data model and conducts coordination analyses. Next, the basic functionalities of the proposed model checking system are presented.

5.1 *Rule interpretation*

The discussed approach distinguishes between three types of rules. There are rules that can be checked against the existing data model and rules that require knowledge derived from the model.

For the first type of rules, SPARQL queries are sufficient to formulate them. These queries can be entered directly or reused with parameters. The advantage of these rules is that they can be extended easily without adding and deploying further code. If a new context is to be checked, a new SPARQL query can be created and executed for checking. This corresponds to the second strategy according to (Pauwels & Zhang 2015), *"rule checking by querying"*.

The second type of rules are those rules where code has to be executed to enrich the model with knowledge. The challenge for this type is to find a structure that is understandable for users, on the one hand, and flexible enough to promote the code's reusability to derive information for exchangeable contexts, on the other hand. For collision checks, for example, the derived information refers to the geometrical relationship of two objects, while the objects themselves have individual characteristics for different AECOO domains.

Therefore, it is useful to let the user define the context by selecting objects that relate to the geometric relationship via SPARQL queries, see Figure 5. The SPARQL query returns the URI of matching objects and bundles the dependency to the discipline-specific ontology so that it does not have to be implemented as part of the application. Using the defined objects of the SPARQL queries, the geometric relationship between the two objects can be checked.

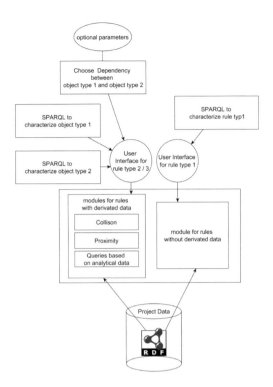

Figure 5. Architecture of the proposed flexible model checking application.

5.2 Model preparation

To prepare the building model, geometric information needs to be derived from relevant objects via SPARQL queries. The Open CASCADE-Brep representations of the objects are used in combination with the BOT ontology to derive geometric relations. Thereby, implicit dependencies in the implementation exist for these two considered data models, only. Further specialist ontologies can be aligned freely.

The interface for the model preparation requires the transfer of the relevant objects' URIs, which can be used to retrieve the corresponding geometric representations. The functionalities to derive geometric relations include collision queries based on aligned bounding boxes and distance queries with the BREPExtrema function of the Open Cascade kernel. Various types of geometric relationships should be available, for example to define safety distances via self-configurable envelopes.

The model preparation reaches its limits when the geometric representation for checking the collision is based on the used specialist ontology. For example, if it is to be checked whether a pipe intersects a wall at a position of high structural stress. In this case, the wall has to be divided into specific stress areas and separate geometric representations need to be defined for each area. In the implementation of this functionality, an implicit dependency on the data model of the ontology of structural analysis needs to be included.

5.3 Rule execution

For executing of the first type of rule, a module for execution that applies SPARQL queries to the data model is required. For the two other types of rule, a separate software module has to be defined for each derivation of building information (e.g. for collision queries, distance functions and definitions of enveloping bodies). After enriching the model, such rules can be checked by SPARQL queries, as well. It is also possible to check the rule directly without integrating the generated information into the data model, but such an integration may be useful if the generated information is to be used efficiently for checking multiple rules.

5.4 Reporting

A report in the form of a tabular list of the checked model instances should be available as a basis. A textual description of the rule is displayed for each instance, showing whether the rule passed or failed. It should be possible to manually combine several rule violations into one issue, similar to the existing planning software, and export issues as BCFXML (BIM Collaboration Format).

The issues and the associated information could also be integrated into the data model and reference the relevant model instances. Planners could be made aware of issues that affect them, when they interact with their planning data on the project platform. When checking geometric properties, graphical reporting is useful for better identification. This can be realized, as in proprietary applications, by marking the relevant components in color and displaying them at a certain camera angle. Visualizations tailored to the rule are useful. These specialized visualizations can be combined with derived information of the rule check and stored in the database. Especially for rules that are based on analytical data of domain models, the visualization concepts should be considered during rule creation.

6 IMPLEMENTATION PLAN

In order to discuss and evaluate the approach for Semantic Web-based model checking, various rules are considered. For this purpose, IFC files serve as a data basis, which are converted manually into a BOT-compliant data model. The Open CASCADE-Brep-objects for geometric representation are created with the IFCOpenShell ("IfcOpenShell" 2020) and referenced to their BOT representations. Regarding the analytical data, a simple example model of a wall with a recess is calculated in Sofistik and the calculation results are extracted via the tool's API. These results could be integrated into the ontological data model in accordance to the data model's definition of Section 4. The queries and program logic are to be implemented in Python and rely on the rdflib and pythonOCC libraries.

Upcoming, examples for each rule type are discussed, including their SPARQL queries and execution workflow.

6.1 Rule type 1: Model checking without geometric properties

The first rule type is suitable for checking completeness and detecting modeling errors. As an example, a query is formulated to recognize whether each wall is assigned to a zone, see Listing 1. For this rule type, reporting in tabular form is suitable.

Listing 1. SPARQL query to check if every wall is assigned to a zone.

SELECT ?object ?passed

WHERE {

?object **a** bot:Element;

gr:hasMakeOrModel ?bpoModel.

?bpoModel **a** bpo:Product;

ptbpo:hasBSDDGUID "0GSQP$JDv5PALB8 zok7r9q".

BIND(EXISTS {

?z **a** bot:Zone;

bot:hasElement ?object.

} **AS** ?passed)

}

6.2 Rule type 2: Model checking with geometric properties

The second rule type is based on knowledge that is derived from the BOT data model and geometry descriptions. For example, the model could be checked for whether walls of a certain thickness collide with pipes of a certain diameter. Therefore, two SPARQL queries are defined, see Listing 2: One to determine walls that fulfill the specification and another to determine the pipes with diameters equal to or larger than the given parameter. Both queries return URIs of the associated objects. Using these URIs, the Brep representations of all relevant objects can be checked for collisions with Axis Aligned Bounding Boxes. Next to the tabular reporting, a graphical reporting is desirable, which is implemented by changing to the appropriate camera angle and coloring the affected objects, see Figure 6.

Listing 2. SPARQL queries to determine the ventilation pipes (above) and walls (below).

SELECT ?object

WHERE {

?object **a** bot:Element;

gr:hasMakeOrModel ?bpoModel.

?bpoModel **a** bpo:Product;

bpo:hasBSDDGUID "1i7kUWjuz45e TW5lnnM0PM";

bpo:hasAttribute ?attribute.

?attribute

bpo:hasBSDDGUID "1uP1j0pmv008tAo1 be_pqe";

schema:value ?value .

FILTER (?value > "130.0"^^xsd:double)

}

SELECT ?object

WHERE {

?object **a** bot:Element;

gr:hasMakeOrModel ?bpoModel.

?bpoModel **a** bpo:Product;

bpo:hasBSDDGUID "0GSQP$JDv5PALB8 zok7r9q";

bpo:hasAttribute ?attribute.

?attribute

bpo:hasBSDDGUID "06fezDNjn7BOOzEWG $4Nh_";

schema:value ?value.

Figure 6. Visualization of a rule type 2 - model check.

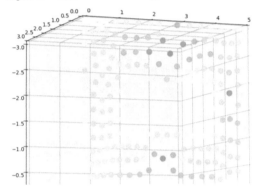

Figure 7. Geometric representation of analytical data.

```
    FILTER ( ?value < "20.0"^^xsd:double)
}
```

6.3 Rule type 3: Model checking with geometric properties and analytical data

For this rule type, multiple domain models are addressed. For example, a model is checked on whether pipes intersect a part of the wall with a certain stress range. This example rule is based on the BOT data model, geometry descriptions and knowledge that is derived from the structural analysis discipline.

The geometric representation of the wall is, therefore, broken down into sub-elements in consideration of the individual stress that is experienced by the wall parts, as is calculated by the structural analysis tool Sofistik. Subsequently, the collision query is performed on these sub-elements instead of the entire wall, see Listing 3. This model preparation is dependent on the data model of the structural analysis tool, i.e., the Sofistik CDB. The visualization's implementation, see Figure 7, can most probably not be reused for other rules.

Listing 3. SPARQL to determine the nodal forces of the walls:

```
    SELECT ?cnode , ?x , ?y , ?z , ?pz
    WHERE {
    ?object a bot:Element;
        gr:hasMakeOrModel ?bpoModel;
        ex:hasSofistikRepresentation ?cnode_gr.
    ?bpoModel a bpo:Product ;
        bpo:hasBSDDGUID "0GSQP$JDv5PALB8zok7r9q".
    ?cnode ex:isPartOf ?cnode_gr.
        ex:hasXCood ?x.
        ex:hasYCood ?y.
        ex:hasZCood ?z.
    ?cnode_L ex:references ?cnode;
        ex:hasSingleForceZ ?pz.
}
```

7 CONCLUSION

The aim of the paper is to show how an application can be designed on the basis of an extendable ontological data model, which is capable of performing model checks that can be configured flexibly. The decisive step is the model preparation, i.e. the derivation of knowledge from the data model. As soon as knowledge needs to be derived, an implicit dependency on the data model has to be implemented in the preparation tools. If no derived knowledge is required (rule type 1), rules can be defined as SPARQL queries and the application can be extended without changing the implementation.

In AECOO, derived information can also refer to geometric properties of the objects (rule type 2). Therefore, a dependency on the geometric data model in the code is sufficient to derive necessary properties for this rule type. The context information for identifying objects between which geometric relationships exist can be based on SPARQL queries. Those queries can, at the same time, bundle dependencies to data models of multiple disciplines.

As soon as a derivation of knowledge based on data models of individual disciplines is required, as is the case for rule type three, such dependencies need to be mapped in the code. Especially for proprietary data models, existing mappings are likely not reusable for other data models, e.g., from competing software vendors, resulting in complex and time-consuming adaptations of the source code for each software application that needs to be considered during model checks.

The effort for integrating singular software applications and their results into model checks can be reduced by unifying data models for abstracting access to the proprietary data models. An example for this, with the structural analysis discipline in mind, would be the TDY ontology for structural models (Thiele et al. 2020). Nonetheless, abstract ontologies face similar problems as other unified interfaces: information may be lost during conversion and, thus, the essential advantage of using ontologies is reduced.

In reporting, similar dependencies arise, because the information that is extracted via rules is also based on the used data models. The gained knowledge of this evaluation should form a basis for a design of a flexible program architecture that allows users to define rules in RDF and associated technologies as far as possible by themselves. A modular structure of the application should also allow easy extensibility of the rules in the program code, similar to the approach for a modular project platform based on microservices that is investigated in *SCOPE*. The evaluated model checking application should subsequently be implemented as a web service in a microservice architecture and complete existing services. Each web service offers various functions for project processing, which are oriented towards the essential tasks in AECOO. This includes conversion tools for importing and exporting from and into ontological databases ("projekt-scope" 2020) as well as tools for product search and administration of product data catalogs (Hoffmann et al. 2019).

ACKNOWLEDGEMENTS

This work is part of the research project *EnOB: SCOPE*, founded by the German Federal Ministry for Economic Affairs and Energy (BMWi). The concepts of the ontology and the microservice were developed in collaboration with Fraunhofer Institute for Solar Energy Systems ISE and Ed. Züblin AG.

REFERENCES

Cruz, I., Xiao, H., Lab, A., 2005. The Role of ontologies in data integration. J. Eng. Intell. Syst. 13.

Eastman, C., Lee, Jae-min, Jeong, Y., Lee, Jin-kook, 2009. Automatic rule-based checking of building designs. Autom. Constr. 18, 1011–1033. https://doi.org/10.1016/j.autcon.2009.07.002

Euzenat, J., Shvaiko, P., 2013. Ontology matching, Second edition. ed. Springer, Heidelberg New York Dordrecht London.

Helm, P. van den, Böhms, H.M., Berlo, L.V., 2010. IFC-based clash detection for the open-source BIMserver.

Hepp, M., 2011. GoodRelations [WWW Document]. URL http://www.heppnetz.de/ontologies/goodrelations/v1 (accessed 3.31.20).

Hoffmann, A., Wagner, A., Huyeng, T., Shi, M., Wengzinek, J., Sprenger, W., Maurer, C., Rüppel, U., 2019. Distributed manufacturer services to provide product data on the web.

Larsen, E., Gottschalk, S., Lin, M., Manocha, D., 2000. Fast Proximity Queries with Swept Sphere Volumes.

Lin, M., Manocha, D., 2008. Overview on Collision and Proximity Queries, in: Otaduy, M. (Ed.), Haptic Rendering. A K Peters/CRC Press, pp. 181–203. https://doi.org/10.1201/b10636-12

Nour, M.M., 2016. BIM Based Clash Detection Applications: Potentials And Obstacles.

Pauwels, P., Zhang, S., 2015. Semantic rule-checking for regulation compliance checking: an overview of strategies and approaches.

projekt-scope [WWW Document], 2020. URL https://github.com/projekt-scope/scope-data-service (accessed 4.1.20).

Rasmussen, M.H., Pauwels, P., Lefrançois, M., Schneider, G.F., 2020. Building Topology Ontology [WWW Document]. URL https://w3c-lbd-cg.github.io/bot/ (accessed 3.31.20).

SCOPE – Semantic Construction Project Engineering [WWW Document], 2020. URL https://www.projekt-scope.de/ (accessed 3.30.20).

SOFiSTiK AG, Oberschleissheim, Germany, 2018. CDB Interfaces Manual, Version 2018-14.

Solibri Workflow [WWW Document], 2020. URL https://www.solibri.com/learn/solibri-workflow-animation-film (accessed 3.30.20).

Supporting Applications [WWW Document], 2019. github.com/buildingSMART. URL https://github.com/buildingSMART/BCF-XML/wiki/Supporting-Applications (accessed 3.30.20).

Sydora, C., Stroulia, E., 2019. Towards Rule-Based Model Checking of Building Information Models. Presented at the 36th International Symposium on Automation and Robotics in Construction, Banff, AB, Canada. https://doi.org/10.22260/ISARC2019/0178

Thiele, C., Huyeng, T., Wagner, A., Shi, M., Hoffmann, A., 2020. Interlinking geometric and semantic information for an automated structural analysis of buildings using semantic web. Presented at the European Conference on Product and Process Modeling 2020.

Wagner, A., Bonduel, M., Pauwels, P., 2019a. OMG: Ontology for Managing Geometry [WWW Document].

Wagner, A., Möller, L.K., Leifgen, C., Eller, C., 2019b. BPO: Building Product Ontology [WWW Document]. URL https://www.projekt-scope.de/ontologies/bpo/ (accessed 3.31.19).

Interlinking geometric and semantic information for an automated structural analysis of buildings using semantic web

T.-J. Huyeng, C.-D. Thiele, A. Wagner, M. Shi, A. Hoffmann & U. Rüppel
Technische Universität Darmstadt, Institute of Numerical Methods and Informatics in Civil Engineering, Germany

W. Sprenger
Zentrale Technik, Digitalisierung & Softwareengineering (DS) · Digitalisation & Software Engineering BIM 5D®, Ed. Züblin AG, Germany

ABSTRACT: The advancing digitization in the building industry highlights weak points in the digital infrastructure. Due to heterogeneous software landscapes, cross-application data exchange is a frequently criticized process in particular, for which no satisfying solution exists so far. Open and application-independent data formats are necessary and therefore the current research project *SCOPE* focuses on developing such formats by applying Semantic Web Technologies. This article proposes an approach that enriches a graph with relevant information for structural design and provides the enhanced description as a calculation basis for structural analysis software. A novel ontology, the TDY-Ontology, is designed to connect geometry descriptions with structural-specific semantic information, like the standard of the analysis to be executed, bearing arrangements, bedding of floor slabs or the definition of joints and loads.

1 INTRODUCTION

The research project *SCOPE* (Semantic Construction Project Engineering) investigates an approach for storing AEC (Architecture, Engineering and Construction) data in an open, Semantic-Web-based format and processing distributed data in a uniform manner by relying on microservices. The research project, with a time-span from 2018 to 2021, consists of the project consortium of the company Ed. Züblin AG, the Fraunhofer Institute for Solar Energy Systems ISE and the Institute of Numerical Methods and Computer Science in Civil Engineering of the TU Darmstadt and further associated partners. The research project focuses on application-independent data management, which should facilitate interdisciplinary cooperation between different project partners ("SCOPE – Semantic Construction Project Engineering" 2018).

In terms of data management, the *SCOPE* project implements a graph-based data format for both the geometric description and the enrichment and linking with semantic information. Regarding the software applications, application independency is achieved by introducing a microservice architecture, which provides an API gateway for access control (Figure 1). Various services have already been developed by the project participants, but the range can be extended as required by additional services.

The structure of the so-called *SCOPE Data Service* ("BIM.5D-SCOPE" 2019) is based on a triplestore (an optimized database for storing data in the form of triples), which can be accessed via the *Fuseki-Service*. Another (optional) service is the *Render-Service* for visualization of the graph and geometric objects. Within the visualisation, users can perform queries (SPARQL) to the data in the triplestore to explore the available data. The *Revit-Service* converts the BIM model, which is created in the authoring system Autodesk Revit, into the data model required for the *SCOPE Data Service*.

Lastly, the *Teddy-Service* is introduced and presented within this paper to support structural design engineers. This paper shows that the geometry created in Revit can be converted into the graph-based *SCOPE* data architecture and automatically applied in the structural analysis software application *SOFiSTiK*.

By linking the information in the graph, changes that may occur during the planning process by other involved planners can be incorporated directly into the structural calculations. This is possible, because, e.g., the material parameters are not only defined in the respective software applications of the planners, but are recorded directly in the graph and are, thus, available to other software applications involved in the planning process.

2 RELATED WORK

In the AEC industry in general, but also increasingly in the field of structural design, the automation

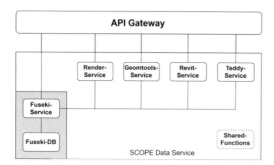

Figure 1. Microservice structure of the *SCOPE data service* (Huyeng et al. 2020).

Figure 2. Ontology structure in the research project SCOPE (Wagner 2020).

of recurring, manual and time-consuming tasks is becoming more and more significant. This is clearly demonstrated by the increasing use of tools such as *Dynamo*, which can be used to parameterize and simplify design and planning processes (Autodesk University & Fudala 2019). However, all these tools are limited to one authoring system and are, therefore, not interdisciplinary.

For an overview of *SCOPE*'s research results so far, see www.project-scope.de. In particular, the combination of geometric and semantic information as well as the developed microservice architecture (Huyeng et al. 2020) is used as a basis for the following concept.

The software application *SOFiSTiK* from the *SOFiSTiK AG*, which is well established in structural design, is a modularly FEM software solution for structural analysis with BIM support. It can be used by a graphical user interface or text editor with its proprietary data structure ("FEM, BIM und CAD Software für Bauingenieure | SOFiSTiK AG" n.d.). A further software company which is widely used in the AEC sector with a diversified portfolio is *RIB Software SE*. The company doesn't focus on structural analysis only, but much more on solutions for the entire planning and construction process. A well-known product is *RIB ITWO* ("RIB Software" n.d.) which covers practically every aspect from construction management up to facility management. *RIB* develops its own data structures as well, *CPIXML* for geometric and *CPIXML-SE* for structural description. Both approaches are suitable for the extension of the SCOPE Data Service as considered in detail in section 3.

The company *Dlubal*, publisher of the software *RSTAB* and *RFEM,* that are well-known in structural design, offers an API or COM interface for both products. According to the developer, the API or COM interface allows to control the calculation externally, to create or manipulate input data and to access results such as internal forces, deformations and bearing reactions (Dlubal Software 2015). Since no free licenses are provided for research and the authors have more experience in using *RIB Software SE* and *SOFiSTiK* (in fact, the implementation of structural design is not the focus of the *SCOPE* project), the underlying data format and the possibility of integrating the interface into the *SCOPE Data Service* could not be investigated in the paper. It is nevertheless intended to analyse the choice of Dlubal software products and integrate it in *SCOPE*. Therefore, the focus in this paper is mainly on the applications of *RIB Software SE* and *SOFiSTiK*.

In conclusion, there are already some approaches on automation in the field of structural analysis. However, the *SCOPE* research project not only allows for an interdisciplinary involvement of all actors, it also offers an application independent data management approach, and provides services based on the *SCOPE Data Service*. This includes the automation approaches taken up in this paper (meant as tools to support, not as a substitute) for the structural engineer.

3 INTERLINKING GEOMETRIC AND SEMANTIC INFORMATION IN SCOPE

As described in (Huyeng et al. 2020), the combination of geometric and semantic information by Semantic Web methods is one of the core areas of the *SCOPE* research project and forms the basis of the approach presented below. The so-called core ontologies, which provide the structure for mapping the required information on different levels, are shown in Figure 2.

The geometric information is structured according to the *Ontology for openCASCADE* (Nothstein et al. 2019). This ontology is based on the open-source geometry kernel openCASCADE. With the help of this ontology, geometric objects can be described precisely and unambiguously.

The *Building Product Ontology* (Wagner & Rüppel 2019) is implemented to specifically describe building products. The BPO makes it possible to not only describe one product but also to specify each component of the product and the relation between them. The *Ontology for Managing Geometry* (Wagner et al. 2019) is used as a link between the information of the BPO, possible further ontologies that describe the product semantically, and the geometric description.

The *buildingSMART Data Dictionary* (bSDD) is used to uniquely characterize products and their individual parts. This is realized by a relationship of the BPO, bpo:hasbSDDGUID. In the long term, this

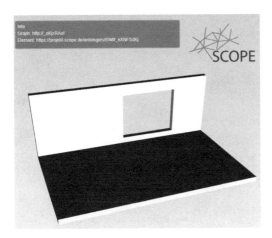

Figure 3. Example structural system: a wall on a continuously bedded ground slab, which is visualized by the *Render-Service*.

taxonomy should be replaced by a separate ontology of the bSDD in *SCOPE*. Currently, products are more and more often modelled parametrically to be able to automatically adapt to boundary conditions. Parametric description of building products is also possible using the core ontology *Ontology for Parametric Systems* (OPS). The ontology allows the mapping of variables as well as equations, which allows to perform calculations depending on other parameters.

4 THE TDY-ONTOLOGY

As illustrated in section 3, the current state of the *SCOPE Data Service* allows the user to store geometric data as well as semantic data in a triplestore, to link them with each other and, consequently, to set them in machine-interpretable relationships to each other. These basic functions are to be extended within the scope of the research project by further application-related scenarios in the area of AEC. One of these interdisciplinary use cases is the derivation of context-related data from the graph for structural design. To store data required for structural analysis without redundancy, an additional ontology is proposed besides the already implemented ontologies. This ontology is explained in more detail below using an example.

For a first approximation of the problem at hand, simple BIM models are generated, such as the continuous elastic supported slab with an upright wall shown in Figure 3. This model is then calculated with the software application of *SOFiSTiK* and eventually analyzed within the *SCOPE Data Service*. The process was also validated with the currently not yet published calculation application *iTWO structure fem* from *RIB Software SE* ("BIM Tragwerksplanung - RIB Software" 2020).

SOFiSTiK as well as *iTWO structure fem* are well suited for the integration to the *SCOPE Data Service*

because the applications provide the possibility to enter all required data for the calculation via the *SOFiSTiK Text Editor* (*Teddy*) or in case of *iTWO structure fem* via a proprietary data format, which is currently developed by *RIB Software SE* (*CPIXML-SE*, see Listing 1). The *CPIXML-SE* schema includes, among others, the material parameters used in the calculation, geometric information such as points, structural lines, structural surfaces and their meshing, supports, loads and load cases, superposition of load cases as well as definitions for the selected reinforcement.

Listing 1. Excerpt of the *RIB Software SE* data schema *CPIXML-SE*. Example definition of Points, Lines and Surfaces.

```
<Rem:Geometry>
  <Rem:Points>
    <Rem:Point Number="1" X="-1" Y="8" Z="0"/>
    <Rem:Point Number="2" X="-5" Y="8" Z="0"/>
    ...
  </Rem:Points>
  <Rem:Curves>
    <Rem:Straight Number="1" Start="1" End="2"/>
    <Rem:Straight Number="2" Start="2" End="3"/>
    ...
  </Rem:Curves>
  <Rem:Surfaces>
    <Rem:PlaneSurface Number="1" Curves="3 1 5"/>
    ...
  </Rem:Surfaces>
</Rem:Geometry>
```

The *SOFiSTiK Text Editor* provides basic functionality via a simple scripting language, which allows the definition of geometries by parameterized points and loops. The text editor generates (usually after the input of data) a file in DAT format, which saves the input as non-compiled plain text. With the microservice described in section 5, the data for the DAT file can be generated directly via the API (see Figure 1 and Listing 2). In earlier versions of *SOFiSTiK*, the input and definition of the system had to be processed manually via the text editor *Teddy*, but this can also be conducted via the *SOFiSTiK Structural Desktop* (SSD) that is equipped with a graphical user interface by now.

In order to be able to store the structure-specific data graph-based, the TDY ontology (Thiele et al. 2020), named after the eponymous input format of *SOFiSTiK Teddy*, is proposed to extend the existing ontology conglomerate accordingly. The TDY ontology, thus, provides the data structure and interconnections of information, which are required for structural design analyses.

Subsequently, the TDY ontology is explained in detail based on the example structural system shown in Figure 3. The parameters for the static calculation of the example model are selected as follows:

– A continuously elastic bedded ground slab with a bedding modulus of 10 MN/m^3.
– The wall and the slab are rigidly connected to each other.

– As acting load, the dead weight (automatically determined by *SOFiSTiK*) and a wind load of 1 kN/m^2 act on the wall (values are characteristic loads).

The TDY ontology defines the description rules at the structural level (T-Box). With currently six classes, five object properties and four data properties that are aligned to the schema.org ontology ("schema.org," n.d.), it is possible to map information with different degrees of complexity for structural design.

Listing 2. *SOFiSTiK-Teddy* file with information derived from the graph.

```
+PROG AQUA
KOPF
NORM DC EN 1992-2004
BETO NR 1 ART C 20
STAH NR 2 ART B GUET '500B' TMAX 32
ENDE

+PROG SOFIMSHC
KOPF
SYST 3D GDIV 10000
STEU MESH 2
STEU HMIN 1/5
SAR 1 MNR 1 T 250 CB 10000 CT 1000 BEZ
 "eXNFSdKj"
SARB Aus
SLNB X1 -10, 8, 0 X2 -5, 8, 0
SARB Aus
...
ENDE

+PROG SOFILOAD
KOPF
LF 1 TYP G FAKG 1 BEZ 'LF selfweight'
LF 2 TYP Q BEZ 'LF wind'
AREA SAR '2' TYP PXX -1.0
ENDE

+PROG ASE
LF ALLE
ENDE
```

All relevant information for the calculation is summarized using a calculation node. Among other things, the underlying calculation standard and the individual objects (walls, floor slabs, etc.) are connected to this node.

Figure 4 shows the connections using the example model. The connection of the products with the node of the calculation is implemented with the tdy:belongsToCalculation property.

According to the core ontologies, the geometry description and other material parameters are linked to the individual objects. The required parameters for the calculation, such as concrete quality, bedding, etc., are structured according to the TDY ontology and can, therefore, be extracted automatically via SPARQL queries. For the ground slab, the relevant parameters are shown as an example in Figure 5. The geometry description according to the OCC is linked to the node of the product but is waived for the

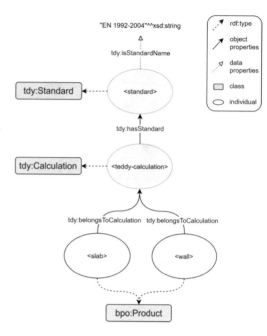

Figure 4. Example wall and slab connected with the TDY ontology via the calculation node.

sake of clarity. The properties tdy:isCStrengthName and tdy:isCESupportNumber are subproperties of the property schema:value ("schema.org," n.d.). This alignment increases the interoperability and reusability of the TDY ontology.

The decisive forces and moments of each element (i.e., the respective MIN and MAX values) are stored in a result node as literals. Additionally, the resulting reinforcement is stored there, as well. The node is connected to the calculation node and the respective element (see Figure 6). In addition to the result, the node *Result* is also provided with a time stamp. This is implemented by using the prov:generatedAtTime property of the PROV ontology ("PROV-O: The PROV Ontology" 2013). It is recommended that the individual geometry descriptions are also assigned a timestamp of their creation using this relationship. A comparison of these two timestamps can, for example, automatically start a new calculation when the geometry is changed.

The flexible structure of the TDY ontology allows covering further cases which are relevant for structural design. For example, more *SOFiSTiK Teddy* attributes describing the material properties can be added or further calculation cases can be covered. New, not yet added attributes can be described using the superclass tdy:Attributes.

Currently, only the load case of wind load is implemented by means of the tdy:Load node (see Figure 7). The value of the wind load is linked via the tdy:isValue property and the direction of the load is described according to *SOFiSTiK* standard. All values described

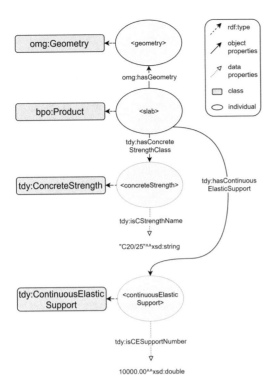

Figure 5. Example slab and the tdy attributes which are needed for the structural analysis.

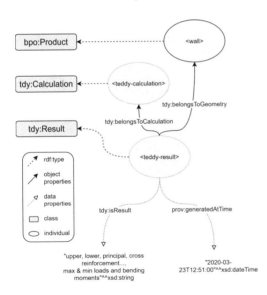

Figure 6. Result node for the example calculation.

with the TDY ontology are currently saved in the respective default unit of *SOFiSTiK*.

The export of the graph-based information into an executable format of the respective software using a service as well as the possible automation steps are shown in the subsequent section.

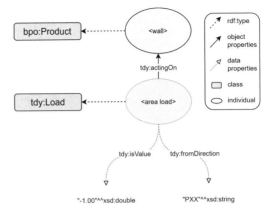

Figure 7. Area load on the example wall (wind load with 1 kN/m^2).

5 THE SCOPE TDY-MICROSERVICE

To perform repeated structural calculations automatically via an API, a demonstrator is implemented and validated. The results are presented below.

To efficiently use the data structure and approaches, as explained in section 4, in the *SCOPE Data Service*, a new microservice is proposed that can accumulate geometric and non-geometric data, which are stored in the graph and relevant for the structural design. Moreover, the service processes this data for it to be used by a calculation program. The service is based on a *Flask* app (micro web framework, "Flask" n.d.) and is implemented in Python. According to the core services of the *SCOPE Data Service*, the service can be accessed via a Rest-API. The service connects to various triplestores via their respective *Fuseki-Services* and by using SPARQL-Queries.

The *Teddy-Service* can automatically create DAT files based on the data stored in the graph. The geometry, which is available as a 3D solid in the graph, is converted into profile center lines and added to the DAT file as structural lines. Material parameters and the underlying standard are extracted according to the TDY ontology described in section 4. The service also automatically triggers the calculation via the *SOFiSTiK Calculation* application, after the DAT file has been generated. This can be implemented with the command

$$C:\backslash...\backslash Sofistik \backslash wps.exe C:\backslash input.dat - b$$

and started, i.e., as a command line call without further user interaction (Felkner 2016) or even with sps.exe for a complete silent calculation. The command executes the various calculation modules that subsequently create and mesh the geometry, perform the calculation, superpose load cases and, eventually, output the required amount of reinforcement of the individual components as a result of the static calculation. The result is stored in the proprietary database (CDB) and can be viewed using the application *SOFiSTiK Graphic* (formerly *WinGraf*). The animated representation of

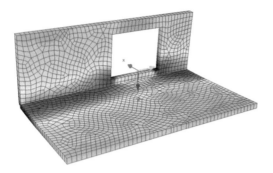

Figure 8. Calculated system and visualization in SSD under wind load.

the deformation of the system can be viewed in the *SOFiSTiK Structural Desktop* or in the *SOFiSTiK System Visualization*. In the long run, the results should be read from the database and written back to the graph.

By activating the *Teddy-Service* with an API request, the service generates a calculation node in the graph, if not already existing, to which the corresponding information and objects are connected, as described in section 4. It is also possible to add a timestamp to the calculation node. Thereby, it can be reconstructed, at which time and with which linked elements the calculation was conducted. For a correct version history and to allow for analysis and change of the DAT, it is possible to save the content of the file as a literal in the graph.

Although the microservice does not implement writing results from the *SOFiSTiK* database back into the graph yet, the TDY ontology already considers this use case. The results can be stored as literals in a result node for the corresponding element and the corresponding calculation, as described in Figure 8. Based on this result node further use cases for the service can be derived. For example, it would be possible that the service checks the integrity of the calculations by comparing the timestamp of the results and the geometric description daily. Subsequently, a change of the geometry description, e.g., by the architect, would trigger a new calculation. If certain parameters are out of range, a message can be sent automatically to the responsible processor.

6 CONCLUSION

Engineering understanding, knowledge and the responsibility that comes with it cannot be replaced by machines in the foreseeable future. However, by providing support for the numerous manual activities involved in the planning process, machine automation can be helpful and holds great potential.

By integrating the TDY ontology and the *Teddy-Service* into the *SCOPE Data Service* and data architecture, it is shown that, in addition to the existing achievements of the research project, such as the linking of semantic and geometric information, this database can also be used for the field of structural analysis and thus contribute to supporting the planners. Thanks to the provision of an API, such as that of the *Teddy-Service*, with which programs like *SOFiSTiK* and *iTWO structure fem* (RIB Software SE) can work directly with the interdisciplinary data stored in the graph, a step has been taken in the direction of further networking of all those involved in the planning process in the building industry. The automatic calculation and replay of the results into the graph allow them to be processed by further participants. Up to now, however, only a few load cases are automatically recognized based on the TDY ontology with the *Teddy-Service*. In the long term, the aim is to extend the service and the ontology so that all important cases can be covered. By providing the documentation of the TDY ontology (Thiele et al. 2020), an individual extension and adaptation by experts is possible. There are also further alignments planned. Especially the alignment to the Quantity-Unit-Dimension-Type ontology ("The NASA Quantity - Unit - Dimension - Type Ontology" 2010) is an important step to make the TDY ontology user-friendly by allowing users to choose the applied units of their data.

The presented service as well as the ontology follow the basic approach of clear data separation as represented in *SCOPE*. The functionalities are provided by services, but the data itself and, thus, the data sovereignty remains with the user.

ACKNOWLEDGEMENTS

This work is part of the research project *EnOB: SCOPE*, founded by the German Federal Ministry for Economic Affairs and Energy (BMWi). The concepts of the ontology and the microservice were developed in collaboration with Fraunhofer Institute for Solar Energy Systems ISE and Ed. Züblin AG and implemented by the authors of this paper.

REFERENCES

Autodesk University, Fudala, T., 2019. Design Automation for Structural Engineering.

BIM Tragwerksplanung - RIB Software [WWW Document], 2020. Digit. TRAGWERKSPLANUNG - ITWO Struct. Für Model. Im Ingenieurbau. URL https://www.rib-software.com/loesungen/tragwerksplanung/innovation/bim-tragwerksplanung (accessed 3.19.20).

BIM.5D-SCOPE [WWW Document], 2019. GitLab. URL https:// gitlab.itc-engineering.com/scope (accessed 3.19.20).

Dlubal Software, 2015. RFEM-/RSTAB-Zusatzmodul RF-COM/RS-COM [WWW Document]. Dlubal. URL https://www.dlubal.com/de/produkte/rfem-und-rstab-zusatzmodule/sonstige/rf-com (accessed 3.23.20).

Felkner, J.M., 2016. Multi-objective design balancing structural, environmental, and architectural considerations (Doctoral Thesis). ETH Zurich. https://doi.org/10.3929/ethz-a-010689078.

FEM, BIM und CAD Software für Bauingenieure | SOFiSTiK AG [WWW Document], n.d. URL https://www.sofistik.de/ (accessed 3.26.20).

Flask [WWW Document], n.d. Pallets. URL https://palletsprojects.com/p/flask/ (accessed 3.25.20).

Huyeng, T.-J., Thiele, C.-D., Rüppel, U., Sprenger, W., 2020. An approach to process geometric and semantic information as open graph-based description using a microservice architecture on the example of structural data. Presented at the European Group for Intelligent Computing in Engineering, accepted for publication, Berlin.

Nothstein, Christian, Sprenger, Wendelin, 2019. OCC: Ontology for openCASCADE [WWW Document]. OCC Ontol. OpenCASCADE. URL https://www.projekt-scope.de/ontologies/occ/ (accessed 2.11.20).

PROV-O: The PROV Ontology [WWW Document], 2013. URL https://www.w3.org/TR/prov-o/ (accessed 3.23.20).

RIB Software [WWW Document], n.d. URL https://www.rib-software.com/home (accessed 3.26.20).

schema.org [WWW Document], n.d. URL https://schema.org/ (accessed 3.25.20).

SCOPE – Semantic Construction Project Engineering, 2018. URL https://www.projekt-scope.de/ (accessed 1.30.20).

The NASA Quantity – Unit – Dimension – Type Ontology [WWW Document], 2010. QUDT Ontol. URL http://www.qudt.org/qudt/owl/1.0.0/qudt/index.html (accessed 3.24.20).

Thiele, C.-D., Huyeng, T., Wagner, A., 2020. Teddy Ontology [WWW Document]. URL https://w3id.org/tdy (accessed 3.18.20).

Wagner, A., 2020. Linked Product Data: Describing Multi-Functional Parametric Building Products using Semantic Web Technologies. https://doi.org/10.13140/RG.2.2.17992.88327.

Wagner, A., Bonduel, M., Pauwels, P., Uwe, R., 2019. Relating geometry descriptions to its derivatives on the web. Proc. 2019 Eur. Conf. Comput. Constr. 1, 304–313. https://doi.org/10.35490/ec3.2019.146.

Wagner, A., Rüppel, U., 2019. BPO: The building product ontology for assembled products. CEUR Workshop Proc. 2389, 106–119.

A BIM to BEM approach for data exchange: Advantages and weaknesses for industrial buildings energy assessment

M. Del Giudice, M. Dettori, S. Magnano & A. Osello
Politecnico di Torino, Turin, Italy

ABSTRACT: Energy assessment has become an important and argued topic in recent years due to climate changes. For this reason, the building industry is crossing a digital transition period. In this context, the development of 3D parametric models as a digital twin is due to the energy assessment of private and public buildings. The purpose of this study is to analyze Building Information Model to Building Energy Model process, with an essential focus on data transfer, which plays a key role to maintain all the information required for energy simulations. Starting from the evaluation of the most used standard exchange formats for energy simulation, this contribution highlights the challenges that occur during the export/import process finding out the lack of data that make the process not error-free. The selected case studies allowed to test the proposed process and analyze the main steps for building energy simulation.

1 INTRODUCTION

1.1 Background

Energy modelling is one of the most important research topics of the last years in line with environmental and economic conditions. Recently, an amendment of the directive on the Energy Performance of Buildings (EPBD) encourages each EU Member State at accelerating the cost-effective renovation of existing buildings, both public and private, into a highly energy efficient and decarbonised building stock by 2050, facilitating the cost-effective transformation of existing buildings into nearly zero-energy buildings (2018/844/EU).

The evaluation of buildings energy performance is a current research field, especially since the regulation EN ISO 52016-1 has introduced the new hourly calculation method for building energy assessment.

In this context, various research focused on energy analysis evaluation related to residential and public buildings, but this topic has not been sufficiently detailed in terms of industrial buildings. Data management optimisation for industrial buildings has to refer to both mass production and energy management to achieve thermal comfort in relation to the activities carried out within industrial environments. Thus, Information and Communication Technologies (ICTs) can give an important contribute for digitalizing data plants with relevant processes and occupancies towards the building data optimization.

For this reason, energy dynamic simulation represents an essential tool to replicate their virtual behaviour, starting from a Building Information Modeling (BIM) methodology. It consists in a set of activities based on a 3D parametric model that contains information about the different features of construction elements, such as measures, materials, costs, physical data, manufacturer or any other data that the designer considers important for the project in addition to geometric data (Osello 2012). Based on a 3D parametric model, graphical and alphanumerical information, it is useful to develop a virtual prototype of the building for simulating its energy behaviors. In addition, the production of an informative model needs to be based on the definition of model uses and objectives. As industrial facilities are particularly demanding in terms of design, due to different thermal requirements of various internal spaces (office, production, storage), regulations of mobility and accessibility and interferences of various Mechanical, Electrical and Plumbing (MEP) systems with building components (Gourlis et al. 2017), data management optimization requires informative models to achieve these objectives.

The purpose of this study is to analyse BIM to Building Energy Model (BEM) process (Kamel et al. 2019), with an essential focus on interoperability, which plays a key role to maintain all the information required for energy simulations. As, in the last years the BEM development focuses on different orientations such as energy consumption (Pezeshki et al. 2019), one of the main challenges consists of developing of a BEM model, starting from a 3D parametric one. In this context, many researchers investigated the topic of Building Energy Performance Simulations (BEPS) to evaluate strength and weaknesses of the BIM to BEM method, developing innovative methods and tools, preserving data quality (Lilis et al. 2018).

One of the main advantages of the connection of BIM authoring platform to BEM tools is the automation of the process for energy simulation of buildings (Kamel et al. 2019). Many researchers investigated this topic to facilitate data sharing, using a specific application for energy simulations. An important step in this procedure is to evaluate the proper standard exchange format that best suits the aims of the energy simulation (Senave et al. 2015). On this matter, using interoperability allows facilitating this step to improve data sharing, developing a unique dataset for specific applications such as thermal comfort and energy consumption. For this reason, the choice of the proper standard exchange format that preserves geometrical (i.e. MEP objects) and alphanumerical (i.e. thermal properties) information, limiting data loss, is fundamental for optimizing data transfer in the target application.

Currently, Green Building eXtensible Markup Language (gbXML) (gbXML 2007) and Industry Foundation Classes (IFC) (BuildingSMART 2008) represent the two communication languages in the building industry used for the BIM to BEM methodology. Several research goals aimed at assessing and comparing the building data transfer via IFC and via gbXML format into Building Performance Simulation (BPS) (Ivanova et al. 2015, Dong et al. 2007). Clearly, interoperability and data transfer from BIM to BEM is paramount in order to reduce lack of data that implies a waste of cost and time in the whole building process (Bahar et al. 2013). With the aim of examining the reliability of the proposed method, IFC and GbXML have been investigated, with particular attention to geometrical surfaces and thermal properties. The accuracy of the energy models object-based (Kim et al. 2016) is evaluated in this contribution, highlighting the possibility to link a BIM authoring platform with an energy analysis tool, monitoring data loss that occurs during data exchange.

1.2 Case study

Currently, the BIM to BEM approach is not widespread for industrial buildings that generally require a high amount of energy generating discomfort and CO2 emissions. For this reason, an industrial existing building was selected as a case study for this contribution. Developed in the 1940s, it is located in Turin (Italy) and it is represented by a huge open space designed for production activities (100.000 m^2).

The building envelop is characterized by prefab walls, typified by two layers of brick with an air chamber between them, a significant number of skylights on the roof, and several glazed surfaces composed by low-performing glasses, in accordance with North Italy buildings in the 1940s.

Whereas two sides of the building border with spaces at different set-point temperatures because of other manufacturing activities, the other two have exterior walls with piers for loading and unloading goods. As regards the plant features of the compound, the energy demand during the heating season is satisfied by an HVAC system, which consists of about 50 Air Handling Units (AHU) and few more than 100 air heaters supplied with both external and internal air. Moreover, since it is a complex case study, a simplified model (BESTEST) was used to verify the efficiency of the BIMtoBEM method. The case study "BesTest600" consists of a single room 8 m wide x 6 m long x 2,7 m high, two openings on one wall and a single thermal zone. The proposed dimensions and shapes have been chosen in consideration of the ANSI/ASHRAE 140-2001 standards (Henninger et al. 2004).

2 METHODOLOGY

2.1 Adopted workflow

In the last years, several researchers investigated the topic of Building Performance Simulation (BPS), as the use of software to predict the energy performance of a building (Altan et al. 2016). In these terms, the use of computer simulation to represent a certain building should be based on an informative model of reality. In this way, it is possible to use various type of data to optimize its energy behavior evaluating energy usage, CO$_2$ emissions and related costs.

The proposed BIMtoBEM methodology summarized in Figure 1 aims to describe the main steps for the development of building simulation using ICTs. The schematic workflow starts from the need to represent the process of data transfer from the real world to the digital one. It consists of five mains steps: i) Data collection; ii) BIM data restitution; iii) export/import step for data transfer; iv) BEM model integration; v) outputs generation. As the case study belongs to the existing building stock, the first step of the followed methodology regards data collection. The majority of the collected documents consisted of 2D drawings and technical documentation related to the building components. In particular CAD drawings, time and activity schedules, and HVAC system data have been analyzed for the development of the 3D parametric model. Through data collection, a great amount of information, related to geometrical, structural, mechanical and thermal properties were overlapped to generate an indispensable database, useful for continuous monitoring of the building throughout its life cycle. Although this procedure represents an important step of the process, to find information because of companies' policy is not easy. The next step consists of creating a parametric model to replicate the investigated building in terms of geometrical, physical and thermal properties. At the beginning, several BIM authoring platforms (e.g. Autodesk Revit, Graphisoft Archicad, Nemetschek Allplan) were evaluated for BIM model creation.

For this study, Autodesk Revit was selected as a BIM platform to develop the 3D parametric model.

Figure 2 highlights the difference between the selected case study and the BesTest in terms of

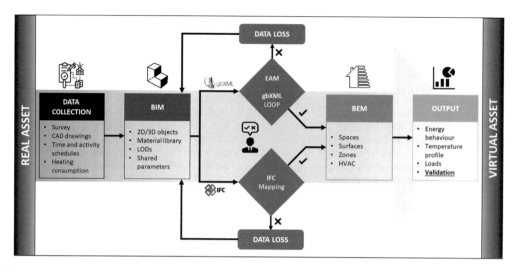

Figure 1. BIMtoBEM workflow.

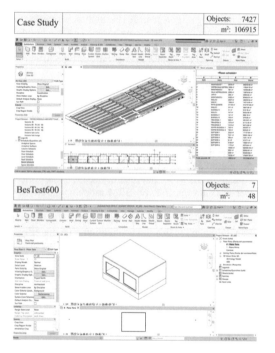

Figure 2. Axonometric view of both the Case Study and Bestest.

objects and square meters. A single comprehensive model was developed, containing architectural, structural and HVAC data, evaluating the Level of Detail/Development (LOD) (AIA 2013) of the modeling process for the energy simulation usage.

According to the need to perform an energy simulation at a later stage, the information detail was increased within the geometrical one focusing on material properties, because of their main role in the building energy behaviour. Rooms have been placed into the model splitting the environment into conceptual subdivision according to heating/cooling loads or structural elements, making the interoperability process be more efficient.

Determining precisely thermal zones in Revit, avoiding overlapping and voids between volumes, allows proceeding towards the BEM model minimizing errors and establishing each room bounding before the simulation.

2.2 BIM to BEM data transfer

At the beginning a literature review was followed to compare the BIMtoBEM data transfer between different BIM authoring platforms and BEM tools, discovering the most used standard exchange formats for energy calculation (Fig. 3).

Currently, this procedure can be performed thanks to two main exchange formats: the gbXML and the IFC. In addition, as regulatory updates are focusing attention on the dynamic hourly calculation, some BEM tools have been selected considering the national context in which the case study is located.

BEM tools have been evaluated based on a literature review concerning their energy engine, Graphical User Interface (GUI), input and output data.

	BIMtoBEM tools	BIM Authoring platform		
		Revit	Archicad	Allplan
BEM tools	Ecodesigner Star	IFC	Integrated	IFC
	Edilclima	IFC	IFC	IFC
	Design Builder	gbXML	gbXML	/
	Termolog	IFC	IFC	IFC
	OpenStudio	IFC/gbXML	IFC/gbXML	IFC
	Green Building Studio	gbXML	gbXML	/
	Termus	IFC	IFC	IFC

Figure 3. BIMtoBEM matrix for standard evaluation.

```xml
<Construction id="aim0193">
    <U-value unit="WPerSquareMeterK">0.1054195</U-value>
    <Absorptance unit="Fraction" type="ExtIR">0.1</Absorptance>
    <Roughness value="VeryRough" />
    <LayerId layerIdRef="aim0197" />
    <Name>Basic Wall: Esterno - Mattone su mont. met.</Name>
</Construction>
<Layer id="aim0197">
    <MaterialId materialIdRef="aim0198" />
    <MaterialId materialIdRef="aim0205" />
    <MaterialId materialIdRef="aim0212" />
    <MaterialId materialIdRef="aim0219" />
    <MaterialId materialIdRef="aim0226" />
</Layer>
<Material id="aim0198">
    <Name>Pannello in gesso (1): 0 [m]</Name>
    <R-value unit="SquareMeterKPerW">0.02</R-value>
    <Thickness unit="Meters">0.013</Thickness>
    <Conductivity unit="WPerMeterK">0.65</Conductivity>
    <Density unit="KgPerCubicM">1100</Density>
    <SpecificHeat unit="JPerKgK">840</SpecificHeat>
</Material>
```

Figure 4. Sample of gbXML code.

In this contribution Design Builder has been selected for GbXML strategy as well as Termolog EpiX 10 has been chosen for IFC one.

2.3 *The gbXML strategy*

GbXML has been chosen as the first exchange format since its development is linked to the improvement of data transfer from CAD-based building information models to disparate building design and analysis software tools. (gbXML 2007) It has been selected particularly to investigate the capability to preserve thermal properties of building components, avoiding to fill in several times the same information.

The hierarchical schema is organized according to a tree structure with different levels of nesting, starting from a root element, which is the parent of all other ones. Depending on the export process and BIM tool the elements may vary. Adopting Autodesk Revit as the BIM tool and the gbXML exporter, Elements are at least: i) Campus; ii) Construction; iii) Layer; iv) Material; v) DocumentHistory.

In detail, Campus element contains all geometrical information and it is composed of other different branches (i.e. Building) that store Rooms and spaces, essential for the export process.

For this study, it is important to focus on the Layer element, that lists all used materials, and on the Material element. Thermal properties are exported correctly in branch Material, that is connected to the other Elements with a specific Id (i.e. MaterialId) (Fig. 4) (Provera 2017).

Starting from Autodesk Revit the Energy Analysis Model (EAM) has been adopted to transfer information into the target application using gbXML.

Currently, EAM is an abstraction of the overall form and structure of a building in a computational network able to effectively capture every main paths and process of heat transfer within the building (Autodesk 2013). However, it does not contain any information about the mechanical systems, nor occupancy schedules, heat gains and weather files information. Then, possible issues (i.e. air gaps) were identified to avoid lack of data that affect the correct creation of analytical surfaces. Design Builder has been chosen as the BEM interface for energy analysis, which is one of the Graphical User Interface (GUI) for Energy Plus, the energy simulation engine that performs the analysis (Kamel et al. 2019).

In these terms the BIM model has been imported into Design Builder, preserving all the surfaces and thermal zones inserted as Rooms as well as thermal properties associated with materials.

The HVAC system has been modelled directly in DesignBuilder because, during the exporting process, HVAC information is lost. Finally, the information about weather file, occupancy and HVAC schedule has been filled in the BEM tool.

2.4 *The IFC strategy*

Industry Foundation Classes (IFC) is an international standard defined in ISO 16739-1: 2018. The functioning is based on the definition of a data modelling language, identified in EXPRESS (ISO 10313-11) and a precise data coding, commonly known as STEP or Clear text encoding of the exchange structure (ISO 10303-21). In order to scan a STEP-file, IFC Class can be displayed using a common notepad application. Nowadays exists a lot of software that makes the control of an IFC file faster and more schematic, however, it is also true that although they have a high level of precision, this software could also be subject to malfunctions.

The first step to rate the quality of the data flow from BIM to BEM is to choose the right tool according to the targets. Various tests have been made which led to the choice of the Termolog EpiX 10 software. This is due to: i) Status of a tool that is already prepared for the new set of regulations, therefore the results obtained in addition to complying with the legal requirements can be certified; ii) Interoperability value between BIM and BEM. Nowadays, to find a BEM tool capable of reading 100% of the information contained in an IFC file is arduous.

Usually, the tools read the file and use only the IFC Classes for performing technical analysis. However, it is much more difficult to find a tool that is able to optimally read all the IFC Classes, preserving HVAC system in an automated way. For this reason, various tests were concentrated on the correct data import relating: i) Geometric shape; ii) Thermal and stratigraphic properties of the elements; iii) Thermal zones.

The first tests, executed using the case study, have shown critical issues in the data flow for its huge number of objects. The BIM model has been exported from the Autodesk Revit in IFC format using the IFC4 Design Transfer View scheme and then imported into the Termolog EpiX 10. Although, the model has been correctly imported, except for all thermal properties. Figure 7 shows the preserved IFC classes (i.e. IfcMaterial) in grey, but also that one available (i.e. IfcMaterialProperties) in white, that unfortunately are

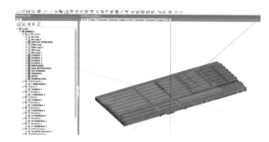

Figure 5. Axonometric view of the imported BEM model.

Figure 6. GbXML strategy transfer.

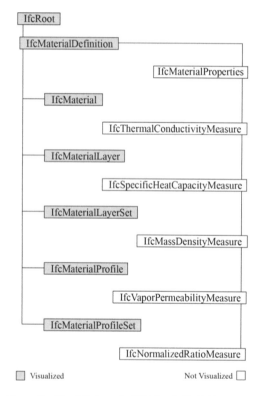

Figure 7. Simplified graph of IfcMaterialDefinition.

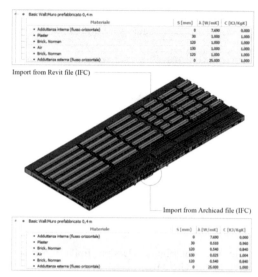

Figure 8. Comparison of IFC files imported into Termolog.

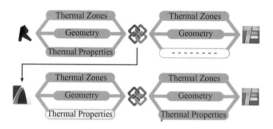

Figure 9. IFC strategy transfer.

3 RESULTS

3.1 *BIM to BEM advantages and weaknesses*

Several tests have been conducted in this contribution to examine BIM to BEM methodological workflow, focusing on the differences between specific exchange formats used for the import/export process, gbXML and IFC. A specific analysis has been carried out to examine the IFC and gbXML code, in particular advantages and critical points of each schema, focusing on their structure.

The obtained results have been synthesized in Figures 10–11 in order to evaluate the strengths and weaknesses of the two strategies. In this way, some data transfer has been used to examine the main steps that characterize the BIMtoBEM process. Although at the beginning information such as location or thermal properties were filled in a BIM platform, however, this kind of data had to be added manually in the target tool. In the study, the missing data (i.e. HVAC system, schedules, internal loads) has been added manually to Design Builder and Termolog.

The two tables show how data exchange efficiency depends on the export/import capabilities of each tested tool. The advantages deriving from BIMtoBEM strategy consist in automating some steps that in the

not compiled automatically. To trace the reason for this issue, another test has been performed with Archicad (Graphisoft) using the Bestest model, exported via IFC from Revit. Even in this case, the model has been correctly imported but without thermal properties.

These one have been filled in manually and the model re-exported in IFC format, using the previously used scheme.

The gbXML Strategy	BIM Input	BIM to gbXML	gbXML to BEM	BEM Input	Comments
Location	X	Transfer	X	-	Weather file is necessary to performs energy simulations
Weather data	-	-	-	X	
Building geometry	X	Transfer	X	-	Only analytical surfaces come from EAM
Rooms and Spaces	X	Transfer	X	-	Rooms objects are required
Building Construction	X	Transfer	X	-	Sometimes thermal properties need to be filled in manually
Internal loads	-	-	-	X	BEM manual input
Operation schedule	-	-	-	X	BEM manual input
HVAC system	X	Transfer	-	X	Data loss and manual input is required

Figure 10. Data transfer evaluation for the gbXML strategy.

The IFC Strategy	BIM Input	BIM to IFC	IFC to BEM	BEM Input	Comments
Location	X	Transfer	-	X	Weather file are integrated and based on location.
Weather data	-	-	-	X	
Building geometry	X	Transfer	X	-	Some objects need to be customized
Rooms and Spaces	X	Transfer	X	-	Rooms objects are correctly imported
Building Construction	X	Transfer	X	-	Thermal properties have to be filled in manually
Internal loads	-	-	-	X	BEM manual input*
Operation schedule	-	-	-	X	BEM manual input
HVAC system	X	Transfer	-	X	System data loss; only geometrical data is preserved

Figure 11. Data transfer evaluation for the IFC strategy.

traditional process would mean a data duplication with a consequent waste of time and resources.

Moreover, graphical and informative contents for building performance simulation require the development of a BIM model oriented to that achievement. According to the tests carried out, some parts of the process (i.e. HVAC system modelling) cannot be fully automated. In addition, the tools available on the market today do not guarantee error-free data exchange, in order to ensure efficient interoperability between BIM and BEM platforms.

From the tests carried out, the approach to performing BPS appears relatively tricky and difficult to validate. Whilst the process shows a number of advantages related to the BPS, it still has some limitations in the data update, which is not bidirectional yet.

3.2 Tools

The results of this analysis become extremely important in order to select which input is necessary to reach the aim of the research: BIM model for energy assessment. The development of a digital building registry has to consider the expected outputs to establish data entry. Indeed, the extension of the industrial building represented an initial limitation for the tests carried out which justified the use of a simplified model, as the BesTest, for this research.

Figure 12 shows the data flow necessary to develop an energy model in an innovative way, starting from specific input data (e.g. building survey, thermal and HVAC system characteristics, time and activity plant schedules). Several BIMtoBEM processes have been highlighted to enhance the importance of the data exchange procedures between different tools selected for this contribution. To overcome the data losses occurred during the import/export process, two further steps in the process were required, as already highlighted in Figure 1: i) gbXML loop; ii) IFC mapping. While for the first step, the action was done on BIM model in order to simplify its geometries, in the second one the IFC file has to be customized due to its frame. The diagram demonstrates, despite gbXML format has been developed with the aim of energy simulation, that the adoption of openBIM format is growing to reach the same goal.

In conclusion, it shows that the calculation engine can be adopted by several BEM tools able to filter BIM-based information with different ways and exchange formats. As an example, the open source engine Energy Plus, is used by the various GUIs that read different exchange formats to transfer data, performing simulation.

4 CONCLUSIONS

In the last years, the Architectural Engineering and Construction (AEC) industry is crossing a transition period oriented to the digitalization process. Therefore, the BIM methodology applied to energy modelling is one of the actual challenges due to economic and environmental condition. For this reason, the BIM to BEM strategy has to be considered as an innovative approach to data management considering that industrial building stock consumes large amounts of energy compared to the entire architectural heritage.

Through this contribution, the investigated process was studied according to the owner requirements establishing a hierarchy for the data sharing in the energy saving field for building performance simulation. The development of BIM models oriented to energy efficiency implies that some data may not be present within the BIM model at the beginning if not required. On the other hand, as the BIM method is based on a set of activities and processes based on the development of a 3D parametric model starting from data collection step, the model uses definition

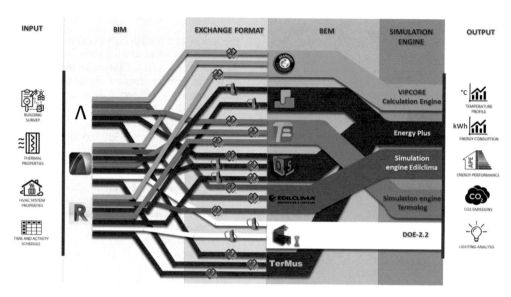

Figure 12. A resulting workflow, comparing the available standard exchange formats in the BIMtoBEM process.

is one of the main tasks to achieve specific objectives (i.e. energy assessment). For this reason, data loss can also be read as model optimization that is adapted to a certain target application. From this point of view interoperability and standard exchange formats facilitate to filter the model, according to the desired purpose, although this is not perfect today. Unfortunately, this data sharing is still characterized by issues and challenges pertaining to the prevailing design processes and available technological approaches (Farzaneh 2019).

Hopefully, in the future the use of standard languages will be developed to reduce errors due to interoperability, improving the alphanumeric and graphical representation of data, making the whole exchange process more stable.

5 AUTHORSHIP

The authors have agreed on the contents, the methodological approach and the final considerations presented in this research. Specifically, Matteo Del Giudice introduced the contribution in Paragraph 1 and the method in Paragraph 2.1, Monica Dettori described the gbXML strategy in Paragraph 2.2, 2.3 and 3.2, Salvatore Magnano described the IFC strategy in Paragraph 2.4, and Anna Osello the Paragraph 3.1.

REFERENCES

Altan H., Padovani, R. & Hashemi, A. 2016. Building Performance and Simulation. In ZEMCH: Toward the Delivery of Zero Energy Mass Custom Homes, *Springer Tracts in Civil Engineering*, 2016: 311–338.

Bahar Y.N., Pere C., Landrieu J. & Nicolle C. 2013. A thermal simulation tool for building and its interoperability through the Building Information Modeling (BIM) platform. *Buildings* 2013; 3(2): 380–398.

BuildingSMART. 2008. https://www.buildingsmart.org/ Web site retrieved on March 2020.

Directive (EU) 2018/844 of the European Parliament and of the Council.

Dong B., Lam K. P, Huang Y. C. & Dobbs G. M. 2007. A comparative study of the IFC and gbXML informational infrastructures for data exchange in computational design support environments. *Proceedings from Building Simulation 2007*: 1530–1537.

gbXML, 2007. https://www.gbXML.org/ Web site retrieved on March 2020.

Gourlis, G. & Kovacic, I. 2017. Building Information Modelling for analysis of energy efficient industrial buildings – A case study. *Journal of Renewable and Sustainable Energy Reviews* 68: 953–963.

Henninger, R.H. & Witte, M.J. 2004. *EnergyPlus Testing with ANSI/ASHRAE Standard 140-2001 (BESTEST)*, California, Ernest Orlando Lawrence Berkeley National Laboratory Berkeley.

Kamel, A. & Memari, A. M. 2019. Review of BIM's application in energy simulation: Tools, issues, and solutions. *Automation in Construction, Elsevier* 97: 164–180.

Kim, H, Shen, Z., Kim I., Kim, K., Stumpf A. & Yu, J. 2016. BIM IFC information mapping to building energy analysis (BEA) model with manually extended material information. *Automation in Construction, Elsevier* 68: 183–193.

Ivanova, I., Kiesel, K. & Mahdavi A. 2015. BIM-generated data models for EnergyPlus: A comparison of gbXML and IFC Formats. *Proceedings from Building Simulation Applications BSA 2015*. Bolzano, 4–6 February 2015.

Lilis, G.N., Giannakis G., Katsigarakis K. & Rovas, D. V. 2018. A tool for ifc building energy performance simulation suitability checking. *Proceedings of the 12th European Conference on Product and Process Modelling, ECPPM 2018, 12–14September, 2018*. Copenhagen, Denmark: CRC Press.

Osello, A. 2012. *The Future of Drawing with BIM for Engineers and Architects*, Palermo, Dario Flaccovio Editore.

Pezeshki, Z., Soleimani, A. & Darabi, A. 2019. Application of BEM and using BIM database for BEM: A review. *Journal of Building Engineering* (23): 1–17.

Provera, F. 2017. *Implementation of BIM methodology for district data visualization,* Master degree thesis, Politecnico di Torino.

Senave, M. & Boeykens, S. 2015. Link between BIM and energy simulation. In Building Information Modelling (BIM) in Design, Construction and Operations, *WIT Transactions on The Built Environment*, Vol. 149: 341–352.

Occupant-centric ontology as a bridge between domain knowledge and computational applications

A. Mahdavi, V. Bochukova & C. Berger
Department of Building Physics and Building Ecology, TU Wien, Vienna, Austria

ABSTRACT: In the past, building information modeling (BIM) in general and building performance simulation (BPS) in particular have developed and deployed fairly detailed representations of building geometry, fabric, construction, and technology. Relatively less attention has been paid to the representations of building users. Thereby, one of the key challenges concerns the matching between the nature and level of building-related performance queries on the one hand and the required or appropriate resolution of applied occupant models. This paper addressed this challenge via a two-fold path: A "top-down" path examines the ontological scope of comprehensive, theory-driven representations of building occupants. Such representations could be presumably condensed so as to match the informational requirements of specific performance queries. On the other hand, a "bottom-up" path involves the reverse-engineering of typical building performance simulation tools, focusing thereby on the implicit schema rooted in the tool's input modalities vis-à-vis occupant information. The top-down path can inform the definition of a common schema for occupant-related information. The bottom-up reverse-engineering can provide a reality check in view of the ontological depth of common occupant representations in analysis tools. The pursuit of these paths is suggested to highlight a gap between behavioral domain knowledge and default occupant representation in BIM and the analysis of this gap can facilitate the derivation of related ontologies.

1 INTRODUCTION

In the past, the representation of building users in building information modeling in general and building performance simulation in particular has been arguably rather reductionistic. Accordingly, a number of recent research efforts have been concerned with the resolution of representations of building users in simulation models. However, there is still a lack of conclusive understanding of the scope and format of adequate occupant-related computational representations. Whereas it is generally agreed that the resolution of occupant-related representations must match the nature of the performance queries, there are not robust methods available to identify the fitting levels of resolution (Mahdavi & Tahmasebi 2016). To systematically address the related challenges, a broader discourse is needed that entails three essential layers. These layers may be briefly characterized as follows:

i) The layer of foundational theories represents the necessary source of knowledge regarding those aspects of human perception and behavior that are relevant to building performance. People's expectations and requirements mandate specific indoor-environmental conditions, and these very conditions are influenced by occupants' actions. Foundational theories explain and hence facilitate the understanding of the semantics of perceptual and behavioral processes (Mahdavi 2020).

ii) The level of ontology represents the necessary conditions for a shared representation and operationalization of domain knowledge underlying computational models of people's perception of and behavior in the built environments. Given the syntax character of the constitutive elements of the ontology, they may be developed and deployed in a manner that would be accommodating of a variety of theoretical expressions of the pertinent domain knowledge.

iii) The level of application pertains to the actual computational implementation of knowledge-based, ontologically organized models of occupants' presence, perception, and behavior in buildings. Despite the wide scope of performance-based queries in building simulation, the respective implementations may all revert back to a common ontology, albeit with different takes in terms of coverage and detail.

The present contribution pursues the above discourse via a two-fold path, which we refer to as "top-down" and "bottom-up". The top-down path examines the ontological scope of comprehensive theory-driven representations of building occupants. Theoretically, such representations could be variously truncated so as to match the informational requirements of specific performance queries. On the other hand, the bottom-up path involves the reverse-engineering of typical building performance simulation (BPS) applications, focusing thereby on the occupant-specific input requirements. This is meant to reveal the implicit scheme rooted in the application's input modalities vis-à-vis occupant information.

The pursuit of this two-fold path is suggested to expose a discontinuity between the ontological implications of theoretical domain knowledge (on human behavior) and current input modes (for occupant-related information) in technical applications. The top-down path can support the work on defining a common schema for occupant-related information (i.e., position and movement, physiological and cognitive processes, control actions, social interactions). The bottom-up reverse-engineering can render a reality check concerning the ontological depth of common occupant representations in analysis tools. The pursuit of these paths is suggested to highlight a gap between behavioral domain knowledge and default occupant representation in BIM. The analysis of this gap can in turn facilitate the derivation of related ontologies.

Note that somewhat comparable observations concerning the discontinuity of the approaches in sociology and engineering approaches have motivated some previous efforts in this area. For instance, an ontology was previously proposed to "represent energy-related occupant behavior in buildings" (Hong et al. 2015a,b). At the theoretical level, a related effort (D'Oca et al. 2017) proposed an "interdisciplinary framework for context and occupant behavior in office buildings". These efforts, while well-intentioned and useful, display also a number of limitations at both theoretical and ontological levels. Neither the choice of theories, nor the logic behind their synthesis are apparent. Rather, the framework leads to a questionnaire-based assessment of a fairly large number of variables suspected to influence occupants' adaptive actions. The conceptual and terminological haziness, already present to some degree in the original theories adopted, is further aggravated in the synthesized framework. A theory-driven ontology that is expected to effectively support high-resolution (e.g., agent-based) modeling of human behavior in buildings arguably needs to be grounded on a more solid basis.

This is a formidable challenge and the current paper is not claimed to provide a conclusive solution. Rather, the objective is to offer the contours of an ontological approach for bridging the gap between high-level behavioral theories in human sciences and occupant representations in engineering applications.

2 THEORIES OF BEHAVIOR AND THEIR APPLICATIONS

2.1 Introductory remark

High-resolution representations of occupants in computational building models require more than the consideration of basic physical factors, such as occupants' emission of sensible and latent heat and indoor air pollutants. Likewise, detailed representations would have to do more than simple rules and schedules in order to model occupants' interactions with buildings' environmental control systems. Especially the latter circumstance implies the need for consulting pertinent theories on human behavior in general and their control-oriented actions in particular. A recent paper (Heydarian et al. 2020) attempted a review of a large number of behavioral theories assumed to be potentially relevant to people's behavior in buildings. However, given the broad scope of this effort, it could not go into the details of the elements, logic, practical application potential, and ontologically relevant implications of each and every paper reviewed. Hence, rather than a longitudinal review, we focus here on four studies, each of which includes references to a distinct behavioral theory. Thereby, a key point of inquiry concerns the theories' potential to inform a shared ontological framework.

2.2 Energy behavior in offices

Lo et al. (2014) studied office energy-saving behaviors in four different organizations in the Dutch provinces Zuid-Holland and Limburg. Toward this end, they used an extended model of the Theory of Planned behavior (TPB, Ajzen 1985; Ajzen 1991; Fishbein & Ajzen 2010), with perceived habit as an additional construct (Figure 1). The idea is that a repeated behavior eventually becomes a habit and no longer involves an evaluation process. The authors tested the reliability of TPB constructs and the relevance of the organizational context for predicting energy-saving behavior. They examined actions by office workers regarding lighting and shading (i.e., switching off lights) and usage of appliances and electronics (i.e., printing and switching off monitors). The participants took part in an anonymous online survey concerning their energy consumption. The TBP considers attitude, perceived norm, and perceived behavioral control as determinative constructs for behavioral intention, which results in behavior itself. A previous experience of a performed behavior and the expected outcome of it form the individual's evaluation of a particular behavior, i.e., the individual's attitude. The individual's moral evaluation of a behavior forms their personal norm, which is considered, by some, as a separate construct of the theory. The perceived behavior control mirrors the individuals' judgment whether they are capable of performing an action given resources, knowledge, and skills. The influences of the social environment on the individual are expressed through the perceived norm, which combines other people's judgment of the individual's behavior and their behavior in the past. The particular study of Lo et al. (2014) combines the TPB constructs with habit and physical context to evaluate the collected survey data from the participants. Their findings suggest that everyday office energy-saving behavior is significantly influenced by the physical context, e.g., organization and availability of operational devices and electronic appliances.

2.3 Energy behavior in social housing

DellaValle et al. (2018) aimed to explain the gap between expected and actual energy performance of social housing buildings. To this end, a pre-retrofit survey was conducted among occupants. The collected data was analyzed through the lenses of Social Practice Theory (Reckwitz 2002; Shove et al. 2012) and Neoclassical Economic Theory (Simon 1957, 1995) in order to identify the behavioral and social levers that would increase retrofit effectiveness (Figure 2). Neoclassical Economic Theory claims that individual's decisions are based on rational choice and are influenced by the cognitive biases created in the momentary context. Preferring short-term benefits to long-term energy performance displays a type of myopic behavior, manifested, for instance, in the choice of electricity suppliers or devices. In addition to momentary social context, individual's moral norms can also predict behavior. This applies to considered sources of information and what they deem to be socially appropriate. According to the Social Practice Theory, the material and cultural context plays a key role in decision making. Examining the context where individuals grew up (long deep socialization) and where they now live (actual context) helps to understand their energy-related behavior and the potential to influence it. Housing energy use results from different practices (e.g., showering, cooking, heating), which satisfy the individual's preferences and bring about personal comfort. Understanding and investigating these everyday social practices is suggested to provide guidance toward more effective retrofit solutions.

2.4 Households, behavior, and energy use

Al-Marri et al. (2018) examined the energy consumption behavior in Qatari households and the residents' views on renewable energy and sustainability. Two different methods were used to conduct the experiment: Quantitative data was collected from a survey among a large number of occupants and qualitative data was collected through interviews with energy experts. The participants' self-reported actions regarding windows opening and ventilation together with lighting and shading behavior were investigated. To analyze and interpret the results, reference was made to the Self-Determination Theory (Ryan & Deci 2000) and Maslow's Hierarchal Theory of Needs (Maslow 1943) (Figure 3). The Self-Determination Theory suggests that individuals could be externally motivated to engage in a specific behavior. They could be encouraged into a specific behavior using mechanisms involving rewards, benefits, or penalties. Maslow's Hierarchal Theory of Needs postulates that individuals prioritize their needs and are more likely to choose short-term comfort over the long-term impact of their behavior. However, both theories suggest that social engagement, awareness, and especially education, as well as motivation and encouragement could influence people's environmental behavior and result into more sustainable energy consumption choices.

2.5 Behavioral changes in university buildings

Matthies et al. (2011) conducted an interventional experiment among university staff in 15 public university buildings in Germany. Different types of data were collected, including energy consumption, self-reported behavior, and behavioral observation. Using the Norm Activation Model (NAM) (Schwartz 1977; Schwartz & Howard 1981), the information was analyzed, an intervention program was developed, and eventually behavioral changes regarding energy consumption were observed. The NAM differentiates four stages of forming a behavior (Figure 4). The attention stage consists of the individual's needs and awareness of behavioral consequences, along with the perceived behavior control. The second motivation stage mirrors the individual's personal norms together with the established social norms and moral. The third evaluation stage concerns the individual's judgement of the outcome of a behavior. The final fourth stage results in the specific behavior or in its denial. The NAM, as its name suggests, concentrates on activation and influence of personal norms. This was achieved, in the case of the specific experiment discussed here, through provision of information on environmental behavior and its impact, together with rewarding techniques. NAM explains the conflict between personal norms and the social context. This sheds light on the question of why, in this case, fulfilling the expectations of colleagues and superiors leads to behavioral changes.

2.6 Reflections

The examination of the above instances of behavioral theory applications in building-related domains underline the previously voiced concerns. It could be argued that the theories themselves have not converged at a unified conceptual framework. This is in part understandable, as their development has been triggered by different contextual settings and different problem statements. Efforts to synthesize some of these theories into some form of a unified framework have not been based on consistent and transparent steps. Rather, they seem somewhat ad hoc and eclectic (D'Oca et al. 2017). It thus should not come as a surprise, when the applications of the theories, as exemplified in sections 2.2 to 2.5 create a similar impression of disunity in concepts and constructs, and deviating layers of postulated causal relationships.

These observations corroborate the perceived gap that motivated the top-down inquiry formulated at the outset of this contribution. The high-level behavioral theories and their applications in building-related settings have not resulted in comprehensive, consistent, and versatile ontologies toward shared representations of building occupants.

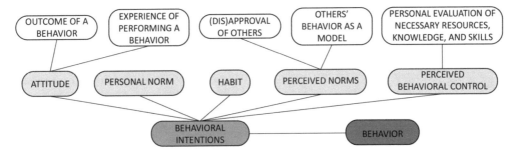

Figure 1. Schematic illustration of the application of the Theory of Planned Behavior (based on Lo et al. 2014).

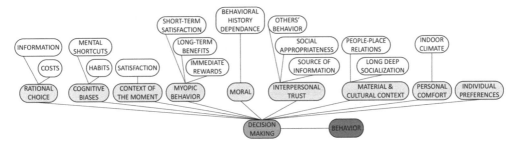

Figure 2. Schematic illustration of the application of Social Practice Theory and Neoclassical Economic Theory (based on DellaValle et al. 2018).

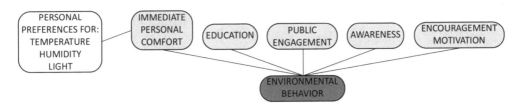

Figure 3. Schematic illustration of the application of Self-Determination Theory and Maslow's Hierarchal Theory of Needs (based on Al-Marri et al. 2018).

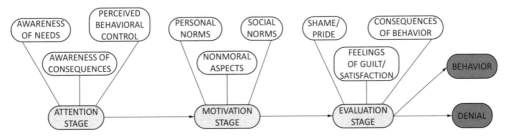

Figure 4. Schematic illustration of the Norm Activation Model (based on Matthies et al. 2011).

3 IMPLICIT TOOL-EMBEDDED SCHEMATA

As alluded to above, the bottom-up path involves the reverse-engineering of common building BPS applications. Thereby, the focus is on input requirements with regard to occupant-specific information.

Table 1 summarized core elements of occupant representations in typical BPS applications (Hong et al. 2018; Ouf et al. 2018). Aside from some basic information concerning the occupants' state (location, activity, clothing), these elements separately address passive and active effects of occupants on the indoor environment (Mahdavi 2011). Passive effects mainly refer to the emission of sensible and latent heat, carbon dioxide, water vapor, and other substances. Active effects refer to occupants' interaction with the

Table 1. Schematic illustration of occupant-specific input in common BPS applications.

Basic state attributes of the occupants	Occupants' effects on the indoor environment	
	Passive effects	Active effects
• Presence • Metabolic Rate • CLO Value	• Sensible heat • Latent heat • CO_2/Pollutants • H_2O	• Schedules • Rules

buildings' indoor-environmental control devices and systems.

This brief observation reveals, from the ontological perspective, a number of gaps and inconsistencies in conventional occupant-related representations in common BPS tools. As such, different representational strategies are pursued with regard to passive and active effects. Passive effects are mapped to simulation zones. This is done, for instance, by specification of the number of occupants in each zone. To give an example, in this case, the people-related internal sensible heat gains are often expressed in area-related terms.

For instance, people-related sensible heat gains are expressed in units of power per zone floor area (e.g., $W.m^{-2}$). Alternatively, people-related internal heat gains are computed by multiplication of the number of occupants with a (default) per-occupant power term (i.e., $W.person^{-1}$).

When representing active behavior, even this rudimentary link to representation of occupants as individuals is absent. Instead, the operation of windows, blinds, and luminaires are captured in term of either schedules or rules associated with such devices. The resulting thermodynamic effects are, in case of thermal simulation, assigned to thermal zones. The schedules and rules may have been initially derived based on data or assumptions concerning occupants' behavior. However, the implementation does not include necessarily an explicit ontological representation of the occupants as individual active agents.

4 NECESSARY CONDITIONS FOR A THEORY-DRIVEN ONTOLOGY

Having looked at *i*) instances of building-related studies involving references to behavioral theories, *ii*) an implicit schema in typical computational building performance modeling tools, and *iii*) recent efforts in ontology proposals for representations of building occupants in energy analysis applications, we must reconfirm our initial observation at the outset of this contribution: As far as representations of occupants are concerned, there is still a gap between high-level behavioral theories in human sciences and technical applications in engineering domains. And there is still a need for enhanced ontological efforts to bridge this gap.

As a possible step in this direction, consider a concrete application scenario that could help with the formulation of the necessary theoretical underpinnings of an occupant-centric ontology. Thereby, the explanatory theory must not necessarily represent an ultimate scientific understanding of the salient aspects of the human behavior, which is – despite advances in the relevant fields of psychology, cognitive science, and sociology – still outstanding. Rather, the applied theory needs to provide a plausible conceptual scaffolding that can serve as the basis of an occupant-centric ontology. To exemplify such a theoretical instance, a rather simple schema is outlined in the following, which takes inspirations from both previous forays in theoretical biology and human ecology (Knötig 1992; Mahdavi 1998a,b, 2016), cybernetics (Ashby 1956; Wiener 1948), and more recent work in cognitive neuroscience (Damasio 2010).

The conceptual constituents of this scheme (Figure 5) may be described as follows. The overall discourse domain is divided into the "individual" (in our case, a human agent) and the "surrounding outside world". The latter includes both physical entities and processes as well as social settings and relationships. The concept of "ecological valency" refers to the totality of the surrounding world's attribute (resources, opportunities, risks, etc.) as relevant to the individual. As applied to indoor environment, the existence of devices (and respective interfaces) for passive and active environmental control can be viewed as proxies of the environment's ecological valency. On the other hand, the concept of "ecological potency" refers to the totality of the individual's physical (sensorimotor) and cognitive capabilities in dealing with the surrounding world. Note that, attributes such as the individual's age, sex, health condition, education, training, experience do not constitute ecological potency as such, but may be regarded as its first-order proxy candidates.

Notwithstanding epistemological qualms, we conveniently assume that both ecological valency and ecological potency are objective properties of the world. What is more relevant to a behavior theory, is their cognitive representations. We suggest that the surrounding world is mapped in terms of a primary representation as the individual's "environment". The surrounding world's ecological valency is mapped in terms of its "affordance", which we can understand as the perceived ecological valency (e.g., recognized opportunities regarding nutrition, shelter, social inclusion, as well as risks and hazards). Likewise, the mapping process includes also the representation of the individual's "self". This implies a kind of meta-mapping process, leading to a cognitive model of the "self within the environment". The meta-mapping is assumed to be accompanied by the capacity to imagine and anticipate future states of the individual through time and space.

Focusing on the cognitive domain associated with an individual, we can make a number of additional assumptions. For one thing, the individual's mind is arguably not a tabula rasa, but entails a memory-based (historical) reservoir of "experience and knowledge". This repository can be assumed to inform the perception of the affordance and contribute to the anticipatory evaluation of behavioral options. Moreover, the individual can be assumed to be – at least to some extent – guided or conditioned by a set of "beliefs and norms", which can narrow down the space of principle behavioral opportunities, for instance to those deemed admissible or proper on, say, ethical grounds.

The above conceptual reflections facilitate the conception of a basic explanatory model of control-oriented behavior. We assume that the individual's control-oriented or regulatory behavior is guided by the outcome of a process that involves the value-driven assessment and evaluation of its current state in view of possible distance to states that would be preferable. The preferable or desired state is the one that is – at the most basic (biological) level of values – oriented toward the individual's "survival". Seeking nutrition and shelter are instances of such behavior. Higher-level expressions of control-oriented behavior may be motivated by desire for elevated levels of physical or intellectual values associated with "satisfaction, comfort, and pleasure". Note that we are concerned here more with behavioral manifestations geared toward short-term and mid-term regulatory functions, rather than behavior with complex cognitive background targeting long-term planning agenda.

Prior to execution, behavioral options may be assumed to be virtually enacted in terms of "action models". Thus, the potential of "planned actions" in achieving the desired state can be assessed in an anticipatory fashion. Actions are executed if such pre-screening promises success and reconsidered or revised otherwise. The entire process is, as alluded to before, informed and supported by the affordance (perceived ecological valency of the surrounding world), the memory-based repository of knowledge and experience, and the filter-function of belief systems and norms.

Actions that have been repeatedly successful in the past may become part of the repository of experience in terms of "habits" and be executed without prior explicit assessment of their ramifications. Note that "habitual behavior" must be distinguished from "reflexive behavior". Whereas the former entails "automated" versions of previously conscious behavior, the latter denotes primarily biologically driven responses to specific stimuli and do not involve higher level cognitive (consciously planned) actions.

Notwithstanding its schematic nature, this model (see Figure 5) is suggested to embody the minimum conceptual repertoire for the formation of an ontology that would address core aspects of human behavior as relevant indoor-environmental applications (e.g., interactions with physical elements and interfaces constituting the environment's affordance).

It is important to reiterate that this model is not suggested to be a physiologically or psychologically detailed, accurate, or validated model of the human control-oriented behavior. What is suggested that it

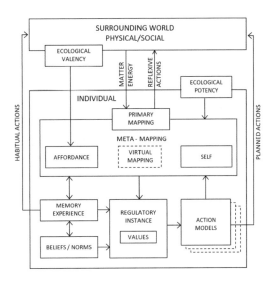

Figure 5. Schematic illustration of the structure and elements of a behavioral model for occupant-related ontological developments (Mahdavi 2020).

entails, as opposed to existing solutions, the minimum ingredients necessary for a general-purpose ontology versatile enough to support the implementation of occupant behavior models for engineering applications concerned with supporting the design and operation of built environments.

5 CONCLUSION

In this paper, we argued that there is a need to advance the state of art concerning building-related occupant behavior ontologies. The bottom-up reverse-engineering approach displays the limitations of the implicit occupant-related schemata in computational building performance analysis tools. The top-down look at the applications of behavioral theories in building-related domains (e.g., energy use) display the absence of consistent – let alone unified – ontological views of the relevant phenomena and processes. Consequently, already implemented occupant behavior schemata in the building performance domain do not appear to be built on solid foundations. We argued that closing the gap in this area requires a consistent high-level behavioral framework. A possible instance of such a framework was presented. We are currently exploring the applicability and coverage of an ontology built upon this framework. This ontology is intended to inform a versatile agent-based representation of occupants' presence and behavior in built environments.

ACKNOWLEDGMENT

The development of the content of this paper including the entailed concepts, ideas, and proposals benefitted from the authors' participation in the IEA EBC Annex 79.

REFERENCES

Ajzen, I. 1985. *From Intentions to Actions: A Theory of Planned Behavior BT – Action Control: from Cognition to Behavior.* In: Kuhl, J. & Beckmann, J. (eds.): 11–39. Berlin, Heidelberg: Springer Berlin Heidelberg. doi: 10.1007/978-3-642-69746-3_2.

Ajzen, I. 1991. The theory of planned behavior. *Orgn. Beh. Hum. Decision. Proc.* 50: 179–221.

Al-Marri, W., Al-Habaibeh, A. & Watkins, M. 2018. An investigation into domestic energy consumption behaviour and public awareness of renewable energy in Qatar. *Sustain. Cities Soc.* 41: 639–646. doi: 10.1016/j.scs.2018.06.024.

Ashby, W. R. 1956. *An Introduction to Cybernetics.* London: Chapman & Hall.

Damasio, A. 2010. *Self Comes to Mind: Constructing the Conscious Brain.* New York: Pantheon books. ISBN 978-0-307-47495-7.

DellaValle, N., Bisello, A. & Balest, J. 2018. In search of behavioural and social levers for effective social housing retrofit programs. *Energy Build.* 172: 517–524. doi: 10.1016/j.enbuild.2018.05.002.

D'Oca, S., Chen, C.-F., Hong, T., & Belafi, Z. 2017. Synthesizing building physics with social psychology: An interdisciplinary framework for context and occupant behavior in office buildings. *Energy Research & Social Science* 34: 240–251. doi: 10.1016/j.erss.2017.08.002.

Fishbein, M. & Ajzen, I. 2010. *Predicting and Changing Behavior: The Reasoned Action Approach.* New York: Psychology Press.

Heydarian, A., McIlvennie, C., Arpan, L., Yousefi Jordehi, S., Syndicus, M., Schweiker, M., Jazizadeh, F., Rissetto, R., Pisello, A. L., Piselli, C., Berger, C., Yan, Z. & Mahdavi, A. 2020. What drives our behaviors in buildings? A review on occupant interactions with building systems from the lens of behavioral theories. *Building and Environment.* 179. 106928. doi: 10.1016/j.buildenv.2020.106928.

Hong, T., D'Oca, S., Turner, W. & Taylor-Lange, S. 2015a. An ontology to represent energy-related occupant behavior in buildings. Part I: Introduction to the DNAs Framework. *Building and Environment* 92: 764–777. doi: 10.1016/j.buildenv.2015.02.019.

Hong T, D'Oca S, Taylor-Lange SC, Turner WJN, Chen Y, & Corgnati, S.P. 2015b. An ontology to represent energy-related occupant behavior in buildings. Part II: Implementation of the DNAS framework using an XML schema. *Building and Environment* 94: 196–205. doi: 10.1016/j.buildenv.2015.08.006.

Hong, T., Chen, Y., Belafi, Z. & D'Oca, S. 2018. Occupant behavior models: A critical review of implementation and representation approaches in building performance simulation programs. *Building Simulation*, 11: 1–14. doi: 10.1007/s12273-017-0396-6.

Knötig, H. 1992. Some essentials of the Vienna School of human ecology, in: Proceedings of the 1992 Birmingham Symposium; Austrian and British Efforts in Human Ecology, Archivum Oecologiae Hominis, Vienna, Austria.

Lo, S. H., Peters, G.-J., Breukelen, G. & Kok, G. 2014. Only reasoned action? An interorganizational study of energy-saving behaviors in office buildings. *Energy Efficiency* 7. doi: 10.1007/s12053-014-9254-x.

Mahdavi, A. 1998a. Steps to a general theory of habitability. Summer 1998, *Human Ecology Rev.* 5(1): 23–30.

Mahdavi, A. 1998b. The human ecology of the built environment. *J. Southeast Asian Architect.* 3(1): 25–44. September 1998. ISSN: 0128–9593.

Mahdavi, A. 2011. The Human Dimension of Building Performance Simulation. Driving better design through simulation. Proceedings of the 12th Conference of The International Building Performance Simulation Association, Sydney, Australia. V. Soebarto, H. Bennetts, P. Bannister, P.C. Thomas, D. Leach (eds): K16 – K33. ISBN: 978-0-646-56510-1.

Mahdavi, A. 2016. The human factor in sustainable architecture. in: Khaled A. Al-Sallal (ed.), *Low Energy Low Carbon Architecture: Recent Advances & Future Directions (Sustainable Energy Developments)*: 137–158. London: Taylor & Francis. ISBN: 978-1-138-02748-0.

Mahdavi, A. & Tahmasebi, F. 2016. The deployment-dependence of occupancy-related models in building performance simulation. *Energy and Buildings* 117. doi: 10.1016/j.enbuild.2015.09.065.

Mahdavi, A. 2020. Explanatory stories of human perception and behavior in buildings. *Building and Environment* 86. doi: 10.1016/j.buildenv.2019.106498.

Maslow, A. H. 1943. A Theory of Human Motivation. *Psychological Review* 50(4): 370– 396.

Matthies, E., Kastner, I., Klesse, A. & Wagner, H.-J. 2011. High reduction potentials for energy user behavior in public buildings: How much can psychology-based interventions achieve? *Journal of Environmental Studies and Sciences* 1: 241–255. doi: 10.1007/s13412-011-0024-1.

Ouf, M.M., O'Brien, W. & Gunay, H.B. 2018. Improving occupant-related features in building performance simulation tools. *Build. Simul.* 11: 803–817. doi: 10.1007/s12273-018-0443-y.

Reckwitz, A. 2002. Toward a theory of social practices: A development in culturalist theorizing. *Eur. J. Soc. Theory* 5: 243–263.

Ryan, R. M. & Deci, E. L. 2000. Self-determination theory and the facilitation of intrinsic motivation, social development, and well-being. *American Psychologist* 55(1): 68–78. doi: 10.1037/0003-066X.55.1.68.

Schwartz, S.H. 1977. Normative influences on altruism. In: Berkowitz, L. (ed) *Advances in experimental social psychology.* New York: Academic, New York: 221–279.

Schwartz, S.H. & Howard, J.A. 1981. A normative decision-making model of altruism. In: Rushton, J.P. & Sorrentino, R.M. (eds), *Altruism and helping behavior.* Hillsdale: Erlbaum: 189–211.

Shove, E., Pantzar, M., & Watson, M. 2012. *The Dynamics of Social Practice: Everyday Life and How it Changes.* London: SAGE Publications Ltd.

Simon, H.A. 1957. *Models of man; social and rational.* New York: Wiley.

Simon, H.A. 1995. A behavioral model of rational choice. *Q. J. Econ.* 69(1): 99–118.

Wiener, N. 1948. *Cybernetics, or Control and Communication in the Animal and the Machine.* Cambridge: MIT Press.

Application of ontologically streamlined data for building performance analysis

D. Wolosiuk & A. Mahdavi
Department of Building Physics and Building Ecology, TU Wien, Vienna, Austria

ABSTRACT: Buildings are increasingly equipped with data monitoring infrastructures to collect multiple layers of dynamic data pertaining to the states and events related to systems' performance, indoor-environmental conditions, occupants' location, movement, and control-oriented actions representation of dynamic building related data. Efficiently utilized, this data could not only enhance the quality and effectiveness of buildings' operational regimes, but also enrich the knowledge base for building design decision support. However, to exploit the potential of this data effectively, seamless data transformation processes are needed, starting from raw monitoring data and ending in adequately structured and visualized building performance data. In the present contribution, we provide a detailed blueprint of a well-defined and ontologically supported instance of such a transformation process. To illustrate this process, we focus in this paper on a specific class of building performance queries that require information on buildings' visually relevant boundary conditions. Such queries pertain to, for example, the intensity of incident solar radiation of buildings' envelope components, the daylight availability and distribution in interior spaces, and the expected output of building-integrated solar energy harvesting systems.

1 INTRODUCTION

The developments in the last decades have shown that architecture, engineering, and construction (AEC) industry can benefit from well-defined data schemas (such as those offered by International Foundation Classes and green Building XML). Incorporating these schemas into the Building Information Modelling (BIM) software has been suggested to facilitate better communication between parties involved and to improve the overall efficiency of the building design, construction, and operation processes. However, the primary focus of most common schemas lies in the representation of primarily static building attributes, including geometry and semantic information on building components and systems. Thus, there is a potential to further enhance the existing schemas toward a more comprehensive coverage of dynamic processes – particularly those relevant to the operation phase of buildings. Specifically, buildings are increasingly equipped with sophisticated data monitoring infrastructures to collect multiple layers of dynamic data pertaining to the states and events related to systems' performance, indoor-environmental conditions, occupants' location, movement, and control-oriented actions representation of dynamic building related data. Exponentially growing volume and availability of building-related performance data enabled by wireless sensor networks and low-power microcontrollers translate into massive quantities of data on occupancy, indoor-environment, and energy systems' performance. Efficiently utilized, this data could not only enhance the quality and effectiveness of buildings' operational regimes, but also provide an empirically-grounded knowledge base for building design decision support. However, to exploit the potential of this data effectively, seamless data transformation processes are needed, starting from raw monitoring data and ending in adequately structured and visualized building performance data. In the present contribution, we provide a detailed blueprint of a well-defined and ontologically supported instance of such a transformation process. Toward this end, we first describe a recently proposed ontology for inherently dynamic building-related data, including multiple variables expressing the state of building systems, indoor environments, and occupants (position, movement, perceptual clues, behavioral manifestations). Collected raw data from buildings' monitoring infrastructure can be structured and streamlined via projection of such data into the structure of this ontology. Data thus ontologized, can then be accessed from a host of analysis-oriented applications, including data visualization, performance simulation, and optimization. To illustrate the scope and applicability of this ontology and the developed data transformation process, we focus in this paper on a specific class of building performance queries that require information on buildings' visually relevant boundary conditions. Such queries pertain to, for example, the intensity of incident solar radiation of buildings' envelope components, the daylight availability and distribution in interior spaces, and the expected output of building-integrated solar energy harvesting systems. Computational applications for building design and operation support typically deploy predefined or standardized formats for data input necessary for the execution of the internal algorithms. In this context, ontologically guided data processing facilitates the development of interfaces between ontologized data and performance analysis applications depending on standardized input formats. Given the generalized and scalable schema captured in the ontology, the corresponding processed data acquires a highly structured and widely deployable characteristics. This results in increased opportunities to use the same set of ontologized data for multiple purposes and by multiple stakeholders.

2 BACKGROUND

Building related ontologies have been known for quite some time. These became a foundation for primarily "static" data schemes for description of building systems and construction. There is an increasing need for mapping "dynamic" data into building information models, specifically for maintenance and control purposes. This development is related to the growing importance of energy-efficient design and the rising volume of available monitored data.

2.1 Building ontologies and "dynamic" data

Recently, there have been a few attempts to address the need for an ontology and data schema for dynamic building related data for performance assessment processes and applications. Instances of such efforts include an ontology for building monitoring (Mahdavi & Taheri 2017), and a building performance indicator ontology (Mahdavi & Taheri 2018, Mahdavi & Wolosiuk 2109a, b). The ontology for building monitoring intended to capture the aspects of data streams coming from various monitoring systems. Based on some of the former efforts in that area (e.g., Mahdavi et al. 2011, 2016) main categories (occupants, indoor environmental conditions, external environmental conditions, control systems and devices, equipment, and energy flows) and pertaining sub-categories were defined based on numerous instances of monitored variables. In addition, a data schema was developed that captures necessary attributes concerning diverse monitored variables.

Building performance indicators (frequently included in standards, technical literature, and simulation applications) play a major role in evidence-based evaluation of building design and operation. There have been very little attempts to capture the characteristics of this domain toward formulating an explicit BPI ontology. The BPI ontology proposed in Mahdavi (2018, 2019a) was based on a review of a large number building performance indicators pertaining to different domains. Five relevant categorical groups were identified, namely energy and resources, indoor air quality, thermal performance, acoustical performance, and visual performance. Moreover, essential data attributes were considered that are needed to comprehensively capture the complexities of various performance indicators. The most recent development combined these two ontologies into a Building Performance Data ontology.

2.2 Building performance data ontological schema

Over the course of development of BPI ontology, it became clear that that ultimately most performance indicators are compounds of measured data, or they can be traced back to a single measured variable. This observation indicated that attributes used to describe monitored data should be mostly covered by attributes domain of BPI data. For example, measured indoor temperature or CO_2 concentration level can function BPIs, for instance in thermal performance and indoor air quality categories. Compound BPIs can be derived from multiple measured variables. For instance, daylight factor (an indicator in the visual performance domain) is derived by dividing measured or computed indoor illuminance by the simultaneously obtained outdoor illuminance. It was therefore logical to slightly modify the elaborate indicators schema to include monitored data, arriving thus at a comprehensive Building Performance Data (BPD), that is an ontological schema that would encompass both previously mentioned ontologies (Mahdavi & Wolosiuk 2019b).

The general schema for BPD is largely similar to the BPI schema proposed in Mahdavi (2018). Table 1 presents the main features of the BPD schema. Each considered variable falls under a specific main performance category and a sub-category, and has a unique name. Each variable is of a certain type (discrete or continuous), has a magnitude, possibly a direction (in case of vector-type variables), and a unit. Properties included in the spatial domain allow to associate a variable with a position in a Cartesian space or a certain topologically relevant location. Those included in the temporal domain provide necessary (time stamp) and supplementary (duration, time step, or aggregation method) details on variable's temporal features. Properties in the Frequency domain help to specify details relevant for specific performance variables (for example in the acoustics domain).

The Agent ID property is relevant if a variable is attributed to a building user or other relevant agent. The properties included in notes help to provide further details, annotations, and specifications concerning the captured data.

3 FROM RAW DATA TO APPLICATIONS

Structured transformation of raw data (in the context of this work mostly understood as a numerical measure of a performance variable) into an organized ontological schema involves a number of challenges and

Table 1. General BPD schema (modified based on Mahdavi 2018).

Category	Sub-category	Name		
Variable	Value	Type		
		Magnitude (size)		
		Direction (vector)		
		Unit		
		Spatialdomain	Point	
			Plane	
			Volume	
			Topological ref.	
			Aggregation method	
			Grid size	
		Temporaldomain	Time stamp	
			Duration	
			Time step	
			Aggregation method	
		Frequencydomain	Range	
			Band (filter)	
			Weighting	
			Aggregation method	
	Agent	ID		
	Notes	Data sources	Category	
			ID/name	
		Derivationmethod	Details (formula, link, etc.)	

questions at each step. Figure 1 presents an overview of the proposed process, where primary "low-level" performance related data undergoes various processing stages to finally be utilized in various "high-level" applications.

Starting with data sources, one has to address the multiplicity of possible source data structure and formatting. There are multiple standards of raw data storage (e.g., tabular files, database files, spread sheet files etc.) and formatting (e.g., date formatting, number formatting, header structure, measurement unit, etc.). These are related, in part, to manufacturers' standards (e.g., for sensors), performance application output standards, or individual users' needs or preferences. To prepare the data for further processing, frequently a universal or automated extraction of measured data is not feasible. Rather, customized approaches are often necessary. Other preprocessing steps such as temporal aggregation/segmentation and quality check can be performed depending on the data specifics. The need for aggregation or segmentation depends on data source's sampling type. Fixed temporal intervals in ontologized dataset could increase usability in future applications (e.g., in statistical analyses). Quality check is important especially when data comes from physical sensors. These are prone to power outages or communication interruptions.

Finally, the ontologization process – categorization and attribution of variables – can also become a challenging task. This step is highly dependent on the nature of the variable in question. More specifically, in many cases the properties pertaining to a variable cannot be scraped from the header of a data source file. More often they need to be identified or provided manually, making this step potentially prone to errors that could be hard to detect in the final semantically enriched dataset.

When choosing a deposition format, a list of requirements should be fulfilled and technical specifications considered. First and foremost, such a format must capture and store variables and variables' metadata (of a various type) in a hierarchical manner. Secondly, given the nature of building performance data and possible large size volumes, provision of effective support for queries is of a high importance.

4 PBD ONTOLOGY IMPLEMENTATION

The proposed ontology and the data schema were tested as a part of a structured transformation workflow (see Figure 1), where performance data collected by multiple sensors monitoring indoor and outdoor environment were put through the transformation steps to form ontologically structured dataset.

The primary data source type in this implementation was multiple database files containing sensor readings supplemented with a timestamp. The source files' standardized naming convention facilitated the identification of the performance variable and its relevant (e.g. temporal, spatial) attributes. To facilitate the attributes supplementation and deposition process, all variables and their attributes were aggregated in a single csv file. The content of this file followed the design specifications of conversion algorithm. Some of the variables required individual approaches in preprocessing or conversion (e.g. sky scanner, luminance camera) due to non-standard data source formats.

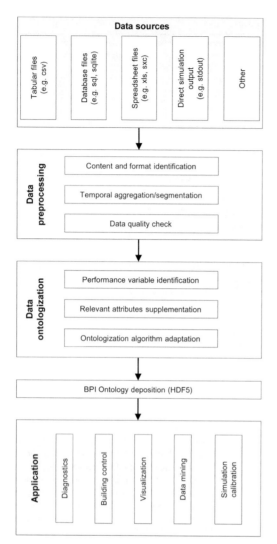

Figure 1. Schematic process overview for transformation (preprocessing, ontologization, storage) of performance-data for use in various down-stream applications (modified based on Mahdavi & Wolosiuk 2019b).

Python programming language was used in the process of performance data transformation. Several functions for conversion, quality check, attributes supplementation, and serializations were written in a HDF5 file (The HDF Group 2019). HDF5 is a primarily scientific data format capable of storing various data type structures. It is capable of storing these structures in hierarchical manner and it makes it possible to assign and accommodate complex metadata to the elements of the structure. Given these features, the HDF5 file format appeared to provide a suitable foundation for testing the BPD Ontology and the proposed data schema. Note that we do not suggest that this is the only or the ultimate implementation solution. Another implementation approach, for instance, could be based on Semantic Web technologies.

The semantically enriched dataset was then tested in a series of applications (see Mahdavi & Wolosiuk 2019a, 2019b). The following section presents an illustrative instance of an advanced application.

4.1 *Illustrative example*

In this illustrative example a task specific interface between ontologized data, stored in an HDF5 file, and Ladybug Tools was created. The interface relies on defined terms and data structure of BPD Ontology.

Ladybug Tools (Roudsari 2013) integrates the potential of well-known performance simulation engines such as the EnergyPlus (Crawley et al. 2001) or Radiance (Ward 1994), with the Rhino 3D (McNeel 2019) modelling software. It is a collection of small applications that couple these simulation engines with a 3D modelling and visualization potential of the Rhino software. This linking is enabled through Grasshopper – a visual programming environment built-into the Rhino software.

4.2 *From BPD ontology to Grasshopper*

Grasshopper quickly grew beyond initial 3D algorithmic modelling and parametric design platform. This was enabled by giving the community a possibility of creating custom components and component packages that could be shared online. The Ladybug tools is an instance of such component package development. Grasshopper supports a number of programming languages for creating new components. These components can be of a universal nature (simple mathematical operations on input) as well as complex or task specific nature, such as calling external software for output generation. The latter is the case of this implementation, where several custom components written in Python language were created to enable and test interfacing between BPD ontology serialized in a HDF5 file and elements of Ladybug Tools. In general, the created custom components take HDF5 file input, and some input options to extract specific performance variable in a desired quantity and form that is consistent with particular requirements of Ladybug's component input. Other basic scenarios that were tested involved browsing HDF5 content for specific performance variable selection (based on attribute filtering) for data visualization in Rhino 3D.

4.3 *Location-specific daylight studies*

Many Ladybug elements take EnergyPlus Weather Files (EPW) as an input. EPW is a tabular-style file containing detailed (hourly) typical meteorological year (TMY) weather data for a specific location. One of the elements of the EPW file is the information on global, diffused horizontal, and direct normal solar radiation. This data is used as an input for some of the daylight study simulations provided in Ladybug. Specifically, it is used for generation of sky model for the Radiance simulation software.

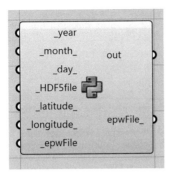

Figure 2. A custom component for modification of Energy-Plus weather file, for use with climate-based sky generator in solar radiation studies.

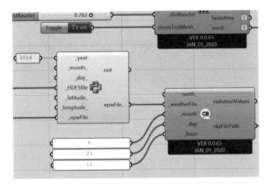

Figure 3. Custom BPD interfacing component integrated into simulation setup.

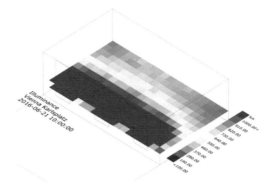

Figure 4. Visualization of the indoor illuminance simulation results based on local historical data for Vienna.

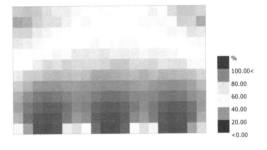

Figure 5. Daylight autonomy studies based on local data extracted from BPD stored in a HDF5 repository.

In this scenario, an application specific component was created to take measured radiation data from the HDF5 file and to replace TMY values in the original EPW file. Values recorded in a specified year (or a specific period within a year) are *i)* aggregated in terms of hourly values, *ii)* direct normal component is calculated from global and diffused horizontal radiation (if not provided), *iii)* latitude and longitude information is replaced (if provided), *iv)* and finally a new EPW file is generated, stored locally, and provided as an input for the *Generate Climate Based Sky* component. Figure 2 shows this component with the required and optional inputs on the left and outputs on the right side. Being able to modify parts of an existing EPW allows for seamless integration of localized environmental data into the standard Ladybug design or analysis workflow.

The created BPD interfacing component was tested by integration in two illustrative simulations, involving snapshot-type Illuminance and annual Daylight Autonomy analyses. The test case for the created component involved modified sample studies provided by the creators of Ladybug software.

The illuminance simulation results are based on selected diffuse horizontal and direct normal irradiance values obtained from modified EPW file generated by our custom component. The new EPW file contains radiation values for the entire year as recorded by pyranometers in 2016 in Vienna city center. The selected point in time for the simulation is 2016-06-21 10:00:00. Figure 3 presents the integrated BPD interfacing component into Ladybug components-based simulation setup (only a small part of the setup canvas is visible here). The resulting daylight illuminance distribution values are visualized in Figure 4 in terms of a color scale.

In a similar manner, an annual Daylight Autonomy for the same space was performed. This time the local historical solar radiation data from the entire year 2016 was used to visualize and analyze Daylight Autonomy (DA) based on a default office type occupancy schedule. The simulated results represent the percentage amount of occupancy time when the illuminance is above the given threshold of 300lux. Again, results are visualized as a color-mapped grid representing value threshold for a given analysis grid tile (see Figure 5).

5 CONCLUDING REMARKS

This paper discussed the recent version of the BPD ontology and its data schema. We discussed the process and challenges concerning the standard ontological data transformation process. We further presented instances of advanced application of semantically structured datasets. Toward this end, "ontologized" dataset was deployed, which is based long-term daylight monitoring data from our Department's microclimatic observatory. We illustrated how this dataset was used to support building performance analyses and visualization pertaining to indoor illuminance distribution and Daylight Autonomy.

As such, the experiences thus far confirm the promising potential of the proposed approach. Thereby, multiple streams of monitored or computed building performance data, once structured according to a standardized ontological schema (BPD) can be made available for seamless utilization in multiple application scenarios, involving not only building design, but also building operation and management support.

REFERENCES

Crawley, DB., Lawrie, L K., Winkelman, F.C. et al. 2001. EnergyPlus: Creating a new-generation building energy simulation program. *Energy and Buildings*, 33(4), 319–331.

Mahdavi, A. 2011. People in Building Performance Simulation. In Hensen, J.L.M., Lamberts R. *Building Performance Simulation for Design and Operation.* Taylor & Francis Group.

Mahdavi, A., Glawischnig, S., Schuss, M., Tahmasebi, F. & Heiderer, A. 2016. Structured Building Monitoring: Ontologies and Platform. *Proc. of 11th ECPPM Conference.*

Mahdavi, A., Taheri, M. 2017. An Ontology for Building Monitoring. *Journal of Building Performance Simulation.* 10(5–6), 499–508.

Mahdavi, A., Taheri, M. 2018. A building performance indicator ontology. *Proc. of 12th ECPPM Conference*, 385–390.

Mahdavi, A., Wolosiuk D. 2019a. Integration of operational data in building information modelling: From ontology to application. CLIMA 2019 Congress. *E3S Web Conf.Volume 111.*

Mahdavi, A., Wolosiuk D. 2019b. A Building Performance Indicator Ontology: Structure and Applications. *16th International IBPSA Conference.*

McNeel, R. & Associates 2019. Rhinoceros Version 6. Available from https://www.rhino3d.com/

Roudsari, M.S. & Pak, M. 2013. Ladybug: a parametric environmental plugin for grasshopper to help designers create an environmentally-conscious design. *Proc. of the 13th International IBPSA Conference.*

The HDF Group. Hierarchical Data Format Version 5 (HDF5) 2019. Available from https://www.hdfgroup.org/

Ward, G.J. 1994. The RADIANCE Lighting Simulation and Rendering System. *Proc. of the 21st SIGGRAPH Conference.*

Microservice system architecture for data exchange in the AEC industry

G. Šibenik & I. Kovacic
Institute of Interdisciplinary Building Process Management, TU Wien, Vienna, Austria

T.-J. Huyeng & C.-D. Thiele
Institute of Numerical Methods and Informatics in Civil Engineering, TU Darmstadt, Darmstadt, Germany

W. Sprenger
Zentrale Technik, Direktion Digitalisierung und Software-Engineering (DS) - BIM.5D, Züblin, Stuttgart, Germany

ABSTRACT: Realization of inter-domain model-based data exchange in the AEC industry has not yet managed to fully satisfy the end user's needs. While microservices are getting more present on the market, its application for the data exchange in the building planning process has not been considered. This paper investigates the potential of microservices in realizing the data exchange within planning and construction. System architecture of an existing monolithic software tool is converted to a microservice system architecture with the help of literature review, analysis of Docker system and expert knowledge. The results show multiple possibilities to containerize a monolithic application with diverse advantages, differing in effort to be containerized. While the microservices do not offer a smooth transition of system architectures from the existing data exchange solutions, containerization offers the flexibility for non-standardized processes which could be tailor-made organized and easily replaced. Although the orchestration between the services is not easy to achieve and the implementation requires significant effort and a paradigm shift in the planning process, the advantages of microservices can already be used in the data exchange process.

1 INTRODUCTION

Multiple building models are used during a building life cycle, some having significant semantical and geometrical differences, but still describing the same real-world building. Interpretation of building models during a data exchange takes place in three ways: in the native tool, importing tool or in the third-party solutions, and all of them still have drawbacks such as misinterpretations or missing information (Sibenik & Kovacic 2020; Turk 2020). Non-proprietary third-party solutions which use exchange standards like industry foundation classes (IFC) show greatest potential for seamless exchange in the heterogeneous AEC industry (Lee et al. 2020; Ramaji & Memari 2016). Interpretations are numerous and vary depending on the planning workflow. The data exchange based on the exchange of files did not deliver sufficient usefulness to end users and the existing web- and cloud-based solutions need to be investigated (Shelden et al. 2020). Researchers emphasize the opportunities which new technologies offer across the industry (e.g. Woodhead et al. 2018), but particular applications for single tasks are still understudied.

Previously, we developed a third-party solution that facilitates interpretations between architectural design and structural analysis models as a monolithic application (Sibenik et al. 2020). A system of loosely coupled services contributing to the business process could show advantages compared to standard monolithic system architectures, however to our best knowledge it has not been investigated for the data exchange purposes in the AEC industry. In this paper, we aim to explore the applicability of microservices for data exchange purpose, specifically between architectural design and structural analysis.

A monolith represents a software application whose modules cannot be executed independently whereas the microservice application is a distributed application made of modules that perform partial services (Dragoni et al. 2017). A monolithic application has all its functionalities in one place and can be scaled by reproducing the same application. It can get large in size, hard to manage and in order to contribute to the application, each developer needs to understand the logic behind the whole application. On the other hand, microservices can be scaled by breaking them up into smaller pieces, the functionalities are provided independently, they can be based on different technologies and it is easier for new developers to join the project. In this paper, we investigate possible applications of microservice system architecture to facilitate the data exchange in the planning process.

In our work, an existing monolithic application, as a non-proprietary third-party solution, developed to support the data exchange, will be redefined with microservices. A new system architecture is created by containerizing an existing application, whereby the application transitions from .NET framework to multiple Docker containers that form a loosely coupled system.

In the section 2 we present the literature review, following with the methodology in the section 3, and presentation of the possible conceptual solutions in section 4. Finally, in the section 5 we will describe the opportunities identified in the concepts, and the implementation challenges.

2 LITERATURE REVIEW

2.1 Application of microservice architecture

Some existing applications of microservice technology in the AEC industry are Speckle, BIMSWARM, and SCOPE (Huyeng et al. 2020). However, research which relates the technology with the data exchange in the planning phase lacks. Krylovskiy et al. (2015) describe the use of microservices for Internet of Things (IoT). Their platform DIMMER, in the early stage of development, facilitates the interoperability between GIS and BIM models. Li et al. (2020) use microservices to facilitate infrastructure smart services. They recognize that software tools are mostly developed as monolithic applications in the infrastructure, as well as in the AEC industry, and as the application functions grow and perform more tasks, more errors tend to occur.

Hoffmann et al. (2019) conceptualize a system to provide product data from the manufacturing industry via microservices. Heterogeneous one-time organizations comprised of multiple domain-specific stakeholders brought together by a single project are typical in the building planning process. The microservice system architecture could be suitable for technical realization of project-specific applications since both the organizations and microservices form a loosely coupled system. Di Orio et al. (2019) also describe the approach to data exchange problems in the manufacturing industry and use microservices in order to address them. They say that the manufacturing industry is inert towards the digitalization, and that it does not manage to support changing market and customer needs. This problem is even more present in the AEC industry, where no product is equivalent.

Huyeng et al. (2020) investigate data exchange possibilities with microservices. They use ontologies and the concept of linked-building-data to deal with geometrical and non-geometrical information, and the system architecture is realized with microservices and a triplestore database. They aim to interrelate manufacturing with AEC industry and connect product models with BIM. Munonye & Martinek (2020) researched the data storage possibilities with the microservice system architecture. Since each service has its own data storage, they recognize a challenge in retaining consistency and integrity of data while providing a loose-coupling principle of microservices.

The above presented research shows that the microservices are investigated for various applications, including the data exchange, but there has not been research which directly refers to the data exchange between domains involved in the building planning process. The reasons for that could lie in the error-prone and insufficiently reliable exchange practices. Some of these issues might be resolved with the technologies which have not yet been considered.

2.2 Decomposing a monolith

Li et al. (2020) describe a popular three-layered system architecture including presentation, business logic and data access layer. Gos & Zabierowski (2020) similarly divide it in web, service and repository layer.

Baskarada et al. (2020) identify opportunities and challenges when moving from a monolith to a microservice system architecture. Baskarada et al. (2020) state there are no clear directives if and when an application should be monolithic or microservice. Decomposition of software architectures to microservices happens more often than the creation of microservices from the scratch. The organizations meet the problems with their existing monolithic application such as scalability and rapid development and have to detect a way to switch their application to microservices (Tapia et al. 2020).

Ponce et al. (2019) performed a rapid review of migration procedures from monolithic to microservice architecture. They conclude that model-driven approaches which include business aspects of the software tool and data flow diagrams are the most commonly used, alongside static analysis of the existing code and dynamic analysis of the runtime functionalities. Kecskemeti et al. (2016) migrated to a microservice system architecture for the ENTICE project and describe primarily technical implementation of various images and dependencies with image synthesis and image analysis approaches. In the work of Sarkar et al. (2018) advantages of a migration from a Windows-based application is described, as well as the additional advantages when the application is further modularized.

Krause et al. (2020) propose the use of static and dynamic methods while decomposing a monolithic application to microservices. They list bounded context of the domain-driven design, static code analysis and refactoring based on non-functional requirements as common techniques used for conversion to microservices, and add dynamic analysis to the list. Chen et al. (2017) use dataflow-driven approach for decomposing a monolithic application. They propose purified and decomposable Data Flow Diagrams (DFD) next to traditional DFDs which could help with the transition.

In our case, we are familiar with the existing application and its development plan. In this work we aim at using a model-driven approach to propose new concepts with the help of DFDs. We also consider the flexibility and scalability of the future system architecture which is not directly tackled by the existing research.

3 METHODOLOGY

3.1 Existing monolithic application

Numerous problem-specific solutions to overcome data-exchange issues in forms of algorithms, software applications or platforms can be found in the literature for various domains in the planning process (e.g. Ladenhauf et al. 2016; Lee et al. 2014; Ramaji & Memari 2016). These all can be considered for a conversion of the system architecture. The AEC industry is heterogeneous which is reflected in software tools, business practices and interactions. The digitalization in the industry is still on the low level compared to the other industries (Agarwal 2016), but new applications for various purposes are continuously being developed with the potential to be candidates for microservice applications.

The existing monolithic application used for data exchange between architectural design and structural analysis will be used as a starting point for the implementation of microservices. For analysis of implementation potential within this data exchange process we use the application created previously by the corresponding author. The application is part of a data-exchange framework implementation focused on supporting varying business processes and the scalability was recognized as a potential issue in the case of extended use.

Within the application, domain-specific building models are described as open building models, based on IFC taxonomy extended with missing terminology, structured in JSON format and additional geometrical representation according to the Open Cascade (OCC) kernel. Models are read, interpreted and recreated with the. NET application called redDim, with the help of geometry kernel methods.

The interpretations in redDim consist of dimensionality reduction and reconnecting elements which are specific for the data exchange between architectural design and structural analysis. These two interpretations can further be decomposed into three methods each. The methods can vary depending on the experience and intuitive knowledge of structural engineers. Additional methods can be developed for additional geometry definitions. The same approach can be implemented for other domains as well, especially if the data exchange involves geometry interpretations, but even if solely semantical interpretations take place. The additional data exchange considerations would require new interpretation methods. Therefore, due to its potential in being scaled in various directions, we consider this software tool a suitable candidate for the consideration of a microservice application.

We define redDim as a three-tier architecture (Li et al. 2020) and analyze its services. redDim consists of a presentation, a business logic and a data access layer. The presentation layer is an OCC GUI, which is in the current system architecture realized with Windows Frames. The business logic layer involves the communication between the GUI and the database which is realized via the MongoDB API, and all the methods which edit the model between the initial and final model state developed with C# and the OCC wrapper. The data access layer is a MongoDB database, where the two models, initial architectural design and final structural analysis open model, are stored. The monolithic model also uses the working memory which includes C# objects during the run.

3.2 Decomposition methodology

The existing monolithic application will be decomposed to a microservice application. Therefore we researched the literature which is mostly available in other research fields, available instructions for Docker and migration of. NET applications to Docker (Docker 2020a; Docker 2020b; Microsoft 2020a; Microsoft 2020b) and expertise of authors involved with the research project SCOPE (SCOPE 2020).

We focus on the model-driven approach, as the application functions are developed by the authors and the application is not yet widely used. Static and dynamic analysis of the application could deliver insignificant results in the case of future optimization of code of the monolithic application, which would be required for the wider application in the industry.

4 IMPLEMENTATION AND RESULTS

Based on the literature review, provided instructions and authors' expertise we propose four different concepts, where one is equivalent to the initial application (see Figure 1). The representation does not distinguish the difference between an inter-container communication and an API, nor if a service is realized with Docker or other technologies. Partial tests were performed in order to realize the intended system architectures. Parts of redDim concept were created as microservices, as well as simplified interpretation methods. Our different concepts to containerize the existing monolithic system architecture are identified as:

a) MongoDB is already a self-contained module, therefore this concept is equivalent to the initial monolithic application. There is a possibility to containerize it and use it with Mongo Express. This was the easiest approach to containerize with Docker one part of the existing application. However, it does not bring any significant advantages, except the ones accompanying Mongo realized with Docker.

b) In this case the presentation layer containing the GUI and the business process with interpretations are separated. This means, that the GUI can be realized both as a container or remain in the existing Windows Form Application and communicate to a

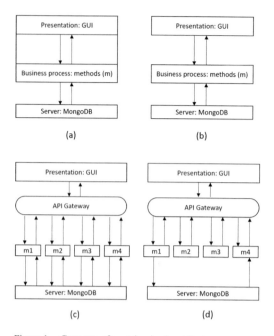

Figure 1. Concepts of containerized architectures.

JSON format is used. In this case a serialization of OCC objects is needed, but the realized system gains on flexibility.

5 DISCUSSION

The proposed containerized solutions have different advantages and disadvantages compared to the monolithic architecture. Containerizing the MongoDB server does not represent a significant logical change compared to the monolithic architecture which connects to the individually running server via the MongoDB API. This can already be considered a microservice architecture regarding the database – redDim connection. The separation of presentation layer and interpretations makes the modules smaller and more task-specific, however, the current use of C# OCC wrapper makes it more complex. The OCC wrapper can only be used as a Windows Forms application, therefore it is necessary to use Windows containers with Docker. This is a significant drawback compared to Linux containers which are technologically more advanced (Sarkar et al. 2018). Another drawback is that the OCC classes used with the wrapper are not part of the open geometry kernel. However, the business process layer could be redefined with C++ architecture, where the original OCC core can be used, or Python OCC wrapper which is open and more advanced, and communicate with a GUI in any form required. In that way it would remain open and be more flexible for additional business processes in the complete framework. The third option (c) provides a further modularization of the program; however, the constant communication of business process layer with the database requires implementing the conversion methods before and after each of the interpretation methods. This approach may end up being computational expensive compared to the monolithic architecture, but it provides the possibility of flexible constellation of interpretation methods which is desired for various heterogeneous processes. The final system architecture concept avoids the requirement of constant communication with the database, but it requires a significant amount of information transfer between the containers in the form of objects, which could also lead to high computational effort. The advantage of both methods with individually containerized methods is that the methods can be separately edited, new ones can be added or existing ones removed, without distorting the whole system. This means as well that a different database, GUI or geometry kernel can easily replace the existing ones if needed. More attention must be paid to communication processes between the services.

The full potential of containerization could be used when the API Gateway is able to transfer more data with a similar performance as a monolithic system. This approach could deliver various business processes which were not possible until now. The single containers could be setup in various ways to respond to

containerized central application. The communication can send HTTP(S) queries, which describe a state of the model, however, it reads the model in a working memory, therefore the responses are serialized in the form of OCC objects. This approach does not offer light communication between the presentation layer and the GUI, which is one of the principles when using a containerized architecture. However, it separates the GUI from the interpretation methods, but due to the serialized responses they are highly dependent on each other.

c) In this concept the business process layer is divided into multiple containers. Each container represents an individual method used for data exchange. In this case each container communicates with the database. Therefore, it is necessary to convert JSON objects on the MongoDB during each request to OCC classes which will be used by the separate interpretation containers. Methods represent interoperations as in the monolithic application, namely reduction of dimensionality and reconnecting elements, however, can be further dissolved to e.g. punctualization, linearization and planarization. The conversion of OCC objects into JSON format and back can be realized as separate containers or in each container alongside the main method.

d) The final concept has the same containerization as in (c), however, the communication does not take place with the server container after each step but is realized as cache memory between the GUI and containers in business process layer. This approach deals with OCC classes except for the communication with the database, taking place only during the initial import and final export method, when the different data-exchange needs of end users. The other forms of containerizing monolithic systems could also be useful, however, the microservice architecture provides more flexibility and usability for the interoperation methods. The microservices as single methods can deliver new business processes not only limited to the data exchange process.

Hoffmann et al. (2019) find that combining independently smaller applications which can be individually replaced and reorganized would make great advantage to support a heterogeneous business process and provide new technical solutions, but also deliver new problems, as coordination, which already poses a challenge for microservices. This can also deliver new business processes as the services can be individually software independently developed and provided to the end user. Since the considered concepts exchange serialized data, problems could occur if the containers are realized with different programming languages.

Complex systems and changing workflows in the AEC industry could benefit from the distributed systems provided on the small scale: greater possibilities to integrate efforts in the community; partial solutions can be used together to achieve a needed solution, even if the underlying technologies differ; it can lead to an IoT ecosystem which is considered to be the industry's future (Woodhead 2018), not limited to the data exchange. Additionally, containers have multiple advantages regarding the performance, portability, more convenient deployment and sharing. However, the focus of this paper is on the scalability and how the containerized system architecture could serve the heterogeneous workflows. New system architecture provides a basic framework for the data exchange and adds to it interpretations for the specific data exchange. Since the majority of interpretations are not standardized, they might need editing, or completely different interpretation to support a specific planning workflow. Different workflows can involve multiple interpretations for the same building elements, requiring tailor-made solutions to serve varying needs of the industry. The tailor-made solutions are created by combining multiple containers and adding the new ones instead of the development from scratch. This research fills the gap which relates the micro-service system architecture with highly variable process within the AEC industry.

6 CONCLUSION

In this work we analyzed possible advantages to the inter-domain data exchange practices in the building planning process. The AEC industry is stepping behind the other industries in the digitalization process, therefore it is difficult to recognize the benefits which could come with the up-to-date technological solutions. This work considered the use of microservices with the process-supporting software solution that can be used for model-based data exchange between architectural design and structural design teams. The migration to microservices would primarily offer more flexibility, from the technical side, as well as from the business logical side. The drawback that is recognized is, that the services are tightly coupled and exchange significant amount of geometrical and non-geometrical information which defies in a certain amount the logic of microservices. However, the flexible organization of modular services could be a technical counterpart of heterogeneous processes existing in the AEC industry.

The existing solutions are still conceptual and the monolithic application has not been completely realized, only the parts recognized by the authors as crucial. The initial application which is used to define microservice concepts is developed as part of the research project, whereby the data exchange practices in the industry are still not digitalized or automated on that level. Besides these limitations we tried to contribute to the digitalization boost that will inevitably reach the AEC industry

As the next steps we aim to fully develop the solution with microservices and test the prototype. Therefore, we will connect multiple proprietary tools to the already created central database, and eventually test the scalability and flexibility when the microservices are used.

ACKNOWLEDGEMENTS

The authors wish to express their gratitude to Strabag SE and its subsidiary Züblin for the help with the realization of this cooperative work. Furthermore, the authors would like to acknowledge the exchange with the research project SCOPE, funded by the German Federal Ministry for Economic Affairs and Energy (BMWi).

REFERENCES

Agarwal, R., Chandrasekaran, S. & Sridhar, M. 2016. Imagining construction's digital future (Blog post, June). https:// www.mckinsey.com/industries/capital-projects-and-infrastructure/our-insights/imagining-constructions-digital-future

Baskarada, S., Nguyen, V. & Koronios, A. 2020. Architecting Microservices: Practical Opportunities and Challenges. *Journal of Computer Information Systems* 60(5): 428–436. doi: 10.1080/08874417.2018.1520056

Chen, R., Li S. & Li, Z. 2017. From Monolith to Microservices: A Dataflow-Driven Approach. *Proceedings of 24th Asia-Pacific Software Engineering Conference (APSEC)*: 466–475, Nanjing, 4 – 8 December 2017. doi: 10.1109/APSEC.2017.53.

di Orio, G., Malo, P. & Barata, J. 2019. NOVAAS: A Reference Implementation of Industrie4.0 Asset Administration Shell with best-of-breed practices from IT engineering. *Proceedings of IECON 2019 - 45th Annual Conference of the IEEE Industrial Electronics Society*: 5505-5512, Lisbon, Portugal, 14 - 18 October 2019. doi: 10.1109/IECON.2019.8927081.

Docker. 2020a. Docker overview. Docker docs. https://docs.docker.com/

Docker. 2020b. mongo. https://hub.docker.com/_/mongo

Dragoni, N., Giallorenzo, S. Lafuente, A. Mazzara, M. Montesi, F. Mustafin, R. & Safina, L. (2017) Microservices: Yesterday, Today, and Tomorrow. In Mazzara M., Meyer B. (Eds.) *Present and Ulterior Software Engineering*. Springer International Publishing AG. https://doi.org/10.1007/978-3-319-67425-4_12

Gos, K. & Zabierowski, W. 2020. The Comparison of Microservice and Monolithic Architecture. *Proceedings of XVIth International Conference on the Perspective Technologies and Methods in MEMS Design (MEMSTECH)*: 150-153, Lviv, Ukraine, 22 - 26 April 2020, doi: 10.1109/MEMSTECH49584.2020.9109514.

Hoffmann, A., Wagner, A., Huyeng, T., Shi, M., Wengzinek, J., Sprenger, W., Maurer, C. & Rüppel, U. 2019. Distributed manufacturer services to provide product data on the web. In Geyer, P., Allacker, K., Schevenels, M., De Troyer, F. & Pauwels, P. (eds.), *Proceedings of 26th International Workshop on Intelligent Computing in Engineering (EG-ICE 2019)*, Leuven, Belgium, 30 June - 3 July 2019. http://ceur-ws.org/Vol-2394/paper23.pdf

Huyeng, T.-J., Thiele C.-D., Wagner, A., Shi, M., Hoffmann, A., Sprenger, W. & Rüppel, U. 2020. An approach to process geometric and semantic information as open graph-based description using a microservice architecture on the example of structural data. In Ungureanu, L. C. & Hartmann, T. (Eds.) *Proceedings of Workshop on Intelligent Computing in Engineering (EG-ICE 2020)*, Berlin, 1-3 July 2020, Berlin: Universitätsverlag der TU Berlin, doi: 10.14279/depositonce-9977.

Kecskemeti, G., Marosi, A.C. & Kertesz, A. 2016. The ENTICE approach to decompose monolithic services into microservices. *Proceedings of International Conference on High Performance Computing & Simulation (HPCS)*: 591-596, Innsbruck, Austria, 18 – 22 July 2016, doi: 10.1109/HPCSim.2016.7568389.

Krause, A., Zirkelbach, C., Hasselbring, W., Lenga, S. & Kröger, D. 2020. Microservice Decomposition via Static and Dynamic Analysis of the Monolith. *Proceedings of 2020 IEEE International Conference on Software Architecture Companion (ICSA-C)*, Salvador, Brazil, 16-20 March 2020, 9-16, doi: 10.1109/ICSA-C50368.2020.00011.

Krylovskiy, A., Jahn, M. & Patti, E. 2015. Designing a Smart City Internet of Things Platform with Microservice Architecture, *3rd International Conference on Future Internet of Things and Cloud*: 25-30, Rome, 24 - 26 August 2015. doi: 10.1109/FiCloud.2015.55.

Ladenhauf D., Battisti K., Berndt R., Eggeling E., Fellner D.W., Gratzl-Michlmair M. & Ullrich T. 2016. Computational geometry in the context of building information modeling, *Energy and Buildings* 115: 78-84. https://doi.org/10.1016/j.enbuild.2015.02.056

Lee J., Jeong Y., Oh M. & Hong S.W. 2014. A Filter-Mediated Communication Model for Design Collaboration in Building Construction, *The Scientific World Journal* 2014. http://dx.doi.org/10.1155/2014/808613.

Lee, Y.-C., Eastman, C.M. & Solihin, W. 2020. Rules and validation processes for interoperable BIM data exchange. *Journal of Computational Design and Engineering*, qwaa064, https://doi.org/10.1093/jcde/qwaa064

Li, X., Xi, Y., Zhu, H., Ling, J. & Zhang, Q. 2020. Infrastructure Smart Service System Based on Microservice Architecture. In Correia, A., Tinoco, J., Cortez, P. & Lamas, L. (eds.), *Information Technology in Geo-Engineering. ICITG 2019*. Springer Series in Geomechanics and Geoengineering. Springer International Publishing. doi: 10.1007/978-3-030-32029-4_12

Microsoft. 2020a. Tutorial: Containerize a. NET Core app. https://docs.microsoft.com/en-us/dotnet/core/docker/build-containertabs=windows

Microsoft. 2020b. Create a web API with ASP.NET Core and MongoDB. https://docs.microsoft.com/en-us/aspnet/core/tutorials/first-mongo-appview=aspnetcore-3.1&tabs=visual-studio

Munonye, K. & Martinek, P. 2020. Evaluation of Data Storage Patterns in Microservices Archicture. *Proceedings of 15th International Conference of System of Systems Engineering (SoSE)*: 373-380, Budapest, Hungary, 2-4 June, 2020, doi: 10.1109/SoSE50414.2020.9130516.

Ponce, F., Márquez G. & Astudillo, H. 2019. Migrating from monolithic architecture to microservices: A Rapid Review. *Proceedings of 38th International Conference of the Chilean Computer Science Society (SCCC)*: 1-7, Concepcion, Chile, 4 - 9 November 2019, doi: 10.1109/SCCC49216.2019.8966423.

Ramaji, I.J., & Memari, A.M. 2016. Interpreted Information Exchange: Systematic Approach for BIM to Engineering Analysis Information Transformations. *Journal of Computing in Civil Engineering* 30(6): 230–246. https://doi.org/10.1061/(ASCE)CP.1943-5487.0000591.

Sarkar, S. Vashi, G. & Abdulla, P.P. 2018. Towards Transforming an Industrial Automation System from Monolithic to Microservices. *Proceedings of 23rd International Conference on Emerging Technologies and Factory Automation (ETFA)*: 1256-1259, Turin, 4-7 September 2018. doi: 10.1109/ETFA.2018.8502567.

SCOPE. 2020. SCOPE Semantic Construction Project Engineering. https://www.projekt-scope.de/en/home-en/

Shelden, D., Pauwels, P., Pishdad-Bozorgi, P., & Tang, S. (2020). Data standards and data exchange for Construction 4.0. In A. Sawhney, M. Riley, & J. Irizarry (Eds.), *Construction 4.0: An Innovation Platform for the Built Environment*: 222-239, Taylor and Francis Ltd.. https://doi.org/10.1201/9780429398100-12

Sibenik, G. & Kovacic, I. 2020. Assessment of model-based data exchange between architectural design and structural analysis, *Journal of Building Engineering* 32, https://doi.org/10.1016/j.jobe.2020.101589.

Sibenik, G., Kovacic, I. & Petrinas, V. 2020. From Physical to Analytical Models: Automated Geometry Interpretations. In Ungureanu, L. C. & Hartmann, T. (Eds.) *Proceedings of Workshop on Intelligent Computing in Engineering (EG-ICE 2020)*, Berlin, 1-3 July 2020, Berlin: Universitätsverlag der TU Berlin, doi: 10.14279/depositonce-9977.

Tapia, F., Mora, M.Á., Fuertes, W., Aules, H., Flores, E. & Toulkeridis, T. 2020. From Monolithic Systems to Microservices: A Comparative Study of Performance. *Applied Sciences* 10(17), https://doi.org/10.3390/app10175797

Turk, Ž. 2020. Interoperability in Construction – Mission Impossible. *Developments in the Built Environment* (In Press). https://doi.org/10.1016/j.dibe.2020.100018.

Woodhead, R, Stephenson, P. & Morrey, D. 2018. Digital construction: From point solutions to IoT ecosystem. *Automation in Construction* 93: 35–46, https://doi.org/10.1016/j.autcon.2018.05.004

Analysis of design phase processes with BIM for blockchain implementation

M. Srećković, G. Šibenik & D. Breitfuß
Institute of Interdisciplinary Building Process Management, TU Wien, Vienna, Austria

T. Preindl & W. Kastner
Institute of Computer Engineering, TU Wien, Vienna, Austria

ABSTRACT: The increasing digitalization and thus evidently advancing change in the architecture, engineering and construction (AEC) industry, requires new business models, processes and strategies. Blockchain (BC), smart contracts and decentralized applications (DApps) are still underused in AEC. BC and its potential of inclusion into the communication between project stakeholders has shown that it is not just a technology that is ready to use, but requires a thorough insight into the design process of domain-specific stakeholders, their interests and their collaboration workflows for a holistic Building Information Modeling (BIM) and BC-supported solution for the design phase. This paper introduces process modeling of BIM-workflows in the design phase. We propose a conceptual framework for the implementation of a design process with BC based on the integration of three underlying theories: design theory, configuration theory and task-technology fit. The main assumption is, before we can capture processes (1) we need to understand them (design theory) in order to re-engineer them for distributed ledger technologies (DLT) (2) we need to adapt them to changing requirements (configuration theory), and finally (3) continually re-adjust Information Technology (IT) and processes interdependence (task-technology fit).

1 INTRODUCTION

Underlined by recent publications (Hunhevicz & Hall 2020; Nawari & Ravindran 2019; Pradeep et al. 2019) there is great need for empirical research to investigate Building Information Modeling (BIM)-workflows and BIM-models and the requirements they would necessitate in order to be linked and used with distributed ledger technologies (DLT) such as blockchain (BC) and decentralized applications (DApps). It also needs to be explored how BIM-workflows could benefit from the decentralized and trust-independent characteristics of DApps. Hence, due to the presence of these technological advances, which are making their way into the AEC industry, there is a research gap to investigate the implementation of these technologies for building design. It is important to find out their advantages and disadvantages and to test their applicability.

Empirical research of process modeling in the design phase is still lacking as well as best practices of process research or the post hoc evaluation of process modeling in planning; primarily because it is very difficult to get accurate and transparent data concerning workflows in building design, even with the implementation of BIM. However, for the inclusion of DLT (blockchain and smart contracts) into the BIM-workflow, it is necessary to analyze and define generic processes in the design phase suitable for the implementation of these technologies.

We argue that the concept of process analysis and modeling needs an integration of multiple theoretical paradigms in order to meet the research challenges of increasing system complexity in building design. Therefore, in this paper we propose an integrative conceptual framework using three theories: design theory, configuration theory and task-technology fit. The main assumption is, before we can capture a process we need to understand it (design), further in order to re-engineer it for DLT, we need to adapt it to changing requirements (configure) and continually adjust IT and business processes interdependence (task-technology fit). The proposed framework is based on our research within the project BIMd.sign.

This paper is organized as follows: In chapter 2 we develop our research framework based on the three theories, in chapter 3 we present our methodology and use case, followed in chapter 4 by the discussion and conclusion.

2 CONCEPTUAL ANALYSIS FRAMEWORK

The implementation of digital technologies and increasing system complexity through digitalization

Table 1. Integration of 3 theories.

Theory	Main idea	Point of our analysis
Design	Prescriptives for design and action	Design workflow
Configuration	Alignment of structure, process, environment	Process modeling Information processing
Task-technology fit	Fit between IT and business processes	BIM, Blockchain, DApps, data exchange & transferability, data formats

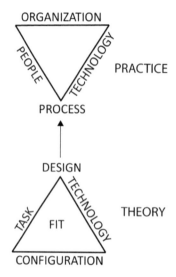

Figure 1. Conceptual framework.

necessitate an integration of multiple theoretical paradigms, with different points of analysis. In this paper, we focus on process modeling of BIM-workflows in the design phase and propose a conceptual framework for the analysis of a design process based on BC and DApps grounded in three theories. This novel combination of the underlying theories - design, configuration and task-technology fit – is the basis for our framework (Table 1).

Our conceptual framework connects theory and practice (see Figure 1). The point of our analysis is - in order to capture processes it is necessary: (1) to understand how they are designed in the complex domain of people, organization and technology interaction; (2) how they need to be configured for BIM and DLT in human-machine systems; and further (3) how to continuously re-adjust the fit or interdependence between IT and organizational structure.

2.1 Design

The design chapter focuses on the workflow analysis using design theory. First the design theory is presented, followed by the application of theory for the analysis of the design workflow in the AEC industry.

2.1.1 Design theory

The paradigm of design science is rooted in engineering, architecture and the sciences of the artificial (Simon 1996), and has found its way into the Information Systems (IS) discipline (Walls et al. 1992). Essentially, it is a problem-solving paradigm seeking "to create innovations that define ideas, practices, technical capabilities, and products through which the analysis, design, implementation, management, and use of information systems can be effectively and efficiently accomplished" (Hevner et al. 2004 p.76).

The IS discipline explores the use of information-technology-related artifacts in human-machine systems (Gregor & Hevner 2013) in the complex realm of people, process, organization and technology interaction. The five classes of theory relevant to IS are (Gregor 2006): (1) theory for analyzing, (2) theory for explaining, (3) theory for predicting, (4) theory for explaining and predicting, and (5) theory for design and action. The focus of theory for design and action is on explicit prescriptions *how to design* and develop an artifact, whether it is a process, technological product or a managerial intervention (Gregor & Jones 2007; Simon 1996).

Design theory is considered to be prescriptive knowledge as opposed to descriptive knowledge (Walls et al. 1992) which encompasses the other types of theory in the taxonomy of Gregor (2006). Design theory applies in a certain design context, defined by the nature of the system, its size, the design phase, the type of technology, the type of users or designers (Walls et al. 1992, 2004). Design theory's scope and purpose is also dependent on environmental requirements such as capabilities and conditions linked to the principles of form and function of the artifact (Spagnoletti et al. 2015).

Hence design research needs to address "wicked-type problems" in planning (Kunz & Rittel 1972; Rittel & Webber 1973) characterized by: (1) unstable requirements and constraints based upon ill-defined environmental contexts, (2) complex interactions among subcomponents of the problem and its solution, (3) inherent flexibility to change design processes as well as design artifacts, (4) a critical dependence upon human cognitive abilities (e.g. creativity) to produce effective solutions and (5) a critical dependence upon human social abilities (e.g. teamwork) to produce effective solutions.

2.1.2 Design workflow

Under the term workflow, one can refer to a business process, specification of a process, software that implements and automates a process, or software that supports the coordination and collaboration of people that implement a process (Georgakopoulos et al. 1995).

Design is both a product and a process and thus design theory must include both aspects (Walls et al. 1992), meaning i.e. the design workflow and the building model itself. The design process is a sequence of expert activities that produce an innovative product (i.e. the design artifact or building model) (Hevner et al. 2004) in the course of designing, planning an action in advance or during the action, including reflection in action (Aken 2004). The evaluation of the artifact (process, building model) then provides feedback information and a better understanding of the problem in order to improve both the quality of the product and the design. This build-and-evaluate loop is typically iterated a number of times before the final design artifact is generated (Hevner et al. 2004), meaning rework and iteration is an essential part of the design phase.

For the purpose of our research, we define design workflow as the flow of information, deliverables, specifications, and other design resources between the project stakeholders (Hattab & Hamzeh 2016), *which are included in numerous processes in the building design phase (e.g. information process, data management process, BIM workflow).* In practice, these processes are not linear or rigid, but rather dynamic and very complex. Hence, workflow is more than a technique to model a process. It is a method to analyze and improve a process, including its modeling.

2.2 Configuration

The configuration theory is used to introduce organizational peculiarities of building construction workflows as well as other artifacts relevant for the design phase. First, the configuration theory is presented; it is followed by the resulting process modeling concept proposed for the building design phase.

2.2.1 Configuration theory

"Configuration refers to any form of organization that is consistent and highly integrated and where all pieces fit neatly together…there is internal consistency, synergy among processes, fit with the external context." (Mintzberg 1991 p.54). This configuration ('gestalt', 'archetype', 'generic type') (Miller 1986) or observable characteristics or behaviors which appear to lead to a particular performance outcome, success or failure (Ward et al. 1996), of an organization, are interrelated. The coalignment or fit of multiple variables and organizational elements such as alignment of strategy, systems, or processes, is reflected in observable patterns in practice (Flynn et al. 2010). This means that the design and structure of an organization and its business processes should match or fit characteristics of certain variables both inside and outside the organizational system (Tushman & Nadler 1978) or in the case of AEC, the system of project-based organization.

According to Tushman & Nadler (1978 p.634), "organizations are information processing systems facing external and internal sources of uncertainty". As systems, their organizational structure should create and enable the most appropriate configuration of work units (as well the linkages between these units) to facilitate the effective collection, processing and distribution of information (i.e. plans, work standards, budgets, feedback on performance etc.).

Information processing is an essential feature of design workflows. Designers use information as raw material (Tribelsky & Sacks 2011), where further processing and flow of accurate and timely information enables an efficient and successful project performance (Pradeep et al. 2019). This conjectures a need to explore current patterns of information processing in the design and BIM-workflow, as well as the interdependence between project stakeholders and organizational structure. Furthermore, it necessitates a configurational fit between process models and information processing requirements, especially for the implementation of DLT.

2.2.2 Process modeling

Process modeling is a way of capturing the operations of organizations in real-world domains (Recker 2009). It is widely used within organizations as a method to increase awareness and knowledge of business processes, and to deconstruct organizational complexity (Bandara et al. 2005). It is an approach for describing how businesses conduct their operations and typically includes graphical depictions of at least the activities, events/states, and control flow logic that constitute a business process (Curtis et al. 1992; Davenport 2005). Process models may also include, among other things, information regarding the involved data, organizational/IT resources, and potentially other artifacts such as external stakeholders and performance metrics (Scheer 2000).

From an IS perspective, information processes relate to automated tasks (i.e., tasks performed by programs) and partially automated tasks (i.e., tasks performed by humans interacting with computers) that create, process, manage, and provide information. Database, transaction processing, and distributed systems technologies provide the basic infrastructure for supporting information processes (Georgakopoulos et al. 1995). Information processes are rooted in an organization's structure and/or the existing environment of information systems, which corresponds to the before presented configurational approach, where organizations are viewed as information processing mechanisms and their organizational structure should reflect that.

Processes are relationships between inputs and outputs, where inputs are transformed into outputs using a series of activities or tasks which add value to the inputs (Aguilar-Savén 2004 p.140). One or more software systems, one or a team of humans, or a combination of these can perform a task.

Human tasks include interacting with computers closely (e.g., providing input commands) or loosely (e.g. using computers only to indicate task progress).

Examples of tasks include updating a file or database, generating a blueprint. In addition to a collection of tasks, a workflow defines the order of task invocation or condition(s) under which tasks must be actuated, task synchronization, and information flow (Georgakopoulos et al. 1995). Task complexity and interdependence are sources of uncertainty. Depending on their degree of interdependence or degree of complexity, their information processing requirements can be minimal or very demanding.

2.3 Technology

Finally, the task technology fit theory is used to relate the process models with their practical implementation. Therefore, the theory itself is presented as well as core technologies: BIM, BC and DApp.

2.3.1 Task Technology Fit (TTF)

Task-technology fit (TTF) is the degree to which a technology supports the performance of tasks, where task requirements, individual abilities and the functionality of technology are in accordance (Goodhue & Thompson 1995). In the context of TTF theory, technology has to match business processes (Karim et al. 2007), enabling a tight coupling of IT function, business strategy and the organization's information needs (Strnadl 2006).

IT artifacts extend the boundaries of human problem solving and organizational capabilities by providing intellectual as well as computational tools (Hevner et al. p.76). IT artifacts are broadly defined as constructs (vocabulary and symbols), models (abstractions and representations), methods (algorithms and practices), and instantiations (implemented and prototype systems) (Hevner et al. 2004). Many IT artifacts have some degree of abstraction but can be readily converted to a material existence; for example, an algorithm converted to operational software (Gregor & Hevner 2013).

In this paper, we use the term *artifact to refer to a thing that has, or can be transformed into, a material existence as an artificially made object (e.g., model, instantiation) or process (e.g., method, software)* (Goldkuhl 2002 p. 5; Gregor & Hevner 2013).

2.3.2 BIM

In the last decades Information and Communication Technology (ICT) has been widely used as a facilitator of AEC collaboration (Lee & Jeong 2012), with the implementation of BIM - a modeling technology (Eastman et al. 2008) respectively a joint digital knowledge domain supporting activities of all stakeholders in AEC; based on various data models with geometrical and/or non-geometrical information; allowing data generation, exchange and processing within the life cycle of built structures (Sibenik & Kovacic 2019). Nevertheless, successful implementation of evolving digital technologies, such as BIM, and furthermore the generation of innovation processes in the digital economy require changes in traditional organizational processes, a dynamic strategic fit and the development of adequate organizational capabilities for competitive advantage.

2.3.3 Blockchain and Decentralized Applications (DApps)

The literature review of blockchain and its potential of inclusion into model-based communication (Nawari & Ravindran 2019; Pradeep et al. 2019) has shown that it is not just a technology that is ready to be used, but requires a thorough insight into the design process of domain-specific stakeholders, their interests and their collaboration workflows in order to find a holistic BIM-BC-supported solution for the design phase.

In an industry in which collaboration is based on expert knowledge and a high degree of trust, the potential of DApps should be examined, as these could enable innovative forms of collaboration between project members and teams in segments of the value chain. They would be expressed automatically (running on a BC network), especially if this could save costs and time for administrative work, reporting, control, monitoring of responsibilities and risk transfer. For this purpose, it is also necessary to examine the role of intermediaries and to understand them better, and what added value they create at what cost for the project and the design process, respectively.

Potentials of making BIM processes and design procedures in building-design more transparent, traceable, more consistent, more efficient, more cost-effective and cheaper with BC and DApps remain unused so far. The implementation of these technologies would also result in the possibility of real-time communication in the model and compliance checking.

In general, BC and DApp technologies make it possible to determine a consensus on the current state of a workflow in a decentralized fashion. The state of the individual BIM artefacts (model) could be fixed with links in the BC that uniquely reference the artefacts content at every step of the BIM-workflow. Additionally, any new version of such an artefact could reference the previous one, which creates a distinct trace of the workflow progress. Such a clear history of the design process would make it possible to determine the responsibilities for individual steps retrospectively. Smart contracts included in such a DApp offer possibilities to determine the roles during the workflow and pass responsibility for the next step as well or approve the completed steps. The potential benefits could however be much greater than these examples.

3 DESIGN PHASE PROCESSES WITH BIM FOR BLOCKCHAIN

In our research project BIMd.sign (BIM digitally signed with blockchain) we are analyzing the design

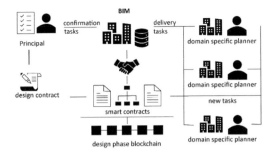
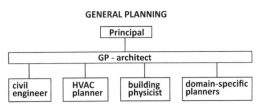

Figure 2. Interaction of stakeholders, BIM and smart contracts.

Figure 3. General planner procurement model.

workflow from the initial information search to execution design (phases according to the HOAI) in the design phase. Empirical research on process modeling in the design phase is missing, as well as research on the implementation possibilities of BIM with DLT in building design. In our project, we aim at closing this gap. Therefore, the main premise of our research is to first analyze stakeholders, processes and data flow in the design workflow and next propose a conceptual framework for the analysis of process modeling for BC, smart contracts and DApps.

A further step includes proposing a conceptual process model for BC implementation. The exploration is grounded in our conceptual analysis framework integrating three theories – design, configuration and task-technology-fit. The point of departure is - in order to capture processes it is necessary, first to understand how they are designed, second how they need to be configured for BIM and DLT and third, how to continuously adjust the fit between these technologies and business processes. Concisely said, in this research step, we are exploring the system interdependence between processes, stakeholders (people) and data flow in the design phase, based on our presented conceptual framework.

Figure 2 shows the general setup of the proposed concept for BIM and DLT integrating four elements: stakeholders/blockchain actors, BIM model, blockchain, smart contracts/DApps.

3.1 Use case

The selected case study is based on an Austrian architecture company offering general planning services in the design phase (see Figure 3). The General Planner (GP) is the lead consultant and appoints all the domain-specific planners, in disciplines i.e. architecture, structural engineering, building services engineering, building physics, fire protection engineering and landscape design The GP delivers all the usual specialist services required for a project, is a single contractual partner for design and engineering and therefore takes on overall responsibility in the design phase. Predominantly, the architect assumes the role of the GP and appoints sub-planners to undertake work in the other disciplines for which they have signed a contract. The GP undertakes a range of coordination tasks, as well as managing and coordinating his specialist design consultants' works. He carries responsibility for all of the services assigned to them, particularly in respect to design, program and costs. The GP is free to choose own sub-consultants, and is therefore able to influence the quality of the overall project design. The principal has thus only one contractual design partner. The GP assumes responsibility and liability for the individual design services, and provides the principal with a guarantee that the individual design services, including all interfaces, are correct. The GP owes the principal a model as planned and contractually agreed.

In Figure 4 we can see the configuration of the workflow in the design phase, including information processing of data (data-flow) and coordination of tasks and activities between the different project stakeholders (process flow). Each step in the process flow has stakeholders responsible for their own domain-specific tasks and the appropriate fulfillment of those. As mentioned, the GP is coordinating and organizing the timing of the tasks and is acting as the interface between the principal and the domain-specific planners.

At the end of each process in the design phase, the GP presents the results to the principal, who either approves to move forward or requires changes, which end in iteration loops until the revision is approved. The sub-process "execution of domain-specific tasks" reflects the data exchange between the different stakeholders during a design task showing the complexity and interrelatedness in the information processing of different data-formats.

Figure 5 shows a conceptual process model for BC implementation. Due to the data stored in the blockchain, the smart contract monitors the status of progress as well as gives permission of further processing the data, if certain requirements are met. A simplified example would be an architect, who develops a conceptual design and forwards it to the structural engineer, who is responsible to check its functionality. If not approved, a list of required changes is transferred back to the architect, who will adapt the design and again send the model back to the structural engineer. If approved by the structural engineer, the architect then authorizes further steps. In this case, the smart contract would be able to track the changes and responsibilities of the involved parties, as well as give clearance for further steps, when all approvals are

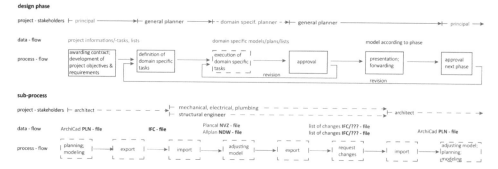

Figure 4. Workflow & sub-processes in the design phase.

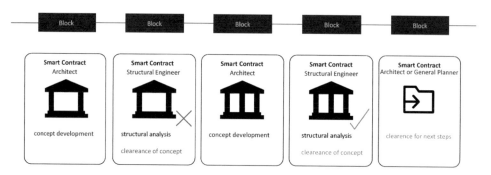

Figure 5. Blockchain schema in the design phase.

met. This means that individual BIM artefacts (e.g. model, drawing) could be fixed with links in the BC that uniquely reference the artefacts content at every step of the BIM-workflow. In addition any new version of the BIM Model could reference the previous one, which creates a distinct trace of the workflow progress.

4 DISCUSSION AND CONCLUSION

In this paper, we introduced a conceptual framework for the analysis and process modeling of BIM-workflows in the design phase. We argue that for the inclusion of DLT (blockchain and smart contracts) into the BIM-workflow it is necessary to understand and capture the entire design workflow (encompassing numerous processes/stakeholders/data formats). In conclusion, this means: (1) to understand how processes are designed and who the participating stakeholders are (2) how processes need to be configured and aligned for the implementation of these technologies and finally (3) how IT needs to be continuously adjusted to fit the organizational structure and processes in the design value chain.

This work serves as a guideline for the incorporation of BC implementations in the design phase. Further steps in the project will involve a framework development for the design phase processes for which BC shows the greatest potentials for the model-based communication.

ACKNOWLEDGEMENT

The research projects BIMd.sign - BIM digitally signed with blockchain in the design phase, Grant No. 873842; and BIMCHAIN/FMCHAIN, Grant No. 873827; are funded by the Austrian Research Promotion Agency (FFG), Program ICT of the Future and Austrian Federal Ministry for Transport, Innovation and Technology (BMVIT). The authors are grateful for the support.

REFERENCES

Aguilar-Savén, R.S. 2004. Business process modelling: Review and framework. *International Journal of Production Economics* 90(2): 129–149.

Bandara, W., Gable, G.G. & Rosemann, M. 2005. Factors and measures of business process modelling: model building through a multiple case study. *European Journal of Information Systems* 14(4): 347–360.

Erri Pradeep, A.S., Yiu, T.W. & Amor, R. 2019. Leveraging Blockchain Technology in a BIM Workflow: A Literature Review. In *International Conference on Smart Infrastructure and Construction 2019 (ICSIC)*. 371–380.

Georgakopoulos, D., Hornick, M. & Sheth, A. 1995. An overview of workflow management: From process modeling to workflow automation infrastructure. *Distributed and Parallel Databases* 3(2): 119–153.

Goldkuhl, G. & Ågerfalk, P.J. 2002. Actability: a Way to Understand Information Systems Pragmatics. In

Coordination and Communication Using Signs: Studies in Organisational Semiotics. Boston, MA: Springer US, 85–113.

Goodhue, D.L. & Thompson, R.L. 1995. Task-technology fit and individual performance. *MIS Quarterly* 19(2): 213.

Gregor, S. 2006. The Nature of Theory in Information Systems. *MIS Quarterly* 30(3): 611–642.

Gregor, S. & Hevner, A.R. 2013. Positioning and Presenting Design Science Research for Maximum Impact. *MIS Quarterly* 37(2): 337–355.

Gregor, S. & Jones, D. 2007. The Anatomy of a Design Theory. *Journal of the Association for Information Systems* 8(5): 312–335.

Flynn, B.B., Huo, B. & Zhao, X. 2010. The impact of supply chain integration on performance: A contingency and configuration approach. *Journal of Operations Management* 28(1): 58–71.

Hanseth, O. & Lyytinen, K. 2010. Design theory for dynamic complexity in information infrastructures: the case of building internet. Journal of Information Technology 25(1): 1–19.

Hattab, M.A. & Hamzeh, F. 2016. Analyzing Design Workflow: An Agent-based Modeling Approach. *Procedia Engineering* 164(510–517.

Hevner, A.R., March, S.T., Park, J. & Ram, S. 2004. Design Science in Information Systems Research. *MIS Quarterly* 28(1): 75–105.

HOAI, die Honorarordnung für Architekten und Ingenieure, 2013. https://www.hoai.de/online/HOAI_2013/HOAI_2013.php (accessed 24.5.2020)

Hunhevicz, J.J. & Hall, D.M. 2020. Do you need a blockchain in construction? Use case categories and decision framework for DLT design options. *Advanced Engineering Informatics* 45(101094).

Karim, J., Somers, T. & Bhattacherjee, A. 2007. The Impact of ERP Implementation on Business Process Outcomes: A Factor-Based Study. *J. Manage. Inf. Syst.* 24(1): 101–134.

Kunz, W. & Rittel, H.W. 1972. Information science: on the structure of its problems. *Information Storage and Retrieval* 8(2): 95–98.

Miller, D. 1986. Configurations of Strategy and Structure: Towards a Synthesis. *Strategic Management Journal* 7(3): 233.

Mintzberg, H. 1991. The Effective Organization: Forces and Forms. *Sloan Management Review* 32(2): 54.

Nawari, N.O. & Ravindran, S. 2019. Blockchain technology and BIM process: Review and potential applications. *Journal of Information Technology in Construction* 24(12): 209–238.

Recker, J., Rosemann, M., Indulska, M. & Green, P. 2009. Business process modeling-a comparative analysis. *Journal of the association for information systems* 10(4): 333–363.

Rittel, H.W.J. & Webber, M.M. 1973. Dilemmas in a general theory of planning. *Policy Sciences* 4(2): 155–169.

Sibenik, G. & Kovacic, I., 2019. Automation of software independent data interpretation between architectural and structural analysis models. *In Proceedings of the 36th International Conference of CIB W78, Newcastle-upon-Tyne, UK*. 810–820.

Simon, H.A. 1996. The Sciences of the Artificial. *MIT Press Books*.

Spagnoletti, P., Resca, A. & Lee, G. 2015. A design theory for digital platforms supporting online communities: a multiple case study. *Journal of Information Technology* 30(4): 364–380.

Strnadl, C.F. 2006. Aligning business and IT: The process-driven architecture model *Information Systems Management* 23(4): 67–77.

Tribelsky, E. & Sacks, R. 2011. An Empirical Study of Information Flows in Multidisciplinary Civil Engineering Design Teams using Lean Measures. *Architectural Engineering and Design Management* 7(2): 85–101.

Tushman, M.L. & Nadler, D.A. 1978. Information processing as an integrating concept in organizational design. *Academy of Management. The Academy of Management Review (pre-1986)* 3(3): 613–624.

van Aken, J.E. 2004. Management Research Based on the Paradigm of the Design Sciences: The Quest for Field-Tested and Grounded Technological Rules. *Journal of Management Studies* 41(2): 219–246.

von Rosing, M., Laurier, W. & M. Polovina, S. 2015. The BPM Ontology. In *The Complete Business Process Handbook*. Boston: Morgan Kaufmann, 101–121.

Walls, J.G., Widmeyer, G.R. & Sawy, O.A.E. 1992. Building an Information System Design Theory for Vigilant EIS. *Information Systems Research* 3(1): 36–59.

Ward, P.T., Bickford, D.J. & Leong, G.K. 1996. Configurations of Manufacturing Strategy, Business Strategy, Environment and Structure. *Journal of Management* 22(4): 597–626.

Digital traceability for planning processes

D. Breitfuss, G. Šibenik & M. Srećković
Institute of Interdisciplinary Building Process Management, TU Wien, Vienna, Austria

ABSTRACT: The increasing complexity of communication and collaboration processes due to the use of building data models and object-based change management, create the need for efficient ways of transparency and traceability within planning processes. The growing presence of BIM (Building Information Modelling) digitalises the AEC industry and produces numerous assets in digital form, but their added value is not completely exploited. Blockchain (BC) and Smart Contracts (SC) might deliver improvement to communication and collaboration in the design phase. Based on a framework for assessing BIM based workflows, processes are analysed according to three sub-categories: project-stakeholder involvement, data-flow and single actions, which form a process-flow. The information gained with the application of this framework, grounded in a use case, is further explored in a conceptual model for SCs in the design phase of a building project. The focus hereby lies on existing processes which can be translated into a SC, and not on the alternation of currently applied workflows. SCs will be used as a tool, for supporting mandatory actions by each project stakeholder and to create a digital reference of changes in a BIM model on a BC.

1 INTRODUCTION

With the increasing application of BIM in the AEC industry, a number of challenges have arisen, hitherto many of them have remained unresolved (Erri Pradeep et al. 2019; Nawari & Ravindran 2019). One of the main challenges in BIM workflows is the traceability of changes within a BIM model and closely coupled with it, the accountability for clearances and the sharing of model information (Coyne & Onabolou 2018). In order to understand BIM-workflows and find possible intersection points with the newly arising technologies Blockchain (BC) and Smart Contracts (SC) within the Architecture, Engineering, Construction (AEC) industry, an extensive exploration of process participants, data exchanges and process-steps is imperative. SC are mainly implemented in financial industries, although recent examples (Fridgen et al. 2018) show an effort in process modelling with SC as well.

In our research project BIMd.sign, we investigate if SCs and BC can be implemented in BIM-based planning and if they can deliver sufficient support as project management tools, and in a further step automatize certain aspects of the planning processes. In this paper we focus on the research question, if process-based SCs could contribute to a BIM-workflow, by enabling traceability and transparency through the entire design phase of a building project. The analysis of a scenario from a use case and the development of a conceptual model which will be presented here, form the base layer for further steps in our research. We argue that SCs could possibly serve as a key element, to enforce certain pre-defined process-sequences, as well as enable a revision-safe database for the enforcement of traceability throughout an entire design phase of a building project.

This paper is organized as follows: In section 2 the state of the art will be presented with the focus on processes in BIM workflows, SC and their application in the AEC industry. Section 3 demonstrates a framework for process analysis and section 4 presents a possible implementation of SC in the planning process followed by a conclusion in section 5.

2 STATE OF THE ART

2.1 Design phase processes and BIM

According to Chi et al. (2014) building projects can be mainly divided into three phases: conceptual development, structural detailing and construction. In all of those phases modelling, analyzing and optimizing plays a significant role in the progress of the project. Conceptual development and structural detailing have an extensive impact on the life-cycle cost of a building, due to the importance of delivered design decisions (building as planned). Hence, it is important to mention, that during the construction phase changes in design are significantly more expensive, due to the lack of flexibility and increasing rework (Singh & Ashuri 2019). Thus, creating a better environment for collaboration and cooperation in the design phase is imperative for planning efficiency.

The development of computer aided design (CAD) and later building information modelling (BIM) and

3D CAD has greatly contributed to a higher flexibility in modifications and efficiency for planners in the AEC industry. BIM is especially delivering advantages in process efficiency and in collaboration in general. Nowadays, BIM is one of the most widely used planning methods, but its full potential is still not exploited, where compatibility of BIM and organizational culture (Son et al. 2015) are mentioned as some of the reasons for that problem. BIM basically fosters the exchanges of data and information during design, construction and operation of buildings. The innovation to more traditional methods is, that the information is mostly directly linked with model data and can be accurately detailed according to the project phase (Chi et al. 2014).

2.2 Smart contracts

Smart Contracts are computer protocols, able to digitally facilitate, verify and enforce contracts between two or more parties (Wang et al. 2019). The term contract in this context is often misleading, as a SC basically enforces an agreement and is not a contract in a legal sense. Literature suggests that SC are either code, which represent legal contracts or act as software agents, which simply fulfill their pre-defined duties (Clack et al. 2016). The latter definition is also the focus of this paper, as a beneficial use of SC is seen in a process-based implementation.

SC were first mentioned in the mid-1990s, but firstly implemented with the BC technology, which offers a decentralized environment to ensure disintermediation of third parties and equality between participants. Transactions which are conducted through a BC are sent directly to the receiver and therefore can't be controlled by a third party (Zheng et al. 2017). Within a BC context, SC can also be seen as computer codes and scripts, which are stored on the BC. SCs are triggered, when they are addressed by transactions and then they execute a pre-defined task (Christidis & Devetsikiotis 2016).

As SC are closely linked to BC technology, they are also profiting from its unique properties. (Wang et al. 2019). A BC is a revision-safe distributed ledger where all transactions are verified through so called consensus mechanisms. These mechanisms serve as a fundamental agreement between all participants, as they are vital to the network (Nguyen & Kim 2018) and creating new blocks. As SCs are stored in the blocks of a BC, they also need to be verified through the consensus mechanism and therefore from each stakeholder who takes part in the verification process. Once stored and deployed on the BC they can't be tampered with, due to the unique "chaining" of the blocks and the distribution of copies over the entire network. Each block references its predecessor with a so-called HASH reference and each HASH reference is the product of all data in the block being hashed (Nawari & Ravindran 2019). If a SC or any other data in the BC is changed, the copy wouldn't be identical to all other copies. Therefore, it is almost impossible to tamper with a BC, if not all copies are changed at the same time.

SC & BC in the AEC industry

Smart Contracts could streamline the potentials of BC to an efficient process progression, by automating certain tasks, saving enormous amounts of time and money and enforcing actions on the BC. According to McNamara & Sepasgozar (2018) a set of pre-defined rules dictates the progression of a process. If- and Then-conditions which resemble the business agreements can be programmed and embedded in the transaction database and indicate a suitable procedure (Nawari & Ravindran 2019).

SCs are seen to automate performance by monitoring all incoming data and managing it accordingly to pre-set rules to settle the stipulations. This could help to prevent disputes, as possible friction points between participating parties can be avoided upfront by an automated protocolling mechanism enabled by SCs. Additionally, emerging disputes can be resolved easier, due to the tamper-proof records. The records ensure a proof of provenance from the data recorded (Erri Pradeep et al. 2019), which allows to track the author of changes in the context of a BIM model, as they reflect all incoming data to the blockchain and enforce different actions if triggered (McNamara & Sepasgozar 2018).

Furthermore, SC could possibly help with a standardization (Mason & Escott 2018) of processes, which can lead to higher quality in the building design, as well as faster and more efficient process design, due to an easy adaption for each project through modular sets of SCs. The information filtered from a BIM model could help to automatize certain tasks during a project. This automatization requires a step-by-step description (Bore et al. 2019) of each process and possible outcomes to fully use all potentials. An overall automatization is not seen at this point, as there will be a need for human input to start and direct certain processes (Mason 2017).

On the other hand, what is problematic in the context of SC is that, once they are deployed on the BC, it is hard to change them. Wrongly coded SCs could therefore affect entire projects and cause tremendous amounts of damage (Mason 2017). Additionally, frequent changes in workflows are seen to be problematic. Changing a SC after its deployment is difficult and linked with a vast amount of work. A workflow standardization, as mentioned before, might help to build up a SC system easier, but also makes the whole workflow less flexible and more difficult to react to unexpected issues (Fridgen et al. 2018). Therefore, a standardization demands a comprehensive and broad preliminary planning of a project.

3 METHODOLOGY

The methodology used is first based on the review of existing literature on the state-of-the-art application of SC and BIM in the AEC industry (section 2). Further, it is grounded in a conceptual framework (Table 1) that

Table 1. Integration of 3 theories.

Theory	Main idea	Point of our analysis
Design	Prescriptives for design and action	Design workflow
Configuration	Alignment of structure, process, environment	Process modeling Information processing
Task-technology fit	Fit between IT and business processes	BIM, Blockchain, DApps, data exchange & transferability, data formats

Figure 1. Conceptual framework.

was developed in a previous step in our research project for the analysis of planning processes. And finally, it includes the exploration of a use case scenario for the conceptual model development (section 4).

An implementation of SC and BC in the increasingly complex system of BIM-based planning requires an analysis from different points of view. In order to investigate a BIM workflow during the design phase of a building project, and keep the focus on process modelling, a framework was developed, based on three underlying theories – design, configuration and task-technology fit as presented in Table 1 and Figure 1.

The conceptual framework showed in Figure 1 connects theory and practice. The analysis framework constitutes the connection between people, organization and technology interaction. Hence, the design of a process with a task-technology-configurational fit requires an adaptive interaction of implemented processes (through people), organization (actions and delivery order) and technology (software and data) (Sreckovic et al. 2020).

At first, for the analysis of our processes the information delivery manual (IDM) was considered. IDM is a standard, which *"provides help in getting the full benefit from a BIM"* (ISO 29481-1, 2016). ISO 29481-1, (2016) and ISO 29481-2 (2012) describe a methodology to identify and describe processes within the context of BIM to support use cases by providing the information in a satisfactory quality at the required time. It is another component, which forms a fundamental part of our analysis framework. Nevertheless,

for our research purposes, a deviation from IDM was necessary - where the focus is in the detail on the comprehensive procedures of the complete project; whereas our framework connects aspects from a theoretical point of view and a practice point of view, delivering an integrative configurational fit of task-technology-people, embedded in an organizational structure resp. environment. Additionally, the focus in our presented conceptual model development is on smaller and more scaled process-scenarios, which furthermore are the entry points for a SC implementation.

4 CONCEPTUAL MODEL DEVELOPMENT

4.1 *Conceptual framework*

The conceptual framework was used to analyze a scenario from a use case in the design phase of a medium sized-office building. The aim was to explore relevant data and define the interaction of people (project-stakeholders), technology (data-flow and software) and tasks (process-flow) for the creation of a conceptual model of SCs.

To describe this conceptual model, we chose the following scenario as base-layer: the architect makes a necessary change in design to a load-bearing wall. The next step includes a needed verification and clearance from the structural engineer, as the aforementioned change includes a structural element.

Figure 2 shows this process with an implementation of SC. The proposed system is divided into four columns: Action, Blockchain, Database and Smart Contract. The first column describes the actions, which are taking place during this process. The framed actions are additionally indicating a SC implementation. The blockchain hosts the SCs and the model references, the database mainly contains the model data in an exchangeable format.

A SC is basically addressed through two different approaches. Firstly, a manually triggered input is needed. This could be for instance the upload of a new model-version to the system. In this case, the SC is responsible for creating a reference and storing it on the BC. Secondly, a SC could be triggered by another SC. When the beforementioned SC creates the reference, it parallelly can monitor the changes in the model-version.

Due to an object-based file (see 4.2 Data Management), object parameters could be one possible indicator for an automatization, such as the parameter load-bearing indicates a verification from the 2structural engineer, who then will be automatically informed. The project manager in this case, will only be notified of the progress and will not be bothered to intervene.

4.2 *Data management*

Within the scenario that we analyzed, data management in the context of BIM-workflow has been file-based. However, to achieve object-based change

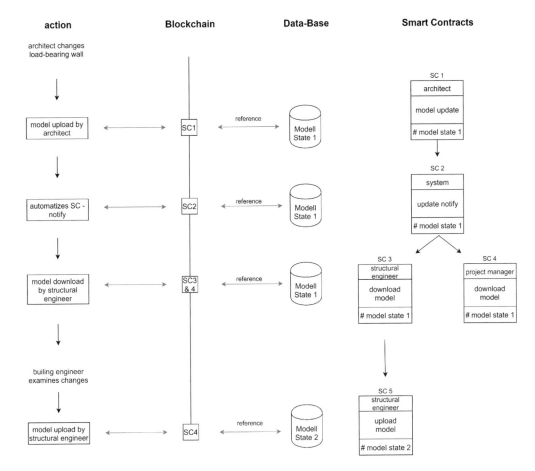

Figure 2. Conceptual model.

management a data-centric approach is required. The potential of data-centric management has not been fully exhausted within the AEC industry (Chassiakos & Sakellaropoulos 2008). In order to provide a suitable data management concept, we reformatted the industry foundation classes (IFC) models to a JSON format. In that way the IFC models can be used on a database such as Mongo DB (Sibenik & Petrinas 2020).

IFC is the most widely used standard to define neutral building models. This format can be used for the file-based data exchange which is still burdened with many problems (Sibenik & Kovacic 2020). Lack of standards, heterogeneous building model representations and slow digitalization makes this task highly complex. As a first step, we use the available IFC models, but the issues as inconsistent building element IDs and non-synchronous link with the database prevent it from being suitable for all data management aspects which are required. The existing solution provides a way to reference a building element from the SC, but it does not fully correspond BIM-based workflows form the technical side. To further improve the data management, the aim is to directly connect proprietary software tools with the central database.

5 DISCUSSION

The conceptual model shows mainly three factors, why BC and SC can be considered to deliver benefits to a BIM process if implemented:

– Documentation: Traditionally, documentation was not an issue, as each document was stamped and in paper version (Singh & Ahsuri 2019), but due to BIM-methods, the exchange rate of data increased drastically, which often leads to data loss, unused information (Erri Pradeep et al. 2019).
– Transparency: Due to the process design conducted through SC, a new level of process integrity can be reached, which is transparent to all project-stakeholders and agreed upon, before the project starts. This could foster and strengthen a new way of trust between the project stakeholders (Li & Kassem 2019).
– Traceability: Documentation and Transparency create the base layer for traceability in the design phases. SC can enable automated reference-making on a BC and therefore create a revision-safe database. With a BIM element-based system, SCs facilitate the traceability of each change in the

model. Author and date of a change, can be tracked on the BC. The traceability enforces a responsible decision-making process and can help avoid legal disputes upfront, as well as minimize the restraint of using BIM in an interorganizational setting.

6 CONCLUSION

Although BIM has become an important method within the AEC industry, it still faces a lot of challenges. In this paper we introduced a conceptual model for the implementation of SCs in the conventional BIM processes, with the aim to a) enforce a more regulated process-flow, b) create a digital traceability of alterations to an interoperable BIM and c) track changes of actions/decisions made in the planning phases of a building project (BIM as planned) and amendments to design during construction (BIM as built). In conclusion we argue, that the implementation of process based SCs could foster a better environment for BIM based projects, due to a standardization of these processes, which would also deliver benefits in the form of efficiency and speed.

Limitations in this paper are the small amount of analysed processes so far as well as a lack of standardized actions (in the context of a taxonomy necessary for a standardized use of SC in the planning processes). Hence, the identification of intersections between people (project-stakeholders), technology (data-flow and software) and tasks (process-flow) has not yet been fully adjusted to the needed requirements for SC.

In this paper we established the guidelines for the implementation of SC with BIM through the development of a conceptual model. The next steps will include an exploitation of further scenarios and use cases, a standardization of terms for SCs (taxonomy) and the development of a practice-based Proof of Concept within our research project.

ACKNOWLEDGEMENT

The research projects BIMd.sign - BIM digitally signed with blockchain in the design phase, Grant No. 873842 is funded by the Austrian Research Promotion Agency (FFG), Program ICT of the Future and Austrian Federal Ministry for Transport, Innovation and Technology (BMVIT). The authors are grateful for the support.

REFERENCES

Abanda, F. H., Vidalakis, C., Oti, A. H., & Tah, J. H. 2015. A critical analysis of Building Information Modelling systems used in construction projects. *Advances in Engineering Software*, 90, 183–201.

Bore, N., Kinai, A., Mutahi, J., Kaguma, D., Otieno, F., Remy, S. L., & Weldemariam, K. 2019. On Using Blockchain Based Workflows. *Paper presented at the 2019 IEEE International Conference on Blockchain and Cryptocurrency (ICBC)*.

Chassiakos, A. P., and Sakellaropoulos, S.P. 2008. A web-based system for managing construction information. *Advances in Engineering Software*, 39, pp. 865–876. doi:10.1016/j.advengsoft.2008.05.006

Chi, H.-L., Wang, X., & Jiao, Y. 2015. BIM-Enabled Structural Design: Impacts and Future Developments in Structural Modelling, Analysis and Optimisation Processes. *Archives of Computational Methods in Engineering*, 22(1), 135–151. doi:10.1007/s11831-014-9127-7

Christidis K, Devetsikiotis M. 2016. Blockchains and Smart Contracts for the Internet of Things. IEEE Access, 4, 2292–2303

Clack, C. D., Bakshi, V. A., & Braine, L. 2016. Smart Contract Templates: essential requirements and design options. *arXiv preprint arXiv*:1612.04496.

Coyne, R., & Onabolu, T. 2018. Blockchain for architects. *Architecture Research Quarterly*, 21(4), 369–374. doi:10.1017/S1359135518000167

Erri Pradeep, A.S., Yiu, T.W. & Amor, R. 2019. Leveraging Blockchain Technology in a BIM Workflow: A Literature Review. *In International Conference on Smart Infrastructure and Construction 2019* (ICSIC). 371–380.

Fridgen, G., Radszuwill, S., Urbach, N., & Utz, L. 2018. Cross-organizational workflow management using blockchain technology-towards applicability, auditability, and automation. *Paper presented at the Proceedings of the 51st Hawaii International Conference on System Sciences.*

ISO, International Organization for Standardization, ISO 29481-1:2016: Building Information Models - Information Delivery Manual - Part 1: Methodology and Format, ISO, Geneva, Switzerland, 2016.

ISO, International Organization for Standardization, ISO 29481-2:2012: Building Information Models - Information Delivery Manual - Part 2: Interaction Framework, ISO, Geneva, Switzerland, 2012.

Li, J., Greenwood, D., & Kassem, M. 2019. Blockchain in the built environment and construction industry: A systematic review, conceptual models and practical use cases. *Automation in Construction*, 102, 288–307. doi:https://doi.org/10.1016/j.autcon.2019.02.005

Mason, J. 2017. Intelligent contracts and the construction industry. *Journal of Legal Affairs and Dispute Resolution in Engineering and Construction*, 9(3), 04517012.

Mason, J., & Escott, H. 2018. Smart contracts in construction: views and perceptions of stakeholders. *Paper presented at the Proceedings of FIG Conference*, Istanbul May 2018.

McNamara, A., & Sepasgozar, S. 2018. Barriers and drivers of Intelligent Contract implementation in construction. *Management*, 143, article number 0251-7006.

Nawari, N.O. & Ravindran, S. 2019. Blockchain technology and BIM process: Review and potential applications. *Journal of Information Technology in Construction* 24(12): 209–238.

Nguyen, G.-T., & Kim, K. 2018. A Survey about Consensus Algorithms Used in Blockchain. *Journal of Information processing systems*, 14(1).

Sibenik, G. and Kovacic I. 2020. Assessment of Model-Based Data Exchange between Architectural Design and Structural Analysis. *Journal of Building Engineering* [In press]. https://doi.org/10.1016/j.jobe.2020.101589

Sibenik, G. & Petrinas, V. 2020. *IFCtoJSON* [Software]. htts://zenodo.org/badge/latestdoi/246248514 (accessed 31.7.2020)

Singh, S., & Ashuri, B. 2019. Leveraging Blockchain Technology in AEC Industry during Design Development Phase. *In Computing in Civil Engineering 2019* (pp. 393–401).

Son, H., Lee, S., & Kim, C. 2015. What drives the adoption of building information modeling in design organizations? An empirical investigation of the antecedents affecting architects' behavioral intentions. *Automation in Construction*, 49, 92–99. doi:https://doi.org/10.1016/j.autcon.2014.10.012

Sreckovic, M., Sibenik, G., Breitfuß, D., Kastner, W. & Preindl, T. 2020. Analysis of Design Phase Processes with BIM for Blockchain Implementation. *Available at SSRN 3577529*.

Succar, B. 2009. Building information modelling framework: A research and delivery foundation for industry stakeholders. *Automation in Construction*, 18(3), 357–375. doi:https://doi.org/10.1016/j.autcon.2008.10.003

Wang, S., Ouyang, L., Yuan, Y., Ni, X., Han, X., & Wang, F. 2019. Blockchain-Enabled Smart Contracts: Architecture, Applications, and Future Trends. *IEEE Transactions on Systems, Man, and Cybernetics: Systems*, 49(11), 2266–2277. doi:10.1109/TSMC.2019.2895123

Zheng Z, Xie S, Dai H, Chen X, Wang H. 2017. An overview of blockchain technology: Architecture, consensus, and future trends.: *IEEE*, 557–564.

*Models, methods and tools for design, engineering,
construction and maintenance*

Polyhedral space partitioning as an alternative to component assembly

V. Galishnikova
Peoples' Friendship University of Russia, Moscow, Russia

W. Huhnt
Technische Universität Berlin, Germany

ABSTRACT: Building information models are conventionally assembled from individually constructed components. For communication and collaboration, data are mapped to the model using standards like the Industry Foundation Classes (IFC). The specification of the topology of a component is shape-dependent. Interoperability of software systems of different vendors has not been achieved with the IFC concept. Polyhedral space partitioning is proposed as an alternate approach to the solution of the interoperability problem. Some fundamental features of the approach are presented in this paper. Unbounded space is partitioned by splitting one component at a time. Polygons and dihedral cycles suffice to model the topology. Collisions and gaps are detected by testing only the component that is being split. Handling of the geometric user surface is simple because the topology is treated by a hidden core model. The model structure supports semantic expansion for physical, functional, economic, social, administrative and other applications.

1 INTRODUCTION

Building Information Modeling (BIM) is becoming more and more the state of the art in AEC projects. Software tools such as Autodesk Revit, Allplan or ArchiCAD are examples for modeling tools used in daily practice. Different kinds of information are specified; the most relevant kinds are semantics and geometry. Semantics are terms which classify components such as beam, slab, column, door or stairway including terms to name properties and values for properties. This paper focuses on geometry.

Software companies usually do not publish details of their implementations. This is true to data structures and algorithms. Contrary to this, theoretical approaches are published and taught at universities all over the world. The most frequently used theories are based on boundary representations (BRep) or Constructive Solid Geometry (CSG). In addition, the shape of components can be described by parameters (Mäntylä 1988).

Existing modeling tools map their internal data onto standardized models for data transfer between different software tools. The most frequently used standard are the Industry Foundation Classes (IFC) (IFC 2020a).

The IFC concept, its implementation and its widespread application have contributed very significantly to the benefits of BIM for AEC. Some of the initial expectations associated with the introduction of the concept did, however, not materialize. When IFC was first presented, one of the main targets was to achieve interoperability between the products of different vendors by standardization.

Unfortunately, this goal has not been achieved. In the place of interoperable systems of independent vendors, we observe acquisition and integration of specialized systems into large general purpose products. Interoperability is mainly limited to the product spectrum of each of the major vendors, complemented by selected strategic partnerships. The disadvantages of this situation are obvious. There may be important commercial reasons for the observed development, but the essential reasons lie much deeper in the IFC concept itself, as discussed below.

IFC are based on ISO standard 10303 and use the above mentioned theoretical approaches for boundary representation of shapes, constructive solid geometry or parametric modeling, treating topology as a property defined for specific shapes as standardized e.g. in the IfcTopologyResource (IFC 2020b).

The most important prerequisite for the interoperability of two systems in the field of geometry is the unhindered transfer of topology between the systems, because topology defines the neighborhood relations of their objects. It is not sufficient to transfer the internal topology of each individual object. A much more difficult task is the transfer of the topological relations between a large number of independent objects of many different classes with a wealth of shapes. The IFC concept does not handle the complexity of this problem adequately at present (Huhnt 2018).

The authors of this paper are convinced that space partitioning can handle topology much more

DOI 10.1201/9781003191476-19

efficiently than component assembly. A main target of polyhedral space partitioning is to solve the topology aspect of the interoperability problem with a shape-free generalized topology treating all objects as one coherent topological system independent of standardization.

A fundamental feature of existing and broadly used approaches is the construction of models by assembly of individually constructed components. Considerable precautions must be taken to avoid collisions and unintended gaps between the objects of a model with serious consequences for the construction process. There is a fundamental difference between the assembly of components by positioning in space and defining objects by the partitioning of space itself. If an object is added to an existing set of objects, it must be tested against each of these objects to avoid collisions and gaps. Ingenious methods have been devised to reduce the scope and effort of this task, but the basic problem persists (Kraft 2014). If space is partitioned by splitting one object at a time, the checking can be restricted to the object that is being split. Very significant gains in simplicity, accuracy and efficiency are achieved.

Partitioning of space is not a new concept (Mäntylä 1988). Orthogonal space partitions with quadtrees and octrees are well known, as are binary space partitions with half-spaces utilizing BSP-trees. If it is sufficient to bound objects with bounding volumes, then axis-aligned bounding boxes AABB or oriented bounding boxes OBB and related concepts can be employed (Ottmann 2002). While these concepts have many important and successful applications, for example for collision detection in computer games, they are not suitable to describe explicitly the complex polyhedral character of buildings which are only in part rectangular and possess many concave and multiply connected spaces. Other concepts of partitioning have therefore been investigated specifically for constructed facilities.

This paper presents first studies by the authors of an alternate approach to space partitioning. The general introduction in section 1 is followed by a review of the state of the art in section 2 and a presentation in section 3 of alternate concepts which distinguish the approach of the authors from the state of the art approaches. Section 3.1 treats differences between partitions of bounded and unbounded space. Reasons for the separation of the partition model into a user and a core model are given in section 3.2. The user specification of a partition is presented in section 3.3, its automated mapping to the core model in section 3.4. Section 3.5 illustrates navigation in the partition model. In section 3.6 the basic model is expanded to models with enhanced semantics. The paper ends with conclusions and an outlook in section 4.

2 STATE OF THE ART

The theory of polyhedra is a well-established branch of mathematics dealing with neighborhood relations. However, the complexity of the polyhedral bodies in building information models significantly exceeds that of typical polyhedra in mathematics because buildings consist of thousands of elements in irregular arrangements.

The faces and cells of polyhedral objects in technological applications are frequently concave or even multiply-connected. In conventional approaches faces and cells are subdivided into convex elements by triangulation (de Berg 1997). A standard tessellation language (3D Systems Inc. 1989) has been developed for triangulations. Triangular meshes for building models are provided by software tools, e.g. by Autodesk Revit (Tammik 2018).

Conventional index tables for topology are not advantageous for digital models due to a variable number of entries per row (buildingSMART 2018a,b,c; Bungartz 2002; Steel 2012). Advanced data structures replace index tables by winged edges (Baumgart 1972), half-edges (Mäntylä 1988) and quad-edges (Guibas 1985). Half-edges are replaced in three-dimensional polyhedral objects by half-faces (Alumbaugh 2005; Bilchuk 2012; Boguslawski 2010, 2011; Feng 2013). The use of dual half-edges for three-dimensional polyhedra is treated in (Boguslawski 2016). Holes and cavities in cell complexes are handled by connecting the inner and outer face loops by artificial face bridges which must be added to the model by the user. The model construction is incremental.

Detection of collisions, gaps and contact faces is a current subject of research (Huhnt 2018; Kraft 2014, 2016). Such errors have proved to be difficult to detect and to correct. The alternate approach of subdividing space to create polyhedral objects consisting of nodes, edges, faces and cells has been considered in principle by OGC (2016).

The largest possible degree of separation of topological and geometrical attributes of polyhedral objects is advantageous. One approach is to specify the geometry of the object and to compute its topological relations from its specified boundary (Borrmann 2009a,b; Daum 2014). Another approach is to construct the topological model and then to specify the geometry by fixing the nodal coordinates (Pahl 2012).

The theory of polyhedral topology was enriched by the local pyramid approach of Nef (1978), Bieri (1998). A Nef polyhedron is defined by a set of sphere maps at its vertices. For the construction of algorithms, a Nef complex (SNC) representation is defined. The concept has been implemented at the Max-Planck-Institut für Informatik at Saarbrücken (CGAL 2019; Granados 2003; Hachenberger 2006).

After reviewing the state of the art, the authors came to the conclusion that the complexity of the Nef polyhedra and the problems in handling multiply connected spaces in (Boguslawski 2016) could be avoided by basing polyhedral partitions on the concept of embedment established by Bernhard Riemann. They became aware that only polygons and dihedral cycles at edges are required to define the structure of polyhedral

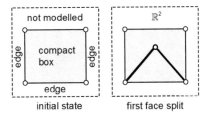

Figure 1. Conventional model in a compact box.

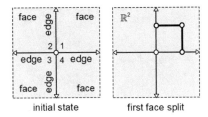

Figure 2. Partitioning of unbounded space.

partitions. Polyhedra need not be constructed and modified by Boolean operations on half-planes as in conventional constructive geometry, but can be constructed easily by splitting and merging domains such as edges, faces and cells.

3 POLYHEDRAL SPACE PARTITIONING

3.1 Partitions of unbounded space

Most of the conventional methods of partitioning do not treat an unbounded space, but replace it by a compact box which contains the model as shown in Figure 1. The box approach leads to a model which is not a true partition of the unbounded space because the part of space outside the box is not mapped to the model. This approach avoids the use of unbounded components, but does not permit the formulation of a consistent theory. Special algorithms must be employed at the boundary of the box.

The approach of the authors uses a global coordinate system in unbounded 3D space which is subdivided into 8 octants. The concept is illustrated in Figure 2 with 2D space subdivided into 4 quadrants. The model is constructed by splitting domains one at a time. Some of the edges, faces and cells of the model are bounded, others unbounded.

3.2 Separation of user and core model

Practicing engineers do not think in topological terms, but are well familiar with geometric concepts. The partition model is therefore split into a geometric user model and a topological core model.

The user is aware only of the surface and functions of the user model with which she or he specifies data, controls functions and evaluates results. The user model is automatically mapped to the core model,

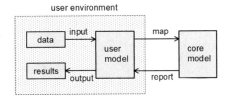

Figure 3. Separation of partition user and core model.

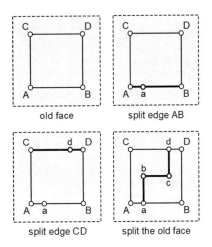

Figure 4. Split of face ABCD into two faces.

whose methods check data validity, construct the topological model and operate on the model to provide requested results.

3.3 Construction of the user model

The partition user model is constructed stepwise. In each step an existing edge, face or cell is split to form 2 objects of the same type. An edge is split by adding an internal node, a face by adding internal polyedges and a cell by adding internal polyfaces. The coordinates of the nodes define the shape of the new objects.

The principle of construction is illustrated in Figure 4 where face ABCD is split into two faces in three steps. In step 1 node a splits edge AB. In step 2 node d splits edge CD. In step 3 polyedge abcd splits the old face into two new faces.

3.4 Construction of the core model

The core model is constructed in phase with the user model according to a set of concepts of which the user is not aware. Some of these concepts are presented in this section.

3.4.1 Interior and exterior of a domain

Conventionally a curve in a plane is regarded as the boundary of the bounded domain. In the alternate concept the curve is regarded as the boundary of both the bounded domain and the unbounded domain. One of the two domains is explicitly chosen as the interior

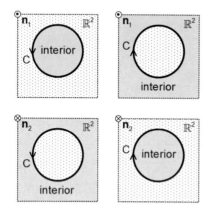

Figure 5. Interior of a domain in a plane.

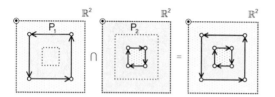

Figure 6. Construction of a multiply-connected facet.

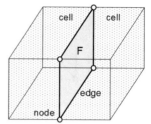

face F in the conventional model

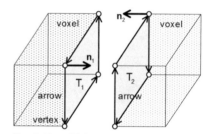

Facets T_1 and T_2 in the alternate model

Figure 7. Twin arrows and twin facets.

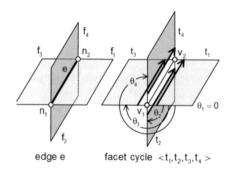

Figure 8. Dihedral cycle at edge e.

of a shape, the other as its exterior. The distinction is made by orienting the curve with its direction relative to the normal vector of the plane. Figure 5 shows that 4 combinations are possible.

3.4.2 Multiply-connected domains

In conventional models multiply-connected domains are treated as special cases. In some algorithms they are decomposed into convex domains, in others they are reduced to simply-connected domains with user-defined components.

The partition concept makes it unnecessary to treat the topology of a multiply-connected face or cell as a special case. Artificial user-defined components and user intervention are not required.

The multiply-connected domain is defined by the intersection of domains. The concept is illustrated in Figure 6 with the intersection of polygons P1 and P2. Arrow polygon P1 is oriented such that it is bounded, arrow polygon P2 such that it is unbounded. The intersection of P1 and P2 is a multiply-connected facet.

3.4.3 Twin arrows and twin facets

In conventional models the direction of the normal vector of a common face of two cells is not specified. The edges of the face are not directed. The face is not assigned to one of the cells. In a core model the face is mapped to two oriented facets as illustrated in Figure 7. A facet consists of the same point set as a face, but the direction of its normal vector is defined. Each of the twin facets is assigned to one of the cells such that the normal vector points to the outside of the cell. Each edge is mapped to two oriented arrows with identical point sets but opposite directions. Each facet is bounded by its own arrow polygons whose direction depends on the orientation of the facet.

3.4.4 Dihedral cycles

The three-dimensional structure of a partition is described with dihedral cycles as illustrated in Figure 8. The faces which are incident at edge e are mapped to twin facets, their edges to twin arrows.

The arrows which point from vertex v_1 to vertex v_2 form an arrow bundle. The facets whose boundaries contain these arrows form a facet bundle. One of the facets is arbitrarily chosen as reference facet. The angle measured clockwise around the arrows from the reference facet to another facet T in the bundle is called the dihedral angle of facet T. The facets of the bundle are sorted with ascending dihedral angle in a cyclic list which is called their dihedral cycle. Polygons and dihedral cycles are the main tools with which the topological structure of polyhedral partitions is described.

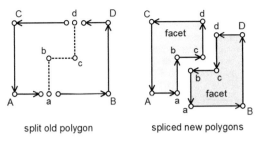

Figure 9. Splitting and splicing of polygons in the core model.

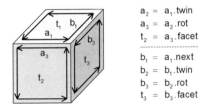

Figure 10. Neighbors t_2 and t_3 of facet t_1.

3.4.5 Split algorithm

A work step split is specified in the user model and executed in the core model. For example, the old face in Figure 4 is mapped to the vertices and arrows of the polygons of two facets, one of which is shown in Figure 9. The added polyedge abcd is mapped to vertices and polyarrows. The polygon and the polyarrow share vertices a and d, at which the closed polygon is split into two open polygons. Each of the open polygons is spliced with one of the polyarrows to form a new facet.

3.5 Navigation and transient data in a partition

The large number of objects of a complex facility necessitates a concise data structure. Only data which cannot be derived with algorithms from other data are therefore stored as persistent data. For example, the following attributes of an arrow a are persistent data:

facet	facet t which contains arrow a
org	vertex at the origin of the arrow
twin	arrow with the same parent edge as a
next	successor of a in the boundary of t
rot	successor of a in its dihedral cycle

Attributes which are not persistent are called transient. A facet t_2 of a voxel x is called a neighbor of a facet t_1 of x if an arrow of t_1 and an arrow of t_2 consist of the same point set. The facets of a voxel x which are neighbors of a given facet t_1 of x are transient attributes of facet t_2.

As an example the neighbors of facet $t_1 = a_1$.facet in Figure 10 are collected. Arrow $a_2 = a_1$.twin is an arrow of facet t_1.twin $= a_2$.facet which is not a facet of voxel x. The successor of a_2 in its dihedral cycle is arrow $a_3 = a_2$.rot which is an arrow of voxel x. Facet $t_2 = a_3$.facet of voxel x is a neighbor of facet t_1. Substitution yields $t_2 = a_1$.twin.rot.facet.

The arrow succeeding a_1 in facet t_1 is $b_1 = a_1$.next. The neighbor of facet t_1 at arrow b_1 is facet $t_3 = b_1$.twin.rot.facet. The procedure terminates when arrow a_1 of facet t_1 is reached and all 4 facets of voxel x which are neighbors of facet t_1 are known.

3.6 Semantic expansion and reduction

The basic partition model contains topological and geometric data which are used to navigate in the topology of the partition and to derive geometric properties such as length, area and volume. The basic model is readily expanded to include semantic data such as physical properties of solids or fluids and functions of spaces. Economic, social and administrative models can be derived, as well as models for specific types of constructed facilities such as bridges and power stations.

Semantic attributes can be assigned to nodes, edges, faces and cells in the user model as well as vertices, arrows, facets and voxels in the core model. The user is aware of the attributes, but not of their assignment to components of the core model.

The basic partition model can also be reduced to a skeleton model for applications such as diagrammatic perspectives. Special values are assigned to some of its geometric properties for this purpose. For example, the thickness of walls, floors and roofs is set to null, and the cross-section of a beam or column is reduced to a point on the axis of the member. The topological model remains untouched.

4 CONLUSIONS AND OUTLOOK

The ongoing research of the authors into polyhedral space partitioning as an alternative to component assembly has confirmed that the embedment concept of Riemann is a sound basis for a generalized treatment of topology without shape-bound standardization. The presented examples of alternative concepts describe some of the fundamental features of the new approach. More comprehensive and detailed descriptions will be presented in additional papers as the work progresses.

ACKNOWLEDGEMENTS

The authors thank Prof. Dr. Peter Jan Pahl for his excellent and extensive contribution to this research. The research is funded by the Russian Foundation for Basic Research (RFBR) – Project Number 20-57-12006, and by the Deutsche Forschungsgemein-schaft (DFG, German Reserach Foundation) – Pro-jektnummer 429900376.

REFERENCES

Alumbaugh, T. J. & Jiao, X. 2005. Compact Array-Based Mesh Data Structures. In: B. W. Hanks (Ed.), *Engineering, IMR '05*, Springer Berlin Heidelberg, pp. 485–503.

Baumgart, B. 1972. Winged edge polyhedron representation. Computer Science Department, Stanford University.

Bieri H. 1998. Representation Conversions for Nef Polyhedra. In: Farin G., Bieri H., Brunnett G., De Rose T. (eds) *Geometric Modelling. Computing Supplement*, vol 13. Springer, Vienna.

Bilchuk I. & Pahl P.J. 2012. Three-dimensional Topological Models of Buildings. Vestnik, *Journal of the Moscow State University of Civil Engineering*, Vol. 69, 2012.

Boguslawski, P. 2011. Modelling and Analyzing 3D Building Interiors with the dual half-edge data structure, PhD thesis, University of Glamorgan/Prifysgol Morgannwg, 2011.

Boguslawski, P. et al. 2010. Modelling and analysing 3D buildings with a primal/dual data structure. In: *ISPRS Journal of Photogrammetry and Remote Sensing*.

Boguslawski, P. & Gold, C. M. 2016. The Dual Half-Edge – A Topological Primal/Dual Data Structure and Construction Operators for Modelling and Manipulating Cell complexes. *ISPRS International Journal of Geo-Information* 2016, 5(2): 19; doi: 10.3390/ijgi5020019.

Borrmann, A. & Rank E. 2009a. Specification and implementation of directional operators in a 3D spatial query language for building information models. *Advanced Engineering Informatics*, 23: 32–44.

Borrmann, A. & Rank, E. 2009b. Topological analysis of 3D building models using a spatial query language. *Advanced Engineering Informatics*, 23: 370–385.

buildingSMART 2018a. IFC Overview. Retrieved from buildingSMART website: http://www.buildingsmart-tech.org/specifications/ifc-overview, accessed on January 27, 2018.

buildingSMART 2018b. IfcFacetedBrep. Retrieved from buildingSMART website: http://www.buildingsmart-tech.org/ifc/IFC4/final/html/schema/ifcgeometricmodelresource/lexical/ifcfacetedbrep.htm, accessed on January 27, 2018.

buildingSMART 2018c. IfcRelFillsElement. Retrieved from buldingSMART website: http://www.buildingsmart-tech.org/ifc/IFC4/final/html/schema/ifcproductextension/lexical/ifcrelfillselement.htm, accessed on January 28, 2018.

Bungartz H.-J. et al. 2002. Einführung in die Computergraphik. Vieweg Verlag, Wiesbaden, 2002. 302 p. ISBN 3-528-16769-6.

CGAL 2019. The Computational Geometry Algorithms Library, Package 3D Boolean Operations on Nef Polyhedra. https://doc.cgal.org/latest/Nef_3/index.html (last visited: 23.2.2019).

Daum, S. & Bormann, A. 2014. Processing of Topological BIM Queries using Boundary Representation Based Methods. *Advanced Engineering Informatics*, 28: 272–286.

de Berg, M. et al. 1997. Computational Geometry- Algorithms and Applications. *Springer Berlin* 1997. 365 p. ISBN 3-540-61270-X.

Feng, X. et al. 2013. Compact combinatorial maps: A volume mesh data structure. Graphical Models. Volume 75, Issue 3: 149–156.

Granados M. et al. 2003. Boolean Operations on 3D Selective Nef Complexes: Data Structure, Algorithms, and Implementation. In: Di Battista G., Zwick U. (eds) *Algorithms – ESA* 2003. ESA 2003. Lecture Notes in Computer Science, vol 2832. Springer, Berlin, Heidelberg.

Guibas, L. & Stolfi, J. 1985. Primitives for the Manipulation of General Subdivisions and the Computation of Voronoi Regions. *ACM Transactions on Graphics*. Vol. 4, No. 2.

Hachenberger, P. 2006. Boolean Operations on 3D Selective Nef Complexes: Data Structure, Algorithms, Optimized Implementation, Experiments, and Applications. Dissertation, Universität des Saarlandes.

Huhnt W. & Galishnikova, V. 2018. Partitioning of space as basis for data structures to describe digital building models. Conference proceedings: *17th International Conference on Computing in Civil and Building Engineering*, ICCCBE 2018, Tampere/Finland: 42–49.

IFC 2020a. https://technical.buildingsmart.org/standards/ifc/ last visited February 28, 2020.

IFC 2020b. https://standards.buildingsmart.org/IFC/DEV/IFC4_2/FINAL/HTML/schema/ifcrepresentationresource/lexical/ifctopologyrepresentation.htm last visited February 28 2020.

Kraft, B. 2016. Ein Verfahren der Raumzerlegung als Grundlage zur Prüfung von Geometrie und Topologie digitaler Bauwerksmodelle, Doctoral Thesis, TU Berlin, Germany, 2016.

Kraft, B. & Huhnt, W. 2014. Geometrically Complete Building Models. In: *Proceedings of the 21th International Workshop: Intelligent Computing in Engineering*. Cardiff, UK.

Mäntylä, M. 1988. Introduction to Solid Modeling. W. H. Freeman & Co. New York.

Nef 1978. W. Beiträge zur Theorie der Polyeder. Herbert Lang, Bern.

OGC 2016. Open Geospatial Consortium. OGC IndoorGML – with Corrigendum, http://docs.opengeospatial.org/is/14-005r3/14-005r3.html, 2016, accessed on March 18, 2020.

Ottmann T. & Widmayer P. 2002. Algorithmen und Datenstrukturen, Spektrum Akademischer Verlag, Heidelberg Berlin.

Pahl, P.J. 2012. Topology of Buildings. Lecture Notes. Chair of Bauinformatik, Technische Universität Berlin, 2012. 124 p.

Steel, J. et al. 2012. Model Interoperabilty in Building Information Modelling. *Software and Systems Modelling, Springer Berlin*, 2012, Issue 1: 99–109.

Tammik, J. 2018. The building coder samples. https://github.com/jeremytammik/the_building_coder_samples/blob/master/BuildingCoder/BuildingCoder/Creator.cs, accessed on March 18, 2020.

3D Systems Inc. 1989. The StL Format. http://www.fabbers.com/tech/STL_Format, accessed on March 18, 2020.

A hierarchical kit library to support content reuse for mass customization

J. Cao & D.M. Hall
ETH Zurich, Zurich, Switzerland

ABSTRACT: Design reuse is an approach to achieve product variety while keeping the low cost and high efficiency of product development found in mass customization. Kit library is an application of virtual prototype reuse. Based on the object-oriented modeling, a framework of hierarchical kit library is proposed. The category, representation and approaches to customize the library content is illustrated in the context of hybrid industrialized construction. The process of hierarchy construction is given to connect the distributed library resources from spatial layout, structural modules and production parts respectively. The strategy facilitates a wider scope of customer-driven design by providing building kits that can support end-to-end building configuration, rather than traditional client participation at a single phase of the project.

1 INTRODUCTION

1.1 Design reuse in industrialized construction

One characteristic found in the recent rise of industrialized construction is the embrace of mass customization (Kolarevic 2015). Many companies put significant effort on product platform development to rapidly respond to customer requirements. Platforms in construction are characterized by the reuse of process and technical solutions (Jansson et al. 2014). In this context, the term *reuse* refers to the act of main stakeholders retrieving data, information, and knowledge from solutions of previous projects and applying them in new projects (Sönmez 2018). The reuse of the knowledge in future projects will lead to continuous improvement of project quality and return of investment made initially inside a single project (Tetik et al. 2019). However, due to the loosely-coupled nature of project-based organizing in construction, design reuse remains an under-utilized source to improve project performance (Dubois & Gadde 2002).

1.2 Virtual prototype reuse: Kit library

Owen et al. classified engineering knowledge representation into five types: 1) pictorial (e.g. sketches, drawings); 2) symbolic (e.g. diagrams, decision tables); 3) linguistic (e.g. textual communication, verbal communication; 4) virtual (e.g. CAD models, simulations) and 5) algorithmic (e.g. computer algorithms, computational procedurals) (Owen & Horváth 2002). Among these various knowledge representations, virtual prototypes are widely used to predict building performance prior to the actual construction (Hiyama et al. 2014). An important application of virtual prototype reuse is kit libraries or product families (Gross 1996). A kit library is "a collection of discrete building components that are pre-engineered and designed to be assembled in a variety of ways to define a finished building" (Howe et al. 1999). BIM (Building Information Models) is used as product digital representations in construction industry. There are public libraries that offer BIM objects, such as Autodesk Seek, BIMObject, RevitCity and Open Source Wood. These libraries enable designers to find similar designs and templates on which to copy or modify. Apart from those public libraries, some industrialized construction pioneers, such as Katerra and Project Frog also develop local kit libraries within their internal platforms through collaboration with manufacturers (Hall et al 2019). A well-defined kit library can meet different design needs while also satisfying the capability of the factory, automating the production process (Bonenberg et al. 2019).

However, the development and application of kit libraries are still limited. Since kit libraries are usually distributed resources, most studies on kit library application is limited in single product development phase (Jin et al. 2008), such as architectural design (Cui et al. 2020; Wikberg et al. 2014), engineering simulation (Hiyama et al. 2014) and part assembly (Liu et al. 2018). A drawback of such single-phase kit libraries is the lack of compatibility between the other trades' systems. This requires extra effort to later modify the kit. For example, industry studies find that multiple classification standards exist even *within* the same suppliers' system (Andersson & Lessing 2017). More importantly, during the early course of project, clients might prefer to compare different element suppliers' products and how they affect the overall building performance (El-Diraby et al. 2017). Without available production kit libraries connected to the abstract design prototypes, it is difficult for stakeholders to make a quick analysis and decision.

To overcome the challenge, this research seeks to provide a framework to build a hierarchical kit library.

The kit library consists of multi-level content satisfying the entire building configuration, from spatial layout to structure composition to detailed production planning. Each level of the kit library is systematically connected with its super system and subsystem. The connection is achieved via automatic data extraction and semantic mapping, so as to avoid unnecessary manual interpretation.

The paper is organized as follows. Section 2 conducts a literature review on the kit library development and application. Section 3 introduces the object-oriented methodology (OOM) and applies it to sketch the general framework of the library. Section 4 elaborates on the proposed framework using the example context of hybrid industrialized construction. Section 5 gave a simple example of how the proposed hierarchical kits library can be modeled in JSON. Finally, section 6 and 7 provide a discussion and future research directions for kit library development.

2 LITERATURE REVIEW

2.1 Development of kit libraries

Design kits or catalog usually comes from internal sources, such as a design firm's in-house libraries. To identify the reusable assets, most research focus on retrieval of design rational of previous projects (Johannesson 2014). Fruchter and Demian developed "CoMem" prototype to help users visualize the design evolution process and understand the project context (Fruchter & Demian 2002). The prototype is relied on design information, data, knowledge and models over time, which might not be recorded clearly. Besides, it requires time and expense to analyze and reconstruct the design rationale due to multiple implicit and intractable influences (Tang et al. 2007). Based on prior project experience and analysis, such as similarity and clustering approaches (Kashkoush & ElMaraghy 2016), the reusable modules which share some preferred sizes or similar structures can be gathered or consolidated. This allows a limited number of sizes of a particular building component available in the market and eases the interchangeability of components to achieve product variability (Singh et al. 2019). For example, Cui et al. put forward a method for prefabricated house design with a hierarchical module library that contains components, spaces and suites (Cui et al. 2020). The modules are designed using modules "2700" and "1800" (millimeters) and their multiples. Similar approach has been done in a prefabricated BIM library (with LOD 300 content) of the Hong Kong Housing Authority (Li et al. 2019), This highlights that the physical and functional information for fabrication is necessary. In general, the richer the library is, the more design variations can be developed.

2.2 Application of kit libraries

Once a kit library is built, it is supposed to be applied to future projects. In this scenario, an intelligent search engine is crucial for retrieving the building objects according to designers' requirements (Li et al. 2020), so as to speed up the access to the large amount of reusable assets. In mechanical product design, Lupinetti et al. defined a multi-criterial similarity assessment to detect the reusable CAD models from existing standard libraries (Lupinetti et al. 2019). The similarity metrics include geometry, structure, attributes and component positions. However, the interrelationship between parts needs to be interpreted with much effort. Costa et al. introduced semantic web to retrieve design context, especially the boundary conditions to adjacent elements. Users could then search for the appropriate components from a production library to suit design requirements (Costa & Madrazo 2015).

However, one limitation of predefined libraries is the use of typical Made-to-Stock (MtS) and Made-to-Order (MtO) components with few Engineered-to-Order (EtO) components (Zhang, Xing & Engineer 2013). Industrialized construction projects contain many ETO components, such as timber or steel frames, precast concrete, façade panels and customized kitchen or bathroom modules, which demand sophisticated engineering and careful collaboration between designers, fabricators, and installers (Tillmann et al. 2015). The design of ETO components require knowledge from many interdependent trades and substantial work in developing virtual models (Liu et al. 2018).

2.3 Research opportunities

From the review above, there are still two difficulties that prevent the wide application of kit libraries. Firstly, the bottom-up logic on how kits or parts could be combined to a whole structure, especially an ETO structure, is less studied in the architectural domain (Retsin 2016). Secondly, due to the segmented nature of project organization, most design kit libraries do not provide a connection to production and construction libraries in which the product manufacturers participate (Costa & Madrazo 2015).

Therefore, there is a need for a hierarchical kit library that a) complies with the existing standards and b) can be reused across different phases of the project for end-to-end project delivery. The proposed approach in this paper is for a library that uses a system design approach to connect flexible design representation during multi-phase of project.

3 METHODOLOGY

3.1 Hierarchical object-oriented modeling

In order to model the construction system in modular fashion, a hierarchical, object-oriented modeling is applied to reflect the structure of buildings. Scott applied the OOM method to model a kit library in three levels, including geometry, assembly and construct (Howe 1997). The object higher in the hierarchy (super-class), such as assembly, contains a set of objects lower in the hierarchy (sub-class) by inheritance. This study extends the model with flexible object modeling to represent diverse virtual prototypes used in reality. For example, a virtual prototype for a floor plan might be a 2D-sketch, which can be modeled using graph structures. Therefore, on each level of hierarchy, an object structure specific to that prototype is implemented. To encode different categories of objects on the same level, a super-class object is a specification or map of the interconnection of its sub-class objects. The map correlates to the quantity and types of sub-class components. The correlation is built up via metadata representing the objects rather than the file itself. The configuration of kits then becomes the process of hierarchy construction by which these maps are defined (Luna 1992).

4 PROPOSED FRAMEWORK

4.1 *Hybrid industrialized construction*

The proposed hierarchical kit libraries are targeted at a hybrid industrialized construction. A hybrid construction combines 3D-modules for the highly serviced and integrated units, such as kitchens and bathrooms, and 2D-panels, such as walls and floors, for more open spaces. The approach provides design flexibility to optimize the standardization-variability balance in housing (Lawson & Ogden 2008). The kit libraries are built in accordance with the structure of the construction, including three levels: super-system level, system level and sub-system level (Piroozfar et al. 2019). Each of library contains specific knowledge for the corresponding level of products' views. The super system in this context is the building spatial layout. The system level contains 3D-modules and 2D-panels. The sub-system level consists of components of the system-level parts, such as structural supports. Figure 1 shows a conceptual example of a hierarchical kit library and how it is mapped to the multi-phase building design, including spatial layout, structural design and building component production. Customers can participate the design through the selection and configuration of each level of libraries. Meanwhile, library managers emerge as a new role in this context to take charge of the standard definition, system maintenance and other coordination work. To illustrate the reusability of the kit libraries, this study focuses on the following perspectives in each level of library:

– Categories of kits
– Representations of kits
– Approaches to customize kits

4.2 *Super-system library*

The super-system library consists of spatial layouts of houses. Typical layouts include one-direction slicing,

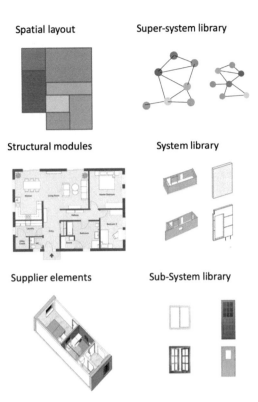

Figure 1. From top to bottom: architectural super-system library (left) for spatial layout (right), structural system library (left) for structural design (right), manufacturing sub-system library (left) for building components (right).

two direction slicing, three direction slicing, heteromorphic slicing and multiple splicing (Cui et al. 2020). Each layout object is stored as an adjacency graph in the database. The graph representation has been widely used in building spatial design (As et al. 2018; Sharafi et al. 2017; Wang et al. 2018), where each node denotes single space while edge denotes the connection between spaces. Both nodes and edges can be assigned with attributes to represent space properties. Interrelationship rules between spaces can be associated with the graph operations. The operations include node addition and deletion, node type specification, and edge addition and deletion. This also helps customers without sufficient knowledge and skills to participate the design process.

4.3 *System library*

The 3D-modules and 2D-panels in the system level library are classified by structure and functionality. For example, standard structures of 3D-modules include four-sided modules, open-ended modules and corner-supported modules (Lawson et al. 2014). New types of structures can be added to the library by designers. Apart from the standard structure of modules, there are modules designed for specialist functionalities, including stairs, balconies, rooftop and exhibition spaces.

The modules are built in parametric-driven BIM applications and constraints on object dimensions are corelated to parameters' definition (Khalili-Araghi & Kolarevic 2020). The exact values of parameters are determined by clients' specification and building regulations, which are further formulated as mathematic functions of building performance. For example, the thickness of walls could be customized according to building insulation requirement. Other performance metrics that drive the design variability include loading capacity, fire resistance, and energy performance. In this way, the kits can be customized parametrically by adjusting the standard of performance.

4.4 Sub-system library

The design content of the sub-system level library is categorized by element suppliers. Suppliers for industrialized construction usually include window and door suppliers, piping and ducting suppliers, lighting suppliers, furnishes suppliers and structural components suppliers. To overcome the heterogenous representations (e.g. PDF, text files, Excel sheets) of above products, a unified structure is defined in the document-oriented database. The document content encodes five important types of attributes, including dimensions, constraints, materials, costs and identity data (Hamid et al. 2018). The Suppliers have access to the database to update their product information via SQL-like query language. More importantly, general contractors can leverage the library to make detailed design for ETO products and complete bidding processes at early stage of projects (Andersson & Lessing 2017). In this case, the clients can compare the project cost and duration of design alternatives to support them in selection of suppliers.

4.5 Hierarchy construction

The hierarchy construction process consists of three steps. Take the connection between super-system and system as an example (shown in Figure 2). The first step is object definition. A planned space layout is determined and transformed into a graph model stored in the database. Assuming that a customer wants to make an interior design within a single space, the space is selected, as well as the corresponding node of the graph model in the backend. The node object contains the space-related information, such as space name, space area and space boundary. The object also contains a method to enable customers to assign other requirements, such as sound insolation, construction cost range, to the space.

The second step is component searching. The space attributes and attached user inputs are encapsulated as a request to the system library in order to retrieve the structural components. The objects in the system library contain methods defined via API to extract parameters in BIM files. The retrieval is completed by mapping request metadata with extracted parameters, in terms of geometry (e.g. dimensions), property (e.g.

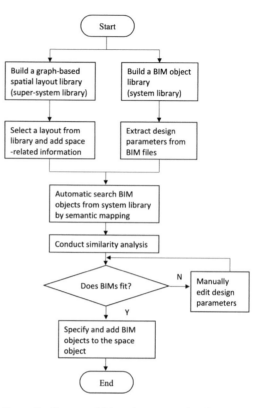

Figure 2. Process of hierarchy construction between a super-system library and a system library.

material), and performance (e.g. structural capacity). The third step is component addition. The customer is able to select the structural components from the response list and place them on the space. Meanwhile, the extracted parameters can be adapted to the new project context manually. The space objects contain a method to add selected components. In this sense, each space object is connected to the structural objects. A similar process can be implemented to link the system library with the sub-system library.

5 IMPLEMENTATION

In this section, we applied JSON (JavaScript Object Notations) to model the objects in each library. JSON has been widely used in web applications. It is easy to read and write, independently with the programming language. Besides, most document-oriented databases store and query data as JSON-like formats. Hence, JSON is an ideal syntax for information storage, processing and exchange. In object-oriented design, JSON can be treated as a unified data model to support object identification and communication. Figure 3 shows examples of a floorplan object in JSON (left) and 3D-module objects in JSON (right). Assuming that a user is configuring the bedroom in the floorplan, the system will conduct semantic mapping between

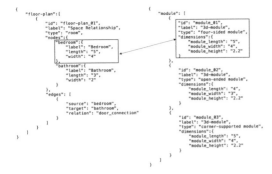

Figure 3. A floorplan object in JSON (left) and 3D-module objects in JSON (right).

the bedroom properties and module properties. The first module can be selected as it is the fittest one in dimensions. Finally, the selected 3D-module's GUID is identified and added to the bedroom node for the next-step configuration.

6 DISCUSSION

The hierarchical kit library proposed in this study not only provides a pipeline to reuse kits from distributed sources, but also highlights how those kits could be connected to satisfy a multi-phase project development. Instead of enforcing a unified data structure between different trades, we consider various data representation approaches, including graph structure, parameter-based models, table-based structure, for corresponding product detailing. Then, a hierarchical object-oriented modeling is applied to map the sub-level objects to supper-level ones. For customers, a benefit for a hierarchical library is to enable customers to visualize different degrees of design freedom and participate in the design following a guided process. For a project team, the hierarchical system helps them connect to the resources stored in other stakeholders' libraries.

However, there is still need for access between the libraries of independent parties to be linked via cloud services. This is not discussed in the paper but could be an area of future research. There is need to study how the use of cloud services can encapsulate the distributed libraries in a centralized platform. In that case, customers can request different libraries from different third parties, including design firms and element suppliers, to fulfill their housing customization.

7 CONCLUSION

The reuse of design knowledge has been emphasized in product mass customization. The most useful design knowledge is represented in the virtual prototypes. A trend in the AEC world is to integrate life-cycle project information in BIM, so-called nD-BIM. However, most research on virtual prototypes reuse is still limited in the single product development phase, which causes a high threshold for third parties to be involved in the early customer decision process. This paper proposes a hierarchical framework to reuse virtual prototypes, namely kit libraries. Compared with existing research on kit library reuse, we highlight the importance to build a hierarchical kit library to connect stakeholders' distributed resources. The development and integration of each hierarchy is achieved by object-oriented modeling. Within each hierarchy, an appropriate representation is leveraged to encode the product information and a set of operations is suggested to customize the product. The kit libraries lay a foundation to build a housing customized system.

REFERENCES

Andersson, N., & Lessing, J. 2017. The Interface between Industrialized and Project Based Construction. *Procedia Engineering*, 196: 220–227.

As, I., Pal, S., & Basu, P. 2018. Artificial intelligence in architecture: Generating conceptual design via deep learning. *International Journal of Architectural Computing*, 16(4): 306–327.

Bonenberg, W., Wei, X., & Zhou, M. 2019. BIM in Prefabrication and Modular Building. Advances in Human Factors, *Sustainable Urban Planning and Infrastructure* 788: 100–110.

Costa, G., & Madrazo, L. 2015. Connecting building component catalogues with BIM models using semantic technologies: An application for precast concrete components. *Automation in Construction*, 57: 239–248.

Cui, Y., Li, S., Liu, C., & Sun, N. 2020. Creation and Diversified Applications of Plane Module Libraries for Prefabricated Houses Based on BIM. *Sustainability*, 12(2): 453.

Dubois, A., & Gadde, L. E. 2002. The construction industry as a loosely coupled system: Implications for productivity and innovation. *Construction Management and Economics*, 20(7): 621–631.

El-Diraby, T., Krijnen, T., & Papagelis, M. 2017. BIM-based collaborative design and socio-technical analytics of green buildings. *Automation in Construction*, 82: 59–74.

Fruchter, R., & Demian, P. 2002. CoMem: Designing an interaction experience for reuse of rich contextual knowledge from a corporate memory. *Artificial Intelligence for Engineering Design, Analysis and Manufacturing: AIEDAM*, 16(3): 127–147.

Gross, M. D. 1996. Why can't CAD be more like lego? CKB, a program for building construction kits. *Automation in Construction*, 5(4): 285–300.

Hall, D. M., Whyte, J. K., & Lessing, J. 2019. Mirror-breaking strategies to enable digital manufacturing in Silicon Valley construction firms: a comparative case study. *Construction Management and Economics*, 0(0): 1–18.

Hamid, M., Tolba, O., & El Antably, A. 2018. BIM semantics for digital fabrication: A knowledge-based approach. *Automation in Construction*, 91: 62–82.

Hiyama, K., Kato, S., Kubota, M., & Zhang, J. 2014. A new method for reusing building information models of past projects to optimize the default configuration for performance simulations. *Energy and Buildings*, 73: 83–91.

Howe, A. S., Ishii, I., & Yoshida, T. 1999. Kit-of-Parts: A Review of Object-Oriented Construction Techniques. In *Proceedings of the 16th IAARC/IFAC/IEEE International Symposium on Automation and Robotics in Construction*: 165–171.

Jansson, G., Johnsson, H., & Engström, D. 2014. Platform use in systems building. *Construction Management and Economics*, 32(1–2): 70–82.

Jin, B., Teng, H. F., Wang, Y. S., & Qu, F. Z. 2008. Product design reuse with parts libraries and an engineering semantic web for small- and medium-sized manufacturing enterprises. *International Journal of Advanced Manufacturing Technology*, 38(11–12): 1075–1084.

Johannesson, H. 2014. Emphasizing Reuse of Generic Assets Through Integrated Product and Production System Development Platforms. In *Advances in Product Family and Product Platform Design: Methods and Applications*. New York: Springer

Kashkoush, M., & ElMaraghy, H. 2016. Product family formation by matching Bill-of-Materials trees. *CIRP Journal of Manufacturing Science and Technology*, 12: 1–13.

Khalili-Araghi, S., & Kolarevic, B. 2020. Variability and validity: Flexibility of a dimensional customization system. *Automation in Construction*, 109: 102970.

Kolarevic, B. 2015. From Mass Customisation to Design "Democratisation." *Architectural Design*, 85(6): 48–53.

Lawson, M., Ogden, R., & Goodier, C. 2014. *Design in Modular Construction. Design in Modular Construction*. London: CRC Press.

Lawson, R. M., & Ogden, R. G. 2008. "Hybrid" light steel panel and modular systems. *Thin-Walled Structures*, 46(7–9): 720–730.

Li, N., Li, Q., Liu, Y. S., Lu, W., & Wang, W. 2020. BIMSeek++: Retrieving BIM components using similarity measurement of attributes. *Computers in Industry*, 116: 103186.

Li, X., Shen, G. Q., Wu, P., & Yue, T. 2019. Integrating Building Information Modeling and prefabrication housing production. *Automation in Construction*, 100: 46–60.

Liu, H., Singh, G., Lu, M., Bouferguene, A., & Al-Hussein, M. 2018. BIM-based automated design and planning for boarding of light-frame residential buildings. *Automation in Construction*, 89: 235–249.

Luna, J. J. 1992. Hierarchical, modular concepts applied to an object-oriented simulation model development environment. In *Proceedings of the 24th conference on Winter simulation*: 694–699.

Lupinetti, K., Giannini, F., Monti, M., & Pernot, J. P. 2019. Content-based multi-criteria similarity assessment of CAD assembly models. *Computers in Industry*, 112: 103111.

Owen, R., & Horváth, I. 2002. Towards Product-Related Knowledge Asset Warehousing in Enterprises. *International Symposium on Tools and Methods of Competitive Engineering, April 22-26, Wuhan, China*: 155–170.

Piroozfar, P., Farr, E. R. P., Hvam, L., Robinson, D., & Shafiee, S. 2019. Configuration platform for customisation of design, manufacturing and assembly processes of building façade systems: A building information modelling perspective. *Automation in Construction*, 106: 102914.

Retsin, G. 2016. Discrete Assembly and Digital Materials in Architecture. *Ecaade 2016*, 1: 143–151.

Scott Howe, A. 1997. A Network-based Kit-of-parts Virtual Building System. In *CAAD futures 1997*: 691–706.

Sharafi, P., Samali, B., Ronagh, H., & Ghodrat, M. 2017. Automated spatial design of multi-story modular buildings using a unified matrix method. *Automation in Construction*, 82: 31–42.

Singh, M. M., Sawhney, A., & Borrmann, A. 2019. Integrating rules of modular coordination to improve model authoring in BIM. *International Journal of Construction Management*, 19(1): 15–31.

Sönmez, N. O. 2018. A review of the use of examples for automating architectural design tasks. *CAD Computer Aided Design*, 96: 13–30.

Tang, A., Jin, Y., & Han, J. 2007. A rationale-based architecture model for design traceability and reasoning. *Journal of Systems and Software*, 80(6): 918–934.

Tetik, M., Peltokorpi, A., Seppänen, O., & Holmström, J. 2019. Direct digital construction: Technology-based operations management practice for continuous improvement of construction industry performance. *Automation in Construction*, 107: 102910.

Tillmann, P. A., Viana, D., Sargent, Z., Handling, S. A., & Tommelein, I. 2015. BIM and LEAN in the design – production interface of ETO. In *23rd Annual Conference of the International Group for Lean Construction*: 331–340.

Wang, X., Yang, Y., & Zhang, K. 2018. Customization and generation of floor plans based on graph transformations. *Automation in Construction*, 94: 405–416.

Wikberg, F., Olofsson, T., & Ekholm, A. 2014. Design configuration with architectural objects: Linking customer requirements with system capabilities in industrialized house-building platforms. *Construction Management and Economics*, 32(1–2): 196–207.

Zhang, J., Xing, Z., & Engineer, S. 2013. An IFC-based semantic framework to support BIM content libraries. In *Proceedings of the 30 th International Conference of IT in Construction (CIB W78)*: 847–854.

Archi-guide. Architect-friendly visualization assistance tool to compare and evaluate BIM-based design variants in early design phases using template-based methodology

K. Jaskula, A. Zahedi & F. Petzold
Technical University of Munich, Germany

ABSTRACT: Most of the Building Performance Simulation (BPS) tools are rarely used by architects as they are considered as too complex and cumbersome, as they are not compatible with architects' needs (Attia et al. 2009). The goal of this work was to develop an architect-friendly template-based assessment tool for variant evaluation in early design phases. Based on a literature review, both qualitative and quantitative criteria were selected and simplified into three categories: social, economic and environmental factors. The main objectives for developing the tool were user-friendliness, flexibility and integration of a knowledge base. As proof of concept, a visualization tool called Archi-guide was put forward and tested with a sample project. The tool includes multiple panels giving freedom in terms of exploring the building variants' data. In the end, the tool was evaluated in an online user study by architects and architecture students.

1 INTRODUCTION

1.1 Problem statement

Decisions taken in the beginning of the design process have the biggest impact on the performance and costs of the building, while in the same time the costs of changes are the lowest. Early decisions are crucial to achieve sustainable solutions (Schlueter & Thesseling 2009).

In early design phases architects create many design variants, then compare, select and continuously detail them. Usually architects evaluate variants manually, based on their own judgement and different criteria such as the owners' requirements, building performance, and cost. Comparing different variants might be difficult, when the models are incomplete or in different stages of development (Zahedi et al. 2019).

Building Information Modeling (BIM) is nowadays one of the most important planning methodologies used in Architecture, Engineering, Construction (AEC) industry. It is based on the idea of continuous use of digital building models across the entire lifecycle of a built facility. However, it is still rarely implemented in the very beginning of the design process, where using model-based simulations and analysis tools are rather neglected (Zahedi & Petzold 2018).

BIM offers new prospects to facilitate the creation and assessment of different design variants and their influence on the future building performance. The results of those simulations must be delivered in a clear and helpful way to support the architect's decision.

Building Performance Simulation (BPS) software and tools are being rapidly developed to be more user-friendly and accurate. However, simulation results' visualization and analysis tools are usually still lacking in more user-friendly interfaces (Gadelhak et al. 2017).

Despite the increasing number of building simulation/energy analysis tools in the last ten years, architects and designers are still finding it difficult to use even basic tools. The reason is that most of these BPS tools are not compatible with architects' working methods and needs, as they are considered as too complex and cumbersome. There is a growing gap between architects as users and BPS tools, which are usually developed by technical researchers, building scientist or HVAC engineers. However, building simulation is also a human, psychological and social discipline as it directly involves man-computer interaction and human knowledge processing. Therefore, architects' problems in interacting with such tools must be comprehended, as architects have a different background, different knowledge processing methods and they are visually oriented (Attia et al. 2009).

Design problems are often both multi-dimensional and highly interrelated. Especially architectural design rarely serves only one purpose, as it is frequently necessary to devise an integrated solution to many different requirements. The qualities of good design are difficult to translate into quantitative standards. Both quantitative and qualitative factors increase the complexity of design problems. Designer must be able to balance both of them in their decision-making process (Lawson 2006).

2 STATE OF THE ART

2.1 Architectural design process and evaluation criteria

According to Kibert (2016) sustainability or sustainable development is a foundational principle that covers a variety of efforts to ensure a civilized quality of life for the future generation. The base of sustainability has the following three pillars: environmental, social and economic.

The German Sustainable Building Council (Deutsche Gesellschaft für Nachhaltiges Bauen – DGNB e.V) and the German approval system for sustainable construction (Deutsche Gütesiegel Nachhaltiges Bauen) currently offer the most comprehensive criteria catalogue to certify sustainable buildings in a systematic way. Some of these criteria are already relevant during schematic design, other aspects do not have to be considered until later design phases or the project completion (Fuchs et al. 2013). Based on DGNB criteria a system for sustainability requirements in planning competitions (Systematik für Nachhaltigkeitsanforderungen in Planungswettbewerben – SNAP) was developed. After examining which criteria are relevant for early design phases, 19 criteria grouped into 5 theme blocks were chosen (listed in Table 1) (Bundesministerium für Verkehr & Bau und Stadtentwicklung 2013).

To address the difficulties associated with the inadequacies of information during the early design stages, Abualdienen & Borrmann (2018) presented a Meta-Model for adding the intrinsic vagueness attached to geometric and semantic information involved with individual building elements during these early stages. They introduced a new classification term – Building Development Level (BDL) to describe the maturity of overall building models in five levels.

A main challenge in early design stages is to provide collaboration and communication between different stakeholders and domain-experts, while managing design information properly and avoiding redundancies and inconsistencies. Zahedi & Petzold (2019) introduced a novel minimized BIM-based machine-readable communication protocol, which works based on schematized information exchange requirements (templates) for different use cases.

In 2009, a German research project Mefisto developed a management system for implementing the partnership, process-driven and risk-controlled handling of construction projects. Mefisto integrates model-based information in a distributed multi-model space to support decision making in different disciplines as well as on different management levels of the project organization (Scherer & Schapke 2014).

Another project eeEmbedded (Collaborative Holistic Design Laboratory and Methodology for Energy-Efficient Embedded Buildings) proposed a new design control and monitoring system based on hierarchical key performance indicators supporting the complex design collaboration process. During the project an open BIM-based holistic collaborative design and simulation platform, a related holistic design methodology, an energy system information model and an integrated information management framework were developed (Scherer & Katranuschkov 2017).

Table 1. Design criteria and 15 sustainability criteria relevant in early design phases.

Topic	Nr	Type	Criteria
Design		design	Urban integration
		design	External space quality
		design	Building quality
		design	User and task-specific image
Functionality	01	additional	Accessibility
	02	additional	Public access
	03	additional	Barrier-free access
	04	additional	Spaces for social integration
Comfort and health	05	additional	Safety
	06	additional	Sound insulation
	07	basic	Daylight
	08	basic	Interior climate
Economy	09	basic	Spatial efficiency
	10	additional	Use flexibility
	11	basic	Life-cycle cost
Resources and energy	12	additional	Surface sealing
	13	basic	Materials
	14	basic	Energy demand
	15	basic	Energy demand covering

2.2 Design decision support systems

In recent years multiple Decision Support Systems (DSS) were generated to assist in the problem-solving process by combining quantitative data and qualitative knowledge/perceptions, processing information in order to present, compare, and rank potential alternatives, and, ultimately, selecting the one that meets the established decision criteria (Lu et al. 2007). DSS can improve decision makers' efficiency, productivity and effectiveness. It can also improve the communication between stakeholders and contribute to a quick problem solving (Jalaei et al. 2015).

Due to the lack of efficient and user-friendly visualization tool for BPS, Gadelhak (2017) proposed a single visualization tool, which can tackle a problem of both visualizing and analyzing a large amount of data, understanding the relations between different parts of the model and giving an overview of design variants and their consequences. Visualization dashboard guidelines were presented and finally a prototype dashboard was developed and evaluated with a case study.

There are also some examples of commercial tools supporting design decision making process, such as Autodesk Insight 360 which has a direct plug-in for Revit. Insight provides fast analysis of building energy and environmental performance throughout the building lifecycle and allows a direct comparison between different variants. It can visualize key performance indicators, benchmarks, factors, ranges, and specifications, and therefore helps to make better design decisions.[1] Another tool from Autodesk, Project Refinery, gives users the power to quickly explore, evaluate, and optimize their generative designs, using such visualizations like a parallel coordinate diagram.

2.3 Architect-friendly user interface

Building Performance Simulation (BPS) tools are mainly used by researchers, physicists and experts who value empirical validation, analytical verification and calibration of uncertainty. However, architects using those tools are much more concerned with the Usability and Information Management (UIM) of interface and the Integration of Intelligent design Knowledge-Base (IIKB). Those two factors are principles of architect-friendly user interface (Attia et al. 2009).

The term Usability incorporates better graphical representation of simulation input and output, simple navigation and flexible control. However, according to survey (Attia et al. 2009) graphical representation was a top priority for users, followed by the flexibility of use and navigation. For example, architects would like to see results presented in a concise and straightforward way. A visual format or 3D spatial analysis are preferred to numerical tabulation. Similarly, information management is becoming a growing concern for architects using BPS tools. There is a need for quality control of simulation input and the ability to evaluate alternatives quickly, accurately, and provide complete analysis for a design.

Supporting decision making under risk and uncertainty requires the Integration of Intelligent design Knowledge-Base in the tools. For architects it is important that tools can support sustainability design decisions and make detailed comparisons between different building designs. In order to make design decisions, the designer must have a higher level of knowledge and details. Therefore, it is essential that the simulation tools include an interface that supports such a knowledgebase (Attia et al. 2009).

Another survey from Wolff (2018) proved that most architects and engineers find a simple graphical representation of results as the most important feature regarding user-friendliness of the communication tool. Regarding information management the majority agreed on the creation of comparison overviews for variants of their choice as the most important feature for the communication tool. Finally, the interviewees were asked to point out the most important feature for the communication tool. Most participants agreed on the user-friendliness of the interface.

3 RESEARCH METHODOLOGY

3.1 Evaluation method

To begin developing an assistance tool, criteria for variant evaluation had to be chosen. Based on the SNAP

[1] https://www.autodesk.com/products/insight/overview

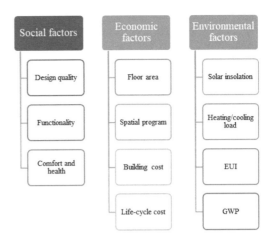

Figure 1. Selected criteria and criteria categories.

criteria and available simulation tools, the relevant criteria for early design phases were selected. To simplify the large number of criteria, they were divided into 3 categories according to the 3 pillars of sustainability. Qualitative criteria, which are as important as quantitative data, are also included in the assistance tool (Figure 1).

The first group of criteria – social factors – includes 3 criteria subgroups: design quality, functionality, and comfort and health, based on the SNAP criteria system. All the criteria in those groups, except for the daylight criteria, are based on subjective opinion and are therefore evaluated by the user. Daylight is represented by the daylight factor simulation outcome and it will be evaluated according to chosen sustainability certificate threshold (DGNB, LEED or BREEAM).

The economic factors include 4 criteria: floor area, spatial program, building cost and life-cycle cost. Floor area is a simple number which can be evaluated as a TOPSIS comparison, with the total floor area required by the client. The spatial program is calculated as a sum off points for each type of room required by the client. Points are given depending on the fulfillment of a minimum area for each type of room required in the spatial program. Building and life-cycle costs are both numeric outcomes and can therefore be compared to the maximum cost set by the client.

Environmental factors include 4 criteria: solar insolation, heating/cooling load, Energy Use (EUI) Intensity and Global Warming Potential (GWP). They are all results of simulations, which are possible thanks to BIM-integrated software Insight 360 or Oneclick LCA. They can all be compared with one another because they can be reduced to a simple number.

3.2 Visualization method

The main questions which must be answered before starting a visualization are what, why and under which circumstances will the visualization happen. The object of a visualization (what?) is defined by input data which are available, the goal (why?) is defined by the task, which must be solved by a user. Generally, the characteristics of the data to be displayed and the goals of the user are the most important influencing factors in the generation of visual representations (Lange et al. 2006).

The input data of the Archi-guide tool is determined by the criteria chosen in the previous chapter and the simulation results needed for each criterion. The goal is defined by the task that the user – an architect – wants to solve, which in this case is variant comparison and evaluation using analysis results. Characteristics of the user such as his cognitive abilities, experiences and preferences influence the selection of a visualization method. In this case, simulation results need to be easy to understand, as the architect is not an expert in every given field of analysis (Zahedi et al. 2019).

Context represents the background of the application, which in this case is the early design phases. As in project models at any design stage there are often elements and assemblies at various LOD, which describe the maturity of the elements inside the building model and not the overall building maturity, this system is not suitable to use in our tool. However, the BDL classification contains the information about quantity and quality regarding the design process of an entire building model. A building model at a certain BDL can include components with diverse LODs. Therefore, the authors of this paper used the BDL concept to illustrate the maturity of different design variants presented in the Archi-guide tool. As an input for our tool, models with BDL 3 and 4 will be used. BDL 1 and 2 are not suitable, as they do not provide enough information to conduct required analysis.

3.3 Objective and structure

The objectives to achieve an architect-friendly tool will be based on the surveys conducted by Attia (2009) and Wolff (2018). As in both studies, the graphical representation was mentioned as one of the most important features for architects, it will be also the main objective of Archi-guide to be a user-friendly tool through a high level of graphical representation and simplicity. In the same time, the tool must be flexible, so that the user can create comparison overviews for variants as well as compare 2 or 3 variants directly and see detailed information about one variant. They should also feel free to control the importance of each criteria to adjust to client's requirements or project context. The last, very important characteristic of an architect-friendly tool is the integration of a knowledge-base. As previously stated, architects cannot be experts in every discipline and need guidelines for example for building codes or rating systems compliance.

The next step determines the structure of the UXD, including interaction design and information architecture. The tool structure is divided into 5 main panels: Decision tree, Criteria overview, Variant comparison, Variant detail and Variant rating panels (Figure 2).

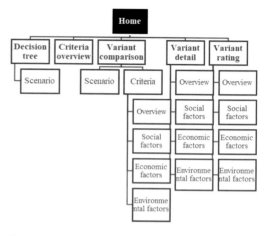

Figure 2. Structure of the tool.

4 IMPLEMENTATION

As a proof of concept, a preliminary prototype was developed. The prototype was built using PowerBI – a program that enables to transform data into visuals, visually explore and analyze data and create interactive dashboards and reports. As a case study, input data of student projects from BIM Seminar at Technical University of Munich in winter semester 2018/19 were used. Each student project was used as a variant in this project. Some of them were developed during this master thesis into the next level. Each variant was saved as an individual Revit model and needed simulations were conducted using Insight 360 and Oneclick LCA plug-ins. All numerical data were exported and combined in a single Excel table. Graphic data was saved in OneDrive and an embedded link to each image was saved in the table. The prototype was saved as a Power BI report and published online. It can be accessed by a link in any browser.

The tool is divided into 5 main panels. The user can switch between panels using the navigation bar located on the left side. Starting from the top, first Home button is located which links to the Decision tree panel. It is followed by buttons to Criteria overview, Variant comparison, Variant detail and Variant rating panels. On the bottom of navigation bar arrow button is located, allowing the user to go back to previous page.

4.1 Decision tree panel

The first panel visible for the user is the Decision Tree panel (Figure 3), as it gives an overview of all design variants and the development of the design process. Including the history of design in the form of a tree or a graph was one of the wishes expressed by professionals that were interviewed by Wolff (2018). From this level the user can switch between different view modes of the decision tree. In the basic view, the name of the variant is visible, and the color indicates which BDL

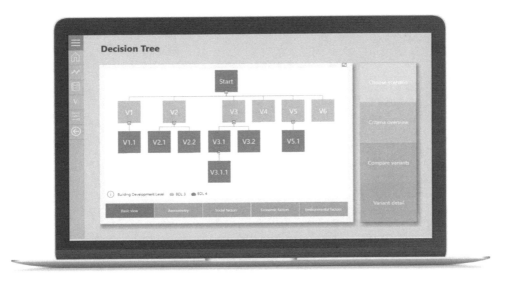

Figure 3. Home panel preview.

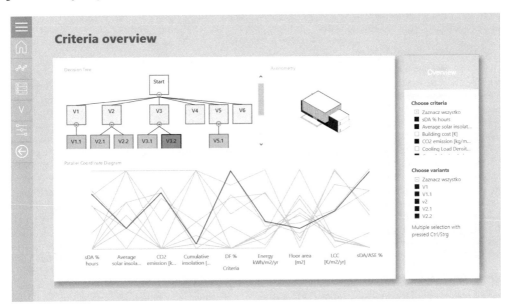

Figure 4. Criteria overview panel.

a variant has. The explanation of what BDL means is shown after clicking the info button in the bottom-left corner. Other view modes include axonometry view, social, economic and environmental factor. The axonometry view is especially important for architects to associate each variant with its design. The last 3 modes are designed to provide a quick overview of all design variants and their results for each criteria category. From the level of the Decision Tree panel, the user can go directly to Choose scenario tab, where he can select on a slider what weight each criteria group has. Unfortunately, the limitation of Power BI is that it is not possible to save user's choices into the source table and therefore it is not possible to show the overall results with real-time changes in the decision tree. Other actions available from the action panel are switching directly to Criteria overview, Compare variants and Variant detail panels.

4.2 *Criteria overview panel*

Detailed information and context are necessary for the decision-making process. The ability to examine and compare several aspects at the same time is also equally important (Gadelhak et al. 2017). For this purpose, a criteria overview panel is created. The criteria overview panel (Figure 4) allows to compare all variants with each other according to selected criteria.

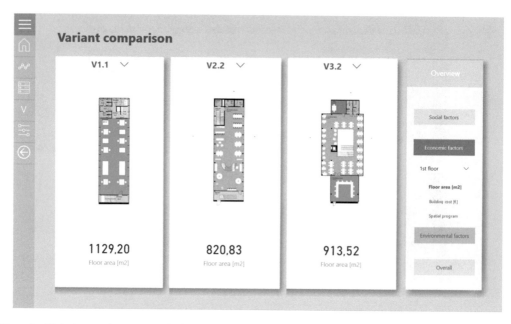

Figure 5. Variant comparison – economic factors.

A parallel coordinate diagram was selected, as it gives an overview of the performance of all variants at once. In contrast to the decision tree diagram, the user can check the exact values of every criteria and variant by simply hovering over it. In the action panel on the right side, the user can select which variants should be compared and which criteria should be shown. On the decision tree graph the user can select which variant should be highlighted. Adding a decision tree in this panel gives a better understanding which variant is selected (in variant V3.2 is selected). Additionally, an axonometry of chosen variant is shown to give a possibility to associate the variant with the respective design.

4.3 Variant comparison panel

Variant comparison gives the opportunity to compare three variants directly with each other. Among the users' wishes, stated by participants of the interview conducted by Carolin Wolff (2018), was also to include so called variant-cards. Therefore, in the variant comparison panel three variants can be compared through variant-cards, which recap and review all the essential info related to each design variant into a card. By using such variant-cards the architect can summarize and sort out his design variants (Sousa et al. 2019). Variant-cards enable direct comparison of both numerical and graphical information, which is essential in design decision-making process. The standard variant-card includes an axonometry of a 3D model and a gauge chart presenting the overall performance of each variant. On top of each variant-card the user can select freely from a dropdown list which variants to compare.

From the Variant comparison panel, the user can switch to the Choose scenario tab or Choose criteria tab. In the Choose scenario tab he can adjust the weights of each criteria category (Figure 6). The result of his choice influences the result of the overall score for each variant visible in the gauge chart. Here the user can also choose which sustainability certificate should be used for benchmarking (DGNB, LEED or BREEAM). In the Criteria tab, the user can select which criteria should be compared with each other. For each criteria category (Overview, Social, Economic and Environmental factors) a separate button is displayed on the action panel. After clicking each button, the criteria list for each group is displayed and the information on each variant card is adjusted. For each criteria category the result for the group is presented in a gauge chart.

The overview button gives access to the floorplans and sections of the building, which are presented as a simple black and white orthogonal drawings. These are the basic architectural means of representing a design and therefore it might be the first information which an architect looks for to understand the building. In the social factors tab, the user can first compare the rating of each criteria group (Design quality, Functionality and Comfort and health). It is also possible to directly compare perspectives of each variant as well as the results of daylight factor analysis presented in the color-coded plan. For economic factors a color-coded use plan is presented (Figure 5). It is possible to switch between different floors of the building using a dropdown list in the action panel. Information about floor area, spatial program and building costs can be directly compared. For environmental factors the results of solar insolation analysis are shown as a color-coded

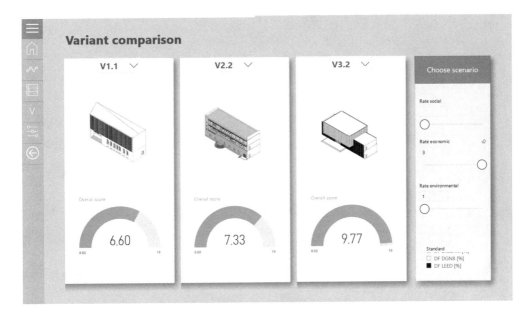

Figure 6. Variant comparison panel – scenario selection.

Figure 7. Variant detail main panel.

axonometry and a number for the average solar insolation. Moreover, Global Warming Potential can be displayed and compared as bar charts.

4.4 Variant detail panel

In the Variant detail panel, the user can see all the available information about one variant. The variant can be chosen from a dropdown menu in the top-right corner. In the action panel the user can select which criteria category should be shown in detail. All the visuals can be zoomed into a full screen mode by clicking an icon in the top-right corner of each visual.

On the first Variant detail panel the user can see a 3D viewer on the left and on the right a score for each criteria category and an overall score displayed as a gauge chart above (Figure 7). The used 3D viewer is a free of charge html viewer downloaded from the Power BI Marketplace. It can display any content provided in an URL link. Therefore, all variants' models were uploaded as IFC files to an online free of charge cloud service called Modelo (www.modelo.io) and

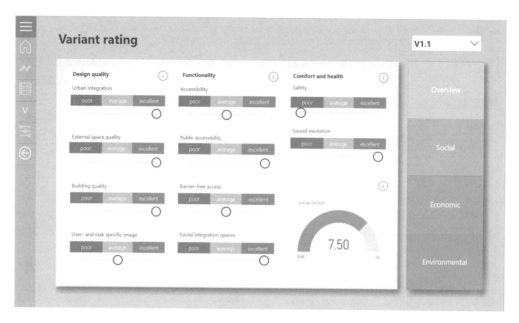

Figure 8. Variant rating panel.

embedded link for each of them was created. The user can load and interact with the model directly in the dashboard. It is possible to freely move the model around, rotate it and zoom in and out. Additionally, the user can switch to a predefined Walk-through mode in the bottom-left corner. Generally, the viewer is compatible with most of the browsers except of Mozilla Firefox.

In the overview tab the basic orthographic drawings are shown instead of a 3D model viewer. On the left a simple floor plan is presented. The user can switch between different floors by using a dropdown list on top of the plan. For better understanding an axonometry drawing is also presented.

The social factor tab shows the results for all qualitative criteria groups as well as simple perspectives and the result of daylight factor analysis. A color-coded floorplan displays how many percent of the time the daylight factor is above the minimum of 2% required by all sustainability certificates. Additionally, a bullet chart shows the overall score for daylight factor with a target required by a rating system, which can be chosen in a dropdown list. An info-button, which is a part of a knowledge-base, displays a pop-up window with information about daylight factor and a legend for the floorplan. The economic factors tab presents all the numerical data, such as floor area, building cost and LCC, as well as graphs representing data of the spatial program. An interactive floorplan, created with Synoptic Panel visual, shows with colors information about different rooms, which are in parallel presented in a donut chart. After hovering over each room on the floorplan, an information about its name and area is shown. Information about the spatial program is shown as a bullet chart with targets specified by client's requirements.

The environmental factors tab shows information about Global Warming Potential, insolation, EUI and heating/cooling loads. All the graphics include an info-button, where user can read about each criterion, as understanding the result requires special knowledge. Additionally, all the numerical data presented in bullet charts or gauge charts include a color scale indicating with green color what is a satisfying result.

4.5 *Variant rating panel*

In the variant rating panel, the user can rate all the qualitative criteria for each variant. Each criteria group (Design quality, Functionality and Comfort and health) has a separate card with an info-button (Figure 8). The user can rate criteria in a 3-grade scale using a slider. For each grade a color was used to indicate the meaning of the result (red for poor, yellow for average and green for excellent). A guideline for the meaning of each grade is given after clicking the info-button. For each criteria a short description and a list of example requirements for each grade is presented. The grades are based on the number of indicators that are fulfilled. As the indicators are mostly qualitative the user should decide how well each variant is fulfilling the requirements. In order to do that he should be familiar with the rules listed in the information pop-up window as well as with the variant itself. That is why he can switch directly from this level to Variant detail panel criteria categories.

The grades the user gives are calculated into the overall grade for social factors and can be visible in the gauge chart in the bottom-right corner. Unfortunately, as Power BI cannot save the choices of a user from the dashboard, it is not possible to use them in other panels.

In order to enable the prototype to work smoothly, all the variants were pre-evaluated.

5 EVALUATION AND FEEDBACK

In the end of the project an online user evaluation was conducted. 20 participants took part including architects, architecture students and engineers working in AEC industry.

In one part of the survey participants had to solve a short task using the Archi-guide tool. In the first task they were advised to use the decision tree panel and afterwards the parallel coordinates diagram. Then they were asked which visualization method gave a better overview of the development of all variants from the beginning of the design process. The answers were almost equally distributed between both visualization methods. It is worth mentioning that our initial presumption was that the architects may prefer the decision tree visualization method for being more abstract and less complex. Nonetheless, apparently many architects are already familiar with the parallel coordinates diagram and prefer to capture all the details in one shot.

In the next 3 tasks participants were advised to use different options of the tool and read some information from it. The purpose of these tasks was to prove, if the users understand how the tool works and if they can read the information correctly. For every one of these tasks there was only one correct answer. The results were quite satisfying with 80% correct answers in the first task, 90% in the second and 95% in the last question.

The feedback results showed that generally users were rather satisfied with the tool. The tool was ranked as very "interesting" and "valuable". Also, "useful", "organized", "innovative" and "creative" got nearly 4 out of 5 points. Interestingly, engineers found this tool as most interesting, as they all rated it with maximum grade. Architects in contrast rated the tool as more organized, understandable, easy to learn, friendly and clear. Students found this tool as the friendliest from all the groups.

Surprisingly we also received feedback from professors and experts suggesting the possibility of using such a tool to present the design process and variants detailing/development to the client and other stakeholders, in order to better justify, explain or legitimize the decisions made by the architect during the design process. This was not an intended purpose of the tool during its development phase, but the experts saw its potential and suggested it for representing and legitimizing the design process.

6 CONCLUSION

There is a growing gap between architects as users and BPS tools, as they are not compatible with architects' needs. The goal of this work was to bridge this gap by developing an architect-friendly assessment tool for variant evaluation. First, the architectural design process, decision-making process and existing decision support systems were presented. In the next part the features of an architect-friendly tool were defined. Criteria relevant for early-design stages were chosen and simplified according to 3 pillars of sustainability into social, economic and environmental factors. Both qualitative and quantitative criteria were included to deliver a holistic result. As proof of concept, a template-based visualization tool called Archi-guide was put forward and tested with a sample project. The tool includes multiple panels giving freedom in terms of exploring the building variants' data. The developed variants in the design process can be visually compared in different ways which supports fundamental decisions. In the end, the tool was evaluated in the online user study. The results of the study demonstrate that the tool supports decision making in early design stages and educates architects in terms of sustainability of buildings, through integration of a knowledgebase in form of info buttons.

Our system concept, approach and the mockup implementation integrate and combine all the design aspects in one tool (regarding criteria from difference origins and essence) to make them all available at once for the architect. Using such a comprehensive tool eliminates the need of switching between different programs. As a proof of success, this work received 2nd Prize in 'Auf IT gebaut' competition which awards the most promising works in the field of digitalization in construction industry.

REFERENCES

Abualdenien, J. & Borrmann, A. 2018. Multi-LOD model for de-scribing uncertainty and checking requirements in different design stages. In: 12th European Conference on Product and Process Modeling (ECPPM 2018).

Attia, S., Beltran, L., Hensen, J.L.M. & Herde, A. de 2009. Architect friendly: A comparison of ten different building performance simulation tools. In: 11th IBPSA Building SimulaTion Conference: 204–211.

Bundesministerium für Verkehr, Bau und Stadtentwicklung 2013. Systematik für Nachhaltigkeitsanforderungen in Planungswettbewerben.

Carolin Wolff 2018. Development of a Computer-aided Framework to Facilitate Collaboration Between Architect and Engineer in Early Design Stages: A Proposal for Integrated Design Planning. Master of Science (M.Sc.) Thesis, Technical University of Munich.

Fuchs, M., Hartmann, F., Henrich, J., Wagner, C. & Zeumer, M. 2013. SNAP Systematik für Nachhaltigkeitsanforderungen in Planungswettbewerben-Endbericht.

Gadelhak, M., Lang, W. & Petzold, F. 2017. A Visualization Dash-board and Decision Support Tool for Building Integrated Performance Optimization. Proceedings of the 35th eCAADe Conference 1: 719–728.

Jalaei, F., Jrade, A. & Nassiri, M. 2015. Integrating decision support system (DSS) and building information modeling (BIM) to optimize the selection of sustainable building components. Journal of Information Technology in Construction (ITcon) 20: 399–420.

Kibert, C.J. 2016. Sustainable construction: green building design and delivery. John Wiley & Sons.

Lange, S., Nocke, T. & Schumann, H. 2006. Visualisierungsdesign – ein systematischer Überblick. In: Simulation und Visualisierung (SimVis 2006): 113–128.

Lawson, B. 2006. How designers think: The design process demystified. Routledge.

Lu, J., Zhang, G., Ruan, D. & Wu, F. 2007. Multi-objective group decision making: Methods, Software and applications with fuzzy set technology. Imperial College Press, London.

Scherer, R.J. & Katranuschkov, P. 2017. eeEmbedded – D10.400 Final Project Report. https://cordis.europa.eu/project/id/609349/reporting

Scherer, RJ & Schapke S-E 2014. Informationssysteme im Bauwesen. VDI-Buch. Springer Vieweg, Berlin.

Schlueter, A. & Thesseling, F. 2009. Building information model based energy/exergy performance assessment in early design stages. Automation in Construction 18:153–163. https://doi.org/10.1016/j.autcon.2008.07.003

Sousa, J.P., Henriques, G. C. & Xavier, J.P. 2019. eCAADe + SIGraDi 2019: Architecture in the Age of the 4th Industrial Revolution, vol. 1.

Zahedi, A. & Petzold, F. 2018. Seamless integration of simulation and analysis in early design phases. In: Proceedings of the IALCCE 2018: The Sixth International Symposium on Life-Cycle Civil Engineering, Ghent, pp 2007–2015.

Zahedi, A. & Petzold, F. 2019. Adaptive Minimized Communication Protocol based on BIM. In: 2019 European Conference on Computing in Construction.

Zahedi, A., Abualdenien, J., Petzold, F. & Borrmann, A. 2019. Minimized communication protocol based on a multi-LOD meta-model for adaptive detailing of BIM models. In: Proc. of the 26th International Workshop On Intelligent Computing In Engineering.

Navigating the vast landscape of spatially valid renovation scenarios

A. Kamari, B. Li & C. Schultz
Department of Engineering, Aarhus University, Aarhus, Denmark

ABSTRACT: Renovation, as a design task, can be defined so as to change building elements – repair, replace, add, remove, refurbish – in order to meet a set of given performance criteria (key performance indicators, KPIs). As a search and optimization task, the set of all ways that a building can be changed to maximize performance criteria is enormous. The aim of this paper is to present a new approach, and prototype search engine, that enables designers to explore aspects of geometric and spatial changes to a building under renovation in addition to a large variety of non-geometric renovation options, e.g. adjusting the dimensions and positions of windows. Our new renovation scenario search engine automatically identifies spatial inconsistencies, such as: windows stretching so far that they overlap each other or go beyond the embedding wall; added walls physically intersecting (clash detection), etc. We evaluate our method and prototype system on a large scale residential building in Denmark.

1 INTRODUCTION

Renovating the existing building stock holds massive potential to meet EU's targets for mitigating climate change, energy efficiency, and large-scale sustainable development (BPIE 2011). The initial aim for renovation is consistently set to fulfill energy-saving measures, due to the many forms of national and international energy policies and subsidies (EU 2016). However, building renovations can substantially influence or improve user comfort, health, aesthetics, safety, and well-being of the residents in a broader perspective (Kamari et al. 2017a).

The renovation design task is as follows: given a built environment, we can repair, replace, remove, refurbish, modify and add building facilities and architectural elements, in order to meet a set of given performance criteria (key performance indicators, KPIs). The particular details of exactly what can be repaired, replaced, etc., and how, varies drastically between renovation projects. Thus, the design team requires a suitable "language" to express the particular renovation choices available, and this language should be formal (i.e. unambiguous) so that software tools can be developed that can read and process such renovation information. In addition, as a search and optimization task, the set of all ways that a building can be changed to maximize performance criteria is enormous (e.g. the case study for our empirical evaluation has a scenario space in the order of 5.7×10^{28}), and thus numerous decision support tools have been developed that aid designers in exploring the space of renovation options.

In this paper, we develop a new approach, and prototype search engine, that enables designers to explore aspects of geometric and spatial changes to a building under renovation in addition to a large variety of non-geometric renovation options, e.g. adjusting the dimensions and positions of windows, removal or addition of certain (non-load bearing) walls, adjusting the height of ceilings, and so on. Specifically, we extend the Renovation Domain Model (NovaDM) to include geometric and spatial changes of building features. NovaDM is a formal (logic-based) domain-specific language for expressing and capturing key concepts for renovation alternatives, for the rapid constraint-based generation of holistic renovation scenarios during the early design stages. Still, it does not currently consider physical alterations to buildings (Kamari et al. 2018a). Our new renovation scenario search engine automatically identifies spatial inconsistencies, such as windows stretching so far that they overlap each other or go beyond the embedding wall, added walls physically intersecting (clash detection), etc.

A key research challenge we address in this paper is how to rapidly explore the enormous search space by efficiently "pruning" regions of the search space that represent spatially invalid options. Within the set of spatially consistent renovation scenarios, our system generates optimal scenarios that maximize/minimize combinations of KPIs within the context of multi-objective optimization. We have implemented our search engine in Answer Set Programming (that has an in-built optimization search feature) that we have extended to natively support fast and efficient spatial reasoning. We evaluate our method and prototype system on a large scale residential building case study in Denmark.

This research builds upon our previous work (Kamari et al. 2019a,b,c) in developing a BIM-based decision support system, named PARADIS (Process integrAting RenovAtion DecIsion Support), which generates optimal and sustainable renovation scenarios, and evaluate them for multiple KPIs (i.e. investment cost, energy consumption, indoor thermal comfort). PARADIS is being developed for rapid constraint-based generation and evaluation of holistic renovation scenarios during the early design stages, tailored towards the renovation of dwellings in Denmark.

2 SPATIAL AND GEOMETRIC CHANGES IN THE CONTEXT OF RENOVATION

Depending on several factors (e.g. building's age, conditions, lifespan, function, client's budget, etc.), renovation scenarios are developed for a given building within four distinct levels:

A) *Urban fabric*, which is relevant to the form of a city, the correlation between public and private regions and areas, and the quality of the built environment;
B) *Building form*, which refers to the shape or configuration of a building;
C) *Architecture/interior layout*, which is the relationships between rooms, spaces, and physical features at one level of a building;
D) *Individual elements*, which are the main components of a building.

For the renovation of dwellings in particular, which is the subject in this paper, changing both the architecture layout (i.e. merging two bedrooms, relocating the kitchen, increasing the toilet dimensions, adding/resizing the balcony) and building elements (i.e. replacing windows, exterior and interior walls) are often suggested, whereas re-designing the urban fabric or building form is not very common because it increases the renovation cost significantly. However, it is observed that changes in urban fabric and building form happens when there is a demand to either change the building function or deep architectural transformation of the built environment. Another factor is the renovation intervention level (renovation *depth*): renovation scenarios can be developed as *Minor, Moderate, Deep,* and *Nearly Zero Energy Building* – NZEB (TECNALIA 2015).

In this paper, spatial and geometric changes are considered related to the categories C and D. In light of this, Figure 1 lists these renovation strategies according to (Kamari 2018; Kamari et al. 2017b,c) based on the study in (Boeri et al. 2015).

2.1 *Spatial and geometric arrangement with respect to existing renovation alternatives*

Spatial arrangement is regarded by discovering likely places and dimensions for a set of interconnected building elements that fulfill all design requirements and increase design performance related to the design intentions. Spatial arrangement is relevant to all physical geometrical design problems, and in the renovation context, it can be regarded as meaningful renovation

A) Replacement of existing elements or their integration with new components aimed at modifying the existing conditions and ensuring a superior level of performance and functionality.

B) *Box in box* approach, where one or more box-shaped volumes inside an existing building are used to indicate building elements (inside a box) where the opportunity to operate is very limited.

C) Building envelope implementation, wrapping up the building in a new envelope.

D) Models of volumetric additions that vary depend on adjusting dimensions and layout concerning the original geometry and features of the existing building envelope and form.

Figure 1. Renovation strategies (A-D) according to (Kamari 2018) based on the study in (Boeri et al. 2015).

alternatives and approaches fulfilled for that purpose. Following the previous work by the authors in (Kamari et al. 2018b) and based on (Baker 2009; Boeri et al. 2015; Burton 2012), a list of 32 renovation alternatives including 139 renovation approaches, and about 3831 renovation actions were identified and presented (the definition of terminologies are presented in Section 3.1). Table 1 presents nine renovation alternatives out of 32, which we choose to work within this paper due to their link with the four spatial and geometric renovation strategies presented in the previous section. In the following, we elaborate on each alternative.

Exterior walls depend on their types (i.e. solid walls or cavity walls) are either being *upgraded* by adding layers of insulation from outside and inside or being *replaced* with new solid or curtain/glass walls, based on the clients' demands.

Interior walls are often *added, removed,* or *replaced* depends on the architectural layout and clients' demands. Adjusting interior walls are highly demanded to create bigger bedrooms and living room, or in case of rearranging interior spaces such as relocating the kitchen. Adjusting interior walls play a significant role in creating new spatial arrangements.

Table 1. Renovation alternatives in terms of spatial and geometric arrangements.

Exterior walls	1
Interior walls	2
Curtain walls	3
Roof	4
Windows	5
Exterior doors	6
Interior doors	7
Balcony	8
Solar shading	9

Curtain walls are usually demanded to keep and *upgraded* in an ordinary renovation. Rarely, they are *replaced* with solid walls.

Roof is usually *upgraded* by adding an insulation layer from inside since it is cheaper and requires less time construction time. However, roofs can also receive insulation from outside, or new tiles and final finish as part of the renovation scenario. Rarely they are *replaced* with new solid or new green roofs.

Windows are often *replaced* with double or triple glazing-highly efficient new windows and different framing materials (uPVC, fiberglass, etc.). Clients highly demand it to relocating and enlarging existing windows in their renovation scenario.

Exterior doors or entrance doors are always *replaced* with new doors of the same size and different materials (e.g. wood, steel, glass, etc.).

Interior doors are usually *replaced* with new doors, and relocated together with the interior walls.

Balcony is a trendy alternative, and given a building, it can be *upgraded* or *added* newly. It is highly demanded to upgrade balconies as volumetric addition to the buildings. Depends on the structure (concrete or steel) and facade types (solid or curtain wall), it can be made by steel and concrete.

Solar shading is often *added* to buildings to keep the sun's solar radiation out of the building during the summer period in order to avoid overheating. Depends on the structure (concrete or steel) and facade types (solid or curtain wall), it is usually made by steel or concrete.

2.2 Spatial and geometric design objectives

There is a wide range of benefits that can be obtained as an outcome of a holistic and sustainable renovation to higher energy performance standards. Many are tangible and possible to quantify, while others are less so and may be difficult to allocate a monetary value. Kamari et al. (2017a) made an effort and addressed sustainability in its full sense for renovation from three categories, including *Functionality, Accountability,* and *Feasibility* (18 sustainable value-oriented criteria, and 118 sub-criteria, have been identified). Along with developing PARADIS (as briefly described in section 1), we have been working on extending a version of an hourly dynamic simulation tool ICE-bear (Purup et al. 2017)

Table 2. Outline for the selecting KPIs for evaluation of the generating renovation scenarios.

Energy consumption (Danish Energy Agency 2017)	Reduction of energy consumption for heating measured in kWh/m2/year [less better]
Discomfort hours above 27 and 28 (°C) (Danish Building Research Institute 2017)	Number of hours [less better]
Indoor Air Quality – IAQ (Dansk Standards 2013)	% out of Class III according to EN 15251 [less better]
DF (daylight factor) (VELUX 2016)	0 < DF < 5 [bigger better]
Daylight requirements according to BR18 (Danish Building Research Institute 2017)	% >= 10 [bigger better]
View-out quality	% of openings area on facade regarding adjacent buildings [**proportional** to opening area; client dependent]
Degree of privacy	% of openings area on facade regarding adjacent buildings [**inversely proportional** to opening area; client dependent]
Degree of Satisfaction (Xu et al. 2018)	% regarding indoor thermal comfort & air quality [bigger better]
Health & Well-being, (Norback et al. 2014)	% regarding Energy improvement, indoor thermal comfort, air quality and their effects on Asthma, Allergy, and Eczema diseases [bigger better]
Investment cost (Molio 2016)	Price of the procurement in DKK (Danish Krone) [less better]

to evaluate several user-specified KPIs on rapidly generating renovation scenarios. Table 1 summarizes the evaluating KPIs for generating renovation scenarios in PARADIS.

In the framework of spatial and geometric arrangement and their relation with the existing renovation alternatives described in Section 2.2, all of the ten listed KPIs in Table 2 will get affected. For example, adjusting the dimension of windows and openings can improve or worsen daylight, indoor air quality, thermal comfort, energy consumption, view-out quality, degree of privacy, etc.

3 METHODOLOGY FOR EXTENDING NOVADM TO INCLUDE GEOMETRIC AND SPATIAL CHANGES OF BUILDINGS FEATURES

In this section, we present the components of our system and elaborate on the novel extensions for

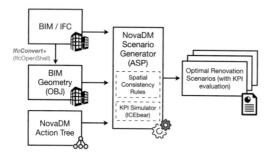

Figure 2. Spatially consistent scenario generation workflow.

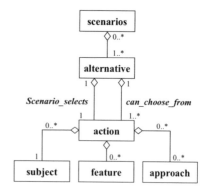

Figure 3. The relationships between concepts in NovaDM.

supporting spatial reasoning. Figure 2 illustrates our workflow and system pipeline.

The geometry of an IFC model is generated into our custom OBJ-like format using our modified version of IfcConvert (IfcConvert+) from the IfcOpenShell project (2020). The IFC model, OBJ geometry and a user's Action Tree describing the set of renovation scenarios (presented in Section 3.1) is given as input to our *NovaDM* optimal scenario generation engine implemented in Answer Set Programming (ASP) that includes spatial consistency rules and an external KPI simulator to evaluate scenarios (ICEbear). A set of n KPI-optimal scenarios is generated along the pareto front (in the context of multi-objective optimization), where the user defines the number of optimal scenarios to generate, n.

3.1 NovaDM – Renovation Domain Model

The *Renovation Domain Model* (NovaDM) is a structured collection of renovation concepts and relationships between those concepts, similar to the notion of a "schema" (Kamari et al. 2018a, 2019a). Our contribution in this paper is we **extend NovaDM to support renovation scenarios that change the geometry of a building**. NovaDM concepts are represented as classes. Figure 3 presents a Unified Modelling Language (UML) class diagram of the six classes in NovaDM that are described below. Using standard UML class diagram notation, boxes represent classes, and lines connecting boxes represent relations between classes. Lines with a diamond symbol represent aggregation (part-whole) relation between two classes with the relative multiplicity of each class to the other annotated on the ends of the arrow. For example, an *action* consists of exactly one subject, and a *subject* is part of zero to many *actions*.

Renovation Subject: the object that is the target of renovation, e.g. windows, doors, etc.

Renovation Approach: the way that the subject will be renovated, e.g. repair, replace, remove, refurbish, or modify.

Renovation Feature: the part of the subject being renovated, e.g. window frame material. Features have types (e.g. material) and values (e.g. fiberglass, uPVC, etc.).

Renovation Action: an instance of a subject, an approach, and a set of feature types and values, e.g. "window frames will be replaced with fiberglass material".

Renovation Alternative: the set of (mutually exclusive) actions for a given subject, e.g. all the ways that windows can be renovated according to the specified approaches and features.

Renovation Scenario: a set of actions such that each action belongs to a different alternative, e.g. one action for windows and one action for floors.

To compactly specify the set of renovation scenarios applicable to a given project, NovaDM defines *Action Trees*. Figure 4 illustrates an example of an Action Tree for windows. Each node in an Action Tree is annotated with zero or more *subjects*, *features* and *approaches*. Action Trees consist of *And* nodes (+) and *Xor* nodes (−), and scenarios are represented by traversing the Action Tree from the root node as follows: if an *And* node is visited then all children must be visited, and if an *Xor* node is visited then exactly one child must be visited. The aspects (features, subjects, approaches) annotating each visited node are collected together, and when there are no more nodes to traverse, then the aspects collected define the renovation scenario. The example tree in Figure 4 defines six scenarios, e.g. one such scenario illustrated in red is: subject = window, approach = refurbish, feature = plastic (material), feature = double (glazing), feature = fixed (mechanism).

As a visual notation shorthand, we illustrate a set of mutually exclusive aspect values as follows, which expands to the three mechanism (feature) children nodes illustrated in Figure 4:

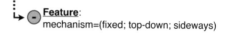

3.2 Geometry-modifying actions in NovaDM

Based on the types of geometry-modifying renovations presented Sections 2 and 2.1, we have extended NovaDM to support the following renovation actions

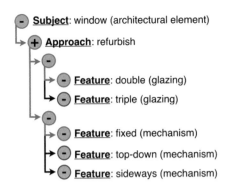

Figure 4. Example of an *action tree* in NovaDM.

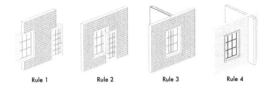

Figure 5. Examples of spatial consistency rule *violations*.

that can apply to building elements that impact the physical structure of a building:

- Modify physical dimensions of an element;
- Modify the position of an element;
- Add/remove an element.

While these modifications can apply to any building element in NovaDM, in this paper we focus on spatial modifications to doors, windows, walls, and balconies based on the cases identified in Section 2 and 2.1.

Renovation Feature: we define three new spatial features *modify_width*, *modify_height*, *modify_thickness* that are assigned a value *delta* specifying the change in the corresponding spatial feature, where the *subject* must be a wall, door, or window element, or group of elements. We define *translate_position* that either (a) translates windows and doors along the walls they are embedded in (i.e. their "host" walls) or (b) translates interior walls by a 2D vector in the horizontal plane, including all embedded elements. These features can only be selected together with the *Renovation Approach* of *modify*. For each of these features, the user specifies a set of discrete spatially modifying options, e.g. [*modify_width* = (0.25; 0.5; 0.75; 1.5)]. A scenario can consist of any combination of these features applied to a single object, and can consist of multiple spatial modifications applied to different objects, e.g. a scenario may increase the dimension of a window by 0.25m and translate it along its host wall by 0.5 m in combination with various other spatial modifications to other windows, doors, and walls.

Renovation Approach: *modify* and *remove* are already supported, although we now extend the semantics of *modify* to include new features for spatial modification as presented above. We also introduce a new approach *add*: when this is selected in the definition of a renovation scenario, the subject and feature described by the corresponding subtree is added to the building. Subjects must be walls, doors, or windows.

3.2.1 *Spatial inconsistencies*
Given a building model and a set of candidate geometry-modifying actions, not all combinations of those actions are physically realizable, i.e. renovation scenarios may now be spatially *inconsistent*: newly added walls might clash with each other or existing walls; windows could be translated beyond the boundaries of their host wall; the dimensions and position of doors and windows could be modified so that they clash with each other, and so on.

Thus, we define *rules* that must be satisfied by renovation scenarios that ensure spatial consistency (see Figure 5). In all prominent BIMs such as IFC windows and doors are modeled by first modeling an *opening* object that *voids* the host wall, and then the door/window component *fills* the opening. Thus, we define spatial consistency rules with respect to their *openings* in order to capture problematic cases for all combinations of doors and windows.

Rule 1. Openings must be spatially *part of* their host elements, e.g. an enlarged window/door cannot exceed the boundaries of its host wall.

Rule 2. Openings must be spatially *disconnected* from all other openings e.g. a relocated door should not intersect with a neighboring window.

Rule 3. Openings must be spatially *disconnected* from all elements, except the element intended to *fill* the opening, e.g. a relocated interior wall cannot overlap with an exterior door.

Rule 4. Walls must not spatially *overlap* with any non-embedded element, e.g. the interior of two walls must not intersect (they can have external contact, i.e. touch at their boundaries).

3.3 *Spatially consistent Scenario generation in answer Set programming*

We have implemented NovaDM scenario generation using the logic programming language Answer Set Programming (ASP) (Brewka et al. 2011). The novelty in this paper is that we extend our implementation to **efficiently prune spatially inconsistent renovation scenarios** from the scenario generation process; this relates to our previous research on extending logic programming to natively support spatial reasoning (Bhatt et al. 2011; Li et al. 2019; Walega et al. 2015; Walega et al. 2017). Furthermore, we have connected ASP to a simulation tool ICEbear that evaluates the Key Performance Indicators for a building listed in Table 2, such as energy performance, thermal comfort, average yearly daylight, and so on. This enables us to use the ASP search engine to generate **optimal spatially consistent scenarios** on the pareto front in the context of multi-objective optimization.

Figure 6. Example action trees with geometry-modifying features.

ASP is a Knowledge Representation and Reasoning programming language that has its foundations in first-order logic (Brewka et al. 2011). Similar to Prolog, ASP has a knowledge base of facts and rules of the form "*Head :- Body*" meaning that, if the *Body* is true, then the *Head* must also be true. Rules with no *Head* are ASP *integrity constraints*, written ":- *Body*" meaning that the *Body* must be false. Choice rules enable one to express that the "*Head*" may, or may not, be deduced, written "{*Head*}:- *Body*", enabling us to define *combinations* of valid deduced facts. ASP reasoning engines are specifically designed to rapidly find combinations of deduced facts that are consistent with all given domain rules (referred to as *answer sets*), and can perform multi-criteria optimization by minimizing cost valuations that are assigned to deduced facts.

We have implemented our renovation domain model NovaDM as ASP rules, integrity constraints, and choice rules such that valid combinations of deduced facts correspond to complete renovation scenarios. We encode project-specific renovation options as ASP facts and implement our spatial consistency rules as ASP rules and integrity constraints. We specifically use the ASP solver Clingo (Gebser et al 2016).

Example. Consider the BIM models illustrated in Figure 5 (Rule3, Rule4) with an exterior wall, a window, and an interior wall, and the Action Tree in Figure 6. If we omit our spatial consistency rules, then our system generates 111 scenarios, each representing a distinctive renovation scenario that includes replacing or repairing windows, and replacing, removing, or repairing interior walls. When we include our spatial consistency rules, then our system generates 28 spatially consistent models involving removing the interior wall or translating both the window and the interior wall so that they are not *externally connected*.

We implement our scenario generator in ASP in the following steps.

Step 1. We identify relevant object types (IfcWall, IfcWindow, IfcOpening) and relationships (IfcRelVoidsElement, IfcRelFillsVoids) from the IFC model and declare them as *facts* in the ASP knowledge base.

   wall(id("0HGvCQAnX4UPtrSTt1ippL")).
   wall(id("2T8Dr8CdD1EBqUHFm0xQ$g")).
   opening(id("0HGvCQAnX4UPtrSSx1ippM")).
   window(id("0HGvCQAnX4UPtrSTt1ippM")).
   fillsVoids(id("0HGvCQAnX4UPtrSTt1ippM"),
      id("0HGvCQAnX4UPtrSSx1ippM")).
   voidsElement(id("0HGvCQAnX4UPtrSSx1ippM"),
      id("0HGvCQAnX4UPtrSTt1ippL")).

Step 2. We extract object geometries into an OBJ-like format (using IfcConvert+) and compute their new representation if they undergo any dimensional or positional changes.

new_representation(Window, NewRepr) :-
   window(Window),
   representation(Window, OldRepr),
   scenario_feature(translate_position, Translation),
   shift(OldRepr, NewRepr, Translation).

Step 3. We check whether the generated renovation scenarios are spatially viable for the given BIM instance with *integrity constraints* in ASP.

%%% Rule 1: openings must be part of host element %%%
   :- opening(Opening), wall(Wall),
   voidsElement(Opening, Wall),
   new_representation(Opening, ReprO),
   new_representation(Wall, ReprW),
   part_of(ReprO, ReprW).

The 28 spatially consistent scenarios that our system generates entail changes in the parameters of the BIM instance (U-value, window dimensions, wall areas), and thus each scenario performs differently with respect to KPI values, which are evaluated using the ICEbear simulator. The KPI outputs from ICEbear are then fed back into the ASP search engine in order to generate **optimal** renovation scenarios that maximize different combinations of KPIs. This is implemented in ASP as follows:

new_parameter(window_U_value, 1) :-
   scenario_feature(window_glazing, double).
   #minimize{Score: kpi(energy_consumption, Score)}.

4 CASE STUDY

In this section, we evaluate our system on a real BIM project, including real geometry-modifying actions that were considered in the renovation project. The purpose of the case study is to demonstrate that our system can rapidly generate and evaluate optimal spatially consistent scenarios within a reasonable amount of time. The selected case (Figure 7) is the renovation of a multifamily residential building block A17 located in Aarhus, Denmark. The building is a part of a dwelling area consisting of nine identical building blocks built-in 1967–1970. A17 is a multifamily residential building block, composed of 32 apartments (8 apartments per floor, and total 4 floors) with a similar layout, total heated floor area of $2700\,m^2$, 340 doors, and 129 windows.

The action tree considered includes 12 renovation subjects, 60 renovation approaches, and 66 renovation features resulting in 5.7×10^{28} renovation scenarios targeting at replacing, refurbishing or modifying walls, roofs, floors, ceilings, and doors. The action tree further introduces 35 geometric changes involving interior doors (translation, width modification) of

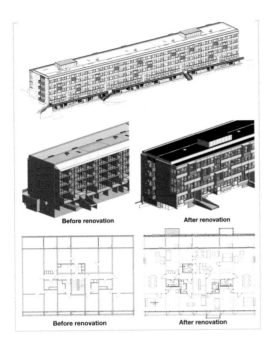

Figure 7. Residential building block A17 in aarhus, denmark used in the case study.

Table 3. Statistics of the BIM used in our case study and scenario generation runtimes.

Number of objects	3936
Number of 3D meshes	886
Generating N optimal scenarios ($N = 1, 3, 10, 20$)	29.7 seconds (average total time for $N = 1, 3, 10, 20$)

which only three are spatially consistent as doors must be contained in their hosting walls.

Our spatial consistency check then only retains renovation scenarios that are spatially viable and uses the derived model parameters to simulate 8 KPIs such as energy consumption, discomfort hours, indoor air quality, daylight factors, etc. Table 3 shows statistics of the extracted first floor of the Danish residential building BIM and runtimes for selecting optimal scenarios in terms of energy consumption.

The results show that our system performs within a practical runtime on large, real-world BIMs with large action trees, by finding the first optimal scenario and subsequent optimal scenarios in an average of 29.7 seconds. To clarify, to generate $N = 20$ optimal renovation scenarios (models) takes approximately 30 seconds. In fact, only 8.6% of all renovation scenarios are spatially consistent, which enables a total of 1680 combinations of model parameters. In our experiments we determined that finding N optimal models using ASP for some N (e.g. N = 1,3,10 or 20), always requires computing 144 ICEbear simulations and each case only varies in the number of optimal models with respect to the chosen KPI. Thus, finding $N = 1$ or $N = 20$ takes a similar amount of time ($\sim$30 seconds), as both cases require 144 ICEbear simulations.

5 CONCLUSIONS AND FUTURE WORK

This paper presented a new approach for supporting renovation decision making by enabling renovation actions that modify the geometry of a building. We extended the Renovation Domain Model (NovaDM) with geometry-modifying actions, and developed a prototype software tool that generates optimal spatially consistent renovation scenarios evaluated against 8 Key Performance Indicators. Our empirical evaluation on a real building model from a renovation project in Denmark demonstrated that our approach could scale to real-world BIMs.

Regarding future work, certain spatial modifications may be physically realizable but nonetheless should still be avoided as they interfere with the intended function of a building. E.g., suppose that an interior wall is added without any embedded door, which makes a part of a room physically inaccessible. Such cases cannot be ruled out in general and must be at the discretion of the user, e.g. perhaps the intention of adding a wall with no door is to create a service shaft that must *not* be accessible for safety reasons.

Therefore, rather than extending *NovaDM* with general rules to capture such cases, in our future work, we are instead extending our renovation scenario search engine to support functional and affordance constraints.

ACKNOWLEDGMENT

The authors of the paper would like to show their gratitude to the Danish Innovation Foundation for financial support through the ReVALUE research project, and to the Independent Research Fund Denmark for their financial support of the project "Intelligent Software Healing Environments" (DFF FPT1).

REFERENCES

Baker N. 2009. *The Handbook of Sustainable Refurbishment: Nondomestic Buildings*. London: Earthscan.

Bhatt M., Lee J. H., & Schultz C. 2011. "CLP (QS): a declarative spatial reasoning framework." In International Conference on Spatial Information Theory, Springer, Berlin, Heidelberg: 210–230.

Boeri, A., Antonin, E., Gaspari, J. & Longo, D. 2014. *Energy Design Strategies for Retrofitting: Methodology, Technologies, Renovation Options and Applications*. Southampton: WIT Press.

BPIE 2011. Europe's buildings under the microscope. Retrieved from http://bpie.eu/wp-content/uploads/2015/10/HR_EU_B_under_microscope_study.pdf

Brewka, G., Eiter, T., & Truszczynski, M. 2011. Answer set programming at a glance. Commun. ACM 54(12): 92–103.

Burton, S. 2012. *The Handbook of Sustainable Refurbishment: Housing*. Abingdon: Earthscan.

Danish Building Research Institute 2017. 'Be18'. Retrieved from https://be18.sbi.dk/be/.

Danish Energy Agency 2017. 'Energimærkning af huse', Energimærkning boliger. Retrieved from https://sparenergi.dk/forbruger/boligen/energimaerkning-boliger/huse.

Dansk Standard 2006. DS/EN ISO 7730 Ergonomi inden for termisk miljø – Analytisk bestemmelse og fortolkning af termisk komfort ved beregning af PMV- og PPD-indekser og lokale termisk komfortkriterier.

Dansk Standard 2007. DS/EN 15251 Input Parameters for Design and Assessment of Energy Performance of Buildings – Addressing Indoor air Quality, Thermal Environment, Lighting and Acoustics. Retrieved from https://webshop.ds.dk/en-gb/standard/ds-en-152512007

Dansk Standard 2013. DS 447 Ventilation i bygninger – Makaniske, naturlige og hybride ventilationssystemer.

EU [European Union] 2016. Boosting Building Renovation: What potential and value for Europe? Retrieved from www.europarl.europa.eu/RegData/etudes/STUD/.../IPOLSTU(2016)587326EN.pdf.

Gebser, M., Kaminski, R., Kaufmann, B., Ostrowski, M., Schaub, T., & Wanko, P. 2016. Theory solving made easy with clingo 5. In OASIcs-OpenAccess Series, Informatics 52, Schloss Dagstuhl-Leibniz-Zentrum fuer Informatik.

IfcOpenShell 2020. IfcOpenShell project. Retrieved from http://www.ifcopenshell.org/

Kamari, A. 2018. A multi-methodology and sustainability-supporting framework for implementation and assessment of a holistic building renovation: Implementation and assessment of a holistic sustainable building renovation [PhD thesis]. Aarhus: Dept. of Eng., Aarhus University.

Kamari, A., Corrao, R. & Kirkegaard, P.H. 2017a. Sustainability focused Decision-making in Building Renovation. *International Journal of Sustainable Built Environment*. 6(2): 330–350. doi:10.1016/j.ijsbe.2017.05.001

Kamari, A., Corrao, R., Petersen, S. & Kirkegaard, P.H. 2017b. Sustainable Retrofitting Framework: Introducing 3 levels of Integrated Design Process Implementation and Evaluation. In PLEA 2017, Edinburgh, UK.

Kamari, A., Jensen, S.R., Corrao, R., & Kirkegaard, P.H. 2017c. Towards a holistic methodology in sustainable retrofitting: Theory, Implementation and Applications. In WSBE 2017 conference, Hong Kong, China.

Kamari, A., Schultz, C. & Kirkegaard, P.H. 2018a. NovaDM: Towards a formal, unified Renovation Domain Model for the generation of holistic renovation scenarios. In ECPPM 2018 conference, Copenhagen, Denmark: 197–205.

Kamari, A., Corrao, R., Petersen, S. & Kirkegaard, P.H. 2018b. Towards the development of a Decision Support System (DSS) for building renovation: Dependency Structure Matrix (DSM) for sustainability renovation criteria and alternative renovation solutions. In SER4SE 2018 conference, Catania, Italy (ISBN: 978-88-96386-56-9): 564–576.

Kamari, A., Schultz, C. & Kirkegaard, P.H. 2019a. Constraint-based renovation design support through the renovation domain model. *Automation in Construction* 104: 265–280.

Kamari, A., Schultz, C., & Kirkegaard, P.H. 2019b. Towards a BIM-based Decision Support System for rapid generation and evaluation of holistic renovation scenarios. In CIBW78 2019, Northumbria University, Newcastle, UK: 244–254.

Kamari, A., Schultz, C. & Kirkegaard, P.H. 2019c. Unleashing the diversity of conceptual building renovation design: Integrating high-fidelity simulation with rapid constraint-based scenario generation. In SimAUD 2019 conference, Atlanta, USA: 29–36.

Li, Beidi, Bhatt M. & Schultz C. 2019. "lambdaProlog (QS): Functional Spatial Reasoning in Higher Order Logic Programming (Short Paper)." In 14th International Conference on Spatial Information Theory (COSIT 2019). Schloss Dagstuhl-Leibniz-Zentrum fuer Informatik.

Molio 2016. Molio Price data. Retrieved from https://molio.dk/molio-prisdata/prisdata-footer/brug-m

Purup, P.B., & Petersen, S. 2017. Rapid simulation of various types of HVAC systems in the early design stage, *Energy Procedia* 122: 469–474.

TECNALIA 2015. Intervention criteria & packaged solutions for buildings renovation towards a NZEBR. Retrieved from http://www.nezerproject.eu/download/18.343dc99d14e8bb0f58bb3b/1440579936965/NeZeR_D2_3_NZEBR%20criteria.pdf.

VELUX 2016. Architecture for Wellbeing & Health | The Daylight Site: 1–19. Retrieved from http://thedaylightsite.com/architecture-for-well-being-andhealth/

Xu, H., Huang, Q., & Zhang, Q. 2018. A study and application of the degree of satisfaction with indoor environmental quality involving a building space factor. Building and Environment, 143: 227–239. doi:10.1016/j.buildenv.2018.07.007.

Wałęga, Przemysław A., Bhatt, M., & Schultz, C. "ASPMT (QS): non-monotonic spatial reasoning with Answer Set Programming modulo theories." In International Conference on Logic Programming and Nonmonotonic Reasoning, pp. 488–501. Springer, Cham, 2015.

Wałęga, Przemysław A., Schultz, C., & Bhatt, M. "Non-monotonic spatial reasoning with Answer Set Programming modulo theories." Theory and Practice of Logic Programming 17, no. 2 (2017): 205–225.

Approaching the human dimension of building performance via agent-based modeling

C. Berger & A. Mahdavi
Department of Building Physics and Building Ecology, TU Wien, Vienna, Austria

ABSTRACT: This paper provides an overview of recent research efforts regarding the application of agent-based modeling (ABM) of building occupants in the context of buildings' energy and indoor-environmental performance assessment. Toward this end, we focus on the methods used in ABM for the representation of occupants' behavior and their environment. Our review suggests that occupant-centric ABM applications frequently rely on sparse domain knowledge and limited theoretical foundations. The representations of the built environment, on the other hand, involve lesser challenges. With regard to coupling techniques (between models of occupants and models of environment), there is a lack of consistent and scalable approaches. However, the main challenge of ABM lies in the paucity of empirically-validated knowledge concerning processes related to occupants' perception, evaluation, and behavior. Future research must thus pursue broad collaborative efforts to collect and utilize observational data from both field and laboratory studies of occupants' requirements, attitudes, and behavior.

1 INTRODUCTION

Building information modeling environments, enhanced with simulation capability, can provide high-resolution dynamic digital models that can be used for a large range of performance queries. Thereby, it is important to include the human dimension in the equation, given the central standing of building users, not only as recipient of services, but also given the fact that their presence and behavior influence buildings' energy and environmental performance.

It is thus important to examine the representational practices of building occupants, particularly in simulation applications for buildings' energy and indoor-environmental performance. Recent efforts have targeted the development of detailed occupancy-related data input and processing for building performance simulation. Specifically, the agent-based modeling (ABM) paradigm appears to offer a promising tool toward capturing the complex patterns of occupants' patterns of presence and behavior in buildings. Accordingly, the present contribution examines a number of ABM-related research and development efforts that relate to the representation of occupants in pertinent computational applications. As such, ABM had the potential capacity to model occupants' behavior, including socially relevant interactions and learning processes.

ABM has been generally deployed in many domains (e.g. economics, medicine, social sciences, population dynamics, geographical systems, and transportation). There have been also applications more closely related to the built environment (e.g. pedestrian's movement, evacuation scenarios). In this paper we mainly concentrate on ABM implementations that concern occupants in the context of buildings' energy and indoor-environmental performance simulation (i.e. energy demand, adaptive thermal comfort, visual comfort, acoustic comfort, and air quality, water consumption, HVAC system design and operation). Such implementations must include representations of both occupants and their context (environment). Whereas the former are almost always implemented in terms of autonomous agents, the latter is commonly captured in terms of conventional (i.e. typically numeric procedural) codes. Therefore, specific coupling techniques must be adopted to provide runtime information exchange between the agent-based occupants' representation and the procedural context representation.

In the present paper, we review the aforementioned efforts in view of the methods used for the representation of occupants' behavior and their environment.

2 BACKGROUND

Agent-based modeling (ABM) uses decision-making entities that are referred to as agents. The occupants of buildings, for example, can be represented in terms of computational agents. Agents are conceived to have the capacity to assess certain aspects of their context (Bonabeau 2002). As such, ABM makes it possible to model how agents learn and behave under complex contextual settings (Wilensky & Rand 2015). In view of specific technical advantages of ABM, Bonabeau (2002) points to a number of technical opportunities ABM can provide, specifically its aptitude to map emergent occurrences, and to represent natural systems.

ABM has been applied in many areas, including social sciences, economics, medicine, population dynamics, geography, and archeology (Folcik & Orosz 2006; Gilbert & Troitzsch 1999; Heppenstall & Crooks 2016; Macy & Willer 2002; Pablo-Marti & Santos 2013; Singh et al. 2016; Tesfatsion 2002; Rindfuss 2008; Wurzer et al. 2015).

General programming languages (e.g. Python (2019) or Java (2019)), software tools (e.g. spreadsheets) and specific ABM software environments (e.g. Repast (2019), Swarm (2019), MASON (2019), AnyLogic (2019), NetLogo (2019)) have been employed toward implementation of ABM applications (Macal & North 2008).

3 ABM APPLICATIONS IN THE BUILDING PERFORMANCE DOMAIN

Broadly speaking, there have been many ABM applications relevant to the general built environment. Instances of such applications pertain to, for example, mobility and transportation, pedestrians' movement and evacuation (Hu et al. 2012; Huynh et al. 2013; Zheng et al. 2019). However, the present treatment focuses on some twenty relatively recent ABM-related publications that focus on representation of occupants in simulation models of buildings' energy and indoor-environmental performance. Thereby, the respective use cases, modeling methods, and implementation approaches are examined.

3.1 Use cases

ABM has been used to simulate occupants' impact on buildings' energy use (Azar & Menassa 2010; 2012; Azar & Papadopoulos 2017; Chen et al. 2016). Likewise, occupants' behavioral adaptation toward achieving thermal comfort has been addressed via ABM (Alfakara & Croxford 2014; Langevin et al. 2014; Lee 2013; Lee & Malkawi 2014; Thomas et al. 2016). Other topics included occupants' visual and auditory comfort (Andrews et al. 2011, Barakat & Khoury 2016), air quality (Jia et al. 2017, 2018, Jia & Bharathy 2018), water consumption patterns (Linkola et al. 2013), and HVAC (Heating, Ventilation, Air-conditioning) design and operation (Putra et al. 2017).

3.2 Modeling methods

Theoretically, ABM methods could be applied to all constitutive aspects of an occupant-centered building performance simulation scenario. As Figure 1 schematically illustrates, ABM-based formalisms could be deployed to model agents' internal (cognitive) processes (instead of adopting simple rules), and interactions amongst agents, interactions between agents and features of the indoor environment (including building control systems). Even indoor and outdoor

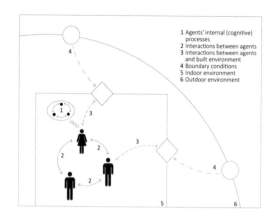

Figure 1. Schematic illustration of processes involving occupants and buildings. Whereas occupants decision-making processes is typically captured via ABM, indoor and outdoor environments are typically represented using numeric simulation.

environmental conditions could be presumably represented in terms of computational agents.

In practice, however, application of ABM formalism is restricted to behavioral manifestation associated with occupants. Ideally, the corresponding knowledge base (be those captured in terms of rules or other programming techniques) should come from relevant disciplines such as psychology, physiology, and sociology. Computational representations of the context (typically indoor environment or building systems) are, however, not agent-based. Rather, the context is emulated using common numeric codes for simulation of thermal, visual, and acoustical processes. This means that coupling codes must mediate between agent-based representations of occupants and numeric representations of their context.

3.2.1 Modeling of the behavior

There is not a unique source of the requisite knowledge for the definition of agents' profile and their behavior in ABM applications. Some efforts rely on interviews and surveys (Andrews et al. 2011; Jia et al. 2017, 2018; Jia & Bharathy 2018; Langevin et al. 2014; Luo et al. 2017; Putra et al. 2017, Schaumann et al. 2017; Wang et al. 2016; Zhang et al. 2016). Others adopt information in literature, including technical papers, standards, and guidelines. To arrive at a basis for the formulation of behavioral rules, some efforts refer to theories of occupants' perception and behavior such as the Reasoned Action Model (Fishbein & Ajzen 2010; Lee 2013; Lee & Malkawi 2014), OB DNAS (Chen et al. 2016, 2018), hierarchical decision-making model (Alfakara & Croxford 2014), Theory of Planned Behavior (Ajzen 1991; Andrews et al. 2011; Linkola et al. 2013), the Belief-Desire-Intention (BDI) framework (Andrews et al. 2011; Linkola et al. 2013; Rao & Georgeff 1998).

Some efforts differentiate occupants in view of their assumed energy use habits (Azar & Menassa

2010, 2012), categorized as "Low", "Medium", and "High". They also try to take cross-individual energy-relevant behavioral influences into account. In case of Jia et al. (2018), occupants' actions (e.g. making choices amongst a number of building systems control options) are geared toward improving their perceived state. A "Goals, Standards, and Preferences" (GSP) tree was used by Jia et al. (2018) to encapsulate agents' mindset. Occupants' response to perceived (thermal, visual, etc.) discomfort (i.e. operation of environmental control systems) is guided by GSP.

Langevin et al. (2014) refer to Perceptual Control Theory (PCT) (Powers 1973) and define five adaptive behaviors toward achieving thermal comfort. These include adjustment of clothing, use of a heater, fan, or thermostat, and window operation. They utilized a one-year field study in an office building to test this approach (Langevin et al. 2015). In Lee & Malkawi (2014), occupants' beliefs (behavioral, control-specific, and normative) in the context of control-oriented comfort-seeking actions are adopted from the Reasoned Action Model (Fishbein & Ajzen 2010). Papadopoulos & Azar (2016) refer to ASHRAE (2019) in view of acceptable comfort ranges, which guide occupants' energy use attributes (i.e. preferences for thermostat set points, lighting usage, and plug loads).

Putra et al. (2017) take the differences in occupants' preferences concerning visual and thermal conditions into account. They use a multi-attribute utility function to arrive at agents' adaptive actions. A number of efforts (Schaumann et al. 2017; Wang et al. 2016; Zhang et al. 2016) rely on interviews and observations as the informational basis for the definition of agents' behavioral repertoire, for instance, with regard to occupants' light switching behavior.

3.2.2 Models of the environment

The majority of the ABM work relevant to indoor environment focus on actual or case study (model) office buildings (Andrews et al. 2011; Azar & Menassa 2010; Azar & Papadopoulos 2017; Barakat & Khoury 2016; Jia et al. 2018, Langevin et al. 2014, 2015; Lee & Malkawi 2014; Papadopoulos & Azar 2016; Putra et al. 2017; Thomas et al. 2016; Wang et al. 2016). A smaller number of studies concerns other building types, such as residential buildings (Alfakara & Croxford 2014; Linkola et al. 2013), hospitals (Schaumann et al. 2017), and schools (Zhang et al. 2016).

3.3 Implementation approaches and tools

The following treatment of the adopted representational approaches in the reviewed studies is structured in terms of *i)* occupants, *ii)* context, and *iii)* occupant-context coupling.

3.3.1 Occupants

Multiple studies deploy the open-source java-based tool NetLogo (2019) to model occupant agents (Andrews et al. 2011; Linkola et al. 2013; Putra et al. 2017). Other deployed tools include AnyLogic (2019) (Azar & Menassa 2010, 2012; Barakat & Khoury 2016; Papadopoulos & Azar 2016; Zhang et al. 2016), the web-based Occupancy Simulator (2019) (Chen et al. 2016, 2018; Luo et al. 2017), the open-source ABM library Repast Symphony (2019) (Alfakara & Croxford 2014), MATLAB (2019) (Langevin et al. 2014, 2015; Lee & Malkawi 2014), PMFserv (Silverman et al. 2006) (Jia et al. 2018; Jia & Bharathy 2018), and the video game engine Unity 3D (2019) (Schaumann et al. 2017).

3.3.2 Context

The context in which the agents act is typically not modelled via ABM. Rather, reviewed studies rely on numeric simulation applications to model buildings' energy use and the thermal and visual indoor-environmental conditions (Alfakara & Croxford 2014; Andrews et al. 2011; Azar & Menassa 2010; 2012; Langevin et al. 2014, 2015; Lee & Malkawi 2014; Linkola et al. 2013; Jia & Bharathy 2018; Jia et al. 2018; Putra et al. 2017). Instances of such numeric simulation applications and associated tools include EnergyPlus (2019), OpenStudio (2019), DesignBuilder (2019), eQuest (2019), TAS (2019), and Radiance (2019).

3.3.3 Coupling agents with their context

Some of the past studies computationally couple agents (ABM) with their context (building energy models) (Azar & Papadopoulos 2017; Langevin et al. 2014, 2015; Lee & Malkawi 2014; Thomas et al. 2016). Others involve the "manual" coupling of agents' behavior and their environment (Jia & Bharathy 2018; Jia et al. 2017, 2018).

Azar & Papadopoulos (2017) couple the ABM (defined in MATLAB) and the BEM (simulated in EnergyPlus). MATLAB is used both as a coupling engine and as a surrogate model for training and validation. Langevin et al. (2014) use BCVTB (Building Controls Virtual Test Bed) (2019) to couple the ABM in MATLAB (2019) and the BEM in EnergyPlus (2019). Lee & Malkawi (2014) deploy MLE Legacy (2019) and BCVTB (2019) to connect agents with their context, whereas Thomas et al. (2016) couple the BEM and the ABM using LCM based communication (Lightweight Communications and Marshalling) (LCM 2019).

4 CONCLUSION

We reviewed a number of recent research efforts that rely on ABM to represent occupants' behavior in built environments. A number of beneficial features have been attributed to ABM. It facilitates the representation of the perceptual tendencies, comfort-related preferences, and behavioral habits in a systematic manner. Moreover, the behavior and interactions of computational agents can evolve over time without the need for explicit procedural coding. These attributes render ABM highly useful, especially when dealing with complex questions, such as the responsiveness of occupants to energy saving appeals and campaigns.

The potential and capabilities of ABM notwithstanding, the reviewed efforts reveal also a number of non-trivial shortcomings. The ABM challenges do not necessarily pertain to the technical implementation matters. There is of course much to be done to enhance the computational dexterity and usability of ABM platforms.

However, the main factor that occupant-centric ABM applications for building design and operation support face is not a purely technical one. Rather, the main problem lies in the paucity of broad and reliable empirical knowledge concerning processes related to occupants' perception, evaluation, and behavior.

As such, the majority of the reviewed efforts are based on rather limited and statistically non-representative sources of occupant-related information. As a corollary, the results of ABM-based analysis are hardly ever validated against detailed information from actual environments.

Occupants' behavior requires domain knowledge from different disciplines, such as psychology, physiology, and sociology. One of the main findings of our contribution is the documentation of the relatively sparse (frequently limited to few buildings and populations) deployed domain knowledge in most occupancy-centric ABM implementations pertaining to buildings. Likewise, our finding point to the limitation of the repertoire of underlying behavioral theories deployed in such theories.

To further highlight these issues, Figure 2 displays the fraction of the reviewed studies that, in order to establish the agents' behavioral repertoire, relied (solely or in combination) on the following resources: information included in common professional codes, standards, and guidelines (C), some measure of collected original data (M), and some reference to applicable theoretical underpinnings of the relevant behavioral traits (T). As this figure illustrates, few studies use (original) empirical information toward the development and validations of their ABM-based findings. In fact, none of the efforts reviewed utilized the kind of empirical information that could be considered to be statistically representative for the building type studied.

The representations of the built environment, on the other hand, involve lesser challenges. But, as mentioned before, this is because these representations rely more on conventional numeric modeling techniques (e.g. thermal and visual simulation). With regard to coupling techniques, our study reveals a lack of consistent and scalable approaches. In certain cases, the coupling has to be instantiated manually, reducing the effectiveness of the ABM's overall applicability.

To summarize, it is one thing to develop computationally advanced formalisms and platforms. But it is an entirely other thing to obtain and to communicate real insights in occupant-environment interaction patterns and processes. To be truly effective, ABM-based analyses of building performance (and occupants' influence thereupon) must be supplied with relevant and detailed domain knowledge. As alluded to before, the technical implementation issues (including coupling methods) may not be the main challenge concerning ABM deployment. Rather, the main problem lies more in the paucity of empirically-validated knowledge concerning processes related to occupants' perception, evaluation, and behavior. Thus, future research efforts must thus not only seek the embellishment of ABM techniques, but also engage in broad – preferably collaborative – efforts to collect, analyze, and structure observational data from both field and laboratory studies of occupants' requirements, attitudes, and behavioral patterns.

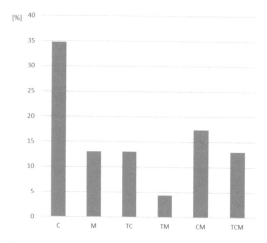

Figure 2. Fraction of the reviewed studies, which rely on codes and standards only (C), include some original empirical information (M), include both theories and codes/standards (TC), both theories and empirical data (TM), both codes/standards and empirical data (CM), as well as all three ingredients, i.e. theories, codes/standards, and empirical data (TCM).

REFERENCES

Ajzen, I. 1991. The theory of planned behavior. Orgn. Beh. Hum. Decision. Proc. 1991, 50, 179–221.

Alfakara, A. & Croxford, B. 2014. Using agent-based modelling to simulate occupants' behaviours in response to summer overheating. *Simulation Series*, 46(7), 90–97.

Andrews, C. J., Yi, D., Krogmann, U., Senick, J. A. & Wener, R. E. 2011. Designing Buildings for Real Occupants: An Agent-Based Approach. IEEE Transactions on systems, man, and cybernetics, 1–15.

AnyLogic 2019. https://www.anylogic.com/. Accessed 22.11.2019.

ASHRAE 2019. https://www.ashrae.org/about. Accessed 22.11.2019.

Azar, E. & Menassa, C. C. 2010. A conceptual framework to energy estimation in buildings using agent based modeling. *Proceedings of the 10th Winter Simulation Conference*, 3145–3156. https://doi.org/10.1109/WSC.2010.5679007.

Azar, E. & Menassa, C. C. 2012. Agent-Based Modeling of Occupants and Their Impact on Energy Use in Commercial Buildings. *Journal of Computing in Civil Engineering*, 26(4), 506–518. https://doi.org/10.1061/(asce)cp.1943-5487.0000158.

Azar, E. & Papadopoulos, S. 2017. Human Behavior and Energy Consumption in Buildings: An Integrated Agent-Based Modeling and Building Performance Simulation Framework. *Proceedings of the 15th IBPSA Conference*, San Francisco, USA.

Barakat, M. & Khoury, H. 2016. An agent-based framework to study occupant multi-comfort level in office buidlings. *Proceedings of the 2016 Winter Simulation Conference*, 1–13.

BCVTB 2019. https://simulationresearch.lbl.gov/bcvtb/. Accessed 22.11.2019.

Bonabeau, E. 2002. Agent-based modeling: methods and techniques for simulating human systems. *Proceedings of the National Academy of Sciences of the United States of America*, 99 Suppl. 3, 7280–7287. https://doi.org/10.1073/pnas.082080899.

Chen, Y., Luo, X. & Hong, T. 2016. An Agent-Based Occupancy Simulator for Building Performance Simulation. *2016 ASHRAE Annual Conference*.

Chen, Y., Hong, T. & Luo, X. 2018. An agent-based stochastic Occupancy Simulator. *Building Simulation*, 11(1), 37–49. https://doi.org/10.1007/s12273-017-0379-7.

DesignBuilder 2019. https://designbuilder.co.uk/. Accessed 22.11.2019.

EnergyPlus 2019. https://energyplus.net/. Accessed 22.11.2019.

eQuest 2019. http://doe2.com/equest/. Accessed 22.11.2019.

Fishbein, M. & Ajzen, I. 2010. Predicting and Changing Behavior: The Reasoned Action Approach, *Psychology Press*, New York.

Folcik, V. & Orosz, C. 2006. An agent-based model demonstrates that the immune system behaves like a complex system ad a scale-free network. *SwarmFest 2006*. South Bend, IN.

Gilbert, N. & Troitzsch, K. G. 1999. Simulation for the social scientist. *Buckingham UK: Open University Press*.

Heppenstall, A. & Crooks, A. 2016. Agent-based modeling in geographical systems. *AccessScience*, 3, 1–9. https://doi.org/10.1036/1097-8542.YB160741.

Hong, T., D'Oca, S., Turner, W. & Taylor-Lange, S. 2015. An ontology to represent energy-related occupant behavior in buildings. Part I: Introduction to the DNAs Framework. *Building and Environment*. 92. 764–777. 10.1016/j.buildenv.2015.02.019.

Hu, Y., Liu, X., Wang, F. & Cheng, C. 2012. An overview of agent-based evacuation models for building fires. *Proceedings of 2012 9th IEEE International Conference on Networking, Sensing and Control*, ICNSC 2012, 382–386. https://doi.org/10.1109/ICNSC.2012.6204949.

Huynh, N., Namazi-Rad, M., Perez, P., Berryman, M. J., Chen, Q. & Barthelemy, J. 2013. Generating a Synthetic Population in Support of Agent-Based Modeling of Transportation in Sydney. *20th International Congress on Modelling and Simulation*, 1–6. https://doi.org/10.13140/2.1.5100.8968.

Java 2019. https://www.java.com/. Accessed 22.11.2019.

Jia, M. & Bharathy, G. 2018. Exploring the validity of occupant behavior model for improving office building energy simulation. *Proceedings – Winter Simulation Conference*, 3953–3964. https://doi.org/10.1109/WSC.2018.8632278.

Jia, M., Srinivasan, R., Ries, R., Bharathy, G., Silverman, B. & Weyer, N. 2017. An Agent-Based Model Approach for Simulating Interactions between Occupants and Building Systems. *IBPSA Building Simulation* 2017, 2407–2413.

Jia, M., Srinivasan, R. S., Ries, R. & Bharathy, G. 2018. A Framework of Occupant Behavior Modeling and Data Sensing for Improving Building Energy Simulation. https://doi.org/10.22360/simaud.2018.simaud.015.

Langevin, J., Wen, J. & Gurian, P. 2014. Including Occupants in Building Performance Simulation: Integration of an Agent-Based Occupant Behavior Algorithm With Energyplus. *Building Simulation Conference*, Atlanta GA, (2010), 417–424. https://doi.org/10.1108/sr.2011.08731bad.003.

Langevin, J., Wen, J. & Gurian, P. 2015. Simulating the human-building interaction: Development and validation of an agent-based model of office occupant behaviors. *Building and Environment*, 88, 27–45. https://doi.org/10.1016/j.buildenv.2014.11.037.

LCM 2019. https://lcm-proj.github.io/. Accessed 22.11.2019.

Lee, Y. S. 2013. Modeling multiple occupant behaviors in buildings for increased simulation accuracy: an agent-based modeling approach, Dissertation.

Lee, Y. S. & Malkawi, A. M. 2014. Simulating multiple occupant behavior in buildings: An agent-based modeling approach. *Energy and Buildings* 69, 407–416.

Linkola, L., Andrews, C. J. & Schuetze, T. 2013. An agent based model of household water use. *Water*, 5(3), 1082–1100. https://doi.org/10.3390/w5031082.

Luo, X., Lam, K. P., Chen, Y. & Hong, T. 2017. Performance evaluation of an agent-based occupancy simulation model. *Building and Environment*, 115, 42–53. https://doi.org/10.1016/j.buildenv.2017.01.015.

Macal, C. M. & North, M. J. 2008. Agent-based modeling and simulation: ABMS Examples. *Proceedings of the 2008 Winter Simulation Conference*. Miami, Florida.

Macy, M. & Willer, R. 2002. From factors to actors: computational sociology and agent-based modeling. *Annual Review of Sociology* 28, 143–166.

MASON 2019. https://cs.gmu.edu/~eclab/projects/mason/. Accessed 22.11.2019.

MATLAB 2019. https://www.mathworks.com/. Accessed 22.11.2019.

MLE Legacy 2019. https://www.mathworks.com/help/simulink/slref/legacy_code.html. Accessed 22.11.2019.

NetLogo 2019. https://ccl.northwestern.edu/netlogo/. Accessed 22.11.2019.

Occupancy Simulator 2019. http://occupancysimulator.lbl.gov/. Accessed 22.11.2019.

OpenStudio 2019. https://www.openstudio.net/. Accessed 22.11.2019.

Pablo-Marti, F. & Santos, J. L. 2013. An agent-based model of population dynamics for the European regions. *Emergence: Complexity & Organization*, 17 (2), 1–19. https://doi.org/10.emerg/10.17357.2d9791afc384e40b47f02d089976c627.

Papadopoulos, S. & Azar, E. 2016. Integrating building performance simulation in agent-based modeling using regression surrogate models: A novel human-in-the-loop energy modeling approach. *Energy and Buildings*, 128, 214–223. https://doi.org/10.1016/j.enbuild.2016.06.079.

Powers, W. T. 1973. Behavior: The Control of Perception. New Canaan, CT: Benchmark.

Putra, H. C., Andrews, C. J. & Senick, J. A. 2017. An agent-based model of building occupant behavior during load shedding. *Building Simulation*, 10(6), 845–859. https://doi.org/10.1007/s12273-017-0384-x.

Python 2019. https://www.python.org/. Accessed 22.11.2019.

Radiance 2019. https://www.radiance-online.org/. Accessed 22.11.2019.

Rao, A. S. & Georgeff, M. P. 1998. Decision procedures for BDI logics. *J. Logic Comput.* 8 (3), 293–342.

Repast 2019. https://repast.github.io/repast_simphony.html. Accessed 22.11.2019.

Rindfuss, R., Entwisle, B., Walsh, S., Li, A., Badenoch, N. & Brown, D. 2008. Land use change: Complexity and comparisons. *Journal of Land Use Science*, 3(1), 1–10. https://doi.org/10.1080/17474230802047955.

Schaumann, D., Breslav, S., Goldstein, R., Khan, A. & Kalay, Y. E. 2017. Simulating use scenarios in hospitals using multi-agent narratives. *Journal of Building Performance Simulation*, 10(5–6), 636–652. https://doi.org/10.1080/19401493.2017.1332687.

Silverman, B., Johns, M., Cornwell, J. & O'Brien, K. 2006. Human Behavior Models for Agents in Simulators and Games: Part I: Enabling Science with PMFserv. Presence. 15. 139–162. 10.1162/pres.2006.15.2.139.

Singh, K., Sajjad, M. & Ahn, C. W. 2016. Simulating Population Dynamics with an Agent Based and Microsimulation Based Framework, Icassr 2015, 335–339. https://doi.org/10.2991/icassr-15.2016.90.

Swarm 2019. http://www.swarm.org/. Accessed 22.11.2019.

TAS 2019. https://www.edsl.net/. Accessed 22.11.2019.

Tesfatsion, L. 2002. Agent-based computational economics: growing economies from the bottom up. *Artifical Life 8* (1), 55–82.

Thomas, A., Menassa, C. C. & Kamat, V. R. 2016. Distributed simulation framework to analyze the energy effects of adaptive thermal comfort behavior of building occupants. *Proceedings of the 2016 Winter Simulation Conference*, 3225–3236. https://doi.org/10.1109/WSC.2016.7822354.

Unity 3D 2019. https://unity3d.com/. Accessed 22.11.2019.

Wang, C., Yan, D. & Ren, X. 2016. Modeling individual's light switching behavior to understand lighting energy use of office building. *Energy Procedia*, 88, 781–787. https://doi.org/10.1016/j.egypro.2016.06.128.

Wilensky, U. & Rand, W. 2015. An introduction to agent-based modeling: modeling natural, social, and engineered complex systems with NetLogo.

Wurzer, G., Kowarik, K. & Reschreiter, H. 2015. Agent-based Modeling and Simulation in Archaeology. *Advances in Geographic Information Science*. Springer.

Zhang, T., Siebers, P.-O. & Aickelin, U. 2016. Modelling Office Energy Consumption: An Agent Based Approach. Srn, 1–15. https://doi.org/10.2139/ssrn.2829316.

Zheng, L., Peng, X., Wang, L. & Sun, D. 2019. Simulation of pedestrian evacuation considering emergency spread and pedestrian panic. *Physica A: Statistical Mechanics and Its Applications*, 522, 167–181. https://doi.org/10.1016/j.physa.2019.01.128.

BIM-based cost estimation in a road project – proof of concept and practice

D. Fürstenberg
Norwegian University of Science and Technology (NTNU), Trondheim, Norway
COWI, Bergen, Norway

T. Gulichsen
COWI, Bergen, Norway

O. Lædre & E. Hjelseth
Norwegian University of Science and Technology (NTNU), Trondheim, Norway

ABSTRACT: Digital transformation can improve productivity and reduce uncertainty in AEC projects. BIM-based workflows play an important role in this transformation. The IFC format has emerged as exchange standard in road projects and allows BIM to be exported from design software to cost estimation software. Therefore, we tested a workflow in a real-life road project from Norway. We show how BIM-based cost estimation can be made with commercial software and uncover related challenges. We used a Design Science Research approach and developed an artefact in the form of an IFC based link between design and cost estimation software. IFC entities were expanded with coded properties based on standard specification texts. We concluded that BIM-based cost estimation improves productivity, reduces uncertainty and eliminates random human errors by automating repetitive and time-consuming tasks. Further research with different software and specification systems in more countries are necessary.

1 INTRODUCTION

Digital transformation can improve productivity and reduce uncertainty in AEC projects (Bertschek et al. 2019; Borrmann et al. 2018). Building Information Modelling (BIM) is central in this transformation (Almeida et al. 2016) and ways of implementing BIM into existing workflows are theoretically and practically explored. BIM research has mainly focused on buildings, and research on BIM for infrastructure is lagging (Shou et al. 2015). However, governmental initiatives in Norway (Norwegian Public Roads Administration 2015), Germany (Bundesministerium für Verkehr und digitale Infrastruktur 2018), the Czech Republic (Ministry of Industry and Trade 2017) and the UK (UK Roads Liaison Group 2018) have enforced the use of BIM for infrastructure and thereby promoted a change in workflows. Cost estimation has traditionally been a manual and repetitive task prone to human error (Firat et al. 2010), something which has reduced its reliability for the contractor. However, according to Fürstenberg and Lædre (2019) contractors need exact and reliable information about the planned assets in a project, especially for cost drivers like constructions and masses for road layers. Traditionally, the estimate and the specification of work is produced by quantity take offs derived from 2D drawings in PDF format and/or generic 3D models. These quantity takeoffs are presented as Excel tables and the quantities are manually copy-pasted to a cost estimation software. If the drawings and/or the generic models are changed, this work must be manually redone. Figure 1 illustrates the difference between the traditional and a BIM-based workflow.

The automation of quantity takeoff is regarded as one of the most useful tasks in BIM-based AEC projects (Hartmann et al. 2012; Monteiro & Martins 2013). Discipline models in the open source format IFC have become the standard in many infrastructure projects. These models contain quantity information and can be imported into commercial cost estimation software. Still, the practical research on BIM-based cost estimation on infrastructure projects is limited.

Therefore, this paper addresses the followings research questions:

- How can BIM-based cost estimation for infrastructure projects be developed to enable automatic quantity takeoff?
- What are the challenges in developing an applicable solution based on commercial software?

2 METHOD AND CASE DESCRIPTION

We used Design Science Research (DSR) as method like described in (Vaishnavi & Kuechler 2015).

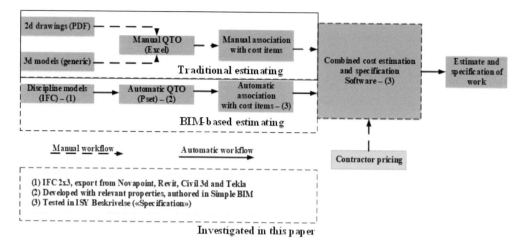

Figure 1. Overview of traditional vs. BIM-based estimating, adapted and simplified from Eastman et al. (2011).

A recent study (Fürstenberg in press) confirmed that DSR is one of the most used research methods on information management in AEC projects. These results are also applicable to the case study presented in this paper. It gave priority to development of an artefact and testing this artefact under real life conditions. This pragmatic approach is one of the characteristics with Design Science Research. Our artefact was a file-based link between design and cost estimation software through IFC (2x3). We had several loops and refinements.

The case was a real-life design-bid-build road project in Norway, consisting of 7 km highway and 3 km secondary roads. The client, the Norwegian Public Roads Administration ordered a "model-based" project with discipline models delivered in open source formats like defined in their design manual V770 (Norwegian Public Roads Administration 2015). The designer produced 50 discipline models in IFC (2x3) format for all disciplines (road, construction, water & sewer, electrical, landscape and earthworks). The case was a pilot project with the aim of reducing the number of drawings as far as possible. Therefore, only a zoning plan and some overview drawings for the constructions were demanded. This project was at the crossing between BIM level 2 and 3, as defined by the UK government's 2011 Construction Strategy (Cabinet Office UK 2011). In short, BIM level 2 involves extracting and integrating data from models while BIM level 3 goes further and involves automation of work processes.

We had workshops with the discipline leaders where we screened Norwegian standard specification texts used for road projects. The goal was to identify cost items suitable for automatic quantity take offs. Then, we identified the corresponding entities in the discipline models and added coded properties to a specific property set according to the Norwegian Public Roads Administration's general specifications. As a next step we imported IFC files into a cost estimation software where cost items based on general specification texts were automatically created. We verified several IFC sample imports against manually created bill of quantities. No disagreements between the IFC files and the manually created bill of quantities were found.

Different design software from Trimble – Novapoint and Tekla Structures – and Autodesk – Civil 3D and Revit – was used to produce IFC files with a coded property set. Simple BIM from Datacubist was used as an authoring software. ISY Beskrivelse ("Specification") from Norconsult Informasjonssystemer was used as cost estimation software. Traditional manual quantity takeoffs could not be used because drawings were not created as part of the digitalization strategy.

3 THEORETICAL BACKGROUND

3.1 Standard specification text (R761/R762)

The client, The Norwegian Public Roads Administration, has standard specification texts for all works which are mandatory to use. These specifications consist of two parts, General Specifications 1 (R761) and General Specifications 2 (R762) which are published in two handbooks, R761 (Norwegian Public Roads Administration 2018a) and R762 (Norwegian Public Roads Administration 2018b). In short, the specification texts include prescribed units. This is similar to the German Standardleistungskatalog, STLK (FGSV 2020) and Swedish BSAB96 (Svensk Byggtjänst 2020). In the UK the practice is different, since *"the specification generally does not include information on quantities"* (National Building Specification 2017). In detail, a specification text in R761/R762 consists of a numeric code, a title, a descriptive text and a specific unit. Thus, it is both a specification and classification system. If the standard specification texts do not describe all involved work processes sufficiently, the designer can create a user-defined specification text with a corresponding user-defined code and unit. Following is one example of a standard specification

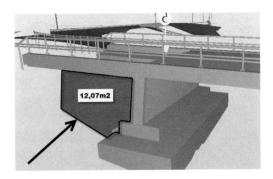

Figure 2. Cost item 71.16 in the construction discipline model (unit: m$^{2''}$).

Table 1. Property set ISY Beskrivelse with properties for one IFC entity.

Property	Value
ISY Unit	m2;m3;m2
ISY Quantity	12,07;1,45;12,07;
ISY Code	71.16;71.14;71.15
ISY Title	Sorting, delivery and masonry; Concrete sole; Geotextile
ISY Localizer	S15;S15;S15

text from R761. For the convenience of the readers it was translated by the authors.

"71.16 Sorting, delivery and masonry

a) Includes sorting, uploading and transport, as well as delivery, if applicable, of natural stone from storage or siding, including masonry. Also includes grouting if specified in the specification.
b) Quantity is measured as projected area. The lower boundary of the area is the bottom edge of the lowest stone of the wall. The upper boundary of the area is the top of the wall at the front edge. At the top of the wall, areas that are horizontal, inclined or rounded back from the front edge of the wall are not measured.

In design-bid-build projects for the Norwegian Public Roads Administration designers produce a combined estimate and specification of work based on R761 and R762. The contractors use it as basis for their bidding and insert their specific unit prices.

3.2 Cost estimation software ISY Beskrivelse

At first sight ISY Beskrivelse might resemble other cost estimation software e.g. VICO Office. One main difference is that ISY Beskrivelse solely extracts the quantities as numeric values from a pre-defined specifically named property set and not from the IFC entities themselves. This makes it possible to attach several cost items to the same IFC entity without decomposing it in the design software. For instance, the above illustrated wall of natural stone (17.16) has two cost items attached since it has a concrete sole (71.14) and the backfill is protected by a geotextile (71.15). Instead of modelling these two cost items separately, their values are appended to the values of the wall. By separating the values with a semicolon ";" ISY Beskrivelse creates 3 independent cost items from this single IFC entity. There is just one property set in the IFC file (Pset_ISY Beskrivelse) which holds all necessary properties. The properties have several values which are interpreted independently by ISY Beskrivelse. Table 1 illustrates the content of the property set for one IFC entity. ISY Beskrivelse expects the Norwegian names for the properties. However, for the convenience of the readers, they were translated by the authors.

The order of the values must be identical for all properties for the correct creation of the cost items. ISY Beskrivelse extracts the correct information from each property and sums up the numeric values of ISY Quantity per cost item. The numeric code, the title and the unit are checked against an internal database and the corresponding specification text is loaded. If user-defined codes and titles were applied ISY Beskrivelse creates user-defined cost items but the specification texts must be generated manually.

The specification of work is created and contains a numeric code, a title, a specification text, a unit and a quantity per cost item but no unit prices. The specification of work is exported as PDF, XML or in several generic formats by the designer. In the procurement phase the contractors insert their specific unit prices to complete the cost estimate for the tender.

3.3 IFC road

BIM for infrastructure, especially road infrastructure has no specific IFC extension, yet. This makes the IFC format less user-friendly for road projects than for building projects. Therefore, buildingSMART (2020a) initiated the IFC Road Project which is supported by national road authorities from Sweden, Finland and Germany and other industry experts. Phase 1 was completed in 2018 with a report and road map. Phase 2 began in 2019 and will end up with the definition of IFC 4x3. This is expected to happen in 2021 (buildingSMART 2020b).

4 FINDINGS

This paper presented a possible automated workflow for BIM-based cost estimation based on commercial software answering the following questions:

– How can BIM-based cost estimation for infrastructure projects be developed to enable automatic quantity takeoff?
– What are the challenges in developing an applicable solution based on commercial software?

4.1 BIM-based cost estimation

In the investigated case a BIM-based cost estimation was made up by creating a dynamic link between design and cost estimation software. This dynamic link was an IFC file with a coded property set based on properties from Norwegian standard specification texts (R761/R762) for road projects. We screened these specification texts for cost items with units that were directly available in the discipline models (m, m^2, m^3, piece). Depending on the design software we excluded most cost items with units that were only indirectly available (lump sum, ton, week, hour). Next, we expanded the relevant entities with a set of specifically named properties containing the R761/R762 data. These properties became a custom user-defined property set (Pset_ISY Beskrivelse) when exported as IFC (2 × 3) files. A unit (ISY Unit), a quantity (ISY Quantity), a numeric code (ISY Code), a title (ISY Title), and an internal localizer (ISY Localizer) was used to map BIM entities to cost items in the specification of works. Depending on the geometry of each BIM entity, it either was identical to a cost item or had to be decomposed or attached to others.

4.2 Challenges when using commercial software

We were able to map approximately 70% of all cost items to entities and extract their quantities in the discipline models. Some cost items involved entities that were not modelled, like geotechnical entities. The quantities of these entities had to be estimated and were added manually in the cost estimation software. Other entities were not modelled either e.g. earthing of light poles. However, they could still be associated automatically to cost items because they are attached to already modelled entities.

The cost estimation software only accepted either manually or automatically imported quantities for the same cost item. It was not possible to manually override quantities created from an IFC import without breaking the link to the discipline models. The checked samples from the IFC files resulted in the same quantities in the cost estimation software as those manually copy-pasted from Excel sheets.

The three design programs allowed different levels of automation for assigning the necessary R761/R762 property set. In Novapoint the decomposition of the entities to the respective cost items was done automatically by applying filters to the entire discipline model. In Revit this was automated as well through the visual programming interface Dynamo and in Tekla through the visual programming interface Grasshopper. In Civil 3D it was done manually by filtering the layers. The authoring software, Simple BIM, was necessary to edit some properties to a format that the cost estimation software could interpret. While this program can automate all of its actions by generic scripts, it was only half automatic in our case. Many functions use filters based on the IFC entity. Since road specific IFC entities are yet to become official buildingSMART standards, extra manually defined filters were needed.

5 DISCUSSION

We clustered our observations on 1) the software, 2) the workflow, 3) the standard specification texts (R761/R762) and 4) the IFC standard.

Some entities correspond to several cost items with different units. E.g. the sub-base course corresponds to one cost item with m^2 and one cost item with m^3. In Novapoint it is not possible to assign the correct codes to both the surface and the volume of the same entity nor to use formulas to convert the numbers. However, in Revit/Dynamo, Tekla/Grasshopper and to a limited extend in Civil 3D it is possible. It is similar for cost items with weight units. Revit and Tekla Structures calculate the weight automatically, Novapoint does not. Half-automatic workflows are therefore necessary for discipline models created in Novapoint.

Some cost items involve temporary installments or transport. Cost items for temporary installments, e.g. scaffoldings, are not modelled but are created by appending their values to existing model entities e.g. bridge deck. This is possible in all the design software used. However, it is different for cost items involving transport, e.g. blast stone. The quantity has several dependencies. If the whole quantity cannot be used at once it must be transported for storage. If the quality is right for being used as raw material, it is transported within the site otherwise, it is dumped. If the quantity does not match the needs on-site, materials must be transported to or from the site. Rulesets must be applied to the quantities for the subtraction and the quality factor. Then the values can be appended to already existing model entities like in the above-mentioned example for the scaffolding. Only Dynamo and Grasshopper have this functionality.

The cost estimation software extracts the values from the IFC file as text and numbers from predefined properties. On the one hand, this gives the designer freedom to extract either area or volume from the same entity. On the other hand, it is not possible to check if the quantities corresponded with the expected units, they just appeared as numbers. Therefore, the estimators have to trust that quantities appearing as m^3 really are m^3.

The presented automated workflow shifts the source of error from random human mistakes to systematic mistakes made by machines, something which makes it much more reliable and repetitive. Manual copy-paste of numbers gives the estimators a misleading impression of control and reliability, as the process depends on individuals. Whereas automated workflows require more effort the first time they are set up (because of the necessary filtering and coding), they make quantity takeoffs and cost estimation more efficient over time and are less dependent on individuals. Already at the first revision the estimators save time when updating the quantity takeoff. The only necessary action is to update the IFC files before re-importing them to the cost estimation software. The client demanded discipline models in an open format and the designer decided to use the IFC format. Therefore, it was easy to expand the entities with the necessary additional properties to automate the workflow from the design software to the cost estimation software.

One major observation is that BIM-based cost estimation could turn the traditional workflow upside down. Traditionally, estimators set up the different cost items first and supply them with quantities when the models have reached the necessary maturity. With an automated workflow, designers can map cost items to BIM entities directly in the design software and export IFC files. Estimators import the IFC files into the cost estimation software and get the cost items and bill of quantities at the same time. The effort and entailing cost of updating the specification of work is lowered and allows an overview over the quantities already in the early phases of the project.

The standard specification texts were set up long before BIM was introduced to AEC projects in Norway and were originally adapted to a drawing-based workflow with manually copy-pasting numbers into the cost estimation software. Estimators still use specific cost items with corresponding units – often out of old habits – without reconsidering their validity in BIM-based projects. Our case study showed that the standard specification texts open for cost items better suited for a BIM-based workflow. E.g. normally the cost item *41.1 Open ditches in ordinary soil* is used. This cost item is measured in meters and cannot directly be mapped to a BIM entity in the used design software. However, there is a sub cost item *41.11 Digging* which is measured in m^3 that can be mapped directly to a BIM entity.

Cost items with a lump sum appear in two alternatives. In the first alternative, they appear only once per discipline model and can be attached to another single entity cost item. This is especially relevant for construction discipline models since there is one IFC file per construction. E.g. cost item *84.11 Design* (of scaffolding) can be attached to cost item *71.16 Sorting, delivery and masonry* (of a wall). The cost item 84.11 can be created automatically by leaving the quantity empty and the contractor enters a specific unit price. In the second alternative, the cost items are made up by several sub cost items with the same code multiplied by different unit prices. E.g. cost item *44.31 Cable drawing pipes* consists of drawing pipes with 110mm, 75mm and 40mm diameter, each with a specific unit price. The total price is generated manually from an Excel sheet with the quantities for each diameter multiplied by its corresponding unit prices. While it is technically possible to extract the quantities from each sub item, all cost items with a lump sum must be delivered as a total cost in Norwegian kroner. Manual workflows are therefore necessary.

Through a critical revision of the cost items used in the presented case, we could have increased the 70% share of automatically mapped cost items, especially cost items with a lump sum. The client prefers lump sums because the risk attached to uncertainty in the cost estimation is pushed to the contractor. However, the use of lump sums restrains a complete automation of the workflow which in return could reduce uncertainty for the client. So, a revision of either the use of cost items on project level or of the handbooks R761 and R762 in general will increase the benefits of BIM-based cost estimation.

Missing specific IFC entities limits the level of automation to some extent, especially for the road discipline model. However, the IFC 4x3 standard which is in its final stage before publication can mitigate this limitation.

6 SOFTWARE USED IN THIS STUDY

6.1 *Design software*

Civil 3D 2019 by Autodesk was used for designing landscape architecture and lighting discipline models, assigning the coded property set and exporting the models to IFC (more info on https://www.autodesk.com/products/civil-3d/overview).

Revit 2019.2 by Autodesk was used for designing some of the construction discipline models, assigning the coded property set and exporting the models to IFC. The designers could choose to use either Revit or Tekla for designing their construction models (more info on https://www.autodesk.com/products/civil-3d/overview).

Dynamo 2.0.2 by Autodesk was used for parametric design of constructions and automatization of Revit workflows (more info on https://www.autodesk.com/products/dynamo-studio/overview).

Novapoint 21 by Trimble was used for designing road, water & sewer, electrical and earthwork discipline models, assigning the coded property set and exporting the models to IFC (more info on https://www.novapoint.com).

Tekla Structures 2018 by Trimble was used for designing some of the construction discipline models, assigning the coded property set and exporting the models to IFC. The designers could choose to use either Revit or Tekla for designing their construction models (more info on https://www.tekla.com/products/tekla-structures).

Rhino 6.23 by McNeel, including Grasshopper was used for parametric design of constructions and automatization of Tekla workflows (more info on https://www.rhino3d.com).

6.2 *Authoring software*

Simple BIM 8.1 by Datacubist, in Scandinavia distributed as Naviate Simple BIM 2019 by Symetri, was used for authoring property sets in the IFC files. Properties for additional cost items were added by using a native, customizable script (more info on https://simplebim.com).

6.3 Cost estimation software

ISY Beskrivelse 11 by Norconsult informasjonssystemer was used for creating cost items based on the Norwegian standard specification texts (R761 and R762) by importing IFC files with specific coded property sets. The program is available in a 32bit and 64bit version. We used the 64bit version which could handle IFC files larger than 200MB (more info on https://www.nois.no/en/products/specification-and-project-management/isy-beskrivelse).

7 CONCLUSION

This paper presented a possible automated workflow for BIM-based cost estimation based on commercial software answering the following questions:

- How can BIM-based cost estimation for infrastructure projects be developed to enable automatic quantity takeoff?
- What are the challenges in developing an applicable solution based on commercial software?

The short answer to research question one is that BIM-based cost estimation with commercial software works in practice. BIM-based cost estimation reduces the need for drawings and manual specification of work.

When it comes to the second research question, we clustered the restraints on 1) the software, 2) the workflow, 3) the standard specification texts (R761/R762) and 4) the IFC standard. Rulesets must be applied to quantities but not every investigated software is able to do this. A BIM-based workflow for cost estimation shifts the source of error from random human mistakes to systematic mistakes made by machines, something which makes it much more reliable and repetitive. The definition of some cost items in the standard specification texts, especially those with a lump sum, are not adapted to a BIM workflow. Missing specific IFC entities limits the level of automation to some extent but the upcoming IFC 4x3 standard can mitigate this problem.

Automated workflow plays an important role in the digital transformation of the AEC industry. It improves the productivity by automating repetitive and time-consuming tasks and reduces uncertainty by eliminating random human errors. Automated workflows are not flawless, either, but instead of random errors they imply systematic errors which are easier to uncover.

When it comes to further work, there are two limitations in this presented case: the sole use of commercial software and that we investigated only a Norwegian case. Both limited the possible outcome. While the results can be transferred to BIM-based projects in other countries where specification of work is done systematically, the concept must be tested and improved further. On a national level, a critical revision of the general specifications and the mapped cost items are necessary. On an international level, different software and different specification systems need to be tested with BIM-based cost estimation. Moreover, we have to investigate if it is the client, the designer or the contractor who benefits most from a BIM-based cost estimation, and if the delivery method (design-bid-build, design-build, integrated project delivery) plays a role.

REFERENCES

Bertschek, I., Niebel, T. & Ohnemus, J. 2019. Zukunft-Bau. Beitrag der Digitalisierung zur Produktivität in der Baubranche. Endbericht. Mannheim: ZEW – Leibniz-Zentrum für Europäische Wirtschaftsforschung GmbH Mannheim.

Borrmann, A., König, M. & Koch, C. 2018. Building Information Modeling, Springer.

buildingSMART. 2020a. IFC Road [Online]. Available: https://www.buildingsmart.org/standards/calls-for-participation/ifc-road/ [Accessed 03-24-2020].

buildingSMART. 2020b. IFC Road Expert Panel: Project Conclusion, Online Meeting, open to the public. Organized by buildingSMART. Date: 2020-03-25.

Bundesministerium für Verkehr und digitale Infrastruktur 2018. Umsetzung des Stufenplans "Digitales Planen und Bauen".

Cabinet Office UK 2011. Government construction strategy. Cabinet Office London.

Eastman, C. M., Teicholz, P. & Sacks, R. 2011. BIM handbook: a guide to building information modeling for owners, managers, designers, engineers, and contractors, Hoboken, John Wiley & Sons.

FGSV. 2020. Standardleistungskatalog – STLK [Online]. Available: http://www.STLK.de [Accessed 03-11-2020].

Firat, C. E., Arditi, D., Hamalainen, J.-P., Stenstrand, J. & Kiiras, J. 2010. Quantity take-off in model-based systems.

Fürstenberg, D. Information Management in AEC Projects: A Study of Applied Research Approaches. *18th International Conference on Computing in Civil and Building Engineering*; 37th International CIB W78 Conference, in press São Paulo.

Fürstenberg, D. & Lædre, O. 2019. Application of BIM Design Manuals: A Case Study. *Proc. 27th Annual Conference of the International Group for Lean Construction (IGLC)*, 2019/07/03 2019 Dublin, Ireland. Dublin, Ireland, 145–156.

Hartmann, T., Van Meerveld, H., Vossebeld, N. & Adriaanse, A. 2012. Aligning building information model tools and construction management methods. Automation in construction, 22, 605–613.

Ministry of Industry and Trade 2017. BIM Implementation Strategy in the Czech Republic. Ministry of industry and trade.

Monteiro, A. & Poças Martins, J. 2013. A survey on modeling guidelines for quantity takeoff-oriented BIM-based design. *Automation in Construction*, 35, 238–253.

National Building Specification. 2017. An Introduction to Specification Writing [Online]. Newcastle-upon-Tyne. Available: https://www.ribacpd.com/articles/nbs/5340/an-introduction-to-specification-writing/200002/ [Accessed 03-05-2020].

Norwegian Public Roads Administration 2015. V770; Modellgrunnlag, Krav til grunnlagsdata og modeller, Oslo, Vegdirektoratet.

Norwegian Public Roads Administration 2018a. R761; Prosesskode 1; Standard beskrivelse for vegkontrakter; Hovedprosess 1–7, Oslo, Directorate of Public Roads and Transport Department.

Norwegian Public Roads Administration 2018b. R762; Prosesskode 2; Standard beskrivelse for bruer og kaier; Hovedprosess 8, Oslo, Directorate of Public Roads and Transport Department.

Rodrigues de Almeida, P., Bühler, M., Gerbert, P., Castagnino, S. & Rothballer, C. 2016. Shaping the future of construction – a breakthrough in mindset and technology. In: *World Economic Forum in Collaboration with the Boston Consulting Group* (ed.). Cologny/Geneva: World Economic Forum.

Shou, W., Wang, J., Wang, X. & Chong, H. Y. 2015. A Comparative Review of Building Information Modelling Implementation in Building and Infrastructure Industries. *Archives of Computational Methods in Engineering*, 22, 291–308.

Svensk Byggtjänst. 2020. BSAB96 [Online]. Available: https://bsab.byggtjanst.se/bsab/om [Accessed 03-11-2020].

UK Roads Liaison Group 2018. BIM guidance for infrastructure bodies. Department for Transport.

Vaishnavi, V. K. & Kuechler, W. 2015. Design Science Research Methods and Patterns: Innovating Information and Communication Technology, CRC Press.

Automatic detection of construction risks

Q. Cui & A. Erfani
Department of Civil and Environmental Engineering, University of Maryland, USA

ABSTRACT: Risk detection and allocation becomes increasingly important to successful project delivery. This is particularly true for mega infrastructure projects, where technical and institutional complexity increases the risk and challenges for collaboration. While early studies developed various methods and tools for risk detection and allocation, industry practice remains experience-based and focuses on opinions and discussions from subject matter experts. This paper will examine the effectiveness of existing methods to identify construction risks and then presents a novel approach to risk detection using case based reasoning and text mining techniques. The method is built on a large project risk database and features semantic inquiry and automatic generation of risk register according to specific project characteristics. I-495/270 managed lanes project from the state of Maryland will be used to demonstrate the process of automatic risk detection.

1 INTRODUCTION

Construction projects are characterized by carrying a high level of uncertainty and complexity. There are numerous risks and uncertainties laden in every construction project from the beginning phase to completion (Siraj & Fayek 2019). This is particularly true for mega infrastructure projects, where technical and institutional complexity increases the risks and challenges for collaboration. Therefore, appropriate risk detection is necessary to ensure successful project delivery.

While early studies developed various methods and tools for risk detection and allocation, industry practice remains experience-based and focuses on opinions and discussions from subject matter experts. However, the most experienced experts have limited personal experience, and their judgment under uncertainty suffers the cognitive limitation. On the other hand, using conventional risk detection workshops and other questionnaire survey methods is time consuming and expensive for construction companies (Gondia et al. 2020).

In response to these limitations, this paper proposes an automatic approach to detect construction risks using an objective source of data. The model is built on a large project risk database and features semantic inquiry. The main objective of the research is to generate a risk register according to specific project characteristics automatically. To do so, this study uses case based reasoning (CBR) and text mining techniques. CBR and text mining techniques provide a great opportunity to extract useful knowledge and solutions by utilizing numerous data that are mainly left unattended or unutilized in the construction industry currently.

2 BACKGROUND AND LITERATURE REVIEW

2.1 Risk detection in the construction industry

Risk detection is the most crucial step of the risk management process because it provides the basis of the following stages. Risk assessment and allocation are performed on the potential detected risks in the first step (Hwang et al. 2013). Therefore, comprehensive risk identification through project life cycle becomes increasingly important to successful project delivery.

Early studies in the construction industry developed different methods and tools for risk identification. These various tools include brainstorming, checklist analysis, literature and documentation review, workshops, Delphi technique, questionnaire survey, root cause analysis, cause and effect diagrams, SWOT (Strength, Weakness, Opportunity, and Threat) analysis and fault tree analysis (El-sayegh & Mansour 2015; Siraj & Fayek 2019; Tavakolan & Mohammadi 2017)

While these tools and their combinations have been used extremely, industry practice remains experience-based and focuses on opinion and discussion from subject matter experts. This situation poses several key limitations. First, these tools rely on a limited domain of experts' knowledge and experience; second, these methods are time-consuming and too expensive for construction companies. To overcome these limitations, researchers recently utilize objective sources of data instead of subjective data to perform risk detection. Diao et al. in 2020, propose a novel method to detect litigation risks in construction projects using accessible social media data.

Although much effort has been expanded on risk detection, the lack of automatic risk detection approach still remains in the construction industry.

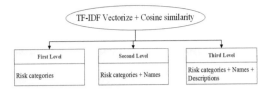

Figure 1. TF-IDF text similarity process for risk registers document.

To fill this gap, we propose a novel approach to risk detection using case based reasoning and text mining techniques. The model is built on a large project risk database and feature semantic inquiry to automatic generation of risk register according to specific project characteristics.

2.2 Effectiveness of existing methods

A risk register is an effective tool that generally records and documents the identified risks within a project. The purpose is to detect, categorize, quantify, and finally track risks in order to improve project performance regarding its objectives (Patterson & Neailey 2002).

Different methods and tools as explained in section 2.1 are applied to prepare risk registers. However, in practice, the prepared risk registers in construction projects are commonly the same. Usually, these risk registers follow a same list of risks. To support this claim, we gathered detailed risk registers for a large number of highway projects in the US over the past 20 years.

After that, text similarity techniques are applied to compare and measure similarities between these documents to each other. Measuring the similarity between two text documents is a broad topic that is interesting for a large number of researchers. There are different methods of comparing similarity using frequency terms, semantic meaning, and word embedding models. The difference between these methods is out of the scope of this research. We apply a common and simple "TF-IDF" technique to measure the similarity between risk registers. Figure 1 illustrates the process of comparing risk register documents for highway projects in the US.

TF-IDF is a simple natural language processing method that calculates the similarity between two documents based on the normalized frequency of words and consideration of the different words. TF (term frequency) explains the frequency of the word in each document in the corpus. It is the ratio of the number of times the word appears in a document compared to the total number of words in that document. It increases as the number of occurrences of that word within the document increases. IDF (inverse data frequency) is used to calculate the weight of rare words across all documents in the corpus. The words that occur rarely in the corpus have a high IDF score. The idea behind this method is when a word repeats again and again in a document it gets more important; however, when

this increases a lot it shows that the word is a common and cannot help to distinguish two documents (Sidorov 2019). Finally, each word in a document based on TF-IDF should obtain a weight and finally a vector will represent the document to calculate similarity. To do so, cosine similarity has proven to be a robust metric for scoring the similarity between two strings, and it is increasingly being used in complex queries (Fan & Li 2013).

Table 1. Risk registers samples.

Project	Name
A	I-95 Express (Phase 3)
B	SR 202L South Mountain Freeway
C	South Capitol Street Corridor Project
D	Three Mile Pensacola Bay Bridge
E	SR 429- Wekiva Parkway
F	I-94 North-South
G	I-75/SR826 Managed Lanes

Project	Jurisdiction	Delivery Method	Contract value Million $
A	FL	DB	1004
B	AZ	DBM	1773
C	DC	DB	669
D	FL	DB	509
E	FL	DBB	684
F	WI	DBB	1625
G	FL	DB	1024

$$TF(w_{ij}) = \frac{\text{number of words i in document j}}{\text{total number of words in document j}} \quad (1)$$

$$IDF(w_{ij}) = Log\left(\frac{\text{number of words i in document j}}{\text{total number of words in document j}}\right) \quad (2)$$

$$Weight(w_{ij}) = TF(w_{ij}) \times IDF(w_{ij}) \quad (3)$$

$$Similarity(Doc1, Doc2) = Cos(V, W) = \frac{V \cdot W}{||V|| \times ||W||}$$
$$= \frac{\sum_{i=1}^{k} Vi Wi}{\sqrt{\sum_{i=1}^{k} Vi^2} \times \sqrt{\sum_{i=1}^{k} wi^2}} \quad (4)$$

A sample of risk registers from available project information is selected to calculate similarity (see Table 1). After that, a python script is developed to calculate the TF-IDF cosine similarity between risk register files based on the process explained in Figure 1. The stop words like "and", "the", etc. removed from data. The similarity result is provided in Table 2 and Figure 2. The result indicates there is a significant similarity among the risk registers. The results show the existing methods of risk detection do not bring new information usually. The average similarity in different risk register comparison is 0.82 in level 3 comparison.

Table 2. Similarity comparison result in third level.

	A	B	C	D	E	F	G
A	1						
B	0.853	1					
C	0.851	0.865	1				
D	0.667	0.728	0.685	1			
E	0.860	0.898	0.856	0.722	1		
F	0.875	0.915	0.881	0.694	0.892	1	
G	0.852	0.903	0.871	0.683	0.860	0.860	1

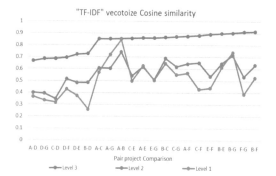

Figure 2. Text similarity comparison among project risk register level by level.

The second point is by adding more detail, risk description, the measure of similarity increased. This supports the issue that in some cases, using different language for risk names does not make a significant difference in risk items.

Figure 2 illustrates the increase of similarity by adding more detail in level 2 and level 3 in risk register pairwise comparison in comparison to comparing just risk categories. There are different examples to support the reason behind the increases of similarity in level 2 and level 3 comparison. For example, "Traffic management" in project F and G is a risk category. However, project B considers "Traffic management" as an individual risk item under the "Design" category. Project E considers "Traffic management" as an individual risk item under the "Construction" category.

Another example, project D includes individual risk item "New permits or new information required" under the "Environmental and Hydraulics" category. Project E includes "Permitting agency requesting something different" and project B contains risk item "Environmental permits". Without consideration of risk description, the text similarity is lower than when the risk description is added and defines that these individual risk items talk about the same context.

2.3 CBR and text mining application in construction

Case based reasoning (CBR) generally is the process of using previous knowledge and solutions to solve a new similar problem (Cheng & Ma 2015). CBR has been applied in various domains of construction industry problems. Ahn et al have utilized CBR to estimate the cost of a new project based on similar projects in 2014. Koo et al. have applied CBR to predict construction project duration in 2010. Dispute management is another topic that researchers have exploited CBR to solve construction dispute cases (Fan & Li 2013; Liu et al 2019). Safety hazard identification is another application of CBR in the construction industry (Goh & Chua 2010). However, there is not any significant research in the literature that pays attention to applying CBR in construction risk detection. On the other hand, text mining and natural language processing provide a powerful tool to extract knowledge inside a large amount of document data (Zhang et al. 2019). In the construction industry, this topic has got more attractive recently in accident analysis and contract analysis (Lee et al. 2020). In this research, case based reasoning and text mining techniques will help to propose an automatic detection of construction risks.

3 DATA AND RESEARCH METHODOLOGY

3.1 Data source and preprocessing

The basis of an acceptable CBR model is to collect appropriate cases. In this research, the risk library is developed by using the project database of major transportation projects in the US. The database includes various types of transportation projects in different states of the US over the past 20 years. Detailed risk registers are extracted from these projects. These risk registers do not follow the same structure usually. To consider this issue, a comprehensive structure for risk library developed to contain all these various unstructured data. And then, all the risk registers data stored in the risk library under the same structure.

3.2 Research design

Aamdot and plaza introduced a classical CBR model involve five phases of case representation, retrieval, reuse, revision, and retention in 1994. The proposed model in this research as shown in Figure 3 is adapted by the classical CBR and adding text mining techniques in the case reuse phase. The details of the proposed model are as follows.

3.2.1 Case representation

The purpose of this step is to define a structure for case storage in the risk library as well as describe a new project to perform risk detection. Each construction project can be described by multiple project features. However, there is not any detailed research in the literature to investigate the project features having a critical impact on the project risks. We extract project attributes based on the risk management

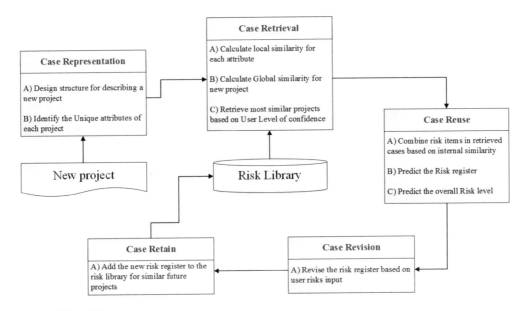

Figure 3. CBR model.

literature and CBR application in other construction domain. Project type, project location, project delivery method, owner type, contract type, contract value are selected to describe project features (Liu et al. 2019).

After that, each case is presented in a vector space based on its attributes. Equation 5 shows how each project is represented in the vector space for further algebra calculation.

$$C_i = \begin{bmatrix} \text{Project type} \\ \text{Project location} \\ \text{Project delivery method} \\ \text{Owner type} \\ \text{Contract type} \\ \text{Contract value} \end{bmatrix} \quad (5)$$

3.2.2 Case retrieval

The purpose of this step is to search among cases in the risk library and retrieve the most similar cases based on similarity calculation in the vector space. The new project will be represented as Equation 1. When all cases represented in vector space, it is easy to calculate similarity by using different methods.

In this research, we follow the local-global method to measure similarity (Cheng & Ma 2015). When the new project is represented as $N = \{n_1, n_2, \ldots, n_6\}$, similarity is calculated for each attribute in the first step. To do so, we use some predetermined rules to calculate cosine similarity between texts. Word embedding models are used to calculate similarity for each attribute like project type, location, delivery method, etc. word embedding models are pertained neural network models that define a different word in a vector space.

Local similarity $(L_{sm})_i$ is calculated for each attribute by Equation 6.

$$L_{sm}(N1, C1) = L_{sm}(\text{Project type}_{newcase}, \text{Project type}_{storagecase}) = \cos inesimilarity(text1, text2) \quad (6)$$

where $N =$ new case; $C =$ storage cases; and $attribute =$ project type.

In word embedding models, word2vec, as explained each word will transform into a vector. Then, cosine similarity between these words is calculated by using Equation 4 (Fan & Li 2013).

After calculating the local similarity for each attribute, the global similarity (G_{sm}) is calculated based on the average of local similarities by Equation 7. The G_{sm} shows how similar the new case is to each case in the risk library.

$$G_{sm}(N, C) = \frac{\sum_{i=1}^{6} L_{sm}(N, C)}{6} \quad (7)$$

Finally, all cases in the risk library are sorted based on the G_{sm} value. The user confidence level will determine the minimum value of G_{sm} for a case to be selected for the next phase. All the cases with at least G_{sm} value based on the user confidence level will be retrieved for the next phase.

3.2.3 Case reuse

The purpose of this step is to predict the risk register for the new project based on similar retrieved cases. In this step, text mining techniques are applied to compare internal similarity between risk items. To do so,

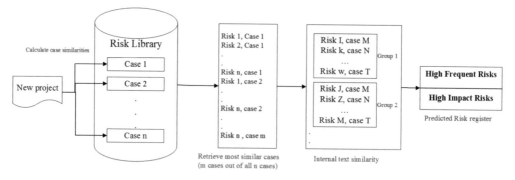

Figure 4. Case reuse process.

as explained in section 2.2, the TF-IDF text mining method is applied to calculate the internal similarity between all risk items in retrieved cases. Each risk name and risk description is converted to a text and then text similarity among individual risk items is calculated. The process of the case reuse step to obtain high frequent and high important risks is shown in Figure 4.

By calculating the internal similarity among retrieved risk items, those risks that have high frequency among similar cases are detected. On the other hand, it is possible that a risk item not repeated in other projects; however, that risk has a high impact on project objective based on that specific risk register. Those risks should consider in the predicted risk register as well. Therefore, the final predicted risk register will include those high frequent risk items as well as high impact risks from similar retrieved cases.

3.2.4 Case revision

The purpose of this phase is to modify and add any new risk item by the project team to the risk register. Because, it is possible that any specific risk item affects the new project which is not considered in the previous projects, this phase considers a way to revise and improve the output of the model.

3.2.5 Case retain

The last step in the CBR cycle is the case retain phase. The purpose of this phase is to improve the quality of the risk library by adding new cases to it. The predicted risk register will be added to the risk library to use for similar projects in the future.

4 CASE STUDY

I-495/270 managed lane project is selected as the case study to demonstrate how the process of automatic risk detection will occur. I-495/270 is a remarkable program to reduce congestion along the entire length of I-495 (Capital Beltway) as well as the entire length of I-270 (Dwight D. Eisenhower Memorial Highway) up to I-70 in Frederick County in Maryland State.

The purpose of the project is to improve trip reliability and multimodal mobility and connectivity on I-495 and I-270. First, to provide the risk register by using the model, the characteristics of the project should be represented in a vector. The project type is "Managed lanes", project location "Maryland State", project delivery method is "Public Private Partnership (p3)", and the owner is "Government". The "contract type" and "contract value" can be determined in different values based on user preferences.

Case similarity measurement will be calculated based on predetermined rules. Most similar cases based on the user confidence level will be retrieved from the risk library. Finally, the text mining techniques applied to the shortlist of similar cases. A long list of frequent and high impact risks will be the output of the model.

5 CONCLUSION

Including multiple stakeholders, having a dynamic environment, and influencing external factors caused a high level of risks and uncertainties in infrastructure projects. Therefore, risk detection becomes increasingly crucial to assure successful project delivery. Previous studies rely on involving subject matter experts to develop project risk registers that is a time consuming and inadequate approach. This paper develops a data based approach to detect construction risks automatically. By using case based reasoning and text mining techniques, this study investigates an opportunity to detect high frequent and high impact risks based on a large database of previous similar projects. With a case study of I495/270 managed lanes in Maryland State, the paper demonstrates the process of automatic risk detection. This approach represents enhancement over other existing methods that have been used in the construction industry.

REFERENCES

Aamodt, A. & Plaza, E. 1994. Case-based reasoning: Foundational issues, methodological variations, and system approaches, *AI communications, 7.1)*: 39–59.

Ahn, J. Ji, S. H. Park, M. Lee, H. S. Kim, S. & Suh, S. W. 2014. The attribute impact concept: Applications in case-based reasoning and parametric cost estimation, *Automation in construction, 43*: 195–203.

Cheng, J. C. & Ma, L. J. 2015. A non-linear case-based reasoning approach for retrieval of similar cases and selection of target credits in LEED projects, *Building and Environment, 93*: 349–361.

Diao, C. Liang, R. Sharma, D. & Cui, Q. 2020. Litigation Risk Detection Using Twitter Data, *Journal of Legal Affairs and Dispute Resolution in Engineering and Construction, 12(1)*: 04519047.

El-Sayegh, S. M. & Mansour, M. H. 2015. Risk assessment and allocation in highway construction projects in the UAE, Journal of Management in Engineering, 31(6): 04015004

Fan, H. & Li, H. 2013. Retrieving similar cases for alternative dispute resolution in construction accidents using text mining techniques, *Automation in construction, 34*: 85–91.

Goh, Y. M. & Chua, D. K. H. 2010. Case-based reasoning approach to construction safety hazard identification: adaptation and utilization. *Journal of Construction Engineering and Management, 136*(2), 170–178.

Gondia, A. Siam, A. El-Dakhakhni, W. & Nassar, A. H. 2020. Machine Learning Algorithms for Construction Projects Delay Risk Prediction, *Journal of Construction Engineering and Management, 46(1)*: 04019085.

Hwang, B. G. Zhao, X. & Gay, M. J. S. 2013. Public private partnership projects in Singapore: Factors, critical risks and preferred risk allocation from the perspective of contractors, *International Journal of Project Management, 31(3)*: 424–433.

Lee, J. Ham, Y. Yi, J. S. & Son, J. 2020. Effective Risk Positioning through Automated Identification of Missing Contract Conditions from the Contractor's Perspective Based on FIDIC Contract Cases. *Journal of Management in Engineering, 36(3)*: 05020003.

Liu, J. Li, H. Skitmore, M. & Zhang, Y. 2019. Experience mining based on case-based reasoning for dispute settlement of international construction projects, *Automation in Construction, 97*: 181–191.

Patterson, F. D. & Neailey, K. 2002 A risk register database system to aid the management of project risk, *International Journal of Project Management, 20(5)*: 365–374.

Sidorov, G. 2019. Vector Space Model for Texts and the tf-idf Measure, *In Syntactic n-grams in Computational Linguistics*: 11–15. Springer: Cham

Siraj, N. B. & Fayek, A. R. 2019. Risk identification and common risks in construction: Literature review and content analysis, *Journal of Construction Engineering and Management, 145(9)*: 03119004.

Tavakolan, M. & Mohammadi, A. 2018. Risk management workshop application: a case study of Ahwaz Urban Railway project, *International Journal of Construction Management, 18(3)*: 260–274.

Zhang, F. Fleyeh, H. Wang, X. & Lu, M. 2019 Construction site accident analysis using text mining and natural language processing techniques, *Automation in Construction, 99*: 238–248.

Optimization method for choosing a set of means for probability of failure reduction of critical infrastructures

O.S. Burukhina
Ural Federal University, Yekaterinburg, Russia

A.V. Bushinskaya & S.A. Timashev
Science and Engineering Center "Reliability and Safety of Large Systems and Machines" of Ural Branch Russian Academy of Sciences & Ural Federal University, Yekaterinburg, Russia

ABSTRACT: Various measures can be implemented to improve mechanical system reliability. Choosing the optimal set of measures is a difficult task. It can be solved in three steps: forming a full set of possible combinations of safety measures, assessing the reliability increment increase and the cost of each combination of sets; selecting the champion set (the maximal reliability increase), or by optimization. The main difficulty in posing such a problem is in describing how each of the used safety module interacts with the protected system; assessing what reliability increase will the safety means deliver (its cost known as an initial data), for a specific potentially dangerous object of known configuration, in the context of a specific incident scenario or normal operation. This paper describes a new universal matrix-based algorithm that solves the increment increase reliability problem due to the implementation of the safety means for complex systems.

1 INTRODUCTION

The mechanical system is a holistic set of the interconnected elements separated from its environment. Systems are simple if they can be presented as a chain of sequentially or parallel connected elements. Failure of at least one of sequentially connected elements or all parallel connected elements leads to the system's failure. In this paper disunited complex systems (DCS) are considered. Such complex systems are characterized by possessing a combined structure, when each of its subsystems (and their reliability) can be considered separately, interdependently from the basic system. This paper addresses a complex system with elements, whose reliability is an independent event defined in advance.

Reliability block diagram is a widely used method of system reliability analysis (Timashev 2016). It is used if it's possible to split the complex system into subsystems, for each of which the probability of success or failure can be defined. It is assumed that when in operating, the object can be only in a workable or failure state. Hence, the element is characterized by the probability of success p_i or the probability of failure q_i.

In this paper an innovative method for reliability assessment of mechanical systems is considered. It is based on the reliability matrix, which is the difference between the characteristic matrix and the connection matrix, which are formed using the structural scheme of the analyzed complex system. Probability of failure (PoF) or probability of success (PoS) of the system is equal to the determinant of the reliability matrix.

The innovative use of the proposed method significantly reduces the time of analysis and decreases the risk of errors, when working with multi-element mechanical systems (Lin et al. 2018). Regarded method can be used universally if it is possible to interpret the task of the corresponding subject area in the mechanical system form. Eventually the method can be used for automated modeling, as a means to reduce time and labor costs when analyzing the reliability of complex objects models.

2 APPLICATION OF THE RELIABILITY MATRIX TO ASSESS THE SYSTEM RELIABILITY

The existing matrix reliability analysis method is based on the logical-probabilistic method, where Boolean functions are used to express analytic conditions of systems operability and their strict interconnections with probability functions. A matrix of incompatible states is formed to enumerate all possible states of the system. Each state in the matrix is checked by the logical condition for operability (Chernov et al. 2004). This method is not wide-spread, since it is quite laborious and inconvenient method even with the matrix application.

According to the method based on graph theory and Boolean functions, mechanical system is represented

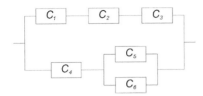

Figure 1. Reliability block diagram of a mechanical system.

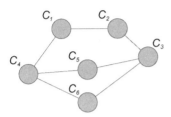

Figure 2. System graph for the mechanical system.

as a graph, its elements are nodes of the graph, and the connection line between the elements is indicated by a link (Tang 2001). Figures 1, 2 show a reliability block diagram of an arbitrary mechanical system and its graph correspondingly.

A complete expression for system reliability is given by three matrices: the components physical interconnection matrix Ω, the components reliability matrix R, and the system reliability matrix Γ.

The component reliability matrix R of a complex system comprised out of n components is diagonal:

$$R = \begin{pmatrix} p_1 & 0 & \ldots & 0 \\ 0 & p_2 & \ldots & 0 \\ \ldots & \ldots & \ldots & \ldots \\ 0 & 0 & \ldots & p_n \end{pmatrix} \quad (1)$$

where p_i = reliability of i-th element.

The system reliability matrix Γ of the system is also a diagonal matrix and is defined as:

$$\Gamma = R - \Omega \quad (2)$$

where Ω – is the components connection matrix Ω.

The system reliability is then calculated as a determinant of the system reliability matrix

$$P_s = \det(\Gamma) \quad (3)$$

In the proposed method the elements and sets of elements connected in series are characterized by their (PoS), and the elements and systems of elements connected in parallel are characterized by their PoF.

Before analyzing reliability of complex systems, they should be reduced to simple ones by replacing each of their subsystems by elements with the same PoS and PoF. This is a crucial moment, as it helps avoiding mistakes. For complex systems (the number of elements greater than 20), it will be difficult or not possible to repeat such a procedure for all subsystems. In order to avoid their difficulty we represent the system reliability matrix Γ as the difference between the component reliability matrix R and the components connection matrix Ω

These matrices are determined on the basis of the reliability block diagram, but could also be formed using Boolean functions or Bryant trees (Bryant 1986)

The components reliability matrix R is always a diagonal matrix. It is formed in such a way that the PoF values or reliability of elements are placed on the matrix main diagonal. The values of q_i and p_i are determined as follows:

– The reliability block diagram of mechanical system is parsed into simple elements and subsystems;
– Subsystems are considered as separate systems, and are sequentially parsed, in their turn, into elements and subsystems until only simple elements remain;
– All simple elements are associated with PoS, if they are connected in series with the elements of their subsystem, and with the PoF, if they are connected in parallel with the elements of their subsystem.
– The connectivity matrix Ω is formed as follows:
– If elements C_i and C_j ($j = i + 1$) are connected in series, then -1 is placed in the ω_{ij} cell; if they are connected in parallel, then 1 is placed in the ω_{ij} cell (the matrix is filled above the main diagonal);
– If elements C_i and C_j ($j \neq i + 1$) are the subsystem initial and final element of identically connected elements, then -1 is placed in the ω_{ij} cell (the matrix is filled below the main diagonal). This doesn't hold true if the given chain of elements represents the whole system;
– If the chain of elements contains complex elements, then the above rules hold true for the last element of the first subsystem and to the first element (if the matrix is filled above the main diagonal) and to the last element (if the matrix is filled below the main diagonal) of the next subsystem.

If the initial system is ultimately presented as a set of series connected elements, the matrix Γ determinant module is interpreted as PoS of the whole system; in case of parallel connection it is interpreted as PoF.

Restrictions of the method:

– The analyzed mechanical systems are considered as hierarchical systems with: finite number of subsystems, definite structure and scheme of interconnections of its subsystems and elements, and, hence, the reliability block diagram;
– Each subsystem is considered as consisting of its own separate subsystems and also meets the described above requirements;
– For each element and the whole system only two possible states are allowed: working state or failure state;
– Failure of one element of the system does not affect the performance of any other element of the same system.

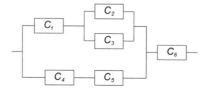

Figure 3. Reliability block diagram of the mechanical system.

Table 1. The reliability values of considered mechanical system elements.

Element	C_1	C_2	C_3	C_4	C_5	C_6
Reliability	0.98	0.8	0.8	0.96	0.99	0.99

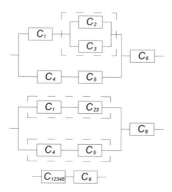

Figure 4. Decomposition stages of the considered mechanical system.

3 A SHOW-CASE APPLICATION OF THE METHODOLOGY

Consider the system shown in Figure 3, where elements C_4, C_5 could be considered as a security subsystem for the remaining system (C_1–C_3, C_6). Start reliability analysis by selecting all its subsystems. Elements C_2 and C_3 are connected in parallel, they can be represented by subsystem C_{23}, its characteristic is PoF. Elements C_1, C_{23}, and C_4, C_5 are serially connected elements, their characteristic is PoS. These elements are represented by the subsystems C_{123} and C_{45} with the corresponding characteristics. Subsystems C_{123} and C_{45} are connected in parallel, they can be represented by subsystem C_{12345}.

In turn subsystems C_{12345} and C_6 are serially connected and represent the system in general (Figure 4); hence, the determinant module of matrix Γ is interpreted as the system PoS.

The components reliability matrix R will take following form:

$$R = \begin{pmatrix} p_1 & 0 & 0 & 0 & 0 & 0 \\ 0 & q_2 & 0 & 0 & 0 & 0 \\ 0 & 0 & q_3 & 0 & 0 & 0 \\ 0 & 0 & 0 & p_4 & 0 & 0 \\ 0 & 0 & 0 & 0 & p_5 & 0 \\ 0 & 0 & 0 & 0 & 0 & p_6 \end{pmatrix} \quad (4)$$

Now construct the components interconnection matrix Ω. Elements C_2 and C_3, connected in parallel are a subsystem, hence $\omega_{23} = \omega_{23} = -1$. Elements C_4, C_5 connected in series, are a subsystem, hence $\omega_{45} = \omega_{54} = 1$. Element C_1 is connected in series with element C_2 and C_3, hence $\omega_{12} = \omega_{31} = 1$. Elements C_1, C_2, C_3 are connected in parallel with C_4 and C_5, hence $\omega_{34} = \omega_{51} = -1$. Elements C_1, C_2, C_3 C_4, C_5 are connected in parallel with C_6. They represent the system as a whole, hence $\omega_{56} = 1$. For all other elements of matrix Ω $\omega_{ij} = 0$ (no correlation exists)

$$\Omega = \begin{pmatrix} 0 & 1 & 0 & 0 & 0 & 0 \\ 0 & 0 & -1 & 0 & 0 & 0 \\ 1 & -1 & 0 & -1 & 0 & 0 \\ 0 & 0 & 0 & 0 & 1 & 0 \\ -1 & 0 & 0 & 1 & 0 & 1 \\ 0 & 0 & 0 & 0 & 0 & 0 \end{pmatrix} \quad (5)$$

Therefore, the reliability matrix Γ will be:

$$\Gamma = R - \Omega = \begin{pmatrix} p_1 & -1 & 0 & 0 & 0 & 0 \\ 0 & q_2 & 1 & 0 & 0 & 0 \\ -1 & 1 & q_3 & 1 & 0 & 0 \\ 0 & 0 & 0 & p_4 & -1 & 0 \\ 1 & 0 & 0 & -1 & p_5 & -1 \\ 0 & 0 & 0 & 0 & 0 & p_6 \end{pmatrix} \quad (6)$$

Let us introduce the values of reliability and PoF of the elements for numerical calculations.

Substitute these values in the reliability matrix Γ. Now the module of the matrix Γ determinant can be calculated, which would be the system PoS, since the system is formed by two subsystems, connected in series:

$$P_s = |\det(\Gamma)| = 0{,}9889987 \quad (7)$$

The system reliability can be also directly and independently assessed by the following expression

$$P_s = \{1 - [1 - p_2 \cdot (1 - q_3 \cdot q_4)] \cdot (1 - p_5 \cdot p_6)\} \cdot p_7$$
$$= 0{,}9889987 \quad (8)$$

Hence, the proposed algorithm is correct and could be used for assessing reliability of complex, multi-element systems and infrastructures of different nature, provided they fit the stated above classification of complex systems and restrictions of the method

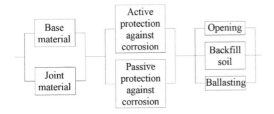

Figure 5. Reliability block diagram of linear part of pipelines.

4 APPLICATION

Reliability block diagram is often used to analyze the reliability of various complex systems, electrical connection diagrams, operation diagrams of electrical substations and switchgears, pipelines, as well as building services systems (Rudachenko & Baikin 2008; Tarasik 2013; Voronin 2008). For instant, Figure 5 presents one of the possible approaches to the analysis of the failure-free operation of linear part of pipelines, which is applicable to oil and gas pipelines and delivery main pipeline of services systems.

The reliability block diagram consists of series-connected blocks with parallel connection of elements within the blocks. The first block represents the reliability components of the pipe elements themselves. Depending on the used rolled pipeline cross section, the block can be supplemented with a spiral weld or a weakness zone, because practice shows that pipelines failures caused by the elements are independent events. The second block presents active and passive protection of pipes against corrosion, the third block unites the elements directly mechanically connected to the pipeline. If each element is assigned a PoF or PoS value, then the reliability analysis of the linear part of the pipeline can be reduced to the method described above. It should be noted that when composing a model, its elements are combined into blocks that have a similar nature of links.

Also, the proposed matrix method considered in the article has the potential to be implemented in the form of a program code and included in CAD software.

5 CONCLUSION

The method proposed in this paper:

- contributes to solving the curse of dimensionality for the considered class of complex systems (such as civil and industrial infrastructures) and can significantly reduce the computer time needed for conducting design calculations;
- permits conducting low cost comparative design of competing structural systems and potentially dangerous objects in terms of their reliability/safety/security;
- permits compilation of the R, Ω and Γ matrices using the reliability block diagram, as well as Boolean functions and Bryant trees).

A system with 20 elements can be approximately in 1,000,000 states. The Boolean approach and Bryant trees approach would have to account for all of them. Our method represents the system as 20 × 20 matrix, consisting only 400 elements.

REFERENCES

Bryant, R.E. 1986. Graph-based algorithm for Boolean function manipulation *IEEE Transactions on Computers* C-35(8): 677–691.

Chernov, V.Y. 2004. *Reliability of Aviation Instruments and Measuring and Computing Systems*. Saint Petersburg: Saint-Petersburg State University of Aerospace Instrumentation.

Lin, S., Wang Y. & Jia L. 2018. System reliability assessment based on failure propagation processes. *Complexity* 2018: 1–19.

Rudachenko, A.V. & Baikin, S.S. 2008 *Operational reliability of pipeline systems*. Tomsk: Publishing House of Tomsk Polytechnic University.

Tang, J. 2001. Mechanical System Reliability Analysis Using a Combination of Graph Theory and Boolean Function. *Reliability Engineering and System Safety* 72: 21–30.

Tarasik, V.P. 2013 *Mathematical modeling of technical systems*. Minsk: New Knowledge.

Timashev, S.A. 2016. *Infrastructures* (*Reliability. Longevity vol. 1*). Yekaterinburg: Scientific and Publishing Council of the Ural Branch of the Russian Academy of Sciences.

Voronin, A.V. 2008. *Mechatronic Systems Modeling*. Tomsk: Publishing House of Tomsk Polytechnic University.

A new approach for delay analysis process

H.A. Dikbas & C. Durmus
Istanbul Medipol University, Istanbul, Turkey

ABSTRACT: Float time is defined as a delay of early start activities, which is not causing a delay of the entire project. Nowadays, one of the most unknown answer around the construction sector is who owns the float and how can it be used. Especially, during large scale and complicated projects, reckless usage of float times by the name of 'first come first served', causes unexpected disputes as a result of change of the critical path. In this study parameters of delays such as float concepts were stated. Relations between float usage and delay analysis types were investigated through an example. Importance of an objective float sharing was emphasized due to giving confidence to parties by avoiding leaving an open field that will cause conflict between them. Depending on these subjects, the main purpose of this study is to clarify why it is needed to search for a new, fair and analytical delay analysis and also a hypothesis was mentioned.

1 INTRODUCTION

Critical path method (CPM) is the commonly used progress to plan the project schedule. Through this method, from the beginning to the end, line of activities, their early and late initial and final start dates as well as their float times can be calculated. Thus, parties can plan their workflow during the construction process.

Float can be described as one or more activities delayable time, which does not extend the entire project end date. Float times are sometimes used due to a predecessor activity's delay or some other requirements of parties in their own way. However, neither laws nor private sector or academy have any mathematical or scientific common idea about float ownership. Thereby, 'first come first used' philosophy has been taken part in the float usage progress. This is not a scientific philosophy, so it lets parties use these limited float times in their own way. Anysz (2019) said that, when the float time is consumed, task becomes critical. Thus, parties start to have delay clashes between each other. In brief wastage of float times bring undesired processes for the project. However, todays delay analysis methods (DAM) do not satisfy parties due to their fairness and nonanalytical calculations. Especially the most important delayable times called 'float' are ignored in these methods. In this paper, delay analysis methods were discussed, new iteration was suggested, and an analysis hypothesis was developed.

2 LITERATURE STUDY

It was stated in the study conducted by Yang (2016) that four kinds of float types were calculated with traditional CPM. These float types were described with their formulas. In the formulas, the dates of early start, early finish, late start, late finish were used.

2.1 Total float (TF)

Total float (TF): The most used float type. The amount of time an activity can be delayed without affecting the overall project's end time. When A is the predecessor activity of B

$$TF_A = LF_A - EF_A = LS_A - ES_A$$

2.2 Free float (FF)

The amount of delayable time for an activity without effecting the early start date of it's successor activity

$$FF_A = ES_B - EF_A - (Lag\ time)$$

2.3 Interfering float (INTF)

The amount of delayable time for an activity retarding the early start date of it's successor activity however not effecting the late star date of it's successor activity

$$INTF_A = TF_A - FF_A$$

2.4 Independent float (INDF)

The amount of delayable time for an activity without delaying the early start date of it's successor activity and without restricting the predecessor activity as well.

When A is successor activity of C and predecessor activity of B

$$INDF_A = FF_A - INTF_C = FF_A - (TF_C - FF_C)$$

In conclusion, according to the Yang, only the total float is used among the floats types specified in the last sixty years in planning and delay analysis. Total float accuracy is decreasing hence the difficulties of planning and moving away from the original planning as the projects are developing.

Garza & Prateapusanond used sharing method instead of "first come first served" approach to define the float ownership. Total float time was divided equally between the employer and contractor in their case study. In addition, it was emphasized that float time division must be defined in the contract.

Referring to Spanos (1984), if employer has the absolute float ownership, he could control the time of all non-critical activities without responsibility. However, if contractor has the float ownership, he could claim damages for every non-critical delay. For these reasons, float ownership must be stated in the contract.

Float types, which are contained in the New Engineering Contract (NEC), written by Institution of Civil Engineers (ICE) were paraphrased by Evans. According to him, total float, by the "first come first served" basis, can be used by both employer and contractor to decrease delay damages. Free float ownership was given to the contractor in the NEC as long as it is not exceeded. Consequently, total float usage belongs to both employer and contractor however free float belongs to the contractor.

According to Roux (2017), employer owns the float time if an existing delay is non-critical besides contractor cannot claim any extension of time. Nevertheless, contactor can claim damages for an employer's risk event.

Float ownership was primarily elaborated in Society of Construction Law (SCL 2017)'s latest protocol. According to the Clause 8.1, contractor may insist that he owns the float because works may not advance as planned so he would like to use the float time. Also, if a non-critical delay happens by an employer risk event, contractor can have an extension of time. In the other hand, employer may not like to give an extension of time when an employer risk event is not critical. Eventually, SCL Protocol doesn't identify the float ownership clearly and the project owns the float. In addition, Clause 8.2 indicates that float ownership should be fairly explained in the contract.

Trauner (2009) mentioned that, a wise contract writer should give the float ownership to the employer because, relying to the float time, a cunning contractor can change activity schedules and cover-up his mistakes. Also mentioned that float ownership is ignored in contracts, so project owns the float.

To make an accurate delay analysis, primary delay parameters such as float ownership, concurrency and criticality must be clarified by parties in the contract first. These parameters can have assumptions in reasonable level but mainly they should be mathematical and be closed for comments. Therefore, the analyst can do the calculation without dilemmas.

3 DAM CALCULATIONS IN A CASE STUDY

During this section, different delay analysis methods were evaluated in a simple program. Relations between float times and float ownership according to delay analysis methods were evaluated. Oracle Primavera P6 v.18 was used to create and to analyze the program. Activity list and float times were given in Table 1 and Table 2.

It was calculated that total planned project duration is 26 days. However, six delay events happened during the process and the project was delayed for 5 days, therefore its actual duration was increased to 31 days. During the analysis, employer's delay events were categorized as 'Excusable-Compensable'(EC) in the other hand, contractor's delay events were categorized as 'Nonexcusable- Noncompensable'(NN). According to Table 3, employer's total delay is 10 days

Table 1. List of activities.

Activity	Activity Time	Predecessor Activities
A	3	–
B	3	A
C	3	B – G
D	6	C
E	4	A
F	4	E
G	4	F
H	3	A
I	5	H
J	2	I
K	5	J
L	4	K – M
M	5	J
N	2	D – L

Table 2. Times.

Activity	Total Float	Free Float
A	0	0
B	9	9
C	0	0
D	0	0
E	0	0
F	0	0
G	0	0
H	2	0
I	2	0
J	2	0
K	2	0
L	2	2
M	2	0
N	0	0

Table 3. List of delayed activities.

Delay Number	Delayed Activity	Liability Type	Delay Time (Day)
1	A	NN	2
2	B	EC	5
3	B	NN	3
4	F	EC	2
5	J	EC	3
6	L	NN	2

and contractor's total delay is 7 days. Thus, it can be seen that only 5 days of these delays were critical.

Referring to the 'as planned' schedule (APS) critical path of the project was A-E-F-G-C-D-N activities way but after delays critical path was switched to A-H-I-J-K,M-L-N activities way. To analyze this, in this study, five delay analysis methods were used to identify who was responsible and how much did float times become useful to decrease the damage. Used delay analysis methods are 'as planned' vs 'as built', 'as planned but for', 'collapsed as built' (CAB), 'impacted as planned' (IAP) and 'window slice analysis'(WSA).

It was seen that methods were used with different perspectives during the delay analyze process. For example, when working on the 'as planned' vs 'as built' method, Kamandang et al. (2017), calculated the global delay time between 'as planned' end date and 'as built' end date. However, Braimah (2013) individually calculated delay times for both employer and contractor. In this study Braimah's way was used because of more detailed results.

3.1 As planned vs as built

'As planned' program is 26 days. 'As built' program is 31 days. $\Sigma NN = 7$ days and $\Sigma EC = 10$ days so concurrent delay is 7 days. Total EC delay is $10 - 7 = 3$ days. Total NN delay is $5 - 3 = 2$ days. It can be said that, according to the 'as planned' vs 'as built' method, float times are ignored and weren't used mathematically during the analyze process.

3.2 As planned but for

In this method there are two ways to calculate. Depend on choice, contractor's or employer's delays are added on the 'as planned' schedule. The difference between chosen party and 'as built' schedule (ABS) results other party's delay time.

Contractor's perspective: $APS + NN = 26 + 7 = 33$ days. Employer's delay time $= (ABS) - (APS + NN) = 31 - 33 = -2$ days.

Employer's perspective: $APS + EC = 27 + 10 = 36$ days. Contractor's delay time $= (ABS) - (APS + EC) = 31 - 36 = -5$ days.

Referring to these results, if there are many activity paths, this method may not give accurate results. In addition, float times and usage were not taken any role in this analyze method.

3.3 Collapsed as built

In this method there are also two ways to calculate. Depend on choice, contractor's or employer's delays are removed from the 'as built' schedule. The difference between chosen party and as planned schedule results other party's delay time.

Employer's point of view: Employer's delay events were removed from the 'as built' schedule. In this case, $(ABS) - (EC) = 28$ days. Employer's delay time $= 28 - 26 = 2$ days and Contractor's delay time $= 31 - 28 = 3$ days.

Contractor's point of view: Contractor's delay events were removed from as – built program. In this case, $(ABS) - (NN) = 28$ days. Contractor's delay time $= 31 - 28 = 3$ days and Employer's delay time $= 28 - 26 = 2$ days.

Eventually it was seen that, float ownership had not become a parameter in this solution. Also, two perspectives of an analysis may be the frontier of new disagreements.

3.4 Impacted as planned

The first delay has affected the end date by two days. This was a NN delay.

The second delay has occurred during the noncritical B activity. This five days old EC delay did not affect the project due to its nine days of float time. However, by the name of first come first served basis, employer left four float days.

Although third and fourth delays have occurred chronologically at the same time, they took place in the different activity ways. Third delay has occurred in activity B and the fourth one in the F activity. Because of activity F's critical two days of delay, it extended the projects end for two days. Thus, float time of activity B was increased to six days so fourth activity did not extend the project duration.

Fifth delay has occurred on the activity J. Before the fifth delay, previous delays increased the float time from two days to four days. The fifth delay which was caused by employer did not affect the project schedule but decreased the float time to one day which can be considered as a warning.

The last delay has occurred during the activity J. Two days of NN delay have changed the critical path and effected for one day due to the fifth delays float consumption. Three of four days of float time were used by employer, contractor's delay has changed the critical path.

Despite the 'Impacted as planned' method considers float times, its fairness is contested because even the contractor caused lesser delay than the employer, contractor became responsible for critical delay.

3.5 Window slice analysis

The project was sliced weekly into five sections. Every delay was added to its window chronologically.

Table 4. Delay analysis results.

Analysis Method	EC (Day)	NN (Day)	Concurrent Delays (Day)
As planned vs as built	3	2	7
As planned but for			
– Employer's vision	10	−5	–
– Contractor's vision	−2	7	–
Collapsed as built			
– Employer's vision	2	3	–
– Contractor's vision	2	3	–
Impacted as planned	2	3	–
Window slice analysis	2	3	–

In the first window, NN delay has delayed the project end for two days due to its criticality.

In the second window, delay in the activity B did not extend the project end because of its nine days of float time.

In the third window, delay in the critical activity F extended the project duration for two days however, delays in D and I activities did not extend the project duration due to their sufficient float times.

There was no delay in the fourth window, so project duration was not affected.

Last and fifth window had two days of critical delays in the activity J which impacted the project due days for one day.

Although window slice analysis considers float times like IAP, WSA also ignores float ownership. As it was similarly seen in the IAP, despite the contractor caused lesser critical delay than the employer, the delay in the fifth window had made contractor became responsible for the critical delay.

According to Table 4 each delay analysis result can be seen individually. Referring these results, both selection method and accuracy of results are open for new disputes.

4 METHODOLOGY

The example given in Table 1 was applied in the hypothesis. Firstly, delay parameters were identified. The free float ownership totally belongs to the contractor. Delays which happen during the free float will not be counted as a delay. Also interfering float will be equally shared between parties. If one party depletes its own interfering float time, can use other parties float time by a written permission. The party who cedes its float time, will not claim its float time back and will be responsible for its own future delays. Parties can also make a consensus about unusable float time to ensure safety. By this clause, certainty was aimed so parties will be more careful and determined when they make a decision.

According to the SCL 2017 Protocol, explanation of true concurrent delay is the occurrence of two or more delay events at the same time, one an employer risk

Table 5. Free float and total float durations.

Activity	Total Float	Free Float
A	0	0
B	9	9
C	0	0
D	0	0
E	0	0
F	0	0
G	0	0
H	0	2
I	0	2
J	0	2
K	0	2
L	2	2
M	0	2
N	0	0

Figure 1. Impacted and nonimpacted cumulative delays.

event, the other a contractor risk event and the effects of which are felt at the same time. In this hypothesis, concurrency of a delay depends on its effect on the due date of the project.

Free and total float times for activities were shown in Table 5.

Secondly, delays are sorted by their criticality in their own activity way. In this example, there are three activity ways. In the first activity way, only critical delays will exist, and two other ways will have noncritical delays. Whichever delay effects the project duration will be added to critical way line.

According to the example, first and fourth delays have happened in the original critical path.

As shown in Figure 1, D1 and D4 have happened in the critical path directly. D2 and D3 have happened in a noncritical activity way. Total sum of eight days, these delays had never affected the project duration because of their nine days of float time. On the other hand, D5 also occurred on the second noncritical work path. However, one day of D6 was noncritical, second day became critical due to D5 usage of float time.

Three activity paths of this example assemble in the last activity which is activity N so any impacted delay in these paths will affect the last activity and the Project schedule. For this reason, "one is an employer risk event, the other is a contractor risk event, and the effects of which are felt at the same time" of the SCL concurrency perspective can be accepted.

Parties will not be responsible for delays in the first noncritical delay path because none of those delays had any effect on the project duration. After

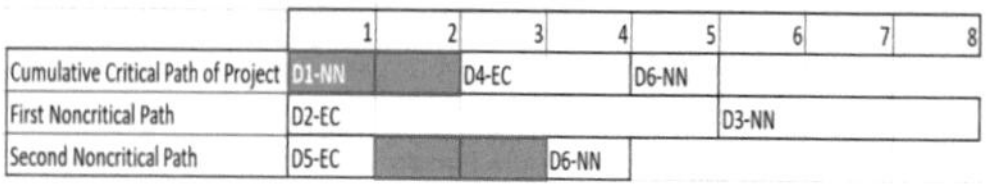

Figure 2. Concurrency between parties and paths.

Table 6. Hypothesis comparison with existing methods.

Analysis Method	EC (Day)	NN (Day)	Concurrent Delays (Day)
As planned vs as built	3	2	7
As planned But For			
– Employer's vision	10	−5	–
– Contractor's vision	−2	7	–
Collapsed as built			
– Employer's vision	2	3	–
– Contractor's vision	2	3	–
Impacted as planned	2	3	–
Window slice analysis	2	3	–
The Hypothesis/Model	3	0	2

this elimination, there are two paths which remain to be calculated in this method.

Thirdly, the second noncritical path which includes D5 and D6, has the key position. It does not seem to be fair that, three of four days of float usage by the employer and remained only one day of float time for future delays. If employer would use two days instead of three days, contractor would not be responsible for one impacted day.

According to the original plan, D5 and D6 had no free float but they have $2 + 2 = 4$ days of total float times. Thus, if float times were equally shared both parties had two days. Nevertheless, the employer used one extra day which cannot be acceptable. In addition, due to the change of criticality the second noncritical path will be compared with the actual critical delay path to identify the concurrency.

Two days long of D1-NN and two of three days of D5-EC will be accepted as concurrent delay because of their effect on the end date. Finally, as shown in Figure 2, concurrent delay days are two days, employer is responsible for three days and contractor has no individual delay responsibility due to employer's extra float usage. The comparison with existing methods can be seen in Table 6.

In the 'as planned' vs 'as built' and 'as built but for' methods, responsible delay times are more than the actual delay time. However, in the hypothesis, parties are responsible for total 5 days which was the actual delay time. In addition, some existing methods can be calculated for employer's and contractor's perspectives separately. This situation may give different results which must be avoided for not to open a new field of dispute. The hypothesis has only one result for every perspective.

Finally, the main purpose is not to accuse or absolve a party for delay so delays within the float times are not counted as responsible delay as long as it was abused by a party. In the example employer used one extra day unfairly so he became responsible. None of other methods takes responsibility on this case. By these results, the hypothesis will prove its fairness in time.

5 CONCLUSION

According to the literature study, float ownership has not substantially taken role in the contracts so far. Even terminologically, only total float has a small role, but other float types are ignored. Nonclaused contracts and uncertain division of float times would cause future conflicts. For these reasons, float concepts must be identified clearly in contracts. This would be more trustful between parties and more efficient for the progress.

The method to identify the float ownership must prove that it is trustworthy. To do this, delay analyzer must avoid subjective perspectives and formulize the solution. The float splitting formulas must be related with mathematical parameters instead of %50-%50 division assumptions.

This case study demonstrated that delay analysis methods are not related with float types and float times primarily. Although IAP and WSA methods considerate float times, their capability is limited with total float only. Moreover, these methods float usage depends on "first come first served" philosophy so even if results are correct, their fairness should still be questioned. As it was seen in the IAP method, despite the contractor's delay time was shorter than the employer, contractor caused change of critical path due to the employer used the float time. This example is very common for delay disagreements in the construction sector.

Awareness of float ownership and float types should be increased. Float ownership must be divided in related to Float time-Delay event-Activity and be mathematical formulized objectively. Considering that, parties that have certain information about their own float times would decrease the amount of disagreements. This will result in decrease of undesirable expenses such as, lawyers, arbitrators and courts also increase efficiency of time managements.

REFERENCES

Anysz, H. 2019. Managing Delays in Construction Projects Aiming at Cost Overrun Minimization. *IOP Conference Series: Materials Science and Engineering*: 603 032004.

Braimah, N. 2013. Construction Delay Analysis Techniques: A Review of Application Issues and Improvement Needs. *Buildings*, ISSN 2075–5309.

Evans, S. N.A.. NEC Questions Who Owns the Float. http://www.stevencevans.com/float/

Garza, J.D.L. & Prateapusanond, A. 2003. A Comprehensive Practice of Pre-allocation of Total Float in the Application of A CPM Based Construction Contract. Virginia Polytechnic Institute and State University, Blacksburg, VA 24061-0105.

Kamandang, Z.R. et al. 2017. An Empirical Study on Analysing Schedule Delays of Construction Project in Indonesia. *34th International Symposium on Automation and Robotics in Construction*. Taipei. ISRAC 2017.

Roux, K.L. 2017. It's Official. Employer Owns the Float. *https://www.linkedin.com/pulse/its-official-employer-owns-float-kobus-le-roux*

Spanos, P. 1984. Be Cautious on Who Controls the Float Time on Some of Those Critical Path Jobs. *Construction Dimensions, January 1984.*

The Society of Construction Law 2017. Delay and Disruption Protocol. *The Society of Construction Law.* Oxfordshire

Trauner, T.J. et al. 2009. *Construction Delays: Understanding Them Clearly, Analysing Them Correctly.* Elsevier Butterworth-Heinemann. ISBN 13: 978-1-85617-677-4. USA.

Yang, J.B. 2016. Reviewing Construction Schedule Float Management. *The Open Construction and Building Technology Journal.* J-11-1.

Requirements analysis for a project-related quality management system in the construction execution

S. Seiß & H.-J. Bargstädt
Chair of Construction Engineering and Management, Bauhaus-University Weimar, Weimar, Germany

ABSTRACT: Quality problems in the execution of construction work were and are a common problem. At present, various support systems for quality control have been developed. However, these existing systems mostly deal with company-related quality management or focus on quality control. Therefore, no requirements for project-related quality management have been considered. This article aims to analyse and to define requirements to design and evaluate a project-related quality management system for the construction execution. The requirements for a project quality management system are derived by a literature review in combination with a system analyses using the system theory and based construction process analyses.

1 INTRODUCTION

1.1 Problem definition

Construction defects are an international problem (Ahzahar et al. 2011), and project participants accept them as an integral part of the construction industry (Mills et al. 2009). Despite the advancing digitalisation of society, quality in construction is a huge problem. In recent years, further developments were introduced in construction operations, particularly concerning building information modelling (BIM), but this has not improved quality on construction sites. For example, a study by the Institute for Building Research shows that between 2009 and 2016, the defects increased by 89% for one and two-family houses in Germany (Helmbrecht et al. 2018). According to a report by the VHV (Vereinigte Hannoversche Versicherung) insurance, 70% of the insured events can be traced back to the execution of construction work and only 10% of the damages can be attributed to planning services. The removal costs of the defects can be up to 12% of the construction costs. The reasons for building damages are especially time pressure, the lack of controls by the construction supervision, the inadequately skilled workers and the cooperation and coordination of the building participants (Böhmer et al. 2020). Similar studies can be found on an international level. For example, an Australian research shows that 14 defects per building on average exist (Johnston & Reid 2019). A survey in England found that 97% of homeowners reported a problem (e.g. snag or defect) to their developers (Home Builders Federation 2020).

Furthermore, quality management is still mostly document-oriented in analogue or digital form, representing an enormous organisational effort. Particularly under time pressure and on huge construction sites, quality-related verifications are not sufficiently performed, or their documentation and tracking become a challenging task.

Therefore, a digital information system for quality management needs to be available for all project participants to meet the continually increasing complexity and higher demands on buildings. Such a system must provide transparency, trust, synchronous information exchange and quality evaluation between contractor and client. This system must also be compatible with and complementary to project management or lean management approaches (Böhmer et al. 2020).

For these reasons, the presented paper deals with the analysis of the current project-related quality management. The analysis highlights the contents of quality management and derives requirements from it. The requirements can be used for further research to design a new concept or evaluation of existing research.

1.2 Project quality management

An exact and sufficient description of quality requirements can often be a difficult task. Quality has different meanings in our language. In general, quality can be understood as the fulfilment or compliance of requirements. Quality represents the divergence between desired properties and the actual properties of a product. According to DIN EN ISO 9000:2015 11, quality is understood to be the "degree to which a set of inherent characteristics of an object fulfils requirements".

Quality management serves the management of quality and is divided into the four areas of 1) quality planning 2) quality assurance 3) quality control and 4) quality improvement (DIN EN ISO 9000:2015 11). The construction quality describes the fulfilment of the building compared to a described condition. A building should correspond to the general state of the art and needs to be free of defects. In terms of high

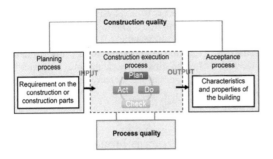

Figure 1. Construction process quality and construction process quality (Weyhe 2005, Masing 1997).

Figure 2. Four elements of a digital quality management model (Marsden 2019).

quality, the building complies with the requirements of the owner.

The quality can be separated into product and process quality. On the one hand, product quality reflects the components' individual qualities and the building's overall quality. On the other hand, the process quality relates to the building processes necessary for the construction of a building. According to DIN EN ISO 9000:2015, a process is defined as the "set of interrelated or interacting activities that use inputs to deliver an intended result". The construction product quality is, therefore, the result of the process quality. This fact becomes evident in Figure 1. It shows that the construction quality results from the process qualities of planning and acceptance of construction. The process quality is linked to the execution process. It is defined as the fulfilment of process-related requirements, e.g. usage of right material or procedure. The execution phase is extended by a Plan-Do-Check-Act-Cycle (PDCA-Cycle). A PDCA-Cycle try continuously to meet the requirement by checking and improvement of qualities. (Swamidass 2000). The process-oriented approach is becoming increasingly more important due to the process-oriented quality management in DIN EN ISO 9000:2015.

The view of quality has been changed over time. Until 1985, the concept of quality was primarily understood as quality control. At a later stage, the term quality assurance becomes established. The idea of process-oriented quality management has become generally accepted since 2000. This full-time approach focuses on the processes relevant for product quality (Walder & Patzak 1997).

In terms of a project, quality management means that all project participants ensure the quality of the structures and construction processes. Together they pursue a quality policy and quality objectives. Project quality management (PQM) represents multi-organisational quality management (Girmscheid 2014). The present paper focuses on PQM in the construction execution.

1.3 Quality information model

Quality information must be shared with all related project participants who take part in the construction process. In this case, especially quality professionals like site manager, construction supervisor and building surveyor must be provided with information about machines, work activities, and workers' information. According to Marsden, digital quality information models have to handle four information management elements people, materials, processes, and machines. Managing these topics by information management, the overall performance can be optimised (Marsden 2019) (see Figure 2).

The method of BIM is one major step to handle qualities related to construction. BIM describes a digital method to manage building-related data over the whole life cycle of a building. By now, BIM is well adapted for the planning phase. In this, the quality of information is described by the BIM execution plan (Marsden 2019) and is checked to comply with customer requirements, codes, and standards. From the perspective of the construction phase, there are special requirements in terms of model definitions and data exchange. The application of BIM in the construction phase is focused on the building, costs and schedules. The reason for this are missing exchange formats for the construction execution, which can handle controlling related information. For example, the data exchange of Industry Foundation Classes (IFC) is specialised for the exchange of building, cost and time information across project participants.

None of the exchange formats does standardise the exchange of information about machines, people, reports, process data, which are necessary to exchange quality-related information between the participants. Especially for small and medium enterprise dominated construction industries are standardised and quality-related data exchange formats necessary. The missing standards lead to quality data management, which is based on isolated applications, instead of project collaboration. Therefore, process and product quality reports are stored in a document-oriented way and are not exchanged in the context of a BIM-Model. By using BIM, data management is raised to a new level, and the building model can become the single source of truth (Marsden 2019).

The BIM Collaboration Format (BCF) standardises the way of communication, including an object-oriented digital model. Defects and plan changes can be clearly described and controlled by exchanging messages with defined parameters. The BCF allows exchanging messages and pictures connected to a BIM-Object by a Globally Unique Identifier (GUID) (buildingSMART International, Ltd. 2020a.). Another solution is to use and connect ontologies of different domains to provide engineers, with quality-related data of different domains. For this, the IFC is translated into the ifcOWL and can be easily linked to other data like geoinformation, machine, sensor data (buildingSMART International, Ltd. 2020b.).

In addition to the execution phase, the handover phase is also critical regarding quality, due to the fact of limited data exchange standards like the Construction Operations Building Information Exchange. The building's documented quality must be handover as an As-built model to use the quality-related data in the execution phase. An As-built model contains all geometric and alphanumeric information of the building's actual condition (Borrmann et al. 2018).

By now, a quality information model for the construction industry does not exist. The introduction of a consistent quality information model will transfer the document-centric quality management approach into data-based quality management (Marsden 2019).

2 RESEARCH METHOD AND SCOPE

This research aims to provide a system analysis of the PQM in the construction execution. In the first step, the actual quality management in the execution phase has to be analysed. Secondly, the requirements are created based on the real system by which an evaluation of existing research and design of a future concept can be done.

A general literature review analyses the critical factors for PQM. Also, a work process analysis based on standards and example projects was done. The aim is to extract the process of quality assurance from the work processes and classify it concerning the planning of quality proof and quality controls. Due to the high diversity of construction processes and the associated documentation and verification requirements, three different process categories are distinguished as the basis for analysis:

1. Machine-dominated work process
2. Logistics-dominated work process
3. Cooperation-dominated work process

The categories represent the basic construction process techniques and are therefore ideal for describing the real system of quality management in construction. The categorisation is based primarily on the performance-determining component. For the process analyses, an exemplary work system for each category was chosen to examine the quality requirements. These requirements describe in a very abstract way what the user expects from a system, which functions a system has to provide and how it should behave on specific inputs (Sommerville 2011). The work system analysis is developed on event-based process chains (EPC). Based on this process analysis, the quality management is described according to the system theory.

3 PROJECT QUALITY MANAGEMENT IN THE CONSTRUCTION EXECUTION

3.1 Critical factors for project quality management in the construction execution

Different participants have different points of view on a project. In the centre of the construction execution is the controlling task, which is divided into 1) costs, 2) quality, 3) time 4) contracts and 5) services. The components are connected and influence each other. The information needed to assess the quality is related to the overall controlling (Girmscheid 2014). Thus, the quality cannot only be considered from one point of view; it must be linked with different construction execution domains.

The control and acceptance of construction work are one of the most important tasks of PQM. From a legal perspective, quality management can be differentiated into private law quality records and public law quality records. Private law certificates are defined based on the service contract. Public quality certificates are defined and required by the state. This includes, for example, proof of the proper installation of fire protection systems. The authorities do not carry out any legal construction supervision or acceptance. However, these authorities try to achieve "preventive averting of danger" by monitoring limited random samples. State commissioned inspectors can carry out the proofs. If such proof is not available for acceptance, authorities can, for example, forbid the building's usage (Weyhe 2005).

On the other hand, a look at quality certificates at the end of a construction project is not enough. Quality-related data must be created, collected, shared and especially smart analysed for the entire construction work execution. This requires a process that includes all project participants as a team (GSA 2019).

The need for a team-based and transparent quality assurance becomes clear by the three different categories of quality proofs. These can be divided into 1) site office inspection, 2) third party inspection, and 3) customer-oriented inspection (Kim 2008). The site office inspection is understood as the self-monitoring by the executing company. The third-party inspection is the so-called external inspection. In this case, building products of a certain class, e.g. concrete with compressive strength, must be inspected by an approved inspection agency. This area also includes building inspection authorities. Customer-related inspections can be carried out by the customer himself or by a construction supervision agency commissioned and authorised by him. The different forms of supervision, as well as obligations, are secured in laws and contracts. Overall, the different supervisions check the construction work several times, and the examination of qualities as well as the acceptance of construction works are carried out according to the 4eye principle (Weyhe 2005).

The participants must continuously carry out the checks or at frequent intervals, especially to counteract large quantities of defects (Kim 2008). On the one hand, this can be done by the site management itself and by the construction supervision commissioned by the client. On the other hand, the follow-up contractors have to check the preliminary work of the previous contractors on which they have to build up. If the contractor does not prevent this, he is responsible for the resulting damages (3 U 814/14 and §4 No.3 VOB/B, VOB/C). Thereby the executing person becomes the inspector himself (Weyhe 2005). The associated continuous checking of the process quality enables the direct correction of defects (Kim 2008).

Furthermore, the successive checks provide the opportunity to verify the different results for plausibility. The results of the plausibility check can be used to conclude the quality of quality control. Also, executing companies can improve their self-control. In general, quality management supports conflict resolutions (Schwerdtner 2005). Additionally, a transparent project quality management system could avoid conflicts.

3.2 Analysis of the project quality management for the construction execution based on the system theory

It is necessary to conduct a system analysis, better to understand the relationships and structures in project management. These properties are typical for a PQM-system of construction projects (Kochendörfer et al. 2018):

1) Open, because the system is related to the environment. For example, data from the construction site and controlling are essential
2) Time-variant and dynamic, since the system variables change over time
3) Socio-technical, since sociological components such as the project participants and technical components such as machines occur in the project
4) Cybernetic, because the system controls itself

We can distinguish four system typologies for further analysis of a system: 1) target system, 2) action system units, 3) action system, 4) product system. The target system focuses on the desired final state of an action result. The action system describes the tasks to achieve for the desired goal. The action system carriers represent the organisations or persons who carry out the tasks or activities. The product system describes the object of action on which tasks are carried out, taking into account the target system. We analyse these four system components in more detail concerning PQM in construction. This analysis enables a detailed

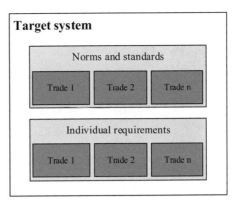

Figure 3. Target system.

presentation of the overall system PQM in subsystems (Patzak 2014; Kochendörfer et al. 2018).

Target system: The overall system aims to ensure that the project meets the required quality standards. The target system is subdivided into the subsystems norms and standards as well as individual requirements (Figure 3). The goals defined for quality can be contradictory (Lucht 2019), to prevent this, public law goals, represented by norms and standards, are above the individual goals. The requirements and standards have a direct influence on the construction processes and components. The target system can relate the processes and components to different trades.

Action system units: Various project participants interact during the construction phase. Figure 4 shows the interaction of different project participants in the construction process in terms of quality. The interaction is orders in three hierarchical layers. The participants of the first layer define requirements, the parties in the second layer control the quality of the building, and the third group of participants do the operations to fulfil the requirements. The actors within the PQM are very diverse and are not limited to the contractor, planners and construction companies. The diversity is justified on the fact that the consideration of construction quality is a domain-spanning task. For example, contractor warranty periods must be considered, which are demanded by the client and the tenants after the acceptance. It becomes clear how important a functioning data exchange between the project participants is, especially concerning the quality of construction. From a quality point of view, it is equally important to define which actors have to respect which construction qualities within the project. Project platforms or a common data environment can support the exchange of data between the participants. By now, common data environment still focuses on document orientated way of information exchange.

Action system: The participants must carry out various activities in order to implement a structural facility. The action system maintains the final states defined in the target system by actions on the product system. In this case, quality assurance actions must

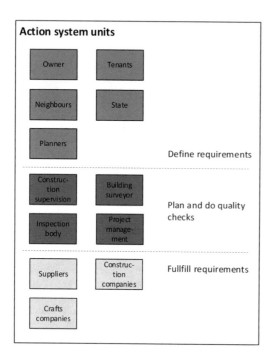

Figure 4. Action system units.

ensure the quality of the construction processes and the structural system. A generic process model of PQM can represent these quality assurance actions. The action system is based on analysing the three presented performance-related processes (machine-dominated, logistics-dominated and cooperation-dominated work process). The action system is set up via an EPC diagram by using scenarios and is transferred into an abstract process diagram (see Figure 5). The process is described in a process map. The illustrated action system is designed generically. This enables the user to use the action system in different work processes and project constellations. It can also be applied to quality management of one component and on all quality activities in a construction project. The diagram is divided into four phases 1) planning, 2) work preparation, 3) construction and 4) operation.

The planning process defines the requirements for the construction project. The requirements are defined in plan, norms and standards, handbooks or project descriptions.

The requirements must be checked for completeness and correctness in the work preparation. Thereupon the planning of the execution, as well as the procurement of materials, personnel and machines take place. Afterwards, the quality-related proofs and examinations resulting from the building activities can be specified. In this phase, a foresighted analysis is also possible. This analysis can point out at which construction units or building activities it frequently comes to quality problems. Thereupon the companies can consider these in the planning and execution of construction or preventive measures can be developed to fulfil the quality.

The construction is understood as a cybernetic system (Proporowitz 2008). The controlling Plan-Do-Check-Act loop mode is always initiated for construction activity and is repeated until it is completed. The loop starts with the recording of qualities on the construction site. This quality will get checked according to plausibility and compliance. The results of the checking process will be recorded in a report. If the quality does not fulfil the requirements, a problem reasoning, as well as a solving process, can be started. After that, the controlling measures can be initialised. After successful construction activity, the final proofs are collected for acceptance and a handover of the service. Finally, companies can use the data to evaluate problems. In this context, it is of interest to find out why quality errors occur. The evaluation results can be used for future projects to avoid future mistakes or to analyse in which process steps quality defects can occur.

The last phase is the guarantee of the product quality. For example, for German companies, this must be four years in terms of contracts has been agreed according to VOB (§13 No.4 VOB/B). Related to a project related manner, this shows that the quality activities do not end with the handover. Construction quality-related activities are within the life cycle of a building.

Product System: In the product system the quality is focused on both: the process of creation and the resulting final product. Thus, the product system is represented by the building and its sum of building components.

3.3 Compilation of the requirements

The following chapter derives the requirements for a PQM system in the construction execution from the previous analysis. As mentioned in the problem definition, this PQM system aims to reduce or avoid quality problems in the construction execution phase. One central idea is to support all project participants with quality-related information by an open, time-variant, social-technical and cybernetic PQM-system. Based on the aim of project quality management system and the analysis, the following three core requirements are defined:

1. The system should interlink data across multiple domains to provide quality management with the necessary data.
2. The system's schema should be open, and the exchange of data should be according to a standard to increase the usability and interaction of information.
3. The system should be able to define rules and queries to plan and run quality checks.

We collect further requirements in Table 1. As mentioned before all requirements are functional user-requirements.

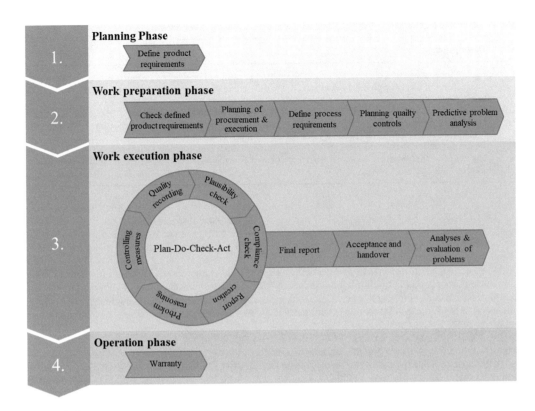

Figure 5. Action system.

Table 1. Extracted requirements by the literature review and the use cases.

No.	Description	Phase
1.	Check planning requirements for completeness, correctness, and fulfilment. (Weyhe 2005)	work preparation
2.	Differentiate requirements into public and private legal quality certificates in order to be able to define the resulting claims	work preparation
3.	Derive requirements from the execution activities (Kim 2008)	work preparation
4.	System defined rules for planning and testing of quality certificates must be defined transparently, according to current standards and system independent.	Work preparation and work execution
5.	Contractors check the services of the previous contractor	Work execution
6.	Repeated controls are executed with existing control results (no blind test)	Work execution
7.	Supervision by the client and/or by a construction supervisor commissioned by the client (4-eye principle)	Work execution
8.	Third-party monitoring by accredited testing agencies (4-eye principle)	Work execution
9.	Construction companies monitor themselves (self-monitoring) (4-eye-principle)	Work execution
10.	Quality proofs are based on each other (e.g. multi-layer soil compaction)	Work execution
11.	Quality documentation and proofs must be recorded synchronously and continuously to construction activities.	Work execution
12.	Tracking, investigation and statistical analysis of quality problems (Kim 2008)	Work execution
13.	Direct feedback to the site engineer and workers (Kim 2008)	Work execution
14.	Integration into the controlling. Controlling factors time, cost and quality interact with each other. (Battikha 2002)	Work execution
15.	Support of communication and cooperation between the construction parties related to quality; accessibility for all project participants as a basis for communication and data exchange (Kim 2008)	overall process
16.	Transparent data management	overall process
17.	Differentiation in process qualities (construction quality) and product qualities (component quality)	overall process
18.	Change of quality proofs by monitoring systems during construction (IoT)	overall process
19.	Handling dynamic changes in the construction process. Especially with regard to the controlling. (PDCR) (Chen & Luo 2014)	Work execution
20.	QM system must be able to handle nested quality certificates	overall process

(*Continues*)

Table 1. Extracted requirements by the literature review and the use cases (Continued).

No.	Description	Phase
21.	System schema of PQM in construction execution must be designed openly and system independent. It has to be possible to integrate it into different platforms or software solutions. (E.g. by linked data)	overall process
22.	Consideration of the responsibilities and competencies of the individual project parties	overall process
23.	Quality certificates contain tolerances that allow a subjective evaluation (e.g. exposed concrete)	overall process
24.	Principle of subsidiarity – independent action of the building participants	overall process
25.	Integration and support of different project management methods like lean management	overall process
26.	Quality related information are stored related to different domains like the BIM and can be used warranty and operation phase	warranty

4 CONCLUSION

The paper offers an approach to understand project-related quality management. It could be shown that quality management in construction execution is a multifaceted process that is not limited to the quality control of the building component. Instead, it is a continuous process from execution planning to handover and its warranty. System theory is an adequate tool to classify the different system aspects of PQM in construction. Furthermore, it was possible to show essential features within the four system typologies and highlight critical factors of PQM. Finally, 26 requirements were found, based on the literature review and the system theory. The collected requirements show how diverse quality management is. The list of requirements cannot claim to be complete but reflects the main requirements resulting from the analyses of the three process categories.

The work represents an essential basis for the development of a project-related quality management system. In further work, already existing systems in practice and research can be evaluated based on the derived requirements. Therefore, it allows identifying development needs for existing systems in technology and science. In a subsequent step, digital PQMsystems can be further developed, or new systems can be designed. The broad term information system made it possible to collect the requirements as unbiased as possible.

An unusual approach to PQM-systems is offered by the DICO project (Digital Construction Ontologie), which deals with ontologies in various domains including quality. By now, it does not deal with the derivation of quality proofs and quality testing. Nevertheless, the ontology represents an important step in development and requires further improvements (Törmä 2019). Other research projects that could influence a digital PQM-systems are Kroqui, Innovance, BIM2TWIN as well as SDac. Also, multiple quality checking systems for construction are available. Especially in the area of information exchange and of linking information will be a need for development.

Based on the defined gaps, a concept and an information system can be developed. Due to the development of the Internet of Things, the increased collaboration and communication between construction parties must be developed towards an open and domain-spanning ontology. It becomes clear that digital methods, such as Building Information Modelling, linked data, as well as the Internet of Things, can contribute to a transparent and high-quality construction. Sciences should not apply these methods in isolation, but rather interlinked to fulfil the requirements for a preventive and accompanying PQM.

REFERENCES

Ahzahar, N., Karim, N.A., Hassan, S.H. & Eman, J. 2011. A Study of Contribution Factors to Building Failures and Defects in Construction Industry – *The 2nd International Building Control Conference* 2011: S.249–255.

Battikha, M. G. 2002 QUALICON: Computer-Based System for Construction Quality Management. *Journal of Construction Engineering and Management* Volume 128 Issue 2 April 2002: 164–173.

Borrmann, A., König, M., Koch, C. & Beetz, J. 2018. *Building In-formation Modeling – Technology Foundations and Industry Practice* Springer International Publishing.

buildingSMART International, Ltd. 2020a. https://technical.buildingsmart.org/standards/bcf/

buildingSMART International, Ltd. 2020b. https://technical.buildingsmart.org/standards/ifc/ifc-formats/ifcowl/

Böhmer, H., Brinkmann-Wicke, T., Sell, S., Simon, J. & Tebben, C. 2020. *VHV Bauschadensbericht*. Hannover VHV Allgemeine Versicherung.

Chen, L. & Luo, H. 2014. A BIM-based construction quality management model and its applications. *Automation in Construction* Volume 46: 64–73.

Törmä, S. & Zheng, Y. 2019 *DICO – Digital Construction Ontologies*. https://digitalconstruction.github.io/

Girmscheid, G. 2014 *Bauunternehmensmanagement – prozessorientiert Band 2*. Berlin Heidelberg Springer-Verlag.

Helmbrecht, H., Simon, J. & Böhmer, H., 2018. *Analyse der Entwicklung der Bauschäden und der Bauschadenkosten – Update 2018* Hannover: Institut für Bauforschung e.V.

Home Builders Federation and National House Building Council 2020. https://www.hbf.co.uk/documents/9690/R173-HBF2020Brochure-v7.pdf

Johnston, N. & Reid, S. 2019. https://www.griffith.edu.au/__data/assets/pdf_file/0022/831217/Examining-Building-Defects-Research-Report-S-Reid-N-Johnston.pdf

Kochendörfer, B., Liebchen, J. H. & Viering M. G. 2018. 5. Auflage. *Bau-Projektmanagement – Grundlagen und Vorgehensweisen*. Wiesbaden: Springer Vieweg.

Lucht, D. 2019. *Theorie und Management komplexer Projekte*. Wiesbaden: Springer Fachmedien.

Marsden, P. 2019. *Digital Quality Management in Construction*. London: Routledge.

Masing, W. 1994. *Handbuch der Qualitätssicherung*. 3. Auflage. München Wien: Carl Hanser Verlag (1994).

Mills, A., Love, P. E. & Williams P. 2009. *Defect Costs in Residential Construction. Journal of Construction Engineering and Management* Vol. 135, Issue 1 2009: 12–16.

PBP Planungsbüro professionell 2015. Vorlage von Konformitätsbescheinigungen und Übereinstimmungserklärungen richtig managen Würzburg: IWW Institut für Wissen in der Wirtschaft Ausgabe 12/2015.

Patzak, G. & Rattay, G. 2014. *Projektmanagement: Leitfaden zum Management von Projekten, Projektportfolios und projektorientierten Unternehmen*. Wien: Linde Verlag Ges.m.b.H.; 6., überarbeitete Auflage.

Proporowitz, A. 2008. *Baubetrieb– Bauwirtschaft*. Leipzig: Carl Hanser Verlag.

Schwerdtner, P. 2005. *Qualitätsmanagement in der Bauwirtschaft als Mittel zur Konfliktbewältigung*. Braunschweig: Intitut für Bauwirtschaft und Baubetrieb 2005.

Sommerville, I. 2011. *Software engineering 9th ed*. Boston: Pearson Education, Inc., publishing as Addison-Wesley.

Swamidass, P.M. (eds) 2000. *Encyclopedia of Production and Manufacturing Management*. Boston: Springer.

U.S. General Services Administration 2019. https://www.gsa.gov/real-estate/design-construction/3d4d-building-information-modeling/guidelines-for-bim-software/document-guides/quality-control-reports.

Walder, F. P. & Patzak, G. 1997. *Quality Management and Project Management*. Wiesbaden: Vieweg+Teubner.

Weyhe, H. 2005. *Bauschadensprophylaxe als Beitrag zur Qualitätssicherung während der Bauausführung*. Weimar: Professur Baubetrieb und Bauverfahren 2005.

Applications of artificial intelligence

Metadata based multi-class text classification in engineering project platform

M. Shi, A. Hoffmann & U. Rüppel
Technical University of Darmstadt Institute of Numerical Methods and Informatics in Civil Engineering, Darmstadt, Germany

ABSTRACT: Enterprises usually establish digital platforms to store and share projects as well as business data. Those files in digital platforms usually have metadata that gives information about the life of a document. Metadata is useful to retrieve useful information about the document. They can be generated fully automatically or given manually by users. Sometimes those manually given metadata can be left out, insufficient, or even wrong. Those metadata stay unaware until the files are being searched for a purpose later on. If the number of files with insufficient metadata is enormous, the filtering of files from platform according to specific metadata becomes difficult. Manually correction can be tedious and enormously time intensive. This paper introduces a method for correcting faulty metadata based on other correct metadata with text classification, natural language processing technique, and linguistic synonymy information in the engineering project platform. The proposed method is evaluated on a real-world dataset. Even though the available information from metadata is limited, the method still gives a promising result and has the advantage of its high computing speed.

1 INTRODUCTION

Transforming documents from non-digital, analog form into digital form has been bringing enormous opportunities and efficiency in the AEC industry. Enterprises usually establish digital platforms to store and share projects as well as business data. Engineers can transfer and share documentation with the digital platform quickly and easily across geographical distance. Digital documents often associate metadata with them. Metadata is "data about data", which provides information about the digital document, such as file types, creation dates, authors, categories, etc. Metadata is useful to build editing process, find a specific document, and retrieve useful information about the document. Digital platforms usually use metadata to filter files. In the case of Microsoft Office documents, when searching for a document in Microsoft Windows, the user can choose to search for a document by the author, the creation date, the datatype, and so forth (Daniel & Daniel 2012). The searching process works fine if the metadata of the searched file is documented correctly. However, metadata can be wrong if it is given by people manually. Dependent on applications in which files are created, some metadata should be edited by the user instead of generated automatically. Some fields have predefined text from which users can choose, some areas support free text, and some fields accept both of them, for the case that the file belongs to none of the given ones. The freedom of editing metadata manually also brings out problems. If a field allows both free text and preset text, it may end up with different descriptions for the same category. Even though a preset text may already exist, users tend to give free text if they are not familiar with the given categories. The faulty metadata hampers the filtering process and the finding of the target file from the data server. Therefore, it is necessary to develop a mechanism for correcting and unifying the value of metadata to improve the filtering process.

When it comes to retrieving information from a large dataset, machine learning is usually an option. In the AEC industry, researches on using machine learning-based text classification already exist in various areas to reduce human efforts. Most of them use textual documents as training data, which makes the training process computationally expensive. Other than that, only a few of them add semantic information to the training process. This article uses natural language processing (NLP) methods and machine learning-based text classification, combined with synonym semantic information to complete faulty metadata and to improve file filtering. Section 2 presents the real-world problem and research motivation. A brief review of related work is given in section 3. Section 4 focuses on the process of classifying with metadata. Several text mining algorithms and different hyper-parameter combinations are being studied.

The results are presented and discussed in section 5. The last section provides the conclusion of this paper and an outlook on future work.

2 MOTIVATION

The German "Digital Administration of Technical Documents" (DVtU) is a central documentation platform for technical drawings, construction plans, static calculations, expertises, approval procedures etc. It is used Germany-wide by the Federal Waterway and Navigation Administration (WSV), who is responsible for the maintenance, expansion, and construction of federal waterways and associated facilities such as locks and ship lifts. DVtU currently has more than 2500 users and is considered as one of the essential IT systems within the WSV ("Wasserstraßen- und Schifffahrtsverwaltung des Bundes," n.d.).

The DVtU enables users to create, edit, and archive project relevant documents through the project lifecycle. DVtU is in the project work comparable with organizing digital files in a folder structure. The difference is that the documents here are called "Technical Documents" and that the user should give extra information, the metadata, while storing the technical documents (Borstelmann & Beier 2018). The metadata includes information like project ID, object name, keywords, the processing status, editor, usage category, etc. Some of this data is generated automatically by the system, e.g., details of the editor. A part of the metadata must be entered by the employee when storing the documents in DVtU.

The use of metadata has the advantages that after all the information entered, users can search and filter documents with their metadata. This makes searching and filtering efficient. The precondition of efficient searching and filtering is that the metadata was given correctly in previous stages. However, since some metadata must be entered manually, users may make mistakes. There are three different types of manually entered metadata: the selection of a drop-down list, free text, a combination of selection, and free text. This paper focuses on the third type metadata with the example "Verwendung" (Engl. "Usage Category" UC), which allows user to select from a predefined list as well as enter free text if the user thinks none of the categories from the drop-down list describe the document correctly. With the gradual increase of technical documents, the diversity of UC also grows dramatically from dozens to more than four hundred. Through a near look in the categories, the authors commit there are many redundancies in the added free text to the original given selection list. Taking the preset UC "Statische Berechnungen" (static calculations) as an example, users also give descriptions with same meaning such as "Statik" (static), "Entwurfsstatik" (design static), "statische Berechnung" (static calculation), "Statische Berechnungen", "Statk" as UC. Some categories have the same words but differ in capitalization; some of them are synonyms of the predefined categories; some words are even spelled wrong (e.g., Statk). Other than redundancy, the author also observed that from the given 141 435 datasets, 101 686 of them share the same UC "Bestandsunterlage" (inventory document), which gives no specific information on the technical document. Since the filtering process in DVtU bases on keywords, which do not understand semantic and synonyms, certain documents are left out if their metadata does not include the same word as in the filtering field. There is a need to unify the categories with the same meaning, correct the wrong labels, and replace the general labels with more specific ones.

Given the fact that the number of datasets in the DVtV system is too large, manually correcting is a time-consuming and tedious process. The good news is that there are 34 metadata areas aligned to technical documents except for the UC. Other than that, 6 485 documents are given with correct metadata after preprocessing. Thus, the authors propose to use machine learning-based text classification methods combined with synonymous information to train a classification model from the correct metadata dataset and use it to rearrange the faulty UC.

3 RELATED WORK

Machine learning-based text classification is widely used in various research fields to reduce human efforts. Caldas Carlos H. et al. (2002) improved information organization and access in inter-organizational systems by using machine learning algorithms, which automatically classify construction project documents according to project-related components. Chi et al. (2016) proposed using TC to quickly identify safety standards by providing the types and violation causes. They have also introduced a methodology of evaluating the strength of text classification categories to suggest well-defined TC categories. Al Qady and Kandil (2014) used a hybrid approach of unsupervised and supervised learning methods to classify construction project documents. With unsupervised learning, they could automatically cluster documents based on text similarities. Then they used the developed core categories to classify documents. Salama and El-Gohary (2016) presented a semantic, machine learning-based text classification algorithm for classifying clauses and sub-clauses for supporting automated compliance checking in construction. Zhang et al. (2019) used TM and NLP techniques to classify the causes of accidents on construction sites and to analyze the construction accident reports. Zhou and El-Gohary (2016) suggested an ontology-based TC that improves the performance of automatically classifying construction regulatory documents by utilizing semantic features on the text. All those researches focused on textual documents in homogeneous form, which is not the case in this paper. Since digital files in DVtV have different formats: .pdf, CAD plans, files created by Microsoft office, pictures, etc., the authors propose to use the limited metadata to classify categories of the documents.

4 CLASSIFICATION APPROACH

Like the normal text classification process, the proposed classification method for metadata consists of following steps: data collection, pre-processing, classifier selection, model training and testing.

4.1 Data collection

For research purpose, the authors get the raw metadata from the engineers of BAW in two excel files, one about the weir system, the other about the water navigation lock. The metadata is exported directly from DVtV with a mixture of corrected labeled "UC" and faulty assigned "UC". There are no format differences between those two files. Both of them have 35 metadata groups that provide information about the document they describe, such as document state, modify datum, ID, comment, object name, the art of document, to name a few. Some metadata groups are numerical, and others are textual. On the whole, there are 141 435 datasets from the two excel files.

4.2 Pre-processing

Usually, raw data cannot directly be used as input data for training models since there might be many irrelevant and redundant data. This process is called pre-processing. Pre-processing includes data cleaning, normalization, transformation, feature extraction, among other things. The output of data pre-processing is the training set. Thus, pre-processing decides the training set's quality that, in turn, affects the performance of the training model significantly. After having a close look into the raw data, the authors take the following steps to pre-process the metadata:

4.2.1 Selecting features and training data

The DVtV system provides 35 metadata groups altogether. However, not every metadata group is relevant to classify the document's usage. The authors choose seven groups as features (exclusive the label) based on experts' (engineers from WSV) knowledge. As said in section 4.1, not all records have correct UCs as a label. Thus, only records with preset UCs from DVtU are considered as training data. There are nine UCs from the DVtU platform. Thus, this is a multi-class classification problem with nine classes. After filtering, only 6 485 from the original 141 435 datasets left. The consideration that some documents might be assigned incorrectly by accident is neglected since it only makes a tiny percentage.

4.2.2 Cleaning data

Due to the reason that some metadata are not mandatory, some values might be missing (so-called NoneType values) in the features. In order to not reduce the size of training data dramatically, the empty metadata is assigned with a reasonable value according to the content of the metadata group.

4.2.3 Stemming, lemmatization, removing stop words

The results of stemming and lemmatization with python packages (nltk.cistem and spacy, designed for general domains) are not ideal. One reason is that terms used in DVtU system are domain-specific. The second reason is that there are many compounds in German, which can not be well converted by the existing opensource stemmer. Removing German stopwords (provided by nltk.stopword) does not bring better performance, because the metadata already has minimal words, and even some stop words contribute meaningful information in this case.

4.2.4 Transforming and vector space model

Since most machine learning algorithms are mathematic models, the text value should be transformed into numerical values to be processed by ML algorithms. On option is to use the vector space model. A vector space model represents a set of documents as vectors in a common space. Term-frequency-inverse document frequency (TFIDF) is one of the most used vector space models (Manning et al. 2009). Through combining term frequency and inverse document frequency, the space vector model can produce a composite weight for each term in each document, in this context a record of metadata, so that the relevance of each word in each document can be considered. This paper uses TFIDF to transform the textual metadata to numeric.

4.3 Classifier selection

Derived from the no free lunch theorem (Wolpert & Macready 1997), there are no general-purpose algorithms that behave the best for all ML use cases. Each classifier has its advantages and limits and is use-case dependent. The authors experimented with ten different classifiers to gain an overview of the performance. Based on the results, three classifiers with the best accuracy, recall, and F1 score are selected for a more in-depth study. The three classifiers are multinomial naïve Bayes (MNB) classifier, linear support vector classifier (Linear SVC), and stochastic gradient descent (SGD) classifier.

4.4 Model training and testing

After pre-processing, there are 6 485 metadata records with seven metadata groups as features and one metadata group "Usage Category" as the label. The data are trained with algorithms from the scikit-learn python package (Pedregosa et al. 2011). The pre-processed metadata is split into training and test data with the proportion of 2:1. The classifier is trained on 4 344 records and tested with 1 390 records. Inner the training data, this paper also applies cross-validation and hyper-parameter tuning with the grid search to find the optimal hyper-parameter combination. The results and reasons will be discussed in the next section.

Other than purely statistical ML-based classification, this research also aligns synonymy semantic information in the classification process. For that, authors create synonym lists for the nine classes (preset UCs) and features that only include free text metadata, respectively. Then each synonym list of the record's features is compared with each class's synonym list to compute a match rate, which shows the similarity rate r between the features and the classes:

$$r = \frac{n}{l} \quad (1)$$

where n = number of same words occurred in both lists; l = length of the shorter list.

The class with the highest match rate, which is more significant than 0.85 at the same time, is selected as the potential class of a record. The threshold of 0.85 derives from several experiments that provide the best performance. At last, this proposed method compares the predicted classes from ML algorithms with the ones selected by the synonym match rate. If the class suggested by the synonym match rate is different from the one generated by ML-algorithm, the former one is considered as the final result.

5 EVALUATION AND DISCUSSION

This paper uses four metrics to evaluate the classifier's performance: accuracy, precision, recall, and F1 score. Accuracy is the fraction of the correct classification in a total number of predictions made by the classifier (Manning et al. 2009). Accuracy only gives rational information about performance if the distribution of the classes is equal or nearly equal. Precision is the ratio of the correct predictions of one class to the total predicted number of this class. Recall, also called sensitivity, gives the proportion of correctly classified data records in the total number of records belongs to the respective classes. F1 score is the harmonic mean of precision and recall. F1 takes both false positive and false negatives into account. F1 supplement the disadvantage of accuracy if the class distribution is not even.

As written in section 3.5, this paper compared the performance of these three classifiers: MNB classifier, Linear SVC, and SGD classifier. The test results are shown in Figure 1 to Figure 6.

As can be seen from all figures, the accuracies of all used classifiers are over 70%, which is good considering the limited information from metadata and the numerous classes (the accuracy of randomly guessing the class is 11%). Comparing Figure 3 (accuracy 70%) and Figure 4 (accuracy 76%), or comparing Figure 5 (accuracy 77%) and Figure 6 (accuracy 78%), we conclude that using synonym information can improve classification performance. Nevertheless, the improvement is not as much as expected. The reason is

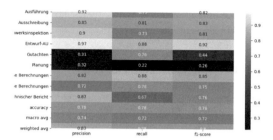

Figure 1. Classification report of Linear SVC.

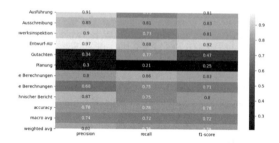

Figure 2. Classification report of Linear SVC with synonym information.

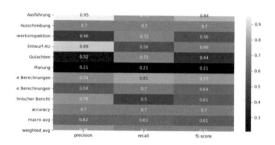

Figure 3. Classification report of MNB classifier.

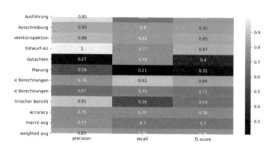

Figure 4. Classification report of MNB classifier with synonym information.

the compound words in the German language, which is the case in our training data. The used stemmer cannot stem compound words into words that comprise it. Also, those compounds are not included in the OpenTheaurus database. Thus, many words with the same meaning cannot be detected. Both problems hamper adding correct synonym information to classifying.

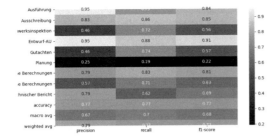

Figure 5. Classification report of SGD.

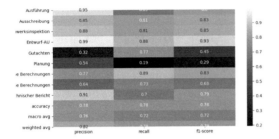

Figure 6. Classification report of SGD with synonym information.

6 CONCLUSION AND FUTURE WORK

This paper focuses on classifying digital documents with metadata. Though limited information from metadata, the testing results provide relatively good performance. Other than that, document classification based on metadata has the advantage of computing speed, since the training data is tiny compared to classification with digital files themselves. The aligning of synonym semantic information improves the performance of a purely ML-based classifier. However, due to the limitation of the existing stemmer and synonym database, the improvement of performance is not as much as expected.

Researches in the future can combine experts' knowledge in the form of ontology with statistical text classification algorithms. Also, semantic and linguistic information such as synonyms can be added to the training process instead of only added to the classification results. Since there is no out-of-box domain-specific symbolic and statistical NLP language toolkit for the German language currently, there is a need for a toolkit that can better deal with compounds, lemmatization, and word-stemming. Also, the same classification process can be experimented on the technical documents instead of only on the metadata to compare performance. Research can also be done in finding documents' concepts with algorithms like LDA and finding relationships between different words with word embeddings such as word2vec. With the generated concepts and relationships from digital files, computers can "understand" and interpret more complicated inquiries. In such wise, TM and NLP algorithms can also produce more specific results when searching for appropriate documentation.

REFERENCES

Al Qady, M., Kandil, A., 2014. Automatic clustering of construction project documents based on textual similarity. Autom. Constr. 42, 36–49. https://doi.org/10.1016/j.autcon.2014.02.006.

Borstelmann, H.-H., Beier, A., 2018. DVtU–Digitale Prüf- und Genehmigungsprozesse.

Caldas Carlos H., Soibelman Lucio, Han Jiawei, 2002. Automated Classification of Construction Project Documents. J. Comput. Civ. Eng. 16, 234–243. https://doi.org/10.1061/(ASCE)0887-3801(2002)16:4(234)

Chi, N.-W., Lin, K.-Y., El-Gohary, N., Hsieh, S.-H., 2016. Evaluating the strength of text classification categories for supporting construction field inspection. Autom. Constr. 64, 78–88. https://doi.org/10.1016/j.autcon.2016.01.001

Daniel, Larry E., Daniel, Lars E., 2012. Digital Forensics for Legal Professionals. Elsevier. https://doi.org/10.1016/C2010-0-67122-7

Manning, C.D., Raghavan, P., Schütze, H., 2009. Introduction to Information Retrieval 569.

Salama, D.M., El-Gohary, N.M., 2016. Semantic Text Classification for Supporting Automated Compliance Checking in Construction. J. Comput. Civ. Eng. 30, 04014106. https://doi.org/10.1061/(ASCE)CP.1943-5487.0000301

Wasserstraßen- und Schifffahrtsverwaltung des Bundes [WWW Document], n.d. URL https://www.contact-software.com/en/industries/logistics-infrastructure/wasserstrassen-und-schifffahrtsverwaltung-des-bundes/ (accessed 5.7.20).

Wolpert, D.H., Macready, W.G., 1997. No free lunch theorems for optimization. IEEE Trans. Evol. Comput. 1, 67–82. https://doi.org/10.1109/4235.585893

Zhang, F., Fleyeh, H., Wang, X., Lu, M., 2019. Construction site accident analysis using text mining and natural language processing techniques. Autom. Constr. 99, 238–248. https://doi.org/10.1016/j.autcon.2018.12.016

Zhou, P., El-Gohary, N., 2016. Ontology-Based Multilabel Text Classification of Construction Regulatory Documents. J. Comput. Civ. Eng. 30, 04015058. https://doi.org/10.1061/(ASCE)CP.1943-5487.0000530

Using topic modeling to restructure the archive system of the German Waterways and Shipping Administration

A. Hoffmann, M. Shi & U. Rüppel
Technical University of Darmstadt Institute of Numerical Methods and Informatics in Civil Engineering, Darmstadt, Germany

ABSTRACT: The German Waterways and Shipping Administration (WSV) is responsible for a large number of technical documents in its archive system. These include the design process in accordance with its administrative regulations (VV-WSV), which covers the entire planning cycle from basic evaluation to implementation planning. In the process of planning, construction and operation of objects of the hydraulic engineering infrastructure, a large and varied number of documents is being accumulated at the responsible authorities. Hierarchical filing systems provided with metadata are often not sufficient to search the documents in a targeted manner. The object of research is therefore machine learning methods that generate new classification systems on the basis of the given document stock and can integrate the existing documents into them. The filing is object-related and the clerk specifies various descriptive attributes. Of interest are now procedures that automatically generate topic models on the basis of the specified texts in the metadata documents in order to assign the documents to them. For this study, the words in the metadata attributes were combined into so-called bag of words and latent Dirichlet allocation (LDA) was applied to automatically find word groups that belong together. With the topic models generated in this way, documents can be searched according to topic composition or, in the case of a keyword search, documents can be displayed which do not contain the keyword but which match the topic. Due to the high number of topics that overlapped within the planning data and the few words per document, the algorithm found it difficult to generate unambiguous topics that could be easily interpreted by humans. In order to generate such topics, so-called Seeded LDA was used. Here the generation of topics can be influenced by setting seed words per topic. With Seeded LDA it is possible to fix certain topics while the algorithm decides others freely and finds new topics.

1 INTRODUCTION

During the life cycle of a building, a large number of documents accumulate. These extend beyond the planning stage with its approval phases, the actual construction and maintenance. The documents are often only accessible decentrally at the offices directly involved. Nevertheless, even with a centralized filing system, the question of structured storage arises to guarantee the granularity and context necessary for information retrieval (Demian & Balatsoukas 2012).

For example, the queries can refer to claim management, quality control and other investigations after the construction phase (Fan et al. 2015). The individuality of the buildings also affects the planning documents and other related documents, so that comparability and classification into a higher structure are difficult to handle manually. The structure would be useful in a different design depending on the user's point of view and expertise. The resulting requirements are often not fully known when setting up such filing systems and change over time as new processes are introduced or older ones are dropped. Usually, a group of experts deals with the existing documents and tries to identify topics with a qualitative approach. The topics and interrelationships can be recorded with an ontology (Kuo 2019).

However, the larger the amount and the more individual the documents, the more difficult the extraction process becomes and the more necessary automated procedures become to support it. In general, finding context, semantic concepts, and relationships from a set of documents is still a major challenge, especially in a domain-specific application like construction (Fan et al. 2015).

2 MOTIVATION

The present study was developed in the context of a preliminary investigation of the document filing system of the German Waterways and Shipping Administration (WSV) (Borstelmann & Beier 2018). The documents refer to ship locks and weir systems and were filed object-related. Furthermore, they refer to all life phases of the structure: They contain, for example, plans, static calculations, proof of economic efficiency and are created mainly in the process of drafting the administrative regulation of the Federal Waterways and Shipping Administration (e.g. VV-WSV 2107, 2016). A glance at the guideline already shows that all possible types of documents (calculations, plans, reports) may occur in the individual phases and, depending on the phase, may have to be submitted again. When filing the documents, a data sheet must be filled out, which contains both selection fields and free text options that serve to describe the document in question (metadata).

The selection fields partly refer to the document types of the VV-WSV, but do not map them analogously. The acceptance of the selection fields does not seem to be high, which has the effect that the most general option from the selection fields is usually chosen to categorize the document. For example, in the "Usage" field, the "Inventory document" option was selected for almost 70% of the approximately 110000 documents referring to the object type "ship lock". The other more specific selections for "Usage" were only used for 2% of the documents. For the remaining 28% of the documents, none of the selection options were chosen, but separate free text categories were defined. The free text field "Detail" was usually filled with some sentences and keywords to describe the document.

In addition, there were also several fields with classic metadata such as format, clerk and various ordinal numbers, which play a subordinate role in finding higher-value concepts. The goal of the investigation was to implement suitable methods for an automated analysis of the existing document inventory. Based on the analysis new input fields and selection field options shall be defined, which better reflect the topics in the filing system and thus find a higher acceptance among users, both when entering and later searching for documents.

3 RELATED WORK

In the literature, there are various examples for solving the problem of information retrieval from unstructured construction documents. In various papers the vector space model is used (e.g. (Zou et al. 2017)). Thereby each document is transformed into a vector. The transformation is done, for example, by assigning each word of a defined dictionary a dimension in the vector. One way to form the vector for each document is to write the number of occurrences of a word in the document to the position of the word in the vector. However, more common are sophisticated variants such as the TF-IDF measure, which considers the length of a document and the fact that terms that occur in few other documents are more meaningful for characterizing the document.

By using the vectors formed in this way, documents can now be compared with each other and with search queries using similarity measures such as the cosine of the spanned angle of two vectors. In (Fan et al. 2015), the vector space model is implemented as part of a search system and validated with claim management documents from a large Hong Kong building. The results of the vector space model are used through user feedback to train supervised learning algorithms that replace the vector space model.

An interesting approach can be found in (Nedeljković & Kovačević 2017). The researchers use the overall corpus to determine various information-rich word pairs based on Pointwise mutual information (PMI) and other measures based on PMI to characterize the documents. Again with a vector space model based on the word pairs a graph structure is derived, visualized and examined for coherent word groups to find concepts.

A construction-related approach, which is related to the latent Dirichlet allocation (LDA) based Topic Modeling performed in this study, is found in (Kuo 2019). Here, latent semantic analysis is used to find concepts and relationships in different problem contexts. The already mentioned vector space model is combined with a Singular Value Decomposition (SVD). The document matrix, which is composed of the vectors of the vector space model, is decomposed into three matrices, the middle one being a diagonal matrix with singular values. By reducing low singular values, the dimensionality of the vector space is reduced, too. Only the most meaningful dimensions (Topics) remain. The two remaining reduced matrices form the document-topic matrix and the term-topic matrix.

Probabilistic Latent Semantic Analysis (pLSA) replaces the SVD with probabilistic modeling of the distribution of documents and words across the topics (Hofmann 1999).

Latent Dirichlet allocation (LDA) then extends this approach again by assuming a Dirichlet distribution as an a-priori-distribution for the distribution of the topics per document (Blei 2012). LDA has already been used for exploring the thematic structure of construction activities (Lai & Kontokosta 2019) and in various other scientific disciplines such as political science, biomedicine or geography (Jelodar et al. 2019). Possibly technically comparable to this work due to the problem boundary condition of short texts within the documents are analyses of Twitter messages (so-called "tweets") (Sanandres et al. 2018).

4 TOPIC MODELING

4.1 *LDA*

The basic assumption of LDA is that the documents are created by a generative process (Blei et al. 2003). Documents consist of a fixed number of topics and topics of different words. For each document the topic composition is drawn and from each topic different words according to the composition (see Figure 1).

Mathematically, the process is formulated in such a way that the parameters for the multinomial distributions to determine the topic distribution over a document and the word distribution over the topics are drawn from Dirichlet distributions. Additionally a number N of the document's words are drawn

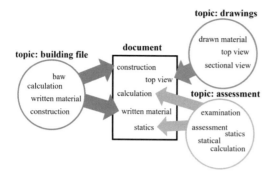

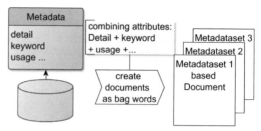

Figure 2. Creating the documents.

Figure 1. Illustration of the generative process.

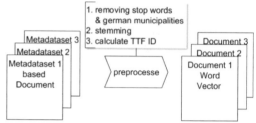

Figure 3. Preprocessing.

from a Poisson or other more appropriate distribution. For each N, a topic is then selected and a word is drawn from that topic according to the multinomial distributions defined first.

In the training process, the topics for each word in each document are iteratively approximated by Gibbs sampling. With this information, the remaining distributions parameters can be calculated (Griffiths & Steyvers 2004). Intuitively, for each word it is found out with which other words they are in a document and which topics these words are assigned to. The word is then assigned the most dominant of these topics. In the beginning the assignment is random. But if the process is repeated often enough for each word, the generated Markov chain converges and a representative sample from the searched common distribution is generated.

Input data from LDA are the words of each document, vectorized as bag of words or via other sophisticated measures such as TF-IDF. In addition, the number of topics to be found must be specified. The hyperparameters alpha and beta are part of the definitions of Dirichlet distributions. Alpha controls how strong the topic blending should be per document. Small alphas increase the chance that only a few dominant topics are found per document, higher ones have the opposite effect. Beta similarly controls the word distribution for each topic. The lower the beta, the more dominant are certain words in a topic. The result is the distribution of topics per document and the distribution of words per topic. New documents can also be evaluated on existing topics based on the words in the document.

4.2 Seeded LDA

In Seeded LDA sets of seed words can be defined to guide the topic creation. Each set will make a topic converge in that direction. Seeded LDA (Jagarlamudi et al. 2012) consists of two combined model extensions of the LDA. On the one hand, additional seed-topic distributions are introduced, which allow only words from the respective seed set to be generated. For each word it is then randomly determined whether it is drawn from a regular distribution according to the standard LDA or from the seed set. In the second extension, the Dirichlet distribution to determine the topic composition in the document is preceded by another Dirichlet distribution, which depends on a binary vector. The binary vector represents which seed set contains words in the document and sets the Dirichlet distribution's mean value. Seeded LDA can also be trained via Gibbs Sampling. The application of Seeded LDA is similar to LDA with one decisive modification. While LDA finds a user defined number of topics in an unsupervised fashion, Seeded LDA allows the data analyst to integrate his own knowledge about these topics. The data analyst can prescribe the model for each topic a list of words that shall appear in it. Seeded LDA will create the topics taking into account this predefined word lists (seed sets).

5 IMPLEMENTATION

For the exemplary analysis all metadata of the object type "ship lock" were selected. The preprocessing of the documents was implemented with the NLP framework Gensim (Řehůřek & Sojka 2010). In Gensim the ldamallet wrapper was integrated to train the model. Depending on the requirements, the attributes from the filing system's metadata records were combined into a bag of words (see Figure 2).

The functions available in Gensim were used to remove so-called stop words, words that are not important for the characterization of the documents and are most common in the German language. In addition, the names of German municipalities were removed as far as they occur in the metadata and were spelled correctly (see Figure 3).

In that way the topics can be formed independently from specific location-based projects. Furthermore, the words were reduced to their stem (stemming).

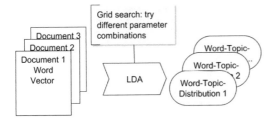

Figure 4. LDA execution.

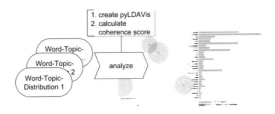

Figure 5. Analyzing the results.

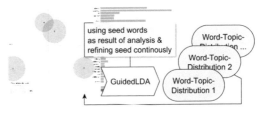

Figure 6. Applying Seeded LDA.

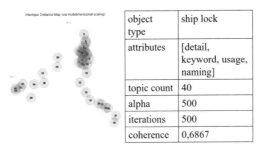

Figure 7. Visualization of the LDA-Topic Model (topic count = 40).

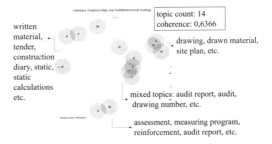

Figure 8. Visualization of the LDA-Topic Model (topic count = 14).

For finding the optimal configuration a grid search was performed, in which different attribute combinations and hyperparameters were systematically tested in different LDA executions (see Figure 4).

For analyzing the results pyLDAvis was used, based on the methods and metrics described in (Sievert & Shirley 2014). The found topics are plotted on a plane where the Jensen-Shannon divergence between their distributions is calculated and projected in two dimensions by principal component analysis (see Figure 4).

The frequency of the topics is represented by the circular area. Depending on the topic, the words are displayed in descending order of saliency and relevance. Saliency evaluates the information contribution of a word to the overall model and relevance corresponds to the information contribution for the specific topic.

For estimating the quality of the trained models the coherence measure c_v was calculated, which is an indirect combination of several other coherence measures like the cosine measure, NPMI and the boolean sliding window (Röder et al. 2015). Intuitively, coherence is to be understood so that words that strongly define a topic often appear together (Syed & Spruit 2017).

Subsequently, the Seeded LDA implementation GuidedLDA (Singh 2017) was used to test different seed word combinations (seed sets) based on the knowledge of the previously executed standard LDA. The seed sets were systematically improved by adding new words to them. In the process more robust and human comprehensible topic models were created (see Figure 6).

6 DEMONSTRATIVE ANALYSIS

A demonstrative analysis was done with the German words of the filing systems documents. These are translated to English for this paper to make the methodology easier to understand. The highest coherence score of 0.69 was found with the configuration that can be seen in Figure 7.

The maximum number of topics and the high alpha speaks for the fact that in the metadata despite the relatively small word count per document a high number of concepts and mixing of these concepts exists. For a data analyst, who has no knowledge or previous experience in the practical handling of the documents, it is however difficult to interpret these topics correctly. The most salient words, i.e. the most essential words for the determination of the topic, correspond to the human understanding (e.g. drawing, assessment, draft, statics, calculation etc.). Therefore, in the next step the topic count is systematically reduced, in otherwise same configuration of hyper parameters (see Figure 8).

With a topic count of 14 a more interpretable division can be recognized in the topics, while the coherence value with 0,64 is still quite high. On the left, there are topics with the words "written material", "tender", "construction diary", "statics", "static

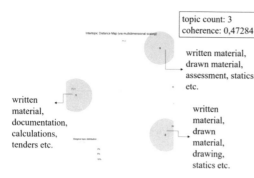

Figure 9. Visualization of the LDA-Topic Model (topic count = 3).

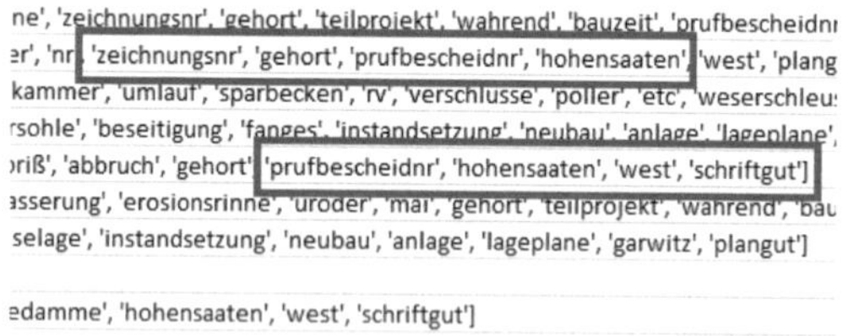

ne', 'zeichnungsnr', 'gehort', 'teilprojekt', 'wahrend', 'bauzeit', 'prufbescheidnr
er', 'nr', 'zeichnungsnr', 'gehort', 'prufbescheidnr', 'hohensaaten', 'west', 'plang
kammer', 'umlauf', 'sparbecken', 'rv', 'verschlusse', 'poller', 'etc', 'weserschleu:
rsohle', 'beseitigung', 'fanges', 'instandsetzung', 'neubau', 'anlage', 'lageplane',
oriß', 'abbruch', 'gehort', 'prufbescheidnr', 'hohensaaten', 'west', 'schriftgut']
isserung', 'erosionsrinne', 'uroder', 'mai', 'gehort', 'teilprojekt', 'wahrend', 'bau
selage', 'instandsetzung', 'neubau', 'anlage', 'lageplane', 'garwitz', 'plangut']

edamme', 'hohensaaten', 'west', 'schriftgut']

Figure 10. Documents with the highest score in one of the mixed topics from Figure 8.

Table 1. Sets of seed words for Seeded LDA-topics.

Topic	Seed Sets
1	[assessment, expert's, inventory, records, opinion, federa, written material]
2	[drawings, layout plan, views, overall plan, top view, sectional drawing, drawn material, draft]
3	[technical, calculations, statical, statics, analysis, written material, draft]

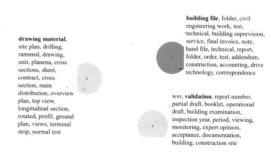

Figure 11. Visualization of the Seeded LDA-Topic Model (seeded topics=3, unseeded topics = 0).

calculations" etc., which probably refer to the written documentation in the construction and planning phase. On the right are topics, in which the words "drawn material", "drawing" and "site plan" are strongly represented, i.e. words that refer to drawn drafts. On the right, the words "assessment", "measuring program", "reinforcement" and "audit report" are strongly represented. These are words, which probably refer to the inspection of existing buildings. If the number of topics is further reduced to three (see Figure 9), the coherence score drops noticeably to 0,47.

As expected, the topics on the left, which refer to the construction and planning phase, merge into one. For topics below, which refer to assessments, and the topics on the right, which refer to drawing material, there is not a clear division. Words like drawing and plan material, which instinctively belong to the topic on the right, are found in the "assessment"- topic, too. To find out why this is the case, another look at the LDA run with 14 topics is needed (see Figure 8). There mixed topics with words from both of the two groups on the horizontal axis can be seen. To understand, which topics cause this effect, documents are looked up, where this topic is most strongly represented. With the model generated in the LDA documents can be searched for existing topics.

The documents, where one of the mixed topics is most dominant, are detected (see Figure 10).

By looking into the documents, it becomes clear that the audit report belonging to an assessment is attached to drawings and plans (see framed words in Figure 10). Both kinds of documents (written material and drawings, German: "Schriftgut" and "Zeichnung") are referring to an "audit report number" (German: "Prüfbescheidnr."). This explains the generated mixed topic and shows that the topic was generated by LDA for human comprehensible reasons.

The mixed topic analysis motivates the use of Seeded LDA. The idea is to incorporate an own interpretation of a topic, such as drawings, without being overlaid by other dominant topics. The seed sets (see Table 1) were determined iteratively by first applying some terms from the experience of the analysis with the standard LDA.

Based on the results, terms that appear in the calculated topics but are not assigned according to one's own interpretation are added to the seed words of the topic, which fits better to the previous assumption. Based on the seed words a new topic model is created, that can be again refined by adding new seed words. After some iterations the topic model created this way shows a clear structure (see Figure 11).

All documents related to "drawings" are in the topic on the left, while "assessment" and "inspection" related files are in the topic below. Words that relate to the construction process are in the topic on the right. This segmentation could be part of a process-oriented filing system.

The guided topic model can now be extended by allowing additional free topics that are not based on seed words. One additional topic is allowed. Close to the "assessment"-topic a further topic has now been created (see Figure 12), which consists of terms like "documentation", "validation", "partial draft" and "operational draft".

The assessment related topic seems to have become more precise again: One interpretation would be,

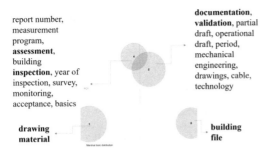

Figure 12. Visualization of the Seeded LDA-Topic Model (seeded topics=3, unseeded topics = 1).

that there are assessments that refer to the structural inspection and acceptance of the building, while other documents with similar terms are also used for the technical documentation and validation of the functionality of the ship lock system (electronics etc.). Here it becomes clear once again that for a more indepth interpretation it is necessary to continuously consult experts in hydraulic engineering to understand the topics and their classification in the processes. The next step could be to analyze subgroups of documents that belong to a certain topic (establish hierarchical topic structures).

7 CONCLUSION

Unstructured filing systems lead to information silos, as the documents required to solve a problem are difficult to find. Particularly in the construction industry, where a large number of different documents accumulate, methods for restructuring document inventories by classifying them into process-oriented meaningful topics are therefore necessary. The paper showed a way to semi-automatically detect concepts and topics in unstructured text data. The probabilistic model latent Dirichlet allocation (LDA) was applied. A method, which is common in other disciplines, but which had been sparsely applied in the construction related studies according to the literature research, carried out here. The present paper tries to contribute to close this gap elaborating how unstructured text data can be analyzed and examined for concepts using the practical example of the filing system of the German Federal Waterways Engineering and Research Institute. For this purpose the standard LDA was used to get an overview of distribution and topics in the document inventory. In order to find the right number of topics and the best configuration of the algorithm, different combinations were systematically tested in a grid search and evaluated with the coherence score. Based on the findings of this analysis, different seed sets were then determined to provide the Seeded LDA with topics and to further specify them iteratively based on the own interpretation.

This way certain topics can be fixed, while by allowing further topics to be added, new insights can be gained into the examined document inventory.

The Seeded LDA allows an iterative semi-supervised approach, which has a great potential in the opinion of the authors of this paper, since it helps to explore the document set automatically, while at the same time systematically placing one's own interpretation and previous knowledge in the model with seed words. In addition to structure-giving topics with their associated dominant words, the algorithm also provides models, with which documents can be examined for topics occurring in them through their word composition. With the found topics an appropriate structure for a document filing system could be defined. Subsequently, the old documents could be automatically inserted into the newly defined structure by applying the models created during the analysis and placing them according to the most dominant topic.

The created topic models can be useful for the information retrieval task, too. On the one hand, documents with a similar composition of topics as the words in the search query can be displayed (concept search). On the other hand, a normal key word search can be improved by adding words from the topics that dominate the key words of a document in addition to the words, that occur in its text (Cheema et al. 2016).

Besides the algorithms demonstrated in this study, there are several other variants of Topic Modeling (e.g. Labeled LDA, Hierarchical Topic Modeling), which are worth to be considered and can extend the methodology shown here.

ACKNOWLEDGEMENT

This work is part of a study for the preparation of a cooperation project with the German Federal Waterways Engineering and Research Institute, which provided the data, the study is based on. The objective of the project would be the development of a construction kit for the systematic restructuring of document inventories and filing systems.

REFERENCES

Blei, D.M., 2012. Probabilistic topic models. Commun. ACM 55, 77–84. https://doi.org/10.1145/2133806.2133826

Blei, D.M., Ng, A.Y., Jordan, M.I., 2003. Latent Dirichlet allocation. J. Mach. Learn. Res. 3, 993–1022.

Borstelmann, H.-H., Beier, A., 2018. DVtU–Digitale Prüf- und Genehmigungsprozesse.

Cheema, M.F., Jänicke, S., Blumenstein, J., Scheuermann, G., 2016. A Directed Concept Search Environment to Visually Explore Texts Related to User-defined Concept Models: in: Proceedings of the 11th Joint Conference on Computer Vision, Imaging and Computer Graphics Theory and Applications. Presented at the International Conference on Information Visualization Theory and Applications, SCITEPRESS – Science and Technology Publications, Rome, Italy, pp. 72–83. https://doi.org/10.5220/0005727400720083

Demian, P., Balatsoukas, P., 2012. Information Retrieval from Civil Engineering Repositories: Importance of Context and Granularity. J. Comput. Civ. Eng. 26, 727–740. https://doi.org/10.1061/(ASCE)CP.1943-5487.0000229

Fan, H., Xue, F., Li, H., 2015. Project-Based As-Needed Information Retrieval from Unstructured AEC Documents. J. Manage. Eng. 31. https://doi.org/10.1061/(ASCE)ME.1943-5479.0000341

Griffiths, T.L., Steyvers, M., 2004. Finding scientific topics. Proceedings of the National Academy of Sciences 101, 5228–5235. https://doi.org/10.1073/pnas.0307752101

Hofmann, T., 1999. Probabilistic Latent Semantic Indexing, in: Proceedings of the 22nd Annual International ACM SIGIR Conference on Research and Development in Information Retrieval, SIGIR '99. Association for Computing Machinery, New York, NY, USA, pp. 50–57. https://doi.org/10.1145/312624.312649

Jagarlamudi, J., Daumé, H., Udupa, R., 2012. Incorporating Lexical Priors into Topic Models, in: Proceedings of the 13th Conference of the European Chapter of the Association for Computational Linguistics, EACL '12. Association for Computational Linguistics, USA, pp. 204–213.

Jelodar, H., Wang, Y., Yuan, C., Feng, X., Jiang, X., Li, Y., Zhao, L., 2019. Latent Dirichlet allocation (LDA) and topic modeling: models, applications, a survey. Multimed Tools Appl 78, 15169–15211. https://doi.org/10.1007/s11042-018-6894-4

Kuo, V., 2019. Latent Semantic Analysis for Knowledge Management in Construction (G4 Monografiaväitöskirja). Aalto University publication series DOCTORAL DISSERTATIONS; 45/2019. Aalto University; Aalto-yliopisto.

Lai, Y., Kontokosta, C.E., 2019. Topic modeling to discover the thematic structure and spatial-temporal patterns of building renovation and adaptive reuse in cities. Computers, Environment and Urban Systems 78, 101383. https://doi.org/10.1016/j.compenvurbsys.2019.101383

Nedeljković, Đ., Kovačević, M., 2017. Building a Construction Project Key-Phrase Network from Unstructured Text Documents. J. Comput. Civ. Eng. 31, 04017058. https://doi.org/10.1061/(ASCE)CP.1943-5487.0000708

Řehůřek, R., Sojka, P., 2010. Software Framework for Topic Modelling with Large Corpora, in: Proceedings of the LREC 2010 Workshop on New Challenges for NLP Frameworks. ELRA, Valletta, Malta, pp. 45–50.

Röder, M., Both, A., Hinneburg, A., 2015. Exploring the Space of Topic Coherence Measures, in: Proceedings of the Eighth ACM International Conference on Web Search and Data Mining, WSDM '15. Association for Computing Machinery, New York, NY, USA, pp. 399–408. https://doi.org/10.1145/2684822.2685324

Sanandres, E., Llanos, R., Madariaga Orozco, C., 2018. Topic Modeling of Twitter Conversations.

Sievert, C., Shirley, K., 2014. LDAvis: A method for visualizing and interpreting topics, in: Proceedings of the Workshop on Interactive Language Learning, Visualization, and Interfaces. Presented at the Proceedings of the Workshop on Interactive Language Learning, Visualization, and Interfaces, Association for Computational Linguistics, Baltimore, Maryland, USA, pp. 63–70. https://doi.org/10.3115/v1/W14-3110

Singh, V., 2017. GuidedLDA, URL: https://guidedlda.readthedocs.io/en/latest/ (Accessed: 15.11.2020).

Syed, S., Spruit, M., 2017. Full-Text or Abstract? Examining Topic Coherence Scores Using Latent Dirichlet allocation, in: 2017 IEEE International Conference on Data Science and Advanced Analytics (DSAA). Presented at the 2017 IEEE International Conference on Data Science and Advanced Analytics (DSAA), IEEE, Tokyo, Japan, pp. 165–174. https://doi.org/10.1109/DSAA.2017.61

Verwaltungsvorschrift der Wasserstraßen- und Schifffahrtsverwaltung des Bundes (VV-WSV) Entwurfsaufstellung VV-WSV 2107, Fassung 06/2016. ed, 2016. Bundesministerium für Verkehr und digitale Infrastruktur (BMVI).

Zou, Y., Kiviniemi, A., Jones, S.W., 2017. Retrieving similar cases for construction project risk management using Natural Language Processing techniques. Automation in Construction 80, 66–76. https://doi.org/10.1016/j.autcon.2017.04.003

Applying weak supervision to classify scarce labeled technical documents

M. Shi, A. Hoffmann & U. Rüppel
Institute of Numerical Methods and Informatics in Civil Engineering, Technical University of Darmstadt, Darmstadt, Germany

ABSTRACT: The digitalization in the construction industry, the number of project relevant documents. It becomes a challenge to organize documents in a searchable manner by classification. The German Waterways and Shipping Administration (WSV) is one of the organizations facing this problem. Manually classifying is due to the considerable expense nearly impossible. In parallel, text classification with machine learning increasingly draws attention. Classification belongs to supervised machine learning, where large labeled data samples are needed. In the filing system used in WSV, only a small amount of data with ground-truth labels are available. It is tedious and expensive to annotate manually. To solve the shortage of training data, we propose applying weakly supervised learning, where noisy and inexact labels can be used in the training process. In this study, we inject the domain knowledge in the training process with weak supervision framework Snorkel to construct a labeling model that programmatically annotates data. We then trained classifiers on the original dataset together with the dataset annoted by the labeling model. The results show that even though the programmatically annoted dataset is noisy, it can still train a generalized classifier and improve the classifiers' performance.

1 INTRODUCTION & MOTIVATION

The digitalization in the construction industry let the number of accessible project relevant documents rise enormously. Documents from the past projects are of high value concerning intercompany knowledge management. Besides general text documents, construction-specific documents such as drawings and planning are also essential information sources. However, because of those documents' unstructured nature, it is very time-consuming to organize them into meaningful groups and retrieve useful information from the big document hub, let alone knowledge extraction.

The German Waterways and Shipping Administration (WSV) is one of the organizations facing this problem. The main tasks of WSV comprise the maintenance, operation as well as the upgrading and construction of the federal waterways[1]. WSV has many technical documents in the digital document platform, the German "Digital Administration of Technical Documents" (DVtU), covering the entire planning cycle from essential evaluation to implementation planning. However, most of the documents are not categorized or wrongly categorized in the platform, making information search and extraction very difficult. In this context, text classification algorithms are usually used to allocate documents in different groups automatically. Text classification, which can process large scale text documents, is a fundamental task in machine learning and natural language processing for decades. Text classification belongs to the supervised learning method. A supervised learning process requires many training data with ground-truth labels to thrive a well-performed and robust classifier. However, in many use-cases, like in the case of WSV, ground-truth labels are scarce due to the time- and cost-intensive manual labeling process. Ground-truth labels refer to accurate labels provided by empirical evidence. When it comes to specific tasks where experts' explicit, implicit knowledge, and experience are needed, the data labeling process can be costly. Another problem with using machine learning in the construction industry is that the training data was labeled for a specific application. When the training goal changes with the business process, data need to be newly labeled from zero. Due to the lack of subject-matter expertise, finance, and time, the insufficient amount of labeled data becomes a significant obstacle for applying machine learning in the construction industry.

The lack of ground-truth labeled data is also a common concern in many other application areas, such as medicine or law. Research that deals with scarce labeled data has shown up in the last years and has given promising results. Weakly supervised learning is a state-of-the-art method that uses noisy, limited, imprecise labels instead of strong ground-truth as training data. The strength of weak supervision is that users can get a reliable predictive model even with those "weak labels" (contrary to ground-truth). These weak labels can be generated programmatically by introducing an external knowledgebase, predefined

---

[1] https://www.bmvi.de/SharedDocs/EN/Articles/WS/federal-waterways-and-shipping-administration.html; Accessed on 01.01.2021

patterns, or by crowd-sourcing labels. In any case, weak labels are quicker and more economical to generate than manually label the data by subject matter experts on a large scale. Another advantage of weakly supervised learning is that users can inject their background knowledge in the training process, which enhances data-driven methods semantically. To ease the access to weakly supervised learning for developers, researchers from Stanford University brought out the Snorkel framework, which can label training datasets programmatically. Snorkel can easily model multiple label sources without access to the ground truth. With the Snorkel labeling functions interface, users can flexibly combine different types of weak supervision strategies, such as pattern matching and heuristic rules (Ratner et al. 2017).

The lack of labeled training data is a mature challenge for application of machine learning methods in the construction branch. In this study, the primary goal is to determine whether we can use background knowledge with the weak supervision framework Snorkel to generate a large number of labels in a short time, andimprove the following text classification's performance. This paper is structured as follows: Section 2 summerized current research work in text classification in construction and infrastructure engineering. Section 3 explains weakly supervised learning and its application with Snorkel framework. The following section 4 describes the use case and research scope of this work. Section 5 shows our concept of applying Snorkel with background knowledge to programmatically label documents in the DVtU platform and train a classification model. We then discuss the results in section 6 and conclude our research in section 7.

2 RELATED WORK

Research on applying experts' knowledge and data-driven system in text classification already exists. In the 1970s and 1980s, the most approachs were based on rule-based expert systems. Expert systems were brittle and fault vulnerable when the system became large. Besides, the rule-based experts' system also has a shallow representation level of achieved knowledge, such that it is challenging to be semantically accessed and interpreted by machines (Niu & Issa 2015) Later on, upcoming statistical machine learning methods were used. Caldas Carlos H. et al. (2002) developed a supervised learning-based program using standard machine learning algorithms to classify construction project documents (such as specifications, meeting minutes, field reports, among others) into multiple topics. In the following years, different machine learning methods, such as Support Vector Machines, Rocchiio, Naïve Bayes, and K-Nearest Neighbor were researched (Caldas & Soibelman 2003; Mahfouz 2012). Later on, approaches to boost the machine learning algorithms' results by integrating knowledge from other sources were introduced. Salama and El-Gohary (2016) proposed a hybrid semantic and machine learning-based method to classify contract clauses for supporting automated compliance checking in construction. They used different classifiers, preprocessing techniques, and feature engineering methods in the implementation. Chi et al. (2016) added the evaluation of category strength in the text classification process for supporting construction field inspection. Recently, deep learning methods such as convolutional neural networks and Bidirectional Transformers for Language Understanding were used to classify Chinese building quality complaints (Fang et al. 2020; Zhong et al. 2019)

3 BACKGROUND

Current automatic text classification is usually based on supervised learning methods where a large amount of training data with ground-truth labels is necessary for satisfying results. Due to the shortage of labeled data in practical applications, weakly supervised learning is gaining more attention in machine learning research.

Weakly supervised learning is a variety of methods that uses incomplete, inexact, and inaccurate labels to construct predictive models (Zhou 2018). Incomplete labels refer to situations where only small amounts of labeled data are given. Inexact labels are less precise labels, while inaccurate labels are not ground-truth (Zhou 2018). Regarding the process of creating weak labels, data pairs can be generated from several sources: distant supervision, where external knowledge bases are heuristically aligned with data points; crowd-sourced labels; rules and heuristics for labeling data, and others. In practice, analysts tend to combine results from different weak supervision data sources to cover as much information as possible (Ratner et al. 2017). However, sources may overlap and conflict.

Snorkel enables the data analyst to use labels from all available weak supervision sources by learning accuracies and correlations over all of them without ground truth (Ratner et al. 2017). It unifies the label sources by producing a probabilistic label distribution over every data point. According to Ratner et al. (2017), Snorkel outperforms distant supervision, one of the most popular forms of weak supervision, by 132% on open-source datasets. Snorkel has also shown its advantages in industrial-scale applications (Bach et al. 2019).

4 USE CASE

The DVtU is a central filing system for technical drawings, construction plans, static calculations, approval procedures, and other technical documents. DVtU currently has more than 2500 users and is considered one of the WSV's essential IT systems[2].

---

[2] https://www.contact-software.com/de/branchen/logistik-infrastruktur/wasserstrassen-und-schifffahrtsverwaltung-des-bundes/; Accessed on 01.01.2021

This plattform enables users to create, edit, and archive project relevant documents through the project lifecycle. DVtU organizes digital project files ("technical documents") in a hierarchical structure. Users provide descriptive information by completing a metadata form, while uploading the technical document (Borstelmann & Beier 2018). The metadata includes information like project ID, object name, keywords, descriptions, comments, the processing status, editor, usage category, to name a few. Some of this data is generated automatically by the system, e.g., details of the editor.

However, since some metadata must be entered manually, mistakes appear with the growth of projects. There are three different types of manually entered metadata: selecting a drop-down list, free text, a combination of selection and free text. This paper focuses on the third type of metadata, specifically on the categorical field "Usage", where users can select from a predefined list or enter free text if they think none of the categories from the drop-down list describe the document correctly. With the gradual increase of technical documents, the diversity of the field "Usage" also grows from dozens to more than four hundred. Many redundancies exist between the added free text and the original given selection list. Taking the preset "Usage"-category "static calculations" as an example, users also give descriptions with the same meaning, such as "static," "design static," "static calculation," and more. Some categories have the same words but differ in capitalization; some are synonyms of the predefined categories; some words are misspelled. Other than redundancy, the author also observed that from the given 141,435 datasets (the total number of documents in the platform is much more than that), 101,686 share the same Usage category "inventory document," which gives no specific information on the technical documents. The unspecific and partially wrong categories hinder the searching and filtering of specific information in the platform. Since the filtering process in DVtU is based on simple keywords, certain documents are left out if their metadata does not include the same keywords as in the filtering field. Thus, there is a need to classify the documents into the nine preset categories, correct the wrong labels, and replace the general labels with more specific ones.

Given that the number of datasets in the DVtV system is extensive and manual classification is a time-consuming and tedious process, the documents' classification process shall be supported by applying machine learning methods to their metadata. There are 34 other metadata areas aligned to technical documents except for the "Usage"-field, of which six have meaningful information. However, after embedding texts in vectors, the number of features reached 14,220 after preprocessing. Due to the unbalanced distribution, only 221 datapoints per "usage"-category, in total 1989 samples, are available as training and testing data, which is not big enough to conduct nonlinear algorithms (such as support vector machines, naïve Bayes, decision tree). Thus, we propose to

```
@labeling_function()
def keyword_schalung(x):
    if ("schalung" in x.Einzelheit.lower().split(" ") ) or \
            ("schalung" in x.Bemerkung.lower().split(" ")):
        return Ausfuhrung
    return ABSTAIN
```

Figure 1. Labeling function code example.

automatically create more training data by integrating the users' background knowledge with the Snorkel framework.

5 INJUNCTION DOMAIN KNOWLEDGE IN THE LABELING FUNCTIONS

The process of creating weak labels with background knowledge for training text classifiers is based on the three steps mentioned in the paper of Ratner et al. (2017):

1. Writing Labeling Functions: users can express patterns, heuristics, external knowledge bases, and others by writing labeling functions.
2. Learning a generative model for labeling functions without supervision: Snorkel learns from labeling functions' agreements and disagreements to estimate their accuracy and correlations. This step provides the quality of labeling functions and returns a set of probabilistic labels in a re-weighted combination.
3. Training a discriminative model: The generated probabilistic labels of step 2 can be used to train supervised machine learning models. The discriminative model generalizes beyond the re-weighted combination of user-defined labeling functions.

5.1 *Writing labeling functions*

People who annotate labels usually use their background knowledge or experience. Snorkel provides the interface, *labeling function* for users to express their knowledge in patterns, heuristics, or more complex forms like preprocessors (Ratner et al. 2017). A labeling function is a snippet of python code, which accepts an unlabeled datapoint, applies the hand-defined logic on it, and returns a decision, whether a specific label should be applied or not. In the case of the BAW filing system, specific words and their synonyms usually appear in the metadata of some classes more often. Besides, words that belong to a domain usually indicate the document's class when they appear in the metadata. For instance, we can infer that if words like "formwork," "reinforcement," or "construction site" exist in the "description" metadata, it may indicate that this document belongs to the class "execution." We wrote 34 labeling functions for all nine classes. Some of them are keywords matching, some are regular expressions, and some are synonyms. One example of the keyword matching labeling function can be seen in Figure 1.

This labeling function checks if the word "framework" (german: "Schalung") exists in the metadata

Table 1. Label distribution of the generative model and majority vote model.

Clas	Labels distribution from generative mode	Labels distribution from majority vote
Ausfuhrung	10,608	12,060
Ausschreibun	58	23
Bauwerksinspektion	73	73
Entwurf-AU	16	15
Gutachte	39	38
Planun	1727	1569
Technische Berechnunge	17	18
Technischer Berich	12	11
Statische Berechnungen	2154	2128
Tota	31,693	31,693

Table 2. Class distribution in datasets.

Clas	Dataset 1	Dataset 2	Dataset 3	Dataset 4 (DS1+DS2)	Dataset 5 (DS1+DS3)
0*	22	20	20	42	42
1*	22	58	20	27	42
2*	22	20	20	42	42
3*	22	20	20	42	42
4*	22	20	20	42	42
5*	22	20	20	42	42
6*	22	20	20	42	42
7*	22	19	18	41	40
8*	22	20	20	42	42

*0 Ausfuhrung
1 Ausschreibun
2 Bauwerksinspektion
3 Entwurf-AU
4 Gutachte
5 Planun
6 Technische Berechnunge
7 Technischer Berich
8 Statische Berechnungen

fields "detail" or "comment". Both of them have document information in free text. The labeling function returns "execution" (german: "Ausführung") as a label if the word exists, "ABSTAIN" if not.

5.2 Learning the generative model for the labels functions

Since we defined labeling functions from different knowledge sources, and each source cover only a small part of datasets, it is unavoidable that the output labels are inaccurate, overlapped, or conflicting. Snorkel provides the generative model to integrate noisy labels from a set of labeling functions (Ratner et al. 2017). This generative model calculates statistical dependencies among labeling functions and produces a single set of probabilistic-weighted labels without access to the actual label (Bach et al. 2017; Ratner et al. 2018).

We applied the labeling functions on 118 290 data samples, and 31,693 of them got at least one label from the labeling functions set. The result from Table 1 shows that the class distribution in the programmatically labeled data is not balanced. It is assumed that this issue is more related to the distribution of documents in the whole dataset and less to the labeling function logic. After the labeling process, we then trained a generative model with the "LabelModel"[3] class of Snorkel. To see whether the trained generative model gives reliable results, we tested it with the ground truth datasets mentioned in section 4. With an accuracy of 65.1% compared to a random guess of 11.1% accuracy, the results justify further investigation.

Snorkel also provides another model for unifying labeling function results, the majority vote model. As its name already suggests, it assigns labels by calculating the majority vote across all labeling functions. The majority vote model leads to increased accuracy of 79.2%. The distribution of labels is listed in Table 1.

We used both models to label the data and compared the end predictive results in section 6.

5.3 Training a discriminative model

Since this work focuses on programmatically creating labels and using weak supervision to improve classification models, we will not dive into the standard training process details of text classification. In a nutshell, after data collection, we preprocessed them with data cleaning, stemming, lemmatization, and transforming text into vectors with the Term-frequency-inverse document frequency (TFIDF).

Now more data pairs are available after procreating labels with Snorkel. We trained the same classifier with five different datasets to see if the generated data can improve the classification performance. Table 2 lists the detailed data distribution of each dataset.

1. The first dataset is the original training data with ground-truth labels. 2/3 of it was used as training data; the other 1/3 data was separated from it as test data for all three datasets.
2. Dataset 2 is composed of data points from the generative model. Since the class distribution in the created dataset is exceptionally unbalanced (see Table 1), we round up the weakly represented classes to 200 by adding ground truth data, which were not used in dataset one but existed in the original data samples.
3. Dataset 3 is derived from the majority vote model with the completion of data with ground truth labels. The completion shares the same process as dataset 2.
4. Dataset 4 is a combination of datasets 1 and 2.
5. Dataset 5 is a combination of datasets 1 and 3.

---
[3] See the details in the online-document at https://snorkel.readthedocs.io/

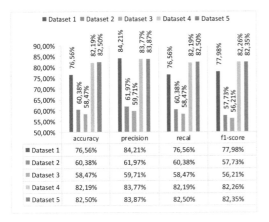

Figure 2. Classification result of SVM without SGD.

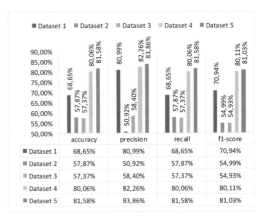

Figure 3. Classification result of SVM with SGD.

6 EVALUATION AND DISCUSSION

The five datasets were trained in three machine learning models: support vector machine with and without stochastic gradient descent (SGD), and the artificial neural network model (ANN), with the same set of parameters. The implementation was programmed in python and used scikit-learn framework (Pedregosa et al. 2011).

This paper used four metrics to evaluate the classifier's performance: accuracy, precision, recall, and F1 score. Accuracy is the fraction of the correct classification in the classifier's total number of predictions (Manning et al. 2009). Accuracy only gives rational information about performance if the distribution of the classes is equal or nearly equal. Precision is the ratio of one class's correct predictions to the total predicted number of this class. Recall, also called sensitivity, gives the proportion of correctly classified data records in the total number of records belonging to the respective classes. F1 score is the harmonic mean of precision and recall, which takes both false positives and false negatives into account. F1 supplement the disadvantage of accuracy if the class distribution is not even.

As can be seen from Figure 2 to Figure 4, dataset 1, the original ground-truth data pairs reached an accuracy of around 68% to 76% in all three classification models. Datasets 2 and 3, even though they are created programmatically with domain knowledge, could reach an accuracy of 60.37%. In datasets 4 and 5, where the programmatically created labels are combined with the ground-truth data pairs, the results show an improvement up to 12.93% in the accuracy compares to dataset 1 (see Figure 3 difference between dataset 1 and 5). Other matrics share the same tendency.

By contrasting results from datasets 2 and 3, as well as datasets 4 and 5, we conclude that the generative model performs similarly to the majority vote model with the supervised learning methods. The results indicate that the first model can better generalize the data,

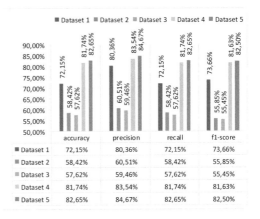

Figure 4. Classification result of ANN.

even though they showed a difference of 14.1% in classification without discriminative classifiers, where generative model has 65.1% while the majority vote reached 79.2% accuracy.

This study shows that adding data with weak labels to a small-sized ground-truth dataset can improve the classification performance. We consider the weak supervise learning a promising method in dealing with text classification with little training data size. By bringing more training data into the training process, the model becomes more generalized by seeing more features, which can prevent overfitting problems.

7 CONCLUSION AND FUTURE WORK

In this paper, knowledge of crowd workers and domain experts were directly injected with Snorkel into the machine learning process. Knowledge such as predefined patterns and heuristics about specific document categories were encoded in labeling functions, which are modular, reusable, and flexible for changes. The labeling functions were applied to unlabeled data, namely, documents without or wrong categories. Then a generative model and majority vote model were then used to learn the labeling functions' accuracy and given weights to their outputs. We then used these artificially created "document-category" pairs to train the commonly used classification models. This concept was implemented in documents from the original filing system of the WSV. The results showed that even noisy and not entirely accurate data could help in training a generalized classifier.

We see the future studies in improving the coverage and accuracy of labeling functions. An external knowledge base shall be connected to the labeling process to inject more domain knowledge into the labeling process. Also, there is need to apply the weak labels on other machine learning classifiers to see if it works on other classification models.

From the study we conclude that knowledge-based weak supervision through labeling functions have the following advantages: datasets can be labeled automatically on a large scale; crowd intelligence and expert knowledge can be combined with the statistical machine learning process; it is modular, reusable, easy to adapt to changes. This concept is adaptable for other engineering application areas such as engineering design with design guidelines, construction site planning, and safety control on construction sites with computer vision, to name a few.

ACKNOWLEDGEMENT

This work is part of a study for the preparation of a cooperation project with the German Federal Waterways Engineering and Research Institute (BAW), which provided the data, the study is based on. The objective of the project would be the development of a construction kit for the systematic restructuring of document inventories and filing systems.

REFERENCES

Bach, S.H., Alborzi, H., Kuchhal, R., Ré, C., Malkin, R., Rodriguez, D., Liu, Y., Luo, C., Shao, H., Xia, C., Sen, S., Ratner, A., Hancock, B., 2019. Snorkel Dry-Bell: A Case Study in Deploying Weak Supervision at Industrial Scale, in: Proceedings of the 2019 International Conference on Management of Data – SIGMOD '19. Presented at the 2019 International Conference, ACM Press, Amsterdam, Netherlands, pp. 362–375. https://doi.org/10.1145/3299869.3314036

Bach, S.H., He, B., Ratner, A., Ré, C., 2017. Learning the Structure of Generative Models without Labeled Data. Proc. Mach. Learn. Res. 70, 273–82.

Borstelmann, H.-H., Beier, A., 2018. DVtU–Digitale Prüf- und Genehmigungsprozesse.

Caldas Carlos H., Soibelman Lucio, Han Jiawei, 2002. Automated Classification of Construction Project Documents. J. Comput. Civ. Eng. 16, 234–243. https://doi.org/10.1061/(ASCE)0887-3801(2002)16:4(234)

Caldas, C.H., Soibelman, L., 2003. Automating hierarchical document classification for construction management information systems. Autom. Constr. 12, 395–406. https://doi.org/10.1016/S0926-5805(03)00004-9

Chi, N.-W., Lin, K.-Y., El-Gohary, N., Hsieh, S.-H., 2016. Evaluating the strength of text classification categories for supporting construction field inspection. Autom. Constr. 64, 78–88. https://doi.org/10.1016/j.autcon.2016.01.001

Fang, W., Luo, H., Xu, S., Love, P.E.D., Lu, Z., Ye, C., 2020. Automated text classification of near-misses from safety reports: An improved deep learning approach. Adv. Eng. Inform. 44, 101060. https://doi.org/10.1016/j.aei.2020.101060

Mahfouz, T., 2012. Unstructured Construction Document Classification Model through Support Vector Machine (SVM) 126–133. https://doi.org/10.1061/41182(416)16

Manning, C.D., Raghavan, P., Schütze, H., 2009. Introduction to Information Retrieval 569.

Niu, J., Issa, R.R.A., 2015. Developing taxonomy for the domain ontology of construction contractual semantics: A case study on the AIA A201 document. Adv. Eng. Inform. 29, 472–482. https://doi.org/10.1016/j.aei.2015.03.009

Pedregosa, F., Varoquaux, G., Gramfort, A., Michel, V., Thirion, B., Grisel, O., Blondel, M., Prettenhofer, P., Weiss, R., Dubourg, V., Vanderplas, J., Passos, A., Cournapeau, D., Brucher, M., Perrot, M., Duchesnay, E., 2011. Scikit-learn: Machine Learning in Python. J. Mach. Learn. Res. 12, 2825–2830.

Ratner, A., Bach, S.H., Ehrenberg, H., Fries, J., Wu, S., Ré, C., 2017. Snorkel: rapid training data creation with weak supervision. Proc. VLDB Endow. 11, 269–282. https://doi.org/10.14778/3157794.3157797

Ratner, A., Hancock, B., Dunnmon, J., Sala, F., Pandey, S., Ré, C., 2018. Training Complex Models with Multi-Task Weak Supervision. ArXiv181002840 Cs Stat.

Salama, D.M., El-Gohary, N.M., 2016. Semantic Text Classification for Supporting Automated Compliance Checking in Construction. J. Comput. Civ. Eng. 30, 04014106. https://doi.org/10.1061/(ASCE)CP.1943-5487.0000301

Zhong, B., Xing, X., Love, P., Wang, X., Luo, H., 2019. Convolutional neural network: Deep learning-based classification of building quality problems. Adv. Eng. Inform. 40, 46–57. https://doi.org/10.1016/j.aei.2019.02.009

Zhou, Z.-H., 2018. A brief introduction to weakly supervised learning. Natl. Sci. Rev. 5, 44–53. https://doi.org/10.1093/nsr/nwx106

Defeasible reasoning for automated building code compliance checking

B. Li & C. Schultz
Cyber Physical Systems Group, Aarhus University, Aarhus, Denmark

J. Dimyadi
Compliance Audit Systems Limited, New Zealand
School of Computer Science, University of Auckland, Auckland, New Zealand

R. Amor
School of Computer Science, University of Auckland, Auckland, New Zealand

ABSTRACT: We present a new approach and a prototype software engine for defeasible reasoning to support automated building code compliance checking. The challenge with formalising defeasible rules in a purely deductive reasoning system is the resulting exponential blow up of the rule set due to the need to explicitly state numerous combinations of exception cases. This leads to a rule set that is brittle, complicated, difficult to understand, and obscures the central point of a given normative provision. Moreover, assessing compliance, to any degree, in the absence of complete information about a building, down to the smallest detail, is very limited in a purely deductive setting in which nothing can be assumed. We empirically evaluate our prototype reasoning engine on an IFC model of a building in Christchurch, New Zealand, for compliance with accessibility criteria from the New Zealand Building Code.

1 INTRODUCTION

Compliance checking has a crucial role in every aspect of a project lifecycle in the architecture, engineering, and construction (AEC) domain. The compliance checking process ensures new and refurbishment work satisfies safety, well-being, and accessibility criteria stipulated by all governing legislations.

There is a growing demand for reliable and effective automated compliance checking systems to take advantage of the open standard digital Building Information Models (BIM) that has gained popularity in the AEC domain. This would save the need to flatten the data-rich BIM model and produce a set of paper-based 2D drawings to suit conventional manual compliance checking practice, which is inefficient and error-prone (Boken et al 2009; Gallagher et al 2004).

Legislation and subsidiary legislation such as building regulations and building codes contain normative requirements conveyed in natural language. Thus, one key challenge to automatically check if a given BIM satisfies a particular set of norms is how certain qualitative natural language terms should be interpreted without ambiguity, e.g. the Building Code of New Zealand (NZBC) Acceptable Solution D1/AS1 document (NZBC 2017) paragraph 1.1.1 states that *"Accessible route shall be provided to give direct access to the principal entrance to the building where practical."*. Here, "accessible route", "principal entrance", and "direct access", and "where practical" are open to interpretation to varying extents. That is, a software tool often needs to support multiple, alternative interpretations and thus compliance is *with respect to* a specific interpretation.

A second key challenge when attempting to formalise natural language norms is handling *defeasibility*, i.e. where the requirement to satisfy a given norm can be *defeated* under certain conditions. Consider paragraph 1.3.1 of NZBC D1/AS1:

> *"Except in 'household units' or where permitted by Paragraph 1.3.2, a single isolated step shall not be permitted but the change of level shall be constructed as a ramp complying with Paragraph 3.0."*

We can deconstruct this rule as follows:

- By default (on authority) this rule applies wherever there is a *"change of level"*
- *"a single isolated step"* is a violation
- *"a ramp"* is compliance
- an exception (that *defeats* the necessity to satisfy the rule) is if the building is a *"household unit"* or satisfies Paragraph 1.3.2.

On the other hand, BIM provides a formal, precise way of describing a building, and while prominent BIMs such as Industry Foundation Classes (IFC) provide object classes for common building elements such as walls, doors, openings, spaces, they lack many classes that, with respect to semantics, more heavily relate to human behaviour and experience that are essential for analysing building code compliance, e.g.: *access routes* (movement behaviours), *accessibility routes* (movement behaviours for particular occupant groups), *building entrance* (transition between outside to inside), a *principal entrance* (occupant experience and intended building use).

Finally, an instance of a BIM itself may be missing particular details that make it impossible to automatically determine whether it complies or violates a given building code, e.g. as we review in our case study in Section 4 the swing direction of hinged doors may not have been specified in a BIM, where certain combinations of door swing directions violate accessibility regulations by not affording sufficient *clear space*.

Together, these four properties present two major challenges for automated compliance checking:

1. *Methodological challenge of providing meaningful information to the user:* given the complexity in formalising natural language codes into software, compliance reports must provide users with meta-information that explains and justifies the underlying logic of analysis. This drives the need for the formal ruleset itself to have certain qualities inspired from requirements engineering, namely *transparency* and *traceability*.
2. *Computation challenge of combinatorial explosion:* reasoning across multiple, alternative interpretations of qualitative terms and multiple alternative assumptions that "fill in" important missing information from a BIM. As a search task this can be formulated as finding combinations of interpretations and assumptions where a rule applies (or not) and is complied with, violated, or defeated by satisfying at least one of the exceptions.

Research contributions in this paper:

- (Section 3) we present a new approach for operationalising defeasible building code rules based on new kinds of BIM objects called *spatial artefacts* (Bhatt et al. 2012a; Bhatt et al. 2012b), and the logic programming language Answer Set Programming (ASP) that we have extended to natively support geometric processing and spatial reasoning (Li et al. 2020; Walega et al. 2017);
- (Section 4) we present our prototype software tool for automated building code compliance checking and empirically evaluate it by checking a NZBC code on a large, real-world IFC model.

2 BACKGROUND

We represent and reason about building codes within the formal framework of defeasible reasoning. Defeasible reasoning is a form of non-monotonic reasoning where the inferred consequents ("Q is true") are contingent on various assumptions and are thus subject to change if new information becomes available: *"by default (on authority), assume proposition Q is true unless we can prove Q is false"*.

We distinguish the following concepts (Eastman et al. 2009; Solihin et al. 2017): Code *formalisation* is the task of representing or expressing the natural language code in a precise, unambiguous way so that it can be interpreted by software (Zhang & El-Gohary 2016). Code *execution* is the computational task of actually checking whether a BIM satisfies the code, and (optionally) modifying an uncompliant BIM in order to satisfy the code.

High quality rulesets: When developing automated code compliance software, building codes are often formalised as software algorithms (an *imperative* approach). In this approach, *formalisation* and *execution* are deeply entangled, having been combined into one step by going directly from natural language code to executable algorithm, complicating the formal representation of the code and thus significantly reducing *transparency*. An alternative approach is to formalize a code as *rules* in a formal logical system such as first order logic (a *declarative* approach) and use general purpose reasoning engines for compliance analysis (Eastman et al. 2009). We adopt this approach with one enhancement of integrating geometric processing natively within the reasoning engine (Borrmann & Rank 2009) to keep the formalised rules as simple and comprehensible as possible.

Inspired by advances in requirements engineering, our aim is to develop a methodology so that the body of algorithms and rules (the knowledge base), has the following qualities:

1. *Transparency:* When code compliance software delivers information to the user, it is of little value to simply report *complies/violates* without presenting the logic and rationale behind the analysis. The formal code representation must as closely as possible reflect the intention and lettering of the natural language code.
2. *Traceability:* The decisions about how qualitative terms relating to human behaviour and experience are disambiguated must be based on research literature in psychology and ergonomics (evidence-based), not ad hoc decisions and magic numbers. The ruleset must accommodate multiple competing, alternative interpretations. Users must therefore have access to the link from the result of rule *execution* back to research literature that provides the rationale justifying the disambiguating definition.

To address the above, our workflow consists of a series of carefully crafted transformations (see Figure 1) that explicitly trace back to the original building code and the providence of research literature used to define ambiguous terms, namely (1) LegalRuleML (LRML), (2) spatial artefacts, (3) reasoning engines that natively support spatial reasoning.

2.1 The ACABIM approach

A commercial automated compliance audit framework known as ACABIM (ACABIM 2020; Dimyadi &

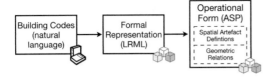

Figure 1. Our workflow translating natural language building codes into operational rules in ASP.

Amor 2018) is used in this work to test the methodology in a real-world implementation. ACABIM has an underlying human-guided automation philosophy in which human experts specify exactly what to check and for the checking engine to execute the procedure accurately. The ACABIM approach argues that human experts should not be tied up in manually undertaking tasks that are trivial but tedious for humans to perform, such as checking if every door in a multi-storey building swings in the right direction to satisfy the requirements for means of escape from fire.

The ACABIM framework fully supports open standards and comprises three main input components, namely BIM, CAP (Compliance Audit Procedure), LKM (Legal Knowledge Model). Additionally, it also supports supplementary human inputs as well as simulations. CAP is the human-guided component of the ACABIM approach where it represents the checklists used by the compliance assessor or building consent authorities. LKM is intended to support open legal knowledge interchange standards such as the emerging standards LegalDocML and LegalRuleML (Dimyadi et al. 2017). LKM is where the current research presented in this paper attempts to augment and add value.

ACABIM checks compliance of a given design aspect in BIM strictly in accordance with the specification in CAP, which is tasked with extracting information from various sources (e.g. IFC/MVD (Model View Definition), LKM, and other supplementary data sources or simulation outputs) and performing calculations using the extracted data.

2.2 LegalDocML and LegalRuleML

The current implementation of the ACABIM approach supports LegalDocML (LDML) and LegalRuleML (LRML), which are two complementary standards that are intended to operate in conjunction. LDML and LRML are both being standardised by OASIS (Organisation for the Advancement of Structured Information Standards). LDML is intended to represent the structure and literal content of a document. LRML is intended to represent the semantics and logical content of the provisions in that document. LRML is built on top of the open standard RuleML, which is a standard developed and maintained by OMG (Object Management Group). LRML extends RuleML with formal features that support norm modelling such as deontic operators (i.e. obligation, prohibition, permission, etc.). LRML supports alternative interpretations as well as defeasible logic.

2.3 Spatial artefacts in BIM

The concept of spatial artefacts was developed by Bhatt et al. (Bhatt et al. 2012a,b) as a way of modelling the semantics of human behaviour and experiences as *objects* that consist of regions of *empty space*, on the same ontological level as walls, doors, and so on. Our contribution in this paper is that we use spatial artefacts to *operationalize* defeasible rules represented in ACABIM LRML, bridging between the qualitative, ambiguous vocabulary of building terms and logical relations, and executable logic programming rules in Answer Set Programming.

Example: In order to use a washbasin, an occupant must be *positioned* within a certain region of *empty space* surrounding the washbasin. To model this, we create a new class of object, called a *functional space* (a subclass of *spatial artefact*), and create an instance of this class for each washbasin that is added to a BIM with a geometric representation as a "blob" of empty space surrounding the corresponding washbasin. We can then formally reason about perceptual, functional and locomotive affordances of the building, and code compliance, by analysing the relationship between this spatial artefact and other objects and spatial artefacts in the environment.

In our empirical evaluation for this paper we define three spatial artefacts: *entrances*, *access routes*, and *accessibility routes*. The geometry of each spatial artefact region is defined based on research literature in psychology and ergonomics (Li et al. 2020). We note that, in our framework, multiple alternative geometric representations can be defined so that compliance of a building code is (traceably) with respect to a particular interpretation.

Spatial artefacts are modelled in BIM on the same ontological level as any other object in the product model, e.g. spatial artefacts are integrated within the Industry Foundation Classes (IFC) as a subclass of *IfcSpace*, and as *IfcSpace* derives from *IfcProduct*, they have an associated geometric representation, i.e. a physical spatial presence in the BIM.

2.4 ASP enhanced with spatial reasoning

Answer Set Programming (ASP) is a logic programming language developed within the field of knowledge representation and reasoning (Brewka et al. 2011).[1] We use ASP by encoding the BIM as ASP facts, encoding building code compliance rules as ASP rules and constraints, and code compliance checking is implemented via ASP answer set search.

In our on-going research (Bhatt et al. 2011; Li et al. 2019; Walega et al. 2017) we have been

---

[1] ASP has a knowledge base of facts and rules of the form "*Head :- Body*" meaning that, if the *Body* is true, then the *Head* must also be true. Rules with no *Head* are ASP *integrity constraints*, written ":- *Body*" meaning that the *Body* must be false. ASP search engines are designed to rapidly find combinations of deduced facts that are consistent with all given domain rules.

extending the base language of ASP beyond propositions so that a set of consistent facts must also be *spatially* consistent. In this paper we exploit ASP with spatial reasoning to achieve fast building code *execution* on large, real-world BIMs while omitting clauses from the formal *representation* of building codes that only serve to speed up execution. E.g. in order to determine if two polygons intersect, a rule checker can first test whether their bounding boxes intersect (a fast test that is a necessary but not sufficient condition for polygon intersection). Rather than adding clauses in the code *representation* for implementing such tests (which reduces *transparency*), they now take place automatically inside the ASP search engine whenever certain spatial relations are evaluated. We have developed a general framework for encoding spatial data structures in ASP (Li et al. 2020).

3 REPRESENTING AND EXECUTING BUILDING CODES

In this section we present our domain models of building code *representation* and *operational form*.

Representation: In LRML, each rule consists of an *IF* and *THEN* element. Each *IF* and *THEN* element consists of an expression *Expr* that we define recursively:

Expr is <*Atom*/Subject, *Operator*, *Data*/Object>
Expr is (*Expr* OR *Expr*)
Expr is (*Expr* AND *Expr*)
Expr is (NOT *Expr*)

IF elements can include an expression of the form (NOT Expr) that corresponds to an exception that defeats the rule. We refer to these expressions as an *EXCEPTION* condition. Each *Atom* consists of "variable" (an entity in IFC) and an optional "relation" (a property in IFC). Operators are logical operators that relate a *Subject* to an *Object* (an IFC entity) e.g. *is*, *equal*, *has*, *by*.

Operational Form: A *building object* consists of a type, e.g. *IfcDoor*, and a unique identifier, e.g. a GUID. A *geometric object* is a geometric representation of a building object, e.g. *IfcRepresentation* including 2D/3D surface representations etc. An *n*-ary relation has a *type*, e.g. *IfcRelVoidsElement*, and is assigned to *n* objects, meaning that the relation holds between those objects.

Given an instance of a BIM denoted M that consists of a set of objects and relations between those objects, the next step is to execute the aforementioned rules to determine compliance. One might be tempted to treat each rule R as a material implication: let P_R, Q_R, E_R be predicates ranging over subsets of objects and relations that express the IF, THEN and EXCEPTION conditions (respectively) of rule R:[2]

For all O in Powerset(M):

$$(P_R(O) \land \neg E_R(O)) \to Q_R(O)$$

Full building code compliance of the BIM instance M then requires that all rules (implications) in the ruleset evaluate to *True* on M. However, this approach is inadequate as the user is not given any insight into the reasoning behind compliance or lack thereof, e.g. a rule might be trivially satisfied if no objects apply ($\neg P_R(O)$) or if every set of object satisfies an exception condition ($E_R(O)$). The user should be able to query the results of compliance analysis directly. Thus we introduce the following three predicates that capture meta-information on which objects and relations O are used to determine scope, exceptions and compliance of a rule R:

Applies(R,O):
Conditions defining the scope of the rule R
Exception(R,O):
Rule R applies to objects O, but is defeated
Compliant(R,O):
Conditions that satisfy a rule that applies

Default reasoning: In practice it is not always possible to determine *Applies*, *Exception* and *Compliant* for a given set of objects. E.g. in NZBC D1/AS1 "household unit" is an exception, but if a BIM is not annotated with its building functional type then we need to supply rules for deciding if a BIM is a household unit or not, and it is reasonable to anticipate that any such a set of rules is incomplete. Thus, in the absence of complete information we should be able to **assume** certain default properties that may later be retracted with new information.

Our operational framework therefore works within a **three-valued logic** setting: when a predicate P is evaluated on a set of BIM objects and relations O in M, it can be either *True* (denoted $P(O)$), *False* (denoted $\neg P(O)$), or neither *True* nor *False*, interpreted as *Undetermined*. We make use of a *not* operator as "failure to prove" with the following definition:

not $P(O)$:
P evaluates to either *False* or *Undetermined* on O
not $\neg P(O)$:
P evaluates to either *True* or *Undetermined* on O

Formally this is expressed via the following axioms A1–A4. Our system determines if a set of objects and relations O apply to a rule:

A1: For all O in Powerset(M):
$P_R(O) \to$ Applies(R,O)

In case O applies, then our system determines if O satisfies the conditions that defeat the rule:

A2: For all O in Powerset(M):
(Applies(R,O) $\land$ $E_R(O)$) $\to$ Exception(R,O)

Importantly, in both the above cases our system does **not** claim to prove that each O does *not* apply, $\neg$Apply(R,O), or that an exception definitely

---

[2] Given a set S, Powerset(S) is itself a set containing all possible subsets of S. The quantifier "For all O in Powerset(M)" means "For all subsets O of M …".

does *not* hold, ¬Exception(R,O). Finally, our system determines compliance (**A3**) and violation (**A4**):

A3: For all O in Powerset(M):
(Applies(R,O) ∧ (not Exception(R,O)) ∧ $Q_R(O)$)
→ Compliant(R,O)
A4: For all O in Powerset(M):
(Applies(R,O) ∧ (not Exception(R,O)) ∧ ¬$Q_R(O)$)
→ ¬ Compliant(R,O)

In this formulation, building codes that are inferred to apply are assumed, by default, to not be defeated (not Exception(R,O)) unless the system can deduce otherwise. If instead we used "…∧ ¬Exception(R,O) ∧…" then there would be the additional burden of needing to prove that the objects and relations do not satisfy the exception in order to reason about compliance. This provides a powerful semantic layer for users to query compliance analysis, for example consider paragraph 1.3.1 of NZBC D1/AS1 presented in Section 1:

Query: *Identify objects that violate some rule:*
Find R, O such that: ¬Compliant(R,O)
E.g.: a BIM instance M with a single isolated step (O) violates the building code (R).
Query: *Identify redundant rules that never apply:*
Find R such that for all O: not Applies(R,O)
E.g.: there is no change of level in a BIM instance M, and thus the code (R) trivially does not apply.
Query: *Identify defeated rules and objects involved:*
Find R,O such that: Exception(R,O)
E.g.: a BIM instance M has a change of level (O), but is also inferred to be a household unit (O), defeating the code (R).
Compliance can be *Undetermined* due to missing information, i.e. where there exists O such that Applies(R,O) but $Q_R(O)$ is *Undetermined*.
Query: *Identify undetermined rule compliance:*
Find R such that: Applies(R,O) ∧ (not Compliant(R,O)) ∧ (not ¬Compliant(R,O))
E.g.: a BIM instance M has a change of level (O), and a ramp object connected to both levels (via two *connectsTo* relations) is present in M but its geometric representation detailing slope and width is missing.

Hypothetical reasoning: In cases where compliance of rule R is *Undetermined*, it is useful to ask under what additional conditions **C** = {$C_1, \ldots, C_n$} can compliance (or its negation) be deduced – i.e. what information is missing that, if added to the BIM instance, will determine compliance of R. We formulate this as a search: given rule R and set of objects and relations O, find C in Powerset(**C**) such that (M ∪ C) satisfies the building code ruleset, axioms **A1**–**A4**, and the following added rule:
  (not Compliant(R,O) ∧ not ¬Compliant(R,O)) → *False*

That is, this added rule makes the inference of *Undetermined* compliance and violation of R on O a logical inconsistency. Examples of conditions include: an assignment of spaces to a particular activity and function type/classification, e.g. "work office space"; topological information about which windows belong to

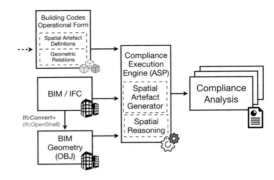

Figure 2. Pipeline for code compliance *execution* using our prototype system.

which spaces; geometric information on the openable and fixed portions of a window; geometric information on door swing direction. In our empirical evaluation we focus on this last case where door swing direction was not included in the real-world BIM (Section 4).

3.1 NZBC to LRML

In 2019, the New Zealand government funded a research under the National Science Challenge project to undertake the translation of 15 documents from the NZBC to LRML (Dimyadi et al. 2020). The digitisation process has three steps, namely document preparation, knowledge extraction, and formalising/encoding.

A software tool was used in the 1st step to capture the structure and literal content of the document. This is followed by domain experts manually translating the text of each provision into logical rules, which are then processed by another software tool for formal encoding into LRML.

The digitisation process of the 15 documents has identified approximately 600 building vocabularies (*buvo*), which, in practice, are entities representing building elements and their properties. Additionally, the process has also identified approximately 50 logical vocabularies (*lovo*), which are logical operators used in the rules in LRML, as well as approximately 20 functional vocabularies (*fuvo*), which are used in conjunction with *buvo* and *lovo*.

3.2 LRML to operational form in ASP

Next, we manually implement building codes represented in LRML into a set of ASP rules, using our operationalisation framework that we presented in Section 3. Spatial artefacts are used to capture concepts of human behaviour and perception as BIM objects. In order to *execute* a building code compliance check (see Figure 2), we first extract the geometry of the BIM using a IfcConvert (IfcOpenShell 2020). Our system then derives spatial artefacts (along with their corresponding geometric representations) and inserts them back into the BIM. Geometric relations that appear in the ASP rules are natively reasoned about within the ASP search engine.

Predicates for *Applies*, *Exception*, *Complies* (Section 3) are then inferred by the ASP search engine, taking values *True*, *False* and *Undetermined* that comprise the compliance analysis report.

Example: Consider the NZBC code provision stating that *"Accessible routes shall be provided to give direct access to the principal entrance to the building where practical"*. The rule is rendered in LRML as a set of *IF-THEN* expressions (ruleml:Expr) composed of Atom/Subject (ruleml:Atom), Data/Object (ruleml:Data), and Operator (ruleml:Fun), specifically: Building *has* Entrance; Entrance *is* Principal; Accessible Route *is* Practical; Entrance *has* Direct Access and *is served by* Accessible Route. Complying with the code thus relies on the interpretation of qualitative terms (*entrance, accessible route*) and their properties (*principal, direct, practical*) that need to be inferred from existing model views.

To implement this rule in an operational form, we augment the building with three types of spatial artefacts, namely, *entrances*, *access routes*[3], and *accessible routes*[4]:

An *entrance* is the threshold between the exterior and the interior.

An access route is a space having a door that is intended for humans to carry out primary activities (in contrast to services and maintenance).

An accessible route is an access route that permits people with disabilities[5] to traverse under normal circumstances. In IFC this is represented as a space where the Pset_SpaceCommon property HandicapAccessible is set to *True*.

The rules are implemented in ASP as follows:

```
entrance(Entrance) :-
door(Entrance),
  connectsTo(Entrance, interior),
  connectsTo(Entrance, exterior).
accessRoute(AccessRoute) :-
  space(AccessRoute), hasFloor(Access
Route),
  area(AccessRoute, Area), Area >
minimum Area,
  functional_activity(AccessRoute, _),
  door(Door), connectsTo(AccessRoute,
Door).
accessibleRoute(AccessibleRoute) :-
occupied_space(AccessibleRoute),
hasProperty(handicapAccessible,
AccessibleRoute).
```

---

[3] Defined in NZBC D1/AS1 as: a continuous route that permits people and goods to move between the apron or construction edge of the building to spaces within a building, and between spaces within a building.

[4] An access route usable by people with disabilities.

[5] People whose ability to use buildings is affected by mental, physical, hearing or sight impairment.

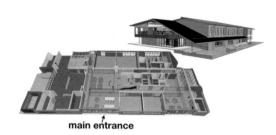

Figure 3. (upper right) Model of building in Christchurch seen from outside; (bottom) Ground floor typology.

We then derive properties from the enhanced building model, including: minimum width/clear opening of an accessible route; whether the entrance is principal; whether the entrance gives direct access. However, such properties are highly context-dependent, and thus require tangible assumptions to be made in the case that no further information is available. For example, consider an accessible route that, by default, must not contain any fire-retardant doors. Such restrictions can be "defeated" if the door is specially designed for a wheelchair user to negotiate unaided. In case of missing information about door operational mode, the *clear space* between consecutive opening doors also needs to be re-evaluated when new information comes to light.

The following example illustrates default reasoning in the context of determining whether an accessible route complies with the minimum unobstructed opening width (760 mm) required in NZBC. As handrails and other minor obstructions are permitted to intrude into this width, they raise an exception that exempts the rule execution. The ASP rules are:

```
hasProperty(clearOpening, Accessible
  Route, Width) :-
accessibleRoute(AccessibleRoute), door
  (Door),
on(Door, AccessibleRoute), width(Door,
  Width).
exception(AccessibleRoute) :-
accessib'leRoute(AccessibleRoute),
handrail(Handrail), on(Handrail,
  AccessibleRoute).
compliant(AccessibleRoute) :-
hasProperty(clearOpening, Accessible
  Route, Width),
not exception(AccessibleRoute), not
  Width< 760.
```

4 EMPIRICAL EVALUATION

We now demonstrate the operational aspects of the proposed framework with a real building in Christchurch, New Zealand (see Figure 3). The model contains 6796 objects, and is divided into two stories and a roof deck.

We identify *access routes* as continuous routes composed of occupied spaces that comply with minimum height clearances (see Figure 4).

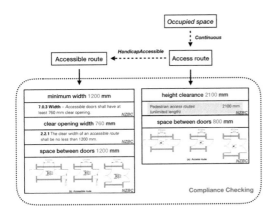
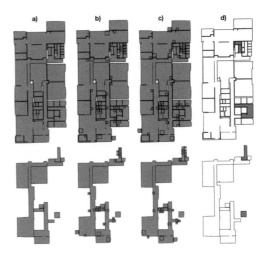

Figure 4. Workflow of deducing compliant access routes and accessible routes (codes from NZBC D1/AS1).

Table 1. Runtime statistics for parsing the IFC model used in our case study, generating different types of spatial artefacts, and checking code compliance.

Model preparation	Runtime (seconds)	Constraint checking	Runtime (seconds)
parsing building into ASP facts	0.98	access route height clearance	0.21
generating object geometry	72.68	access route spaces between doors (facing)	12.40
deriving access routes	0.21	accessible route minimum width	0.29
deriving accessible routes	0.27	accessible route clear opening width	0.27
deriving accessible doors	0.29	accessible route spaces between doors (facing)	8.37
deriving operational spaces of doors	3.65	accessible route spaces between doors (side-by-side)	8.54
deriving access doors	0.27		

Accessible routes are subsets of access routes that:

– have design intent `HandicapAccessible`, and
– comply with accessible route requirements (minimum width, clear opening width, spaces between doors).

Analysing the ground floor, our system first identifies 2831 objects of which 14 slabs, 387 walls, 69 spaces, and 53 doors, and 1454 polyhedral meshes. Our system then derives 61 access routes, 12 accessible routes, 53 access doors, 31 accessible doors, and 84 operational spaces of hinged access doors that are *missing door swing direction* (see Table 1). Our system produces the compliance report concluding that 3 access routes and 2 accessible routes are **not** in compliance with NZBC if all hinged doors swing to the left, and 2 access routes and 2 accessible routes are **not** compliant if all hinged doors swing to the right (see Figure 5) via hypothetical reasoning.

Finally, we identify 6 entrance doors that are accessible from street level, and we define the principal entrance as having the maximum opening width (3.38m). We then assume an accessible route gives practical, direct access to the main entrance if: (1) The main entrance is connected to the accessible route; (2) The accessible route is compliant with NZBC; (3) The accessible route does not contain access control devices such as revolving doors.

Figure 5. Derived spatial artefacts: a) Access routes (top row) and accessible routes (bottom row); b) Access routes and accessible routes with operational spaces (in *red*) of left-swing doors; c) Access routes and accessible routes with operational spaces (in *red*) of right-swing doors; d) Non-compliant access routes and non-compliant accessible routes in case of left-swing doors (*red regions*).

In ASP this is implemented as follows:

```
%% main entrance is the largest entrance
%%
:- main_entrance(D), width(D, W),
entrance(D1), width(D1, W1), W < W1.
%% rule defining practical accessible
route %%
property(AccessibleRoute, practical):-
main_entrance(Entrance),
connectsTo(Entrance, AccessibleRoute),
compliant(AccessRibleRoute),
revolving_door(RevDoor)
not on(RevDoor, AccessibleRoute).
```

The results in Table 1 show that our system generated all 241 spatial artefacts within 5 seconds, and that code compliance reasoning took a total of approximately 19 seconds. This demonstrates that our approach is practical and can scale to real-world BIMs. The main bottleneck is generating 3D mesh data for the original BIM objects (OBJ format, via IfcConvert) which took ~73 seconds.

5 CONCLUSIONS

We have presented a comprehensive methodology for formally: (1) *representing* building codes using LegalRuleML; (2) expressing the formalised codes in a new *operational* form based on defeasible reasoning that supports default reasoning, hypothetical reasoning, and code compliance querying; (3) *executing* the codes as rules using Answer Set Programming extended to natively support spatial reasoning. An empirical evaluation of our prototype code compliance reasoning engine demonstrated that our approach is practical and can be employed on real-world scale BIMs. For future work, we are now in the process of implementing a larger fragment of New Zealand Building Code D1/AS1 in our new operational form.

ACKNOWLEDGMENT

The authors gratefully acknowledge the Independent Research Fund Denmark for their financial support of the project "Intelligent Software Healing Environments" (DFF FPT1).

REFERENCES

ACABIM (2020). Accessed on May 2020 from: http://www.acabim.com/

Bhatt, Mehul, Jae Hee Lee, & Carl Schultz (2011). "CLP (QS): a declarative spatial reasoning framework." In International Conference on Spatial Information Theory, pp. 210–230. Springer, Berlin, Heidelberg.

Bhatt, Mehul, Joana Hois, & Oliver Kutz (2012a). "Ontological modelling of form and function for architectural design." Applied Ontology 7, no. 3: 233–267.

Bhatt, Mehul, Carl Schultz, & Minqian Huang (2012b). "The shape of empty space: Human-centred cognitive foundations in computing for spatial design." In 2012 IEEE Symposium on Visual Languages and Human-Centric Computing (VL/HCC), pp. 33–40. IEEE.

Boken, P., & Callaghan, G. (2009). Confronting the Challenges of Manual Journal Entries.

Borrmann, Andre & Rank, Ernst. (2009). Topological analysis of 3D building models using a spatial query language. Journal of Advanced Engineering Informatics. 23. 370–385.

Brewka, G., Eiter, T., & Truszczynski. M., (2011). Answer set programming at a glance. Commun. ACM 54(12), 92–103

Dimyadi, J., Governatori, G., & Amor, R. (2017). Evaluating LegalDocML and LegalRuleML as a standard for sharing normative information in the AEC/FM domain. In F. Bosche, I. Brilakis, & R. Sacks (Eds.), Proceedings of the Joint Conference on Computing in Construction (JC3) (Vol. I, pp. 637–644). Heriot-Watt University, Edinburgh, UK.

Dimyadi, Johannes, & Robert Amor (2018). "BIM-based Compliance Audit Requirements for Building Consent Processing." Proceedings of EC-PPM: 465–471.

Dimyadi, J., Fernando, S., Davies, K., & Amor, R. (2020). Computerising the New Zealand Building Code for Automated Compliance Audit. In 6th NZ Built Environment Research Symposium (NZBERS2020) (Vol. 6, pp. 39–46).

Eastman, C., Lee, J.-M., Jeong, Y.-S., Lee, J.-K. (2009). Automatic rule-based checking of building designs, Automation in Construction, 18, 1011–1033.

Gallaher, M., O'Connor, A., Dettbarn, J., & Gilday, L. (2004). Cost Analysis of Inadequate Interoperability in the U.S. Capital Facilities Industry. Report. Gaithersburg, Maryland: US Department of Commerce.

IfcOpenShell (2020). Accessed on May 2020 from: http://www.ifcopenshell.org/

Li, Beidi, Mehul Bhatt, & Carl Schultz (2019). "lambdaProlog (QS): Functional Spatial Reasoning in Higher Order Logic Programming (Short Paper)." In 14th International Conference on Spatial Information Theory (COSIT 2019). Schloss Dagstuhl-Leibniz-Zentrum fuer Informatik.

Li, Beidi, Johannes Dimyadi, Robert Amor, & Carl Schultz 2020. "Qualitative and Traceable Calculations for Building Codes" In CIB W78 2020 Conference (accepted for publication).

NZBC [Building Code of New Zealand] DS1/AS1 (2017). Retrieved from: https://www.building.govt.nz/building-code-compliance/d-access/d1-access-routes/

Solihin, W., Dimyadi, J., Lee, Y.C., Eastman, C., Amor, R. (2017). The Critical Role of Accessible Data for BIM Based Automated Rule Checking System, LC3 Joint Conference on Computing in Construction.

Wałęga, Przemysław Andrzej, Carl Schultz, & Mehul Bhatt. "Non-monotonic spatial reasoning with Answer Set Programming modulo theories." Theory and Practice of Logic Programming 17, no. 2 (2017): 205–225.

Zhang, J. & El-Gohary, N. (2016). "A prototype system for fully automated code checking." 16th Int. Conf. on Computing in Civil and Building Engineering, 535–542.

An AI-based approach for automated work progress estimation from construction activities using abductive reasoning

K.W. Johansen, R.O. Nielsen, J. Teizer & C. Schultz
Department of Engineering, Aarhus University, Aarhus, Denmark

ABSTRACT: Current standard approaches for monitoring site progress for lean construction favor weekly or bi-weekly meetings. All trade and construction site management representatives meet to synchronize the forthcoming schedule. Up-to-date information is often not available, causing poor coordination and resulting in delays, rework and waste of monetary resources. Furthermore, infrequent updates on work performance impact scheduling of critical activities. This paper investigates the possibility to automate some tasks in progress monitoring by applying an AI-system with abductive reasoning on real-time localization sensing data (RTLS) and domain expert knowledge. The work proposes a framework, consisting of three modules (data preparation, processing, and update) that utilize abductive reasoning. An experiment was conducted on previously collected data Teizer et al. (2013) to compare progress inferred from the framework with actual progress recorded. The preliminary results indicate the framework is able to reason about progress with high degree of similarity to the paper of Teizer et al. (2013), however, solely based on RTLS data and without any manual input. The future of the framework is promising since it supports the analysis of time series, allowing it to be applied nearly simultaneously to data collection, and thereby significantly increasing the update rate for information.

1 INTRODUCTION

1.1 Problem statement

The continuous task of construction site monitoring is a critical factor in construction management as it describes on-schedule-, on-budget- and safety-parameters in the construction process (Hogam 2019). Nevertheless, progress monitoring is also a difficult task as it involves a sub-processes and manual work (Mccrea et al. 2002). The solution today is a (bi-)weekly meeting where all the trades and the construction management team get synchronized. However, this leads to a relatively low resolution (i.e., missing, inaccurate or errors in temporal assessment), which results in waste factors common in construction. Reports state 38% waste in schedule, which is high compared to manufacturing at only 12% (Aziz & Hafez 2013).

In recent years, the Building Information Model (BIM) method made it possible to plan very comprehensive digital models of a construction site, including 3D and scheduling information. Even though comprehensive models enable construction management to compare the as-performed vs. the as-planned-states, the as-performed state estimation is still mostly a purely manual process. This entails a lower update rate which means that issues are discovered late, after causing an observed negative impact on a project's performance (i.e., schedule, budget, quality, and safety).

1.2 Motivation for research

To achieve a higher resolution in work progress monitoring, and a higher success rate in project performance, the monitoring process must be automated. Existing studies in the domain of monitoring automation have so far been using approaches such as image-based data (i.e., from still or video cameras, or mounted on unmanned aerial vehicles), laser scanning, radio frequency identification (i.e., Ultra Wideband), global positioning satellite systems (Alizadehsalehi & Yitmen 2018). Common to all of them is that data is processed using comprehensive analysis tools, and often relies on intensive computer processing. Computer vision, for example, includes very large models that have to be created before the system works properly. As construction sites are rarely identical, these models are often made for a specific construction site. Furthermore, these systems can only provide information about *visible* progress, and is thus more suited to outdoor construction work environments (Braun et al. 2020).

The purpose of this project is to investigate the possibility of using Ultra Wideband (UWB) data (Teizer et al. 2007) in combination with domain expert knowledge. The way that this project varies from earlier approaches is that the applied AI-based software does not rely on a trained model from large data sets, but instead on domain expert knowledge and common-sense reasoning about action and change,

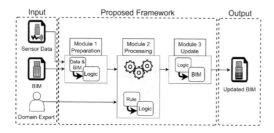

Figure 1. Proposed framework for progress monitoring.

where specified rules allow the system to generate a probabilistic hypothesis of the progress of a construction project.

Figure 1 shows a simplified overview of the approach used in this study. The main idea is to process UWB data through Answer Set Programming (ASP), which produces "most likely" hypotheses from data inputs based on if-then rules created from domain expert knowledge.

2 BACKGROUND

2.1 Localization in lean construction

Lean construction is the concept of optimizing a construction process in terms of time, price, quality, safety, and waste (Womack & Jones 1997). Unlike lean manufacturing, lean construction is project-based, and thus the factors that determine how processes can be optimized are highly project-specific. Moreover, measuring process quality is harder and more expensive to carry out. Lean construction's progress measures often include weekly or bi-weekly meetings, where the state of as-performed is determined and compared to as-planned. This is more time consuming and less agile than the automated process in the manufacturing-world, where measurements are taken and analyzed continuously.

To collect information about the as-performed state of a construction site at a higher temporal resolution beyond only weekly updates, we seek to automate progress monitoring. In recent years, considerable research effort has focused on using localization to support lean construction, for example, as a secondary input to computer vision or wireless tracking. Vision, as one approach, is used to automatically interpret video streams collected from the construction site at regular intervals and translate the semantic content into construction project progress. However, limited field-of-view and big data can limit the applicability of the approach (Bügler et al. 2017; Park & Brilakis 2012; Shahandashti et al. 2011; Teizer 2015; Yang et al. 2011). Ultra-Wideband (UWB) technology (Cheng et al. 2010; Cheng et al. 2011) offers as alternative approach. Like any other RTLS technology, UWB has limitations (Golovina et al. 2016; Vasenev et al. 2014). Johansen et al. (2020) presented a study using UWB data as the primary input to reason about construction activities based on formalized domain expert knowledge with the use of abductive reasoning. Thus, we intend that our system, based on UWB and abductive reasoning, is complemented with other approaches in practice. Posture recognition, for example, helps to determine, what high-level activity a worker is carrying out (Ray & Teizer 2012; Ryu et al. 2020; Valero et al. 2016). However, posture and RTLS data require fusion, as shown successfully before (Cheng et al. 2013; Migliaccio et al. 2013). Another important source of data is three-dimensional information (3D). UAV (Siebert & Teizer 2013) and laser scanning are often used to record 3D as-performed information (Bosché et al. 2015; Braun et al. 2020). Obtained point clouds are linked to BIM (Brilakis et al. 2010; Golparvar-Fard et al. 2009).

2.2 Construction site status monitoring

The information we seek to supply to project managers is whether certain planned construction tasks have been undertaken. However, we cannot directly 'sense' task completion and instead use sensor data about worker location to infer the evidence-based *plausibility* of work being done. We adopt the common approach of decomposing so called 'high level' tasks into a hierarchy of subtasks that can be more reliably reasoned about based on sensor data (Zöllner et al. 2005).

Given sensor data, our system needs to infer subtasks that are consistent with this sensor data. For example, given a timestamped series of geo-referenced locations associated with a particular worker (e.g., tracked UWB tags attached to personal protective equipment (PPE)) our system may infer that worker A was present in work area B, where task C according to the schedule should be performed. Moreover, the time that worker A spent may be used to (hypothetically) infer completion of sub-task $C1$, which is required in task C. This results in an increase of the plausibility of task C being completed.

Such inferences are only plausible, tentative explanations of the data: if it were indeed the case that the worker undertook the subtasks, we would expect such sensor readings. However, many other scenarios would also result in these sensor readings – what if the PPE with the tags were accidentally swapped between workers? Given this new information our system would need to be able to revise its previous inferences. This kind of explanation based hypothetical reasoning is a form of *abduction* (in contrast to deduction in which inferences are never retracted).

The challenge in abductive reasoning is to efficiently produce and select a small number of useful hypotheses that are justified from the data, out of the enormous multitude of hypotheses that are also consistent with the data available. To do so, we apply the concept of weak constraints, which penalizes models where hypotheses are not justified by evidence. The framework is then seeking the most plausible solution through minimization of the penalty connected to the model. To put our research on a formal footing, consider the fundamental equation of inference (Michalski 1993):

$$BK \land H \models Obs$$

Background Knowledge (*BK*) is a set of *IF-THEN* rules between propositions, and facts that we know to hold in the current setting such as the 4D BIM of the construction project, the up-to-date worker roster, current time of day, construction project plan and task schedules, etc.; **observations** (*Obs*) are represented as symbolic facts annotated with probabilities, derived directly from sensor data; **hypotheses** (*H*) are facts that, when combined with *BK*, entail the observations. **Deduction** is the process of inferring *Obs* given *H* and *BK*. **Abduction** (Thagard & Shelley 1997) is the process of inferring *H* given *BK* and *Obs*, such that background knowledge alone is not enough to explain the observations ($BK \not\models Obs$), the hypothesis is not inconsistent with background knowledge ($BK \land H \not\models$ false) and the hypothesis is sufficient to explain the observations together with background knowledge. Our abductive reasoning module is implemented using **Answer Set Programming (ASP)** (Gebser et al. 2007; Marek & Truszczynski 1999).

ASP is a declarative logic programming language that is used to represent and reason about semantic information in a given application domain (such as 4D BIM and safety) in the form of facts and rules and has an in-built search engine for finding models (combinations of deduced facts) that follow from the given premises. In the context of logic programming, the rules in BK that provide the basis for abductive reasoning are a set of Horn clauses of the form: $h \leftarrow b1, \ldots, bn$, where proposition h is true (the rule head) if propositions $b1, \ldots, bn$ are all true (the rule body). We represent a 4D BIM in ASP as ASP facts, background knowledge about construction sub-tasks is formalized as ASP rules, and valid hypothetical explanations are encoded as ASP models discovered by the ASP search engine. Rules in BK are not allowed to be violated unless these are described as *weak constraints*, which intuitively is used to tell the engine that the rule is preferably not violated, but in the case that the rule is violated, a penalty is added to the model. The concept is further explained in (Perri et al. 2003). The role of abduction in construction progress analysis is further explained in the next section.

3 METHODS

3.1 *Preparation*

The first module in the framework has the responsibility of preparing RTLS data and the BIM for abductive reasoning in the second module. The preparation results in observations and export of the BIM, which is used by the processing module.

3.1.1 *BIM input*
From the BIM, the goal is to retrieve the individual building elements, called IfcProducts, and convert them into JSON objects for better compatibility later in the process. Depending on which IfcProduct is being handled, such as IfcWall, IfcSpace, IfcDistributionPort, different actions are needed to generate the elements in BIM JSON. In the case of the IfcWall, a wall may contain a door, which is modelled as an opening voiding the wall, which is then filled by the door. Therefore, the properties of the door are handled with the wall because the door and the wall share the same value for direction, and the location is a reference from the location of the wall.

Every IfcProduct extracted from the IFC BIM, has five custom added properties (1) PlannedStartDate (2) PlannedEndDate (3) ActualStartDate (4) ActualEndDate (5) Status. The Planned-StartDate and -EndDate represent the times at which the management team expects construction has started and ended on a building element, while Actual-StartDate and -EndDate are the times at which the framework has reasoned the actual start and end of construction on the building element. Status represents at which point in the construction process the building element is currently in, either (a) completed (b) possibly completed, and (c) not performed (aka. built).

3.1.2 *RTLS data input*
When the RTLS data from the construction site is loaded into the framework it needs to be filtered in a meaningful way because the raw data is typically noisy.

The first step of the filtering is on speed. All UWB tags were placed on either human workers or on machinery that did not move faster than 7 km/h. Therefore, speeds that exceed 7 km/h must be error spikes.

Because the time is correct, speed filtering will allow for tags being far from each. If the data point also is an error, it will still be included. By using displacement, this does not happen. Displacement filtering only allows for two points to be within a range of 1 meter of each other.

After filtering data points, the trajectory line can still be jagged. The line should show the path of a person walking which is smooth. To smooth out the line, a Savitzky–Golay filter is applied.

The last filtering, Ramer-Douglas-Peucker (RDP), is used to make the subsequent calculations faster by removing redundant data. Because some tags were recording with 60 Hz, a substantial part of the data points do not add new value and can therefore be removed. As a results of applying these filtering techniques, the original data set can be reduced to as much as 6.8% of its original size.

3.1.3 *Observation creation*
Having both the JSON formatted BIM and the filtered data, the process of finding observations can begin. The first step is to map at what time a given tag is inside an area. This gives the observations of what elements the tags are near throughout the period. After

the observations have been found, they are then formulated as logic programing statements and given as input to the reasoning module.

3.2 Processing

3.2.1 Hypothetical reasoning problem

In general, a hypothetical reasoning problem consists of three main components: (1) a knowledge base, (2) evidence, and (3) a hypothesis. The goal of a hypothetical reasoning framework is to infer the most plausible hypothesis (abducibles) based on a knowledge base and the evidence.

The knowledge base-file is the logical representation of the domain expert knowledge that describes the necessary evidence (observations) to explain a hypothesis. The single BIM element-file is a logical formulation of the individual elements of BIM JSON. The separation enables reasoning to be performed on individual building elements, which ensures an easily interpretable progress estimation.

This framework operates with two levels of hypotheses. These levels describe matters from different perspectives. A higher-level hypothesis can, for example, explain if a construction site element has been constructed. The low level, in that case, describes more activity-relevant information, for example, whether an ironworker has been active in a work area.

3.2.2 Formal definition of framework

The formal definition is done in a top-down approach, which is initiated with the high-level hypothesis.

Equation 1 shows how the high-level hypotheses are created, where is the high-level activity that comes from a set of all high-level activities A_h. T is a representation of all time intervals over which the hypotheses are inferred. CS_a represents the set of construction elements on the construction site, where CS_a is a subset of the whole construction site CS. A_{a_h} is a subset of the low-level activities A_l and contains the activities a_l that are reasoned about in the lower-level hypotheses for the higher-level hypothesis to exist. From a high-level hypothesis, $Expected_h$ is inferred with the required activities.

$$\forall a_h \in A_h, t \in T, ce \in CS_a : Hyp_h(a_h, t, ce) \rightarrow$$
$$\forall a_l \in A_{a_h} : Expected_h(a_l, t, ce)$$
$$A_{a_h} \subset A_l$$
$$CS_a \subset CS$$

Equation 1. Inference of high-level expectation (Expected_h) from high-level hypothesis (Hyp_h).

With the $Expected_h$ predicates it is possible to penalize the solution if a corresponding low-level hypothesis cannot be reasoned. This is done with the "⤳", which describes a weak constraint that means if the statement is not True a penalization is added to the model. The penalization happens on w_a^* the construction element to motivate the engine to include the activity at least once. w_a^* denotes the weight of the penalization. This is shown in Equation 2.

$$\forall ce \in CS_a : Expected_h(a, t, ce) \leadsto \exists t_a$$
$$\in THyp_l(a, t_a, ce)[w_a^*, Expected_h(a, t, ce)]$$
$$w_a^* \in \mathbb{N}^0$$
$$CS_a \subset CS$$

Equation 2. Inference of low-level hypothesis (Hyp_l) from high-level expectations (Expected_h).

Every low-level activity a_l introduces a set of observations O_{a_l}, and each of those observations, o, must take place in the evidence for the hypothesis to be an explanation. To do so the $Expected_l$ predicate is introduced for each of the observations. This is shown in Equation 3.

$$\forall a_l \in A_{a_h}, t \in T, ce \in CS_a : Hyp_l(a_l, t, ce) \rightarrow \forall o$$
$$\in O_{a_l} : Expected_l(o, t, ce)$$
$$A_{a_h} \subset A_l$$
$$CS_a \subset CS$$

Equation 3. inference of expected observations (Expected) from low-level hypothesis(Hyp_l).

Equation 4 is introduced to ensure that the expected observation for the activity, denoted O_{a_l} also holds in the set of observed data O. The observed data is the observation predicated that is described in the Holds-predicate.

$$True \rightarrow \forall o \in O_{a_l}, t \in T, ce \in CS :$$
$$Expected_l(o, t, ce) \wedge Holds(o, t, ce)$$
$$O_{a_l} \subset O$$
$$O_{a_l} \subset O$$

Equation 4. Secure that the expected observations (Expected) is actually contained in the observations (Hold).

3.2.3 ASP mapping of hypothetical reasoning framework

To go from a theoretical framework to an operational framework, where a reasoning engine can be used, the theory is mapped to ASP (Perri et al. 2003). In the following listing it is shown how the existence of *hypothesis_h* infers *expected_h*. This listing is a mapping of Equation to ASP:

{hypothesis_h(a, t, ce)}.
k_{min}{
expected_h(a1, t, ce);
expected_h(a2, t, ce)
}k_{max}:- hypothesis_h(a, t, ce).

As described in Equation 2 a model should be penalized if an expected low-level hypothesis, *hypothesis_l*, is missing. This is done using weak constraints as shown in the listing below:

:~ expected_h(a1, t, ce),
not hypothesis_l(a, t, ce).
[, abducible_expected_h(ce)]

For each of the low-level hypotheses, *expected_l*-predicates are inferred as shown in Equation 3. This is done similarly to the high-level hypothesis:

{hypothesis_l(a, t, ce)}.
k_{min}{
 expected_l(o1, t, ce);
 expected_l(o2, t, ce)
}:- hypothesis_l(a, t, ce).

As shown in Equation 4 the *expected_l*-predicates describes the necessary observations (evidence) for a hypothesis to be reasonable. The mapping of this is shown in the following listing, where a model is rejected if the evidence does not exist:

:- expected(o, t, ce), not holds(o, t, ce).

3.2.4 Time series window handling

If new observations are continuously appended to the current set of observations the number of predicates will eventually get very big, which negatively affects processing time. Even though this is not preferable the information from earlier time windows is needed as a low-level hypothesis might have been explained there and is then needed in the next time window to explain the high-level hypothesis without penalization. To overcome this, it is necessary to take pertinent reasoning output from previous runs and feed it into the processing engine.

The infeed of important information from previous runs introduces another problem, hence the reasoning engine would keep updating the time of the high-level hypothesis as it is possible to explain the high-level hypothesis in the following runs. This would mean that the actual end time for the construction element would be updated on each run. To overcome this a filtering of the hypotheses is necessary. It is this filtering process that returns the most relevant information.

3.3 Update

The last module has the responsibility to take the information from **Module 2**, in Figure 1, and update the original BIM with the information such that the management team can inspect the updated BIM and follow the progress on the construction site. **Module 3** receives two inputs from the previous modules, and another input from the original BIM.

The process now is to go backward from what **Module 1** did, namely, convert logical formulated statements back into the IFC data format. This is done firstly by extracting the information from the results and converting them into the JSON formatted BIM.

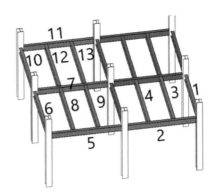

Figure 2. 3D view of ITF with annotated installation order of steel girders to be installed (red) vs. pre-existing (green).

The extracted information is time properties and status on the building elements (1) ActualStartDate (2) ActualEndDate (3) Status. When the values for the properties are extracted, they are updated into the JSON BIM generated in **Module 1**. Lastly, the original BIM is updated with the values from the JSON file, which results in an updated BIM in an IFC-format for every stakeholder in the project to analyze.

4 EXPERIMENTS AND RESULTS

4.1 Construction situation introduction

To establish a fundament from which it is possible to state whether not it is practical to reason about construction progress from BIM, UWB-data, and Domain expert knowledge we run the proposed framework on an experiment. The experiment is based on UWB-data from an Ironworker Training Facility (ITF). The data originates from (Teizer et al. 2013) which studied productivity and safety. The training involves the installation of 13 steel girders in a partial completed work environment. The complete installation of all 13 beams has a duration of approx. 90 min.

The construction situation is illustrated in Figure 2. From this Figure it is possible to determine which of the beams must be installed and in which order. Three types of personnel are involved: 2 connectors installing the girders at height, 1 rigger on the ground floor connecting them to the crane hook, and 1 crane operator in charge of thereof. Figure 3 displays an example of RDP-filtered UWB trajectory data.

UWB data and the BIM is provided as input for the logic-based framework as a formatted JSON file. Along with the transferred data we also assumed for each connecter one *work zone* at both ends of the steel girders (each 0.7 m in radius). This corresponds to half of the distance between two neighboring beams. The BIM is transformed into logic predicates that are interpretable to the reasoning engine. The UWB-data is filtered according to the aforementioned process and afterwards analyzed regarding the work zones. If a tag is observed in a work zone it creates a predicate *holds(in_area(GlobalID, TagID, interval(T1, T2).*

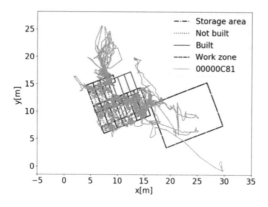

Figure 3. Filtered trajectory data from UWB tag 00000C81 (connector).

Table 1. Description of hypotheses in ITF: Storage area is abbriviated to SA and assembly area to AA.

Hypothesis	Expected	Constraints
High-level Hypothesis		
Beam installed	Beam rigged at T1	T1<T2,
	Beam moved at T2	T2<T3
	Beam connected at T3	
Low-level Hypothesis		
Beam rigged	Crane hook in SA at T1	T1=T2
	Rigger in SA as T2	
Beam moved	Crane hook in SA at T3	T3<T4
	Crane hook in AA at T4	
Beam Connected	Connector 1 in AA at T5	T5 = T6,
	Connector 2 in AA at T6	T6 = T7
	Crane hook in AA at T7	

4.2 Knowledge base formulation

Our approach for gathering relevant construction activity knowledge for the framework's knowledge base was to conduct a domain expert interview. For example, one interview question was: "What is the assembly procedure of steel girders?", to which the expert answered:

1. Connecting a horizontal steel girder requires the columns to be placed beforehand.
2. A rigger connects the girder to the crane hook.
3. Upon signal, the girder is slowly, carefully lifted from the temporary storage area to the assembly location. Small adjustments might be necessary until bolts and nuts on both sides secure the girder in place.
4. The crane hook is detached, and so on.
5. Loads above other workers must be avoided.

Based on the interview we subsequently created a formal definition of the installation task and subtasks. This outcome is shown in Table 1, which refers to the task and subtasks as high- and low-level hypothesis, respectively. This domain knowledge was then encoded as ASP facts and rules.

4.3 Reasoning

Abductive reasoning was done on individual construction elements i.e., girders. The knowledge base infers the high-level hypothesis, which then infers the lower-level ones. The high-level hypothesis is described with the following predicate:

hyp_h(
  beam_installed(GlobalID,
  interval(T_start, T_end)).

T_start and T_end denotes the interval in which the hypothesis is inferred and comes from the times, where connectors are observed to be in a certain work zone. The high-level hypothesis infers the lower-level ones, which contains expectations to the observations contained in the logic instance. If the observations corroborate the expected in the interval, the hypothesis is possibly true.

We perform reasoning on one girder at a time and do this on the two separate parts of data. The separation of data simulates that the analysis is carried out during the day simultaneously to the data collection, described previously by Teizer et al. (2013).

4.4 Results

The output of the above reasoning process is several models containing a high-level hypothesis with corresponding penalization as a measure of missing evidence (i.e. unjustified hypotheses). The engine optimizes the models toward the minimum penalization. These hypotheses are transformed into Figure 4, which shows the interval, mean value, and penalization in comparison to the results presented in Teizer et al. (2013). The figure shows a similarity between the reasoned results (colored dots) and the results presented in the paper (black crosses). Only two of the results from the paper are outside the (relatively broad) intervals. This is a consequence of the girder placement, for example, girder No. 9 is placed opposite to No. 13 which results in a higher number of observations in these work zones. No. 11, which is placed at the outside of the structure, has an interval more similar to Nos. 2, 3, 4, 5, 6, 7 and 8. The intervals are on average covering 16% of the total duration of the work, corresponding to 14 minutes. The average distance between the result from the paper and the reasoned is 5.2 minutes.

Figure 5 shows the mean values from Teizer et al. (2013) in comparison to the inferred values from our framework, along with their trendlines. This figure also emphasizes the similarity, and the R-squared value of the inference being high shows that there is a linear trend in both results. The steeper slope of the "reasoned" trendline (blue) is a result of the broader intervals in No. 9, No. 10, No. 12, and No. 13.

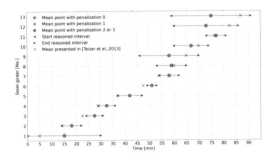

Figure 4. Reasoned intervals and mean values from both parts of data, compared to result presented in (Teizer et al. 2013).

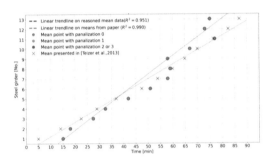

Figure 5. Reasoned mean values of first and last part of data; results presented in (Teizer et al. 2013) along with corresponding trendlines.

Figure 6. Result after reasoning on first (left) and second (right) part of data. Red denotes a penalty of 2 and 3, orange 1, and green 0.

From the reasoned results the JSON representation of the BIM is updated, where penalties of 2 and 3 is interpreted as *Not built*, 1 as *Possibly built*, and 0 to *Built*. The IFC-representation of the BIM is then updated from this, representing the most plausible, evidence-based state of the construction project.

Figure 6 shows the rendering of the IFC model in a 3D-CAD program, where the elements are filtered on the IFC-property *status*, which holds aforementioned information. The output of the different filters is then colored accordingly to status our color scheme i.e., green=*Built*, orange=*Possibly built*, and red=*Not built*.

5 CONCLUSION

Progress estimation in construction is a time-and resource-demanding process as it mainly consists of a manual assessment, which results in a relatively low resolution of updates to the as-performed state. This study proposes a framework that automatically estimates in- and exterior progress based on RTLS- and BIM data, and domain expert knowledge. The latter describes procedures existing in construction processes, which makes it applicable to similar types of projects containing the same tasks.

The proposed framework is presented through formal logic equations and was implemented in Answer Set Programming (ASP), which supports abductive reasoning. We have detailed how the input data is handled, with our simplified approach to multishot reasoning on time series data.

To conclude on the practicality, it was decided to include the full process involved in progress estimation, except for data acquisition. The framework was able to produce plausible hypotheses that were corroborated by evidence in the form of RTLS-data-observations. The BIM that was used as input is updated with the estimation progress to inform construction management about undertaken progress.

Through a comparison of the results from Teizer et al. (2013), it was found that the inferred progress was similar, thus demonstrating the viability of progress estimation using abduction. The practicality of domain expert knowledge formalization is shown in the simple process that was made in our experiment. It was reasonable to estimate progress exclusively from RTLS-data, but in a scenario where tasks are less separable it would most likely require additional sensor inputs.

Thus, this study provides early evidence that abduction can be used to tackle the problem of construction site progress monitoring. In our future research we are conducting follow-up studies on larger and more complex real-world construction projects which will include applications beyond productivity analysis (Li et al. 2020; Schultz et al. 2020).

REFERENCES

Alizadehsalehi, S., Yitmen, I., 2018. A Concept for Automated Construction Progress Monitoring: Technologies Adoption for Benchmarking Project Performance Control. Arabian Journal for Science and Engineering, Vol. 44, pp. 4993–5008.

Aziz, R.F., Hafez, M.S., 2013. Applying lean thinking in construction and performance improvement. Alexandria Engineering Journal, pp. 679–695.

Bosché, F. et al., 2015. The value of integrating Scan-to-BIM and Scan-vs-BIM techniques for construction monitoring using laser scanning and BIM: The case of cylindrical MEP components. Automation in Construction, Volume 49, pp. 201–213.

Braun, A., Tuttas, S., Borrmann, A., Stilla, U., 2020. Improving progress monitoring by fusing point clouds, semantic data and computer vision. Automation in Construction, 116, 103210.

Brilakis, I. et al., 2010. Toward automated generation of parametric BIMs based on hybrid video and laser scanning data. Advanced Engineering Informatics, Volume 24(4), pp. 456–465.

Bügler, M., Borrmann, A., Ogunmakin, G., Vela, P.A., Teizer, J., 2017. Fusion of photogrammetry and video analysis for productivity assessment of earthwork processes. Computer-Aided Civil and Infrastructure Engineering, Wiley, Volume 32, Issue 2, 107–123, http://doi.org/10.1111/mice.12235.

Cheng, T., Yang, J., Teizer, J., Vela, P., 2010. Automated Construction Resource Location Tracking to Support the analysis of lean principles. 18th Annual Conference of the International Group for Lean Construction, pp. 643–653.

Cheng, T., Venugopal, M., Teizer, J., Vela, P., 2011. Performance Evaluation of Ultra Wideband. Automation in Construction, Volume 20, Issue 8, pp. 1173–1184, http://dx.doi.org/10.1016/j.aei.2011.04.001.

Cheng, T., Teizer, J., Migliaccio, G., Gatti, U., 2013. Automating the Task-level Construction Activity Analysis through Fusion of Real Time Location Sensors and Worker's Thoracic Posture Data. International Workshop on Computing in Civil Engineering, https://doi.org/10.1061/9780784413029.079.

Gebser, M., Kaufmann, B., Neumann, A., Schaub, T., 2007. Clasp: A conflict-driven answer set solver. Logic Programming and Nonmonotonic Reasoning, pp. 260–265.

Golovina, O., Teizer, J., Pradhananga, N., 2016. Heat Map Generation for Predictive Safety Planning: Preventing Struck-by and Near Miss Interactions between Workers-on-Foot and Construction Equipment. Automation in Construction, Volume 71, pp. 99–115, http://dx.doi.org/10.1016/j.autcon.2016.03.008.

Golparvar-Fard, M., Peña-Mora, F., Savarese, S., 2009. Application of D4AR – A 4-Dimensional augmented reality model for automating construction progress monitoring data collection, processing and communication. ITcon, Volume 14, pp. 129–153.

Hogan, P., 2019. Top 5 Challenges for Construction Project Managers. Online: https://www.fieldwire.com/blog/top-challenges-for-construction-project-managers/.

Johansen, K.W., Nielsen, R., Schultz, C., Teizer, J., 2020. Non-Monotonic Reasoning for Automated Progress Analysis of Construction Operations. Enabling the Development and Implementation of Digital Twins, Volume 20, pp. 300–311.

Li, B., Schultz, C., Teizer, J., Golovina, O., Melzner, J., 2020. Towards a unifying domain model of construction safety, health and well-being: SafeConDM. Adv. Engineering Informatics.

Marek, V., Truszczynski, M., 1999. Stable models and an alternative logic programming paradigm. The Logic Programming Paradigm, pp. 375–398.

Mccrea, A., Chamberlain, D., Navon, R., 2002. Automated inspection and restoration of steel bridges—a critical review of methods and enabling technologies. Automation in Construction, pp. 351–373.

Michalski, R., 1993. Inferential theory of learning as a conceptual basis for multistrategy learning. Machine learning, Volume 11, Issue 2-3, pp. 111–151.

Migliaccio, G., Cheng, T., Gatti, U., Teizer, J., 2013. Data Fusion of Real-time Location Sensing (RTLS) and Physiological Status Monitoring (PSM) for Ergonomics Analysis of Construction Workers. CIB World Congress, Brisbane, Australia.

Park, M.-W., Brilakis, I., 2012. Construction worker detection in video frames for initializing vision trackers. Automation in Construction, Volume 28, pp. 15–25.

Perri, S., Scarcello, F., Leone, N., 2003. Abductive Logic Programs with Penalization: Semantics, Complexity and Implementation. CoRR, https://arxiv.org/abs/cs/0310047.

Ray, S., Teizer, J., 2012. Real-Time Construction Worker Posture Analysis for Ergonomics Training. Advanced Engineering Informatics, Volume 26, pp. 439–455, http://dx.doi.org/10.1016/j.aei.2012.02.011.

Ryu, J., Alwasel, A., Haas, C., Abdel-Rahman, E., 2020. Analysis of Relationships between Body Load and Training, Work Methods, and Work Rate: Overcoming the Novice Mason's Risk Hump. Journal of Construction Engineering and Management, Volume 146, Issue 8.

Schultz, C., Li, B., Teizer, J. (2020). Towards a Unifying Domain Model of Construction Safety: SafeConDM. European Group for Intelligent Computing in Engineering Conference (EG-ICE), p. 363–372.

Shahandashti, S. et al., 2011. Data Fusion Approaches and Applications for. Journal of Construction Engineering and Management, Volume 137, Isssue 10, pp. 863–869.

Siebert, S., Teizer, J., 2013. Mobile 3D mapping for surveying earthwork using an unmanned aerial vehicle (UAV). Proceedings of the 30th International Symposium on Automation and Robotics in Construction, Montreal, Canada, https://doi.org/10.22260/ISARC2013/0154.

Teizer, J., Lao, D., Sofer, M., 2007. Rapid Automated Monitoring of Construction Site Activities using Ultra-Wideband. Proc. 24th Intl. Symposium on Automation and Robotics in Construction, pp. 23–28, https://doi.org/10.22260/ISARC2007/0008.

Teizer, J., Cheng, T., Fang, Y., 2013. Location tracking and data visualization technology to advance construction ironworkers' education and training in safety and productivity. Automation in Construction, Volume 35, pp. 53-68, http://dx.doi.org/10.1016/j.autcon.2013.03.004.

Teizer, J., 2015. Status Quo and Open Challenges in Vision-Based Sensing and Tracking of Temporary. Advanced Engineering Informatics, Volume 29, Issue 2, pp. 225–238, http://dx.doi.org/10.1016/j.aei.2015.03.006

Valero, E. et al., 2016. Musculoskeletal disorders in construction: A review and a novel system for activity tracking with body area network. Applied Ergonomics, pp. 120–30.

Vasenev, A., Pradhananga N., Bijleveld, F., Ionita, D., Hartmann T., Teizer, J., Dorée, A., 2014. An Information Fusion Approach for Filtering GNSS Data Sets collectd during Construction Operations. Advanced Engineering Informatics, Volume 28, pp. 297–310, http://dx.doi.org/10.1016/j.aei.2014.07.001.

Womack, J.P. Jones, D.T., 1997. Lean Thinking—Banish Waste and Create Wealth in your Corporation. Journal of the Operational Research Society, Volume 48, p. 1148–1148.

Yang J., Cheng, T., Teizer, J., Vela, P.A., Shi Z.K., 2011. A Performance Evaluation of Vision and Radio Frequency Tracking Methods for Interacting Workforce. Advanced Engineering Informatics, Volume 25, Issue 4, 736–747, http://dx.doi.org/10.1016/j.aei.2011.04.001

Zöllner, R., Pardowitz, M., Knoop, S., Dillmann, R., 2005. Towards cognitive robots: Building hierarchical task representations of manipulations from human demonstration. IEEE Intl. Conference on Robotics and Automation, pp. 1535–1540.

Assumption of undetected construction damages by utilizing description logic and fuzzy set theory in a semantic web environment

A. Hamdan & R.J. Scherer
Institute of Construction Informatics, TU Dresden, Germany

ABSTRACT: Within the scope of non-destructive inspections, it frequently happens that damages within the construction structure or at inaccessible areas are not detected. Thus, human experts must assume these damages by evaluating relevant information about the construction and its surroundings, which could become a time-consuming and error prone task. For this reason, a new approach is presented in this paper, in which a knowledge system has been developed utilizing Semantic Web Technologies to automatically infer new information about potential undetected damage areas based on past generalized experiences. Thereby, ontological models for defining information about the affected construction and the already inspected damages are used as data input and reasoned by applying predefined rules formalized in Description Logic. Furthermore, the possibility for occurrence of each assumed damage is determined by utilizing fuzzy set theory.

1 INTRODUCTION

With the emergence of Building Information Modeling (BIM), new methods of planning and controlling the entire life cycle of constructions in a digital and collaborative environment have been developed. Currently, BIM is mainly used for buildings under construction rather than for existing ones, due to the high effort of modelling and converting smart objects from an existing structure (Pocobelli et al. 2018, Volk et al. 2014). Thereby, an important aspect that comes up when working with existing constructions, contrary to those that are newly built, is the detection and assessment of damages to the structure as well as the planning of necessary maintenance measures.

To detect these damages, various inspection methods are applied to the structure, whereby the visual inspection by a human expert is the preferred approach, due to being a non-destructive method as well as requiring no additional measuring devices. However, when applying this method, it frequently happens that damages within the construction structure or in difficult accessible areas are overseen or not detectable by visual inspection. Consequently, the type and approximate position of these undetected damages must be assumed, to apply more and specific inspection methods, which also could be destructive to the structure or require additional measuring devices. To perform an accurate damage assessment, various information regarding the construction and its environment as well as previously detected damages need to be evaluated. Currently, this task is processed manually by human experts. Thus, the evaluation could become highly time-consuming and error prone, depending on the experience of the expert.

In this research a new approach is proposed that utilizes description logic (Baader et al. 2007) in conjunction with the fuzzy set theory (Zadeh 1965) in a knowledge system to automatize the assumption process for undetected damages. Thereby, the system utilizes Semantic Web Technologies to serialize the terminological concepts (TBox) and assertional components (ABox) in a modular ontology. In this regard, the ontology is also used for semantic representation of the existing construction and already inspected damages using the Web Ontology Language (OWL) (Hitzler et al. 2012) as formalization. Description Logic rules for inferring undetected damages are formalized using the Shapes Constraint Language (SHACL) (Knublauch et al. 2017) and the possibility of damage occurrence is determined utilizing the fuzzy set theory. In this paper, the developed methodology as well as the ontological models and rules for assumption of undetected damages are presented.

2 RELATED WORK

Besides multiple methods that utilize machine learning for damage classification and assessment, various approaches have been developed that try to represent damaged constructions in an ontology for subsequent damage assessment.

The Monument Damage Information System (MONDIS) (Cacciotti et al. 2015) allows for representing damages of historical monuments and buildings. Thereby, the focus of the ontology is the semantic description of damage causes and processes. However, an interface to implement more information about the damaged components is missing. Furthermore, concepts for inferring implicit information about the damages or affected structure are not defined, which prevents an automatized assessment.

An ontological knowledgebase for bridge maintenance is proposed by Ren et al. (2019). There, bridge components can be defined and grouped in various zones. Furthermore, additional component information can be defined such as material parameters or previously performed refurbishment measures. Damage objects are assigned either to the bridge component or zone and can be evaluated via rules formalized in Semantic Web Rule Language (SWRL) or Semantic Query-Enhanced Web Rule Language (SQWRL). In this regard, the structural health of the existing bridge is assessed based on a condition index.

A similar principle has been developed by Hu et al. (2019) for suburban tunnels. Besides classes and properties for describing the components, zones and assigned damages, additional environmental objects can be added, e.g. for defining weather or traffic parameters. Furthermore, related persons or organizations as well as resources, e.g. construction equipment or vehicles can be defined and linked to the tunnel representation. In addition, the ontology is designed to be compatible with other corresponding systems such as electricity or sewage systems. The rules applied in this ontology do not only assess the structural health of the tunnel but also the damage causes.

The aforementioned approaches assess only existing damages and construction information. Information about possible undetected damages is not considered. An ontological system that infers information about undetected damages in an existing construction is still missing.

3 ONTOLOGICAL REPRESENTATION

To reason instances of undetected damages, any information that can support the reasoning process, should be collected and stored in an ontology which serves as the primary information input of the system. The ontology used in this research, consists of data about the existing construction as well as detected damages and environmental influences. In this regard, the data gathering as well as the ontology generation has been processed as part of a BIMification, during the anamnesis stage (Scherer & Katranuschkov 2018). A first reasoning can then be performed based on the TBox of the ontology. Although, this approach is aligned towards assessing a damaged bridge as test subject, the principle can also be applied to any other construction types. Thereby, the web ontology for representing the construction type needs to be changed. For an application on buildings the Building Topology Ontology (BOT) (Rasmussen et al. 2017) is recommended due to its compatibility with the damage ontology used in this research. Figure 1 shows the overall ontology structure. Thereby, two core ontologies for representing the existing bridge construction as well as the affecting damages are used. In this regard, the component representations are linked with damage representations through an object property of the damage ontology (brot:component dot:hasDamage dot:Damage). Furthermore, various extensions to these two ontologies were applied to process more specific information, e.g. material or structural parameters. In the following section the web ontologies for representing a bridge and its damage representation as well as their compatible extensions are described.

3.1 Bridge ontology

To create an ontological model of a bridge in its undamaged state, a modular ontology framework for bridge representation is used, which has been developed in a previous work (Hamdan & Scherer 2020). The framework consists of a core ontology named Bridge Topology Ontology (BROT), which provides a TBox for defining a general bridge ontology. Thus, the classes and properties provided by the BROT terminology can be used to describe a basic bridge

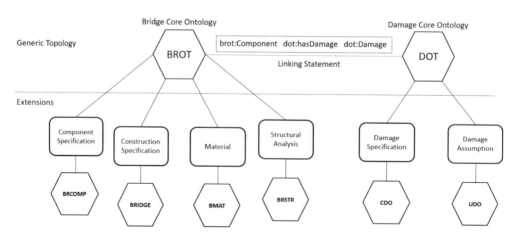

Figure 1. Structure of all used core ontologies and their extensions.

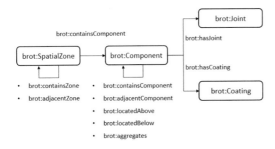

Figure 2. Main classes and object properties of BROT.

representation regardless of its type or used material. Figure 2 shows the main classes and object properties of BROT. The complete ontology can be accessed in a Github repository[1]. The BROT ontology is designed as bridge counterpart of BOT and therefore reuses several concepts of it.

Therefore, in BROT a bridge is represented as a topology of interlinked zone instances (brot:Spatial Zone) that consist of multiple component representations (brot:Component).

However, contrary to BOT, the zone instances are further classified based on common bridge terminology. Thus, a hierarchical zone structure consisting of a site that contains one or multiple bridges can be created as well as the super- and substructure of a bridge. Furthermore, component instances can be complemented with instances that represent joints or coatings, e.g. asphalt. BROT provides multiple types of object properties that are used for describing the various topological relationships between the zone and component representations. This includes properties for containment of zones and components (e.g. brot:containsZone or brot:aggregates) as well as properties for describing adjacencies or height differences (brot:locatedAbove and brot:locatedBelow). Thus, topological relationships, such as the aggregation of reinforcing bars in a concrete beam or higher position of a pier compared to its foundation element can be modeled.

To allow the definition of domain-specific information such as material specific components or the structural behavior of the bridge, the BROT ontology can be extended with compatible ontologies. For instance, BROT does not provide any classes or properties for classifying or characterizing bridge components as these vary often between different bridge types. Therefore, an extension named Bridge Components Ontology (BRCOMP) has been developed, that provides multiple domain-specific classes for component representation, e.g. for defining piers or abutments of a bridge. Furthermore, data properties for defining component parameters such as the installation date or the concrete cover are provided in the TBox. In a similar way, the extension Bridge Classification Ontology (BRIDGE) has been developed for classifying the bridge representation itself and assigning additional data to it. Thus, it enables the use of bridge classes that are dependent on the material, construction design or other characteristics, such as steel bridges or box girder bridges. Examples for possible bridge attributes that are covered in BRIDGE are the construction date or the overall construction length.

The occurrence of damages and their potential characteristics depend often on the structural function of the components. Therefore, an extension for defining structural data has been developed, which is named Bridge Structure Ontology (BRSTR). It allows for classifying load bearing components as well as describing the load transmission between multiple connected components via appropriate object properties.

Besides the structural characteristics, the material parameters of each component are also important for an accurate assumption of undetected damages for which the Building Material Ontology (BMAT) can be used. The web ontology provides various classes for defining building materials as well as corresponding material parameters. At its current state, the extension consists mainly of specifications for reinforced concrete.

The proposed ontology framework does not provide any terminology for defining an accurate construction geometry, since these data are not appropriate and not necessary to be represented through expert knowledge. It is more convenient to limit geometry information to simple geometrical parameters such as the height or length of a component and link the ontology with the BIM-IFC model as Baumgärtel & Scherer (2016) has shown. This can be realized by linking instances from the ontology with corresponding IFC entities in an ICDD Multimodel (ISO 21597-1 2020, Scherer & Katranuschkov 2019). Furthermore, a tool has been developed by Hamdan & Scherer (2020) to generate a bridge topology using BROT and BRCOMP from an IFC model, which can be further semantically enriched by an expert before applying rules for damage assessment.

3.2 Damage ontology

For representing detected damages in the existing construction the Damage Topology Ontology (DOT) is used which has been developed in a previous research (Hamdan et al. 2019). The ontology provides classes and properties for describing damage objects in a generic way, regardless of its specific type or the material it affects. Thereby, a damage topology as well as documentation and structural related data can be defined. Similar to the bridge ontology BROT, DOT is compatible with various extensions that characterize the damage instances further. Since the proposed approach has been applied on a bridge representation made of prestressed concrete, the extension Concrete Damage Ontology (CDO) is used in conjunction with DOT. Figure 3 shows the main classes and object properties for modeling generic damage representations.

---

[1] https://github.com/Alhakam/bridgeOntology

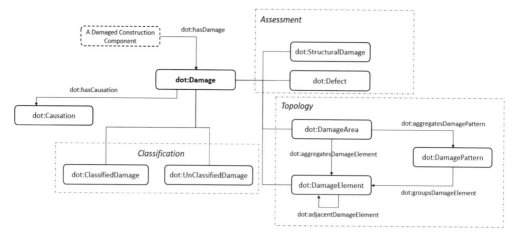

Figure 3. Main classes and object properties of DOT.

The complete ontology can be accessed in a Github repository[2].

DOT provides two ways of modeling damages. Individual damage representations are modeled as *Damage Elements* (dot:DamageElement) and assigned to an affected component that has previously been modeled utilizing a construction topology, e.g. BOT or BROT. Furthermore, a detailed damage geometry, which is ideally stored in a separate geometry file, could be linked to the damage element via a linkset defined in an ICDD Multimodel.

However, modeling geometry of complex damage patterns manually can be a time-consuming process.

Since existing constructions often are affected by areas that contain a huge amount of smaller scaled damages, modeling such complex geometries could lead to intolerable computation times when processing the entire damage model. For this reason, damages representations in the ontology can also be modeled as *Damage Areas* (dot:DamageArea), which could be linked with a more simplified geometry, e.g. a bounding box. Similar to *Damage Elements*, the *Damage Areas* can be assigned to representations of affected components. Furthermore, *Damage Elements* can be embedded in a *Damage Area*, thus a flexible modeling approach at different detail levels is supported. DOT also provides an object property for defining the adjacency relationship between two physically connected *Damage Elements*. This allows for modeling complex damage patterns that can be instantiated in the ontology (dot:DamagePattern) and assigned to a damage area. The impact on the structural capacity of a construction from a damage is described by using two classes. Damages that influence the structural capacity are classified as Structural Damage (dot:StructuralDamage), while damages that are harmless to the structural properties of the construction are classified as *Defect* (dot:Defect).

The CDO extension provides additional terminology for classifying concrete damages, e.g. cracks or reinforcement corrosion. Additional data properties for characterizing the damage instances such as the angle or depth of a crack can be assigned as well.

During the reasoning process new damage instances are created that represent undetected damages. To distinguish these assumed damage instances from those that are known facts, a new auxiliary extension for DOT has been developed, named Undetected Damages Ontology (UDO). It allows for the classification of *Undetected Damages* (udo:UndetectedDamage) and provides a data property to represent the possibility factor of damage occurrence (udo:damagePossibility), which is the entry point for the application of fuzzy set theory methods at a subsequent step.

4 UNDETECTED DAMAGE REASONING

The reasoning of the ontological representation to determine possible undetected damages is processed in multiple consequent steps. First, the ontology is reasoned based on the concepts of the used TBox. In this regard, the concepts are defined in OWL and can be inferred by a reasoning engine. Thereby, mostly superclasses and implicit topological relations of the instances are inferred. Optionally the already detected damages can be classified and assessed according to the approach by Hamdan & Scherer (2019). In the second step, each component is evaluated whether undetected damages could hide in its structure. For each damage type, hidden in the component, a new instance of an undetected damage is created according to predefined SHACL rules. Afterwards, additional classification and property assignments is processed. Finally, the new damage instances are evaluated regarding their occurrence possibility and a fuzzy set member function is attached. In the following section, each reasoning step is described in detail.

4.1 *Preliminary OWL reasoning*

The ontologies for bridge and damage representation provide multiple concepts serialized as OWL2

---

[2] https://github.com/Alhakam/dot

axioms which are based on Description Logic. Besides inferencing superclasses (e.g. each instance that is classified as crack is also consequently classified as damage) topological relations are also inferred. Thereby, two types of rules were relevant for the subsequent damage assessment. First, aggregation or containment relations must be transitive, e.g. that components are also affected by the damages in their subcomponents expressed by eq. 1.

$$\forall x, y, x\ contains(x, y) \cap contains(y, z) \rightarrow contains(x, z) \qquad (1)$$

In OWL this rule can be realized by classifying an object property as transitive property. If the two assertional object properties for containment are different, a property chain axiom is used, e.g. a damage instance that affects a component, which contains the originally affected subcomponent (see eq. 2).

$$\forall x, y, d\ hasDamage(x, d) \cap contains(y, x) \rightarrow hasDamage(y, d) \qquad (2)$$

Second, components that are bearing other components (defined via the property brstr:bears) should be automatically classified as load bearing components (brstr:LoadBearingComponent), since these are often affected by hidden structural damages (see eq. 3).

$$\forall x, y\ bears(x, y) \rightarrow LoadBearingComponent(x) \qquad (3)$$

4.2 Assumption of undetected damages via SHACL

The assumption of *Undetected Damages* involves the generation of new instances (classified as udo:UndetectedDamage) from rules, which are based on expert knowledge. For this reason, SHACL has been used, as it allows for the generation of new OWL instances from rules. Currently, SHACL supports only the inference of one new statement type per rule. Thus, it is not possible to generate new instances and assign properties to them in one process step. Thus, instances need to be generated first before processing rules that assign properties.

Furthermore, SHACL does not directly support the creation of new OWL individuals. Instead, a statement is inferred that links an existing component instance with a generated URI via a property that indicates that the component has an undetected damage (subproperty of udo:hasUndetectedDamage). In this regard, the URI is generated by a SPARQL-based SHACL function that involves the creation of a randomized value. To reason the damage type at a later phase, specific properties that indicate the assumed damage type and undetected status are used. (e.g. udo:hasUndetectedReinforcementCorrosion). In the test case, mainly hidden damages due to reinforcement corrosion were inferred. Therefore, Listing 1 shows the SHACL rule and function for inferring instances of this damage type.

```
urc:DamageAssertion
a sh:TripleRule ;
sh:subject sh:this ;
sh:predicate udo:hasUndetectedReinforcementCorrosion ;
sh:object [
  urc:DamageCreation () ; ] ;
sh:condition [
  sh:and (
    [ sh:property urc:Materialcheck ; ]
    [ sh:or (
      [ sh:property urc:ConstructionYear ; ]
      [ sh:class brcomp:BridgeGirder ; ]
      [ sh:property urc:SeaWater ; ]
      [ sh:property urc:Carbonatation ; ]
      [ sh:property urc:Crack ; ]
      [ sh:property urc:Spalling ; ]
      [ sh:property urc:SpallingExposedReinforcement ; ]
      [ sh:property urc:CoarseGrainAggregate ; ]
      [ sh:property urc:AlkaliSilicaReaction ; ]
    ) ] ) ] .
urc:DamageCreation
a sh:SPARQLFunction ;
sh:select """
  SELECT IRI(CONCAT(STR(NOW()),
  STR(CEIL((RAND() * 10000)))))
  WHERE {}""".
```

Listing 1. Assumption of undetected reinforcement corrosion damages via SHACL.

The rule in Listing 1 consists of multiple conditions that need to be fulfilled in order to generate a new damage instance. One condition that must always be fulfilled is the existence of reinforced concrete in the component (named urc: MaterialCheck in the SHACL file) (see eq.4).

$$\forall c\ \exists m, r\ Components(c) \cap (hasMaterial(c, m) \cap ReinforcedConcrete(m)) \cup (aggregates(r) \cap Reinforcement(r) \leftrightarrow R(c) \qquad (4)$$

The consequence for inferring undetected reinforcement corrosion is the same for every rule used in this research (see eq.5).

$$\forall c \exists u\ hasDamage(c, u) \cap UndetectedDamage(u) \leftrightarrow U(c) \qquad (5)$$

Table 1 shows all rules that were developed for inferring undetected reinforcement corrosion. Besides rules for inferring reinforcement corrosion, additional rules for undetected cracks, sulfate attack as well as stress corrosion damages have been applied, whereby stress corrosion does only occur on prestressed bridge constructions.

4.3 Characterization of undetected damages

Since the generated instances possess no classes or properties, this information must be assigned in

Table 1. Rules for inferring undetected reinforcement corrosion.

Rule in FOL	Possibility-Factor
∀c,b.∃ y R(c)∧ Bridge(b)∧ constructionYear(b,y)∧ y<2012∧ containsComponent(b,c) → U(c)	0.01–0.47 (depending on y)
∀c.∃ d R(c)∧ hasDamage(d)∧ CoarseGrainAggregate(d) → U(c)	0.2
∀c.∃ d R(c)∧ hasDamage(d)∧ Carbonatation(d) → U(c)	1.0
∀c.∃ d R(c)∧ hasDamage(d)∧ Crack(d)∧ crackDepth(d,x)∧ concreteCover(c,y)∧ (y<x) → U(c)	0.5
∀c.∃ d R(c)∧ hasDamage(d)∧ Spalling(d)∧ spallingDepth(d,x)∧ concreteCover(c,y)∧ (y<x) → U(c)	0.5
∀c.∃ d R(c)∧ hasDamage(d)∧ SpallingWithExposedReinforcement(d) → U(c)	1.5
∀c.∃ d R(c)∧ hasDamage(d)∧ AlcaliSilicaReaction(d) → U(c)	0.2
∀c.∃ d R(c)∧ seaWaterContact(c) → U(c)	0.7
∀c.∃ d,bc,bd R(c)∧ adjacentComponent(bc)∧ BridgeCap(bc)∧ adjacentComponent(bd)∧ BridgeDeck(bd) → U(c)	0.3
∀c.∃ d,co R(c)∧ hasCoating(c,co)∧ Coating(co)∧ hasDamage(co,d) → U(c)	0.7

a subsequent step by additional SHACL rules. In SHACL the instances are selected as objects of a specific object property that indicates that the damage is undetected as well as its damage type (e.g. udo:hasUndetectedReinforcementCorrosion). There, every instance is classified as owl:NamedIndividual as well as *Undetected Damage* (udo:UndetectedDamage). In addition, since only *Structural Damages* are considered in this research as *Undetected Damages*, the damage is also classified as dot:StructuralDamage. Furthermore, depending on the damage type that could be inferred a domain specific damage class is assigned and a class that indicates that the damage has been successfully classified (e.g. assigning cdo:Crack and dot:ClassifiedDamage).

Although it would be possible to derive additional properties that characterize the new damage no rules have been developed so far, due to the high level of uncertainty in the current state of this research. Thus, the only property that is inferred for undetected damages is the data property that defines the damage possibility factor, which is needed in a subsequent step to determine the possibility of occurrence for each undetected damage. In Figure 3 a resulting Undetected Damage after processing the described approach is shown.

4.4 Possibility of occurrence for undetected damages

The defined conditions in the SHACL rules, which are based on expert knowledge have different influences on the occurrence possibility of undetected damages. For instance, the existence of spallings in a component where reinforcement is exposed leads to a higher risk that reinforcement corrosion occurs than the existence of cracks. With this knowledge it could be possible to estimate the degree of truth for the undetected damages.

The damage possibility factor has been estimated for each rule based on seven different assumption terms that range from "has no influence on occurrence" to "damage occurrence is highly probable. It must be assumed that the damage is existent". Since

Table 2. Assumption Terms and related Damage Possibility Factor (not perceptual probability).

Assumption Term	Damage Possibility Factor
No influence on occurrence	0.0
Increases probability together with other factors	0.01
Increases probability	0.2
Increases probability greatly. Often solely cause.	0.5
Very probable that damage will occur at some point of time.	1.0
Significantly high occurrence probability. Damage occurrence highly probable.	2.0
It must be assumed that damage is existent.	4.0

the damage possibility factor in this scenario is cumulative for each condition, numerical values have been assigned to these terms. When characterizing the undetected damage instances, a data property is assigned, which holds this numerical value.

Table 2 shows the seven assumption terms as well as the corresponding damage probability levels.

If multiple conditions are true, only one undetected damage of each type and per component is inferred. However, multiple damage possibility factors are assigned to the damage instances. By applying a SPARQL query on the ontology, the sum of all damage possibility factors for each undetected damage can be determined and stored in a new data property, using the CONSTRUCT function of SPARQL. According to the fuzzy set theory by Zadeh (1965) the undetected damages range in a state between 0 to 1, whereby 0 means factual nonexistence and 1 means factual existence. Based on the assumption terms, which are imprecise and non-numerical information, the representative damage possibility factor is applied on a membership function to determine the degree of truth. The membership function is a logarithmic normal distribution function utilizing the error function (see eq. 5 & 6).

Figure 4. Inferred OWL instance that represents an undetected damage.

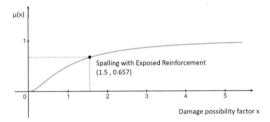

Figure 5. Fuzzy membership function for determining the degree of truth.

Figure 5 shows the graph of this function as well as the determined degree of truth for an undetected reinforcement corrosion, caused by a spalling with exposed reinforcement. Therefore, the estimated possibility for this damage would be 65,7 %.

The function ensures that $\mu(x) = 1$ is not reached, since without checking the hidden damages with appropriate measuring methods it would never be known whether an undetected damage really exists.

$$\operatorname{erf}(x) = \frac{2}{\sqrt{\pi}} \int_0^x e^{-x^2} d\tau \qquad (6)$$

$$\frac{1}{2}\left(1 + \operatorname{erf}\left(\frac{\ln(x)}{\sqrt{2}}\right)\right) \qquad (7)$$

It should be noted that the degree of truth is not equivalent to the probability of damage occurrence. Thus, the value for the degree of truth should only be used as guideline when evaluating the results and validation of the assumed damages via measuring methods is indispensable. Furthermore, by using Fuzzy Logic, it could be possible to reason possible non-refurbishment measures, e.g. setting a speed limit or prohibiting access to certain vehicle types or loads.

5 CONCLUSION AND FUTURE WORK

In this research an approach is proposed for assuming undetected damages hidden in the structure of constructions or inaccessible areas. Thereby, a semantic representation of the construction and already detected damages is used in the form of an ontology serialized in OWL. In this regard, a test subject is used, which is an existing bridge. The TBox is defined by the web ontologies BROT (Hamdan & Scherer 2020) and DOT (Hamdan et al. 2019) as well as compatible ontology extensions. To integrate the ontology in BIM processes, its objects can be linked with corresponding model data, such as an IFC model via an ICDD Multimodel (ISO 21597-1 2020).

The reasoning of undetected damages is performed by using rules based on Description Logic (Baader et al. 2007) in consequent process steps. First, the overall topology and load transmission of the construction and damage ontology is reasoned via OWL axioms. Afterwards, OWL instances that represent the potential undetected damages are inferred through SHACL rules and classified in a subsequent step. Furthermore, one or multiple damage possibility factors are assigned to each undetected damage instance, depending on the rule constraints that are fulfilled by the ontology. Since, the undetected damages are according to the fuzzy set theory (Zadeh 1965) neither factual existent nor nonexistent, the damage possibility factors are used for determining the degree of truth through a membership function. The resulting values for the degree of truth can then be used as guideline for further damage assessment or in an expert system that supports Fuzzy Logic. However, the probability that an undetected damage exists cannot be determined by using these methods.

The assumed damages of the test case still need to be validated by on-site measurements in future research. In this regard, the asserted damage possibility factors and fuzzy membership function can be optimized based on measurement results from the construction site. Furthermore, the approach can be tested and further developed on other construction types, e.g. buildings, tunnels or non-concrete bridges.

It is still subject of future research, how the fuzzy damage objects could be used in subsequent assessment processes. By applying the results to a fuzzy inference engine that uses Fuzzy Logic, decision recommendations could be determined, e.g. possible measures to prevent further degradation of the existing construction. Furthermore, the reasoned damage information can be used in a system identification process as suggested by Lin et al. (2019).

ACKNOWLEDGEMENT

This research work was enabled by the support of the Federal Ministry of Education and Research of Germany through the funding of the project wiSIB (project number 01-S16031C) and EUROSTARS project cyberBridge (project number 01|QE1712C).

REFERENCES

Baader, F., Calvanese, D., McGuinness, D. L., McGuinness, D. L., Nardi, D., & Patel-Schneider, P. F. (2007). *The Description Logic Handbook: Theory, Implementation and Applications*.

Baumgärtel, K., & Scherer, R. J. (2016). Automatic ontology-based green building design parameter variation and evaluation in thermal energy building performance analyses. *Proc. of the 11th European Conference on Product and Process Modelling, ECPPM 2016*, 667–672.

Cacciotti, R., Blaško, M., & Valach, J. (2015). A diagnostic ontological model for damages to historical constructions. *Journal of Cultural Heritage, 16*, 40–48. https://doi.org/10.1016/j.culher.2014.02.002

Hamdan, A.-H., Bonduel, M., & Scherer, R. J. (2019). An ontological model for the representation of damage to constructions. *7th Linked Data in Architecture and Construction Workshop*. http://www.w3.org/1999/02/22-rdf-syntax-ns#

Hamdan, A.-H., & Scherer, R. J. (2019). A knowledge-based Approach for the Assessment of Damages to Constructions. 36th CIB W78 2019 Conference.

Hamdan, A., & Scherer, R. J. (2020). Integration of BIM-related bridge information in an ontological knowledgebase. *Linked Data in Architecture and Construction Workshop*.

Hitzler, P., Krötzsch, M., Parsia, B., Patel-Schneider, P. F., & Rudolph, S. (2012). OWL 2 Web Ontology Language Primer (Second Edition). *http://www.w3.org/TR/owl-primer*

Hu, M., Liu, Y., Sugumaran, V., Liu, B., & Du, J. (2019). Automated structural defects diagnosis in underground transportation tunnels using semantic technologies. *Automation in Construction, 107(August)*, 102929. https://doi.org/10.1016/j.autcon.2019.102929

ISO/TC59/SC13. (2020). ISO 21597-1 Information container for linked document delivery – Exchange specification – Part 1: Container. https://www.iso.org/standard/74389.html

Knublauch, H., Allemang, D., & Steyskal, S. (2017). SHACL Advanced Features. *W3C Working Group Note 08 June 2017*. https://www.w3.org/TR/shacl-af/

Lin, F., Grille, T., Petschacher, M., & Scherer, R. J. (2019). Single / Dual Variation Approach: A Novel Bridge System Identification Method Based on Static Analysis and Parallel Simulation. *CIB W78 2019*.

Pocobelli, D. P., Boehm, J., Bryan, P., Still, J., & Grau-Bové, J. (2018). BIM for heritage science: a review. *Heritage Science, 6*, 30. https://doi.org/10.1186/s40494-018-0191-4

Rasmussen, M. H., Pauwels, P., Hviid, C. A., & Karlshøj, J. (2017). Proposing a Central AEC Ontology That Allows for Domain Specific Extensions. *Lean and Computing in Construction Congress – Volume 1: Proceedings of the Joint Conference on Computing in Construction, July*, 237–244. https://doi.org/10.24928/JC3-2017/0153

Ren, G., Ding, R., & Li, H. (2019). Building an ontological knowledgebase for bridge maintenance. *Advances in Engineering Software, 130(July 2018)*, 24–40. http://doi.org/10.1016/j.advengsoft.2019.02.001

Scherer, R. J., & Katranuschkov, P. (2018). BIMification: How to create and use BIM for retrofitting. *Advanced Engineering Informatics, 38*, 54–66. https://doi.org/10.1016/j.aei.2018.05.007

Scherer, R. J., & Katranuschkov, P. (2019). Context capturing of Multi Information Resources for the Data Exchange in Collaborative Project Environments. *Proceedings of the European Conference on Computing in Construction (EC3 2019)*.

Volk, R., Stengel, J., & Schultmann, F. (2014). Building Information Modeling (BIM) for existing buildings — Literature review and future needs. *Automation in Construction, 38*, 109–127. https://doi.org/10.1016/j.autcon.2013.10.023

Zadeh, L. A. (1965). Fuzzy Sets. *Information and Control, 8(3)*, 338–353. https://doi.org/10.1016/S0019-9958(65)90241-X

An overview of data mining application for structural damage detection in the last decade (2009–2019)

F. Lin, J. Liu & R.J. Scherer
Institute of Construction Informatics, TU Dresden, Dresden, Germany

ABSTRACT: Data mining becomes well-known by the public nowadays. Thanks to the novel updating of programming languages and computer hardware, data mining in the last decade arises rapidly, and shows a board spectrum of varied algorithms and diverse application domains. Focusing on structural damage detection, in which the related concepts e.g. system identification, structural health monitoring, etc. are also involved, this paper investigates frequently utilized data mining algorithms, summarizes the corresponding application scenarios and indicates the tendency of data mining development according to the academic publications from 2009 to 2019. With respect to application objectives (such as stiffness estimation, damage recognition, and sensor placement), algorithms are grouped into three categories: cluster analysis, classification methods and global optimization. In each category, the algorithm principles are introduced, the corresponding applications are systemically elaborated. Finally, the overview of data mining application is summarized and discussed.

1 INTRODUCTION

1.1 Background

Diverse categories of damage incurred by ambient environment, operation and human factors can emerge in structure, such as corrosion, fatigue, creeping, etc. As the number of structure projects increases, the attendant damage in structure along with the service duration is more and more of concern by civil engineers. Conventionally, the structural damage detection is used to be executed through visual inspection, which works as the main reference in structural health estimation. With respect to concrete tasks, structural damage detection or structural health estimation refers to qualitive and quantitive types. The former contains mainly positioning and recognition of damage; the latter covers measurement of damage, assessment of stuffiness reduction, prognosis of damage growth, and others. However, the regular approach to damage identification functions is barely effective, especially in modern complex structures. The rapid developing Data Mining (DM) technologies were brought into consideration to improve the accuracy and quality of structural damage detection, and have been employed in several domains under this topic.

DM, which is nowadays mentioned much frequently, has been under development for decades. It encompasses a series of classical algorithms including cluster analysis, Genetical Algorithms (GA), Artificial Neural Network (ANN) and so on. Benefit from the performance upgrade of modern computers, DM algorithms are capable to be more conveniently for integration in other domains for intensive implementation.

In the year of 2009, Saitta et al. (2009) published the book "Data Mining Application in Civil Engineering", in which the publications till 2008 corresponding to applying DM technologies in Architectural, Engineering and Construction (AEC) industry are studied and demonstrate a comprehensively panorama focusing on two main aspects namely System Identification and sensor placement. Following this research trajectory, this paper summarizes the major advance of data mining application in structural damage detection in the last ten years from 2009 to 2019 through a literature survey, depicts the tendency of utilizing DM algorithms and indicates directions of further research in this field. According to the popularity and the characteristics of DM algorithms, the reviewed papers are classified in to three groups: cluster analysis, classification methods and global optimization.

1.2 Research scope and literature retrieval

There are several correlated concepts, e.g. damage detection, system identification, and Structural Health Monitoring (SHM). Damage detection is interpreted as a sub concept under system identification. It emphasizes seeking for models that are able to reflect the health state of structure, while another related concept SHM is prone to the technologies about structural monitoring acquisition. Both essential parts compose the entire structural damage assessment system.

In order to present a systematic overview for the past decade, publications from 2009 to 2019 are researched and collected in major databases, including Google Scholar, ScienceDirect and IEEEXpore, through key

Table 1. Distribution of publications referring to clustering algorithms.

Algorithms	references*	collected**
Hierarchical methods	6	7
Partitioning methods	4	4
Other conventional methods	3	3
New-developed clustering	2	2

* the number of referenced publications in this paper
** the number of collected publications in entire overview research work

words such as "structural damage detection", "structural system identification", "structural health monitoring", "data mining", "clustering", "genetic algorithms", "ANN", etc. 195 publications (Han et al. 2011) are selected in total, serving as the source of review, in reference solely the necessary ones are listed due to the limited space in this conference paper.

2 CLUSTER ANALYSIS

2.1 General

Cluster analysis or simply clustering is the process of partitioning a set of data objects (or observations) into subsets. Each subset is a cluster, such that objects in a cluster are similar to one another, yet dissimilar to objects in other clusters. The set of clusters resulting from a cluster analysis can be referred to as a clustering. In data mining, clustering is classified as an unsupervised learning method, since the data hereby are labeled automatically through algorithms themselves rather than before feeding models. In this section, the collected 6 papers (see Table 1) referring to application of conventional clustering algorithms in structural damage detection are grouped according to the proposal (Han et al. 2011): hierarchical methods, partitioning methods and other conventional clustering algorithms, including density-based, distribution-based and grid-based clustering.

2.2 Hierarchical methods

As the definition states, a hierarchical method creates a hierarchical decomposition of the given set of data objects. Two approaches, i.e. agglomerative and divisive, can be distinguished by how the hierarchical decomposition is formed. The agglomerative approach starts with each object forming a separate group, and successively merges the objects or groups close to one another, until all the groups are merged into one. On the contrary, the divisive approach starts with all the objects in the same cluster, then a cluster is split into smaller clusters in each successive iteration, until eventually each object is in one cluster, or a termination condition hold (Han et al. 2011).

Modal analysis is a widely used method in structural analysis. The calculated modal parameters can be interpreted as simulation results presenting the bridge behaviors under inherent vibration. Chen et al. (2017) adopted a hierarchical method to deal with the numerous modal parameters, which result from structural dynamic analysis of a bridge model, in order to identify bridge damage and assess the bridge's serviceability. Min et al. (Min & Santos 2017) chose the Euclidian distance criteria as basis of the cluttering analysis procedure for the purpose of modal parameters automatic selection. The effectivity of hierarchical methods in filtering model parameters automatically is also verified by related research work (Marrongelli et al. 2017; Zonno et al. 2018).

This algorithm functions not only in simulation result analysis but also in processing monitoring data. Vertical acceleration data measured by the sensors installed in a bridge can be grouped using hierarchical methods based on the similarity the sensor data values reflect during the monitoring phases (Cho et al. 2010). As an external implementation, the measured data from airplane experiments are also suitable for utilizing this hierarchical method in damage detection due to the resemblance between mechanical and civil structures (MacQueen 1965).

2.3 Partitioning methods

The simplest and most fundamental version of cluster analysis is partitioning, which organizes the objects of a set into several exclusive groups or clusters (Han et al. 2011). 4 related papers are found, among them, one paper elaborates how the k-means algorithm, the most well-known and commonly used partitioning method, exceeded in grouping joints in a bridge and recognizing the abnormal or defect ones (Diez et al. 2016).

In the quired paper collection from the last decade, fuzzy-c-means-algorithm emerges more times than the k-means algorithm, for instance this algorithm is capable to identify the vibration-based damage in a truss bridge (Yu et al. 2013). The kernel fuzzy-c-means-algorithm is derived from fuzzy-c-means-algorithm oriented to the nonlinear input data. Abu-Mahfouz and Banerjee (2017) integrated the kernel fuzzy-c-means-algorithm into a structural crack detection system, according to features in the vibration single. Behnia et al. (2019) combined the kernel fuzzy-c-means-algorithm and the main component method to categorize different damage stages in a reinforced concrete sample.

2.4 Other methods

The previous two methods are more frequently adopted in research work. Besides, other conventional clustering methods demonstrate their feasibility in structural damage detection. A quick search tool, which sets the density-based clustering as the ground idea, is developed to determine the vortex-induced vibration of the bridge deck. The vibration values work as monitoring data of bridge behaviors, and serve for studying the relation between the bridge behaviors and wind velocity (Li et al. 2017). Distribution-based clustering is able to be either utilized along in structural damage (Silva et al. 2017), or merged with such as principle component analysis to manipulate the characteristics of damage sensitivity (Wah et al. 2017).

Facing more complex scenarios and increasing diversity of data types, scientists invent a branch of new clustering methods to overcome the coming challenges in the last decade. Spectral clustering is a typical new-developed method based on graph theory. Alamdari et al. (2017) proposed a methodology, which is based on the clustering of the spectral moments of accelerations, to locate structural and instrumentation anomalies in infrastructures. Adaptive kernel spectral clustering was employed for structural model calibration and structural damage recognition (Langone et al. 2017).

3 CLASSIFICATION METHODS

3.1 General

Unsupervised learning and supervised learning are two categories of Machine Learning (ML), which can be interpreted as implementation though DM techniques. While clustering algorithms introduced in Section 2 are of typical unsupervised learning, this section emphasizes classification methods which are applied usually in supervised learning. Initially, the available data are divided into two parts with different ratios and amount. The entire process contains a training phase and testing phase in the frame of supervised learning. The larger part of data needs to be labeled for feeding into architecture. Training means that the model is established through determining parameters according to the given labeled data. After the establishment of the trained model, the smaller part of data without labels being employed to validate the model are defined as testing. Although a dozen of algorithms are developed (Han et al. 2011). The paper research work shows a relative convergent situation (see Table 2): 80 papers corresponding to classification by supervised learning are identified. 14 focus in regard to Bayes classifier, 12 Support Vector Machine (SVM), 5 Decision tree. It is to be noticed that the number of collected papers about ANN reaches 49 and occupies over half of the collection.

3.2 Bayesian classifiers

Bayesian classifiers based on Bayes' theorem are statistical classifiers, can predict class membership probabilities such as the probability that a given tuple belongs to a particular class. Studies comparing classification algorithms have found a simple Bayesian classifier known as the naive Bayesian classifier to be comparable in performance with decision tree and selected neural network classifiers. Bayesian classifiers have also exhibited high accuracy and speed when applied to large databases (Han et al. 2011).

Jang & Smyth (2017) conceived and implemented a complex combination of methodology, in which a Bayesian model updating scheme is used to improve an agreement between the identified modal properties of the real measured data and those from the FE model for a major long-span bridge. In the condition of data with high noise content, a Bayesian-based method provides reliable modal parameter identification in structural dynamic systems (Ghrib & Li 2017). Ntotsios et al. (2012) presented details on a Bayesian inference framework for the identification of the location and the severity of damage using measured modal characteristics. All the 14 collected papers from the last ten years suggest that Bayesian classification methods mostly either function with integration of other algorithms or mathematical models (Figueiredo et al. 2014; Ghrib & Li 2017), or are applied in a certain component in a complicated research proposal (Arangio & Beck 2012).

3.3 Support Vector Machine

SVM, a method for the classification of both linear and nonlinear data, works as follows. It uses a nonlinear mapping to transform the original training data into a higher dimension. Within this new dimension, it searches for the linear optimal separating hyperplane. With an appropriate nonlinear mapping to a sufficiently high dimension, data from two classes can always be separated by a hyperplane (Han et al. 2011).

Curry & Crémona (2012) calculated the declined eigen frequencies for structural damage identification, in which the structural behaviors can be distinguished with SVM. In order to predict the natural frequency variation of a suspension bridge, Laory et al. executed Support Vector Regression, which is an application of SVM in regression and testified with fewer errors, to separate the monitoring signal effect by environment change from noise.

As an extension approach, Least Squares Support Vector Machine (LS-SVM) delivers a possibility to locate gird failure during the steel bridge inspection (Liao & Lee 2016). Dackermann et al. (2017) explored a multi-label-classifier SVM to classify damage and to value the current state of foundation piles and utility poles.

Table 2. Distribution of publications referring to classification algorithms.

Algorithms	references	collected
Bayesian classifiers	5	14
Support Vector Machine	4	12
Decision tree	3	5
Artificial Neural Network	10	49

3.4 Decision trees

A decision tree is a flowchart-like tree structure, where each internal node (non-leaf node) denotes a test on an attribute, each branch represents an outcome of the test, and each leaf node (or terminal node) holds a class label. The topmost node in a tree is the root node (Han et al. 2011).

Impacts from ambient factors (e.g. temperature, traffic, wind and humidity) on structures are always of a major concern, therefore, this phenomenon that these impacts can modify the nature frequencies is applied by decision three algorithms for the purpose of structural damage detection (Laory et al. 2011; Posenato et al. 2010). Salazar et al. (2017) managed the deformation interpretation and anomalies detection of a dam using Classification And Regression Trees (CART), which described the generation of binary decision trees (Breiman 1998).

3.5 Artificial Neural Network

An Artificial Neural Network (ANN) which is a computational model inspired by networks of biological neurons, wherein the neurons compute output values from inputs (Puri et al. 2016), helps in approximating complex relation between input and output (Ayyadevara 2018). The subsections beneath elaborate shallow neural networks and deep neural networks respectively. The number of former-related paper amounts 43 out of 49 (see Table 2), the latter is studied in the rest.

3.5.1 Shallow neural networks

Shallow neural network is a relative concept against deep neural network, normally it is simply named as ANN. Tadesse et al. (2012) generated the training and testing data from FEA software ABAQUES for an ANN, which is subsequently employed to predict alerting of bridge deflections in a term because of damage. Joghataie and Dizaji (2009) constructed a multi-layer ANN to analyze the non-linear behaviors of a concrete gravity dam caused by earth quake.

As a matter of fact, the proceeding work is solely related to damage detection, it neither located the damage nor assessed the extent of damage. ANN can be combined with particle swarm optimization (PSO) (Liu et al. 2014) or Bayesian methods (Arangio & Beck 2012) to identify damage locations and quantities. In this way, the above mentioned algorithms are integrated into neurons in the ANN to achieve the goal of complementing them. In addition, sensor distribution can be optimized by means of ANN, for example, the Dynamic Artificial Neural Network (DANN) assists an intelligent SHM system to suggest optimal sensor placement (Smarsly & Law 2014).

3.5.2 Deep neural networks

Deep Learning, the sub concept of machine learning based on deep neural network established with more layers, presents a more powerful capability in application than shallow neural networks. A Convolutional Neural Network (CNN), which is nowadays also often reported in media, is also one of the representative deep learning algorithms, becomes a popular and effective solution to image processing. The found publications encompass this topic mainly as well.

With respect to the combination of damage detection und computer vision using CNN, researchers concentrate on damage photo recognition. Yokoyama & Matsumoto (2017) developed an automatic tool to detect concrete cracks using CNN. Beyond cracks detection, the detection accuracy can be further evaluated (Wu et al. 2016). The graphics of steel damage due to metal fatigue can be utilized to estimate its state (Liu & Zhang 2019).

Regions-based CNN (R-CNN) (Girshick et al. 2014) is developed to enhance object recognition in computer version through multi-object classification and rendering the regions of different kinds of identified objects in color. Hence, concrete crack, steel corrosion, bolt corrosion, and steel delamination are successfully located and colored by a R-CNN-based structural visual inspection method (Cha et al. 2018).

4 GLOBAL OPTIMIZATION

4.1 General

Global optimization algorithms are derived from the Monte Carlo method. In the first loop of its procedure, the initial population is in the beginning generated stochastically, sequentially substituted into the fitness function, which implies the mathematic model describing the kernel relationship in study, then adjusted in accordance with the export from fitness function after each iteration loop until the results satisfy the convergence criteria. Several global optimization algorithms are developed with mechanisms of diverse population adjustment, for instance genetic algorithms (GA), PSO, etc. The application domains in damage detection hereby involve model manipulation, damage detection and sensor placement. Totally 62 corresponding publications are collected from 2009 to 2019 (see Table 3). 40 among them are related to GA, PSO is mentioned in another 11 papers, these two kinds of algorithms are mostly applied, the remaining 13 involve other global optimization algorithms. Only the most representative achievements are depicted in

Table 3. Distribution of publications referring to global optimization algorithms.

Algorithms	references	collected
Genetic algorithms	10	40
Particle swarm optimization	4	11
Ant colony optimization	1	6
Monkey algorithms	2	7

this section to show a systematic overview with less redundancy.

4.2 *Genetic algorithms*

Genetic algorithms attempt to incorporate ideas of natural evolution. In general, genetic learning starts as follows. An initial population is created consisting of randomly generated rules. Each rule can be represented by a string of bits. Offspring are created by applying genetic operators such as crossover and mutation. In crossover, substrings from pairs of rules are swapped to form new pairs of rules. In mutation, randomly selected bits in a rule's string are inverted (Han et al. 2011).

Ribeiro et al. (2012) implemented GA to update finite element models of a bowstring-arch railway bridge until the simulation results resulting from the model match the parameter of experiments. Hybrid GA is also usually used in more complicated scenarios, e.g. GA with sequential niche technique performs better in bridge model updating (Jung & Kim 2009). The updating can entail lots of parameters in a model, e.g. geometry, material, mechanical coefficients, etc.

The background behind damage detection using GA is basically identical to model updating. In damage detection, initial populations indicate specifically the stiffness reduction from cracks, which is of a major concern in concrete structural damage detection, or other parameters, which are able to be manipulated due to other kinds of damage. The equilibrium equation of a FE-model serving as the fitness function in GA ought to be constructed. Obey the crack modelling theory. The initial population implies element stiffness or element stiffness reduction. The offspring after iteration steps represent the damage state of certain elements when the convergence trend emerges. This approach is feasible either in determining defect elements in a truss structure (Nobahari et al. 2017), or in identify the cracks in a concrete beam structure (Panigrahi et al. 2009).

As an algorithm inherited from Mont Carlo method, GA owns the uncertainness in random sampling, beyond that people suffer the premature convergence as well. To prevent implementation from these shortcomings, other algorithms or methods are combined with GA. Yu et al. (2019) proposed the hybrid method of GA and fusion of fuzzy-logic, declined the iteration steps to convergency in structural damage identification. The combination with ANN (Betti et al. 2015) or clustering (Silva et al. 2016) presents a more effective performance.

Besides the models and simulation results the previous context in this section narrates, the application domain sensor placement focuses on the monitoring system composed by sensors with various functions. The distribution of sensors is impacted by a series of factors, such as energy consumption, number of sensors, noise and so on. Beygzadeh et al. (2014) enhanced GA to organize an optimal sensor distribution in space structures to minimize the disturbance of noise. Similar contributions are achieved by means of several combination attempts of GA and other algorithms.

4.3 *Other algorithms*

As known and endorsed by the statistics in this paper, GA is one of the most popular and representative global optimization algorithms. Therefore, the cases using GA were comprehensively elaborated. This subsection aims at demonstrating the work referring to damage detection realized through other global optimization algorithms, which are PSO, ant colony optimization (ACO) and monkey algorithms (MA).

PSO is a population based stochastic optimization technique inspired by social behavior of bird flocking or fish schooling. In PSO, the potential solutions, called particles, fly through the problem space by following the current optimum particles. Model updating can be completed through the traditional PSO (Saada et al. 2013) or a hybrid-PSO algorithm (Begambre & Laier 2009). The majority of relevant republications indicate the mainstream that PSO are often implemented in accompany with other methods. Mohan et al. (2013) evaluated the use of Frequency Response Function (FRF) with the help of Particle Swarm Optimization (PSO) technique, for structural damage detection and quantification. Li et al. (2017) employed PSO in acoustic emission study to support the cluster analysis for steel corrosion, so that the corrosion caused by different reasons can be determined.

In ACO, each individual of the population is an artificial agent that builds incrementally and stochastically a solution to the considered problem (Dorigo & Birattari 2010). ACO performs effectively in structural damage evaluation based on the changing eigen frequency (Yu & Xu 2011). However, this algorithm is more suitable to solve the optimal path problem. Monkey algorithm is created by Zhao & Tang 2008), successively implemented by Yi et al. (2012) in a plenty of tasks in SHM.

5 DISCUSSION AND INTERPRETATION

The objectives of damage detection are quite clear namely two parts: 1) estimating the locations of damage and quantities of damage extent; 2) setting optimal plans of sensor placement. The latter could be intuitively interpreted as a single mission. The former contains many topics:

– *In a (reinforced) concrete structure, concrete cracks are of major concern, while corrosion and metal fatigue merely appear in reinforcements or steel structures. Both induce stiffness reduction directly or indirectly, therefore, in these cases, the methods should be used to figure out where the defects occur and how much stiffness of the damaged area remains.*

- Structural monitoring system is installed to collect the data reflecting structural behaviors under certain load cases. Usually, there are noise and missing data in the outcome from monitoring system. The errors resulting from noise are requested to be fixed, the missing data ought to be prognosed.
- The entire procedure of a conceived method functions as a system, which imports inputs and exports outputs after processing. To be aware of the quality and effectivity of a system, sensitivity analysis is executed, too, so that the impacts from inputs (e.g. imperfect monitoring data and approximated load cases) on outputs (e.g. modal parameters and deflections) can be studied.

Researchers from the field of structural damage detection play the role as the middle layer in a sandwich. On one hand, this overview work implies that a variety of methods is developed and explored aiming at structural damage detection. However, the implementation scenarios actually do not emerge in diverse structures, besides a few specific study cases of dams, the majority is about frames, tunnels, or bridges. Even though each kind of these structures can be constructed in a more complex form, they are still relatively convenient to be modelled comparing with really complicated buildings or infrastructures, e.g. a skyscraper or an airport. Thus, the proposed damage detection methods validated in research projects are required to be continuously enhanced for the further practice-oriented application.

On the other hand, data mining is admired yet also controversial. The related DM algorithms are initially created by mathematicians and computer scientists in 1980s. The succeeding work emphasizes mainly in fixing blemishes from these DM algorithms. Brand new ones are barely developed. Certainly, the duty of corresponding researchers is making good use of DM algorithms instead of producing them. It explains why the publications in the last decade focused on assembling classic DM algorithms and other mathematical models together to overcome the inherent shortcomings and to handle knottier situations.

The quantities of collected papers published in corresponding years are demonstrated in Figure 1, where five DM algorithms are chosen as characteristic items, and clustering, GA (the most representative global optimization algorithm) and ANNs are most indicative: after GA reached the pick value of 8 in the year of 2011, fewer and fewer papers employed it; the numbers referring to clustering and ANNs showed a rapid increasement after 2016, where AlphaGo presented human beings the power of artificial neural networks.

Both the reasons of popularity change of DM algorithms and the solution to implementation in more complex civil structures point to the same direction: the improving computing performance. It has leaded to such wide usage of the classic DM algorithms at present, its future updating for more intensive application is worthy of expectation.

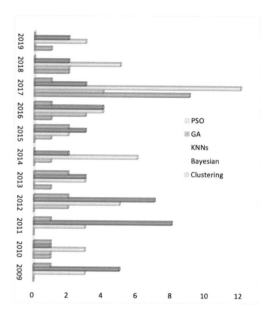

Figure 1. The number of papers corresponding to the chosen algorithms in last decade.

6 CONCLUSIONS

195 papers and several other relative publications from 2009 to 2019 are collated for an overview. Pursuant to a representative selection of 54 among them, this paper summarized the application of data mining algorithms in structural damage detection (involving sensor placement in the broad sense) and developments in this field from the perspective of three major categories: cluster analysis, classification methods and global optimization. Both Features of DM application and the future tendency of structural damage detection methods are discussed based on the literature.

ACKNOWLEDGEMENTS

This research work belongs to an extended study from the projects wiSIB, No. 01-S16031C and cyberBridge, No. 01-QE1712C. The authors would like to acknowledge the support of the Federal Ministry of Education and Research of Germany and to the funding of this research work.

REFERENCES

Abu-Mahfouz, I. & Banerjee, A. 2017. Crack Detection and Identification Using Vibration Signals and Fuzzy Clustering. *Procedia Computer Science*, 114: 266–274.

Alamdari, M.M., Rakotoarivelo, T. & Khoa, N.L.D. 2017. A spectral-based clustering for structural health monitoring of the Sydney Harbour Bridge. *Mechanical Systems and Signal Processing*, 87: 384–400.

Arangio, S. & Beck, J.L. 2012. Bayesian neural networks for bridge integrity assessment. *Structural Control and Health Monitoring*, 19(1): 3–21.

Ayyadevara, V.K. 2018. Artificial Neural Network. In V. K. Ayyadevara, ed. *Pro Machine Learning Algorithms: A Hands-On Approach to Implementing Algorithms in Python and R*. Berkeley, CA: Apress: 135–165. https://doi.org/10.1007/978-1-4842-3564-5_7 1 April 2020.

Begambre, O. & Laier, J.E. 2009. A hybrid Particle Swarm Optimization – Simplex algorithm (PSOS) for structural damage identification. *Advances in Engineering Software*, 40(9): 883–891.

Behnia, A., Chai, H.K., GhasemiGol, M., Sepehrinezhad, A. & Mousa, A.A. 2019. Advanced damage detection technique by integration of unsupervised clustering into acoustic emission. *Engineering Fracture Mechanics*, 210: 212–227.

Betti, M., Facchini, L. & Biagini, P. 2015. Damage detection on a three-storey steel frame using artificial neural networks and genetic algorithms. *Meccanica*, 50(3): 875–886.

Beygzadeh, S., Salajegheh, E., Torkzadeh, P., Salajegheh, J. & Naseralavi, S.S. 2014. An Improved Genetic Algorithm for Optimal Sensor Placement in Space Structures Damage Detection. *International Journal of Space Structures*, 29(3): 121–136.

Breiman, L. ed. 1998. *Classification and regression trees*. Repr. Boca Raton: Chapman & Hall.

Cha, Y.-J., Choi, W., Suh, G., Mahmoudkhani, S. & Büyüköztürk, O. 2018. Autonomous Structural Visual Inspection Using Region-Based Deep Learning for Detecting Multiple Damage Types. *Computer-Aided Civil and Infrastructure Engineering*, 33(9): 731–747.

Chen, Y. & Zhong, Z. 2017. Bridge Structure Modal Parameter Identification Based on Improved DATA-SSI and Clustering Analysis. *Journal of Highway and Transportation Research and Development*, 34(9): 76–85.

Cho, Soojin, Jo, Hongki, Jang, Shinae, Park, Jong-Woong, Jeong, Hyeong-Jo, Yun, Chung-bang & Seo, Ju-Won. 2010. Structural health monitoring of a cable-stayed bridge using wireless smart sensor technology: data analyses. *Smart Structures and Systems*, 6(5–6): 461–480.

Cury, A. & Crémona, C. 2012. Pattern recognition of structural behaviors based on learning algorithms and symbolic data concepts. *Structural Control and Health Monitoring*, 19(2): 161–186.

Dackermann, U., Yu, Y., Niederleithinger, E., Li, J. & Wiggenhauser, H. 2017. Condition Assessment of Foundation Piles and Utility Poles Based on Guided Wave Propagation Using a Network of Tactile Transducers and Support Vector Machines. *Sensors*, 17: 2938.

Diez, A., Khoa, N.L.D., Makki Alamdari, M., Wang, Y., Chen, F. & Runcie, P. 2016. A clustering approach for structural health monitoring on bridges. *Journal of Civil Structural Health Monitoring*, 6(3): 429–445.

Dorigo, M. & Birattari, M. 2010. Ant Colony Optimization. In C. Sammut & G. I. Webb, eds. *Encyclopedia of Machine Learning*. Boston, MA: Springer US: 36–39. https://doi.org/10.1007/978-0-387-30164-8_22 28 March 2020.

Figueiredo, E., Radu, L., Worden, K. & Farrar, C.R. 2014. A Bayesian approach based on a Markov-chain Monte Carlo method for damage detection under unknown sources of variability. *Engineering Structures*, 80: 1–10.

Ghrib, F. & Li, L. 2017. An adaptive filtering-based solution for the Bayesian modal identification formulation. *Journal of Civil Structural Health Monitoring*, 7(1): 1–13.

Girshick, R., Donahue, J., Darrell, T. & Malik, J. 2014. Rich Feature Hierarchies for Accurate Object Detection and Semantic Segmentation. In Proceedings of the IEEE Conference on Computer Vision and Pattern Recognition. 580–587.

Han, J., Pei, J. & Kamber, M. 2011. *Data Mining: Concepts and Techniques*. Elsevier.

Jang, J. & Smyth, A. 2017. Bayesian model updating of a full-scale finite element model with sensitivity-based clustering: Bayesian model updating of a full-scale finite element model with sensitivity-based clustering. *Structural Control and Health Monitoring*, 24(11): e2004.

Joghataie, A. & Dizaji, M.S. 2009. Nonlinear Analysis of Concrete Gravity Dams by Neural Networks. : 6.

Jung, D.-S. & Kim, C.-Y. 2009. FE model updating based on hybrid genetic algorithm and its verification on numerical bridge model. *Structural Engineering and Mechanics*, 32(5): 667–683.

Langone, R., Reynders, E., Mehrkanoon, S. & Suykens, J.A.K. 2017. Automated structural health monitoring based on adaptive kernel spectral clustering. *Mechanical Systems and Signal Processing*, 90: 64–78.

Laory, I., Trinh, T.N. & Smith, I.F.C. 2011. Evaluating two model-free data interpretation methods for measurements that are influenced by temperature. *Advanced Engineering Informatics*, 25(3): 495–506.

Laory, I., Trinh, T.N., Smith, I.F.C. & Brownjohn, J.M.W. 2014. Methodologies for predicting natural frequency variation of a suspension bridge. *Engineering Structures*, 80: 211–221.

Li, D., Yang, W. & Zhang, W. 2017. Cluster analysis of stress corrosion mechanisms for steel wires used in bridge cables through acoustic emission particle swarm optimization. *Ultrasonics*, 77: 22–31.

Li, S., Laima, S. & Li, H. 2017. Cluster analysis of winds and wind-induced vibrations on a long-span bridge based on long-term field monitoring data, Engineering Structures. *Engineering Structures*, 138. https://www.deepdyve.com/lp/elsevier/cluster-analysis-of-winds-and-wind-induced-vibrations-on-a-long-span-Ho0mS6d4a3?key=elsevier 7 July 2019.

Liao, K.-W. & Lee, Y.-T. 2016. Detection of rust defects on steel bridge coatings via digital image recognition. *Automation in Construction*, 71: 294–306.

Liu, H., Song, G., Jiao, Y., Zhang, P. & Wang, X. 2014. Damage Identification of Bridge Based on Modal Flexibility and Neural Network Improved by Particle Swarm Optimization. *Mathematical Problems in Engineering*, 2014: 1–8.

Liu, H. & Zhang, Y. 2019. Image-driven structural steel damage condition assessment method using deep learning algorithm. *Measurement*, 133: 168–181.

MacQueen, J. 1965. Some Methods for Classification and Analysis of MultiVariate Observations. *Proc of Berkeley Symposium on Mathematical Statistics & Probability*. http://www.ams.org/mathscinet-getitem?mr=214227 1 April 2020.

Marrongelli, G., Magalhães, F. & Cunha, Á. 2017. Automated Operational Modal Analysis of an arch bridge considering the influence of the parametric methods inputs. *Procedia Engineering*, 199: 2172–2177.

Min, X. & Oliveira Santos, L. 2017. Dynamic Assessment of the São João Bridge Structural Integrity. *Procedia Structural Integrity*, 5: 325–331.

Mohan, S.C., Maiti, D.K. & Maity, D. 2013. Structural damage assessment using FRF employing particle swarm optimization. *Applied Mathematics and Computation*, 219(20): 10387–10400.

Nobahari, M., Ghasemi, M.R. & Shabakhty, N. 2017. Truss structure damage identification using residual force vector and genetic algorithm. *Steel and Composite Structures*, 25(4): 485–496.

Ntotsios, E., Papadimitriou, C., Panetsos, P., Karaiskos, G., Perros, K. & Perdikaris, P.C. 2012. Bridge health monitoring system based on vibration measurements. *Bulletin of Earthquake Engineering*, 7(2): 469–483.

Panigrahi, S.K., Chakraverty, S. & Mishra, B.K. 2009. Vibration based damage detection in a uniform strength beam using genetic algorithm. *Meccanica*, 44(6): 697–710.

Posenato, D., Kripakaran, P., Inaudi, D. & Smith, I.F.C. 2010. Methodologies for model-free data interpretation of civil engineering structures. *Computers & Structures*, 88(7): 467–482.

Puri, M., Solanki, A., Padawer, T., Tipparaju, S.M., Moreno, W.A. & Pathak, Y. 2016. Chapter 1 - Introduction to Artificial Neural Network (ANN) as a Predictive Tool for Drug Design, Discovery, Delivery, and Disposition: Basic Concepts and Modeling. In M. Puri, Y. Pathak, V. K. Sutariya, S. Tipparaju, & W. Moreno, eds. *Artificial Neural Network for Drug Design, Delivery and Disposition*. Boston: Academic Press: 3–13. http://www.sciencedirect.com/science/article/pii/B9780128015599000016 1 April 2020.

Ribeiro, D., Calçada, R., Delgado, R., Brehm, M. & Zabel, V. 2012. Finite element model updating of a bowstring-arch railway bridge based on experimental modal parameters. *Engineering Structures*, 40: 413–435.

Saada, M.M., Arafa, M.H. & Nassef, A.O. 2013. Finite element model updating approach to damage identification in beams using particle swarm optimization. *Engineering Optimization*, 45(6): 677–696.

Saitta, S., Raphael, B. & Smith, I.F.C. 2009. Data Mining: applications in civil engineering. http://www.researchgate.net/publication/41940444_Data_Mining__applications_in_civil_engineering 1 April 2020.

Salazar, F., Toledo, M.Á., González, J.M. & Oñate, E. 2017. Early detection of anomalies in dam performance: A methodology based on boosted regression trees. *Structural Control and Health Monitoring*, 24(11): e2012.

Silva, M., Santos, A., Figueiredo, E., Santos, R., Sales, C. & Costa, J.C.W.A. 2016. A novel unsupervised approach based on a genetic algorithm for structural damage detection in bridges. *Engineering Applications of Artificial Intelligence*, 52: 168–180.

Silva, M., Santos, Santos, R., Figueiredo, E., Sales, C. & Costa, J.C.W.A. 2017. Agglomerative concentric hypersphere clustering applied to structural damage detection. *Mechanical Systems and Signal Processing*, 92: 196–212.

Smarsly, K. & Law, K.H. 2014. Decentralized fault detection and isolation in wireless structural health monitoring systems using analytical redundancy. *Advances in Engineering Software*, 73: 1–10.

Tadesse, Z., Patel, K.A., Chaudhary, S. & Nagpal, A.K. 2012. Neural networks for prediction of deflection in composite bridges. *Journal of Constructional Steel Research*, 68(1): 138–149.

Wah, W.S.L., Chen, Y.-T., Roberts, G.W. & Elamin, A. 2017. Damage Detection of Structures Subject to Nonlinear Effects of Changing Environmental Conditions. *Procedia Engineering*, 188: 248–255.

Wu, L., Mokhtari, S., Nazef, A., Nam, B. & Yun, H.-B. 2016. Improvement of Crack-Detection Accuracy Using a Novel Crack Defragmentation Technique in Image-Based Road Assessment. *Journal of Computing in Civil Engineering*, 30(1): 04014118.

Yi, T.-H., Li, H.-N. & Zhang, X.-D. 2012. A modified monkey algorithm for optimal sensor placement in structural health monitoring. *Smart Materials and Structures*, 21(10): 105033.

Yokoyama, S. & Matsumoto, T. 2017. Development of an Automatic Detector of Cracks in Concrete Using Machine Learning. *Procedia Engineering*, 171: 1250–1255.

Yu, A., Ji, ajia & Sun, S. 2019. An improved multi-objective genetic algorithm and data fusion in structural damage identification. *International Journal of Security and Networks*, 14(2): 95.

Yu, L. & Xu, P. 2011. Structural health monitoring based on continuous ACO method. *Microelectronics Reliability*, 51(2): 270–278.

Yu, L., Zhu, J.-H. & Yu, L.-L. 2013. Structural Damage Detection in a Truss Bridge Model Using Fuzzy Clustering and Measured FRF Data Reduced by Principal Component Projection. *Advances in Structural Engineering*, 16(1): 207–217.

Zhao, R. & Tang, W. 2008. Monkey Algorithm for Global Numerical Optimization. *Journal of Uncertain Systems*, 2(3): 165–176.

Zonno, G., Aguilar, Boroschek, R. & Lourenço, P.B. 2018. Automated long-term dynamic monitoring using hierarchical clustering and adaptive modal tracking: validation and applications. *Journal of Civil Structural Health Monitoring*, 8(5): 791–808.

Intelligent structural design in BIM platforms: Optimization of RC wall-slab systems

N. Bourahla & S. Tafraout
LGSDS laboratory, National Polytechnic School, Algiers, Algeria

Y. Bourahla
Informatic Department, University of Science and Technology Houari Boumediene, Algeria

ABSTRACT: This paper evaluates the performance of a genetic algorithm (GA) in generating an optimal RC shear-wall slab structure for a given architectural model in IFC (Industry Foundation Classes) format. The concept of the optimization procedure is first outlined. Then a sensitivity analysis is carried out to find out the adequate weights of the multi-objective function to control the torsional eccentricity ratio, the total length of the shear walls, the covered and overlapping area related to the gravity load bearing. The results demonstrate that a proper use of weights is essential in generating optimized layouts which comply with the architectural configurations and satisfy the predefined seismic design criteria for a wide range of building architectural schemes. A repeatability test confirmed the robustness of the procedure. The outcome of this study can help tuning the optimization procedure by providing weight ranges that can be used for different structural and architectural design conditions.

1 INTRODUCTION

A pronounced trend towards a systematic use of the building information modeling (BIM) in the construction industry is being observed. Even though the benefits of using BIM are irresistible, several challenges are still facing the implementation of BIM (Ghaffarianhoseini et al. 2017; Sun et al. 2017). In the context of inter-disciplinary enhancement of the productivity by generative design, significant developments have been achieved in the last decade (Abrishami et al. 2014; Chang & Shih 2018; Chi et al. 2015). A step further towards more efficient and creative design within the BIM environment has been realized through a recent proposal that intelligently generates optimized structural solutions considering the inter-disciplinary conformity between the architectural configuration and the structural engineering requirements (Tafraout et al. 2019). This method is based on applying a genetic algorithm to produce an optimized structural layout from an architectural BIM model extracted in IFC (Industry Foundation Classes) format. This way permits more freedom in the architectural design and explores further the possibilities of the structural design. The performance of this approach in generating optimized reinforced concrete shear-slab structures has been demonstrated using several architectural configurations (Bourahla et al. 2019). However, a challenging problem in multi-objective optimization function is to determine the appropriate weights (Gennert & Yuille 1988). This paper investigates the efficiency of the genetic algorithm in controlling some key structural design criteria and their effect on the optimization procedure. To this end, a sensitivity analysis is carried out by scaling the importance weights of the objective function for the torsional eccentricity, cover-overlapping area, total shear walls length factors for several case studies.

2 CONCEPT OF THE INTELLIGENT STRUCTURAL DESIGN OPTIMIZATION

The concept of the intelligent structural design used here is an optimization procedure based on a genetic algorithm to derive optimized structural layouts that fits into an architectural configuration in IFC format extracted from a BIM model (Tafraout et al. 2019). This concept is applied to generate optimized structural layouts of RC wall-slab systems. Using the architectural model, the genetic algorithm defines first the entire geometric locations where structural elements can be placed at each level with respect to the architectural partitions and the load path continuity. Then, predefined structural criteria are expressed as constraints and a set of performance factors are formulated in terms of a multi-objective function to be optimized using a genetic algorithm. In the present study, the weights of the objective function are varied in order to investigate the effect of each parameter on the seismic performance of the resulting optimized structures.

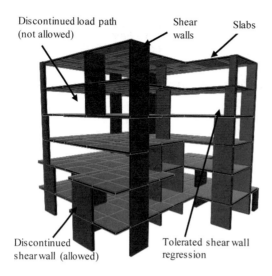

Figure 1. RC shear wall-slab system.

3 CONSTRAINTS AND OBJECTIVE FUNCTION TO OPTIMIZE A RC WALL-SLAB SYSTEM

As mentioned before, the type of structural system used in this study is constituted of RC walls to resist both the gravity and the lateral load with RC slabs as horizontal diaphragms (see Figure 1).

The structural rules and design criteria together with the constraints and the multi-objective function are detailed in (Tafraout et al. 2019). An outline of the main constraints and the objective function for this type of framing are given below.

4 STRUCTURAL CONSTRAINTS

At the conceptual design stage, the engineer usually defines first the structural layout then performs a pre-sizing calculation to determine the structural elements dimensions before proceeding to the analysis and design criteria checking. To automatize this procedure, key controlling parameters are selected and formulated into structural constraints to be implemented into a genetic algorithm to determine an optimized structure. The selected parameters for RC wall-slab systems are:

- Vertical load path continuity to ensure that the loads are safely transmitted from the upper storeys to the foundation.
- Wall rigidity variation decreasing from bottom to top to ensure a uniformity of stress distribution in the walls as the efforts in the latter due to gravity and lateral loadings are more important in the lower storeys.
- Individual shear wall length is bounded within a range for low to medium rise building with a median length as an optimal. The limits are defined on the basis of code recommendations and some technical literature.
- Slab dimensions and supports span are constrained by the ratio of the slab span to its thickness (span/thickness ratio) which varies in the range of 20 to 25 times and the maximum span between two support points in the range of the storey height h to 1.5 h (Tafraout et al. 2019).

5 PERFORMANCE PARAMETERS AND OBJECTIVE FUNCTION

Simplified performance parameters are used to formulate an objective function:

$$f_v = v.w \quad (1)$$

where w is a constant weights vector defined empirically, that is each criterion has a weight proportional to its importance in evaluating a solution. The evaluation vector v corresponds to the following criteria:

- Total length of walls to floor area ratio which expresses the rigidity and strength requirements for low to medium height building. It is based on finite element calculation of simple multi-storey buildings under seismic loading.
- Covered slab area is related to the gravity load bearing capacity represented by a slab contour supported by a shear wall. This criterion ensures a uniform distribution of the RC walls that would respect the maximum span and cover the whole slab area with minimum overlapping.
- Floor torsional eccentricity is a criterion to guarantee a structural regularity to avoid torsional overstressing.
- Torsional radius is a criterion to increase the torsional stiffness. This favor the distribution of the walls towards the periphery of the building.

6 GENETIC ALGORITHM OPTIMIZATION

The optimization is carried out using a genetic algorithm with the above-mentioned constraints and objective function. The input to the GA is an architectural configuration in IFC format which defines the slab contours and the position of partitions of each floor. The output of the GA, after a limited number of iterations, is an optimized RC shear-slab system which best satisfies the constraints and optimizes the objective function. The flowchart in Figure 2 illustrates the main tasks of the GA as defined below:

- Random generation of first population: The algorithm generates an initial population of 80 individuals with different shear wall lengths positioned randomly within the architecture partitions.
- Selection: The selection of the layouts for the next population is simply based on fitness ranking of all the individuals after adding those of the crossover and mutation.
- Crossover function: A new layout is obtained by combining walls from two existing configurations

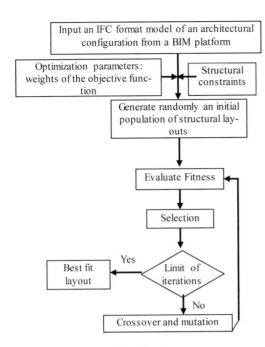

Figure 2. Genetic algorithm flowchart.

so as the new solution inherits features of both parents and it remains in compliance with the predefined constraints.
– Mutation function: Consists in either inserting a new shear wall, removing, shortening or extending an existing shear wall to produce a new layout (individual).

7 EFFECT OF THE CONSTRAINTS AND WEIGHTS OF THE OBJECTIVE FUNCTION ON THE OPTIMIZATION PERFORMANCE OF THE GA

In a shear wall-slab RC system, the shear walls play double role to resist both the total gravity and horizontal (seismic) loads. Therefore, the optimization of the structural system implies that a minimum required total length of the shear walls to be distributed in such a way to cover all the floor area in order to bear the floor gravity load. The shear walls should also be distributed to limit the eccentricity ratio and to maximize the torsional rigidity. All these criteria are sometimes conflicting. Nevertheless, the genetic algorithm has shown superior performance in structural layout optimization (Bourahla et al. 2019; Tafraout et al. 2019). In the present study, a special focus is put towards the capability of the genetic algorithm in controlling some design parameters by varying the constraints and the weights of the objective function related to the covered area ratio (Wco), the eccentricity ratio (Wec) and the total length of shear walls (Wlt).

For this purpose, seven case study samples of architectural configurations have been considered (see Figure 3).

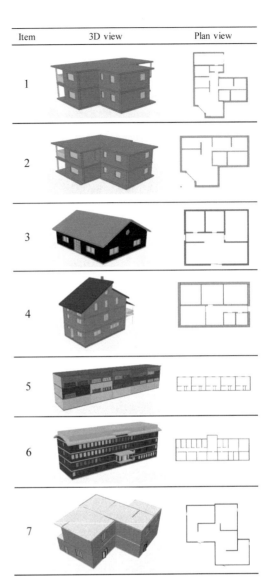

Figure 3. Architectural IFC models.

8 EFFECT OF THE SUPPORT SPAN AND THE WEIGHT OF THE COVERED AREA PARAMETER

Assuming a uniform gravity load distribution, each shear wall can bear an area around its axis delimited by a half span. An optimal distribution of the shear walls should cover with a minimum overlapping the entire slab area (Tafraout et al. 2019). Those the maximum span is used as a constraint and the covered area and overlapping are introduced in the multi-objective function. The performance of the GA is demonstrated by exploring the obtained optimized layouts for different spans and weights of the cover-overlapping parameter Wco. For short spans where it is not possible to cover the complete slab area, the GA tends to distribute the minimum total length of shear walls with minimum

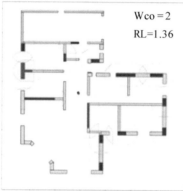

Figure 4. Shear walls distribution for increasing weight of cover-overlapping parameter (short span).

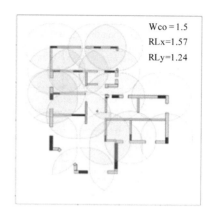
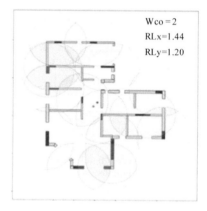

Figure 5. Shear walls distribution for increasing weight of cover-overlapping parameter (typical span).

overlapping area. The total length of shear walls is augmented with increasing values of the weights of the cover-overlapping parameter (see Figure 4).

An important feature of the GA is that once the slab is completely covered, any increase in the weight of the cover-overlapping parameter Wco does not lead to any increase of the total length of shear walls as illustrated (see Figure 5), where RLx and RLy expressed the ratio of the total length of shear walls along the x or y axis respectively, over the minimum required total length of shear walls. This reveals that the GA optimization is dominated by the shear wall length parameter, therefore it is convenient to use higher values of Wco as it does not increase much the total length of the shear walls once the slab is covered.

9 EFFECT OF THE TORSIONAL ECCENTRICITY PARAMETER

In earthquake resistant design, the torsional eccentricity ratio is an important parameter which characterizes the symmetry and regularity of a building.

Table 1. Torsional eccentricity for different buildings with different Wec.

Wec	0.1		1.0		1.5	
Eccentricity Buildings	ex (%)	ey (%)	ex (%)	ey (%)	ex (%)	ey (%)
1	5.7	6.3	2.4	1.0	0.3	0.7
2	8.1	19.8	2.8	6.8	5.3	0.6
3	2.9	37.3	10.0	12.6	0.4	6.3
4	23.1	72.7	0.0	4.8	1.9	4.5
5	6.1	4.9	0.3	12.5	0.1	0.0
6	8.4	35.0	0.3	0.9	1.0	1.3

Table 2. Variation of the total length of shear walls with respect to the weight Wlt.

Buildings	1		2		3	
Wlt	RLx	RLy	RLx	RLy	RLx	RLy
0.1	1.27	1.06	1.22	1.39	**1.71**	1.05
0.5	1.02	1.00	1.03	1.03	1.02	1.00
1.0	1.02	1.03	1.01	1.00	1.05	1.02
1.5	0.99	1.00	1.02	1.03	1.00	1.00
2.0	1.01	1.04	1.00	0.99	1.01	1.00

It has been introduced as a parameter in the objective function of the GA in order to control the structural symmetry of the building. In this study three different values 0.1, 1.0 and 1.5 characterizing low, intermediate and strong importance of the eccentricity parameter are used to investigate its effect on the optimized layouts. The resulting eccentricities of the floors of the optimized layouts for all buildings, obeyed to the variation of the eccentricity weight. Table 1 shows that the torsional eccentricity ratio may exceed 70% for a weight of 0.1 (almost uncontrolled eccentricity) but it can be kept less than 12% for a weight equal to unity and less than 6% for a scale factor weight of 1.5. Taking into account these results, it is recommended for common use a weight in the range of 0.5 to 1.0 to obtain regular structures with an eccentricity less than 15%.

10 EFFECT OF THE TOTAL SHEAR-WALLS LENGTH PARAMETER

The density of shear walls is controlled by a constraint expressed by the minimum needed total length of shear walls in each direction and a weight value in the objective function related to this parameter. An optimum structural layout is obtained for a minimum total length of shear walls. The weight Wlt corresponding to this parameter in the objective function has been scaled to 0.1, 0.5, 1.0, 1.5 and 2.0 while keeping the other parameters equal to unity. It has been noticed that the GA keeps optimizing the total length and converges to the total length limit in most cases except for Wlt = 0.1 where the total length of the shear walls exceeds appreciably the total length limit and the ratio RLx reached a maximum of 1.71 in one case (see Table 2). Figure 6 shows representative layouts of a building with almost constant RLx and RLy for values of Wlt above 0.5. Within the limit of the considered configuration samples, it can be concluded that the GA is efficient in optimizing the total length of the shear walls to its minimum value for a wide range of Wlt values. To diminish the dominance of this parameter, low values in the order of 0.1 have to be assigned to Wlt.

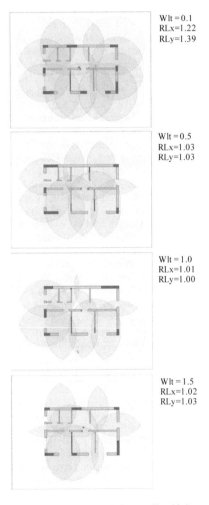

Figure 6. Variation of the total shear walls with increasing weight Wlt.

11 STABILITY AND ROBUSTNESS OF THE GENETIC ALGORITHM

Although, a limited number of seven configuration samples that has been treated, with 16 different sets of weights totalizing 96 cases, all the best solutions output by the GA were consistent with the predefined

Table 3. Comparison of optimized layout characteristics for different solution variants.

Variants	ex	ey	RLx	RLy	OA
1	2.99	2.15	1.0	1.0	2.69
2	2.12	0.64	1.0	1.0	2.53
3	0.82	0.70	1.0	1.0	2.53
4	1.42	1.16	1.0	1.0	2.69
5	3.91	1.63	1.0	1.0	2.63

criteria and the weights of the multi-objective function as demonstrated earlier in this study and in previous works (Bourahla et al. 2019; Tafraout et. al. 2019). Within the limits of these results, the GA was very reliable. To investigate further the stability and robustness of the algorithm, five optimization runs are carried out for two configurations with the same weights. The results of only one configuration is presented here as they were very similar. The obtained layouts are compared in terms of the total length ratios of shear walls (RLx, RLy), eccentricity ratios (ex, ey), overlapping area ratio (OA) and geometric distribution of shear walls. As shown on Table 3, insignificant variations of the eccentricity ratios can be noticed because of the variation of the geometric positions of the shear walls. In all variants, the maximum eccentricity ratio does not exceed 4%. The overlapping area ratio is almost constant and the total length of shear walls is identical for all variants. On the basis of these results, the genetic algorithm performs efficiently and yields stable and reliable optimized layouts with very similar structural characteristics. Because of the non-uniqueness of the solution, the GA outputs best layouts with some differences in the geometric distribution of shear walls (see Figure 7). Continuing research work is being carried out to introduce more construction provisions into the GA which will reduce the variability of the geometric distribution of the shear walls.

12 CONCLUSION

This paper advances a recent approach in generating optimal structural design of buildings within a BIM environment using intelligent methods. The protocol of this approach is first recalled with special focus on the constraints and the multi-objective function of the genetic algorithm used to produce an optimized reinforced concrete wall-slab system in full compliance with an architectural configuration given in IFC format. The challenge in finding the adequate weights of a multi-objective function of an optimization algorithm is addressed for this special case. A sensitivity analysis is carried out by varying the weights of the multi-objective function to control the torsional eccentricity ratio, the total length of the shear walls, the covered and overlapping area related to the gravity load bearing.

The obtained results show that the genetic algorithm is very efficient in optimizing the total length

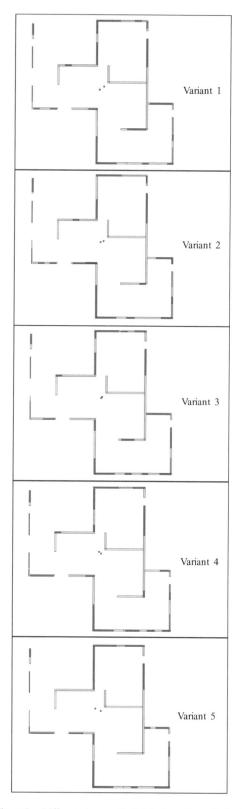

Figure 7. Different shear walls distribution of an optimized layout.

of shear walls to a minimum required value for a wide range of weights and only values as low as 0.1 can lead to higher distribution density of shear walls if needed. The structural regularity of optimized layouts with torsional eccentricity ratios lower than 15% can be achieved for weights of this parameter in the range of 0.5 to 1.0. However, for the weight of the covered-overlapping area parameter, high values of 4 have been used to fully cover the floor area of different building samples for relatively short and long spans without significant overlapping. The main outcome of this study is intended to help tuning the optimization procedure by providing weight ranges that can be used for different architectural configurations.

Finally, by comparing several variants of a layout solution it has been found that the structural characteristics are very similar which confirms the GA stability and robustness.

REFERENCES

Abrishami S., Goulding J. S., Rahimian F. P., & Ganah A. 2014. Integration of BIM and generative design to exploit AEC conceptual design innovation. *J. Inf. Technol. Constr.*, vol. 19, no. September, pp. 350–359, 2014.

Bourahla N., Tafraout S., & Bourahla Y. 2019. Performance Evaluation of an Intelligent Structural Design Process in BIM Environment: case of RC wall-slab Resisting Systems, *International Conference on Recent Advances in Civil Engineering Infrastructure (RACEI)*, Hyderabad 16–18 December 2019.

Chang Y.F., Shih S. G. 2018. Exploring multi-disciplinary communication based on generative modeling at the early architectural design stage. *Preprints 2018, 2018060230. doi: 10.20944/preprints201806. 0230.v1).*

Chi H. L., Wang X., & Jiao Y. 2015. BIM-Enabled Structural Design: Impacts and Future Developments in Structural Modelling, Analysis and Optimisation Processes. *Arch. Comput. Methods Eng.*, vol. 22, no. 1, pp. 135–151, 2015, doi: 10.1007/s11831-014-9127-7.

Gennert, M.A., Yuille, A.L. 1988. Determining The Optimal Weights In Multiple Objective Function Optimization. *Proceedings of the Second International Conference on Computer Vision*, 87–89.

Ghaffarianhoseini A. et al. 2017. Building Information Modelling (BIM) uptake: Clear benefits, understanding its implementation, risks and challenges. *Renew. Sustain. Energy Rev.*, vol. 75, pp. 1046–1053, Aug. 2017, doi: 10.1016/j.rser.2016.11.083.

Sun C., Jiang S., Skibniewski M. J., Man Q., & Shen L. 2017. A Literature review of the factors limiting the application of BIM in the construction industry, *Technol. Econ. Dev. Econ.*, vol. 23, no. 5, pp. 764–779, Nov. 2017, doi: 10.3846/20294913.2015.1087071.

Tafraout S., Bourahla N., Bourahla Y., & Mebarki A. 2019. Automatic structural design of RC wall-slab buildings using a genetic algorithm with application in BIM environment. *Autom. Constr.*, vol. 106, no. July, p. 102901, 2019, doi: 10.1016/j.autcon.2019.102901.

METIS-GAN: An approach to generate spatial configurations using deep learning and semantic building models

H. Arora, C. Langenhan & F. Petzold
Technical University of Munich, Munich, Germany

V. Eisenstadt & K.-D. Althoff
German Research Center for Artificial Intelligence, Kaiserslautern, Germany

ABSTRACT: In order to recommend architects design options, a system was developed which uses artificial intelligence (AI) methods of case-based reasoning (CBR) and deep learning. Since the system uses deep learning, it requires a sufficient amount of data for training, but currently, not enough amount of semantic building data is available publically. In this paper, a Generative Adversarial Network (GAN) is considered to generate the semantic building data to train a Deep Neural Network (DNN) to recommend design options.

1 INTRODUCTION

The development of architecture is characterised by continuous change due to social, ecological and technological conditions. Current as well as future building tasks take place against the background of the change of these framework conditions. For example, the industrialisation of the 19th century and the composite building material reinforced concrete at the beginning of the 20th century changed the formal language of architecture. The resulting approaches of classical modernism in architecture with representatives such as Le Corbusier or the Bauhaus are the starting point of the design methodology movement (Richter 2010). Building on system theory approaches, the first computer-based approaches were developed, which concentrated on design rules (Purcell et al. 1990), but were unable to formalise the complexity of architectural designs in an exhaustive manner. For this reason, the second generation of the design methodology movement in the 1970s, represented by Horst Rittel and others, did not view design procedurally as the fulfilment of requirements, but rather as an individual process that could only be described incompletely (Richter 2010). In the 1980s, mainly the digital approaches of case-based reasoning (CBR) (Kolodner 1993) influenced the building design, which led to AI research in the building industry called case-based design (CBD) in the 1990s.

The search for analogies in references of already built or designed buildings is an established method to examine ideas, to clarify design parameters or to show new ways and options for the design process. Built and planned buildings serve as a knowledge base and includes spatial arrangements as well as solutions for specific architectural characteristics. In order to achieve this task, on the one hand, approaches for the development of inherent knowledge in references are developed, together with the formalization of knowledge in digital semantic building models (BIM/IFC). On the other hand, techniques and methods for finding formal structures as well as the corresponding description and query language for spatial arrangements should be developed for an efficient determination of analogically fitting design patterns.

Most computational search methods available today rely on textual rather than graphical approaches to represent information. To address these shortcomings, (Langenhan & Petzold 2010; Langenhan et al. 2013; Langenhan 2017) introduced a novel approach which facilitates the automatic lookup of reference solutions from a repository using graphical search keys. For the search key, the notion of a building floor plan fingerprint was introduced which describes the main characteristics of a building's design. This forms the basis for assessing the similarity of different reference solutions to a specified problem and serves, accordingly, as an index for the building model repository. A generation of different variations based on these similar designs for the current design problem can enrich the building design database with new designs that can be used for search or demonstration of the temporal evolution of the design. In this paper, we present an approach for such generation that is based on artificial intelligence techniques, such as CBR and deep learning (DL).

2 PROBLEM DESCRIPTION

Over the decades, for the purposes of design development, architects have been using different types of architectural visualisations such as floor plans, sections and elevations as well as perspective presentations and physical models. However, methods of digital recommendation of inspirational designs were rarely or never used for such purposes due to the limitations of the computational methods in terms of available digital and digitalized data. This paper deals with topics in the field of building design, semantic technologies, CBR and DL in order to conceptualize, implement and evaluate a data generation solution for knowledge-supported design systems. More exactly, we propose a computational approach to create rich semantic building data by applying the hybrid CBR+DL model Metis-GAN. Metis-GAN is based on CBR and Generative Adversarial Nets (GAN) (Goodfellow et al. 2014) and can be used to provide the systems with variations of architectural floor plans that are similar to the already available ones. The approach extends the learning step "retain" of the CBR cycle (Aamodt & Plaza 1994) in order to expand the case base and integrate references as auto-completion into the design process as future work.

From today's perspective, the formalization of complex cases in architecture is not sufficiently solved and is referred to as a data acquisition bottleneck. While for the formalization of building information, object-oriented approaches such as product data modelling or BIM (Eastman 1999) have been transferred to the building industry in the 1990s, they were not used to automatically extend the data set and strengthen the semantics together with the methods of artificial intelligence.

To be able to provide architects with good recommendations, a sufficient amount of architectural floor plan data is required to train a deep neural network that can be used in tandem with a CBR manager. While data available in public datasets is scarce, a major issue is that the scientific community in architecture does not apply techniques of deep learning to synthesize floor plan data. Our approach can generate new data using Deep Generative Modelling. One of the main goals of the approach is to initiate the impact of Machine Learning and Deep Learning in the architectural domain, in the same way, these methods have impacted other fields.

3 RELATED WORK

Generally, in the field of architecture, different approaches have been used to generate floor plans. Among them are approaches for optimization and search, physically based algorithms, generative grammars, and probabilistic algorithms. In (Martin 2006) the author suggests a novel method for dynamic procedural generation of a residential unit. According to the author, their method is novel for generation of architectural models. Generally, other methods only consider generating exteriors of buildings with no interior structure or layout being considered. The proposed approach starts with the interior and uses the interior structure to dictate the exterior appearance of the building being generated.

Deep Generative Modelling is used in the field of architecture but not as often as it is applied in other fields. Over a period, different methods have been used to perform machine learning techniques to architectural data but most of the approaches consider the architectural floor plan sketches as images. In (Newton 2019), the author of the paper discusses the potentials and limitations of the GAN-based deep generative model in 2D and 3D architectural design generation tasks. The author suggests a method to predict images of architectural floor plans with specific architectural styles. The author also suggests using data augmentation to train the GAN with small training datasets.

As stated above, it is evident from the literature that data currently available in public datasets is insufficient for training deep learning models. In (Sharma et al. 2017), the authors of the paper created a training set of architectural floor plans known as ROBIN with a different set of floor plans for a different number of rooms in the dataset. While using floor plan images for deep learning methods is a promising approach, we think that the problem of non-coverage of all architectural concepts using the form of an image can occur. In the same work, the authors also proposed an AI system that applies deep learning approaches to automatically analyse building floor plan images and retrieve similar plans from a large-scale repository. The system uses a deep learning framework to extract high and low-level features from the query floor plan. Once the features are extracted, all the features will be compared with the extracted features of the floor plans from ROBIN with the subsequent application of a similarity measure. Our model uses deep learning in tandem with a CBR manager to generate variations of floor plans instead of just checking for the similarity between different floor plans.

As et al. (2018) suggested a novel method to find out a subgraph from each housing graph, this subgraph is selected using an artificial neural network (ANN). The unique ability of this ANN is that it selects a subgraph with the highest target function score. The authors also proposed a method to generate new architectural floor plans using the graphical structure representation, i.e., the rooms are depicted by nodes and edges are the connections between two different rooms. In order to generate new samples, the authors used InfoGAN (Chen et al. 2016). While the approach presented in (As et al. 2018) is quite similar to the approach considered in this paper, the main difference is that our approach considers different types of connections between two different rooms. Huang & Zheng (2018) used a modified version of GAN known as pix2pixHD (Wang et al. 2018) which is a generative method for generating images using image synthesizing and image segmentation. The authors of the paper performed recognition of different sections of floor plans by labelling them first and then generating the same samples by inputting the floor plan to the network. The output of the network is the labelled floor plan. The authors also generated floor plans using the labelled floor plan images.

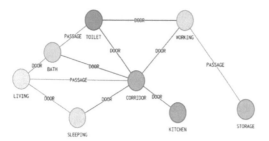

Figure 1. An example of room configuration.

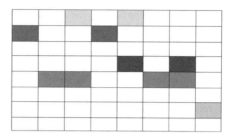

Figure 3. An example of room configuration as an image.

	121		151		
611		672			
			751		731
	561	521		541	531
					392

Connection map

Figure 2. An example of room configuration as connection map.

4 APPROACH

To solve the above-stated problem, we propose a method to generate new architectural floor plans using Deep Generative Modelling, an algorithmic setup that can be used to build Artificial Neural Networks to generate new data based on data provided to the model. The Deep Generative Approach (Goodfellow et al. 2014) is GAN, according to this approach, a generative model is pitted against an adversary, in our case, a discriminative model. The GAN structure is based on a min-max game, in which the generative model aims at generating data as similar as possible to the true data, i.e. minimize the loss between the generated and true data while the discriminator tries to maximize the likelihood of distinguishing between fake and true data points. This approach is used to generate images from the data.

In the second approach, we only considered the 192 samples which we had received after the preprocessing step. This, without resampling, model took 60 iterations to converge to a point where the generator and the discriminator loss were in equilibrium with values 0.69 and 1.38 respectively. The batch size for each iteration was 192 as we wanted to make sure that the model gets trained on the training data as much as possible because of such a small number of training samples. When it comes to the architecture of the generator model (see Figure 4), we specifically used

Layer (type)	Output Shape
dense_21 (Dense)	(None, 400)
batch_normalization_35 (Batc	(None, 400)
leaky_re_lu_36 (LeakyReLU)	(None, 400)
dense_22 (Dense)	(None, 12544)
batch_normalization_36 (Batc	(None, 12544)
leaky_re_lu_37 (LeakyReLU)	(None, 12544)
reshape_12 (Reshape)	(None, 7, 1, 7, 256)
conv_lst_m2d_5 (ConvLSTM2D)	(None, 1, 7, 256)
batch_normalization_37 (Batc	(None, 1, 7, 256)
leaky_re_lu_38 (LeakyReLU)	(None, 1, 7, 256)
conv2d_17 (Conv2D)	(None, 1, 7, 512)
batch_normalization_38 (Batc	(None, 1, 7, 512)
leaky_re_lu_39 (LeakyReLU)	(None, 1, 7, 512)
conv2d_18 (Conv2D)	(None, 1, 7, 512)
batch_normalization_39 (Batc	(None, 1, 7, 512)
leaky_re_lu_40 (LeakyReLU)	(None, 1, 7, 512)
conv2d_19 (Conv2D)	(None, 1, 4, 128)
batch_normalization_40 (Batc	(None, 1, 4, 128)
leaky_re_lu_41 (LeakyReLU)	(None, 1, 4, 128)
flatten_7 (Flatten)	(None, 512)
dense_23 (Dense)	(None, 36)
reshape_13 (Reshape)	(None, 1, 36)
batch_normalization_41 (Batc	(None, 1, 36)
re_lu_5 (ReLU)	(None, 1, 36)

Figure 4. The architectures of the Metis-GAN components: Generator.

Convolutional LSTM (Shi et al. 2015) to help the generator model get some kind of understanding about the structure of the training data.

In the last layer of the generative model, a ReLU activation function was used instead of the

```
Layer (type)                    Output Shape
=================================================
dense_24 (Dense)                (None, 1, 100)

dropout_8 (Dropout)             (None, 1, 100)

leaky_re_lu_42 (LeakyReLU)      (None, 1, 100)

dense_25 (Dense)                (None, 1, 784)

dropout_9 (Dropout)             (None, 1, 784)

leaky_re_lu_43 (LeakyReLU)      (None, 1, 784)

reshape_14 (Reshape)            (None, 28, 28, 1)

conv2d_20 (Conv2D)              (None, 28, 28, 256)

dropout_10 (Dropout)            (None, 28, 28, 256)

leaky_re_lu_44 (LeakyReLU)      (None, 28, 28, 256)

conv2d_21 (Conv2D)              (None, 28, 28, 512)

dropout_11 (Dropout)            (None, 28, 28, 512)

leaky_re_lu_45 (LeakyReLU)      (None, 28, 28, 512)

conv2d_22 (Conv2D)              (None, 28, 28, 512)

dropout_12 (Dropout)            (None, 28, 28, 512)

leaky_re_lu_46 (LeakyReLU)      (None, 28, 28, 512)

conv2d_23 (Conv2D)              (None, 14, 14, 256)

dropout_13 (Dropout)            (None, 14, 14, 256)

leaky_re_lu_47 (LeakyReLU)      (None, 14, 14, 256)

flatten_8 (Flatten)             (None, 50176)

dense_26 (Dense)                (None, 1)

dropout_14 (Dropout)            (None, 1)
```

Figure 5. The architectures of the Metis-GAN components: Discriminator.

LeakyReLU as at the end we wanted the predicted values to be greater than or equal to zero as it was important for the model to make sure that in the end the predicted connection map can again be transformed into a workable aGraphML.

The last layer of the discriminator model (see Figure 5) has a sigmoid as the activation function as we wanted to get the probability of a sample to be fake or true.

5 RESULTS

Given the fact that this is the first implemented version of the Metis-GAN approach, the results (see Figures 6–9) can be seen as promising, however, much improvement of the model is required to achieve better structured and more diverse room configurations. With

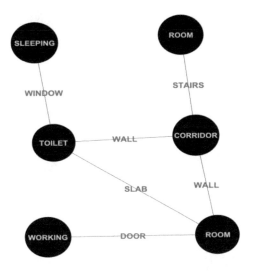

Figure 6. Example of room configurations generated with Metis-GAN.

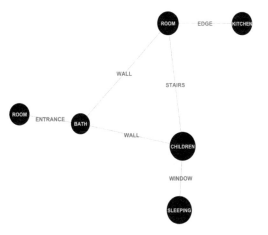

Figure 7. Example of room configurations generated with Metis-GAN.

this approach we intend to initiate a research direction towards a day where the generation of new architectural floor plans can be executed fully automatically using the machine learning methods.

The current approach generates architectural models which have 6 rooms and 6 edges in total. A major drawback is that it generates certain nodes in the room configuration that have the room type as a generic room and some of the edges which connect two different rooms are also sometimes of the generic edge type Edge. While the results achieved with the above stated approach might not be the most optimal, it is still a step forward in the intended direction and the conclusions made from the results need to be considered in the future approaches to make sure that such problems can be avoided as long as possible.

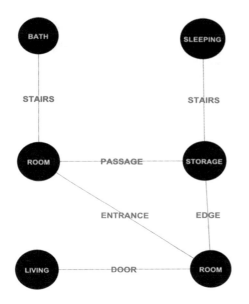

Figure 8. Example of room configurations generated with Metis-GAN.

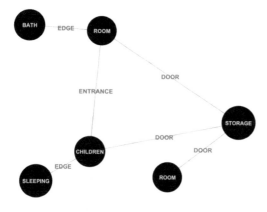

Figure 9. Example of room configurations generated with Metis-GAN.

For the current model, improvement lies in the fact that with just 192 samples we can generate samples which are promising but still don't adhere to architectural constraints.

6 CONCLUSION & FUTURE WORK

According to our understanding of the current problem and the conclusions made from the experiment results, the solution is most probably not to develop a completely new deep learning approach, instead, better data structures and ML-compatible floor plan representations should be developed that we can use to feed the deep learning model.

One possibility would be to use the image representations of the floor plan graphs, and then try to generate new images, for example, using again a Generative Adversarial Network, that was originally developed to generate and evaluate images. The other method that can be used to improve the predictions, is apply the Natural Language Processing (NLP) and Sequence Learning techniques to convert the room configurations in the aGraphML format into a sequence (i.e., sentence) and then try to generate new sentences which can be later converted back to aGraphML. In this approach, we can explicitly share the constraints that are required to be considered to generate architectural floor plans.

ACKNOWLEDGEMENT

This research is part of the research project metis-II, funded by the German Research Foundation (DFG).

REFERENCES

Aamodt, A. & Plaza, E. (1994). Case-based reasoning: Foundational issues, methodological variations, and system approaches. *AI communications*, 7(1): 39–59.

As, I., Pal, S., & Basu, P. (2018). Artificial intelligence in architecture: Generating conceptual design via deep learning. *International Journal of Architectural Computing*, 16(4): 306–327.

Chen, X., Duan, Y. & Houthooft, R. (2016). Info-GAN: Interpretable representation learning by information maximizing generative adversarial nets. In:*NIPS*, https://arxiv.org/abs/1606.03657

Eastman, C.M. (1999). Building Product Models: Computer Environments supporting design and construction. Boca Raton, Fla: CRC Press.

Goodfellow, J. I., Abadie, P. J., Mirza, M., Xu, B., Farley, W. D., Ozair, S., Courville, A. & Bengio, Y. (2014). Generative Adversarial Networks, arXiv:1406.2661 [stat.ML].

Huang, W. & Zheng, H. (2018). Architectural Drawings Recognition and Generation through Machine Learning.

Kingma, P. D., & Ba, J. (2014). Adam: A Method for Stochastic Optimization, Published as a conference paper at the *3rd International Conference for Learning Representations*.

Kolodner, J. L. (eds) (1993). Case-Based Learning (Vol. 10, No. 3). Springer Science & Business Media.

Langenhan, C., Petzold, F. (2010). The Fingerprint of Architecture: Sketch-Based Design Methods for Researching Building Layouts through the Semantic Fingerprint of Floor Plans. *Architecture and Modern Information Technologies*: 4–13.

Langenhan C., Weber M., Liwicki M., Petzold F., Dengel A. (2013). Graph-based retrieval of building information models for supporting the early design stages. *Advanced Engineering Informatics*. Volume 27, Issue 4: 413–426.

Langenhan C. (2017). Datenmanagement in der Architektur. Untersuchung zur Organisation von Entwurfsinformationen in IT-Infrastrukturen und Nutzungsmöglichkeiten in wissensbasierten Systemen. Dissertation – Technische Universität München. ISBN 978-620-2-32014-6.

Martin, J. (2006). Procedural house generation: A method for dynamically generating floor plans. *Symposium on Interactive 3D Graphics and Games*.

Newton, D. (2019). Generative Deep Learning in Architectural Design. Technology |Architecture + Design, 3:2: 176–189. DOI: 10.1080/24751448.2019.1640536

Purcell, P. A. Mitchell, W. J. & McCullough, M. (1990). The electronic design studio: Knowledge and Media in the computer era. Cambridge: MIT Press.

Richter, K. (2010). Augmenting Designers' Memory: Case based reasoning in Der Architektur. Dissertation. Berlin: Logos-Verlag.

Sharma, D., Gupta, N., Chattopadhyay, C., & Mehta, S. (2017). DANIEL: A Deep Architecture for Automatic Analysis and Retrieval of Building Floor Plans. *14th IAPR International Conference on Document Analysis and Recognition (ICDAR)*, Kyoto: 420–425.

Shi, X., Chen, Z., Wang, H., Yeung, Y. D., Wong, W. K. & Woo, C. W. (2015). Convolutional LSTM Network: A Machine Learning Approach for Precipitation Nowcasting, arXiv:1506.04214 [cs.CV].

Wang, T., Liu, M., Zhu, J., Tao, A., Kautz, J. & Catanzaro, B. (2018). High-Resolution Image Synthesis and Semantic Manipulation with Conditional GANs. *2018 IEEE/CVF Conference on Computer Vision and Pattern Recognition:* 8798–8807.

Practical experiences from initiating development of machine learning in a consulting engineering company

T. Alstad & E. Hjelseth
Norwegian University of Science and Technology (NTNU), Gjøvik, Norway

ABSTRACT: There are high expectations for the impact of Machine Learning (ML) in general. However, applicable solutions in the AEC/FM industry are hard to observe. This paper shares experiences from a pilot project in a consulting engineering company utilising existing geotechnical data to develop a ML-based solution based on supervised learning. Even if the case is domain specific, the experiences from the process should be of general interest. This study documents that relatively much can be done by limited investments. Access to enough relevant data for development (training) of algorithms was identified as barrier, in addition to a time demanding process for preparing relevant data. Applicable competency includes both to understanding of the problem, exploring relevant data and developing the digital ML-software solution. This case shows that learning by doing can be a way to develop the relevant competency. ML-projects should therefore be regarded as a learning process to be started promptly.

1 INTRODUCTION

Generally, based on the high expectations on the impact of machine learning (ML), one would expect multiple practical examples in 2020. In principle, the technology is available and is used in other industries like health science, finance/banking and in industries offering services for analysis and prediction as their service.

ML in this study is understood as the learning process combining large numbers of algorithms and large data sets/Big Data (BD). More simply, ML is the intersection of Artificial Intelligence (AI) and BD. There are in general three types of ML: Supervised learning, Unsupervised learning or Reinforced learning (Salian 2018). This study uses supervised learning based on classification of data.

This paper shares experiences from a pilot project in a consulting engineering company that aimed to use existing geotechnical data to develop a Machine Learning (ML) based solution. Even if this study uses geotechnical data as a source, the experiences from developing ML based solutions should be of general interest to the Architects, Engineers, Contractors, and Facility Management (AEC/FM) industry. The industry partner is a lager multidisciplinary consulting engineering company in Norway.

So far has it been hard to find use ML-based solutions in the AEC/FM- industry. The study includes discussion about why ML is not being used as widely as in other industries. The authors practical experiences from the development process is also included in this study.

Development of the ML solution reviewed in this paper was based on Python (Python 2020) programming language, with libraries for data analytics, data exploration, parsing files, data extraction, machine-learning frameworks and the gradient boosting algorithms LightGBM (Ke et al. 2017) and CatBoost (Dorogush et al. 2018).

Utilising ML in the AEC/FM industry is expected to have great potential in terms of productivity, automatization, data driven decision making with more informed and accurate information. This can result in improved productivity, more automatization, data-driven decision making with more informed and accurate information. Dealing with increased complexity and information flow can also enable more fact-based sustainability assessments.

Development of ML solutions can be both beneficial to ongoing developments within BIM, 4D and 5D development (e.g. within time and cost scheduling, analysis and simulations). Generative design and parametric modelling are also a candidate for utilizing ML-based solutions.

The Gartner report by Costello (2019) describes the top challenges in adopting AI for respondents; these comprise of lack of skills (56%), understanding AI use-cases (42%), and concerns with data scope or quality (34%). The most significant were:

"Finding the right staff skills is a major concern whenever advanced technologies are involved".

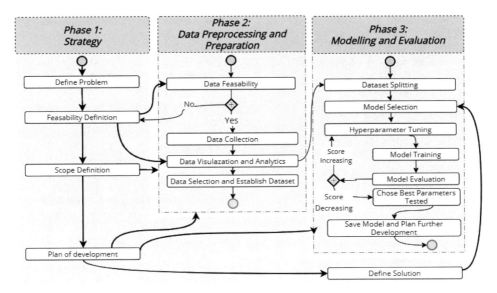

Figure 1. Proposed phases and generalized content of the machine-learning development for this project.

The reports say that the skill gaps can be addressed using service providers, partnering with universities, and establishing training programs for existing employees.

2 METHOD

This study uses combinations of a descriptive approach ("What is done in practice"), a normative approach ("How it should be done ideally") and a constructive approach ("Why is this relevant") to show the importance and potential of ML use in AEC/FM. The objective is to share the authors' experiences and reflections on the ongoing pilot process to identify recommendations.

This study intends to explain the experiences from the development processes. It is not to present a learning book in ML, or to argue for best methods.

Machine learning model development is an iterative process between model improvement and feature development. In the reviewed experience, the chosen processes reviewed in this paper are the following: (1) strategy, (2) data pre-processing and preparation, (3) dataset splitting, modelling and evaluation.

The discussions section is based on the expectation of high priority to invest in development of suitable solutions. This perspective can be a bit critical to the current awareness and the willingness to prioritise and prepare. The discussion section is, therefore, a relatively large part of this paper

The three phases and their connections are visualised in Figure 1. Straight arrows illustrate the processing within each phase, while the curved arrows illustrate how processing of relevant content contributes to the development of the ML-solution findings.

This paper is the first paper in a series of papers within ML. The focus is on the process and only a few of the findings are presented here. The master thesis by the first author will be submitted in June 2020.

2.1 *Phase 1 – Strategy*

The research started when a problem was defined after meeting company leaders in the geoscience department. The meeting was initiated by the realization of vast amounts of data from soil surveys being stored on project file servers. The overall problem definition and agenda were how to utilize the data. It led to the defined problem of whether it was possible to use data from drill rigs and laboratory reports to predict soils or gain a deeper understanding of soil shifts related to construction projects. To further define the strategy, we explored previous projects and how the data from these projects was stored and managed on different file servers. These experiences gave insights to feasibility, scope and problem solution. The collected data files (txt and native Excel formats) were investigated to give insights on how and where information was located inside the files. Furthermore, it also gave insight into which modules were needed in Python to extract information. The next step involved searching and gathering the project root folders manually from company project file servers to a local environment, and this was done to ensure availability and fast prototype problem solution, instead of working through a slow server interface.

2.2 *Phase 2 – Data preparation and pre-processing*

Two modules were created in python based on the findings in the strategy phase. The total number of files collected was 18632. The numbers for files should

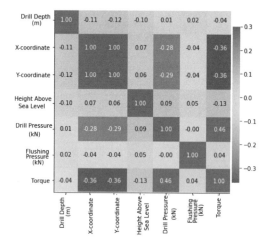

Figure 2. Correlation matrix to visualize the dataset.

indicate that there was enough data to fulfil this project into a practical application. However, when analysing the soil predictions, only 27 percent of the data initially collected was used. Studies by Walker (2012) within big data analytics estimate that only 30 percent of collected data is of value. This elaborates the importance to have a strategic plan for information management involving systems which detail the collection, storage, analysis and distribution of data created in data structures.

To improve the insight in the dataset, a correlation matrix was developed to visualize the dataset, as illustrated in Figure 2.

The use of labelled soil data is based on traditional domain knowledge. How this split is done can influence the results. The algorithms used, LightGBM and CatBoost are of the "gradient boosted" type, their criteria are weighted differently. One benefit with boosted algorithms is that most relevant criteria will get increased weight, and by this created more relevant outcome.

The first Python module created lists of: (1) every folder and file paths, (2) files within the same project referenced by project number given in folder name, (3) corresponding drill data and laboratory analysis files referenced by drill hole number given in file name, (4) incompatible pairs that did not find a reference.

The second module extracted information from the data files (laboratory analysis and drill data); it was programmed to extract the rows and columns containing coordinates, pressure data, torque, drill depths, heights. The dependent variable to be predicted was the soil type. This was housed in an empty data frame made earlier with the before mentioned features. The data extracted was then cleaned from irrelevant data/registrations, which could affect the performance of the algorithm, such as incomplete sets.

Dataset visualisation was conducted to ensure that rare occurrences were identified and removed. At the end of this phase, a total of 137 file pairs were found compatible with the machine-learning modelling phase.

2.3 Phase 3 – Dataset splitting, modelling and evaluation

Based on the final dataset feature (visualization and analytics) it was performed for the training, testing and validation. The amount of data containing the soil type clay accounted for 71% of the data selected, which indicates that the data to be trained on should have over 70% in training, so when the test set not only consists of clay. The split was performed randomly with 70% for training, 30% for testing and validation. The machine learning modules were then implemented.

The case study used supervised learning, which in terms means that the data trained on had labels on the ground truth. The algorithms were adjusted under an iterative process with tuning parameters of the algorithms, where you try out some parameters, and if the prediction results get better you know you are tuning it correct.

The training was then initiated for both algorithms (LightGBM and CatBoost) to evaluate the models' output, given the feature importance as a starting point for adjusting the algorithm and to see if any features should be dropped. The performance of both algorithms was relatively similar related to predict the outcome. One significant difference was the processing time, where LightGBM used approx. 20 min. and CatBoost used approx. 35min. in processing time after each adjustment of the settings. This implies that it is faster, one can do more adjustment of settings with LightGBM than CatBoost. The explanation for the differences in processing time is related to the structure and leaves in the gradient goosing decision tree in the algorithms.

The torque and flushing pressure scored a low importance to the model output, but the prediction accuracy decreased when removed, which advocated the decision to not remove them. Based on the results given a search algorithm called GridSearchCV (GridSearchCV 2020) was initiated to search for the best configurations of the machine learning algorithms.

For both models they scored an accuracy of 97% though the results were suspected of giving partial results (overfitting), which was confirmed by cross-validating the accuracy of 10 random samples, which gave an accuracy of 74% in both algorithms; LightGBM and CatBoost.

Cross-validation (CV) is a technique which involves reserving a particular sample of a dataset on which you do not train the model. Later, you test your model on this sample before finalizing it.

CV is used to evaluate estimation performance. Learning the parameters of a prediction function and testing it on the same data is a methodological mistake: a model that would just repeat the labels of the samples that it has just seen would have a perfect score but would fail to predict anything useful on yet-unseen data. This situation is called overfitting.

To avoid overfitting, it is common practice when performing a (supervised) machine learning experiment to hold out part of the available data as a test set (SCIKIT 2020). The results of 74% is therefore a good indication that the quality of the dataset has improve. The results should be as high as possible. Without CV, it would be easy to be impressed by the 97% score of accuracy which is influenced by overfitting.

2.4 Overview

There were limited differences in the performance of both algorithms; LightGBM and CatBoost. The feature importance was similar and indicates that the data conforms to the required reliability and the features share correlations towards the given problem (dependent variable).

The dataset split should be more varied for further development. It has been recommended to collect more data on other soil types than clay and focus on the ones most critical to identify from a production perspective.

These results illustrate that ML can be regarded as data and algorithms, or $KL = BD + AI$. It is easy to be fascinated by the power in the analysis and how you can train the algorithms to very high accuracy (97% in our test case) by forgetting the influence of the dataset. The CV results indicate a high accuracy due to overfitting. A large dataset is a good start, but one should according to Walker (2015) be aware that only 30% of data will be useful in practice. The dataset can therefore be too low for enabling an applicable solution for trustworthy industrial uses. The total time for training the algorithm in this case study was approximately ten days. Adjusting the settings was based on trial and error to find an optimal value. The optimisation ended when the total performance did not increase. However, identifying relevant types was supported by literature studies and experience.

3 DISCUSSION

This discussion focusses on the experiences of the ML development process using an example from a Norwegian consulting company. The authors question why this type of development has not started before a master's student with mixed competency from the new Master programme in Digital building processes at NTNU (NTNU 2020) gave it a try?

Even if the data has been accessible, it has not been used for systematic studies or development before this pilot was started by the master's student. The geotechnical projects are often small and done by a limited number of practitioners in a short time. These studies are done before large investments are put into the project. Each single project is therefore not critical, and well-established methods can therefore be used. The collected dataset is not prepared for ML, in addition to not knowing if they are useful as they are.

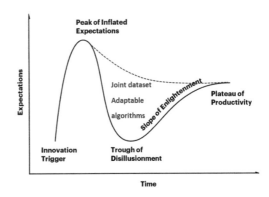

Figure 3. oint dataset and adaptable algorithms for enabling ML (based on Gartner's Hype curve).

The AEC/FM is in general behind other industries when it comes to digitalisation. This position can be an advantage because it enables reuse of relevant types or solutions from other industries. It also puts resources into the process of adapting to the needed appropriate solutions. This does not imply delaying the progress, instead the authors suggest and exemplify that it's time to start with the foundations now.

Large datasets are the foundation for ML, and awareness of this significant aspect can save cost and improve quality.

Experiences show that preparing dataset takes around 80% of the time in developing ML based solutions. It is also essential to be aware of that data preparation for machine learning still requires humans. Our experiences are also confirmed by Walch (2019) and the Cognilytica report (Cognilytica 2020).

In this respect, it is possible to predict that sharing data could be a way to lower the threshold to start exploring and developing ML-based solutions.

The learning aspect is important because it can be hard to know what type of information you need in what format. Practical experiences with exploring ML solutions can give the answer in a better way, as opposed to just traditionally thinking about what is used for manual professional assessment.

The solution could also be seen as concerning characteristic with the AEC/FM industry; there are no dominating actors with significant market shares, which should support new business models where one focus more on economy sharing economy principles, used for manual professional assessment. Data is a recourse that gets value by sharing (Moody & Walsh 1999). This situation is illustrated in Figure 3.

4 SUMMARY OF EXPERIENCES

Developing ML-based solutions should be regarded as a multidisciplinary and multi activity-based learning process. ML is not a production process towards a predefined program. This is one of the most crucial experiences in this pilot study.

This paper explores several possibilities and issues related to enabling ML-based solutions. ML requires large amounts of data (big data) to predict reliably. Even if ML, in principle, can use all types of data, including unstructured data. However, this is a time-consuming process to prepare a processable dataset for training and validation. The preparation task took approximately 80% of the time in developing the ML solution. The preparation of dataset should, in theory, be one of the first things to start with, but the relevant requirements for data are often discovered as part of the development process. The lack of relevant datasets can be addressed as a potential joint effort for companies and organisations in the AEC/FM-industry.

Development of algorithms by use of existing low-cost software solutions like Python and related plug-in for optimisation of algorithms enables new opportunities for the development of ML-based solutions. Development of algorithms requires collaboration between the ML developer and the geotechnics specialist to identify relevant topics in the easiest way.

This is an initial study in a rapidly changing theme, and the authors do not have enough evidence to draw hard conclusions. However, this study has revealed some interesting observations for further discussions and studies, especially within the process of development and expected outcomes.

However, there is no doubt that digitalization in general and ML is unique and has the capacity to change the AEC/FM-industry. This transformation can result in significant impacts. The exciting perspective is whether the AEC/FM industry enables to invest enough to get synergic effects of pervasive digitalization. Increased awareness of the real potential of advanced digitalization is essential to avoid that the development is flattening out because an appropriate solution is reached. We still see much manual solution in both modelling and in file-based exchange of information. This is both time consuming and error prone. In this respect ML stands for a radical change where the interesting perspective is what we can learn the machine (software) to do automatic, and not software we have learn.

It is important to be aware that technology is just an enabler, not a driver for change in itself (Owen et al. 2013). The sense of emergency to identify drivers for change is critical to do something about today's situation.

In this context, preparations for relevant datasets is one of the most important elements that need to be developed. As we experienced, this study used data from 18632 files, but this is not enough to get high enough quality on the algorithms. There are two reasons for this limited usefulness. First is the general one; that only approximately 30 of data is useful (Walker 2015), we reached 27%.

The second one is that quick clay, the fraction of the soil that is of most interest, is only a very limited fraction of all samples of soil types. Therefore, we do not have enough data to train the algorithms to be accurate enough for practical use.

This was not discovered until we started developing the case. However, it was discovered early enough to avoid large investment in further development. The aim of this project will be changed towards other types of soil (or other criteria) which can be useful in practical applications.

The window of opportunity within ML is open and the AEC/FM has now the possibility to start preparing datasets.

ML is also about learning, not only about data set and algorithms. General experiences by collaboration in multidisciplinary teams should therefore be a way to enhance the learning process. However, experiences for just hiring ML-experts from IT-related consultancy companies has not (in these engineering consultancy company) resulted in applicable contributions. These companies are good in selling in the benefits by examples from other industries, by examples (and by this experiences) from AEC/FM related industry is missing. The impression is that they are "solving the wrong" problem, and their solutions is not applicable. Another aspect is of course the cost by hiring these types of companies. These observations do therefor indicate that they need increased domain knowledge – or that ML is a more fundamental learning process then we are aware of.

Understanding of long-term development can be illustrated by Gartner (2017) where the management scheduled a one-year for plan and begin pilot before they launch the solution. How long time it really takes shows that planning and piloting takes three years before launch happens. The technology itself is available, but how to use it in the optimal way on the the cases with most impact takes time to explore.

When developing ML-based solutions, this should be measured after its value for continuous learning, rather than the usefulness of a single predefined solution. Machine Learning supports Human Learning and vice versa in all phases in the learning process. This implies developing internal competency and work in multidisciplinary teams. The attitude for implementing ML could, therefore, be summarised as following:

"Just learn it! – Just do it! – Just learn from doing it!"

REFERENCES

Cognilytica (2020). Data Preparation & Labeling for AI. Document ID: CGR-DLP20. https://www.cognilytica.com/2020/01/31/data-preparation-labeling-for-ai-2020/.

Costello, K. (2019). Gartner Survey Reveals Leading Organizations Expect to Double the Number of AI Projects In Place Within the Next Year. Gartner Newsroom, Press Releases, July 15, 2019 Available at: https://www.gartner.com/en/newsroom/press-releases/2019-07-15-gartner-survey-reveals-leading-organizations-expect-t.

Dorogush, A. V., Ershov, V., & Gulin, A. (2018, October 24). CatBoost: gradient boosting with categorical features support. LearningSys. Retrieved April 7, 2020, from https://learningsys.org/nips17/assets/papers/paper_11.pdf.

Gartner (2017). Annual Enterprise survey, Build the AI Business case, Gartner group.

GridSearchCV (2020). sklearn.model_selection.GridSearchCV. scikit learn. Available at https://scikit-learn.org/stable/modules/generated/sklearn.model_selection.GridSearchCV.html.

Ke, G., Meng, Q., Finley, T., Wang, T., Chen, W., Ma, W., Ye, Q., & Liu, T. (2017). LightGBM: A highly efficient gradient boosting decision tree. In Advances in neural information processing systems, pp. 3146–3154. Retrieved April 7, 2020, from https://papers.nips.cc/paper/6907-lightgbm-a-highly-efficient-gradient-boosting-decision-tree.

Moody, D.L. & Walsh, P. (1999). Measuring the Value Of Information-An Asset Valuation Approach. In ECIS, June, pp. 496–512.

NTNU (2020). Master programme in Digital building processes, Masterprogram (Sivilingeniør) 2-årig, Gjøvik, Byggog miljøteknikk – Studieretning: Digitale byggeprosesser. Available at: https://www.ntnu.no/studier/mibygg/digitale-byggeprosesser.

Owen, R., Amor, A., Dickinson, J., Matthjis, P., and Kiviniemi, A. (2013). Research Roadmap Report CIB Integrated Design and Delivery Solutions (IDDS), CIB Publication 370, International Council for Building. ISBN 978-90-6363-072-0. http://site.cibworld.nl/dl/publications/pub_370.pdf.

Python (2020). Python Software Foundation. Available at: https://www.python.org/.

Salian, I (2018). What's the Difference Between Supervised, Unsupervised, Semi-Supervised and Reinforcement Learning?, NVIDIA, https://blogs.nvidia.com/blog/2018/08/02/supervised-unsupervised-learning/.

SCIKIT (2020). 3.1. Cross-validation: evaluating estimator performance, https://scikit-learn.org/stable/modules/cross_validation.html.

Walch, K. (2019). Data preparation for machine learning still requires humans, Cognilytica, https://searchenterpriseai.techtarget.com/feature/Data-preparation-for-machine-learning-still-requires-humans.

Walker, R. (2015). From Big Data to Big Profits, Success with data and analytics, Oxford University Press, ISBN: 9780199378326.

Digital twins and cyber-physical systems

A cyber physical system for dynamic production adaptation

M. Polter, P. Katranuschkov & R.J. Scherer
Technical University Dresden, Dresden, Germany

ABSTRACT: This paper introduces the Geo Production 4.0 platform, a cyber physical system (CPS) for dynamic production adaptation based on predictions of system development with focus on all types of basic structures such as deep construction pits, tunnels, underground tubes or culverts. The results show that static problems resulting from inaccurate predictions during the construction process can be identified early with the help of CPS and solved inexpensively by timely adaptation of the production process.

1 INTRODUCTION

Soil behavior is difficult to predict and is usually estimated on the basis of soil samples, which creates many problems. The samples are often spatially far apart, which causes great uncertainty due to interpolation between the samples. Changes in the soil parameters due to external influences are also difficult to predict because the samples are also far apart in time.

One attractive solution is the continuous measurement of the ground structure behavior at easy to attach points instead of analyzing the soil material and back calculation of the structural mechanical system changes to identify the soil mechanic system and the related soil material law. In this iterative process, these material laws are continuously adapted to the actual soil system behavior and the accuracy is continuously increased. Due to the high number of unknowns, this inverse method can neither be solved numerically nor analytically, but only by the scenario method, i.e. parameter studies by statically calculating model variants. This large number of system identifications can only be carried out efficiently in real time with a system that has the appropriate IT resources and parallelization options.

The Geo Production 4.0 platform is a cyber physical system (CPS) for dynamic production adaptation based on continuous measurements and simulation-based prediction of the structural-mechanical behavior of the foundation body and subsoil. The digital twin of the examined object is continuously refined, thereby enabling increasingly precise predictions.

The remainder of the paper is structured as follows: First, the Geo Production System is classified in the area of cyber physical systems. Then, in addition to the software architecture (section 3), the individual components of the workflow are considered (section 4). Section 5 is dedicated to the concept and implementation of the digital twin as a collaborative database for the entire process. This is followed in section 6 by technical details of the workflow engine as the central core component of the system. The use of the models and simulation data created during the workflow for the adaptation of the production process as well as an evaluation of the developed composite platform are treated in section 7 and 8. The paper concludes in section 9 with a summary of the knowledge gained and an outlook for future extensions.

2 CYBER PHYSICAL SYSTEMS

With the development of the Internet of Things (IOT), the virtual world of software and services and the physical world of objects are moving closer together (Seiger et al. 2019). With the advent of CPS, physical objects can be automatically converted into digital representations and their state can be compared in real time. The resulting concept of the digital twin, i.e. the digital representation of objects in the physical world, allows the investigation and prediction of the behavior of these objects under changing conditions, often even before they are actually completed.

The inclusion of different views on objects based on sensors and the use of advanced information analytics enable networked systems to interact efficiently with each other at a high level. This connection of the physical "factory" with the cyber computational space forms the basis for the expansion of traditional manufacturing processes to Industry 4.0 (Jiang 2018).

According to Alam and El Saddik (2017) CPS have the following key properties: computation, control and communication (3C). This simple characterization is extended in various fields of science and engineering in a domain-specific manner by additional features (Liu et al. 12017). Lee emphasizes the integration of calculation processes and the control of physical processes, which affect each other vice versa (Lee 2007). In (He 2010) CPS are understood as controllable, credible and scalable networked physical equipment systems, the integration of which is based on environmental perception. Gonzalez et al. (2008) focus on the monitoring and control of physical systems based on the computations of networked engineering systems. Since CPS combine different domains from computer science, mechanical engineering and simulation engineering, they offer good conditions for collaborative software systems for engineers from various fields.

Seiger et al. (2018) have designed a system for self-adaptive workflows, in which inconsistencies between digital and real world objects are automatically corrected in special compensation actions based on sensor and context information. If changes in the physical world are detected by sensors, the compensation actions automatically synchronize the digital representation. In the Geo Production System, the refinement of the digital twin is a main goal and many of the workflow steps serve this purpose only, whereby sensor and context information is also used. Unlike in (Seiger et al. 2019), the workflow is not interrupted by external sensors, but their values are queried at specified points in the workflow and then used to adapt the digital object. Our object representation can therefore not be called a "smart object" because changes in the physical world do not automatically trigger its update in real time as in Seiger et al. (2018).

Since the use of cloud resources is already a central part of many software systems, the integration of the digital twin into the cloud infrastructure creates the actual bridge between the physical layer and the application layer of CPS (Alam & El Saddik 2017). The advantages of integrating cloud technologies into the CPS cyber layer include scalability in terms of storage, computation power and the availability of cross domain communication capabilities. This is where the Geo Production platform comes in: cloud-based remote services continuously update individual components of the digital twin based on simulation (see section 5).

Perera et al. (2014) present a model in which the IoT is described as Sensing-as-a-service (SenAS). In this model, sensors are provided by various actors (owners), who provide the data of the physical environment as sensor services. This perspective plays a crucial role in the Geo Production platform, since the sensor data from third-party providers is integrated into the workflow in a service-based manner.

This overview has shown that the Geo Production platform that will be presented in the following fulfills the characteristics of a CPS. Most systems focus on sensor-based real-time synchronization of the digital twin with the physical object in order to control automated processes. In the system presented here, however, this is not the only focus. In addition to the sensor-based update, the Geo Production platform also continuously refines the digital twin based on simulations, which enables predictions to be made about the physical change in the object. This means that not only the processes directly related to the object (here the excavation model), but also auxiliary processes that run depending on the current state of the object, can be specifically simulated and adjusted in advance.

3 GEO PRODUCTION 4.0 PLATFORM ARCHITECTURE

The Geo Production 4.0 Platform is a web-based CPS for dynamic production adaptation based on predictions of system development. A core function is the implementation of an iterative, simulation-based system identification method for dynamically changing systems, as they result from the construction progress. This allows the high number of safety factors to be reduced while maintaining the static safety level. In addition to the automatic generation of model variants, a continuous comparison with sensor-based measurements is also carried out, whereby the digital twin of the monitored structure is constantly refined and enables ever more precise predictions of system behavior.

The software architecture is component based and encapsulates various services in modules. Each module has a specific task within the platform, such as provision of sensor data, implementation of system identification, management of the digital twin or control of the workflow execution. This architecture enables the interchangeability of individual components with others with the same or similar functionality as well as the integration of remote web services as functional components. Depending on the current configuration, it can also be understood as a composite web platform, similar to the concept of mashups. This allows variants of the workflow with different functionality and/or software to be added, which enables almost unlimited areas of application for simulations in different phases of the building life cycle as well as the adaptation to specific company requirements. The following remote services and partner platforms were integrated via the workflow to a multi-platform (see Figure 1) with individual models, integrated to an interoperable information system through the multimodel approach (Fuchs & Scherer 2017):

– Atena Cloud: Provision of the engine for the generation of model variants, the solver for system identification and connection to cloud resources for parallel execution of a large number of simulations.
– iGM.NET: Provision of a dashboard for remote management of the sensors and an interface for reading the sensor data directly from the construction site
– INERPROJECT: Project management platform for the provision of individual models, project management data and the aggregation of sensor data from multiple sources in a uniform representation.
– FBGuard: Fiber optic sensor system with online connection for strain measurements
– Construction Simulation Toolkit (CST): Construction process simulation tool – this tool is

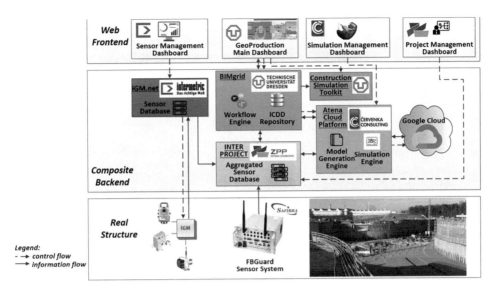

Figure 1. Geo Production CPS platform top level architecture.

desktop-based and is required for the last step of the workflow, the construction process simulation.
- BIMgrid: Workflow management engine, management of the digital twin, provision of the Geo Production dashboard

In the following section, the functionality of the Geo Production platform and the workflow engine as a central component are presented.

4 GEO PRODUCTION WORKFLOW

The aim of the platform is to provide a realistic, static model of an excavation pit, which is used to predict and adapt the construction process. Figure 2 shows the models created during the process and thus the evolution of the digital twin. The sub-processes and corresponding services of the Geo Production workflow are explained below.

Step 1: Sensors on the construction site measure occurring forces and soil deflection parameters at certain points on the ground. These are collected through an online service and made available via corresponding web-based interfaces (iGM.NET). The INTERPROJECT platform aggregates the sensor data from different providers in a uniform model in a specially developed notation language (Geo Production Sensor Markup Language – GPSML) and makes them available to the workflow engine. INTERPROJECT also plays the role of the central project management platform.

Step 2: The architect provides the building information model of the construction pit. The IFC format is provided as standard in the workflow, since this offers the greatest interoperability and flexibility for different tools. Together with the sensor data and all other models created during the workflow, it is added

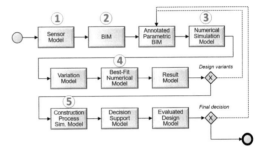

Figure 2. Model-based view on the Geo Production workflow.

to the digital twin of the excavation pit. The executing civil engineer can also annotate the model to give the structural engineer information on the positioning of sensors or the mapping of certain elements in the numerical model.

Step 3: The structural engineer uses the BIM model to create a numerical simulation master model for the simulation of structural changes in the subsoil. The positions of the sensors on the construction site are taken into account and corresponding monitoring points are inserted in the model so that the results of the simulations can be compared with the actual values on site. Material laws are defined as and with variable parameters. The variations of the corresponding soil parameters for the structural analysis are defined in a specially developed variation model, which frees the user from the manual creation of a large number of input models. The Atena Solver is used here as software for structural response calculation.

Step 4: The numerical master model, the sensor data and the variation model are uploaded to the Atena Cloud remote platform, where the model variants for the parameter study are first generated and then calculated in parallel in a cloud-based backend. The aim of the parameter study is to determine the material parameter values as accurately as possible, which most realistically describe the soil conditions of the excavation pit. The Atena Solver for structural analysis calculates the ground deformation under different soil conditions. From the large number of calculated variants, the model is then selected with a best-fit algorithm, which best matches the data measured on the excavation pit. This process can be repeated with additional material laws until the match is satisfactory. Another feature is the possibility to optimize the sensor positions on the construction site.

Step 5: The best fit model, i.e. the real soil system without oversized safety factors, is the starting point for the simulation of geotechnical structure variants and various construction processes. If the results are not satisfactory, the simulations can be repeated iteratively with refined parameter values. With the Construction Simulation Toolkit (see section 7), various construction processes can be simulated, taking into account the knowledge gained, and the most optimal variant can be identified according to various criteria (e.g. costs, completion date etc.). It is also possible to check and demonstrate compliance with security requirements. The decision support model is not strictly defined and, depending on the application, can be, for example, a schedule, an alternative geotechnical structure variant or an annotated model from a previous process step. The evaluated design model bundles all the information required for the final decision about a construction process. These can also differ from case to case.

The Geo Production platform is controlled via a web-based user interface, which controls the individual steps of the above workflow. This simplifies operation and lowers the learning curve as it is designed intuitively for the engineer. The user is guided through the subtasks and always has an overview of where in the overall process he is. If a task has to be carried out on one of the sub-platforms, either its front end is automatically called up or the remote functionality is integrated transparently for the user as a web service. Furthermore, users have the option of downloading the current status of the Digital Twin at any time and continuing to work with the current models in third-party software or reviewing intermediate states.

5 DIGITAL TWIN

The digital twin approach enables the behavior of the object to be simulated before it is actually manufactured, which is not possible with current Building Information Models (BIM). The data is managed in the Geo Production platform using the multimodel approach (Fuchs & Scherer 2017) according to ISO 21597-1 (ISO 2020) compliant information containers for linked document delivery (ICDD). This form of managing and linking documents, formalized

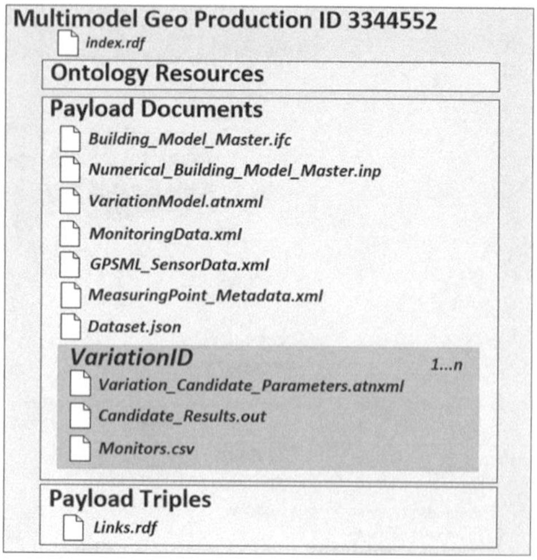

Figure 3. Multimodel container structure in Geo Production.

processes and objects offers the necessary prerequisites for the compact mapping of the monitored structure in a digital twin with all models and data relevant to the workflow. Another argument for using ICDD as a data management format is the possibility of linking models or individual components with meta information. They can be used for instance for advanced filter functions to determine a best fit model in a set of simulation results.

The permanent updating of the digital twin with the physical object representation is an important aspect in order to enable precise predictions in every phase of the object life cycle. The user can download the current status of the digital twin at any time and use it for the documentation of interim results and further simulations.

Figure 3 schematically shows the folder structure of the digital twin multimodel container. The inclusion of a unique ID in the name makes the platform multi-user capable. The file *index.rdf* describes the structure of the container in a human and machine readable way. The payload documents contain models and additional files for certain tools (e.g. *dataset.json* for transferring sensor data formatting descriptions between applications). In addition, any simulation results are stored here. For each simulation run, a folder with the ID of the simulation is created, which contains the simulation models saved in the delta process (file *Variation_Candidate_Parameters.xml*) as well as results of the structural response calculation (*Candidate_Results.out*) and comparison values for the sensor data (*Monitors.csv*). As defined in the ISO 21597-1 standard, Ontology Web Language (OWL) – based cross-model semantic links between elements of the individual models are defined (ISO 2020). These links can also be annotated with metadata and offer the possibility for software-supported semantic filter operations as well as AI-based data mining. Data obtained in this way not only facilitate targeted information retrieval for the user, but also offer potential for self-adaptive workflows. For example, a continuous automatic aggregation of simulation results can decide the number of workflow runs. If the sensor-based counterpart falls below a certain threshold for the deviation of calculated forces, the workflow execution is not restarted and the current model is presented as best fit.

6 WORKFLOW ENGINE

The workflow engine is a central part of the Geo Production web application and implemented in the BIMgrid sub-platform. Its task is to guide the user through a sequence of defined subtasks, to control the data flows and control flows between the individual software components and to synchronize the services of the remote platforms in order to enable the workflow to be processed in a highly automated manner. User input should only take place where it is absolutely necessary, e.g. where automation is not possible or visible, due to engineering responsibility. Data transfers, recurring sub-tasks (such as temporarily storing data) and entering control parameters into the software applications used are almost completely automated and not only relieve the burden on the user, but also enable the necessary fast processing of the workflow of the mass of simulations while simultaneously reducing human input errors.

A workflow is integrated into the system by mapping the individual work steps onto workflow step components. There are four general workflow step types: *Inputs* of user and control information into the system can be done by uploading files or entering strings in a mask. These sub-processes interrupt the workflow and trigger a user dialog. *Calculations/simulations* are executed by third-party software (solvers) or directly implemented algorithms (e.g. for result aggregation). If all input data is a priori made available, these workflow steps can run automatically and make the input data available for subsequent steps. *Data reading, transfer* and *storage* processes consist of several logical and technical sub-processes and can be automated in most cases. *Output* and *visualization* tasks can be represented by downloads or a graphical representation of data and do not necessarily mark the end of a workflow run. Resources are assigned to each workflow step in the form of abstract execution engines (see Figure 4). An execution engine for simulation-based workflow steps defines a certain software and a hardware type for the execution of the task. Hardware types can be, for example, a certain server, but also more abstract constructs such as public cloud or private grid resources as well as sensor interfaces. This enables the engineer to specify that sensitive data remain exclusively within the company's network or to reserve certain resources for dedicated operations. In addition, the respective network topology can be taken into account and

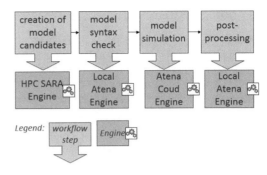

Figure 4. Assignment of execution engines to workflow steps.

data-intensive processes can preferably be carried out on IT resources in the local network with a higher bandwidth. For example, an execution engine called *Atena Cloud engine* defines a group of cloud nodes on which the Atena Solver for structural analysis is installed. If this execution engine is assigned to a simulation workflow step, the workflow engine automatically distributes the simulation jobs to free cloud nodes and ensures ideal load distribution. Syntax check and postprocessing are calculated in the local company network (*Local Atena Engine*) so that the necessary models do not have to be transmitted over the Internet. The generation of model candidates can be resource-intensive, which is why a special High Performance Computing (HPC) server is reserved for this task (*HPC SARA Engine*).

Execution engines are defined by the administrator when installing the platform, but like workflow steps, they can also be added or updated later. This approach allows the free definition of partially automated workflows using third-party software, provided that this supports execution via the command line. The corresponding execution commands are generated dynamically depending on the situation by the workflow engine. It is also possible to integrate physical sensors into the workflow in real time.

With knowledge about the structure of the numerical model, a specially developed variation model approach also allows the engineer to store parameter variations for a large number of model variants centrally in an XML-based file, the variation model. A special software component interprets this and automatically generates all the model instances based on the master model and the parameter values defined in the variation model.

The workflow engine is observer-based, with each workflow step monitoring the status of its predecessors. If status of a step changes to finished, all successor steps start automatically. The workflow processing is only interrupted by necessary user input, predefined alarms or occurring software and hardware errors. Operations such as file conversions, data storage or querying sensor values are automatically performed transparently in the background.

7 CONSTRUCTION SITE SIMULATION

As soon as a model with a satisfactory match of the real situation of the soil system is identified, it can be used as a starting point for simulation-based predictions about possible changes of the construction process to figure out the optimal improvement of the construction process. This follows similar objectives as finding the optimal energy-efficient building design (Baumgärtel & Scherer 2016). The construction site simulation tool is intended to assist the designers in comparing different construction methods and examining variations of key construction parameters such as resource capacity and schedules (see Figure 5). The goal is to carry out various variated process simulations and enable:

– calculation of construction time, resources and costs to provide for informed design decisions,
– comparing different construction methods, resource capacity and schedules and
– optimization of the production process of the construction pit, e.g. the sheet piles and anchor installation.

The main features of the Construction Simulation Toolkit are (Ismail et al. 2014):

– process-based simulation using formal Reference Process Models in BPMN notation
– BIM-SImulation Model (BIM-SIM) integration based on the IFC standard
– modular and extendable for different structure types.

With the early passage of identified variants through necessary approval procedures, there is maximum flexibility for situation-related adjustments of the construction process without delay. The Construction Simulation Tool is a complex tool and was developed as part of several research projects (Ismail et al. 2017).

8 CASE EXAMPLE AND VALIDATION

In our validation scenario (see Figure 6), various construction flowcharts for a section of an excavation pit were created. This section was again divided into 6 sub sections, in which excavation work can take place in parallel. With the help of the Geo Production Platform, various variants of ground protection installations were designed, whereby the number and positions of the anchorages of the protection elements in the ground were varied.

Figure 6 shows a screenshot of the CST. On the right a Model View is shown for creating anchor configurations. The 3D representation of the model is displayed on the left in an external model viewer. The anchors

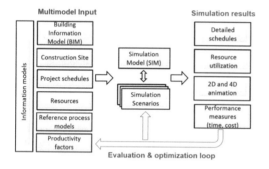

Figure 5. Features of the Construction Simulation Toolkit.

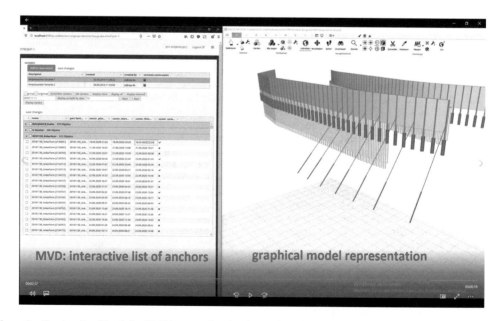

Figure 6. Construction Simulation Toolkit screenshot showing anchors filtered from the multimodel (left) and visualized (right) from the validation scenario.

are displayed in list form and can be activated or deactivated by the user with a click. Since both the CST and the model viewer use an ICDD multimodel container as a common data source, the model representations of the two tools are synchronized in real time if a change has been made in one of the two representations. The construction process of these different variants is simulated with different resources, e.g. different numbers of workers. The simulation showed that anchor elements could be saved, but the construction process as a whole could not be accelerated because of dependencies between the sub-processes. It was also determined that the number of workers did not play a significant role in the construction planning of this scenario due to these dependencies.

This section presented in short the structure and functionality of the Geo Production system and provided an overview on the architecture, the general workflow, the digital object representation and the final construction process simulation. Features such as user, project and backend management – not presented here – are under preparation for publication.

9 CONCLUSIONS

The Geo Production 4.0 web platform was presented. It is designed as a cyber physical system for the dynamic adaptation of construction processes, based on various simulations and real-time integration of sensor measurements. In addition to an intervention in the production process, the platform also enables an almost online reconfiguration option for the basic structure. This not only makes the production process safer, but also more robust against unpredictable events by providing variants. The management of models and simulation results in ISO standardized multimodel information containers for linked document delivery supports a continuous refinement of the digital twin as well as software interoperability and collaborative work of engineers from various fields. The evaluation based on the data of a real construction project showed that not only real-time monitoring but also nearly real-time predictive simulation-based adaptation of processes and models can be realized with the geo Production cyber physical system.

Some technical disadvantages result from the current implementation status of the software. By integrating various third-party software platforms into the workflow, the user has to register with various platforms. However, this can be simplified with a credential delegation mechanism, in which case the user only has to log on to the overall platform once. In addition, the platform currently only offers a programming application interface for adapting workflows. This is sufficient for the current application, but increases the effort for adapting to other scenarios. By providing a graphical user interface or an interface for process modeling languages such as the Business Process Modeling Language (BPMN) to define new workflows, the platform could be expanded more easily for other fields of application. In terms of flexibility, the platform already allows the implementation of the transparent, automated exchange of remote services based on their interface and functional descriptions. This offers potential for further increasing the flexibility of the workflow adaptation with regard to the availability and capabilities of third-party services.

The digital twin of the physical object that arises during the workflow run can become very large, in terms of number as well as size of the managed files. Suitable solutions for data transfer must be found here. For example, individual model candidates for mass simulations are not saved directly, but according to the delta method. If a master model is available, the simulation models can be reconstructed at any time from the parameter assignments saved in simple text files. Nevertheless, the ICDD multimodel container can reach sizes of up to several gigabytes. Appropriate compression methods and the provision of options for filter-based downloading of components of the ICDD could significantly reduce the data traffic.

ACKNOWLEDGEMENTS

This research has received funding as an EUROSTARS project under the Grant Number 01|S17099C ("Geo Production 4.0 – Cyber-physical controlled production process for geotechnical structures") by the European Union. The funding is gratefully acknowledged.

REFERENCES

Alam, K. M. & El Saddik, A. 2017. C2PS: A digital twin architecture reference model for the cloud-based cyber-physical systems. *IEEE access*, 5. Jg., p. 2050–2062.

Baumgärtel, K. & Scherer, R.J. 2016. Automatic ontology-based Green Building Design Parameter Variation and Evaluation in Thermal Energy Building Performance Analyses, In: *Proc. eWork and eBusiness in Architecture, Engineering and Construction*. ECPPM 2016, *7–9 September 2016*. Limassol, Cyprus.

Fuchs, S. & Scherer, R. J. 2017. Multimodels – Instant nD-modeling using original data. *Automation in Construction*, vol. 75, p. 22–32.

Gonzalez, G. R., Organero, M. M. & Kloos, C. D. 2008. Early infrastructure of an internet of things in spaces for learning. in *Proc. 8th IEEE Int. Conf. Advanced Learning Technologies*, Santander, Cantabria, p. 381–383.

He, J. F. 2010. Cyber-physical systems. Commun. China Comput. Feder., vol. 6, no. 1, pp. 25–29.

Ismail, A., Srewil, Y. & Scherer, R. J. 2014. Collaborative web-based simulation platform for construction project planning. In: *Working Conference on Virtual Enterprises*. Springer, Berlin, Heidelberg, p. 471–478.

Ismail, A., Srewil, Y. & Scherer, R. J. 2017. Integrated and collaborative process-based simulation framework for construction project planning. *International Journal of Simulation and Process Modelling*, vol. 12, Nr. 1, p. 42–53.

ISO 21597-1. 2020. *Information Containers for linked document deliery*. [Online]. Available at: https://www.iso.org/standard/74389.html, Accessed: 5 April 2020.

Jiang, J.-R. 2018. An improved cyber-physical systems architecture for Industry 4.0 smart factories. *Advances in Mechanical Engineering*, Nr. 6.

Lee, E. A. 2007. Computing foundations and practice for cyber-physical systems: a preliminary report. Tech. Rep. UCB/EECS-2007-72, University of California, Berkeley.

Liu, Y., et al. 2017. Review on cyber-physical systems. *IEEE/CAA Journal of Automatica Sinica*, Nr. 1, p. 27–40.

Perera, Charith, et al. 2014. Sensing as a service model for smart cities supported by internet of things. *Transactions on emerging telecommunications technologies*, vol. 25, Nr. 1, p. 81–93.

Scherer, R. J. & Schapke, S.-E. 2014. Information Systems in Construction (in German), vol. 1 and 2. ISBN 978-3-642-40883-0.

Seiger, R., Assmann, U. & Huber, S. 2018. A case study for workflow-based automation in the internet of things. In: *2018 IEEE International Conference on Software Architecture Companion (ICSA-C)*. IEEE, p 11–18.

Seiger, R., et al. 2019. Toward a framework for self-adaptive workflows in cyber-physical systems. *Software & Systems Modeling*, 2019, vol. 18, Nr. 2, p. 1117–1134.

A framework for development and integration of digital twins in construction

A. Corneli, B. Naticchia, A. Carbonari & M. Vaccarini
DICEA Department, Polytechnic University of Marche, Italy

ABSTRACT: Generation and sharing of data have always been critical within the Architecture, Engineering, Construction and Operation (AECO) sector due to its fragmentation and it is well recognized that this led to an inefficient and poor integration of processes along the building lifecycle. Despite the introduction of Building Information Modelling (BIM) that enhances interoperability many issues related to the management of information evolution still persist. Management of vast amount of data is complex especially in change management process during construction phase when a real-time processing, integration and immediate feedback are usually necessary. In order to exploit the potential of digital technologies for process efficiency improvement linking directly to BIM models and defining BIM processes is necessary. Product Life-cycle Management (PLM) and Model-Based Systems Engineering (MBSE), commonly applied to other engineering fields, embody the key approach to complex construction management and model-based information use. The challenge is not only to support the various phases of the building's life cycle, but also to capture its multidisciplinary nature which has led to the creation of various frameworks. To this aim, "digital twin" with its dynamic digital representation of the physical system continuously updated with latter's performance, maintenance and health status data can represent a valuable tool. However, current MBSE digital twins face the following challenges: specific domains applications, unrelated modelling theories, different semantic models and uncommon storage representations. All these issues led to seamless integration across disciplines during the whole lifecycle. MBSE toolchain is an emerging technique in the area of systems engineering which is expected to become a next generation approach for supporting complex Cyber-Physical System (CPS) development as the integration system of systems. This paper proposes a framework for development of digital twin integration based on MBSE toolchain technique, providing a semantic integration among layers such as social (integration of stakeholders' views), process (integration of management models for process control and monitoring), technical (integration of domain specific model for predictive co- simulation), information (integration of data, model and tool operations). The framework has two main purposes: to support toolchain development with a Model Based System Engineering approach; to promote interoperability of the whole developed toolchain through a service-oriented approach where services are provided to the other components by application components, through a communication protocol over a network.

1 INTRODUCTION

A recent study reveals that the productivity of the construction industry suffered a not-incremental trend over the last 40 years (McKinsey 2017). Although construction projects are increasingly integrated around Building Information Modelling (BIM) based platforms, this approach is still leaving behind the full exploitation of the digital revolution.

Current standardizations, such as Industry Foundation Classes (IFC), are meant to facilitate software document exchange mainly in the design phase. The current level of BIM maturity is mainly based on federated information models and far away to support flawlessly the knowledge flow between lifecycle processes. Therefore, construction projects suffer from a lack of data integration, from the presence of disconnected documents, and from insufficient data for process simulation and management. This flaw causes a lot of unexpected project management and process integration issues during the whole building lifecycle.

Moreover, Architecture, Engineering, Construction and Operation (AECO) frameworks lack holistic approaches and tools that integrate four fundamental dimensions: Sociality & Sustainability; Process; Information & Service Infrastructure; Technology. In order to make an impact on the overall project management and process interoperability and control, several steps ahead are needed: (i) developing platform-neutral frameworks requiring limited programming skills and limited knowledge of the product model schema for their quick adoption by the construction industry; (ii) developing flexible tools supporting human intervention in management processes; (iii) capturing semantic knowledge for interoperability across stakeholders and lifecycle processes.

Technically, the availability of new digital technologies (e.g. new sensing, miniaturized and robotic tools, IoT), artificial intelligence, big data analytics and enhanced computing power, pave the way to the adoption of advanced management approaches, which are expected to transform current approach to planning, building, integrating and maintaining the built assets.

In this context digital twins can play a decisive role since they can be defined as a semantic representations describing one object, its behaviour, its history, and its possible relationships with the complex context in which it can operate (IEEE 2019). Furthermore, according to their multidisciplinary nature digital twins also embed the mentioned above four dimensions at once.

The Digital Twin (DT) here meant is a support for the whole life-cycle, as an integration of symbiotic autonomous systems (SAS) merging bits with atoms, by taking into consideration an ecosystem where players can potentially interact, make use, and leverage data and semantic information, getting rid of specific domain syntaxes.

Thus in this paper we propose a general, scalable, transdisciplinary methodology for twinning products and processes in AECO sector that conveys model-based systems engineering (MBSE) tools, knowledge management and reasoning models which in form of services allows integration and deployment of DT as toolchains.

2 LITERATURE REVIEW

The proposed framework is inspired by the state of the art from engineering sectors where the digitalization of processes has currently gained a relevance both in terms of technologies and methodologies.

2.1 Digital Twin

AECO activities form a complex system: in them, several agents act independently to reach both personal and shared goals, resulting in emerging behaviours. Furthermore, the final product in the construction industry is built and maintained through the contribution of several different disciplines.

According to latest definition of DT the 'twin' has to be considered as a representation that resembles (rather than mirrors) certain desired behaviours of the physical artifact faithfully enough. In true 'twinness' the 'twin' comes into existence synchronously with the artifact it resembles and it also ceases existence when the artifact ceases to exist. The properties of the 'twin' are also updated synchronously with changes to the physical artifact and vice versa (Tomko & Winter 2019). The digital twin is also defined as a digital representation of things from the real world (Kuhn 2017); a concept with which data and information of atoms are assigned to bits (Datta 2016); a computer-aided model of a tangible or intangible object; a synonym for the Industry 4.0 asset administration shell.

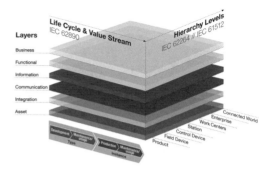

Figure 1. Reference Architectural Model Industrie (RAMI 4.0).

In particular the new reference architecture for industry 4.0, the RAMI 4.0 is gathering a deep attention for its capability to unify the numerous, heterogeneous, and multidimensional aspects of the lifecycle of processes. A new theory of the DT has to be taken into account within the context of RAMI 4.0 and its three dimensions (Deuter 2019): Layers, Life-cycle & Value stream, Hierarchy levels. Hence, the DT becomes an infinite source of information that can be accessed only by means of continuously changing finite points of observation that depend on the observers, from the evolutions of the context, and from the position in the three dimensions of the process reality as defined by the RAMI 4.0 (Figure 1).

SE on the other hand developed methodologies for coping with complex systems. One effective methodology is MBSE, which proved to be a powerful and flexible tool in several fields (e.g. engineering of cyber-physical systems, software engineering, aerospatial industry, etc.), and would fit well in the construction engineering domain. MBSE is defined as the "formalized application of modelling to support system requirements, design, analysis, and verification and validation activities beginning in the conceptual design phase and continuing throughout development and later lifecycle phases" (INCOSE 2011). Thanks to MBSE-based approaches, the team or the organization responsible for the realization of the construction, can monitor its progress; embrace positive emerging behaviours from the stakeholders, and early recognize and reject negative behaviours that undermine the quality, security, and performance of the final product or of the environment in which it participates.

Even in the construction domain, similarly to Industry 4.0, the existence of a virtual replica of a physical asset that keeps in sync with its physical counterpart, promises to optimise operations and maintenance and support better decision making. Lately, the construction industry is starting to invest in digital twin technologies. The National Digital Twin is an ambitious pilot project to bring new efficiencies to the UK's infrastructure (Corke 2019) by means of federated twins. Digital twins can perform linear infrastructures lifecycle management (Tchana 2019), providing synchronized and standardized data flow for intelligently interconnecting all construction processes (Naticchia et al. 2019). The in-progress digitization of the construction artefacts is favored by the exploitation of technologies to acquire information from buildings and construction sites: drones, cameras, laser scanners, GPS, mixed reality, mobile devices, RTLS, etc. A geometric digital twin from labelled point clusters about existing reinforced concrete bridges was presented in (Lu 2019) by proposing a slicing-based object fitting method. RTLS have been adopted (Li 2016) in the form of sensor technologies empowered by artificial intelligence. The concept of modular building performance simulation has been accredited to realize the vision of an integrated building simulation platform for several stakeholders (Mitterhofer 2017). This approach is the one we refer to exploit through the integration of different domain specific models (e.g. energy, structural, fire safety, etc.) to build a digital twin for complex buildings.

2.2 Product Lifecycle Management (PLM)

In order to ensure effective and efficient development in short cycles while maintaining high product quality, the management of data and information throughout the entire life cycle is of great importance (Gilz 2014). In engineering, which is currently subject to constant changes (global markets, distributed development, increasing complexity and digitalization), a central concept, based on methods and IT solutions, has been established under the term PLM (Product Lifecycle Management). PLM manages product-related information among companies, suppliers and customers involved in the realization and usage of the product (Eigner & Stelzer 2009), and has been successfully applied to several industrial sectors.

An approach similar to PLM is also required in the BIM environment. For example, ISO 19650-2 requires a federated common data environment (CDE) to be used as the shared source of information, to collect, manage and disseminate documentation, graphical model and non-graphical data for the entire project team (ISO 2018). Both PLM and BIM require reliable IT hardware and software infrastructures. The rapid evolution of the latter enabled product- and process-related information to be tracked more accurately and more promptly, motivating the definition of Product and Project Lifecycle Management (PPLM). PPLM is a process extending PLM where the focus is shifted to the lifecycle of project around the product (Perelman et al. 2008; Sharon et. al. 2008). This integration between product and process lacks in the construction field where it would be disruptive instead. The knowledge of the building could be structured at a different degree of complexity, and evolving under inputs of humans as well as autonomous agents, from the early stages (e.g. preliminary feasibility, conceptual design) of the project till the end of the lifecycle.

The proposed framework advances state-of-the art of Digital Twin enable construction by:

- defining a model-based methodology for the AEC field, to assist stakeholders when defining the digital-twin as the integration of several viewpoints of the construction execution or management;
- developing an architecture based on the MBSE and tools such as toolchains and toolchain generators;
- applying the methodology to develop reusable ontologies, services, and toolchains that link and share the information along the entire construction phase;
- integrating the digital twin of the product together with the digital twin of the processes involved in the project execution, handover and having an impact on the next phases of the life cycle.

3 METHODOLOGY

The final aim of our approach is to automatize and normalize the development of digital semantic-based technologies for the creation of digital twins in construction, following a MBSE approach. Tools and models integrated in the proposed framework are related to four technical areas: automated work progress assessment and deviation analyses, automated quality assessment, dynamic resource planning, real-time health and safety management.

The work methodology follows the subsequent steps:

1. Preparatory work for data acquisition: this includes identification and implementation of every data source, including preparatory field work, if necessary;
2. Identification/creation of models and process patterns: this step includes automated model creation, and manual completion of the models whenever necessary, which are relevant to the specific application, including the selection of patterns from which process models is derived;
3. Selection of the Digital Twin development tools for the case at hand;
4. Production of the Digital Twin Toolchain: in this step the knowledge base, viewpoints, tools, methods, resources are linked in the form of the DT Toolchain;
5. End-user interfaces: a set of libraries is available to develop web services as interfaces between end-users and the digital twin viewpoints;
6. Validation: this section describes the available mechanisms for effective DT implementation, and how to set up the environment and devices to effectively provide this implementation;
7. Execution: this section describes the mechanisms to apply the DT in the management of construction works.

The framework main services composing a generic toolchain can be still organized by their purpose:

- Data acquisition and information enhancement;
- Process management and prediction;

– Advanced reasoning and assessment;
– Advanced computation and simulation.

The data acquisition and information enhancement unit is in charge of sensing the physical building/infrastructure and feeding the intelligent digital twin in real-time. Customized technology (e.g. AR/MR) for operators can be exploited to make them enjoy the advantages of the digital twin directly on-site.

The process management and prediction unit includes services necessary to manage, assess the performances and monitor business processes, which is the prerequisite to interact with stakeholders, to reach a certain degree of automation in project management, to update the instance of the digital twin according to the actual progress and to timely trigger events.

The advanced reasoning and assessment unit must generate actionable information from diverse data supplied by means of different tools, such as artificial intelligence, probabilistic inference, management of historical data and data mining, decision making models.

The advanced computation and simulation unit performs both real-time and forward looking simulations to support inference, what-if analyses, automatic rule-checking (social rules, constraints and policies, affecting processes, expressed in semantic rule-based languages); basically, a real-time multi-physics simulation environment is used to analyse interaction among systems and the environment, including all physical behaviours (e.g. mechanics, spatial interaction, thermal behaviour, fire simulations).

The digital twin is the result of dynamic instantiation of the physical infrastructure through the set of services managed by the proposed framework. The instantiation may vary according to the specific purpose, e.g. users interacting with the digital twin and progress in the construction process. It includes several capabilities:

– event management and control actions (e.g. interface provided to interact with users and among different users, on-demand trigger of scenario and what-if analysis, inference and simulation services and events such as alerts and notifications, control actions based on effectiveness KPIs);
– visualization and applications (e.g. integration of both visual and non-visual information, real-time and predictive management of key issues, namely quality assessment, resource tracking and optimisation, health and safety management, work progress monitoring, on-site man-machine collaboration).

4 SYSTEM ARCHITECTURE

As depicted in Figure 2, the proposed framework can be seen as a multiple cybernetic loop at different time-scales that aims at controlling and improving the physical reality under focus: a) short-term, a control loop that feeds DT with data from the field (e.g. IoT data, and social IoT) and uses them to elaborate the control actions; b) a mid-term loop, where the DT

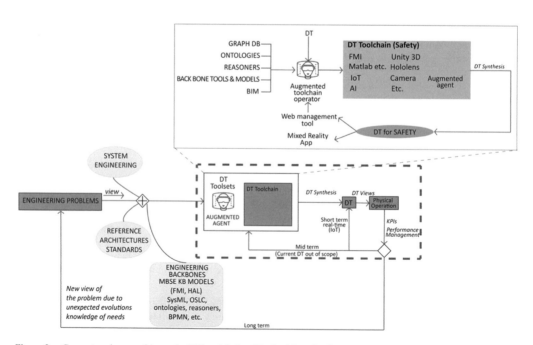

Figure 2. Concept and zoomed in on the DT toolchain of the health and safety management DT.

needs to be updated or regenerated from the toolchains in order to resume effectiveness in the current evolution of the controlled reality; c) a long-term loop, where the information (situational awareness) from physical or digital reality are used to infer new kind of problems or to signal unexpected contexts (possibly due to new requirements), and then possibly steer the purposefulness of the toolchains.

The zoomed-in part of Figure 2 shows an example of an instance and an example of the DT toolchain for the objectives of health and safety in construction as described in (Naticchia et al. 2109). In this case the DT is foreseen as a running mixed reality application developed by components like a game engine, real 3time sensors, mixed-reality headsets, holonic multi-agent procedures, tensor flow AI, cameras, drones, etc. The technologies implied are collected into a backbone of information and services. The process of collecting the tools to compose a convenient DT makes use of all the semantic and ontology-based reasoners, modelling tool, services provided from the technological backbone, along with knowledge graph tools that makes the augmented agent able to automatize most of the tasks and the programming of the toolchain mechanism. Then, the toolchain is executed to program, assemble, compile, deploy and run the DT.

The proposed platform is conceived with a multi-layer architecture where five layers continuously collect and process information of the work that is happening at the physical layer, where the construction as well as the back-office activities happen (Figure 3). The Physical Layer is not part of the architecture but is just assumed. The proposed architecture is built for the purpose of observing actions and behaviours while they happen, to assist the stakeholders along the entire Product and Process Lifecycle Management activities. To bridge this gap, two activities are central for the proposed architecture, viz. continuous Knowledge Extraction and Information Lifecycle Management: only through them one can assure that Digital Twins convey the most updated and coherent picture of the infrastructure and all the processes related to it.

The Data Sources layer collects several sources of data by observing the daily activities. At this layer both technical (more product-related) as well as social (more process-related) information are extracted. This

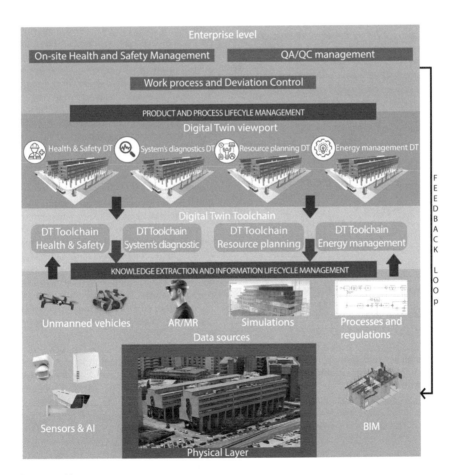

Figure 3. System architecture.

295

layer includes the data standardization tasks, where the extracted information is represented using open or de-facto standards, enabling future integrations and exchanges. The obtained knowledge base is the ground from which a proper Digital Twin of the construction is derived, but the knowledge base cannot be considered a Digital Twin per se, since it is meant to be an optimal representation of the extracted information to be further processed and refined by algorithms and autonomous agents.

In the Toolchain Development layer, the focus is on giving the user the opportunity to "chain together" several pieces of information taken from the common knowledge base that are mutually dependent. At this level, each Digital Twin Toolchain is intended to contain the interfaces of the Digital Twin developed for the semantic dimension of pertinence (viz. health and safety management, dynamic resource planning, work progress control and quality assessment), as well as the tools and services that can be invoked for keeping the Digital Twins synchronized among themselves. The order in which pieces of information should be updated along the toolchain is relevant, since the Digital Twins contain mutually dependent pieces of information, thus the toolchain must be designed to produce a unique, 4coherent, and stable, representation of the world that is changing under the eyes of the stakeholders. The role of the toolchain is to collect every update at the level of the Digital Twins, update back the shared knowledge base (see the grey down-ward arrow). Examples of information coming from the Digital Twin level would be the input provided by the user, or the result of some co-simulation or rule inference algorithm that allows to discover (or predict) new knowledge about the physical twin. The Digital Twin Viewpoints layer contains the Digital Twins specific of each of the four semantic dimensions, providing a focused viewpoint of the current construction for that semantic dimension. Finally, at the Enterprise Level, the architecture provides a set of interfaces through which the final user receives a coherent view of the Digital Twins they must be aware of. Every change the user applies through such interfaces is reflected on the Digital Twin viewpoints below, by means of the toolchain connecting all the DT viewpoints. Note that at the Enterprise Level we find tools and interfaces that are already familiar to the end users and support them along the usual Product and Process Lifecycle Management activities. Examples of such interfaces are:

1. On-site health & safety management, where health and safety information are combined with resource planning information;
2. Work progress and deviations control, where information about resource usage is combined with information coming from monitoring the work progress;
3. QA/QC management, where a view of the work progress is combined with a view of the quality assessment.

The proposed system makes use of currently in force European and International standards and open source platforms, in order to guarantee further interoperability and continuity of the solutions proposed.

5 AN APPLICATION FOR EMERGENCY MANAGEMENT SCENARIO

Among the phases of building lifecycle the management is the one chosen for first tests of the proposed framework. The application already developed in (Naticchia et al. 2019) especially focuses on real-time emergency management when a DT is able to show unforeseen scenarios, giving the possibility to cope with them. A DT with the aim of finding the best path in case of fire emergency was developed exploiting the holonic theory.

The architecture of the developed holonic system, depicted in Figure 4, supports fire emergency management and rescue operation, detecting the most effective way-out. The developed system, utilizing building updated data, hosted in BIM and loaded into a virtual reality platform, is able to carry out a real-time pathfinding to tackle emergency scenarios, unforeseen by the standard emergency plan. The switch from the standard way-out to the alternative one (see Figure 5) occurs automatically, as soon as the fire is detected by sensors and simulated in the virtual reality platform. The implemented algorithm, detecting all the connected floors surfaces (highlighted by purple color in Figure 5), finds rooms internally connected by doors. The alternative path circumvents the zone affected by fire and smoke and leads the endangered users towards the closest stairwell.

This first DT, used for emergency management, improves the current approach, leveraging more updated and significant information. The crucial convenience in developing DT is that a knowledge-based approach is hard to pursue since it should represent every possible situation. Developing twins that mirrors real condition, simulate random situations (determined by a scene definition) let new unforeseen scenarios emerge and enrich the knowledge base for further analysis. Exploiting data from BIM, sensors, AI, and other data sources to feed a virtual reality platform has been demonstrated with this previous work that can definitely help.

The developed system architecture tested in a large mixed-use public building, shows its potentiality through the contribution given to officers in charge of emergency management. More in details, the results of real-time pathfinding, support endangered users as well as rescuers, dealing with urgent decision within a really short deadline.

For future developments the connection between sensors and virtual reality platform represents relevant matters of work in order to automate the functioning of the whole system.

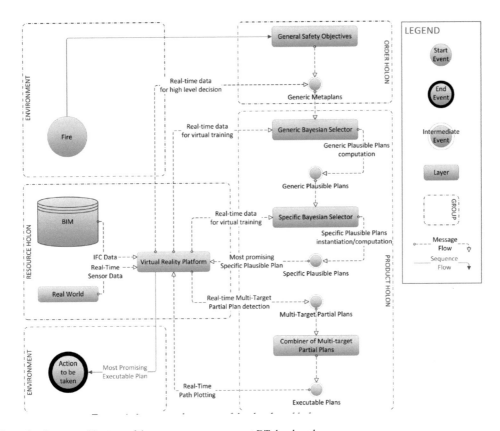

Figure 4. System architecture of the emergency management DT developed.

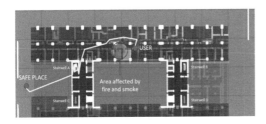

Figure 5. Standard (above) way-out and the alternative way-out (below) found with the DT.

6 CONCLUSIONS

The digitalization of the construction industry is an ongoing process and its need has been deeply discussed in order to try to enhance the sector productivity. Anyway, an organic approach to this change is still missing. Furthermore, the fundamental involvement of processes is not always taken into consideration.

This paper proposed a new approach to this transition gathering other engineering industries experiences in modelling product and processes.

Product life-cycle management and MBSE represent the approaches that combined with building information modeling can drive the AECO sector to its 4.0 step.

Digital twins developed within this framework embody the tool to provide a semantic integration among social, process technical and information layers. The framework proposed is able to support toolchain development with a system engineering approach and to promote interoperability of the whole developed toolchain through a service-oriented approach.

Finally, the first application for emergency management stands as a proof of concept of the development of DT for construction. This is a first use case but the approach can be easily extended to others and it can also be enriched with links to the physical layer (e.g. on-site sensors).

REFERENCES

Ardanza, A., Moreno, A., Segura, Á., de la Cruz, M. & Aguinaga, D. 2019. Sustainable and flexible industrial human machine interfaces to support adaptable applications in the Industry 4.0 paradigm. *International Journal of Production Research*, 57(12), 4045–4059.

Collins, J., Regenbrecht, H. & Langlotz, T. 2017. Visual coherence in mixed REALITY: A systematic enquiry. *Teleoperators and Virtual Environments*, 26(1):16–41.

Corke, G. 2019. Seeding the National Digital Twin. AEC Magazine. https://www.aecmag.com/component/content/article/59-features/1835-seeding-the-national-digital-twin

Datta, S.P.A. 2016. Emergence of Digital Twins. Computing Research Repository (CoRR) abs/1610.06467.

Deuter, A. & Pethig, F. 2019. The Digital Twin Theory, Industrie 4.0 Management 35 (1), p. 27–30.

Eigner, M. & Stelzer R. 2009. Product Lifecycle Management – Ein Leitfaden für Product Development und Life Cycle Management. *Springer-Verlag*, Berlin, Heidelberg.

Fujino, Y., Siringoringo, D. M. & Abe, M. 2009. The needs for advanced sensor technologies in risk assessment of civil infrastructures. *Smart Struct. Syst*, 5(2), 173–191.

Gilz, T. 2014. PLM-Integrated Interdisciplinary System Models in the Conceptual Design Phase Based on Model-Based Systems Engineering. Ph.D. Thesis, Schriftenreihe VPE, Vol. 13, University of Kaiserslautern.

Ham, Y., Han, K.K., Lin, J.J. & Golparvar-Fard, M. 2016. Visual monitoring of civil infrastructure systems via camera-equipped Unmanned Aerial Vehicles (UAVs): a review of related works. Vis. in Eng. 4, https://doi.org/10.1186/s40327-015-0029-z.

IEEE 2019. "Symbiotic Autonomous Systems, White Paper III, November.

INCOSE 2011. Final Report, Model-Based Engineering Subcommittee, NDIA, Feb. 2011.

ISO 2018. Technical Committee: ISO/TC 59/SC 13 – ISO 19650-2:2018 "Organization and digitization of information about buildings and civil engineering works, including building information modelling (BIM) — Information management using building information modelling — Part 2: Delivery phase of the assets".

Kuhn, T. 2017. Digitaler Zwilling. Informatik Spektrum 40, 5, S. 440–444.

Li, H., Chan, G., Wong, J. K. W. & Skitmore, M. 2016. Real-time locating systems applications in construction. *Automation in Construction*, 63, 37–47.

Lin, J., Han K. & Golparvar-Fard, M. 2015. A Framework for Model-Driven Acquisition and Analytics of Visual Data Using UAVs for Automated Construction Progress Monitoring, *Computing in civil engineering*, 156–164, 10.1061/9780784479247.020.

Lu, R. & Brilakis, I. 2019. Digital twinning of existing reinforced concrete bridges from labelled point clusters, *Automation in Construction*, Volume 105, https://doi.org/10.1016/j.autcon.2019.102837.

Martinez-Aires, M. D., Lopez-Alonso, M. & Martinez-Rojas, M. 2018. Building information modeling and safety management: A systematic review. Safety science, 101, 11–18.

McKinsey, 2017. McKinsey Global Institute "Reinventing construction: A route to higher productivity" report, 2017.

Mitterhofer, M., Schneider, G. F., Stratbücker, S. & Sedlbauer, K. 2017. An FMI-enabled methodology for modular building performance simulation based on Semantic Web Technologies. Building and Environment, 125, 49–59.

Naticchia, B., Messi, L. & Carbonari, A. 2019. BIM-based Holonic System for Real-Time Pathfinding in Building Emergency Scenario, *Proceedings of 2019 European Conference on Computing in Construction*, 117–124, DOI: 10.35490/EC3.2019.174.

Park, C. S., Lee, D. Y., Kwon, O. S. & Wang, X. 2013. A framework for proactive construction defect management using BIM, augmented reality and ontology-based data collection template. *Automation in Construction*, 33:61–71.

Pellerin, R., Perrier, N. & Berthaut 2019. F. A survey of hybrid metaheuristics for the resource-constrained project scheduling problem. *European Journal of Operational Research*.

Perelman, V., Sharon, A. & Dori, D. 2008. Towards a Unified Product and Project Lifecycle Model(PPLM) for Systems Engineering. *In Proceedings of ESDA 2008 – 9th ASME Engineering Systems Design and Analysis Conference*, Haifa, Israel, July 7–9.

Pérez, C., Fernandes, L. A. & Costa, D. 2016. A literature review on 4D BIM for logistics operations and workspace management. *Proceedings of the 24nd International Group of Lean Construction*: 486–495.

Sadeghpour, F. & Andayesh, M. 2015. The constructs of site layout modeling: An overview. *Canadian Journal of Civil Engineering*. 42. 10.1139/cjce-2014–0303.

Salehi, S. A. & Yitmen, I. 2018. Modeling and analysis of the impact of BIM-based field data capturing technologies on automated construction progress monitoring. *International Journal of Civil Engineering* 16.12: 1669–1685.

Sharon, A., Perelman V. & Dori, D. 2008. A Project-Product Lifecycle Management Approach for Improved Systems Engineering Practices. *In Proceedings of the Eighteenth Annual International Symposium of the International Council on Systems Engineering* (Utrecht): INCOSE.

Tchana, Y., Ducellier, G. & Remy, S. 2019. Designing a unique Digital Twin for linear infrastructures lifecycle management. *Procedia CIRP*, 84, 545–549.

Tomko, M. & Winter, S. 2019. Beyond digital twins – A commentary. Environment and Planning B. *Urban Analytics and City Science*, 46(2), 395–399. https://doi.org/10.1177/2399808318816992

Vick, S. & Brilakis, I. 2018. Road Design Layer Detection in Point Cloud Data for Construction Progress Monitoring. *Journal of Computing in Civil Engineering*, 32(5), 04018029.

A Digital Twin factory for construction

C. Boje & S. Kubicki
Luxembourg Institute of Science and Technology, Esch-sur-Alzette, Luxembourg

A. Zarli
R2M Solution Roquefort-Les-Pins, France

Y. Rezgui
Cardiff University, Cardiff, UK

ABSTRACT: As our society is becoming increasingly interconnected, the construction industry is faced with unparalleled needs for digital infrastructure as it is beginning to adopt sophisticated cyber-physical integrations, known as Digital Twins. Previous work on BIM and cyber-physical systems within the construction sector as well as from nearby engineering fields have already contributed significantly around the subject of digital twins in terms of their definition and potential uses across several construction application domains. We propose the concept of a Digital Twin Factory within the Architecture, Engineering and Construction sector, which is aimed to support the rapid deployment of construction site digital twins in a more practical sense. Within this article, we outline the initial requirements followed by a potential blueprint on what a Digital Twin Factory should resemble and what its main roles are in creating, hosting and updating the data about the construction site on multiple levels.

1 INTRODUCTION

The architecture, engineering and construction industries are encouraged to follow the latest technological developments which are able to go beyond the scope of Building Information Modelling (BIM) – which is falling behind in terms of connectivity to real world construction assets and the application of Artificial Intelligence (AI). The Digital Twin (DT) paradigm unveils new perspectives on managing the construction, operation and maintenance of the construction site and built assets by ensuring a synchronicity between the physical world with its respective digital models (Grieves 2014). A DT promises a very high degree of fidelity when it comes to representing its Physical Twin (Schleich et al. 2017) by gathering data about the real world construction site and all its assets in real-time. This data would be used for the application of engineering models and artificial intelligence to aid in the lifecycle management of built assets overall.

If the construction industry is to adopt a Digital Twin paradigm however, it is necessary to understand what is expected when deploying such a concept in practice, and how this could be done in a more standardized way, paving the way towards the automation of Digital Twin deployment.

As such, the discourse within this article introduces the novel concept of a "Digital Twin Factory" (DTF) and expands upon several specific aspects which were considered relevant for the construction and operation stages of built environment assets which are governed by DTs. These are narrowed down to: 1) procedural and 2) technical requirements for creating a digital twin; 3) modularity of digital twins; and 4) the inter-digital twin communication. We consider these aspects of importance in understanding how DT would work in practice, and therefore dictate the manner in which a DTF should be considered.

Once these aspects are discussed, an initial framework can be constructed and expanded, which would enable the concept of Digital Twin Factory to be explored and tested in research and development.

In terms of structure, the article highlights several relevant existing studies around the subject of BIM and DTs in section 2. Following this, the key perceived requirements for developing a DT are enumerated and discussed in section 3. Based on the previous sections, the novel concept of a DTF is introduced along with a first proposal for a DTF reference framework in section 4. Finally, the key recommendations and main conclusions are presented in section 5.

2 RELATED WORK

2.1 Digital twins in manufacturing

The DT paradigm is already in use on a daily basis from the aerospace sector being a first adopter (Tuegel et al. 2011), to the car manufacture and constant monitoring (e.g. Tesla has a digital twin for every single car it produces). Nevertheless, the needs for research and innovation around the DT subject are mandated by Strategic Research and Innovation roadmaps e.g. in Europe (ECTP Innovative Built Environment 2019), with the DT being one of the main concepts generally associated to the Industry 4.0 generation. The manufacturing sector is a prime investigator and one of the first adopters of the DT paradigm, with several important studies having described the various applications and technologies for enabling DTs (Grieves 2014; Schluse et al. 2018; Tao et al. 2019). The real-time surveillance of manufactured products is estimated to bring unprecedented benefits to the manufacturing industry, allowing the possibility of virtual testbeds, huge client customization offering or simulating endless scenarios and use-cases (Haag & Anderl 2018; Schluse et al. 2018). These would in return improve the lifecycle of the product in question, while significantly reducing the costs and material waste of testing using real products. DTs within this domain aim to simulate to a very high level of detail the behaviour of the products from structural integrity, maintenance and running costs as well as overall sustainability assessment. However, no common methods or frameworks have been defined to date in terms of which tools, technologies or data schemas should be used across the board.

2.2 Digital twins for construction

The construction industry has similar goals with the manufacturing one, but the scale and complexity of the supply chain and its organization are significantly different. The expected lifetime of built assets is also much more prolonged, where different stakeholders intervene and leave at different lifecycle stages.

It remains unclear how DTs within the construction domain are different from existing Building Information Models and Cyber-Physical systems, which already achieve certain parts of the site sensing and management processes. We believe that a Construction Digital Twin (CDT) differs from BIM because the latter does not traditionally include real-time data collected from the construction site or a building in operation, with an initial framework outlined in (Boje et al. 2020). The DT concept uses tools and technologies to collect and process real-time data and information from devices, components and vehicles on an ongoing construction site and the structures in use, as well as information from dedicated remote services (e.g. weather forecast, geographical information systems, projected logistics etc.). This borders the field of Cyber-Physical Systems, which has already investigated several uses of site material tracking (Srewil & Scherer 2013) or formwork sensing and monitoring (Yuan et al. 2016). However, the key bridge between the two is the BIM, storing the site captured data in a semantic way. With CDTs, as-designed and as-built BIMs are intended to be synchronized, thus allowing stakeholders to continuously monitor real progress against the initial BIM-based planning.

2.3 Construction site monitoring

The more recent methods of scan-to-BIM has been gaining traction on monitoring construction sites, via the use of photogrammetry (Tuttas et al. 2017), laser scanning with point clouds (Turkan et al. 2012) and aerial drone captures (Hamledari et al. 2017). To date, these are the more efficient technologies of capturing highly detailed site information. However, photos, videos and point clouds lack data structure and concise semantics, facing many difficulties when interfacing with BIM virtual objects. Additionally, these captures are taken at regular intervals, rather than in a continuous real-time manner, as it is not always necessary and would require huge data storage capabilities. Their integration in BIM-based information workflows thus requires careful planning and data handling.

In parallel to scan-to-BIM methods, Internet of Things (IoT) devices with BIM integration (Dave et al. 2018) seem to be the logical choice on monitoring infrastructure due to lower data storage and processing requirements (Pargmann et al. 2018), as well as more affordable deployment and maintenance costs. However, implementing IoT devices solely for construction purposes and their potential uses remain largely unexplored. As such, there is a major technology gap between construction and building operation, which needs to be reflected in the CDT transformation from one stage to another. Amongst other characteristics, IoT data flows differs from BIM-based files or databases as they consist in large volumes of dynamic sensed data, usually managed in ad-hoc time-series databases systems.

3 DIGITAL TWIN REQUIREMENTS

The deployment of DTs differs from BIM and cyber-physical systems, as it has to deal with a more comprehensive and interconnected system of systems, supported by semantics. This poses new challenges on both procedural (how can it be done) and technical (what can be used) levels. Additionally, the level of DT granularity (to what level) and its communication with other 'things' and DTs need to be carefully considered.

3.1 Procedural perspective

From a procedural perspective, there is a gap in understanding how a Construction Digital Twin (CDT) can be deployed in practice on physical sites, where its digital components should reside (physical data storage and access) and what precisely are its envisaged uses on and off site when compared to already existing BIM uses.

Methods on instantiating cyber-physical systems using BIM have already been described by Akanmu and Anumba (2015), considering the inclusion of sensor networks. Additionally, devices used to track the construction site with laser scanning and photo/video feeds have become more affordable and user friendly, and are being used more and more often by construction practitioners. The challenge remains in deciding which types of sensors to use during construction and how a combination of stationary simple sensors interact and complement more sophisticated devices such as laser scanners. This is expected to vary depending on the construction site (location, size, site availability during the day), the construction schedule (a balance of costs vs time), the required final level of quality that needs to be monitored, but also the health and safety of construction workers and nearby local communities (monitoring noise levels, pollution, traffic, etc.).

Considering that the deployment of site sensing is an additional cost to either the client or the construction companies, the benefits of using such devices need to be analyzed and potentially divided between client and contractors during procurement stages. Additionally, the types of devices which can be used throughout the entire lifecycle of the future built asset need to be considered at this stage. Unlike in existing buildings where sensing equipment can be stationary for longer periods of time, the dynamic nature of construction sites means that sensing devices themselves are in constant replacement and re-calibration – which equates to more time and labor in adapting the site sensing when required. Where structural monitoring sensors can aid during the construction as well as in the operations of subsequent lifecycle stages (Davila Delgado et al. 2018), sensing equipment attached to things such as formwork (Yuan et al. 2016) need to be constantly re-deployed and has no role in subsequent stages.

3.2 Technical perspective

From a technical perspective, the challenges appear with regard to managing dynamic data from site, interfacing Internet of Things (IoT) devices and linking data with the BIM, as well as ensuring cohesion between the various data sets involved.

There is no doubt that a CDT is seen as a virtual representation of the (real) construction site. However, this poses serious challenges in terms of who deploys the DT in the first place, where is it stored, who can access it (ownership, cyber-security) and in what formats it is kept in. Traditionally with BIM, the virtual models start appearing from the designers/constructors, with some information being transferred on to facility management and operation stages. However, as the DT is envisaged to traverse through all lifecycle stages, the data, its format and physical hosting need to shift from a consultant/design responsibility to the client and the facility owners and/or managers. As a consequence however, the DT is firstly dependent on BIM specific data sets (i.e. Industry Foundation Classes) which are scoped for the construction domain only. Additionally, as the DT is expected to evolve from one lifecycle stage to another, the built asset schematics need to be complete and up-to-date in terms of data and all its semantics, as well as integrated with the operation stage sensing equipment. As such, these requirements and restrictions on data transfers and ownership need to be considered carefully from procurement stages. The access to the DT instance (with all its data) needs to pass from a design/creation phase (overlapping with the physical asset construction stage) to an as-built phase using common semantics and similarly aligned data pipelines.

In terms of data semantics and their flexible interchange, a common solution at this stage is presented in the form of Semantic Web Linked Data (Pauwels et al. 2017), which offers the capability of expressing specific domain knowledge in machine interpretable formats. Formats such as RDF/S (Resource Description Framework/Schema) and OWL (Web Ontology Language) allows more flexibility and extensibility in modelling the semantics. This could allow a more dynamic definition of models at each stage and ways of adapting semantics depending on building type, installed sensors and required DT services at each stage of the lifecycle. Several existing ontological models have already been defined and tested (W3C LBDG 2020). However, most developments are focused on existing buildings and the facility management stage. On the other hand, the mapping between such models and the IFC has been a research subject for some time (Bonduel et al. 2018), bridging the initial gap of information transfer, and even recently applied to construction management (Boje et al. 2019). However, the actual use and integration of semantic web technologies for live construction sites remains un-explored. Additionally, the full extent of the IFC schema semantics remain unapplied during construction stages, in favor of third-party out of the box tools and equipment.

3.3 Modularity for CDTs

The end goal of DTs is to provide different services to help manage and optimize the lifecycle of the physical assets (Tao et al. 2018). Services represent business requirements that are implemented using proven engineering models and methods. These can vary greatly depending on construction tasks on and off site, and are expected to change and evolve during and after the construction stage. (Schleich et al. 2017) refer to a DT as an incorporation of different models, each representing different states of the physical twin. These need to be interoperable (the ability to establish equivalences between different models) and expansible (the ability to add, integrate and remove or replace models).

Considering the complex system that a DT represents, the different parts that make up a CDT should ideally be modular and independent from the whole and from each other. Each module needs to incorporate specific models and to provide a specific service. Additionally, a CDT should be dynamically adaptable to site conditions irrespective of infrastructure

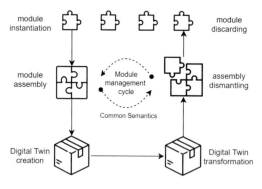

Figure 1. DT modular deployment process (the DTF shop floor in Figure 2).

type, ensuring information cohesion and transferability while not constraining future lifecycle modules.

A modular deployment process as represented conceptually in Figure 1 would be required, underpinned by a common set of semantics specifically designed for digital twin services – their description and the interactions between them using (functional) Application Programming Interfaces (APIs). This would account for both the DT initial creation (or instantiation of initial set of services) and transformation (when its services are added, updated, changed or removed). Thus, each module should be easily replaced and refitted on demand, depending on lifecycle requirements.

Whilst a modular approach seems logical and convenient, the challenge lies in deciding on a common platform with a set of semantics which is able to incorporate widely adopted industry data standards. Overlapping with the technical requirements from the previous section, a common semantic layer ensuring module-to-module communication would allow for a seamless integration of services, data formats, and even DT-to-DT communication and information exchanges.

3.4 Inter-digital twin communication

As a DT is envisaged to connect to real world things and be a digital replica linked to a physical asset, it cannot exist in isolation and therefore its communication to interconnected physical devices and indeed other DTs needs to be considered. While communication between a CDT and its site devices can be established using standard IoT protocols, the communication between multiple DT needs to occur at a different, higher level.

Another degree of complexity arises from the exchanges between DT from different domains (e.g. a CDT with smaller manufactured DTs or on-site construction machinery DTs), and DT of different scales (e.g. a CDT with existing building DTs, or district level DTs). Where the communication with its own incorporated site sensing equipment is at actuation level (data with low semantics), communication with other services, the web or consumers need to encompass a greater level of meaning (data with high semantics). Meeting this requirement would go hand-in-hand with having available a description of a common semantic layer, which would be passed on using popular formats such as XML (Extensible Markup Language) or JSON (JavaScript Object Notation) for typical structured information transfer between web-based services. However, this does not guarantee the alignment or equivalencies of concepts, as it requires re-interpretation and potentially re-development on the receiving end. Conversely, a semantic web linked data paradigm accounts for the transfer of the data as well as the definition of the models themselves (using common vocabularies, taxonomies and ontologies). In such a case, typical RDF/S and OWL can be used to describe resources which can be exchanged using various graph formats (e.g. ".rdfxml", ".ttl", etc.) or even via a hybrid approach using the JSONLD (JSON for Linking Data) format over common internet protocols. This would ensure a more complete transfer of semantics, allowing the receiving-end DT to interpret the information using the same definition, thus avoiding discrepancies or conflicts.

4 DIGITAL TWIN FACTORY CONCEPT

Considering the listed requirements and constraints from the previous, an initial conceptual Digital Twin Factory is proposed for the construction industry, with its simplified workflow depicted in Figure 2.

The main workflow steps of the DTF concept are considered here over multiple lifecycle stages, as suggested in Figure 2. These have been grouped up for simplicity in two different groups, being a) Construction/Deconstruction (assuming a dynamic construction site) and b) Operation/Renovation (assuming a more static built environment, with occasional small renovation works). The DTF workflow needs to adapt to the specific specifications and restrictions at each of these stages, as follows:

1. Requirements analysis – based on the type of a) construction site or b) operated building, a set of requirements is outlined by stakeholders. These requirements need to be considered from the prism of creating and managing a DT (as discussed in section 3) by considering the initial available datasets (BIM models and others), the required business services for the business processes (e.g. construction management, quality control etc.), the preferred sensing types and site equipment, but also the data specifications (format standards, storage types, accessibility, security, etc.). Once the requirements are broadly defined, the DTF can begin defining a DT template of modules based on a common semantics layer.

2. Digital Twin a) creation and b) transformation – once the requirements are met and validated, the DTF can begin the initial instantiation of the

modules and assemble them into a full-fledged Digital Twin (a – creation), which is not yet connected to its Physical Twin. Testing and validation of its services can be carried out at this stage, to assess its suitability before the actual deployment and real-time use. In the case of existing sites with already deployed DT (e.g. in-between lifecycle stages, or temporary renovation), the DTF should be able to operate on existing DTs and adapt them according to new requirement specifications (1-b), thus allowing a modular way of transforming (adding, changing, updating, removing) services.

3. Digital Twin Deployment – this step consists of the actual linking of data from the site physical assets (sensing equipment) to their corresponding virtual objects within the DT of the construction site. Testing the connection and calibrating the sensors are evident steps here, along with confirming to the DTF that the provided services are working as expected and can begin to be managed under the deployed DT. From this point one, the DT becomes its own entity – the live, virtual representation of the construction site. The physical storage of the CDT remains unclear. One possible way is to adopt a BIM like process of a Common Data Environment (CDE) to be shared between the stakeholders.

The common semantics layer represented in Figure 2 is a vital component ensuring the interoperability and standardization of the integral factory process. The DTF can be seen as a generic builder of templates which use common standards and models for expressing the CDT and its data. The conceptualization of semantics within the DTF have to be generic/abstract enough to incorporate any type of service (embedded models), sensing types (observable properties) and equipment (latest tools used on site ranging from sensing equipment to actuators, construction machinery, etc.). Within the specific case of the data requirements part, the DTF should gather and keep track of meta-data about the data specifications (e.g. format, databases, external data sets, external CDEs used, etc.). A basic semantic integration of concepts at the DTF level can allow conceptual integration of modules, which can be extended to represent more complex concepts such as categories of modules by different roles, capabilities, building types. This addition of meta-data object facility would stimulate the growth and constant improvement of the DTF.

In addition to a common semantics layer, which would ensure a unified vocabulary and language, the DTF requires a transactional layer. This would manage and keep track of all transactions related to creating and transforming its outputted DTs, allowing dynamic updates and roll-backs to previous states. Additionally, newly inputted requirement definitions should be flexible enough to allow the registering of new services, sensors and tools which would undergo though this transactional layer, becoming aligned to the DTF semantics. Thus, the DTF would have a prime role in updating the common semantics of each DT as time progresses, being the point of synchronization between multiple DTs and the prime entity that governs the transformation of DTs over time, in a standardized way.

5 CONCLUSION AND FUTURE WORK

This article proposal aims to explore the opportunity to go further with the DT paradigm, allowing fully automated processes and data pipelines from DT conception to its lifecycle management.

If digital twins are the next major implementation in the construction sector as was outlined within this paper, there needs to be a standardized way to deploy and manage them over time. To this end, we have proposed the concept of a Digital Twin Factory which would have the role in creating Construction Digital Twins of built assets in a modular way, using common semantics. The conceptual framework for the DTF was scoped down to the specifics of the architecture and construction engineering sector, but it can be extended to nearby engineering domains which are encouraged to adopt a similar philosophy for DT standardization. This is mainly due to the need for interconnectivity of our society, and the interoperability between DT of different types, scales and application domains.

Future work aims to explore the DTF concept in more detail, looking at more practical ways of implementation (e.g. survey of existing proprietary and open source tools) and the investigation of the suggested transactional layer. Additionally, based on the presented framework, we believe that the interaction of DTs with linked data remains a potential research avenue for years to come. Aspects such as the integration of DT with process models like LEAN construction, or even environmental assessment methods like BREAM/LEED/HOE still need to be considered and

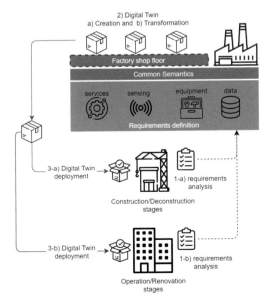

Figure 2. The Digital Twin Factory workflow.

analyzed, unlocking more potential than a BIM-based world view.

ACKNOWLEDGEMENTS

The authors acknowledge the financial support from Fonds National de la Recherche (FNR) Luxembourg, and Agence Nationale de la Recherche (ANR) France to the 4DCollab grant reference: 11237662 (LU)/ANR-16-CE10-0006-01(FR).

REFERENCES

Akanmu, A. & Anumba, C.J. 2015. Cyber-physical systems integration of building information models and the physical construction. *Engineering, Construction and Architectural Management* 22(5), pp. 516–535.

Boje, C., Bolshakova, V., Guerriero, A., Kubickia, S. & Halin, G. 2019. Ontology assisted collaboration sessions on 4D BIM. *Creative Construction Conference 2019, June – 2 July 2019*. Budapest, Hungary, pp. 591–596.

Boje, C., Guerriero, A., Kubicki, S. & Rezgui, Y. 2020. Towards a semantic Construction Digital Twin: Directions for future research. *Automation in Construction* 114.

Bonduel, M., Oraskari, J., Pauwels, P., Vergauwen, M. & Klein, R. 2018. The IFC to Linked Building Data Converter – Current Status. *6th Linked Data in Architecture and Construction Workshop (LDAC), CEUR Workshop Proceedings*.

Dave, B., Buda, A., Nurminen, A. & Främling, K. 2018. A framework for integrating BIM and IoT through open standards. *Automation in Construction* 95, pp. 35–45.

Davila Delgado, J.M., Butler, L.J., Brilakis, I., Elshafie, M.Z.E.B. & Middleton, C.R. 2018. Structural Performance Monitoring Using a Dynamic Data-Driven BIM Environment. *Journal of Computing in Civil Engineering* 32(3).

ECTP Innovative Built Environment 2019. *Strategic Research and Innovation Agenda*.

Grieves, M. 2014. Digital Twin: Manufacturing Excellence Through Virtual Factory Replication. White paper.

Haag, S. & Anderl, R. 2018. Digital twin – Proof of concept. *Manufacturing Letters* 15, pp. 64–66.

Hamledari, H., McCabe, B., Davari, S. & Shahi, A. 2017. Automated Schedule and Progress Updating of IFC-Based 4D BIMs. *Journal of Computing in Civil Engineering* 31(4).

Pargmann, H., Euhausen, D. & Faber, R. 2018. Intelligent big data processing for wind farm monitoring and analysis based on cloud-technologies and digital twins: A quantitative approach. *2018 IEEE 3rd International Conference on Cloud Computing and Big Data Analysis (ICCCBDA)*. IEEE, pp. 233–237.

Pauwels, P., Zhang, S. & Lee, Y.-C. 2017. Semantic web technologies in AEC industry: A literature overview. *Automation in Construction* 73, pp. 145–165.

Schleich, B., Anwer, N., Mathieu, L. & Wartzack, S. 2017. Shaping the digital twin for design and production engineering. *CIRP Annals* 66(1), pp. 141–144.

Schluse, M., Priggemeyer, M., Atorf, L. & Rossmann, J. 2018. Experimentable Digital Twins—Streamlining Simulation-Based Systems Engineering for Industry 4.0. *IEEE Transactions on Industrial Informatics* 14(4), pp. 1722–1731.

Srewil, Y. & Scherer, R.J. 2013. Effective Construction Process Monitoring and Control through a Collaborative Cyber-Physical Approach. *IFIP Advances in Information and Communication Technology*., pp. 172–179.

Tao, F., Cheng, J., Qi, Q., Zhang, M., Zhang, H. & Sui, F. 2018. Digital twin-driven product design, manufacturing and service with big data. *The International Journal of Advanced Manufacturing Technology* 94(9–12), pp. 3563–3576.

Tao, F., Zhang, H., Liu, A. & Nee, A.Y.C. 2019. Digital Twin in Industry: State-of-the-Art. *IEEE Transactions on Industrial Informatics* 15(4), pp. 2405–2415.

Tuegel, E.J., Ingraffea, A.R., Eason, T.G. & Spottswood, S.M. 2011. Reengineering Aircraft Structural Life Prediction Using a Digital Twin. *International Journal of Aerospace Engineering* 2011, pp. 1–14.

Turkan, Y., Bosche, F., Haas, C.T. & Haas, R. 2012. Automated progress tracking using 4D schedule and 3D sensing technologies. *Automation in Construction* 22, pp. 414–421.

Tuttas, S., Braun, A., Borrmann, A. & Stilla, U. 2017. Acquisition and Consecutive Registration of Photogrammetric Point Clouds for Construction Progress Monitoring Using a 4D BIM. *PFG – Journal of Photogrammetry, Remote Sensing and Geoinformation Science* 85(1), pp. 3–15.

W3C LBDG 2020. Linked Building Data Community Group.

Yuan, X., Anumba, C.J. & Parfitt, M.K. 2016. Cyber-physical systems for temporary structure monitoring. *Automation in Construction* 66, pp. 1–14.

Lifecycle oriented digital twin approach for prefabricated concrete modules

M. Wolf, O. Vogt, J. Huxoll & D. Gerhard
Digital Engineering Chair, Ruhr-University Bochum, Germany

S. Kosse & M. König
Chair of Computing in Engineering, Ruhr-University Bochum, Germany

ABSTRACT: Building Information Modeling (BIM) is increasingly used throughout the construction industry as a crucial tool for project management. In order to use BIM throughout the entire life cycle of a building, aspects of production, maintenance and renovation of the structure must not be neglected, especially if the share of industrial prefabrication with corresponding modularization increases. The challenge is also to digitally support new prefabrication and production concepts during the design and recycling process. In this paper, the authors present an approach to transfer digital twin concepts from the field of mechanical engineering to civil engineering and thus extend current BIM approaches by cyber physical components. As a use case, the production of pre-cast concrete modules made of free-form high performance concrete in the construction industry is presented and in particular a concept for the transfer and systematic collection of requirements, relevant information models and interactions that form an administrative shell for digital twins in BIM environments.

1 INTRODUCTION

According to an average forecast of the United Nations, the world population will grow by another 2 billion people in the next 25 years. The demand for housing, infrastructure and utilities is enormous. Concrete structures play a special role here. Concrete has established itself as the undisputedly most used building material due to its free formability, worldwide availability and low material costs. The aim is to use significantly less material, avoid errors, introduce consistent prefabrication with quality assurance and thus achieve the fastest possible construction on site. The key to this is the dismantling of concrete structures into similar individual modules, which are then mass-produced in a fully digitalized production process.

One of the critical factors for the increased productivity and efficiency of today's manufacturing is the use of information and communication technologies (ICT) across the entire product creation process. Development times for new products can be shortened as a result of early and effective communication as well as coordination between design and production. Especially, for the fast and precise production of pre-cast concrete modules made of free-formable high-performance concrete, relevant data on the production process and the current status of the module must be continuously recorded, compiled and made available. Using this digital information, the individual production steps can be more precisely coordinated, and the utilization of the machines more effectively planned. The addressed modules have series character and enable rapid construction. Visual individuality of the composite structure is being preserved, so that scalable, adaptive and over the life cycle alterable modules and suitable production concepts are sought. The modularization starts retrospectively from the structure (see Figure 1). The production of the modules in the

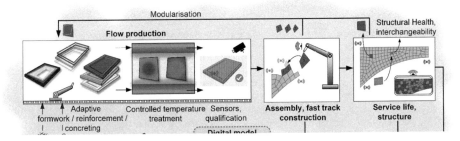

Figure 1. Adaptive modularized constructions made in flux (Source: https://www.ruhr-uni-bochum.de/spp2187).

factory and their quick assembly influence the module design. Production is automated in a linear flow principle with the individual steps of forming, reinforcing, and controlled rapid hardening by heat treatment, qualification of the individual modules and identification by sensors. The sensor technology is used for complete tracking, assembly and as a service life indicator to evaluate the load-bearing capacity or serviceability characteristics of modules over their lifetime. An integrated, digital model controls all processes and interactions.

An important prerequisite for the realization of integrated digital models, also known as digital twins, is the combination of current Building Information Modeling concepts and proven methods from the context of Industry 4.0, which provide for largely self-organized production by integrating digital tools and automated production technology.

Based on the current developments in the field of Building Information Modeling and Industry 4.0, this paper focuses on the development of consistent and adaptable data and interaction models for the industrialized, fault-tolerant rapid production of pre-cast modules. Based on a systematic collection of all relevant information and interactions in the form of ontologies, suitable descriptions for digital twins are developed. A formal and verifiable description of the requirements regarding function and quality, considering the possible uncertainties in the course of production, is essential. For this purpose, information from used materials, systems and processes must be collected, integrated and analyzed. On the basis of interactions with other modules or connections, it is possible to check the requirements continuously. In the sense of Industry 4.0, the modules are supposed to communicate with the respective production machines and other modules during production. Based on the collected data, it can be checked at any time whether the current status of the module still meets the previously defined requirements. In case of deviations, the data can be used to check further use elsewhere or to make automatic corrections. Finally, this approach enables agile control of the planning and production of pre-cast modules. Based on the findings from mechanical engineering and research activities in the context of Industry 4.0, innovative adaptive modular designs with flow manufacturing methods require real-time networking of products, processes and systems based on consistent data.

2 STATE-OF-THE-ART AND RELATED WORK

2.1 Building Information Modeling

Building Information Modeling stands for a cooperative approach using digital building models for various aspects within the life cycle of a building (Borrmann et al. 2018). The focus of international research and development is on supporting the design, construction and operational phases. Different methods have been developed for different use cases to support the participants. These include the management of heterogeneous information sources (Becerik-Gerber & Jazizadeh 2012), the digital exchange of knowledge (Alashwal 2011), automated collision checks (Han et al. 2014), the creation of construction process simulations (Haque & Rahman 2009), the connection of energy analyses (Gupta et al. 2014) and the use of models for operation (Lin & Su 2013).

In the field of planning, production and assembly of pre-cast concrete elements, digital building models are already used in different ways. In the field of planning, the focus lies on the one hand on how precast concrete parts can be described geometrically in a reusable and straightforward way (Sacks et al. 2004) and on the other hand on how the production processes can be implemented optimally under consideration of production-related deviations. Initial approaches for the integration of external information for the control of the production process have also been considered (Zhong et al. 2017).

One of the most important premises for Building Information Modeling is working with three-dimensional geometry. Explicit and implicit methods are used to describe solid models. Another extremely important trend in the construction industry is parametric modelling, which allows geometric models to be provided with dependencies and constraints that it can be adapted to changing boundary conditions quickly and with little effort. Thereby, dimensions of geometric objects can be used as parameters (Borrmann et al. 2018).

The Industry Foundation Classes (IFC) provide a comprehensive and standardized data format for the manufacturer-independent exchange of digital building models. It thus forms an essential basis for the implementation of BIM. Currently, the prefabrication process is not deeply integrated into this development. As prefabricated products become more and more complicated, the demand for a more powerful model exchange interface is increasing.

The IFC4pre-cast project aims to create an extension of the IFC format. However, the focus is on extending the IFC, using certain IFC classes for structuring and defining additional properties. Such developments can be used in the future to improve data exchange. The focus of this paper is on concepts for the holistic establishment of digital twins, rather than individual data exchange formats such as IFC.

2.2 Pre-cast elements

The high-precision rapid construction of the future is based on the reduction of individual, complex parts to basic modules, which can be produced highly scalable with modern flow production methods. The main advantage of this approach is that the majority of the components can be prefabricated in industrial series production and only need to be assembled on-site to form the overall structure. Industrial series production is only possible if the finished parts can be produced with a high repetition rate.

Pre-cast concrete elements are very well suited for high volume series production due to their inherent modular character. However, the modular character also comes with a drawback. Pre-cast concrete elements are unique components that are designed for a specific position in a structure. Therefore, some essential design considerations and careful planning are required, especially in the area of connections and bearing of the components.

The possibilities of connecting pre-cast concrete parts are generally limited to grouting pockets with connecting reinforcement or special installation parts that are filled with concrete at the construction site. Furthermore, welded-on steel plates can be used which are already inserted into the pre-cast concrete element at the right place in the production plant and welded to the connecting steel parts at the construction site.

One way of producing pre-cast parts for bridge construction that fit precisely together is the contact method (also known as match casting). Because one part is used as formwork for the next part, the parts will later fit together very precisely. The finished parts can be pre-stressed with a "dry joint", i.e. without glue or mortar. The demands on the tolerances at the contact points are very high.

Experiences from mechanical engineering show that compliance with strict manufacturing tolerances is essential for the application of flow manufacturing processes. In the construction industry, tolerances typically lie in the range of a few centimeters, while in mechanical engineering, the requirements are much higher, sometimes only a few hundredths of a millimeter. In order to ensure error-free assembly on the construction site, precise planning and production of the prefabricated parts are essential, which goes far beyond the current demands in the construction industry.

Tolerances are applied to deviations from the nominal value which can occur in the form of length and cross-sectional deviations, angle deviations, alignment deviations and deviations from the concrete cover dimension. A distinction must be made between random and systematic deviations. While random deviations can deviate in both positive and negative directions, systematic deviations have a characteristic directional bias. Both types occur with an approximately normal distribution.

The measurement of the deviations is typically carried out by manual measurements with folding rulers or measuring tapes. These methods are not only time-consuming but also inaccurate. New fully automated and digital measurement and evaluation methods must be developed for use in a fully digitized production process. Kim et al. present a method for dimensional and surface quality assessment using 3D laser scanning in combination with BIM (Kim et al. 2014). The presented method can determine the dimensions with an average error of 2.5 mm for simple rectangular surfaces with constant thickness. Defects of the surfaces can be localized with an accuracy of 86.9%. The technique is currently limited to two-dimensional surfaces. In their further work, Kim et al. extended their method so that a comparison between as-designed state and as-built state from 3D laser scans can be performed automatically (Kim et al. 2016). In a BIM-based approach, all data is made available to all stakeholders involved in the building process.

2.3 *Digital twin model representations*

There are many essential terms around the Digital Twin Representation. Those terms are specified in The RAMI 4.0 standard DIN 91345. For this paper, the terms asset and administration shell need a further explanation.

An asset is a uniquely identifiable entity with value to the organization. An asset can be a physical object, but it can also be software, service or design. Every asset has its own lifecycle and a unique identifier, using indicators according to ISO 29002-5 or a URI in the URN namespace according to RFC 3986.

The Asset Administration Shell (AAS) stores and provides all the information about the asset and is a holistic digital representation of the asset. The AAS and asset have a one-to-one relationship, and together in combination, they are an Industry 4.0 (I4.0) component (Plattform Industrie 4.0 2016).

The AAS also has a unique identifier. Therefore, the I4.0-component is uniquely identifiable too. The identifying information about the AAS itself and the asset is stored in the header. The body consists of a various amount of submodels. In those submodels relevant information on specific aspects is clustered. Every submodel contains specific data properties and derived functionalities. With all those submodels the shell can offer specialized services. For automated communication, the submodels should follow the approach of standardization (Plattform Industrie 4.0 2016). Standardized submodels can be used for different AASs, but it is still possible to create unique and individual submodels for every AAS. This concept does not apply only to the submodels, it is also applicable for the AAS itself. A common set of data properties with individual characteristics can be relevant for different I4.0-components. The type-relevant data is stored in a type AAS, the individual AAS of realization or the so-called instance AAS obtains all properties from the type AAS. From a type AAS, many instance AAS can be created when the production of a product starts. Information about the production or test parameters are stored in the AAS. After production, the product is in the usage and maintenance stage. Data from this stage, like sensor data, are also implemented in the AAS. The AAS follows the lifecycle perspective of RAMI 4.0.

Within the scope of Industry 4.0, the automated data transfer is a topic of high interest for any participant of the value chain. Regarding data transfer using AAS, there has not been a profound solution yet. So far, it is still a research topic without an answer about how this transfer will work. Concept ideas propose that AAS can be used in a passive and active mean in order to transfer data in an Industry 4.0 environment. Passive use is achieved by exchanging the AAS like a data sheet between a requesting and a responding instance, for example between two partners along the value chain. The information can, for example, be sent in an XML-, RDF- or JSON-format. In addition, passive usage can be done by the use of an API to request and send data to a shell. The active usage, describing the pro-active exchange between shells, is still under development as of today.

The data exchange formats (XML, RDF or JSON) address the entire live cycle of type and instance administration shells. However, this information needs to link to development or production systems. During the engineering of a product, type and instance information of AAS should be realized using Automation Mark-Up Language (AutomationML). From engineering, these data will be transferred to operational phase within production. In Industry 4.0, OPC-UA has been the framework of choice to enable interoperability in data exchange. Therefore, OPC-UA information models should have access to all information of the administration shell and provide production processes with real-time data.

3 CONCEPT FOR THE ADMINISTRATION SHELL FOR DIGITAL TWINS

For efficient and precise production of flexible modules made of high-performance concrete, all relevant data must be continuously collected, combined and made available. The information can be used for different use cases throughout the life cycle, from design to deconstruction, including recycling. For example, the information is a fundamental basis to coordinate the single production steps more precisely. During production, the module's AAS can communicate with the machines and collect data on product quality. If necessary, machine settings can be tuned mid-process or the integration of additives and other ingredients can be adjusted. Pre-cast modules that do not meet certain tolerances can be exchanged or be used within other structures with lower quality requirements. This enables an automatic rescheduling of the production. In addition, further evaluation can be carried out with regard to the load-bearing capacity of the entire system.

The idea is that all data (process steps, machine data, materials, etc.) is directly assigned to the corresponding module's AAS. Furthermore, requirements that have been defined for an individual module must be integrated in a verifiable manner. This, in turn, requires access to detailed real-time data about the product. In the end, a digital twin is created during product design ("digital twin") and updated during production (when the "physical twin" actually starts its existence). Subsequently, holistic data analysis can be carried out, and conclusions can be drawn as to how such modules can be produced more efficiently and with higher quality in the future using flow production. Such ideas and requirements have already been successfully developed and implemented in mechanical engineering in the context of Industry 4.0.

Within the scope of this paper, consistent and adaptable data and interaction models are developed for the industrialized, fault-tolerant rapid production of adaptive modular products. The main research question, which is being addressed in this paper, focuses on how one improves the quality of individual concrete modules based on the collected data from engineering and production.

To answer this question, asset administration shells, as described by the Platform Industry 4.0, are developed for use in the construction industry. The basic idea is to surround every asset with an asset administration shell that can provide a minimal but sufficient identification and description according to different use cases. The concept of an administration shell in civil engineering is illustrated by the following example.

To establish a more efficient flow production, the building is divided into specific modules during the design phase. In our example, a building in skeleton construction is divided into four different structural modules such as columns, beams, frame corners or junction modules (see Figure 2). Those modules would be produced in a flow production system and put together at the final place of the building's construction. This production system should be capable of

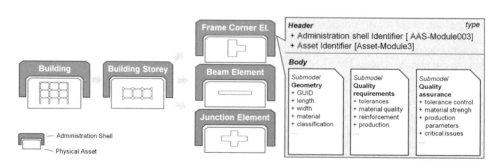

Figure 2. Concept of administration shells for elements in a skeleton building construction.

producing a certain and rather low number of different modules to keep production processes efficient and setup times low. A large number of buildings can be created by putting these modules together. This way, modern and efficient pre-cast processes can be enabled for many kinds of construction projects.

In the context of pre-cast concrete modules, a dedicated asset administration shell is set up for each individual module. The administration shell encapsulates all necessary information and provides interfaces for communication with other administration shells. Other administration shells can be other pre-cast modules, machines for production or quality assurance or also later construction equipment during the assembly.

The asset management shell of a pre-cast module is composed of a number of submodels. These Submodels represent different aspects of the asset concerned. For example, they may contain a description relating to quality requirements, but could also store results for structural analysis or energetic characteristics. Only a partial description is defined and standardized for each aspect to form a template for a submodel.

For example, submodels can be used for the geometric description, the structural design, the requirements for tolerances or the realization of connections and joints. For communication between different components or machines, certain properties can then be assumed or required to exist. For example, a production machine can request information on maximum tolerances directly from the pre-cast module's AAS. In addition, a system for tolerance checking, e.g. a laser scanner, can, in turn, send the results of the geometric target-actual comparison to the pre-cast module.

To represent the complete product lifecycle and the benefit of the AAS concept, a five-staged approach is used. The five stages comprise of the construction design (phase 1), modularization (phase 2), flow production (phase 3), quality control for individual modules (phase 4) and matching the quality requirements measured data (phase 5). The introduced example is used further to clarify this approach.

3.1 Phase 1: Construction design

In the first phase, the administration shell of a single pre-cast concrete element is derived from a 3D model of a building or construction. The 3D model is provided using the Industry Foundation Class (IFC) format, which is established as the standard format for the digital description of building models. The building model contains information about element, properties and geometries, information about the logical structure and the relationship between the building structure and the individual components. The building model follows a hierarchical structure consisting of a project as a root entity. Subordinate to this root entity are entities of the site, building, floor and building elements. Based on the overall model in IFC format, a significant portion of the management shell for this submodel can be created automatically.

3.2 Phase 2: (Automated) modularization

The aim of modularization is to decompose a larger construction element into smaller modules. For this purpose, modules of the same type should be used as far as possible so that a high degree of series production is possible. Consequently, individual modules are created, which in turn are represented by their own administration shell. However, the structure of the administration shell differs from the administration shell of the parent construction element. The individual administration shells can communicate with each other and with the parent building component with the help of standardized interfaces. Thus, the modules know where they are located in the overall building and which other modules are adjacent. If several identical modules can be used for one building element, they can be exposed to different loads due to their position in the building. This can also result in different quality requirements, e.g. that only smaller tolerances are allowed if large forces have to be taken over by connected modules. The geometrical information about the module, its connecting possibilities and requirements for the concrete production must be stored as submodels. Thus, the administration shell contains all necessary information about the module. In the course of production, assembly and also use, further submodels can be added to the individual modules. Also, the administration shell of the parent component gains an additional submodel, which contains information about the modularization process, i.e. under which assumptions and conditions individual modules were formed. If assumptions change, it is possible that the decomposition is no longer useful and has to be adapted. In the context of this paper, however, the focus is first of all on the initial modelling of the administration shell of modules.

3.3 Phase 3: Flow production

The construction of a building is usually performed on a specific construction site outside of modern production factories. This makes it difficult to apply modern production technology to on-site construction processes. Therefore, a stable Industry 4.0 production system needs to be designed for a pre-cast flow production of modules. The modularization of the construction object allows a shift from on-site to factory-site production of the building elements, or more precisely the engineered modules. The assembly of those modules will still be performed on the construction site, but in a time- and cost-saving, more efficient, way. This production system will be able to produce a certain amount of different modules. Modern production technologies like 3D-printing will be adapted for, i.e. carbon concrete printing, in order to produce modules without current limitations. Creating those modules in a stable production environment allows a more controlled and digital supervised value creation. For instance, quality checks will be performed using state-of-the-art sensors to identify anomalies within and after production. The concept of an asset administration shell provides the digital backbone of the information management for the production activities. Each module is individually identifiable and assigned to a specific production order. The AAS integrates the information about what has and what should be done to its asset. In more detail, the production order can be stored as a submodel within the AAS body. This digitally stored data can be used by other AAS, for example, to establish an automated machine-to-machine communication. The module is able to tell a 3D-printer all production parameters before the start of the actual production process. This way, the I4.0-paradigm is applied. The performance parameters of the machine can be added to the production machine's administration shell. The planned and realized requirements are stored in the administration shell of each individual module. This way, the whole production system is able to receive all relevant module information and to realize an efficient flow production.

3.4 Phase 4: Quality control for individual concrete modules

The products manufactured in phase 3 have to be quality controlled in a similar way as in the regular industrial production in mechanical engineering. Since both the production information and the requirements for the targeted geometry are stored in the AAS of the individual concrete modules, they can be physically checked using the pre-planned measurement plans with compatible machines or procedures. The measurement results are then also stored in the concerning submodel of the respective module's AAS.

3.5 Phase 5: Matching quality requirements and measured data

After checking each component for its dimensions and the tolerances achieved compared to the requirements, the 5th phase is carried out. Here the previously defined structures are joined together virtually from the individual concrete modules as a whole. This virtual verification can check on the basis of the individually recorded dimensions and the availability by the AAS whether the structural element, which is composed of the modules, can meet the requirements set for the structural element with regard to the overall dimensions.

If there are deviations from the maximum permissible tolerance, various measures can be initiated to correct the overall structure. These include the reworking of individual concrete modules to match the dimensions of other modules to be connected or the reproduction of individual concrete modules if reworking is not possible for the given instance or batch of concrete modules. The third possibility is to replace a concrete module with another module in the appropriate place that meets the requirement.

All of the mentioned possibilities can be tested algorithmically using the individual AAS of the concrete modules and the AAS of the building element before physical production steps are initiated. In particular, a "faulty", i.e. not requirement-compliant, production of the modules is not first discovered during assembly on the construction site.

4 IMPLEMENTING AN EXAMPLARY ASSET ADMINISTRATION SHELL

As mentioned before, an Asset Administration Shell consists of its manifest, the component manager and several individual submodels. These submodels represent the individual data sources for the relevant activities and interactions enabling flow production in this context like shown in the phases 1 to 5 in the concept section. The following example, depicted in Figure 3, shows an Asset Administration Shell for a building storey, which was created and viewed in

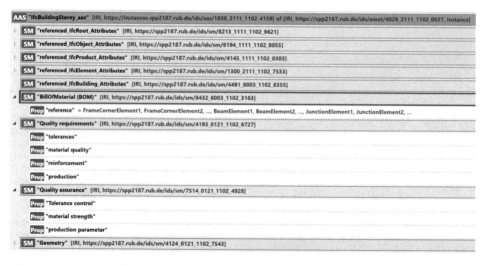

Figure 3. Exemplary administration shell implementation for a single storey instance.

the open source software AASX Explorer (GitHub 2020). For a better understanding this particular AAS shows only the core principles and not every possible submodel for this application.

The *IfcBuildingStorey_aas* has different submodels for the provision of the various information needed for the described processes. The first few submodels starting with *"referenced_"* refer to the hierarchical structure and types of data elements typically found in IFC files. The *BuildingStorey* inherits all attributes and properties from its parent *IfcBuilding,* whereas *IfcBuilding* inherits those from *IfcElement* and so on. Those inherited attributes do not have to be created redundantly. Once they are created, they are inherited by references, meaning that if an attribute of *IfcObject* is changed, the attributes in the *referenced_IfcElement_Attributes* will change too. The described submodels show that the IFC data model can provide both structure and submodels (meaning in this case distinct attributes, which are used in civil engineering use cases) for the Asset Administration Shell concept. In addition to the submodels referring to IFC, the example shows submodels for further information of the storey element.

The next major submodel defines the bill of material (BOM). In this submodel, the references to every single AAS of either generated or manually engineered concrete modules are stored, to gather the "child elements" the storey is to be built of. In this example, the information that the storey is assembled of frame corner elements, beam elements and junction joints is stored in this submodel. Additional to pure references, it is possible to create a submodel for information about the assembly process. This submodel would then store data on how the modules are assembled or where which module belongs in the assembly. For a better overview and understanding, such a submodel is omitted in Figure 3, while it is implemented in the prototype. Besides the submodels for assembly information and organization and IFC structure, submodels for requirements, quality criteria, measurement data, manufacturing information and 3D-models can be created to support the different phases mentioned in the concept.

The shown AAS has a submodel for quality assurance with properties like tolerance control or material strength. More properties can be added for different methods of quality control. The quality requirements submodel consists of properties like tolerances or specific material requirements. Those properties can be specified by adding precise target values, which are later used in measurement procedures following manufacturing. This excerpt of our wholly implemented example shows that IFC and the concept of AAS can be adapted to create an AAS for concrete modules, which, when assembled, form an individual building element.

3As mentioned before, there are many different approaches to how Asset Administration Shells can communicate with each other. In relation to our first implementation, a useful service between multiple AAS could be the exchange of information between the submodel *measurement data* of all **element** modules in the BOM with the submodel *requirements* of the *storey* module. With this relation, it is possible to check whether the requirements, for example, the maximum or minimum dimensions of all modules combined to form the building element, are met or not.

5 CONCLUSION AND OUTLOOK

In this paper, a method to ensure quality control of concrete modules for a modularized flow production by using the concept of the administration shell from mechanical engineering was shown. The basics, the targeted geometries and the exact manufacturing processes are currently being developed in the priority program 2187 (adaptive modularized constructions made in flux) funded by the German Research Foundation in order to be applied in the described processes. The next step for the enhancement of the digital twin for concrete modules is the definition of templates for submodels and relevant properties for the identified processes.

Future work will include the automated generation of AAS from IFC files. While these AAS are certainly not entirely defined by the time all relevant data has been extracted from the IFC files, the granularity and structure of the engineered building elements help to create the desired information structure. These temporary AAS will then be further augmented via assistant systems and relations to production systems and current engineering software.

REFERENCES

Alashwal, A.M., Rahman, H.A., Beksin, A.M. 2011. Knowledge sharing in a fragmented construction industry: On the hindsight. Sci. Res. Essays 6 (7): 1530–1536.

Becerik-Gerber, B., Ku, K., Jazizadeh, F. 2012. BIM-enabled virtual and collaborative construction engineering and management. ASCE J. Prof. Issues Eng. Educ. Pract. 138 (3): 234–245.

Borrmann, A., König, M., Koch, C., Beetz, J. (Ed.) 2018. Building Information Modeling – Technology Foundations and Industry Practice, Springer International.

DIN SPEC 91345:2016-04 2016. Referenzarchitekturmodell Industrie 4.0, Deutsches Institut für Normung e. V.

DIN EN 13306 2012. Instandhaltung – Begriffe der Instandhaltung, Deutsches Institut für Normung e. V.

GitHub repository 2020. aasx-package-explorer, https://github.com/admin-shell-io/aasx-package-explorer.

RFC 3986 2005. The Internet Engineering Task Force. Uniform Resource Identifier.

Gupta, A., Cemesova, A., Hopfe, C.J., Rezgui, Y., Sweet, T. 2014. A conceptual framework to support solar PV simulation using an open-BIM data exchange standard. Autom. Constr. 37: 166–181.

Han, N., Yue, Z.F., Lu, Y.F. 2012. Collision detection of building facility pipes and ducts based on BIM technology. Adv. Mater. Res. 346: 312–317.

Haque, M.E., Rahman, M. 2009. Time-Space-Activity conflict detection using 4D visualization in multi-storied construction project, International Visual Informatics

Conference 2009. Bridging Research and Practice, Visual Informatics: 266–278.

Kim, M., Cheng, J.C.P., Sohn, H., Chang, C. 2015. A framework for dimensional and surface quality assessment of pre-cast concrete elements using BIM and 3D laser scanning, Automation in Construction, Volume 49, Part B: 225–238.

Kim, M., Wang, Q., Park, J., Cheng, J.C.P., Sohn, H., Chang, C. 2016. Automated dimensional quality assurance of full-scale pre-cast concrete elements using laser scanning and BIM, Automation in Construction, Volume 72, Part 2: 102–114.

Lin, Y.C., Su, Y.C. 2013. Developing Mobile-and BIM-based Integrated Visual Facility Maintenance Management System. The Scientific World Journal, Article ID 124249, 10 pages.

Plattform Industrie 4.0 und ZVEI 2016. Struktur der Verwaltungsschale. Fortentwicklung des Referenzmodells für die Industrie 4.0-Komponente, Federal Ministry for Economic Affairs and Energy, Berlin.

Sacks, S., Eastman, C. M., Lee, G. 2004. Parametric 3D modeling in building construction with examples from pre-cast concrete. Automation in Construction, Volume 13, Issue 3: 291–312.

Zhong, R. Y., Peng, Y., Xue, F., Fang, J., Zou, W., Luo, H., Ng, S. T., Lu, W., Shen, G. Q. P., Huang, G. O. 2017. Prefabricated construction enabled by the Internet-of-Things, Automation in Construction, Volume 76: 59–70.

A Digital Twin as a framework for a machine learning based predictive maintenance system

C.-D. Thiele, J. Brötzmann, T.-J. Huyeng & U. Rüppel
Technische Universität Darmstadt, Institute of Numerical Methods and Informatics in Civil Engineering, Darmstadt, Germany

S. R. Lorenzen, H. Berthold & J. Schneider
Technische Universität Darmstadt, Institute of Structural Mechanics and Design, Darmstadt, Germany

ABSTRACT: Due to the advancing digitization in the building industry, improving sensor qualities and growing opportunities of wireless technologies, the potential and possible applications of a digital twin in connection with IoT-Devices (Internet of Things) enlarges. Therefore, the field of implementing digital twins in the AEC (Architecture, Engineering & Construction) sector will be examined in this paper and in relation to the research project ZEKISS. In this project an AI (artificial intelligence) will be developed which is meant to give predictions of the structural health of railroad bridges based on structural FE-models and real-time sensor data. The foundation for the connection of the gathered data and the structural elements of the bridges is the developed digital twin. In the scope of ZEKISS, the definition of a digital twin goes beyond the 3D representation of real objects and the linking of the sensors to the digital elements as known from the BIM method. Furthermore, the digital twin is seen as the whole framework with the AI as the core-element on which any kind of necessary simulation can be performed.

1 INTRODUCTION

An aging infrastructure in large parts of the world and the growing shortage of skilled workers are an increasing challenge in monitoring and maintenance of infrastructures (DB Netz AG 2019). Traditionally, visual inspections, which are carried out at regular intervals and often in combination with the application of measuring systems, play an important role in the detection of defects on the surface of the building element or the structure itself. Nevertheless, this inspection type is labor-intensive, time-consuming and subjective, even for well-trained inspectors (Sun et al. 2020) and can require operational breaks. It also makes it impossible to track the condition of the infrastructure in real time. In addition to the varying operational loads that occur in day-to-day operations, the infrastructure is exposed to a wide variety of environmental influences, this is called environmental and operational changes (EOC) in the structural health monitoring (SHM) community (Chapuis & Sjerve 2018). EOCs include extreme events like earthquakes which can accelerate structural damage and even lead to sudden failure. Real-time monitoring of these structures counteracts this.

Nevertheless, not only the real-time monitoring but also and more importantly predictive maintenance plays a decisive role in avoiding extreme events such as the total failure of structures. With predictive maintenance it is possible to give statements about the lifetime of the structure. Therefore, it can be used as well to rank structures in relation to their lifespan. This makes it possible to tackle infrastructure projects in the order of their importance.

Furthermore, the advancing digitization in the building industry in combination with the rising computational abilities, improving sensor qualities, growing opportunities of wireless technologies and data analysis methods makes it possible to handle and implement novel monitoring and predictive maintenance systems (Araujo et al. 2012). For this purpose, a digital twin as a framework for a machine learning based predictive maintenance system is proposed in the present paper.

Therefore, the field of implementing digital twins in the AEC sector will be examined in more detail in the scope of this paper and in relation to the research project ZEKISS (German acronym, meaning "Condition assessment of railroad bridges and vehicles with AI methods for the evaluation of sensor data and structural dynamic models"). In the research project an AI will be developed which is meant to give predictions of the structural health of railroad bridges based on structural FE-models and sensor data within a digital twin framework.

2 RELATED WORK

In the field of production and manufacturing the principles of digital twins are already widely used (Jones et al. 2020). However, in the AEC sector the trend is just slowly emerging.

In Staffordshire, UK two newly built railroad bridges are equipped with sensors and are connected to a digital twin (Ye et al. 2019). As in ZEKISS, a data-driven and a physic-based data evaluation approach has been chosen. The digital twin provides a connection between both analysis methods and huge amounts of sensor data.

Another example is a research project dealing with an AI based digital twin of an institute building on the Cambridge campus (Qiuchen Lu et al. 2019). Here as well, the focus is on the data-driven prediction of structural failures. In addition, an augmented reality application for visualization purposes has been connected to the digital twin. Qiuchen Lu et al. (2019) describe in detail the challenges concerning the data management, as connecting data from different origins with no unique keys and possibly resulting in duplicates. The basis consists of an IFC building model.

As mentioned, the occurrence of digital twins in the AEC sector is not very common yet. Accordingly, the extent of publications concerning the specific use-case of the combination of AI, IoT and SHM is marginal. Nevertheless, the mentioned publications (Qiuchen Lu et al. 2019; Ye et al. 2019) are showing that they see the digital twin as we do; not only as the mapping of a building in a 3D model but rather as a whole framework respectively as the system architecture. It covers among other things the data gathering and data management up to the analysis of the data. Unfortunately, the used technology of the digital twin as well as the chosen software solution is not described in further detail in any of the mentioned literatures. Additionally, the publications do not use or do not explicitly describe the utilization of the structural model instead of the architectural 3D model.

In the context of digital twins exist many companies which are dealing with developing and offering software solutions. A well-known research project is *Dasher 360* from Autodesk (Autodesk 2011). It focuses on the connection between the sensors and the building element. Also, the visualization of live sensor data is available. There is no information about when *Dasher 360* will be released, but there is a demo application with example data available. There is also no information if open standards like IFC will be supported.

Another software solution is the *Bosch IoT Suite* (Bosch 2020). Many predesigned tools like analytics can be applied in this environment to a common digital twin model. An example for the application in the AEC sector is the construction project of the railroad station Stuttgart 21. A specific column type with a complex shape and unique in every case has been realized with the Bosch IoT Suite (Hill 2019). The limitations of this solution seem to be the junction of custom extensions like the designed ZEKISS-AI. Additionally, the service is not free of charge, even if there is a possibility to test it cost-free with a limited amount of API calls and reduced support.

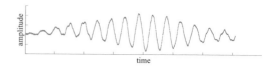

Figure 1. Example sequence of bridge vibration data while operational excitation.

In both solutions, there is no known possibility to use a structural analysis model as a digital twin basis in the investigated software or to use the existing possibilities with custom extensions or microservices like an AI.

Additionally, IoT on railroad bridges is hardly comparable to IoT on buildings. One reason is the difference between simple temperature sensors and the gathered data in ZEKISS. Many existing solutions like Autodesk *Dasher 360* provide a live feed of sensor data and their visualization. In ZEKISS, there is a wide variety of different sensor types (mainly: displacement, acceleration, strain, temperature and wind). Moreover, the connections to other services and databases result in data representations which have to be stored and visualized differently than normal data plots. The main focus is on the measurement of vibrations due to operational excitation (like train crossings). Each excitation generates data as shown in Figure 1. These oscillations will eventually be evaluated by the AI to detect any structural anomalies.

3 GENERAL DEFINITIONS

As an overview of the terms and concepts used in this paper, a brief summary with respective definitions will be given. Figure 2 shows the relation of these terms.

3.1 Big Data

In general understanding Big Data (BD) is a term to describe a large amount of data and its techniques to acquire, store, compute and analyze this data in various ways (Sun et al. 2020). Even if the term per se is not precisely defined, e.g. from which amount of data one speaks of Big Data, in this context and thus in ZEKISS the term Big Data is used especially to reference the techniques that are used.

3.2 Artificial Intelligence

Artificial Intelligence (AI) like Big Data is a not precisely respectively widely defined term and most often used to describe techniques used in the context of its conceptuality. In general, it can be stated that an AI aims to achieve intelligent behavior throughout virtual or physical man-made machines (Salehi & Burgueno 2018). This concept is already known but now its possibilities are enlarging and becoming more applicable with the rising computational power in the recent decades. When designing AI systems, two different approaches can be distinguished: knowledge-based AI

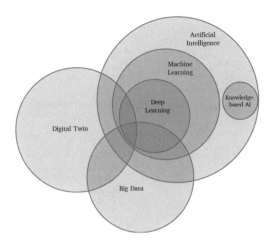

Figure 2. Layers of AI and its relation to DT and BD (based on Avci et al., 2021; Salehi and Burgueno, 2018).

and machine learning. The first one is based on a set of rules that can be translated to "if" and "else" statements. The second one is based on algorithms that enables machines to acquire knowledge or rules from data (Avci et al. 2021). Whenever the term AI is used in this paper, the sub-category machine learning is meant.

3.3 Digital Twin

The term Digital Twin (DT) is used to describe a digital representation of a real object. It was originally introduced by NASA 2010 and primarily used in the production and manufacturing sector (Jones et al. 2020). There are several definitions of a DT in different contexts. Some authors in the structural dynamics community use the term DT as a reference to a surrogate model (Bernt Johan Leira & Jian Dai 2020). In the scope of the ZEKISS project the DT is the whole framework for the machine learning based predictive maintenance system. Sensor and other monitoring data are linked to the DT and update it continuously. Simulations, evaluations and other performances can then be carried out on the digital twin and, depending on the selected method, can evaluate the measured data in real time (Ye et al. 2019). Furthermore, depending on the application, it is also possible to return and update the real object from the digital twin (i.e. a bidirectional data flow). However, this currently mainly affects the production and manufacturing sector, since it can have a direct influence on, for example, the speed of production. At this point, bidirectional data flow plays a subordinate role in SHM (Botz et al. 2020) and the current research project.

3.4 Industry Foundation Classes

An important part of the usability and workability of the 3D digital representation is the interoperability of the model so that it can be used in the context of the framework. Therefore, the IFC-format will be used. The Industry Foundation Classes (IFC) is an open and object-based file format developed by buildingSMART International (bSI). Digital models from different software can be transferred into this format and consequently be exchanged.

4 THE DIGITAL TWIN IN ZEKISS – A FRAMEWORK

The general goal of ZEKISS in the sense of predictive maintenance is to create a tool that gives engineers a resource to decide whether a bridge structure has structural damages which may be invisible or cannot be detected with conventional methods. For this purpose, the ZEKISS-Visualizer serves as a graphical user interface where all information comes together so that the accountable engineer is able to exam all resources, understand the statement of the AI and to make an informed decision. The Visualizer is part of the DT-Framework which is introduced in more detail in section 4.2.

The framework is carried out as a microservice architecture. In the center as in the user interface is a web platform which for now is realized with the python web framework Flask. The structural model of the bridge can come from any software tool which can export to the IFC format and besides using it for the calculations, the user can import the geometry of the model to the Visualizer.

First off, the raw data is acquired by the sensors attached to the bridges. The sensors can then communicate with the framework via *ditto* (Eclipse 2020). Since the paper's focus is the DT-Framework ditto will not be explained in further detail. The acquired data needs to be preprocessed before the AI can analyze it. In this regard, more details will follow in section 4.3. The preprocessed data is then stored in a database from which the AI can use it to learn about the specific bridge. The AI itself is currently in development and therefore not the focus of this paper. All in all, the DT-Framework includes all the mentioned parts like a microservice environment with the AI as its core element. The following figure summons up the framework as above described.

Another key feature of ZEKISS is the reciprocal measurement of all involved parts in the system like train2train, train2bridge, bridge2bridge and bridge2train. This enables a mutual verification of measurement data and sensor data from different sources which can be used and interpreted by the AI. This is not yet implemented in the framework but will be located on the same level as the sensors.

4.1 Structural analysis model as basis

It is not intended to use the ZEKISS-Visualizer as a platform to inform or convince future customers as it is the case by some 3D-Viewers. Rather, the viewer shall serve as a tool for engineers, who decide whether to take an action repairing the regarded bridge or not. Therefore, the viewer needs to provide all relevant

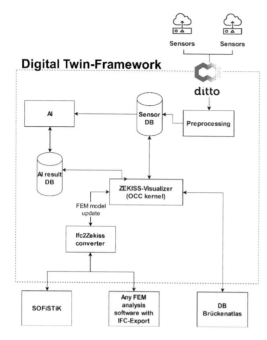

Figure 3. System architecture of ZEKISS.

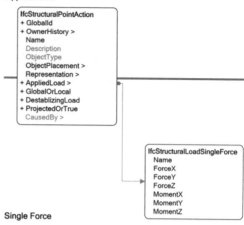

Figure 4. IFC Single Force reference as an example of SAV definitions (Lehtinen & Hietanen, n.d.).

data to evaluate the AI response. Additionally, to viewing functions, the DT acts as the framework including the AI. So, the structural model and its dependencies (like support conditions, results of the calculation, etc.) serve as one pillar for the training of the AI and its conclusive predictions.

In the scope of ZEKISS, the structural analysis model is created among others with the software SOFiSTiK. To ensure a universal connectivity to other structural software solutions, an interface for IFC in the DT-Framework has been created. IFC offers the possibility to export and exchange relevant data via the Model View Definitions (MVD, a subset of the entire IFC schema) according to the respective field of application. In case of structural analysis the *Structural Analysis View* (SAV) 2 × 3 and 4 provides many information of the system, loadings and load cases that can be exchanged via the IFC interface (buildingSMART International 2020). For example, the following information that is used in ZEKISS:

- *IfcStructuralLoadGroup*
- *IfcStructuralPointAction*
- *IfcStructuralLoadSingleForce*
- *IfcBoundaryEdgeCondition*: holds information of the support conditions of an element, i.e. the linear stiffness for each coordinate direction

The definition of the IfcSturcturalPointAction as well as the IfcStrcturalLoadSingleForce can be seen in the Figure 4.

Additionally, SOFiSTiK is able to export the *Coordination View* (CV) beside the SAV. While the SAV is specialized on structural information, the CV holds geometrical information of the structural system. Beside SOFiSTiK, also other structural software like RFEM from DLUBAL can export these MVDs (Dlubal Software 2017).

With the help of the IfcOpenShell (IfcOpenShell 2020) library the geometrical information can be imported into the DT-Framework. For this a python service has been implemented that reads the IFC file, parses it and converts the geometrical information into shapes, accordingly to the OpenCASCADE geometry kernel (using pythonOCC, Paviot 2020). To ensure the correct reference of each object an own data structure has been created that contains the pythonOCC shapes including their reference point which is defined in the IFC file but gets lost when it has been converted with IfcOpenShell. Subsequently, the pythonOCC shapes must be tessellated (disassembled into primitive surfaces) to be prepared for the use within the ZEKISS-Visualizer.

In the future it is planned to connect Ansys to the framework as well. Ansys (Ansys 2020) cannot export IFC but offers the possibility to create STL files which can then be easily integrated into the existing process, though it is only geometrical information. Solely the exchange of the structural information, which is solved in the case of SOFiSTiK via the SAV, has to be reviewed.

Another source of information is the use of an API of structural analysis software if available. In the case of SOFiSTiK, there is no API provided but due to its comprehensive documented SOFiSTiK-Database (CD-Base or CDB, see Figure 5), information like calculation results or natural modes can be retrieved (SOFiSTiK 2020). There are DLL libraries provided in C, VB and Python to access the database, however C compatible data types must be used.

With the access to the calculation results all relevant information can be extracted from the CDB and visualized in the ZEKISS-Visualizer in one place. For example, maximum stresses or an overview about all

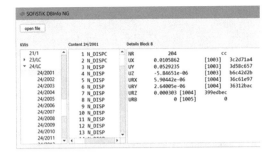

Figure 5. Built-in inspector for a CD-Base of an example bridge.

Figure 7. Example bridge in ZEKISS-Visualizer based on a structural analysis model from SOFiSTiK.

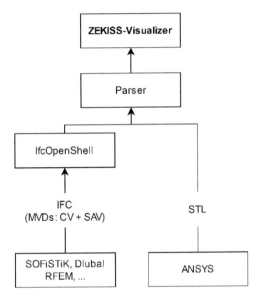

Figure 6. Detailed import options from structural analysis software to the ZEKISS-Visualizer.

connect the *Brückenatlas* (DB Netze 2020), a database of all German railroad bridges with basic information like the location of the bridge, to the DT. This adds more and general information of each specific bridge to the framework. Moreover, the timetable of the train connections could be connected to the DT-Framework, so that the user and the AI even knows which train passed the bridge and produced its oscillation. The services provided by the DB (German railroad company) offer different opportunities to connect to their databases by various APIs (Deutsche Bahn AG 2016).

From a technical perspective the visualizer is separated into a frontend, realized with HTML, JavaScript and Jinja. As geometry renderer the Three.js library was chosen. Tessellated geometries can be interpreted via Three.js and rendered by assigning the corresponding reference point of the geometry. The backend is realized as a python service, using Flask. The backend includes, among other functions, the possibility of importing IFC files respectively reading the information from the SOFiSTiK database. These functionalities will further be outsourced into separated microservices if the range of functions continues to increase.

4.3 Preprocessing sensor data

As previously described, preprocessing the sensor data is an important part in the process of data mining with the aim that the AI can use the processed data for learning and evaluating processes later. First off, the sensors attached to the bridges acquire the raw data. As described, the sensors can communicate with the framework via *ditto*. This raw data needs to be cleaned, for example in form of noise reduction, and must then go through some further data preprocessing steps. These include data integration, data transformation and data reduction so that the data packages can be stored as an object in the database. MongoDB, Couchbase or Redis can be chosen as a database. As a framework of the service FastAPI ("FastAPI" 2020) is used. The stored object then, for instance, represents the data set of one oscillation of the bridge during a train passing with possible additional information

calculated natural modes can be displayed to support the engineer in charge.

Figure 6 gives an overview of the realized or planned connections of the specialist software to the DT-Framework.

4.2 The ZEKISS-visualizer

The ZEKISS-Visualizer, as already mentioned, is a web-based graphical user interface (GUI) and therefore the access point for the accountable engineer. First off, it allows him to visualize the bridge. This is an important aspect as a visualization permits the user to recognize and interact easier with the bridge. This visualization especially helps since the DT-Framework will be used to manage different bridges. In addition to the 3D representation of the bridge the visualizer can show the natural modes and the oscillations of the bridges. The natural modes, for example, can be calculated with SOFiSTiK. Furthermore, it is planned to like the date, the temperature, the train that passed the bridge etc.

Furthermore, the sensors need to have a digital representation of themselves. Therefore, they need to be located on the bridge with a reference point within the 3D model, so the data of the sensors is correlated to the DT and especially to a specific component of the bridge like the bridge itself, the tracks or other parts. The attaching of the sensors to the digital representation of the bridge can be carried out with a graphical interaction, meaning the engineer can use the mouse to place the sensor.

5 CONCLUSION

As shown in the paper, a digital twin framework for an AI-based predictive maintenance system has been proposed and created. Due to IFCs Model View Definitions, especially the Coordination View and the Structural Analysis View, it is possible to import geometric and semantic information from FE software into the framework. IfcOpenShell has proven to be useful for interpreting and converting the IFC files into pythonOCC geometries. To retrieve the semantic information, an own parser has been implemented. The source code of the ZEKISS-Visualizer will be published on GitHub (TU Darmstadt 2020). Consequently, an open BIM approach has been chosen and realized.

The AI, which is seen as another essential part of ZEKISS, is embedded into the framework. Of course, designing, training and applying the AI will still be a challenging task in ZEKISS. However, the focus of this paper is laid on the system architecture and resulting in the proposed digital twin framework. The sensor data will be automatically preprocessed and conditioned to train the AI and to serve as data basis for the further decision making. The next steps developing the AI will be the equipment of the example bridge *HumVib* ("Human-induced Vibration bridge") at TU Darmstadt with sensors. With it, the framework will be tested. Thereby, the aims of the AI must be defined, and the specific AI algorithms will be chosen. Additionally, huge data is already available from different other bridges like the *Schmutter* bridge near Augsburg, Germany, which will be used for further training and verifying of the framework.

Thus, the ZEKISS-Visualizer provides a central web-based platform with all available information about the condition of a bridge to make an informed decision. In the scope of ZEKISS, the digital twin therefore is seen as the overall framework with all necessary parts, assembled in a microservice environment.

One perspective of this work is to realize a bidirectional data flow in the AEC sector, as known from digital twin applications in other areas like manufacturing. A possible use case could be the regulation of loadings by controlling the traffic of a bridge. So, the traffic could be adjusted to the up-to-date structural health state of the construction.

ACKNOWLEDGMENTS

The research project ZEKISS ("Condition assessment of railroad bridges and vehicles with AI methods for the evaluation of sensor data and structural dynamic models", www.zekiss.de) is carried out in collaboration with the German railroad company DB Netz AG I.NPF 22(T), the Wölfel Engineering GmbH and the GMG Ingenieurgesellschaft mbH. It is funded by the mFund (mFund 2020) promoted by the BMVI (Bundesministerium für Verkehr und digitale Infrastruktur).

REFERENCES

Ansys, 2020. Engineering Simulation & 3D Design Software | Ansys [WWW Document]. URL https://www.ansys.com/de-de (accessed 11.11.20).

Araujo, A., Garcia-Palacios, J., Blesa, J., Tirado, F., Romero, E., Samartin, A., Nieto-Taladriz, O., 2012. Wireless Measurement System for Structural Health Monitoring With High Time-Synchronization Accuracy. IEEE Trans. Instrum. Meas. 61, 801–810. https://doi.org/10.1109/TIM.2011.2170889.

Autodesk, 2011. Dasher 360.

Avci, O., Abdeljaber, O., Kiranyaz, S., Hussein, M., Gabbouj, M., Inman, D.J., 2021. A review of vibration-based damage detection in civil structures: From traditional methods to Machine Learning and Deep Learning applications. Mech. Syst. Signal Process. 147, 107077. https://doi.org/10.1016/j.ymssp.2020.107077

Bernt Johan Leira, Jian Dai, 2020. Extreme dynamic response of extended bridge structures subjected to inhomogeneous environmental loading, in: EURODYN 2020: Proceedings of the XI International Conference on Structural Dynamics. European Association for Structural Dynamics, pp. 3481–3495.

Bosch, 2020. Bosch IoT Suite – One open IoT platform for all business domains [WWW Document]. Bosch IoT Suite. URL https://www.bosch-iot-suite.com/ (accessed 11.11.20).

Botz, M., Grosse, C., Große, C., 2020. Monitoring im Rahmen eines digitalen Bauwerkszwillings.

buildingSMART International, 2020. MVD Database [WWW Document]. Build. Tech. URL https://technical.buildingsmart.org/standards/ifc/mvd/mvd-database/ (accessed 11.11.20).

Chapuis, B., Sjerve, E. (Eds.), 2018. Sensors, Algorithms and Applications for Structural Health Monitoring: IIW Seminar on SHM, 2015, IIW Collection. Springer International Publishing. https://doi.org/10.1007/978-3-319-69233-3

DB Netz AG, 2019. Brückenerneuerungen im Streckennetz der DB Netz AG – Größtes Modernisierungsprogramm der Deutschen Bahn.

DB Netze, 2020. Brückenkarte | DB Netze Brückenportal [WWW Document]. URL https://bruecken.deutschebahn.com/br%C3%BCckenkarte (accessed 11.11.20).

Deutsche Bahn AG, 2016. API-Portal – Liste der APIs [WWW Document]. URL https://developer.deutschebahn.com/store/ (accessed 11.11.20).

Dlubal Software, 2017. BIM-Workflow: Datenaustausch mittels IFC-Dateien [WWW Document]. Dlubal. URL https://www.dlubal.com/de/support-und-schulungen/support/knowledge-base/001472 (accessed 11.11.20).

Eclipse, 2020. Eclipse Ditto open source framework for digital twins in the IoT [WWW Document]. URL https://www.eclipse.org/ditto/index.html (accessed 11.11.20).

FastAPI [WWW Document], 2020. URL https://fastapi.tiangolo.com/ (accessed 11.12.20).

Hill, J., 2019. Bosch setzt auf die Economy of Things – Die Schwaben treiben den digitalen Umbau voran. Computerwoche.

IfcOpenShell, 2020. IfcOpenShell [WWW Document]. URL http://ifcopenshell.org/ (accessed 11.11.20).

Jones, D., Snider, C., Nassehi, A., Yon, J., Hicks, B., 2020. Characterising the Digital Twin: A systematic literature review. CIRP J. Manuf. Sci. Technol. https://doi.org/10.1016/j.cirpj.2020.02.002

Lehtinen, S., Hietanen, J., n.d. Structural Analysis View – IFC2x3 Binding.

mFund, 2020. BMVI – mFUND im Überblick [WWW Document]. URL https://www.bmvi.de/DE/Themen/Digitales/mFund/Ueberblick/ueberblick.html (accessed 11.13.20).

Paviot, T., 2020. tpaviot/pythonocc-core.

Qiuchen Lu, V., Parlikad, A.K., Woodall, P., Ranasinghe, G.D., Heaton, J., 2019. Developing a Dynamic Digital Twin at a Building Level: using Cambridge Campus as Case Study, in: International Conference on Smart Infrastructure and Construction 2019 (ICSIC). ICE Publishing, pp. 67–75. https://doi.org/10.1680/icsic.64669.067

Salehi, H., Burgueno, R., 2018. Emerging artificial intelligence methods in structural engineering. https://doi.org/10.1016/J.ENGSTRUCT.2018.05.084

SOFiSTiK, 2020. About CDB — CDB Interfaces 2020 [WWW Document]. URL https://www.sofistik.de/documentation/2020/en/cdb_interfaces/introduction/about_cdb/_about_cdb.html (accessed 11.11.20).

Sun, L., Shang, Z., Xia, Y., Bhowmick, S., Nagarajaiah, S., 2020. Review of Bridge Structural Health Monitoring Aided by Big Data and Artificial Intelligence: From Condition Assessment to Damage Detection. J. Struct Eng. 146, 04020073. https://doi.org/10.1061/(ASCE)ST.1943-541X.0002735.

TU Darmstadt, I., 2020. Github: research project ZEKISS [WWW Document]. GitHub. URL https://github.com/zekiss (accessed 11.11.20).

Ye, C., Butler, L., Bartek, C., Iangurazov, M., Lu, Q., Gregory, A., Girolami, M., Middleton, C., 2019. A Digital Twin of Bridges for Structural Health Monitoring. Proc. 12th Int. Workshop Struct. Health Monit. Stanf. Univ. 2019 Press.

Semantic contextualization of BAS data points for scalable HVAC monitoring

V. Kukkonen
Aalto University, Espoo, Finland
Granlund, Helsinki, Finland

ABSTRACT: Buildings account for approximately 40% of the total primary energy use in the U.S. and EU, while building systems waste a lot of energy due to poor operation and faults. Automated fault detection and diagnostics (AFDD) can be used to reduce the energy waste and ensure occupant comfort, and to support monitoring-based commissioning (MBCx). Applying the methods to existing buildings is cumbersome due to poor data labeling. Scaling the methods to account for different types of buildings requires a flexible way of mapping the data points to relevant monitoring processes. This paper proposes a novel method of using a semantic model to contextualize the data points for downstream analysis applications. The method is evaluated with application examples, and further potential applications are discussed.

1 INTRODUCTION

Buildings consume significant amounts of energy globally, accounting for approximately 40% of the total primary energy use in the U.S. and EU (Cao et al. 2016). Building systems such as heating, ventilation and air conditioning (HVAC) and lighting are major consumers of both electricity and heat. In commercial buildings, as much as 30% of this energy consumption is due to incorrect operation and faults in said systems (Kim et al. 2018).

One approach to reducing the energy waste of buildings is to use automated fault detection and diagnostics (AFDD), which has the potential to save energy and ensure occupant comfort by notifying operators of anomalous or improper behavior of key equipment, such as air handling units (AHUs). Besides detecting faults, AFDD methods are also useful in monitoring-based commissioning (MBCx) for validating energy conservation measures and preserving their effect over time (Gunay et al. 2019). While more and more sophisticated methods of AFDD for HVAC systems have been developed during the past decades (Kim et al. 2018), a key issue of effectively applying the methods in existing buildings and legacy systems remains.

AFDD methods require data about the building systems, which is generally available from building automation systems (BASs). Using the data points requires understanding their content and relations to the other data points. However, BAS data is generally poorly labeled (Gunay et al. 2019) and lacks universal naming conventions or tagging protocols (Harris et al. 2018), due to which applying monitoring in new systems requires a lot of case-by-case work and expert knowledge.

Data points' descriptions are metadata, which can be expressed in many ways. One of the more established approaches in building systems context is the tag-based Project Haystack (Project Haystack 2019). On the other hand, ever since the early 2000s, the use of semantic web technologies has been proposed in the architecture, engineering and construction (AEC) industries (Pauwels et al. 2017). Increasing interest towards semantic web and linked data technologies in the academic and industrial communities of AEC domain is visible in initiatives such as the Linked Building Data Community Group at World Wide Web Consortium and the Linked Data Working Group in buildingSMART International (Rasmussen et al. 2019). In addition to the design and construction phases, ontologies such as the Haystack-inspired Brick (Balaji et al. 2018) have been developed to support data analysis applications in the building operations and maintenance domains.

To support scalable HVAC systems monitoring, this paper presents a novel method for qualifying BAS data points related to HVAC processes by their context using SPARQL queries. To this end, an ontology is developed by extending existing ontologies. The method is explained and evaluated with application examples.

The paper is structured as follows. First, an overview of related work is provided. Second, the general idea of data point contextualization is explained. Third, the used ontology is described. Fourth, an example model is presented, and examples of fault detection rule inputs are formulated as SPARQL queries for relevant data point contexts. Finally, conclusions are presented.

2 RELATED WORK

This section provides an overview of the contemporary research relevant to metadata representations of BAS data points and building monitoring. First, some high-level, generic data models for systems and observations are described. Second, examples of data models focusing on building systems are listed. Third, a selection of applications operating on building data are presented, where those applications involve some building system data models. Finally, the related work is briefly summarized, explaining the novelty of the proposed data model.

2.1 Generic data models

The Sensor, Observation, Sampler, and Actuator (SOSA) and Semantic Sensor Network (SSN) ontologies were developed by a joint working group between the World Wide Web Consortium (W3C) and the Open Geospatial Consortium (OGC) (Haller et al. 2019). SOSA is the more lightweight core ontology, and SSN extends on it. Together the ontologies describe vocabulary for sensors, actuators, features of interest and their properties, among other related things.

The SEAS ontology was developed as a part of the ITEA2 Smart Energy-Aware Systems (SEAS) project (Lefrançois 2017). The ontology describes features of interest and their properties, systems and their connections including subsystem composition, and property evaluations.

2.2 Building system-related data models

Haystack (Project Haystack 2019) uses tags and sets of tags to annotate entities. Entities in Haystack include the data points, but also the equipment. Tags may be simple markers, meaning their information is indicated by their presence or absence, or they may have values. Value types include simple ones such as numbers and strings of Unicode characters, and more complex ones such as references to other entities and collections.

Brick (Balaji et al. 2018) combines semantic models with tags from Haystack. Brick builds on class definitions that follow Haystack tagsets, which form the taxonomy of the data points. As an example, Zone_Temperature_Sensor is a subclass of Temperature_Sensor. Additionally, certain relationships are defined between the entities in a building, including location (hasLocation), composition (hasPart), and basic flow connection (feeds/isFedBy).

CTRLont (Schneider 2019; Schneider et al. 2017) describes automation systems, including building automation systems, and their control logic.

Ploennigs et al. (2014) extend a previous version of SSN with physical processes as connectors between properties of various features of interest.

2.3 Data models in building data applications

Bruton et al. (2015) describe an AFDD tool for AHUs based on expert rules, in which they use a naming convention to relate data points to certain locations and quantities within an air handling unit. The tool uses virtual sensors to establish a type of standard model or a baseline for the available measurements, against which rule-based fault detection is performed. Stinner et al. (2018) describe a more general naming scheme for standardizing monitoring data in buildings.

Schneider et al. (2016) describe an ontology-based tool for configuring and deploying AFDD that focuses on the connection between static building information model (BIM) and dynamic information from building systems. The rules, or "analytics" as termed by the authors, consist of parameterized queries.

Semantic BMS (SBMS) (Kucera 2017; Kucera et al. 2018) builds on and extends a previous version of SSN, viewing the building automation system as a network of sensors and actuators. The purpose of SBMS is to provide a protocol-independent model for intelligent building systems to be used in data analysis. The building systems and automation functions are encapsulated by linking data points as inputs and outputs of "programs" or algorithms.

Delgoshaei (2017) combines multi-domain semantic models with rule-based reasoning to detect and diagnose faults. The equipment descriptions focus on composition where, for example, an AHU may have properties such as hasSupplyAirTemperature and hasCoil.

Further examples not described here in the interest of brevity include (Andriamamonjy et al. 2018; Bourreau et al. 2019; Dibowski et al. 2017; Hammar et al. 2019; Mainetti et al. 2015; Terkaj et al. 2017).

2.4 Summary

In summary, a significant portion of the existing research body has focused on one or more of the following: (1) generic, higher level descriptions of systems, sensors, and such; (2) building automation systems with information about control logic and protocols; (3) high-level views of buildings as collections of spaces and their conditions. In contrast, the perspective of the ontology presented in this paper is a simplified logical model of the HVAC process.

3 SEMANTIC MODELS FOR SCALABLE EXPERT RULES

Following the categorization by Kim et al. (2018), there are three main types of AFDD methods for building systems: process history-based, quantitative model-based, and qualitative model-based. In process

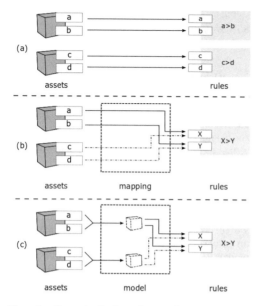

Figure 1. Conceptualizations of approaches to connect asset data points to monitoring rules: (a) direct, per-asset rules; (b) parameterized rules with per-asset mapping; (c) model-based rules.

history-based methods, data from a system is analyzed based on historical values. However, historical data of real buildings rarely has any faults or other significant events annotated, complicating the identification of normal operating conditions. Alternatively, quantitative models, i.e. physical models of varying levels of detail, can be developed that can simulate the expected values of data points and detect deviations from these. However, this generally requires a lot of effort and might be completely infeasible for old systems with unknown parameters. Finally, qualitative models rely on qualitative expert knowledge, which is often encoded as expert rules. These expert rules and their scalability are the running example used in this paper.

Examples of expert rules for HVAC system fault detection include the air-handling unit performance assessment rules (APAR) (House et al. 2001; Schein et al. 2006). The rules define expressions which indicate a potential fault when satisfied. These expressions are defined in terms of variables, such as "return air temperature", which are explained in writing and illustrated in a schematic diagram of an AHU. The challenge with scalable application of the rules stems from this: if the rules were to be applied to an AHU with a different composition, some of them may be completely inapplicable or require creative modifications. The rules assume a certain AHU configuration, but the rules don't explicate these assumptions. While a point name such as "return air temperature" of an air handling unit is quite universally understood, in practice even it is ambiguous due to varying compositions and sensor configurations of AHUs.

A rule in this context is considered as a function from a set of inputs to some type of fault indication. Figure 1 illustrates three general approaches to connecting data points from an asset, such as an AHU, to rule inputs: rules can be defined on a per-asset basis; rules can be parameterized and inputs mapped from each asset to applicable rules; or the assets data points can be mapped to a standard model, with the rules formed against the model. If the rules are always defined on a per-asset basis, they can be specialized to consider the specific characteristics of that particular asset. However, it is likely that certain similarities and abstractions will emerge, leading to a natural tendency to parameterize rules, which leads to the per-rule mapping approach. In this case, the rules encapsulate assumptions about the inputs, which must be upheld when mapping the inputs, introducing a potential for subtle errors. Finally, the assets' data points can be mapped to a model, and the rules can be built on top of this model. This model itself can be a simple naming convention such as used by Bruton et al. (2015), although that has its own limitations in terms of flexibility. Alternatively, the model can be more complex, using a more expressive approach such as Haystack or Brick.

In the proposed method, a type of qualitative semantic model is built, and the rules define their inputs as SPARQL queries. While this model is more complicated and thus requires more effort to build, ideally it reduces effort required later in the application. Additionally, as building systems have limited variance between buildings, it seems feasible that applications for building the semantic models could exploit system and component templates, instead of requiring manual creation from the ground up. In the next section, an ontology for building these models, placing BAS data points in the context of HVAC components, is described.

4 ONTOLOGY DESCRIPTION

In this section, the ontology for annotating BAS data points in the context of HVAC processes is described. The ontology has two aspects: the types and process relationships of the HVAC components, and the types and relationships of BAS data points. The prefixes for the terms related to the aforementioned aspects are hvac: and data:, respectively. The idea of the ontology is to support simplified logical models of HVAC processes and their data points, as opposed to accurate representations of physical equipment or automation control logic.

The HVAC components and relationships are built on the Flow Systems Ontology (FSO) (Kukkonen et al. 2019). The component classes in FSO follow the subtypes of IfcDistributionFlowElement from buildingSMART Industry Foundation Classes (IFC). Here, the hvac namespace introduces some more specific classes relevant to AHUs, which are shown in Table 1. The terms for fluid and heat

Table 1. System and component classes extended from FSO.

Class	Subclass of
hvac:AirHandlingUnit	fso:System
hvac:Coil	fso:EnergyConversionDevice
hvac:Damper	fso:FlowControlDevice
hvac:Fan	fso:FlowMovingDevice
hvac:Filter	fso:FlowTreatmentDevice
hvac:HeatRecoveryWheel	fso:EnergyConversionDevice
hvac:Pump	fso:FlowMovingDevice
hvac:Valve	fso:FlowControlDevice

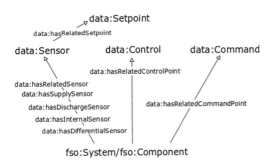

Figure 2. Data point classes and the properties between in-stances of them. Inverse properties are omitted from the figure, but for each "hasX" property there exists an inverse "isXOf".

exchange connections between components and systems defined by FSO (`fso:suppliesFluidTo`, `fso:returnsFluidTo`, `fso:transfersHeatTo` and `fso:transfersHeatFrom`) form the basis of the process model, as will be illustrated in the next section.

The individual sensors, control signals, setpoints, and commands are modeled as instances of `data:DataPoint` and the more specific subclasses `data:Sensor`, `data:Setpoint`, `data:Control`, and `data:Command` as illustrated in Figure 2. Alarms and other potentially available points are omitted from this work for now. Sensors are also a subclass of `sosa:Sensor`.

A sensor is a point that observes a property and can either relate to an actual physical device or be "virtual" and calculated from other observations, but these are not differentiated. An example of a virtual sensor is a sensor that estimates air volume flow from pressure difference over a fan. Setpoints are generally related to properties observed by sensors. Control points refer to control signals for various actuators, generally varying between 0 and 100 percent or normalized between 0 and 1. Command points are related to enumerable commands, such as operating modes or "open/closed" toggles.

Data points are tied to HVAC components: fans, valves, and so on. This is asserted with the `data:hasRelatedPoint` property. The subproperties are `data:hasRelatedSensor`, `data:hasRelatedCommand`, `data:hasRelatedControl`, and `data:hasRelatedSetpoint`, shown in Figure 2. There are specific ways a sensor data point can relate to a piece of equipment: observations can be before or after components – i.e. on the supply or discharge side – or internal to the equipment, such power consumption readings. Additionally, there are sensors that observe a property difference over a component, for example pressure difference. For annotating these, the property `data:hasRelatedSensor` has further subproperties `data:hasSupplySensor` for sensors on the supply or input side, `data:hasDischargeSensor` for sensors on the discharge or output side, `data:hasDifferentialSensor` for sensors observing a difference in property over a component, and `data:hasInternalSensor` for sensors observing a property internal to the component. It is worth noting that a sensor that is on the discharge side of one component may well be on the supply side of another component. That is, being a supply or discharge sensor is not something that is inherent to a point, but it depends on the perspective.

Finally, the properties observed by sensors are modelled with SOSA. More specifically, the properties, such as `data:Temperature`, are modeled as instances of `sosa:Property`, and sensors are tied to the properties with `sosa:observes`.

Although aspects such as units of measurement and their harmonization are important in data analysis applications, they're not considered in the presented ontology for now. While omitted, it should be possible to implement these as an incremental addition to the ontology, and this is one area for future research.

5 EXAMPLES AND EVALUATION

In order to demonstrate and provide means of evaluating the approach, this section discusses examples of fault detection rules using a semantic model built with the presented ontology. The rules focus on air handling units, although similar principles could be used when considering other HVAC equipment.

The first subsection introduces the model of an example AHU. Following that, some illustrative expert rules for fault detection are described, and their inputs are formulated as SPARQL queries. While different expert systems may use different kinds of rules and outputs combined with techniques such as decision/diagnosis trees, the example rules don't go into that much detail. Rather, the reasoning and basic idea of looking at certain data points is given, followed by an illustration of how those data points could be qualified in the semantic model. The formulation of actual rules and combining the results of rule evaluations to arrive to a diagnosis is beyond the scope of this paper. Additionally, the final step of retrieving the data point labels or other identifiers for data retrieval is omitted from the queries.

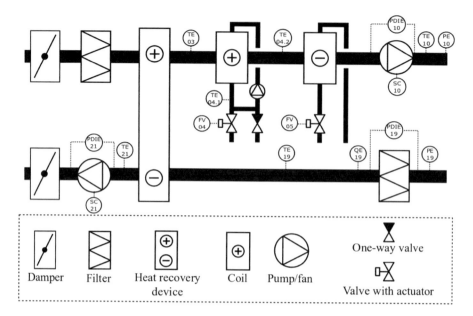

Figure 3. Schematic diagram of an example air handling unit with a subset of the data points relevant to the following examples. Top side is the supply side, where the air flows from left to right, while the bottom side is the return side with air flow from right to left. TE is temperature sensor, PE is pressure sensor, PDIE is pressure difference sensor, QE is CO2 sensor, SC is frequency control signal, and FV is valve actuator control signal.

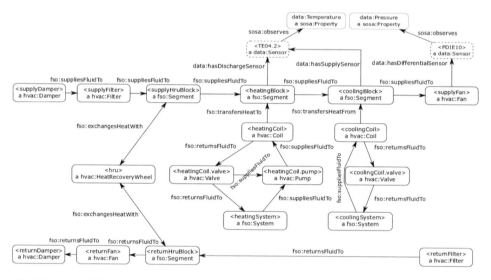

Figure 4. Illustration of the semantic AHU model with a subset of the data points shown as an example. Not shown in this figure are the fso:hasComponent properties connecting the AHU resource to the components.

5.1 Example model

A schematic of the example AHU composition is shown in Figure 3. Beyond the basic fans, filters and dampers, the AHU contains heating and cooling coils, and a heat recovery wheel.

The example AHU was translated to a semantic model, illustrated in Figure 4. All the components in the AHU are marked as such with the fso:hasComponent. On the supply side, the components are connected to the next with fso:suppliesFluidTo, while on the return side the connection is fso:returnsFluidTo, both subproperties of the fso:feedsFluidTo.

It is assumed that each component directly contains only one fluid. If multiple fluids are involved, that is modeled via composition. For example, the AHU coils contain water or some other liquid, which exchanges heat with the air passing through the coil. This is modeled as two components: the coil itself, and a heating or cooling "block" of the ductwork around the coil, as illustrated in Figure 4. This makes it possible to follow fluids through the model.

The heating and cooling coils are connected to the supply heating and cooling blocks with the directional heat exchange subproperties `fso:transfersHeatTo` and `fso:transfersHeatFrom`. Similarly, the heat recovery wheel is connected to both supply and return sides via `fso:exchangesHeatWith`.

The data points are connected to the components with the appropriate properties, with a discharge point of an upstream component modeled as the supply point of the next component downstream, where sensible. As an example, Figure 4 shows a temperature sensor between the heating and cooling blocks and a differential pressure sensor over the supply fan in the upper right quadrant of the figure. Other data points are omitted from the figure.

Overall, the model provides means to qualify data points via their related components, and the components can further be qualified by: (1) composition, being part of a specific system; (2) fluid connections, including logical "supply" and "return" distinctions; (3) heat exchange connections, including directional design intent of transferring heat from / to another component. In the following examples, this information is used to qualify the inputs for some fault detection rules.

5.2 Case 1: Detecting AHU operation mode

As discussed by House et al. (2001), an air-handling unit will generally operate in a few discrete modes, such as heating or cooling. Knowing the operating mode can be used, for example, to narrow down the relevant rules to evaluate. Without going into specifics of the actual modes, the variables required for detecting the mode of the example AHU are supply and return fan controls, heating and cooling coil valve controls, and heat recovery control. An example for retrieving these is shown in Listing 1.

It is worth noting that this query won't return AHUs which don't match the assumptions encoded in the SPARQL query. For incompatible AHU compositions, different rules with different SPARQL queries will need to be used, as is the goal of the method.

5.3 Case 2: AHU supply temperature error

An AHU will generally control the temperature of the air it supplies downstream. For this purpose, there is a sensor measuring the temperature of the air, as well as a setpoint. If the AHU is faulty, it may not be able to reach the setpoint. Multiple potential rules can diagnose different scenarios around this. For example, in heating mode, there may be problems with the heating coil valve or the heating plant producing the heat.

Listing 1. Example SPARQL query for retrieving information about the points required for AHU mode classification. The relevant components are narrowed down, and the data points tied to them are projected to the result.

```
PREFIX fso: <https://w3id.org/fso#>
PREFIX hvac: <https://example.com/hvac#>
PREFIX data: <https://example.com/data#>
PREFIX : <https://example.com/building1/>

SELECT
  ?ahu
  ?supplyFanCtrl
  ?returnFanCtrl
  ?heatingCtrl
  ?coolingCtrl
  ?hruCtrl
WHERE {
  ?ahu a hvac:AirHandlingUnit ;
    fso:hasComponent
      ?supplyFan ,
      ?returnFan ,
      ?heatingCoil ,
      ?heatingValve ,
      ?coolingCoil ,
      ?coolingValve ,
      ?hru .

  ?supplyFan a hvac:Fan ;
    data:hasRelatedControl
?supplyFanCtrl .
  ?returnFan a hvac:Fan ;
    data:hasRelatedControl
?returnFanCtrl .
FILTER EXISTS {
    ?ahu fso:hasComponent ?supply ,
?return .
    ?supply fso:suppliesFluidTo
?supplyFan .
    ?return fso:returnsFluidTo
?returnFan .
  }
  ?heatingCoil a hvac:Coil ;
    fso:transfersHeatTo ?heatingBlock
.
  ?heatingValve a hvac:Valve ;
    fso:exchangesFluidWith+
?heatingCoil ;
    data:hasRelatedControl
?heatingCtrl .
  ?coolingCoil a hvac:Coil ;
    fso:transfersHeatFrom
?coolingBlock .
  ?coolingValve a hvac:Valve ;
    fso:exchangesFluidWith+
?coolingCoil ;
    data:hasRelatedControl
?coolingCtrl .
  ?hru a hvac:HeatRecoveryWheel ;
    data:hasRelatedControl ?hruCtrl .
}
```

For this rule, the assumptions are that there is a temperature sensor and a related setpoint in the supply side of the AHU, and optionally the control of a valve-controlled heating coil upstream of that sensor. If there are multiple temperature sensors that match these assumptions within a single AHU, they will all be returned as separate results.

Listing 2. Example SPARQL query to retrieve relevant points for supply air temperature and setpoint, as well as optional heating coil valve control upstream of the air temperature sensor.

```
PREFIX sosa: <http://www.w3.org/ns/sosa/>
PREFIX fso: <https://w3id.org/fso#>
PREFIX hvac: <https://example.com/hvac#>
PREFIX data: <https://example.com/data#>
SELECT
  ?ahu ?temp ?setpoint ?valveCtrl
WHERE {
  ?ahu a hvac:AirHandlingUnit ;
    fso:hasComponent ?comp .
  ?comp
    data:hasSupplySensor
    | data:hasDischargeSensor
      ?temp .

  FILTER EXISTS {
    ?ahu fso:hasComponent ?upstream .
    ?upstream fso:suppliesFluidTo+ ?comp .
  }

  ?temp data:hasRelatedSetpoint
?setpoint ;
    sosa:observes data:Temperature .

  OPTIONAL {
    ?ahu fso:hasComponent ?coil , ?valve
.
    ?coil a hvac:Coil ;
      fso:transfersHeatTo ?heatingBlock
.
    FILTER EXISTS {
      ?heatingBlock fso:suppliesFluidTo+
?comp
    }
    ?valve fso:exchangesFluidWith+ ?coil
;
      data:hasRelatedControl ?valveCtrl
.
  }
}
```

5.4 Case 3: Heating coil valve leak

A typical example of a detectable fault in an AHU is a fault with a heating coil valve. The basic rule states that if the temperature of the air increases over the coil while the valve is supposedly closed, then the valve may be stuck open or otherwise leaking.

The rule appears simple at first, but varying AHU configurations and instrumentations complicate the matter. In the ideal scenario, there exist air temperature sensors directly before and after the heating coil. However, it is not uncommon that, for example, the AHU composition shown in Figure 3 were to have the latter temperature sensor after the supply fan, leaving not only the heating coil but also cooling coil and supply fan between the sensors. In this scenario, the tolerance for flagging a potential fault would need to be greater, as the air temperature can increase over the supply fan.

An example of considering varying downstream temperature sensors is shown in Listing 3. The query might return multiple results per AHU heating coil. The application could, for example, only consider the one with the smallest tolerance.

Listing 3. Example SPARQL query defining multiple alternative observations with varying characteristics using UNION.

```
PREFIX sosa: <http://www.w3.org/ns/sosa/>
PREFIX fso: <https://w3id.org/fso#>
PREFIX data: <https://example.com/data#>
PREFIX hvac: <https://example.com/hvac#>

SELECT
  ?ahu ?supplyTemp ?dischargeTemp
  ?valveCtrl ?tolerance
WHERE {
  ?ahu a hvac:AirHandlingUnit ;
    fso:hasComponent
      ?heatingBlock ,
      ?coil ,
      ?valve .
  ?coil a hvac:Coil ;
    fso:transfersHeatTo ?heatingBlock .
  ?valve a hvac:Valve ;
    fso:exchangesFluidWith ?coil ;
    data:hasRelatedControl ?valveCtrl .

  FILTER EXISTS {
    ?supplyTemp
      sosa:observes data:Temperature .
    ?dischargeTemp
      sosa:observes data:Temperature
  }

  # supply temperature sensor
  {
    ?heatingBlock data:hasSupplySensor
      ?supplyTemp .
  }

  # discharge temperature sensor
  {
    # sensor directly after the coil
    ?heatingBlock data:hasDischargeSensor
      ?dischargeTemp .
    BIND(0.5 as ?tolerance)
  } UNION {
    # sensor downstream of the coil
    ?heatingBlock fso:suppliesFluidTo+
      ?downStream .
    ?downStream data:hasDischargeSensor
      ?dischargeTemp .
    BIND(2.0 as ?tolerance)
  }
}
```

6 CONCLUSION

The purpose of this paper was to demonstrate a method supporting scalable monitoring of HVAC systems. The presented approach achieves this goal by combining a novel ontology for contextualizing BAS data points related to HVAC processes with a model-based approach for mapping data points to monitoring processes. SPARQL queries are used to qualify the data points as required by the monitoring processes. The method is demonstrated and evaluated with a selection of expert rules for fault detection.

The presented approach allows applications to explicitly qualify inputs for monitoring processes by specifying the data point contexts in terms of HVAC components and connections. This explicitness removes ambiguity related to common HVAC terminology, while enabling the expression of the similarities and differences of system compositions.

While the examples given were limited to expert rules for fault detection, in principle the same type of mapping approach could be used in other applications. As an example, virtual sensors and derivations from the existing data points could be considered monitoring processes with similar input mapping requirements.

For future research, further evaluation of the expressiveness of the proposed data model and queries is required. To this end, more real-world systems need to be modeled and further application examples considered. Additionally, alignments to and integrations of existing ontologies form another future development. These developments will undoubtedly further shape the ontology. A relevant consideration not yet touched on in this paper is the representation and potential harmonization of differing units and scales. Finally, methods and tools for efficiently building the semantic models will require research.

REFERENCES

Andriamamonjy, A., Saelens, D. & Klein, R. 2018. An auto-deployed model-based fault detection and diagnosis approach for Air Handling Units using BIM and Modelica. *Automation in Construction* 96: 508–526.

Balaji, B. et al. 2018. Brick: Metadata schema for portable smart building applications. *Applied Energy* 226: 1273–1292.

Bourreau, P. et al. 2019. BEMServer: An open source platform for building energy performance management. In *Proceedings of the 2019 European Conference for Computing in Construction*. University College Dublin.

Bruton, K. et al. 2015. Comparative analysis of the AHU InFO fault detection and diagnostic expert tool for AHUs with APAR. *Energy Efficiency* 8(2): 299–322.

Cao, X., Dai, X. & Liu, J. 2016. Building energy-consumption status worldwide and the state-of-the-art technologies for zero-energy buildings during the past decade. *Energy and Buildings* 128: 198–213.

Delgoshaei, P. 2017. *Semantic Models and Reasoning for Building System Operations: Focus on Knowledge-Based Control and Fault Detection for HVAC*. PhD Thesis. University of Maryland.

Dibowski, H., Holub, O. & Rojícek, J. 2017. Ontology-based fault propagation in building automation systems. *International Journal of Simulation: Systems, Science and Technology* 18(3): 1.1–1.14.

Gunay, H.B., Shen, W. & Newsham, G. 2019. Data analytics to improve building performance: A critical review. *Automation in Construction* 97: 96–109.

Haller, A. et al. 2019. The modular SSN ontology: A joint W3C and OGC standard specifying the semantics of sensors, observations, sampling, and actuation. *Semantic Web* 10(1): 9–32.

Hammar, K., Wallin, E.O., Karlberg, P. & Hälleberg, D. 2019. The RealEstateCore Ontology. In C. Ghidini et al. (eds), *The Semantic Web – ISWC 2019*. Springer International Publishing.

Harris, N., Shealy, T., Kramer, H., Granderson, J. & Reichard, G. 2018. A framework for monitoring-based commissioning: Identifying variables that act as barriers and enablers to the process. *Energy and Buildings* 168: 331–346.

House, J.M., Vaezi-Nejad, H. & Whitcomb, J.M. 2001. An expert rule set for fault detection in air-handling units. *ASHRAE Transactions* 107: 858–871.

Kim, W. & Katipamula, S. 2018. A review of fault detection and diagnostics methods for building systems. *Science and Technology for the Built Environment* 24(1): 3–21.

Kucera, A. 2017. *Semantic BMS: Semantics-Driven Middleware Layer for Building Operation Analysis in Large-Scale Environments*. PhD Thesis. Masaryk University.

Kucera, A. & Pitner, T. 2018. Semantic BMS: Allowing usage of building automation data in facility benchmarking. *Advanced Engineering Informatics* 35: 69–84.

Kukkonen, V., Kücükavci, A. & Rasmussen, M. 2019. Flow Systems Ontology. Available at: https://w3id.org/fso [Accessed March 22, 2020].

Lefrançois, M. 2017. Planned ETSI SAREF Extensions based on the W3C&OGC SOSA/SSN-compatible SEAS Ontology Patterns. In *Proceedings of Workshop on Semantic Interoperability and Standardization in the IoT, SIS-IoT*.

Mainetti, L., Mighali, V., Patrono, L. & Rametta, P. 2015. A novel rule-based semantic architecture for IoT building automation systems. In *23rd International Conference on Software, Telecommunications and Computer Networks, SoftCOM 2015*.

Pauwels, P., Zhang, S. & Lee, Y.C. 2017. Semantic web technologies in AEC industry: A literature overview. *Automation in Construction* 73: 145–165.

Ploennigs, J., Schumann, A. & Lécué, F. 2014. Adapting Semantic Sensor Networks for Smart Building Diagnosis. In P. Mika et al. (eds), *The Semantic Web – ISWC 2014*. Springer International Publishing.

Project Haystack. 2019. Project Haystack. Available at: https://project-haystack.org/ [Accessed January 12, 2020].

Rasmussen, M.H., Lefrançois, M., Pauwels, P., Hviid, C.A. & Karlshøj, J. 2019. Managing interrelated project information in AEC Knowledge Graphs. *Automation in Construction* 108.

Schein, J., Bushby, S.T., Castro, N.S. & House, J.M. 2006. A rule-based fault detection method for air handling units. *Energy and Buildings* 38(12): 1485–1492.

Schneider, G.F. 2019. *Semantic Modelling of Control Logic in Automation Systems*. PhD Thesis. Karlsruhe Institute of Technology.

Schneider, G.F., Kalantari, Y., Kontes, G., Steiger, S. & Rovas, D.V. 2016. An ontology-based tool for automated configuration and deployment of technical building management services. In *Proceedings CESBP – Central European Symposium on Building Physics/ BauSIM 2016*.

Schneider, G.F., Pauwels, P. & Steiger, S. 2017. Ontology-Based Modeling of Control Logic in Building Automation Systems. *IEEE Transactions on Industrial Informatics* 13(6): 3350–3360.

Stinner, F., Kornas, A., Baranski, M. & Müller, D. 2018. Structuring building monitoring and automation system data. *The REHVA European HVAC Journal*.

Terkaj, W., Schneider, G. & Pauwels, P. 2017. Reusing Domain Ontologies in Linked Building Data: The Case of Building Automation and Control. In *Proceedings of the 8th International Workshop on Formal Ontologies meet Industry*.

Integrating sensor- and building data flows: A case study of the IEQ of an office building in The Netherlands

S. van Gool, D. Yang & P. Pauwels
Information Systems in the Built Environment Group, Department of the Built Environment, Eindhoven University of Technology, The Netherlands

ABSTRACT: The need for integrating building data with other data sources is growing, most notably sensor data. This potential synergy between building and sensor data flows has been researched, but current research is aimed almost exclusively at the automation of transmitting sensor data to building models, although often not focusing on operational and analysis possibilities with the integrated data. This research aims to effectively integrate sensor and building data flows for analysis purposes. Using the developed integration, multiple data sources can be effectively utilized for analysis. Its functionality is proven in this paper on hourly Indoor Environmental Quality (IEQ) analyses of a University Medical Centre (UMC) office building. These results prove that, using building information, measured sensor values can be placed into context, and actionable insights can be derived instead of manually interpreting the sensor results.

1 INTRODUCTION

In the current big data era, research points out that the need for using data gathered in the exploitation phase of a building during the design and construction of new buildings or redevelopment of existing buildings is growing (Rasmussen et al. 2019). To realize this need, the paradigm of using BIM-related technology is increasingly being used. Even though BIM technologies are widely used in the design and construction phase, their adoption in the operations & maintenance and renewal phase is still limited (Heaton & Parlikad 2019). However, such technologies are starting to receive specific interest from property managers that desire to share and utilize information in new ways (Carbonari et al. 2018). Thus, BIM technologies now begin to extend their capabilities over the full building lifecycle, making room for integration of other technologies, and thus effectively constituting Digital Twins for buildings. A promising interaction is sensor deployment. By constantly monitoring a building on multiple fronts, problems can be pinpointed in real-time allowing a building to perform during its whole lifecycle (Deloitte 2018). However, without clear integration of sensors in the building dataflows, it is difficult to use sensor data effectively for multiple purposes (Tang et al. 2019).

The current research on the building and sensor dataflow integration focused almost exclusively on the automation of transmitting sensor data to BIM models (Chang et al. 2018). Therefore, even though the built environment is considered as the most important sector for sensor-related research (Dave et al. 2018), ineffectively integrating building models with sensor data will hinder its possibilities (Chang et al. 2018; Tang et al. 2019). Without information integration and management standards, data analytics with large, heterogeneous datasets is costly and time inefficient (Gerrish et al. 2017).

This research aims to ease this building- and sensor dataflow integration. It focuses on effectively combining these dataflows to enable insights on the Indoor Environmental Quality (IEQ).

This paper shall briefly review current methods related to integrating building-/sensor dataflows in Section 2. After, Section 3 shall propose an integration method and demonstrate its application on a UMC case study (Section 4). Finally, in Section 5, this paper discusses the limitations of this study and future research directions.

2 LITERATURE REVIEW

2.1 Existing integration frameworks

Extensive literature research on building/sensor dataflow integration frameworks has already been conducted. A recent and extensive example is Tang et al. (2019) and Schneider et al. (2018), identifying several main integration framework categories, each having their benefits and limitations. Considering the large variety of options in the identified framework categories, this literature review will narrow its scope to one framework. Following from the categories listed by Tang et al. (2019), the hybrid approach 'semantic web + relational database' seems to be most suitable for IEQ analyses and for the operations & maintenance phase in general. The hybrid approach has the flexibility of linking numerous data types using the semantic web approach, making it possible to extend the project scope. Additionally, it uses the standard query language SPARQL. Lastly, by retaining the sensor time-series data in its native format instead of transforming it to RDF, query performance should be faster. However, it seems too hasty to choose relational databases (SQL-based) without investigating its non-relational counterparts, such as a NoSQL store (e.g. MongoDB). Consequently, this integration will also consider non-relational database options.

2.2 Semantic modelling for building data

The standard Industry Foundation Classes (IFC) is one of the most widely recognized and well accepted models for building information exchange throughout the built environment industry (Linked Building Data Community Group). However, despite its wide adoption, acceptance and open data structure, several challenges exist in using IFC. Firstly, the formal semantics of IFC are defined mostly in EXPRESS, which is a seldom used language for software developers. Moreover, out of the box querying and reasoning is hardly available for this language. Secondly, IFC schemas are not easily extensible in a user-friendly manner and the IFC schema is not flexible when combining different data sources required for new situations. Lastly, the IFC schema is large and complex – the latest version of the IFC schema (IFC4) contains 1200 data types (Pauwels & Terkaj 2016).

Consequently, many relationships and properties within a building that are required in daily operations and maintenance processes are difficult to use in IFC models, and, despite its goal of covering the entire built environment industry of information exchange, combining information from related domains is difficult (Zhang et al. 2018). As an alternative to IFC, ontology-based semantic web modelling is considered promising due to, among others, the common language these technologies rely on as well as its apparent global scale deployment (Pauwels et al. 2017). Semantic web technologies could overcome the challenges in standard data models related to flexible interlinking of data from different domains and scales, and thereby enabling interoperability among systems and actors (Pauwels et al. 2018).

Linking the separate but structured datasets, data should be either represented in RDF or linked to it. Using Uniform Resource Identifiers (URIs) in combination with RDF files, data can be instantly represented on the web of data (McGlinn et al. 2019). Still, standardization and consensus should be achieved among the ontologies to realize industry-wide adoption. The W3C Linked Building Data Community Group (LBD CG) aims for this standardization and consensus on ontologies (Industry Foundation Classes (IFC) 2019).

Central in the deliverables is an intentionally small core ontology that contains the topology of the building, namely the Building Topology Ontology (BOT) (Rasmussen et al. 2017). Several other foundation ontologies have been proposed in literature. For example, the BIM Shared Ontology (BIMSO) was proposed by Niknam and Karshenas (2017), representing a foundation ontology to which BIM Design Ontologies (BIMDO) can be linked. The BIMDOs express the properties of building elements. The idea behind BIMSO seems similar to the idea behind BOT, as defined by the LBD CG. Also, the BRICK ontology is closely related to the BOT ontology, although larger in size as it contains more than only the topology.

The BOT aims to reduce the current redundancy in the building dataflows and is extensible with the other LBD ontologies. The BOT should only define the core building topology using physical and conceptual objects as well as their relations (Rasmussen et al. 2017).

The LBD CG also aims to standardize the way in which products and properties can be formally defined, in combination with the BOT ontology. This resulted mostly in the Ontology for Property Management (OPM) (Rasmussen et al. 2017). OPM defines three levels of complexity: (level 1) no objectification, (level 2) one-time objectification and (level 3) double objectification. PROPS functions similarly to OPM but only contains level 1.

Geometric representation is central in building data models. To conduct IEQ analyses, semantically usable geometry is required. Multiple options were found throughout literature, most notably GEOM, WKT and gbXML.

The GEOM ontology defines the shapes as instances, without having any direct information on absolute coordinates. This is one of the main differences between GEOM and alternative ontologies such as WKT, gbXML, etc., that focus on direct geometry with absolute coordinates (McGlinn et al. 2019). WKT is human-readable, efficient and the semantics related to spatial reasoning are well understood (Pauwels et al. 2017). Using GeoSPARQL, 2D WKT literals can be easily queried from an RDF graph (McGlinn et al. 2019). However, representing 3D geometries remains more of a challenge as GeoSPARQL does not support 3D geometries. Similar to IFC, gbXML contains component tagging such as building, space, surface, material, etc., but the relational structure is not the same as in IFC (Ferguson et al. 2017). Having the XML structure, gbXML is easy to conduct IEQ analyses with.

3 PROPOSED METHOD

This paper aims for standardization and consensus on ontologies in the built environment industry, so it will focus on the deliverables of the LBD CG. Therefore, for the core ontology, it builds forth on the core ontology BOT, as presented in (Rasmussen et al. 2017).

Apart from the core ontology, a property ontology is required. The different complexity levels of the OPM ontology can be very useful, but it is less relevant to this paper. Therefore, a simple property ontology is preferred, for which OPM ontology level 1 seems to be the most suitable.

This paper looks for a semantically rich geometry representation for analysis purposes. Two important criteria in choosing a format are (1) limited complexity and (2) easy querying possibilities. GbXML contains all required geometric components, contains mutual datapoints with the property ontology and is XML-based. Furthermore, the space-oriented data model of gbXML aligns very well with the IEQ analysis aimed for in this paper (space boundaries, areas, volumes, etc.). Therefore, this paper uses gbXML to represent the semantically rich geometry.

Sensors periodically (e.g. 10 min.) write tiny parts of information to the database, after which analyses and visualizations read large parts of data from the database. Therefore, a database should be good at single writes as well as at multiple reads. MongoDB has the performance and scalability of a non-relational database, and it has been around for some time during which it has developed stability. Consequently, MongoDB is chosen as time-series database.

Using the BOT as core ontology, the domain-specific data sources are linked. Material is hereby linked on a data level, rather than on ontology level, in order to maintain ontological independence and value. Using mutual datapoints from the property graph, a link can be established between the graphs, MongoDB and the gbXML geometry. Summing up the proposed method, several steps must be taken during this integration. Figure 1 represents a dataflow diagram showing these steps. The diagram contains circles representing actions and rectangles representing data formats. The steps have been categorized into four categories: data gathering and preparation, data integration, data analyses and communicating results. The case study in the next chapter shall elaborate on the steps taken using these categories as paragraphs.

4 CASE STUDY: ASSESSING IEQ ANALYSES IN A UMC OFFICE BUILDING

To prove the proposed integration, a case study has been conducted on an office building within a Dutch University Medical Centre (UMC). The UMC contains numerous data sources which are desired to be integrated and linked to physical assets, such as rooms. Integrating the building and sensor data sources and analyzing the IEQ forms a solid start to this goal. The integration follows the dataflow diagram shown in Figure 1.

4.1 Data collection and preparation

The first step is collecting and preparing the required data sources. To link the sensor data to physical objects with an own URI in the graphs, the sensors are drawn in Revit as objects. A property that is converted in the IFC/LBD converter has been chosen to add the sensor ID to. This serves as a mutual datapoint.

Additionally, the Revit model must be converted to the required formats. Using the Revit exporter, a gbXML and IFC file have been gathered. The IFC file is then converted to LBD graphs using the IFC to LBD converter.

4.2 Data integration

Semantic modelling aims for, among others, easy data accessibility. Using local Turtle files does not empower this strength. Therefore, GraphDB has been used as a remote graph database (GraphDB Free 9.2 documentation 2020). Using the API of GraphDB, the data is accessed and adjusted. MongoDB has a similar API

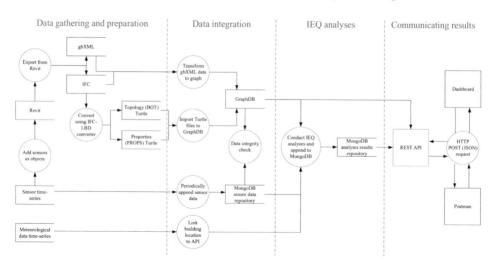

Figure 1. Dataflow diagram of the proposed method.

which is used to query from and post to in a Python script using the Pymongo package.

GbXML still needs to be integrated into the graphs though. This paper only uses a few gbXML datapoints. However, it might be that in a later stage more gbXML elements are to be added to the property graphs. To ease this integration, a dedicated gbXML ontology has been written: Ontology for Managing gbXML spatial data (ogbxml) (Ogbxml: Ontology for Managing GbXML). Supporting possible future use cases, this ontology describes classes and predicates of all gbXML datapoints, not only the ones that are used in this paper.

Having integrated the data formats does not ensure that the datapoints are integrated correctly though. To prevent data issues in the analyses later, a data integrity check has been created. This checks for duplicates in the property graphs, for differences between sensor naming in the property graphs and the sensor database, for wrong sensor allocations and for missing datapoints. Once this check is run through without critical errors, the data was evaluated to be correct and fitting for the IEQ analyses.

4.3 IEQ analyses

Using the developed integration, IEQ analyses can be conducted. The Indoor Air Quality (IAQ), room efficiency and thermal comfort have been assessed. As an example, the mean radiant analysis, part of the thermal comfort, will be discussed as all data sources are required in this analysis.

In the mean radiant temperature calculation the mean of all indoor surface temperatures must be found. For internal surfaces it can be assumed that the indoor surface temperature is equal to the indoor air temperature. For outdoor surface temperatures this is more difficult to assess through. To assess the indoor surface temperature of an external surface, the indoor room conditions must be known, the thermal behavior of the surface must be known and the outdoor conditions must be known. This requires various datapoints from multiple data sources.

Firstly, the bounding surfaces and their thermal properties are found for the selected room in the graphs. The BOT graph is used to find the relevant surfaces and the PROPS graph is used to find the relevant properties of these surfaces. Secondly, the rough geometry, azimuth and tilt of these surfaces is determined through the gbXML file. Thirdly, the outdoor conditions, gathered through the meteorological API, are used in combination with the rough surface geometry to determine the sun's influence based on time of the year for each external surface. Lastly, the indoor air temperature is gathered from the sensor data gathered from MongoDB.

Note that the gbXML file is still required, even though this information could be easily transferred to the property graph. The reason for this is that gbXML represents surfaces differently than IFC, and thereby also the graphs. IFC represents surfaces, such as walls, as one element, while gbXML splits these surfaces per room, enabling easy IEQ analyses per room. Calculating the mean radiant temperature of a room requires all integrated dataflows and is thereby the perfect example of utilizing the integration created in this paper.

4.4 Communicating results

Having building- and sensor data as well as analyses results on a local machine does not make the integration worthwhile. The results must be communicated to external systems to be valuable. Therefore, a REST API was created with HTTP Post requests. However, as a REST API itself is not accessible either, a dashboard was created of which a screenshot is shown in Figure 2. By creating this dashboard, the loop of gathering, interlinking, processing and

Figure 2. An overview of the dashboard.

communicating data has been completed. As shown in Figure 2, the dashboard has three main components: (1) interactive 2D floorplan, (2) room properties and (3) analysis results. The floorplan is gathered directly from the building data graph and the gbXML file, and thereby demonstrates the interaction between different data sources. Additionally, several room properties are shown in the dashboard to demonstrate the link between the room polygons in the dashboard and the room URIs in the graphs. Lastly, the link between the MongoDB collection containing room analyses results is shown by demonstrating the analyses results in line and bar diagrams.

5 CONCLUSION AND DISCUSSION

5.1 Critical evaluation of results

Even though the integration works for the IEQ analyses as this paper intended to, there are some limitations to the integration.

First, this integration is reproducible for a new case study, but errors are probably going to occur related to the data quality and format. When reproducing the integration, it is most important to notice its weak points: the human component involved in linking the sensors with the building data, and the content and frequency of the measured sensor values. If careful attention is spent on these weak points, the integration should be fully reproducible.

Additionally, with the human component involved in the integration steps, labelling of spaces and elements might prove to be a liability. The developed integration links the sensors in the building data to the sensor time-series data using a mutual datapoint. This mutual datapoint is the sensor ID, which is found in the building data through the sensor property 'comments_simple'. Ideally, a more dedicated property should be used that takes in a string as well, but this property must also be converted from IFC to LBD. The IFC to LBD converter considers only several specific datapoints of which the sensor ID must be one. This is essential in choosing the mutual datapoint.

Regarding gbXML geometry, furniture and full object geometry is not included, only spaces and surfaces. The motion sensors used in this case study are placed under desks and, using the developed integration, the desk that the sensor is attached to can easily be identified. However, these desks cannot be automatically drawn into a floorplan as gbXML data does not contain geometry for desk elements. If these desks are to be automatically drawn into a floorplan, another source of absolute coordinates should be found. IFC could be a potential solution as it contains the necessary data, but, as described before, it contains a lot of unnecessary data as well.

5.2 Future research directions

The developed integration is not a goal itself, but rather a means to achieving other goals. Additionally, the developed integration, with the building data as backbone, is suitable for integration of various other sources, as the UMC intends to do, for example social data (Corry et al. 2014). This paper focused on the objective and physical building aspects, but how users interact with and experience a building is just as important.

Effective integration on such a UMC campus requires a scalable integration. In essence the developed integration is scalable, but assessing speed was not part of the scope of this paper. However, speed performance is an essential part to investigate when scaling such integrations to larger projects. This should therefore be researched in future research.

Apart from the research area of integrating building data with various other sources, several related areas can use the results of such data integration in practice.

ACKNOWLEDGEMENT

This paper is part of the master thesis 'Integrating sensor- and building dataflows – A case study of indoor environmental quality assessment of an office building in the Netherlands'. The thesis is conducted at the Technical University of Eindhoven being part of the master program 'Construction Management and Engineering'. The thesis report can be found on the following page: https://www.ofcoursecme.nl/education/master-thesis-database/#.

REFERENCES

Bonduel, M. (2018). Towards a PROPS ontology. Presentation, (March). Retrieved from https://github.com/w3c-lbd-cg/lbd/blob/gh-pages/presentations/props/presentation_LBDcall_20180312_final.pdf

Carbonari, A., Corneli, A., Di Giuda, G., Ridolfi, L., & Villa, V. (2018). BIM-Based Decision Support System for the Mangement of Large Building Stocks. Proceedings of the 35th International Symposium on Automation and Robotics in Construction (ISARC), (Isarc). https://doi.org/10.22260/isarc2018/0049

Chang, K. M., Dzeng, R. J., & Wu, Y. J. (2018). An automated IoT visualization BIM platform for decision support in facilities management. Applied Sciences (Switzerland), 8(7). https://doi.org/10.3390/app8071086

Dave, B., Kubler, S., Främling, K., & Koskela, L. (2016). Opportunities for enhanced lean construction management using Internet of Things standards. Automation in Construction, 61, 86–97. https://doi.org/10.1016/j.autcon.2015.10.009

Dave, B., Buda, A., Nurminen, A., & Främling, K. (2018). A framework for integrating BIM and IoT through open standards. Automation in Construction, 95(August 2017), 35–45. https://doi.org/10.1016/j.autcon.2018.07.022

Deloitte. (2018). Data is the new gold: The future of real estate service providers. Retrieved from https://www2.deloitte.com/global/en/pages/real-estate/articles/future-real-estate-data-new-gold.html

Ferguson, H., Krisnadhi, A., & Cheatham, M. (2017). Semantic-Graph based approach to BIM, resilience, and spatial data interoperability. 1–27.

gbXML. (n.d.). About gbXML. Retrieved from http://www.gbxml.org/

Gerrish, T., Ruikar, K., Cook, M., Johnson, M., Phillip, M., & Lowry, C. (2017). BIM application to building energy performance visualisation and managementChallenges and potential. Energy and Buildings, 144, 218–228. https://doi.org/10.1016/j.enbuild.2017.03.032

GraphDB Free 9.2 documentation. (2020). Retrieved 11 May 2020, Retrieved from http://graphdb.ontotext.com/documentation/free/

Heaton, J., & Parlikad, A. K. (2019). A conceptual framework for the alignment of infrastructure assets to citizen requirements within a Smart Cities framework. Cities, 90(January), 32–41. https://doi.org/10.1016/j.cities.2019.01.041

Industry Foundation Classes (IFC). (2019). Retrieved from technical.buildingsmart.org/standards/ifc/.

Loyola, M. (2018). Big data in building design: A review. Journal of Information Technology in Construction, 23(January), 259–284.

Linked Building Data Community Group. (n.d.). Retrieved from https://w3c-lbd-cg.github.io/lbd/

McGlinn, K., Wagner, A., Pauwels, P., Bonsma, P., Kelly, P., & O'Sullivan, D. (2019). Interlinking geospatial and building geometry with existing and developing standards on the web. Automation in Construction, 103(March 2018), 235–250. https://doi.org/10.1016/j.autcon.2018.12.026

Niknam, M., & Karshenas, S. (2017). A shared ontology approach to semantic representation of BIM data. Automation in Construction, 80, 22–36. https://doi.org/10.1016/j.autcon.2017.03.013

Ogbxml: Ontology for Managing GbXML Spatial Data, Retrieved from sjorsvangool.github.io/ogbxml/.

Pauwels, P., & Terkaj, W. (2016). EXPRESS to OWL for construction industry: Towards a recommendable and usable ifcOWL ontology. Automation in Construction, 63, 100–133. https://doi.org/10.1016/j.autcon.2015.12.003.

Pauwels, P., Krijnen, T., Terkaj, W., & Beetz, J. (2017). Enhancing the ifcOWL ontology with an alternative representation for geometric data. Automation in Construction, 80, 77–94. https://doi.org/10.1016/j.autcon.2017.03.001

Pauwels, P., Poveda-Villalón, M., Sicilia, Á., & Euzenat, J. (2018). Semantic technologies and interoperability in the built environment. Semantic Web, 9(6), 731–734. https://doi.org/10.3233/SW-180321.

Pauwels, P., Zhang, S., & Lee, Y. C. (2017). Semantic web technologies in AEC industry: A literature overview. Automation in Construction, 73, 145–165. https://doi.org/10.1016/j.autcon.2016.10.003.

Rasmussen, M. H., Pauwels, P., Hviid, C. A., & Karlshoj, J. (2017). Proposing a Central AEC Ontology That Allows for Domain Specific Extensions. Proceedings of the Joint Conference on Computing in Construction, 1, 237–244.

Rasmussen, H. L., Jensen, P. A., Nielsen, S. B., & Kristiansen, A. H. (2019). Initiatives to integrate operational knowledge in design: a building client perspective. Facilities, 37(11–12), 799–812. https://doi.org/10.1108/F-02-2017-0021.

Tang, S., Shelden, D. R., Eastman, C. M., Pishdad-Bozorgi, P., & Gao, X. (2019). A review of building information modeling (BIM) and the internet of things (IoT) devices integration: Present status and future trends. Automation in Construction, 101(February), 127–139. https://doi.org/10.1016/j.autcon.2019.01.020.

Terkaj, W., Schneider, G. F., & Pauwels, P. (2017). Reusing domain ontologies in linked building data: The case of building automation and control. CEUR Workshop Proceedings, 2050(March 2018).

Zhang, C., Beetz, J., & De Vries, B. (2018). BimSPARQL: Domain-specific functional SPARQL extensions for querying RDF building data. Semantic Web, 9(6), 829–855. https://doi.org/10.3233/SW-180297.

Abstract life-cycle modeling of cyber-physical systems in civil engineering

D. Legatiuk & K. Smarsly
Bauhaus University Weimar, Germany

K. Lossev & A. Volkov
Moscow State University of Civil Engineering, Russia

ABSTRACT: With recent advancements in embedded sensing technologies for the Internet of Things, cyber-physical systems, instrumented with structural health monitoring and control applications, are increasingly implemented in civil engineering. Several approaches towards metamodeling cyber-physical systems in civil engineering have been proposed in recent years, based on Unified Modeling Language (UML), category theory, and abstract algebra. However, life-cycle metamodeling of cyber-physical systems (CPS) in civil engineering has not yet been addressed in its full generality. The evolution of a CPS during its life cycle typically requires the evolution of the corresponding metamodel, because different components of a system may be added or removed. Therefore, life-cycle metamodeling approaches must provide possibilities to describe temporal behavior of CPS components, which is not supported by current metamodeling approaches utilized in civil engineering practice. Thus, in this study, an abstract modeling concept for integrating temporal evolutions of cyber-physical systems in civil engineering into existing metamodeling approaches is proposed. The integration starts with a detailed description of a typical CPS life cycle in civil engineering, underlying unique features for each life-cycle phase. The features characterizing phases of a CPS life cycle are abstracted and formalized by abstract algebraic constructions, supporting diagram-based modeling approaches, such as UML. Finally, an illustrative example of abstract CPS life-cycle modeling for additive manufacturing of concrete is presented.

1 INTRODUCTION

According to Lee and Seshia (2017), a cyber-physical system (CPS) is a heterogeneous, coupled system integrating computing, networking, and physical processes, incorporating embedded computers and networks that monitor and control the physical process. The coupled nature of cyber-physical systems is reflected on the level of information exchange, since information is exchanged between physical and computational processes, or, in other words, between the "physical domain" and the "cyber domain" (Legatiuk & Smarsly 2018). In recent years, cyber-physical systems have become an indispensable part of engineering fields, such as civil engineering (Huang et al. 2010), construction engineering (Yuan et al. 2016), aerospace engineering (Atkins & Bradley 2013), and aviation (Sampigethays & Poovendran 2013). For details on CPS research, the interested reader is referred to (Letichevsky et al. 2017; Losev & Chulkov 2019; Volkov & Shilova 2019). In civil engineering, the characteristics of cyber-physical systems are related to the engineering subfields, such as structural health monitoring (SHM) and structural control (Bhuiya et al. 2016), additive manufacturing (Smarsly et al. 2020), and smart cities (Mohanty 2016; Volkov 2018).

The broad integration of cyber-physical systems into civil engineering applications has stimulated the evolution of classical engineering structures into sophisticated engineering systems equipped with modern monitoring tools observing and controlling the long-term behavior, referred as "intelligent structures" or "cognitive buildings". According to Korkmaz (2011), intelligent structures maintain and improve the structural performance by recognizing changes in behavior and actions, adapting the structure to meet performance goals, and using past events to improve future performance.

The heterogeneous architecture of cyber-physical systems, thus of cognitive buildings, requires developing new concepts for CPS modeling, because coupling between the cyber domain and the physical domain must be addressed in each step of the modeling process. Moreover, the modeling process in civil engineering nowadays originates from a *conceptual modeling* or *metamodeling* of whole engineering systems, i.e. the heterogeneous CPS architecture must be supported by metamodeling. Metamodeling aims at abstract descriptions of components constituting engineering systems under consideration and at the coupling and the information exchange between the components. Through metamodeling, a general architecture of engineering systems, such as cognitive buildings, may be highlighted and possible sources of design (or modeling) errors can be identified.

Attempts to introduce metamodeling methodologies supporting holistic descriptions of CPS architectures based on different formalisms have been proposed by several authors. For example, set theory and abstract algebraic approaches towards describing CPS components and coupling between the components have been proposed by Simko et al. (2014). Despite the several advantages provided by the flexibility of set theory and abstract algebra, the approach introduced by Simko et al. (2014) cannot be utilized directly in civil engineering because the approach focuses on the computational part of CPS modeling, i.e. on the description of the cyber domain, while neglecting the physical domain, which is an indispensable part of modern cognitive buildings. An alternative to using set theory and abstract relational algebra, building on studies of Nefzi et al. (2015), Simko et al. (2014), and Syarif et al. (2019), has been reported by Legatiuk and Smarsly (2018), where a general architecture of intelligent structures has been analyzed and decomposed into "building blocks" whose composition constitutes intelligent structures.

Apart from set theory-based approaches, diagram-based metamodeling is frequently used in engineering applications. In particular, building information modeling (BIM), based upon object-oriented modeling, is a very popular choice for metamodeling in civil engineering (Borrmann et al. 2018). Using BIM and unified modeling language (UML) for CPS metamodeling has been presented by Fitz et al. (2019) and Smarsly et al. (2019), where a formalization of cyber-physical systems with UML and its practical realization in terms of the Industry Foundation Classes (IFC) have been discussed. Finally, more abstract modeling approaches based on category theory and type theory have been reported by Gürlebeck et al. (2017) and Gürlebeck et al. (2020), respectively. These approaches are not directly related to CPS metamodeling but rather address the general issue of modeling error detection in the overall modeling process in engineering. Moreover, both category theory-based modeling and type theory-based modeling are motivated by classical mathematical modeling related to using partial differential equations. Nonetheless, attracted by the descriptive power of category theory and its support of diagram-based description of models, first attempts towards CPS metamodeling based on category theory have been proposed by Legatiuk et al. (2017).

Although the approaches discussed above may successfully be utilized for metamodeling of cyber-physical systems in civil engineering, a common limitation is obvious, namely, the lack of support of temporal evolutions of civil engineering structures, as addressed to some extend in (Legatiuk & Smarsly 2018; Losev 2019). In other words, the approaches do not address the *life cycle of cyber-physical systems* in civil engineering. Specifically, considering the growing demands associated with digitalization of the construction industry, the complete life cycle of cyberphysical systems, representing vital components of smart city applications, needs to be traceable, thus to be metamodeled along with metamodeling of cyberphysical systems as such. Therefore, metamodeling approaches currently utilized in civil engineering need to be extended to address the life cycle of cyber-physical systems in civil engineering.

Since the life cycle of cyber-physical systems in civil engineering has not been addressed in full generality by metamodeling approaches currently utilized in the field of civil engineering, this paper discusses first steps towards life-cycle metamodeling of cyber physical systems, starting from results achieved in intelligent structures research, which will be generalized towards cyber-physical systems in the remainder of this paper.

2 LIFE-CYCLE METAMODELING OF CYBER-PHYSICAL SYSTEMS IN CIVIL ENGINEERING

When analyzing the life cycle of cyber-physical systems, structuring into life-cycle phases strongly depends on the objectives of the analysis, since different life-cycle phases may be further subdivided to more refined sub-phases, or, in contrary, grouped together to build more general phases. Moreover, even structuring of the life cycle of classical buildings may vary, depending on the objective. For example, from an environmental point of view, the life-cycle phases presented in (Ngwepe & Aigbavboa 2015) include the extraction of raw material and manufacturing, which are typically not considered life-cycle phases in the construction industry. To overcome the problems associated with different views on the life-cycle phases, this paper aims at a more formal perspective on a meta-level (or conceptual level). Therefore, results presented in (Legatiuk & Smarsly 2018) are taken as a basis for the upcoming discussion on the life cycle of cyber-physical systems.

Using abstract algebraic and set-theoretic constructions for conceptual modeling of intelligent structures has been presented by Legatiuk & Smarsly (2018). Therein, temporal evolution of intelligent structure s has been addressed, subdividing the life cycle of intelligent structures into three major phases: (i) design phase, (ii) calibration and validation phase, and (iii) operational phase. In short, these three phases are characterized as follows:

– **Design phase.** The design phase covers definitions of objectives for intelligent structures to be fulfilled during the operational life time, design of experiments to identify locations for placing sensors and actuators as part of the SHM and control systems, and a detailed mathematical modeling of the structure to identify potential critical points, such as stress concentration areas.
– **Calibration and validation phase.** Naturally, this phase addresses the issue of calibration and

validation of all systems of intelligent structures to assure that performance goals defined during the design phase are met at the beginning of the subsequent operational phase. It should be noted that, ideally, performance goals should be met during the whole operational life of intelligent structures, but it cannot be generally guaranteed in advance.

- **Operational phase.** Representing the output of intelligent structures modeling, the operational phase is characterized by continuous analyses of measurements collected by the SHM systems and by validation of the measurements against the models. The models may be mathematical models, created in the design phase and calibrated/validated in the calibration and validation phase, or data-driven models, created and calibrated/validated in the calibration and validation phase.

Taking into account that intelligent structures are specific instantiations of cyber-physical systems in civil engineering, the above subdivision of the intelligent structure life cycle is generalized to cyber-physical systems. However, since this paper aims at discussing principal changes on the conceptual level of CPS modeling in civil engineering, a more refined version of the CPS life cycle is proposed. Moreover, to keep the mathematical constructions general, no specific formalism for CPS metamodeling is assumed in this paper, meaning that all of the presented constructions may be instantiated by any existing metamodeling approach (with appropriate adaptations, if necessary). Figure 1 shows a general schema of a CPS life cycle extended from the intelligent structures life cycle proposed by Legatiuk and Smarsly (2018).

As can be seen from Figure 1, the CPS life cycle is subdivided into five general phases: *planning, construction, testing, operation,* and *demolition*. Moreover, the output of the design phase, i.e. a meta-model, is used as the input for the calibration phase, and so on. To prepare the upcoming discussion of modeling the CPS life cycle on the conceptual level, a short overview of the CPS life cycle phases, introduced in Figure 1, is provided:

- **Planning.** Formulation of the goals for creating a CPS as well as identifying CPS specifications and properties.
- **Construction.** Practical construction of a CPS.
- **Testing.** Performing field tests on the CPS to identify potential changes to the CPS that may be necessary.
- **Operation.** Operation of the CPS during its life time, including maintenance and monitoring activities.
- **Demolition.** Sustainable waste recycling.

Next, the link between the five phase of the CPS life cycle and conceptual modeling level is established. Figure 2 illustrates the five phases and the corresponding objects on the conceptual level together with the abstract mappings $f_i, i = 1, 2, 3, 4, 5$, and $g_j, j = 1, 2, 3$, which are discussed below in detail. For clarity of the mathematical presentations, the mappings f_i are discussed first.

- f_1 maps CPS design specifications into a CPS metamodel. In general, mapping f_1 represents a *formalization process* of modeling assumptions and task specifications in a form of mathematical objects or formal relational constructions. Practical realization of the formalization process may be done through any formal system, such as first order logic or type theory. The use of type theory to formalize modeling process for problems of mathematical physics has been presented in (Gürlebeck et al. 2020). However, only partial differential equations-based modeling has been addressed in the type-theoretic setting so far, and therefore, this type-theoretic approach needs to be extended for a more general metamodeling.
- f_2 is an optional *updating mapping* to a CPS metamodel. During the construction process necessities for model updating can be discovered, and therefore, mapping f_2 is applied to existing CPS metamodel.

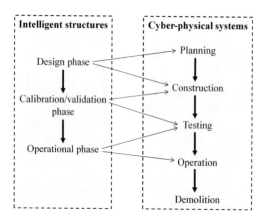

Figure 1. Extension and subdivision of the intelligent structures life-cycle to the CPS life cycle with five life-cycle phases.

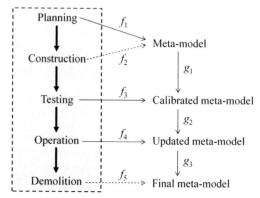

Figure 2. CPS life-cycle phases and corresponding objects on the conceptual level with abstract mappings.

Formally, this mapping can be represented as identity mapping if no changes in the metamodel are necessary, or a modifying mapping, which naturally depends on a particular task and setting, e.g. it may add or remove CPS components throughout the life cycle. The optionality of mapping f_2 is denoted by a dashed arrow.
- f_3 is an *updating mapping* required by any CPS meta-model that incorporates the results of the testing phase, i.e. the CPS is "prepared" for the next phase, the system operation.
- f_4 represents a *mapping of the information* collected by SHM systems, which may be part of the CPS, into the CPS meta-model. The term "updated meta-model", mentioned in Figure 2, will be elaborated in details during the discussion on the mappings $g_j, j = 1, 2, 3$. In short, the updated meta-model represents the process of meta-model updating with information collected by an SHM system installed on a structure.
- Finally, f_5 is an optional *finalizing mapping* devoted to performing last updates to the CPS meta-model with information collected during the demolition phase. Again, optionality of this mapping is indicated by a dashed arrow.

For the discussion on changes in model states on the conceptual level, i.e. changes in the meta-model states, recall the formal definition of an intelligent structure proposed in (Legatiuk & Smarsly 2018):

Definition 1 (Intelligent structure). Consider the following abstract mappings:

- f – mapping of signals to measurements, i.e. this mapping represents sensors
- g – mapping of measurements to a control signal, i.e. this mapping represents a "model" processing measurement and sending a signal to actuators
- h – mapping of a control signal to actions on a structure, i.e. the actions of actuators

Through the abstract mappings, an intelligent structure $\Im$ can formally be considered as the following pointwise noncommutative composition:

$$\Im := h \circ g \circ f \text{ over } \mathbb{R}_+, \qquad (1)$$

where $\mathbb{R}_+$ denotes positive real numbers, i.e. the timeline.

Definition 1 is used for describing cyber-physical systems in civil engineering. Moreover, formula (1) is a general abstract description of a CPS and any practical instantiation of a CPS meta-model by using UML, e.g. as proposed by Fitz et al. (2019), may be abstractly described by (1). Thus, on the conceptual level, a CPS is always described by the noncommutative composition (1). In this case, point evaluations of (1) over the timeline represent states of the CPS along its complete life cycle meaning that the general representation (1) is always valid, but the mappings h, g, and f may vary, depending on the concrete instantiations of meta-models on each life-cycle phase.

The mappings $g_j, j = 1, 2, 3$ shown in Figure 2 represent principal changes on the conceptual level, precisely:

- g_1 indicates that on the conceptual level, a CPS meta-model created according to the design specifications of the planning phase, is modified with the information collected during the testing phase. In short, the image of g_1, i.e. a calibrated meta-model, is used as a basis for CPS life time operation.
- g_2 represents that the calibrated meta-model can be changed during the CPS life time operation, as it has been indicated in the discussion of mapping f_4. In general, g_2 can be the identity mapping, if no changes to the system during the operational phase have been made, or it can be a sequence of modifying maps if changes have been made. For example, if UML diagrams are used as a meta-model, then g_2 would represent adding/removing specific blocks of a diagram.
- g_3 indicates that the latest state of the updated meta-model is used as a basis for the demolition phase. Changes to the final meta-model can be done only by using the (optional) f_5 mapping and only if additional information necessary on the conceptual level is obtained during the demolition phase.

It must be emphasized that the abstract mappings described above will become more concrete once a specific metamodeling formalism is used. Moreover, the abstract mappings may be equipped with additional formal structures, such as compositional semantics and commutative diagrams, the latter case corresponding to a category theory-based approach, where each object on the conceptual level is created by categorical constructions. Among other possibilities, the use of *categorical ontology logs*, or simply *ologs*, proposed by Spivak and Kent (2012) seems to be straightforward to be adapted. When using ologs, each object on the conceptual level will be an individual olog, and then the abstract mappings $g_j, j = 1, 2, 3$, become functors between the ologs.

3 EXAMPE LIFE-CYCLE METAMODELING OF ADDITIVE MANUFACTURING PROCESSES

As a simple illustrative example illuminating the CPS life-cycle metamodeling concept proposed in this paper, a CPS for additive manufacturing of concrete structures, commonly referred to as "concrete printing", is presented in this section. The laboratory setup is shown in Figure 3.

To specify the metamodeling concept presented in Section 2, a UML-based metamodeling approach is used as a formalism to metamodel the concrete printing process. Since the goal here is to illustrate the life cycle metamodeling concept – rather than discussing the concrete printing itself – a simplified UML model, based on the meta-model proposed by Smarsly et al. (2020), is taken as the basis of the formalization. The formalization, according to Figure 2, is realized via mapping f_1, which maps the design specification into components of a UML diagram, by which the metamodel of the concrete printer is obtained. The corresponding UML diagram is shown in Figure 4.

In general, the mapping f_1 is not "just a mapping"; it rather represents the *process* of how design specifications are formalized in the chosen metamodeling formalism, i.e. in terms of blocks in the UML diagram. Clearly, practical realizations and rigorous descriptions of the mapping f_1 strongly depend on the metamodeling formalism that is chosen. More details on modeling formalization concepts are provided in (Gürlebeck et al. 2020).

Once, in this illustrative example, the real-world CPS for concrete printing is constructed and the meta-model is created, the CPS for concrete printing and the corresponding meta-model are passed to the testing phase. During the testing phase, the necessity of adding a temperature sensor to the CPS is observed. Therefore, the CPS meta-model needs to undergo a modification during its life cycle. The modification, according to Figure 2, is realized by two mappings:

– mapping f_3 formalizes the new design specification, i.e. the necessity of adding a temperature sensor, in terms of blocks in the UML diagram, and
– mapping g_1 modifies the meta-model.

It should be noted that mapping f_2 is optional and does not contribute to the example considered here. Again, at each moment of its life cycle, the CPS may be described by (1). The modification mappings, such as g_1, do not change the form of (1) but modify the abstract mappings h, g, and f because the new sensor setup needs to be taken into account. The calibrated meta-model, upon applying f_3 and g_1, is shown in Figure 5.

Next, the CPS for concrete printing and the corresponding meta-model are passed to the operation phase. Updating of the meta-model during the operation phase is controlled by the mappings f_4 and g_2, which, in general, are not singular mappings but collections of several mappings performed during the operation phase. Naturally, the mappings are identity mappings if no changes to the system need to be made; otherwise, the mappings are modifying mappings. The operation of the CPS for concrete printing has shown that no further modifications of the printing system are necessary, i.e. all mappings summarized by g_2 are identity mappings. However, a last inspection indicates that the printhead is not functional and, therefore, the CPS cannot serve its design purpose. Here, two possibilities in the CPS life cycle exist:

– the CPS may be repaired to continue its service, or
– the non-functionality of the printhead means end of service for the CPS.

In the first case, repairing the CPS would be a part of the operation phase, i.e. the corresponding modifications on the meta-model need to be added via mapping g_2 to continue the operation phase. In the second case, which is implemented in the illustrative example, the mapping g_3 represents a final modification of the CPS meta-model, and the finalization of the metamodel is obtained by deleting the block corresponding to the printhead, as shown in Figure 6 (cf. "Printhead" in Figure 5).

Figure 3. A concrete printer serving as a CPS example, source: (Smarsly et al. 2020), modified.

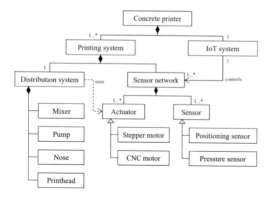

Figure 4. The meta-model of the CPS for concrete printing, denoted in terms of a (simplified) UML diagram.

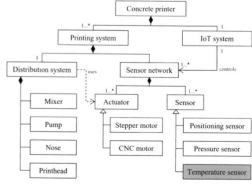

Figure 5. Calibrated meta-model of the CPS for concrete printing.

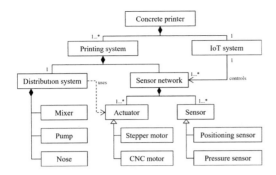

Figure 6. Final meta-model of the CPS for concrete printing.

In summary, a CPS life cycle on the meta-level is a sequence of meta-model states, each state represented by the general form (1), connected with each other by mappings $g_j, j = 1, 2, 3$ shown in Figure 2. Moreover, some of the mappings $g_j, j = 1, 2, 3$, for example $g2$, may further be subdivided into "smaller" mappings to update the CPS meta-model within one particular life cycle phase.

4 SUMMARY AND CONCLUSIONS

The broad integration of modern technologies in the field of civil engineering has promoted the implementation of cyber-physical systems as an indispensable part of smart cities and smart infrastructure. Moreover, analysis of sensor data provided by various sensors installed on a structure supports high performance of cyber-physical systems through their life time operation. Nonetheless, it is rare that a civil engineering structure keeps its state unchanged from the original construction until the final demolition. Considering that, due to complexity of modern civil engineering applications, metamodeling approaches are more frequently deployed to understand and optimize engineering structures, principal changes of cyberphysical systems in civil engineering must also be considered on the level of meta-models, or on the conceptual level. Unfortunately, modern approaches to metamodeling in civil engineering do not address the question of temporal evolution on the meta-level in its full generality. In this paper, first steps towards life-cycle metamodeling of CPS in civil engineering have been presented. The mathematical constructions have been kept very abstract and general aiming to avoid using specific metamodeling approaches probably biasing the generality of the constructions. The evolution of meta-models is described by abstract mappings that may easily be specified in the case of using concrete metamodeling approaches. Finally, an illustrative example of CPS life-cycle metamodeling of additive manufacturing of concrete structures has been discussed.

ACKNOWLEDGMENTS

The financial support of the German Research Foundation (DFG) through grants SM 281/9-1, SM 281/15-1, and LE 3955/4-1 is gratefully acknowledged. Any opinions, findings, conclusions or recommendations expressed in this paper are those of the authors and do not necessarily reflect the views of DFG.

REFERENCES

Aktins, E.M. & Bradley, J.M. 2013. Aerospace cyber-physical systems education. *Proc. of the American Institute of Aeronautics and Astronautics (AIAA) Infotech@Aerospace (I@A) Conference, Boston, MA, USA, August 19.*

Bhuiyan, M.Z.A., Wu, J., Wang, G., Cao, J. & Jiang, W. 2016. Sensing and decision making in cyber-physical systems: The case of structural event monitoring. *IEEE Transactions on Industrial Informatics*, 12(6): 2103–2114.

Borrmann, A., Koenig, M., Koch, C. & Beetz, J. (eds.). 2018. *Building information modeling – Technology foundations and industry practice.* Springer.

Fitz, T., Theiler, M. & Smarsly, K. 2019. A metamodel for cyber-physical systems. *Advanced Engineering Informatics*, 41(2019), 100930.

Grieves, M. 2006. *Product lifecycle management: driving the next generation of lean thinking.* New York, McGraw-Hill.

Gürlebeck, K., Hofmann, D. & Legatiuk, D. 2017. Categorical approach to modelling and to coupling of models. *Mathematical Methods in the Applied Sciences*, 40(3): 523–534.

Gürlebeck, K., Legatiuk, D., Nilsson, H. & Smarsly, K., 2019. Conceptual modelling: Towards detecting modelling errors in engineering applications. *Mathematical Methods in the Applied Sciences*, 43(3): 1243–1252.

Huang, H.-M., Tidwell, T., Gill, C., Lu, C., Gao, X. & Dyke, S. 2010. Cyber-physical systems for real-time hybrid structural testing: A case study. *Proc. of the international conference on cyber-physical systems (ICCPS), Stockholm, Sweden, April 12, 2010.*

Korkmaz, S. 2011. A review of active structural control: challenges for engineering informatics. *Computers and Structures*, 89(23): 2113–2132.

Lee, E.A. & Seshia, S.A. 2017. *Introduction to embedded systems – A cyber-physical systems approach.* Second Edition, MIT Press.

Legatiuk, D., Dragos, K. & Smarsly, K. 2017. Modeling and evaluation of cyber-physical systems in civil engineering. *Proceedings of Applied Mathematics and Mechanics*, 17(1): 807–808.

Legatiuk, D., Theiler, M., Dragos, K. & Smarsly, K. A categorical approach towards metamodeling cyber-physical systems. *Proc. of the 11th International Workshop on Structural Health Monitoring, Stanford, CA, USA, September 12.*

Legatiuk, D. & Smarsly K. 2018. An abstract approach towards modeling intelligent structural systems. *Proc. of the European work shop on structural health monitoring (EWSHM), Manchester, UK, July 10.*

Letichevsky, A.A., Letychevskyi, O.O., Skobelev, V.G. & Volkov, A. 2017. Cyber-physical systems. *Cybernetics and Systems Analysis*, 53(2017): 821–834.

Losev, K. 2019. Buildings life cycle methodology aspects. *The Eurasian Scientific Journal*, [online] 6(11). Available at: https://esj.today/PDF/119SAVN619.pdf (in Russian).

Losev, K. & Chulkov V. 2019. Infographic oriented management model of cyber-physical systems during a building life cycle. In: *Proc. of the International Scientific Conference "Construction and Architecture: Theory and Practice for the Innovation Development" (CATPID-2019), Moscow, Russia, October 1, 2019.*

Mohanty, S. 2016. Everything you wanted to know about smart cities. *IEEE Consumer Electronics Magazine*, 5(3): 60–70.

Nefzi, B., Schott, R., Song, Y.Q., Staples, G.S. & Tsiontsiou, E. 2015. An operator calculus approach for multi-constrained routing in wireless sensor networks. *Proc. of the 16th ACM International Symposium on Mobile Ad Hoc Networking and Computing, June 6, New York, NY, USA.*

Ngwepe, L., & Aigbavboa, C. 2015. A theoretical review of building life cycle stages and their related environmental impacts. *Journal of Civil Engineering and Environmental Technology*, 2(13): 7–15.

Sampigethaya, C., & Poovendran, R. 2013. Aviation cyber-physical systems: foundations for future aircraft and air transport. *Proceedings of the IEEE*, 101(8): 1834–1855.

Simko, G., Levendovszky, T., Maroti, M. & Sztipanovits, J. 2014. Towards a theory for cyber-physical systems modeling. *Proc. of the 4th ACM SIGBED International Work shop on Design, Modeling, and Evaluation of Cyber-Physical Systems, Berlin, Germany, April 14, 2014.*

Smarsly, K., Fitz, T. & Legatiuk, D., 2019. Metamodeling wireless communication in cyber-physical systems. *Proc. of The 26th International Work shop on Intelligent Computing in Engineering (EG-ICE). Leuven, Belgium, June 30, 2019.*

Smarsly, K., Peralta, P., Luckey, D., Heine, S., & Ludwig, H.-M. 2020. BIM-based concrete printing. *Proc. of the International ICCCBE and CIB W78 joint conference on computing in civil and building engineering, Sao Paolo, Brazil, June 2, 2020.*

Spivak, D. & Kent, R. 2020. Ologs: a categorical framework for knowledge representation. *PLoS ONE*, 7(1): e24274.

Syarif, A., Abouaissa, A., Idoumghar, L., Lorenz, P., Schott, R. & Staples, S.G. 2019. New path centrality based on operator calculus approach for wireless sensor network deployment. *IEEE Transactions on Emerging Topics in Computing*, 7(1): 162–173.

Volkov, A. 2018. Smart City: Convergent Socio-Cyber-Physical Complex. *Proc. of the 2018 VI International Scientific Conference "Integration, Partnership and Innovation in Construction Science and Education" (IPICSE-2018), Moscow, Russia, November 14, 2018.*

Volkov, A & Shilova, L. 2019. Cyber-physical systems in construction for sustainable urban development. *Proc. of the 2nd International Symposium on Architecture Research Frontiers and Ecological Environment (ARFEE 2019), Guilin, China, December 20, 2019.*

Yuan, X., Anumba, C.J. & Parfitt, M.K. 2016. Cyber-physical systems for temporary structure monitoring. *Automation in Construction*, 66: 1–14.

Photogrammetry, laser scanning and point clouds

Combining point-cloud-to-model-comparison with image recognition to automate progress monitoring in road construction

A. Ellinger, R.J. Scherer & C. Wörner
TU Dresden/VIA IMC, Dresden/Berlin, Germany

T. Walther & P. Vala
Bauhaus Universität Weimar, Weimar, Germany

ABSTRACT: This paper proposes a method to recognize the current progress of a road construction site by combining geometric comparison of 3D CAD geometry to as-built point cloud data with image-based surface recognition. Recognizing the current status of road construction sites by purely geometric means is error-prone, since the individual road layers are relatively thin compared to possible deviations of the built road from the design geometry and possible inaccuracies in the z-dimension of the as-built point cloud. Recognizing the current state by purely visual means is prone to error if two or more layers look alike or if the appearance of a layer is distorted by e.g. dust. Therefore, we propose to merge the geometric and the visual evaluation into one hybrid probabilistic algorithm. The goal is not only to improve the accuracy of the progress recognition, but also to gain additional information about the confidence of the systems decision.

1 INTRODUCTION

1.1 Motivation

Progress control is a critical task in construction project management (Zhang & Arditi 2013). For successful project control, it is crucial to identify possible undesirable developments at an early stage in order to counteract them in good time (Berner et al. 2015).

Mapping data regarding the current progress for a successful project controlling in a conventional manner is "Achilles' heel" in successful project control (Walther & Bargstädt 2019). Automating this process can help to overcome the limitations associated with presently employed manual data collection and analysis practices (Zhang & Arditi 2013).

In times of digitalization and dynamization of the working world, automated approaches are indispensable in the construction industry due to a continuous need for data tracking and usage. There is still no system that guarantees fully automatic monitoring, analysis, and control of construction site activities (Omar et al. 2018).

1.2 Using UAVs to scan construction sites

Traditionally, construction managers must be personally present at the construction site to examine progress in different activities and understand the present state of the project. This type of progress control is time-consuming, requiring data collection and extraction from construction drawings, schedules, and budget information (Navon 2007).

Measuring the projection of the real world point wise has been performed with surveying technologies like the digital level or tachymeter. With advancing technology, generating point clouds by the means photogrammetry and LiDAR (*light detection and ranging*) is becoming more feasible and accepted within the industry (Qu et al. 2014). Unmanned Aerial Vehicles (UAVs), equipped with RGB camera or LiDAR sensors, can cover several hectares in little time to create a projection of the real world in XYZ-coordinates.

1.3 Sources of errors

When relying on the geometric projection of the real world in XYZ-coordinates generated form data that was captured via remote sensing, several possible sources of errors must be considered.

First, the constructed road layers are not guaranteed to exactly match the design geometry. The acceptable

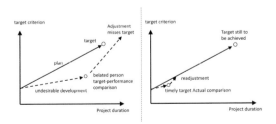

Figure 1. Influence of the time that passes before an undesirable development is detected on the success of a project (Berner et al. 2015).

construction errors from an operational point of view might be higher than the error that is acceptable to recognize the current status by purely geometric means. The error margin is usually defined by contract specifics. The UK Highways Agency for example defines the limit for the vertical error at 6 mm for pavement, 15 mm for base, and 10 mm for subbase levels. The limit for the horizontal error is defined at 25 mm (Vick & Brilakis 2018).

Second, point clouds generated from aerial images using UAVs and photogrammetry are not guaranteed to exactly match the true surface geometry. The process of photogrammetry itself yields many sources for potential errors that must be taken into account in further use of its results, such as errors in stereo plotting and analytical photogrammetry or photogrammetric error propagation (Scherz 1974). Errors following mechanical properties, such as lens distortion, operating errors, motion blur during flight or varying light and weather scenarios, can affect the results in horizontal and vertical accuracy of each given point. Using GCPs (Ground Control Points) can mitigate many of those errors but will not guarantee a perfect projection of the real world (Moser et al 2016).

The margin of vertical error reported in other studies ranges from under 3 cm (Moser et al. 2016) up to 0.5 m (Zulkipli & Tahar 2018).

1.4 Proposed solution

The proposed solution combines image- and point cloud-based information to deal with the problem of uncertainty that comes with the various sources of errors described before.

When uncertainty is a factor, it seems plausible to combine more than one independent source of information to gain a more reliable result. Taking in account results from two independent components allows for analyzing how much the two components "agreed". This kind of information could not be gained by merely looking at the confidence scores of a single component.

The focus of this article lies on the basic structure of the proposed hybrid approach, its basic components, and the fundamental ideas to combine the visual information with the geometric one.

To the best of our knowledge, this is the first article that proposes the combination of image-based surface recognition with geometric point-clout-to-CAD (Computer Aided Design) comparison to address the problem of inaccuracies and uncertainties in automated road construction progress monitoring.

2 RELATED WORK

2.1 Progress monitoring in building construction

Within the building construction industry, attempts to automate progress monitoring based on point clouds and image data are already well advanced compared to the infrastructure construction industry.

2.1.1 Point Cloud to 3D CAD comparison

In 2009, Frederic N. Bosche discusses in detail how as-built point cloud data and 3D as-planned models could be used to automatically recognize construction progress.

His research plays a central role in the method described by Turkan et al. (2012), which aims to automate construction progress tracking by fusing 4D modeling and laser scanning. This method was also reported to be able to increase the accuracy compared to manual methods. The project schedule was updated automatically according to the recognized progress. The core algorithm of this system was based upon the works presented in (Bosche 2009). The entire approach consists of manually coarse registration of the as-built point cloud, automated finetuning of the registration via Iterative Closest Point Algorithm and the final Object recognition via a robust surface-based recognition metric.

Kim et al. (2013) report the feasibility of Principal Component Analysis (PCA) in combination with Iterative Closest Point algorithm and Levenberg–Marquardt algorithm to automatically register as-built point cloud data to the 3D CAD Modell. The study points out that this method is also beneficial for use in project progress monitoring.

An approach presented by Pucko et al. (2018), utilizes a novel method for data acquisition. Small low precision 3D scanning devices that could be placed on the workers protective helmets record the construction site. From this data a constantly updated partial point cloud is generated and registered to a 4D as-built point cloud and compared to the 4D as-paned model. This way, deviations from the time schedule can be reported in real time.

2.1.2 Methods involving image recognition

In 2009, Zhang et al. (2009) demonstrate that computer vision can assist "work in progress measurement" to a certain extend. In combination with 3D design model and a model-based fitting approach, their computer vision module was able to detect the construction state of building components.

Using video-based construction monitoring was proposed by Gong and Caldas (2011). The applicability of the developed reasoning mechanism to various types of construction operations was demonstrated in a case study.

Progress monitoring on an operation-level (e. g. formwork vs. concrete surfaces) is still a challenge, since this requires additional appearance information apart from the raw point cloud geometry. Han & Golparvar-Fard propose to tackle this issue by leveraging the interconnectivity of site images and BIM-registered point clouds. They included formalized knowledge about the process sequence of construction operations as well as a histogram-based representation of possible construction material types. This method is reported to accurately detect the current progress even in the presence of occlusions (Han & Golparvar-Fard 2014).

Golparvar-Fard and Pena-Mora present a method that detects the current construction progress based on 3D BIM as-planned models and unordered daily photographs from site as well as sparse as-built point clouds generated from those photographs. They introduce a machine-learning scheme built on a Bayesian probabilistic model. This model was specifically designed to detect physical progress even in the presence of occlusions (Golparvar-Fard & Pena-Mora 2015).

Mahami et al. (2019) propose a photogrammetric Multi-View Stereo method to improve the accuracy and completeness of as-built point clouds generated from daily site images. This in turn allows for more accurate and detailed determination of the current progress of a construction site.

2.2 Material recognition

According to Dimitrov and Golparvar-Fard, material classification plays an important role in any vision-based system for automated construction progress monitoring. They report an average classification rate of 97.1% when training and testing different classifiers on a on a real-world dataset. (Dimitrov & Golparvar-Fard 2013).

Wieschollek & Lensch (2016) discuss the feasibility of leveraging pretrained convolutional neural networks to improve the performance of material classifiers. They compared the result to methods that use hand-crafted features and conclude that even a simple Convolutional Neural Network (CNN) trained on an opensource image dataset called "Flickr Material Dataset" outperforms all published hand-engineered features when using transfer learning on a CNN pre-trained for object recognition to compensate for the relatively small size of this dataset (100 examples per material category).

2.3 Differences of building- and road construction sites

Although there are many similarities in the methods and technologies used to automate progress monitoring for building construction sites and infrastructure construction sites (e.g. 3D CAD design geometry, point clouds, photogrammetry, machine learning), there also are significant differences to be considered. While building consist largely of distinct objects than can be oriented in space in many ways, roads usually consist only of multiple layers that are always relatively close to being parallel to the XY-plane. In addition, building objects can have basically any shape, ranging from flat disc-like ceilings over approximately linear columns up to arbitrarily shaped elements in architectural sophisticated building envelops. The dimensions of road layers, however, almost always follow the pattern $d_h \ll d_w \ll d_l$, meaning they are extremely thin

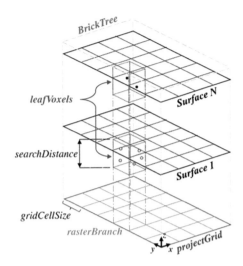

Figure 2. *BrickTree* structure (Vick & Brilakis 2018).

compared to their width and length. Thus, in road construction monitoring, a different way of partitioning space may be advantages to register as-built point clouds to 3D CAD Models.

2.4 Progress monitoring in road construction

Steven Vick and Ioannis Brilakis seem to be the first to describe a way of partitioning Euclidean space specifically optimized for detecting road layers in as-built point clouds (Vick & Brilakis 2018). The geometric component described in chapter 3 of this article is largely based upon their work. The core idea of their method is a hierarchical space partitioning data structure called *BrickTree* which is based upon the idea of *k*-d trees. This method partitions space in the XY-plane in an evenly spaced orthogonal grid. In the Z-dimension, space is not partitioned evenly. The branches of the *voxelTree* consist of rectangular boxes whose height is independent of their length and width. The resulting leaves of this data structure are therefore not necessarily cubical, hence the name *brickTree*.

In addition to the *brickTree* structure, Vick & Brilakis also propose a machine learning-based approach to find the best criteria to decide which layer should be classified as the one represented by the as-built point cloud. Since they lack the necessary training set containing many examples from a variety of different construction sites, they implement their proof of concept using different statistical thresholds like e.g. maximum number of points in a given *voxelLeafe*.

A method to utilize aerial images to recognize progress of linear infrastructure construction sites was proposed by Behnam et al. (2016). This system identifies the current construction stage from multi-temporal satellite images. It combines three different image processing methods: object-based image analysis (OBIA); template-based image recognition and normalized vegetation index images (NDVI).

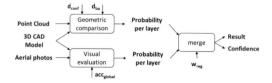

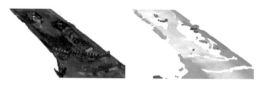

Figure 3. Interaction of the basic components (schematic).

Figure 5. Removing safety barrier, hanger and stockpile using estimated surface normal. Computed and visualized using *Open3D*.

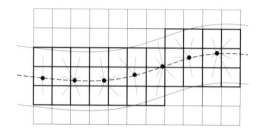

Figure 4. Projecting grid-based results on the alignment curve.

small angel to the global XY-plane, surface normals can be used as a feature to remove points that are part of the none-horizontal parts of the occlusions, e.g. the side faces of a car.

Other methods such as identifying entire point cluster based on their geometric features are capable of removing also the none-horizontal regions of occlusions. Such methods are not implemented in the scope of this article and are subject to further research.

3 PROPOSED HYBRID ALGORITHM

3.1 *Basic idea*

The hybrid method proposed in this article consists of three components. The first two components compute the current status independently, one of them by comparing the as-built point cloud to the design model geometrically, the other one by evaluating the surface of the road at its current state visually and predicting its surface material type. The third component merges the output of those components and returns the final result together with a "confidence score".

The outputs of all three components are given in the form of probability vectors. Each vector element represents the probability that the road layer with the corresponding index is the currently highest one in the evaluated as-built data.

All results are first computed for distinct regions defined by a regular grid of squares in the XY-plane. Those results are then projected on discrete sections of the alignment curve, represented by points sampled from this curve in regular distances (measured in curve length) as shown in Figure 4. Only grid cells whose midpoints lie within XY-boundaries of the as-planed geometry are considered.

3.2 *Removing occlusions*

Removing occlusions such as vehicles, safety barriers or machines from the as-built point cloud is a necessary preprocessing step to prevent distorting the results of the geometric comparison.

One way to automate this process is to remove points based on the geometric features of their local neighborhood. Principal component Analysis (PCA) is a method for geometric features extraction, especially to estimate the normal vectors of surfaces (Hu 2019). Assuming that any small region of a road surface can be approximated by a plane with a relatively

3.3 *Geometric component*

3.3.1 *Rasterized point-cloud-to-point-cloud*

As mentioned in chapter 2, the algorithm that compares the geometry of the as-built point cloud to the geometry of the as-planned layers per grid cell is largely based on the article *Road Design Layer Detection in Point Cloud Data for Construction Progress Monitoring* by Vick and Brilakis (2018), especially the ideas on how to partition space.

The essential difference to their approach lies in not selecting a specific layer as the result for a given grid cell (based on either fixed statistical rules or learned decision boundaries), but rather assigning a probability to each of the design layers. This step is essential to later merge the geometry-based results with the results of the image classification component. The procedure of computing those geometrical probabilities is as follows:

First, from the surface of each layer of the design model, one synthetic point cloud is sampled. These point clouds encode the shape of the as-planned geometry. Together with the as-built point cloud, they are registered in one common coordinate system.

Second, space is partitioned into a regular square grid in the XY-plane. Each grid cell is vertically divided into regions separated by horizontal boundaries that lie in the middle of two adjacent layer surfaces. The height of those boundaries is therefore calculated from the average heights of the adjacent layers with in the XY-boundaries of a particular grid cell (shown as green dotted lines in Figure 6).

To calculate this average height efficiently, voxel down sampling is used to compute one centroid per layer. It stores the averaged coordinates of all points that belong to the respective layer and fall into the respective XY-Boundaries. The projection of the voxel grid onto the YX-plane therefore defines the XY-grid. Using optimized voxel downsampling algorithms is significantly faster compared to naively averaging all the points per grid cell using e.g. nested loops. In this study, *Open3D's* voxel downsampling method was used.

3.3.2 Geometric probability scores

As mentioned before, the key assumption is that the z-value of the as-built point cloud is unlikely to exactly match the height of an as-planned layer surface. It deviates from this height due to inaccuracies and errors in building, scanning and data processing. Thus, we assume that no layer can be detected with absolute certainty, but rather each layer has a certain probability of being the actual currently highest as-built layer.

Since in the scope of this study, not enough data was available to compute the real distribution of the z-error for neither a specific project nor for a more general case, we illustrate the concept based on the assumption that the error is normally distributed. The z-value of the centroid of the as-built point cloud defines estimate for a given grid cell (represented by the red dotted line in Figure 6). We also assume that the confidence interval of the distribution is known or can be sufficiently estimated. Determining a more realistic distribution will be subject of further research.

Given those assumptions, the probability of a layer i corresponds to the area under the distribution function in between the lower boundary $d_{b,i}$ and the upper boundary $d_{b,i-1}$, where the upper boundary lies in the middle of the upper adjacent layer i-1 and the lower boundary in the middle of the layer i. The as-build layer height defines the estimate of the distribution for a given grid cell (see Figure 6). The probability per layer is therefore given by the integral of the distribution function between those boundaries.

$$P(layer\_i) = \Phi(b_{i-1}) - \Phi(b_i) \quad (1)$$

$$\Phi(z) = \frac{1}{\sqrt{2\pi}} \int_{-\infty}^{z} e^{-\frac{1}{2}t^2} dt \quad (2)$$

To prevent the highest and lowest design layer from getting disproportionately high probability scores when a significant portion of the area under the distribution function lies above the highest or below the lowest layer surface, two additional "artificial" boundaries are introduced. They form the upper (respectively lower) boundary for the integral when computing the probability for the highest and lowest design layer (see the highest green dotted line in Figure 6).

The probability per layer for a discrete section along the alignment curve is given by the element-wise mean of the probability vectors of all grid cell that belong to this region. The belonging grid cells per region are computed by finding the closest sample point of the alignment curve for each grid cell as illustrated in Figure 4.

3.4 Visual component

The current implementation of the visual component also follows a rasterized approach. The entire orthomosaic created from the images taken by the UAVs

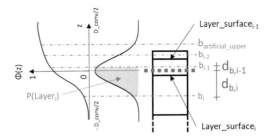

Figure 6. Concept of computing the probability of layer i for a normally distributed error in the z-value (Illustration).

RGB-Sensor during one flight is divided into a grid of regular squared tiles. Geolocation, size, and orientation of each tile is stored. This allows for mapping the result of the classification process back to the area the classified image represents.

Each tile is evaluated by an image classification algorithm that was trained to distinguish between road layer types, e.g. asphalt surface course, gravel base course, and so forth. The resulting vector contains the probability scores per type. Since the image recognition component does not directly output the index of a specific layer, but rather a surface material type, a subsequent function first evaluates which layers belong to this type. The probabilities for each surface type are then assigned to those layers. The material-type-per-layer-information could for example be derived from the BIM CAD model from which the as-planned point clouds were sampled. Since this is rather a problem of interoperability and sematic model enrichment, it is not further discussed in this study.

As classification algorithm a pretrained convolutional neural network (CNN) was chosen. In particular, *PyTorch's* implementation of the *SqueezeNet* architecture, which is pretrained on the *ImageNet* dataset, was leveraged. During the training process, only the last layer of the multi-layer perceptron component of the network was "unfrozen" and trained to learn the new categories. This method is widely used in many fields of computer vision since it speeds up the training process, reduces the amount of necessary training data and in general yields better results than training all parameters of a network on use case specific data from scratch (cf. chapter 2).

The resulting vector that contains the scores per class is normalized and multiplied with the global accuracy of the classifier computed on the test set:

$$\vec{p}_{vis\_tile} = \hat{y} \cdot \frac{1}{sum\{\hat{y}\}} \cdot acc_{glob} \quad (3)$$

3.5 Combining the two results

Finding a valid mechanism to merge the two results appears to be the most challenging part of the proposed hybrid method. The current exemplary implementation simply computes a weighted sum from the two probability vectors. The regularization parameter w_{vis}

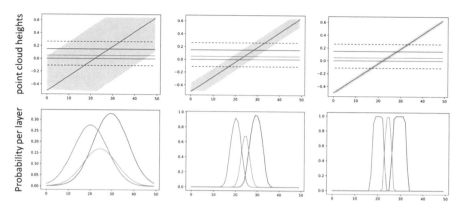

Figure 7. Probability per layer as a function of the position of the as-built point cloud centroid (red line), plotted for 3 different confidence intervals (light red region). Computed using synthetic point clouds representing idealized as-built data and design geometry.

allows to assign more weight to the result of one of the components (see equation 4). While this parameter is useful during the experimental stage, it seems unlikely that one "ideal" value for this parameter can be found that is sufficient in all possible scenarios. Therefore, a more refined combination mechanism is yet to be developed. Some possible concepts are discussed in section 5.

$$\vec{p}_{res} = \vec{p}_{geo} + \vec{p}_{vis\_reg} \cdot w_{vis} \qquad (4)$$

3.6 Computing the confidence scores

The confidence of the system is computed for each discrete section along the alignment curve separately by calculating the absolute difference of the highest probability score of the combined result to the second highest score. Additionally, a "penalty score" is introduced that reduces the confidence value if the two components do not assign their highest confidence value to the same layer.

4 PROOF OF CONCEPT

A prototype of the proposed system was implemented in Python; open source libraries used include the 3D data processing library *Open3D*, the computer vision and deep learning library *PyTorch* and the scientific computing package *NumPy*.

The prototype was tested on both synthetic and real-world data, although synthetical data was only used to validate the geometric component. The point cloud representations of the design geometry as well as the synthetic test data were generated in *Rhinoceros 3D* using the graphical algorithm editor *Grasshopper*.

The real-world data was captured on a two-lane urban road construction site in Leipzig, Germany. At the time of the UAV scan, the entire base course and about one third of the surface course were built in the test section (see Figure 8). The as-built point cloud contained large errors in the z-values of some point

Figure 8. Preliminary results of the image classification component achieved on the test data.

regions. A processing error lead to deviations from the real as-built surface geometry of up to almost half a meter. The confidence interval for the test was therefore set to 0.5 meter, which is far greater than the acceptable construction error specified in most construction contracts. The distance of the artificial boundaries to the surfaces of the outer layers was set to 0.2 meters. This corresponds to the design thickness of the surface course. The entire surface course was regarded as one single layer, the fact that it consists of three consecutive asphalt layers was discarded in the scope or this study due to the large error in the point cloud data.

The image recognition component was tested under simplified "laboratory" conditions; the primary goal was to demonstrate the applicability of the whole pipeline in principal. The number of different material classes, the size of the training set and the conditions under which the training set was produced do not reflect the conditions necessary to validate a production-ready system.

The classifier was trained on images taken on the same construction site as the test data but on a different section (none of the images used for testing were seen by the classifier during training). The three test-classes were *asphalt surface course*, *concrete base course* and *not part of road*, with about one hundred training examples per class.

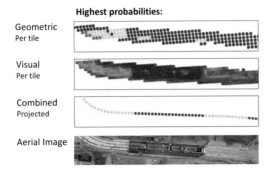

Figure 9. Color coded visualization of the result of the individual components (highest probability per tile/segment) and comparison to the aerial image. Visualized using *Open3D*.

5 RESULTS AND DISCUSSION

The exemplary implementation validates the general functionality of the proposed algorithm. It shows that the algorithm is capable of computing the probability scores from rasterized aerial images as well as from as-built point clouds. It also demonstrates how the raster-grid based scores can be projected on the alignment curve, which is the common reference frame for infrastructure projects. Furthermore, the concept of the merging mechanism could be demonstrated. A visualization of the results of those consecutive steps is shown in Figure 9.

Tests performed on synthetically "as-built" point clouds sampled directly form the design geometry showed that the geometric component is robust when the error in the z-value of the as-built data is smaller than half the vertical distance of the two layer surfaces adjacent to the as-built point cloud and when the confidence interval is chosen accordingly (see e.g. Figure 7).

The tests on real world data shows that the method is applicable to real projects, yet the results concerning the improvement in the overall accuracy are inconclusive. To evaluate the reliability of the method, more tests on different construction sites of various types need to be conducted.

So far, the main benefit of the proposed hybrid approach seems to lie in the additional information gained about the systems confidence. In addition to the individual confidences of the two separate components, the hybrid system is also able to analyze "how much the two components agreed". The drop of the confidences system on the southern ("right") end of the test section corresponds with the increasing error in the z-value of the as-built point cloud (see Figure 10).

5.1 *Alternative combination mechanisms*

One possible approach to deal with the problem of "how much weight to give to which component under which circumstances" could be to primarily rely on the result of only one component. The result of the

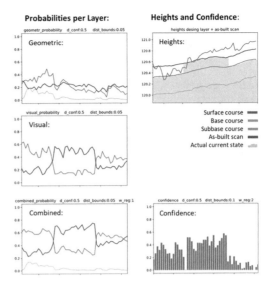

Figure 10. Probability scores per layer and confidence scores of the combined result for the real-world example. Computed and visualized using Python.

secondary component would only be considered if the result of the primary component is ambiguous.

Another approach could be to use the second component solely for additional information about the reliability of the results of the primary component, possibly highlighting sections in which the two components disagree. In this scenario, only one component would be used to compute the actual result. The other component would merely "comment on it".

Ideally, an intelligent mechanism would identify the best combination based on the specific values of each individual result and based on additional circumstantial parameters. Whether such a method can be developed utilizing supervised machine learning algorithms is subject to further research.

5.2 *Applicability to other fields*

The central aspects of this approach seem also feasible for other types of linear infrastructure like railways or airport runways. In general, combining two or more independent components to gain insights about the reliability of a recognition system appears to be advantages for any kind of autonomous progress monitoring system.

6 CONCLUSION

This paper introduced a new approach to recognize the current progress of a road construction site by combining geometric Point-Cloud-to-CAD comparison with image-based surface recognition. The goal was to improve the reliability of the systems results as well as to gain additional information about the systems confidence in each of its decisions.

Tests performed on both synthetic and real-world data validated the general applicability of the proposed algorithm. While there are promising aspects regarding the systems reliability, the results concerning the overall improvement in accuracy under real-world conditions are yet inconclusive due to insufficient test data and the need to develop a more refined combination mechanism.

Currently, the main benefit of the proposed hybrid approach appears to lie in the additional information gained about the systems confidence. In addition to the individual confidences of the two separate components, the hybrid system is also able to analyze "how much the two individual components agreed".

It seems unlikely that autonomous progress recognition systems developed in the near future will be guaranteed to work absolutely flawless. Therefore, proposing a mechanism that provides insights about the reliability of the systems' results appears to be the main practical contribution of this work.

REFERENCES

Behnam, A.; Wickramasinghe, D. C.; Ghaffar, M. A. A.; Vu, T. T.; Tang, Y. H. & Isac, H. B. Md. 2016. Automated progress monitoring system for linear infrastructure projects using satellite remote sensing. *Automation in Construction* 68: 114–127.

Berner F., Kochendörfer B. & Schach R. 2015. *Grundlagen der Baubetriebslehre 3: 59–60*. Wiesbaden: Springer Fachmedien Wiesbaden.

Bosche, F. 2009. Automated recognition of 3D CAD model objects and calculation of asbuilt dimensions for dimensional compliance control in construction. *Advanced Engineering Informatics* 24: 107–118.

Dimitrov, A. & Golparvar-Fard, M. 2014. Vision-based material recognition for automated monitoring of construction progress and generating building information modeling from unordered site image collections. *Advanced Engineering Informatics* 28: 37–49.

Golparvar-Fard, M.; Pena-Mora, F.. and Savarese, S. 2015. Automated Progress Monitoring Using Unordered Daily Construction Photographs and IFC-Based Building Information Models. *Journal of computing in civil engineering* 29(1): 04014025.

Gong, J. & Caldas, C. H. 2011. An object recognition, tracking, and contextual reasoning-based video interpretation method for rapid productivity analysis of construction operations. *Automation in Construction* 20: 1211–1226.

Han, K. & Golparvar-Fard, M. 2014. Multi-Sample Image-based Material Recognition and Formalized Sequencing Knowledge for Operation-Level Construction Progress Monitoring. *Computing in Civil and Building Engineering*: 364–372.

Hu, C.; Pan, Z.; Li, P. 2019. A 3D Point Cloud Filtering Method for Leaves Based on Manifold Distance and Normal Estimation. *Remote sensing* (11): 198.

Kim, C.; Son, H. & Kim, C. 2013. Fully automated registration of 3d data to a 3d cad model for project progress monitoring. *Automation in Construction* 35: 587–594.

Mahami, H.; Nasirzadeh, F.; Ahmadabadian, A. H. & Nahavandi, S. 2019. Automated Progress Controlling and Monitoring Using Daily Site Images and Building Information Modelling. *Buildings* 2019(9): 70.

Moser, V.; Barišic, I.; Rajle, D. & Dimter, S. 2016. Comparison of different survey methods data accuracy for road design and construction. *4th International Conference on Road and Rail Infrastructure*.

Navon, R. 2007. Research in automated measurement of project performance indicators. *Automation in Construction* 16(2): 176–188.

Omar H.; Mahdjoubi L. & Kheder G. 2018. Towards an automated photogrammetry-based approach for monitoring and controlling construction site activities. *Computers in Industry* 98: 172–182.

Pucko, Z.; Šuman, N. & Rebolj, D. 2018. Automated continuous construction progress monitoring using multiple workplace real time 3D scans. *Advanced Engineering Informatics* 38: 27–40.

Qu, T. & Coco, J. & Rönnäng, M. & Sun, W. 2014. Challenges and Trends of Implementation of 3D Point Cloud Technologies in Building Information Modeling (BIM): Case Studies. *Computing in civil and building engineering* 2014: 809–816.

Scherz, J. 1974. Errors in Photogrammetry. *International Journal of Rock mechanics and Mining Science and geomechanics Abstracts* 11(8): 493–500.

Turkan, Y.; Bosche, F.; Haas, C. T. & Haas, R. 2012. Automated progress tracking using 4d schedule and 3d sensing technologies. *Automation in construction* 22: 414–421.

Vick, S. & Brilakis, I. 2018. Road Design Layer Detection in Point Cloud Data for Construction Progress Monitoring. *Journal of computing in civil engineering*. 32(5): 04018029.

Walther T. & Bargstädt H.-J. 2019. Automated data acquisition to show the actual performance during the construction process: In: *International Association for Bridge and Struc-tural Engineering, 20th Congress of IABSE New York City 2019*: 1566–1573.

Wieschollek, P. & Lensch, H. 2016. Transfer learning for material classification using convolutional networks. *arXiv preprint arXiv*. 2016: 1609.06188.

Zhang, C. & Arditi, D. 2013. Automated progress control using laser scanning technology. *Automation in Construction*. 36: 108–116.

Zhang, X., Bakis, N., Lukins, T., Ibrahim, Y., Wu, S., Kagioglou, M., Aouad, G., Kaka, A. & Trucco, E. 2009. Automating progress measurement of construction projects. *Automation in Construction* 18: 294–301.

Zulkipli, M. & Tahar, K. 2018. Multirotor UAV-Based Photogrammetric Mapping for Road Design. *Hindawi International Journal of Optics*, Volume 2018, Article ID 1871058.

Research on BIM and virtual pre-assembly technology in construction management

D. Han, C.N. Rolfsen, B.D. Engeland & H. Hjelmbrekke
Oslo Metropolitan University (OsloMet), Oslo, Norway

H. Hosamo
University of Agder, Gimlemoen/Grimstad, Norway

K. Hu
Chongqing Jiaotong University, Chongqing, China

T. Guo
Southeast University, Nanjing, China

C. Ying
Chongqing Smart City and Sustainable Development Academy (SCSDA), Chongqing, China

ABSTRACT: In the field of architecture, the virtual pre-assembly of prefabricated components (steel structure, some of concrete structure) need to obtain data through 3D laser scanning or other instruments, and then realize by computer virtual imitation. In recent years, the combination of BIM and virtual pre-assembly is more and more used in some large and complex steel structures, especially in the application of technology, and the effect is good. However, for the combination of BIM and virtual pre-assembly, there is little research on management. In this paper, the combination of BIM and virtual pre-assembly in the management of the application of a more in-depth study. Compared with the traditional construction management, the combination of BIM and virtual pre-assembly optimizes the key construction steps such as construction preparation, quality inspection, schedule management, resource coordination, etc., which can serve the whole life cycle of the project. We have applied the combination of BIM and virtual pre-assembly to the construction management of a large complex steel structure. The results show that the integration of BIM and virtual pre-assembly technology has many advantages over the traditional construction management. It can find some problems that traditional management can't find in advance, and allocate resources reasonably.

1 INTRODUCTION

Construction project management defined by Ding Zhang et al. (2011) is the fundamental to ensure the construction progress, project quality and safety construction. The traditional management mainly depends on the rich experience of managers, while the arrangement of materials, labor and general equipment depends on the well prepared construction organization design. Woonseong et al. (2016) proposed a simulation framework integrating building information model (BIM) to predict the productivity dynamics during the construction planning stage. To develop this framework, the researchers examined key factors that affect operational level productivity and then predicted productivity dynamics. Chen et al. (2014) discussed the advantages, organization and process (POP) data definition structure of 4D BIM in high-quality application based on construction specifications by building models in products, and verified it with the construction stage of Wuhan International Expo Center as an example. The efficiency and quality of construction can be guaranteed if the site construction meets the construction organization design. When the site construction situation does not conform to the construction organization design and the situation is not serious, the construction organization design can be modified appropriately to adapt to the actual situation. However, when the actual situation of the site is seriously inconsistent with the construction organization design, the managers have to take a lot of remedial measures, or even rework.

Virtual construction can prevent and solve some key problems in construction management, especially for prefabricated structures Obonyo (2011) said. In virtual reality system, managers can know the construction results of existing structures in advance. The management personnel can optimize the preparation work according to the construction results to reduce the risk in the construction. Of course, it needs simulation results to be reliable. Through the virtual construction demonstration of the future conditions, managers can find the construction hidden dangers and safety hidden dangers in advance. Managers take corresponding measures to control in time to ensure the safety of personnel and avoid unnecessary losses. Through the relevant discussion, managers can find the deficiencies of the construction scheme. They will revise the plan to ensure the quality of the project.

Guo et al. (2018) applied BIM system to carry out virtual pre-assembly of steel structure to improve the quality of construction. Li et al (2019) applied BIM Technology to the construction of a large-scale anisotropic steel structure. By using the BIM model to reduce the collision and simulate the site construction, the fine construction and management of the site steel structure and curtain wall project are realized. Xu et al. (2019) used the function of BIM Technology visualization to deepen the design of long-span truss. This method avoids collisions and ensures the construction progress on site.

Reliable and comprehensive data is the guarantee of virtual construction simulation results. The total station has been proved to be a reliable instrument. However, the total station often needs other media to locate the measured objects, such as reflectors and manual line drawing. The existence of medium has certain influence on feature acquisition. In addition, the rate of the total station limits the number of points, and the limited points have certain limitations in large and complex structures. 3D laser scanning is one of the most effective tools to measure 3D objects and structures. It can capture millions of points per second with linear accuracy in the millimeter range.

In the application of virtual pre-assembly with measuring instruments, Case et al. (2014) used the total station to assemble the bolt hole position of Chernobyl nuclear power plant protective cover project by digital simulation. The principle is the generalized procrustes analysis method proposed by researchers. Qin et al (2019) combined the three-dimensional laser scanning technology with BIM Technology, and put forward the integrated engineering quality control methods of component processing quality inspection, virtual pre-assembly on construction site and real-time monitoring during construction for an arch bridge. Liang et al (2018) introduced a set of virtual pre-assembly steps and methods of bolt connected bridge components, which can be used for multi-faceted splicing of steel truss beams. Liu et al (2020) studied the virtual pre-assembly technology of steel truss beam in combination with BIM and 3D laser scanning, and virtual pre-assembly the truss under three working conditions, and achieved satisfactory results.

According to the related literature, there are few researches on BIM and virtual pre-assembly technology in construction management, which is a topic worthy of more discussion. This paper will summarize a set of technical routes about BIM and virtual pre-assembly technology in construction management, and conduct a more in-depth study on this topic through a case.

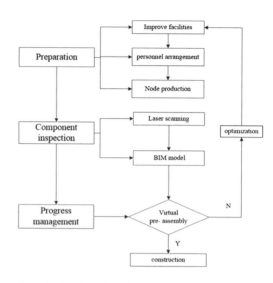

Figure 1. Optimized construction management workflow.

2 BIM AND VIRTUAL PRE-ASSEMBLY TECHNOLOGY IN CONSTRUCTION MANAGEMENT

Construction organization management is divided into three elements by Yang (2019): construction preparation, component inspection and progress management. BIM and virtual pre-assembly technology can be used to optimize the traditional construction management. The following is the optimized construction management workflow.

2.1 Preparation stage

Improving construction facilities is one of the necessary steps before construction. In addition, managers should arrange personnel reasonably. For components, especially prefabricated structural components, nodes are the key elements of the successful construction of components, so high-quality manufacturing nodes are one of the key tasks.

2.2 Component inspection

2.2.1 Data acquisition
The prefabricated components need to be inspected, and the traditional method can use the total station to collect the limited points of components. The measurement equipment determines the data acquisition of the structure object. Among them, the data measured by the total station can't reflect the three-dimensional shape of component splicing comprehensively and intuitively, but the construction of corresponding component in collision inspection is important. The potential risk of construction only depends on the finite point

simulation and prediction, and the hidden danger of safety may not be reflected in time.

We used three-dimensional laser scanning and photogrammetry to detect components and obtain comprehensive and reliable information of components. It is worth noting that node is the key factor of component, and data acquisition is needed. Different 3D laser scanners have different precision, and the distance is an important factor affecting the point cloud, so we should choose the appropriate instrument according to the actual distance and other factors, Cai et al. said (2019).

2.2.2 BIM model

The acquired point cloud data needs to be preprocessed, including the steps of station collage, denoising and down sampling. Among them, the classic algorithms of point cloud station include ICP Gregory et al. (2002) and so on. The feature extraction of components is obtained by corresponding algorithms, such as point cloud objects with plane features can be processed by growing region algorithm (Xue et al 2008) and RANSAC (Yang et al. 2016).

The extracted feature parameters are the basis of the reverse BIM model. The common BIM software includes Revit, Tekla, Catia, etc. Among them, Revit is widely used in housing construction and other fields, Tekla is often used in the construction of steel structures, and Catia surface design is unique. We need to choose different BIM software according to our needs.

2.3 Progress management

2.3.1 Construction progress prediction

In order to predict the construction progress of components through the virtual pre-assembly and to carry out the construction the virtual pre-assembly results must meet the schedule requirements

The goal of traditional alignment algorithm is to minimize the sum of the least squares of the difference between two matrices A and B. In order to ensure the effect of virtual pre-assembly, according to the principle of Procrustes analysis method, the two matrices to be spliced are simulated and aligned. The alignment goal of this method is the principle of minimizing the iterative change of the center of mass of two matrices, which is better than the minimization of the sum of squares of the difference between A and B matrices.

2.3.2 Optimization preparation

If the results of virtual pre-assembly meet the requirements of component splicing, then construction can be carried out. However, when the results of virtual pre-assembly do not meet the requirements of component splicing, managers need to optimize the preparatory work to ensure the next step of construction smoothly. Generally, nodes can be processed to meet the requirements. Such advanced optimization preparation can reduce the risk of subsequent construction and ensure the smooth progress of subsequent construction.

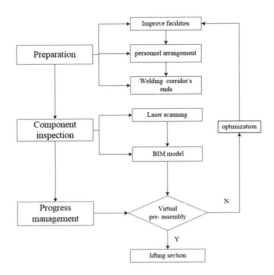

Figure 2. Construction management workflow of Chongqing Raffles.

2.4 Guarantee the construction progress

The follow-up construction steps are also an important guarantee for the construction. Generally speaking, as long as the front work is completed with high quality, the follow-up construction work only needs to be carried out step by step. Of course, managers need to deal with unexpected situations.

3 CASE STUDY

3.1 Project overview

Chongqing Raffles air corridor is more than 200 meters away from the ground. The corridor is a steel structure which is connected at roofs from the ground to the air corridor. There are three lifting sections in total, of which the maximum weight is 1100t. According to the principle of Cartesian coordinate system, there are three possible deviations when it is combined with roof's ends, i.e. axial deviation (X direction), transverse deviation (Y direction) and vertical deviation (Z direction). When a deviation exceeds the limit, it will cause the closure to fail to work normally. The construction personnel have to cut high above the limit. These conditions will greatly increase the closure cycle and the risk of high altitude closure.

If the reliable inspection of components is not carried out, there are three disadvantages of this construction management method:

1. The final state of corridor splicing is unknown, resulting in the inability to accurately organize appropriate construction personnel to cut the end in advance. The cutting equipment cannot be allocated accordingly;
2. Air risk construction is relatively large, which may lead to safety problems;

Figure 3. Scan the ends of the corridor.

Figure 4. Scan the ground lifting section and its ends.

3. The air is affected by wind and other factors, resulting in the construction quality cannot be guaranteed, further affecting the management of construction progress. It can be seen that component inspection is the core step of the construction management route.

Therefore, the virtual pre-assembly must be carried out before the lifting of the ground lifting section. According to the results of virtual pre-assembly, the ground lifting section is cut in advance. The following is a workflow of the application routes of virtual pre-assembly technology in Raffles management, which is a combination of the air corridor BIM and 3D laser scanning.

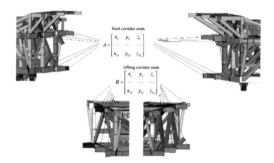

Figure 5. The BIM model of corridor's ends.

3.2 Preparation stage

Managers arrange relevant personnel to improve the equipment before construction, install the crane in place, and check whether the lifting device is in good condition. In addition, managers should focus on organizing personnel to weld the joints of the corridor.

3.3 Component inspection

3.3.1 Data acquisition

Because the ends of the corridor are located in different towers, they are far away. According to the actual situation, we decided to use Leica MS60 whole station scanner to collect the data at the ends of the corridor. The characteristics of this instrument are effective scanning distance and high precision. In addition, for the ground lifting section and its ends, we use the method such as X330 to complete the data acquisition. The instrument has the advantages of fast scanning speed and abundant laser point clouds, and it is suitable for close range scanning.

3.3.2 BIM model

The collected point cloud completed the pre-processing work such as station collage and drying. The corridor and the ground lifting section's ends are key features, which directly determine the results of corridor assembly. In the point cloud processing software Geomagic Control, the corresponding algorithm is used to get the spatial parameters of all the end heads,

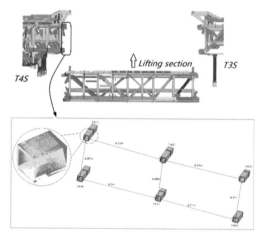

Figure 6. The BIM model of ground lifting section's ends.

which are saved in A and B matrices. The data of A and B will update the BIM model, so that the designed BIM model can be converted into the actual BIM model (Figure 5).

3.4 Progress management

3.4.1 Progress forecast

The result of virtual pre-assembly is the key index of progress prediction. Only when the result of virtual pre-assembly meets the progress requirements can the construction be carried out (Figure 6). Virtual pre-assembly is a way to predict the progress ahead of time.

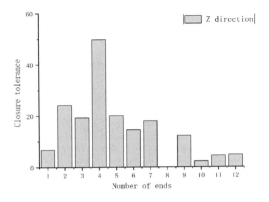

Figure 7. The result of virtual pre-assembly.

It is worth noting that the influence degree of the axial deviation (X direction), transverse deviation (Y direction) and vertical deviation (Z direction) on the structure is different. According to the finite element analysis and simulation, the closure deviation in Z direction has the greatest influence on the structural stress, and the deviation cannot exceed 25 mm.

The results of virtual pre-assembly show that the construction requirements can be met only when the 4 end of the corridor is cut. The managers decided to cut the end 4 to avoid subsequent construction risks.

The following is the result of virtual pre-assembly in Z direction (Figure 7).

3.4.2 Optimization preparation

The managers shall organize relevant personnel to cut the unreasonable end 4, and the cutting requirements shall meet the deviation range. Compared with the previous arrangement of personnel and facilities when closing upstairs, the managers' strategy avoids the arrangement of this operation.

3.4.3 Guarantee the construction progress

As the corridor lifting is strictly controlled according to the pre-assembly results, the corridor construction process is smooth. It not only improves the efficiency of construction, but also ensures the quality of construction (Figure 8).

4 REPRESENTATION OF RESULTS

The three-dimensional laser scanner accurately scanned the end of the corridor and related components, and accurately obtained the key features needed for virtual pre assembly. Virtual preassembly predicted that No. 4 end was unqualified, and managers organized personnel to cut No. 4 end in advance. This measure effectively avoided the situation that the corridor installation was not smooth, and also avoided the risk of high altitude adjustment of the corridor. The manager effectively managed the whole process of corridor construction, improved the efficiency of site construction, and ensured the mechanical performance of corridor.

(a) Lifting equipment

(b) Controller

(c) Closure

(d) Closure completed

Figure 8. Corridor's construction.

5 CONCLUSION

BIM and virtual pre-assembly technology can accurately simulate the construction process of the structure, especially the prefabricated structure, and finally evaluate the simulation results from the mechanical point of view. The results show that the method can obtain reliable simulation results. In this way, the hidden danger of structure manufacturing can be found in advance. If managers can effectively combine the technology, they can achieve the following management functions:

1. Improve the efficiency of building construction. Before the construction of the building, the advanced simulation can be carried out, and the managers can find and solve the problems in advance. This avoids the more likely cost of solving the problem halfway;
2. Reasonable allocation of resources. The simulation results can make the managers optimize the construction preparation steps, arrange the constructors more reasonably, and improve the corresponding facilities;
3. Ensure the quality of construction. This involves the optimization problem, that is, we need the optimal construction scheme according to the actual situation. Managers can use virtual pre-assembly to achieve this goal;
4. Avoid construction risks. The high altitude work risk is very big, the managers through this technology can effectively avoid arranging the construction personnel to carry on the high altitude work, guarantees the construction to carry on smoothly. In addition, due to the high stress requirements of the structure, the structure beyond the acceptance range is unsafe.

BIM and virtual pre-assembly technology can play a great advantage in construction management. Further, how to quantify the contribution index of management in construction project is worth studying.

ACKNOWLEDGMENTS

The related research and funding of this article are supported by Chongqing Smart City and Sustainable Development Academy (SCSDA). The authors would like to acknowledge the support of the SCSDA.

REFERENCES

Case, F., Beinat, A. & Crosilla, F 2014 Virtual trial assembly of a complex steel structure by Generalized Procrustes Analysis techniques. *Automation in Construction*, 37:155–165.

Chen LJ. & Luo H 2014 A BIM-based construction quality management model and its applications. *Automation in construction* 46(oct.):64–73.

Gregory C. Sharp, Sang W. Lee & David K. Wehe 2002. ICP Registration using invariant features. IEEE Transactions on Pattern Analysis; *Machine Intelligence*, 24(1):90–102.

Liang Xiang & Youwei Li. 2018. Virtual pre assembly technology of bolted bridge steel members. *Construction technology*.

Manliang Guo 2018 Virtual pre assembly technology of steel structure. *Architectural technology*, 49 (04): 381–384.

Ningyu Li, Bo Li, Sheng Zhang, Songwu Zhang & Yuchi Zou 2019 Research and Practice on construction application of special-shaped venues based on BIM Technology. *Sichuan building materials*, 45 (10): 137–139.

Obonyo, E.A. 2011 An agent-based intelligent virtual learning environment for construction management. *Construction Innovation*, 11(2):142–160.

Qinghua Yang 2019 Application of BIM Technology in construction management of prefabricated buildings. *Engineering construction and design*.

Tengfei Xu 2019 Application analysis of BIM Technology in the construction of long-span steel truss. *Jiangxi building materials*, (09): 117–118.

JJ Yang & Liangcai Wu 2016. Robust point cloud plane fitting method based on RANSAC algorithm. *Beijing surveying and mapping*, No.127 (02): 77–79 + 83.

Yawei Qin, Wenjie Shi & Mingzhao Xiao 2019 Quality control of bridge steel structure engineering based on BIM + 3D laser scanning technology. *Journal of civil engineering and management*, 36 (4).

Yue Cai, Wenbing Xu & Dan Liang 2019. Impacts of different factors on accuracy of point cloud obtained from Terri all three dimensional laser scanning. *Progress in laser and optoelectronics*, 054 (009): 358–367.

Xiaoguang Liu & Yongjie pan 2018. Application of Liu virtual pre assembly technology in steel truss beam. *Railway Engineering*, 2020.

Wenxue Xu, Tao Jiang & Zhizhong Kang 2008 Plane feature extraction of laser scanning data based on region growth method. *Surveying and Mapping Science*, (S3).

Woonseong J., Soowon C. & Jeongwook, S. 2016. BIM-Integrated Construction Operation Simulation for Just-In-Time Production Management. *Sustainability*, 8(11):1106–1110.

Zhang, D. 2011 Improvement of construction organization design preparation. Science and technology information, 2011 (28): 316–316.

The use of the BIM-model and scanning in quality assurance of bridge constructions

C.N. Rolfsen, A.K. Lassen & D. Han
Department of Civil Engineering and Energy Technology, OsloMet – Oslo Metropolitan University, Norway

H. Hosamo
University of Agder, Gimlemoen/Grimstad, Norway

C. Ying
Chongqing Smart City and Sustainable Development Academy, Chongqing, China

ABSTRACT: For bridges with complex structures and difficult construction, quality is a real issue. The use of the BIM-model is important to seek good quality in the design of the bridge. Furthermore, the model is important for construction and supervision, but it is undeniable that quality assurance after completion could be a problem. The use of a scanner could ease this problem. The bridge completion state mode is collected by 3D laser scanning technology. The theoretical model is compared with the actual state, so that the concrete bridge member is subjected to three-dimensional digital detection such as dimensional deviation and flatness of the surface of the member.

1 INTRODUCTION

In recent years, with the rapid development of China's economy, traffic flow has increased significantly. Some bridges have greatly exceeded the design flow requirements, and in addition some bridges have construction quality problems. Overload transportation, large traffic flow and other direct reasons result in a serious decline in the technical state of bridges at all levels, some sections of the bridge frequently appear early disease, some bridges even occurred structural damage, this situation not only poses a threat to lives of broad masses of people, but also causes high material losses. Therefore, bridge safety has become a major concern, and the quality of the bridge under construction should be increased (Ran 2018).

In the field of engineering, the laser scanner is considered to be the most suitable tool for construction engineering monitoring and construction process control. The BIM model integrates a large amount of data for construction projects, including the location and dimensions of components. Close range photogrammetry is a measurement method to obtain a large amount of physical and geometric information from photos of the measured object in an instant. In this paper, two main routes are combined at the end. The first one is to compare the theoretical BIM model with the measured point cloud model to detect the deviation and determine the location of the deviation. The second route is to collect high-definition photos of components, automatically identify photos with damage, and locate the actual location of photos with damage in the real world, so as to achieve rapid detection of component damage.

Through the defect detection and safety evaluation of concrete bridges, it is of great significance to repair and reinforce the bridges according to the inspection results. In the early stage of the cracks maintenance, reducing the investment in bridge reconstruction and increasing the service life of the bridge is in line with the concept of sustainable development. Facing the need for bridge damage and defect detection in China, it is urgent to develop non-intrusive defect detection technology and devices to improve the level of bridge maintenance and safety guarantees in China.

2 PRINCIPLES AND METHOD

A BIM model is a three-dimensional digital representation which integrates all kinds of relevant information of a construction project (Carvalho et al. 2019). It is the digital expression of all the geometric, physical and performance information of the building. The BIM model could be combined with a laser scanner to do the quality assurance. The three-dimensional laser scanning technique can also be called "the real-life scenery duplication technique" (Xiujun 2007). It uses a laser scanner to scan the surface of the object intensively to obtain point cloud data similar to the point cloud modeled by finite element method. The close range photogrammetry technology collects high-definition images of the structure surface through a high-definition camera, and identifies and

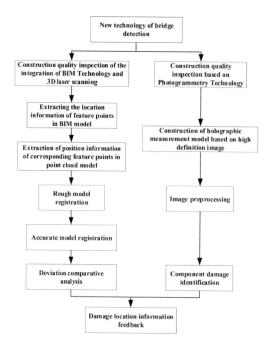

Figure 1. Technical route.

Figure 2. Feature point information extraction.

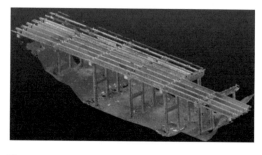

Figure 3. Point cloud model.

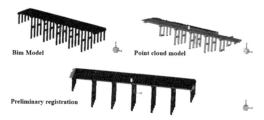

Figure 4. Alignment of BIM model and point cloud model l.

quantifies structure surface damage through image gray information (Hoang 2018).

In this study, BIM technology, 3D laser scanning technology and photogrammetry technology are applied to the quality inspection of construction of completed bridges. Close-range photogrammetry technology is used to detect the appearance defects of components, which are not easy to find, and to accurately identify and locate them, while the BIM model is compared with the actual field model obtained by 3D scanning, so as to detect and evaluate the construction quality of the completed model. The technical route is shown in Figure 1.

3 FUSION OF BIM TECHNOLOGY AND 3D LASER SCANNING FOR QUALITY TESTING

Based on a new bridge project, the technical route is explored and applied, and the construction error of box girder is detected.

Based on BIM technology and 3D scanning technology, the component quality inspection flow is as follows.

3.1 *Extracting feature point position information from BIM model*

By fitting the feature surface, the points intersected by three feature surfaces are taken as the feature points, and the feature points are not less than 3 and the feature points are not coplanar, and the coordinate information of the feature points is extracted for model registration (Shibasaki et al. 2010).

3.2 *Extraction of corresponding feature point position information in point cloud model*

The same method is used to extract the feature point information corresponding to the BIM model in the point cloud model for coordinate transformation (Zhiwei et al. 2016).

3.3 *Rough registration*

Through the manual registration function, the BIM model is preliminarily aligned with the point cloud model. The common part of the two models is selected and the n-point alignment is adopted to make the two models roughly aligned.

3.4 *Accurate registration*

The feature point information input coordinate matrix, which is optimized iteratively by the algorithm, makes the objective function obtain the optimal value, that is, the deviation is minimized.

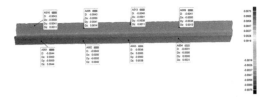

Figure 5. Deviation chromatogram.

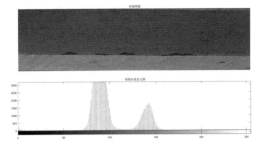

Figure 7. Gray histogram.

Figure 6. Full-bridge photogrammetry model.

Figure 8. Precise location of damage.

3.5 Deviation comparison

Taking the fine BIM model as the reference object and the point cloud model as the test object, the deviation chromatogram is generated, the deviation chromatogram is qualitatively analyzed by the deviation chromatogram, the large deviation area is preliminarily determined, and the deviation value is obtained by annotation in the larger deviation area for quantitative analysis, so as to judge the construction quality (Wang 2015).

4 CONSTRUCTION ENGINEERING QUALITY TESTING BASED ON PHOTOGRAMMETRY

The UAV collects the high-definition image as the data source to identify and locate the component damage. The steps are as follows.

4.1 A holographic photogrammetry model based on HD image photos

The high-definition image data obtained by UAV tilt photography (unmanned aerial vehicles) are calculated in spatial triangle to obtain the location information of each photo, so as to match the image and obtain the structure real scene model (Zhuo et al. 2017).

4.2 Image preprocessing

The concrete surface image collected by the camera is a color image. Although the amount of information in the color image is large, the information in the gray image is enough. In order to improve the operation speed and save the storage space, the collected color image should be converted into the gray image.

4.3 Component damage identification

The lattice matrix is used to collect the structural information of the object, update the information and store it as a digital image. The coordinate system of the digital image is established on the image, the image information is recognized according to the time sequence, the damaged area of the image is extracted, the damaged area of the image is denoised by the Gauss function, and the damaged area of the image is denoised by the wavelet decomposition method for the second time. At the same time, the damage orientation of the object structure is calculated (Jiya & Qian 2017).

4.4 Damage location information feedback

Recording the name of the damaged photo and reading the GPS information (including longitude, latitude, and altitude information), so that it is fed back to the holographic measurement model to accurately locate the damage location.

5 ANALYSIS AND DISCUSSION

As an information carrier, photos or images contain the greatest amount of information of the measured object. Combined with damage identification by imaging technology, the collected massive photos are identified by images to obtain the surface damage of the structure and conduct qualitative and quantitative analysis. Close-range photogrammetry technology requires higher lens pixels of the equipment, and the photographing must follow a certain order, which has poor effect on smooth surface and transparent objects.

In the construction stage, the BIM model used in the field management needs to integrate effective technical means as auxiliary. 3D laser scanning technology can record the complex situation of a construction site efficiently and completely, and compare it with the BIM model, which will bring great help to project quality inspection and project acceptance. The combination of 3D laser scanning and BIM model refers to the comparison, transformation and coordination of BIM model and corresponding 3D scanning model, so as to achieve the purpose of assisting engineering quality inspection, rapid modeling and reduced rework. However, the adoption of new technology also increases the cost to a certain extent, but the long-term economic benefits outweigh this.

6 CONCLUSION

This paper shows how the use of realistic capture technology to collect the appearance, geometry, and deformations of the bridge in the completion stage and performance of big data analysis can map the quality state of the bridge in the physical world in real time and ensure the bridge structure safety. Furthermore, this paper shows that the technology has the following advantages:

(1) The holographic BIM model corresponding to the quality state of the bridge in the physical world is established to realize the timely detection and timely adjustment of the construction quality problems of the bridge under the harsh environmental conditions, so as to significantly reduce the construction risk and the construction period and improve the construction quality of the bridge.
(2) Based on background subtraction, an adaptive threshold segmentation method based on grayscale estimation is proposed to realize the complete extraction of cracks.
(3) The combination of the technologies innovatively integrates 3D scanning, photogrammetry and BIM technology to realize the perception of bridge state depth.

With the further development of 3D laser scanning technology and BIM Technology, their application in project management will be further expanded and deepened.

REFERENCES

Carvalho, J. P., Bragança, L. & Mateus, R. 2019. Optimising building sustainability assessment using BIM. *Automation in Construction,* 102, 170–182.

Hoang, N.-D. 2018. Detection of Surface Crack in Building Structures Using Image Processing Technique with an Improved Otsu Method for Image Thresholding. *Advances in Civil Engineering,* 2018.

Jiya, T. & Qian, L. 2017. Laser damage identification based on digital image processing. *Laser Journal,* 105–108.

Ran, G. 2018. Research on Bridge Detection and Safety Assessment System Based on BIM. *Construction & design for engineering.*

Shibasak, R., Iwat, A. & Kondon, H. 2010. Automatic Large-Scale Point-Clouds Registration By Using 3D Matching of Natural Feature-Points. *International Conference on Power Electronics & Intelli-gent Transportation System.*

Wang, X. 2015. Research on Application of bridge deformation detection technology based on 3D laser scanning.

Xiujun, D. 2007. Research on application of 3d laser scanning technology in acquiring dtm with high accuracy and resolution. *Journal of Engineering Geology,* 15, 428–432.

Zhiwe, L., Kezhao, L. & Leijie, Z. 2016. Implementation of seven parameters coordinate conversion based on Matlab. *Journal of Heilongjiang Institute of Technology.*

Zhuo, X., Koch, T., Kurz, F., Fraundorfer, F. & Reinartz, P. 2017. Automatic UAV Image Geo-Registration by Matching UAV Images to Georeferenced Image Data. *Remote Sens,* 9.

ECPPM 2021 – eWork and eBusiness in Architecture,
Engineering and Construction – Semenov & Scherer (eds)
© 2021 Taylor & Francis Group, London, ISBN 978-1-032-04328-9

Application of phase three dimensional laser scanner in high altitude large volume irregular structure

D. Han, C.N. Rolfsen, E. Erduran & E.E. Hempel
Oslo Metropolitan University (OsloMet), Norway

H. Hosamo
University of Agder, Gimlemoen/Grimstad, Norway

J. Guo & F. Chen
Chongqing Jiaotong University, China

C. Ying
Chongqing Smart City and Sustainable Development Academy (SCSDA), China

ABSTRACT: The precise measurement and positioning of the outline of high-altitude massive special-shaped structure has a guiding role in the installation of its external curtain wall and other ancillary facilities. Traditionally, total station measurement and other technical methods are mostly used, which have the disadvantages of large workload, long time-consuming and low accuracy. This paper takes the high-altitude corridor enclosure structure of Chongqing Raffles Square as the research object, combines the characteristics of Leica Nova MS60 high-precision three-dimensional laser scanning total station which can accurately locate and Faro x330 three-dimensional laser scanner which can scan quickly, and puts forward an excellent combination of two instruments. Based on the point scanning method and the mathematical optimization theory, a high-precision registration formula for point clouds with special-shaped structures is derived. On the premise of meeting the registration accuracy of point clouds, the rapid scanning and global positioning of a large number of special-shaped components are realized, and precise data are provided for the rapid extraction of contours. It has great application prospects and important applications in reverse modeling and deviation analysis.

1 INTRODUCTION

The super-high-rise air corridor is a new type of complex steel structure, which brings great difficulties to the field construction because of its unique structure and complex force characteristics. Taking chongqing Raffles high-altitude view flyover as an example, in order to obtain the true line shape of the enclosure structure and provide the basis for the curtain wall installation, the traditional method uses the total station to carry out single point measurement, and the line of the measured points is connected to obtain the line shape of the enclosure structure.

Now 3D laser scanning technology is introduced, but the data obtained by the ground 3D laser scanner are in the local coordinate system centered on the station, and the data obtained by different stations need to be unified into the global coordinate system. Chen et al. (2014) used a total station to measure the geodesic coordinates of the target, and transformed the data obtained by the scanner through the coordinate system of the target. However, the method is not efficient and the accuracy of the global coordinate system is limited. Zhu et al. (2018) measured the control point as the scanner erection point through the total station instrument. The selection requirements are high, and it is necessary to ensure that there are overlapping parts of the scanning area of the adjacent stations, and the scanning sites are not flexible enough.

In view of the above problems, this paper proposes a new method of combining Leica NovaMS60 high-precision 3D laser scanning total station with Faro x330 3D laser scanner, which gives full play to the advantages of Leica's high-precision global coordinate system positioning total station instrument, and has the function of measuring points of global coordinate system with certain scanning density.

2 NEW METHOD FOR SCANNING HIGH-ALTITUDE LARGE-VOLUME IRREGULAR STRUCTURES

In order to solve the problems of time consuming, difficulty and low precision, this paper proposes a new data acquisition method, combining with the characteristics of two methods of collecting data, a mathematical optimization theory is proposed to derive

DOI 10.1201/9781003191476-50

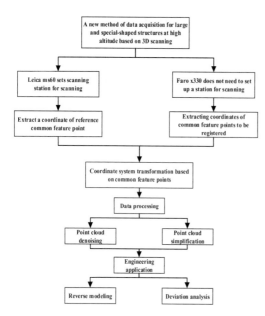

Figure 1. Technical route.

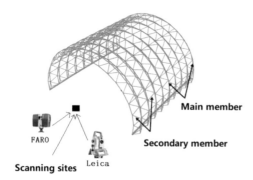

Figure 2. Scanning scheme and site layout.

the point cloud high precision registration formula, which realizes the high efficiency, omni-directional and high precision data acquisition and processing of large volume complex structures.

2.1 Data acquisition and processing

2.1.1 Scanning scheme

According to the coordinates of the two known control points, Leica uses the rear intersection mode to set up the station. The accuracy of the scanning data is controlled at the horizontal interval of 3 mm and the vertical interval of 5 mm, so as to obtain the more dense point cloud data. In order to ensure the accuracy of the later point cloud splicing, faro uses the same site, the same precision way to obtain the point cloud data, and Leica is erected in the same position, which ensures that the collected scanning data is in the same side, the same position, the same distance, and the same scanning accuracy is set by faro scanner to ensure that the two periods of data are scanned at the same accuracy.

2.1.2 Data denoising and streamlining

In this paper, the gaussian filter method is used to denoise the original data (Liang 2019). Gaussian filter method uses the gaussian function to still maintain its characteristics after passing Fourier transform, and locates the designated area weight gaussian distribution to achieve the effect of noise reduction. Although the point cloud after denoising reduces the interference of miscellaneous points, the amount of data is still massive, so it is necessary to streamline and compress the data (Li et al. 2019; Peng et al. 2013; Siddiqui et al. 2007; Wan et al. 2016). In this paper, by curvature sampling, the data is reduced without affecting the surface reconstruction and ensuring certain precision, thus improving the speed of data processing.

2.2 Point cloud registration

Since the point cloud data obtained by Faro scanner is based on the local coordinate system, and Leica is set up by station, the scanned data is based on the field construction coordinate system, so there is a coordinate conversion problem between the two. Using the seven-parameter registration principle based on the common feature points (Xie et al. 2017), at least three pairs of common feature points are obtained in the common part of the point cloud in two phases respectively. The maximum deviation is minimized by iterative optimization of the algorithm, thus the conversion parameters are calculated and the point cloud coordinate system is transformed.

Each selected two-stage point cloud corresponds to three pairs of common feature points. Two-piece point clouds are accurately registered by the algorithm optimization principle (Koguciuk 2017; Quan et al. 2019), where the matrix a represents the faro scan data feature point coordinate information and the matrix b represents the Leica scan data feature point coordinate information.

The point cloud registration rotation matrix is:

$$R_\alpha = \begin{bmatrix} 1 & 0 & 0 & 0 \\ 0 & cos(\alpha) & -sin(\alpha) & 0 \\ 0 & sin(\alpha) & cos(\alpha) & 0 \\ 0 & 0 & 0 & 1 \end{bmatrix} \quad (1)$$

$$R_\beta = \begin{bmatrix} cos(\beta) & 0 & sin(\beta) & 0 \\ 0 & 1 & 0 & 0 \\ -sin(\beta) & 0 & cos(\beta) & 0 \\ 0 & 0 & 0 & 1 \end{bmatrix} \quad (2)$$

$$R_\gamma = \begin{bmatrix} cos(\gamma) & sin(\gamma) & 0 & 0 \\ -sin(\gamma) & cos(\gamma) & 0 & 0 \\ 0 & 0 & 1 & 0 \\ 0 & 0 & 0 & 1 \end{bmatrix} \quad (3)$$

Translation matrix:

$$T_{xyz} = \begin{bmatrix} 1 & 0 & 0 & 0 \\ 0 & 1 & 0 & 0 \\ 0 & 0 & 1 & 0 \\ x & y & z & 1 \end{bmatrix} \quad (4)$$

where α, β, γ are the rotation angles around the X, Y, and Z axes, and x, y, and z are the translational distances along the X, Y, and Z axes.

Post-registration location information:

$$A' = R\alpha \cdot R\beta \cdot R\gamma \cdot TXYZ \cdot A \qquad (5)$$

Registration deviation:

$$D = B - A' = \begin{bmatrix} \Delta x_1 & \Delta y_1 & \Delta z_1 \\ \Delta x_2 & \Delta y_2 & \Delta z_2 \\ \Delta x_3 & \Delta y_3 & \Delta z_3 \end{bmatrix} \qquad (6)$$

The results show that the best registration results can be obtained by using the maximum deviation minimization as the objective function.

The objective function is as follows

$$f(\alpha, \beta, \gamma, x, y, z) \\ = \min\left\{ \max_{i=1,2,3} \left[\sqrt{(\Delta x_i)^2 + (\Delta y_i)^2 + (\Delta z_i)^2} \right] \right\} \qquad (7)$$

2.3 Analysis of splicing accuracy

To test whether the stitching accuracy meets the requirements, it is now verified from the following three aspects: (1) calculating whether the distance determination error between each pair of feature points meets the requirements; and (2) fitting the common part of the circular pipe point cloud to determine whether the diameter of the circular pipe meets the design requirements.

2.3.1 Analysis of stitching accuracy between corresponding feature points

By calculating the distance between each pair of feature points to determine whether the error meets the requirements.

$$d = \sqrt{(x1 - x2)^2 + (y1 - y2)^2 + (z1 - z2)^2} \qquad (8)$$

2.3.2 Comparative analysis of pipe diameter

The point cloud of the same part of Leica and faro scan is fitted respectively, then the diameter of the point cloud of the same part after registration is obtained, and the diameter of the point cloud fitting after registration is compared with the expected value.

Registration error assessment function:

$$W = E - E_p; \qquad (9)$$

$$E = (R_l + R_f)/2; \qquad (10)$$

where
W: registration bias;
R_l: Leica scan point cloud fitting tube diameter value;
R_f: faro scan point cloud fitting tube diameter value;
E: Leica and faro scan point cloud fitted the desired value of the cylinder;
R_P: after registration point cloud fitting tube diameter value.

3 PROJECT CASES

3.1 Engineering profile

Chongqing Raffles "Crystal Corridor" is the first high-altitude view flyover over 200 meters in China, with a total length of about 300 meters and a width of about 30 meters. The cross section of the enclosure structure is arched with a height of about 16 m and a span of 31 m. By the transverse span of the main bar and the longitudinal secondary bar welding composition, all the members of the pipe. The exterior of the enclosure structure is covered with glass curtain wall, so the installation and positioning of the curtain wall should be based on the shape of the enclosure structure. However, the construction of the enclosure structure is complex, and its installation process is the splicing of the upper part of the main structure of the corridor after the ground welding of the local bar, the weight of the bar is affected during the lifting process, and the assembly in the field installation process. Due to the influence of the connection error, the bar has been deformed, so there is a big deviation between the theoretical line and the actual line, which cannot truly reflect the contour and line shape of the enclosure structure, so it is difficult to guide the installation of the external curtain wall accurately.

3.2 Instrument introduction and comparative analysis

The instruments used in the accurate scanning of the enclosure structure of the air corridor in Leica Nova MS60 are phase scanners, which are Leica Nova MS60 high-precision 3D laser scanning total station and Farox330 3D laser scanner. Leica MS60 high-precision 3D laser scanning total station has the functions of total station and scanner, but it is time-consuming. The scanning rate of the instrument can reach 976000 points per second, and the field of vision is wide, but its point cloud data is based on the local coordinate system, so it is difficult to compare with the components in the field construction coordinate system, so there is the problem of coordinate system transformation.

In view of the shortcomings of each of the above two kinds of instruments in the scanning process, a new scanning method is introduced, which combines the characteristics of Leica Nova MS60 and Farox330 fast scanning, and two kinds of instruments are used to scan the same component under the condition of maintaining the same site and the same precision. By splicing the common feature points, the Faro scanning point cloud data to Leica scanning data is spliced together, and the error of multi-point and multi-point registration is tested under the premise of single point and single point registration, so that the local coordinate system can be integrated high-precision and high-efficiency conversion.

Figure 3. Leica (left) and faro (right).

Figure 4. Point clouds: raw (left) and after processing (right).

3.3 Technical applications

3.3.1 Data processing

Using the Gaussian filtering algorithm, the original point cloud data is carefully denoised and the data is reduced by setting the curvature sampling mode to improve the data processing speed.

3.3.2 Data registration and accuracy analysis

Because the enclosure structure is a curved mesh and the members are all round pipes, point cloud splicing is carried out at the node of the intersecting pipe. The central line of the cylinder is obtained by the cylindrical fitting of the circular tube at the node, and the coordinate of the intersection point is obtained by the intersection of the two lines.

The point cloud registration steps are as follows:

1. Acquisition of round tube centerline by feature fitting;
2. Extract the centerline of the intersecting circular tube and obtain the intersection point through the intersection of the straight line;
3. Extract intersection coordinates and input coordinate transformation matrix;
4. By optimizing the algorithm, the objective function is minimized and the conversion parameters are obtained;
5. The point cloud registration is obtained according to the error evaluation function;
6. Error meets required output result, otherwise return to third step.

There are 7 groups of feature points extracted for point cloud registration, coordinate information is extracted and deviations are calculated. The results are presented in Table 1.

The above results show that the X-direction deviation between the seven groups of corresponding

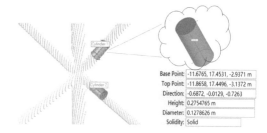

Figure 5. Circular fitting and coordinate information.

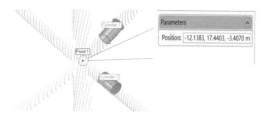

Figure 6. Extraction of intersection information.

Table 1. Corresponding characteristic point deviation.

group indication	Δx/(m)	Δy/(m)	Δz/(m)	d/(m)
1	0.0006	0.0003	0.0007	0.000969536
2	0.0007	−0.0009	−0.0008	0.001392839
3	−0.0007	0.001	0.0006	0.001360147
4	−0.0006	−0.0001	0.0011	0.001256981
5	0.0008	−0.0009	−0.0007	0.001392839
6	−0.0009	0	−0.0009	0.001272792
7	−0.0002	0.0004	0.0007	0.000830662

feature points of two point clouds is between 0.2 mm–0.9 mm, the Y-direction deviation is between 0 mm–1 mm, the Z-direction deviation is between 0.6 mm–1.1 mm, and the calculated corresponding point spacing is 1.4 mm, which meets the precision requirement with this project.

After the registration model is randomly fitted to multiple columns in the common part, its diameter is obtained, and the fitting diameter is compared with the standard diameter to evaluate the point cloud registration accuracy.

The below results show that the registration deviation value is between 0.03 mm and 1.5 mm, and most of them are controlled within 1 mm, and the point cloud registration effect meets the engineering requirements.

By comparing and analyzing the point cloud registration accuracy from the above two aspects, the results all meet the engineering requirements, so this method can realize the fast scanning and accurate positioning of large volume components.

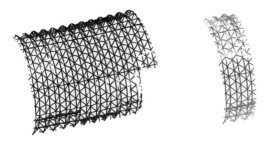

Figure 7. Faro (left) and Leica (right) Scanning Models.

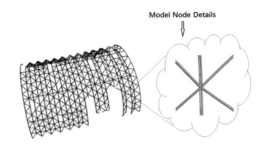

Figure 9. Reverse model of enclosure structure.

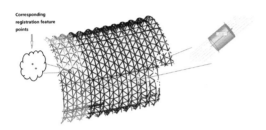

Figure 8. Post-registration point cloud model.

Table 2. Diameter contrast deviation.

Column No.	$R_L/(m)$	$R_F/(m)$	$E/(m)$	$R_P/(m)$	$W/(m)$
1	0.19008	0.19008	0.19008	0.18968	0.00040
2	0.19233	0.19233	0.19233	0.19250	−0.00017
3	0.19450	0.19450	0.19450	0.19388	0.00062
4	0.19165	0.19075	0.19120	0.19129	−0.00009
5	0.19429	0.19491	0.19460	0.19491	−0.00031
6	0.19263	0.19453	0.19358	0.19412	−0.00054
7	0.19456	0.19598	0.19527	0.19673	−0.00146
8	0.19016	0.19148	0.19082	0.19164	−0.00081
9	0.19445	0.19443	0.19444	0.19494	−0.00050
10	0.18671	0.18618	0.18645	0.18653	−0.00009
11	0.19494	0.19498	0.19496	0.19598	−0.00103
12	0.19132	0.19129	0.19131	0.19134	−0.00003
13	0.19298	0.19376	0.19337	0.19488	−0.00150
14	0.19263	0.19453	0.19358	0.19412	−0.00054
15	0.19470	0.19444	0.19457	0.19318	0.00139

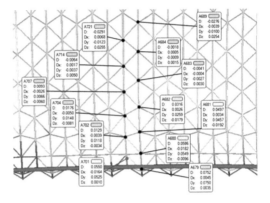

Figure 10. Comparison of deviations.

4 ENGINEERING APPLICATIONS

4.1 Reverse modeling

Through the fusion application of the two instruments, the point cloud data of the enclosure structure are obtained efficiently and completely. Each circular tube centerline is quickly extracted by the algorithm, and the cross section information is given to generate the reverse model (Wang et al. 2017).

4.2 Deviation analysis

The point cloud model obtained by this method can be compared and analyzed. Taking the design model as the reference object and the point cloud model as the test object, the three coordinate axis direction deviation of each node position is obtained by contrast analysis, and it is transferred with the reverse model of the enclosure structure to the site construction side and the curtain wall designer for reference adjustment, so as to play a guiding role in the later curtain wall installation.

5 CONCLUSION

The implementation of high-precision coordinate system conversion algorithm and fast acquisition of point cloud data is decisive for high-efficiency, high-precision and all-directional acquisition of point cloud model with large-scale heteromorphic structure. It provides data support for reverse modeling and deviation comparison analysis, greatly expands the application range of 3D laser scanning technology and the stability, reliability and application range of data, and plays a more efficient role in engineering.

ACKNOWLEDGMENTS

The related research and funding of this article are supported by Chongqing Smart City and Sustainable Development Academy (SCSDA). The authors would like to acknowledge the support of the SCSDA.

REFERENCES

Chen, H.Y., Hu, X.B. & Li, C.G. 2014. Application of ground 3D laser scanning technology in deformation monitoring. *Bulletin of Surveying and mapping* (12), 84–87.

Koguciuk, D. 2017. Parallel ransac for point cloud registration. *Foundations of Computing & Decision Sciences*, 42(3), 203–217.

Liang, J.H. 2019. Research on preprocessing, denoising and registration of scattered 3D point cloud.

Li, J.T., Cheng, X.J., Yang, Z.X. & Yang, R.Q. 2019. Point cloud data compression method based on curvature classification. *Progress in laser and optoelectronics*, 56 (14).

Peng, Y. & Liu, S.G. 2013. Compression and Segmentation Technology of Point Cloud Data in Reverse Engineering. *International Journal of Digital Content Technology and its Applications*.

Quan, S.W., Ma, J., Feng, F. & Yu, K. 2019. Multi-scale binary geometric feature description and matching for accurate registration of point clouds. *International Workshop on Pattern Recognition*.

Siddiqui, R.A., Eroksuz, S. & Celasun, I. 2007. Multi Stage Vector Quantization for the Compression of Surface and Volumetric Point Cloud Data. *Signal Processing and Information Technology*, 2007 IEEE International Symposium on.

Wan, C. H. & He, X.P. 2016. The Compression Method of Point Cloud Data based on the Hypotenuse-height Deviation Method. *Proceedings of the 2016 2nd Workshop on Advanced Research and Technology in Industry Applications*.

Wang, D.B, Yang, H.Y. & Xing, Y.F. 2017. Application of BIM and 3D laser scanning technology in reverse construction of curtain wall project of Tianjin Chow Tai Fook financial center. *Construction Technique*, (23):15–18.

Xie, J., Guo, X.F & Zhao, Y.J. 2017. Coordinate transformation. *Site Investigation Science and Technology*.

Zhu, S.G. & Nie, S.G. 2018. Application of 3D laser scanning technology in building deformation monitoring. *Value engineering*, 37 (31), 267–269.

Automatic detection method for verticality of bridge pier based on BIM and point cloud

D. Han, C.N. Rolfsen & N. Bui
Oslo Metropolitan University (OsloMet), Norway

H. Hosamo
University of Agder, Gimlemoen/Grimstad, Norway

Y. Dong & Y. Zhou
Chongqing Jiaotong University, China

T. Guo
Southeast University, Nanjing, China

C. Ying
Chongqing Smart City and Sustainable Development Academy (SCSDA), China

ABSTRACT: Bridge pier is a building that supports bridge span structure and transmits dead load and live load to foundation. As the main supporting structure of the whole bridge, the perpendicularity of the pier is particularly important. Among them, for the pier with high height, the influence of perpendicularity deviation on the quality of pier is more obvious. In this paper, an automatic detection method of pier perpendicularity is proposed. The 3D laser point cloud is used to obtain the spatial attitude of the pier, and the point cloud of the reliable feature surface of the pier is extracted by combining the automatic recognition algorithm. According to the principle of statistical correlation, the characteristic surface data of piers are processed to get the actual center line and section of piers. At last, we update the BIM model of the designed pier with the center line and section and analyze the 3D deviation between the updated BIM model of the pier and the point cloud. We apply this method to the verticality detection of several bridges. The results show that the method is reliable and robust.

1 INTRODUCTION

With the rapid development of the transportation industry in the world, more and more world-class bridges have come out one after another. As the backbone of the whole bridge, the pier transfers all the loads of the superstructure to the foundation, and its quality directly affects the overall service life of the bridge. With the improvement of construction technology, the construction height of the pier is also increasing. The height of the pier of He Zhang grand bridge, known as "the highest high town in Asia", has reached an amazing 195 m (Yang 2014). Among them, the problem of pier perpendicularity deviation has always been the research focus of many scholars (Liu et al. 2018; Tie et al. 2017). Made a quantitative analysis on the problem of pier perpendicularity deviation by using the finite element model, and the results showed that the pier perpendicularity deviation had a great influence on the bearing capacity of the bridge (Ding et al. 2019). Realized the rapid measurement of pier perpendicularity by using the total station free station non-contact measurement method (Qiu et al. 2018). Improved the method of least square fitting circular curve, fast check the perpendicularity of pier quickly. The main basis of the above research is the limited data points collected by the total station, but as a three-dimensional structure, the method of determining the perpendicularity of the pier by the limited points measured by the total station is not very convincing, in fact, the problem of the detection of the perpendicularity deviation of the pier has not been solved.

As another technological innovation after GPS technology, 3D laser scanning technology has been widely used in the field of deformation monitoring (Deng et al. 2018). 3D laser scanning technology can easily obtain the surface features of the structure, and has the advantages of high data accuracy and easy post-processing. Huang et al. (2012) combined the 3D laser scanning technology with the verticality detection of piers, used the advantages of 3D laser scanning to indirectly measure the verticality of piers, and obtained the verticality deviation of piers. The above research focuses on the feasibility of 3D laser scanning applied to the measurement of pier perpendicularity. The method used to deal with the perpendicularity of the pier is relatively basic, and the calculation method of the perpendicularity deviation is not elaborated in detail. The obtained perpendicularity of the pier is only based on the deviation values of several pier sections, so the perpendicularity is not accurate. Based on the research background of the hollow thin-walled rectangular high pier, starting from all the point cloud data of the pier, the deviation value of the characteristic points of each horizontal section of the pier is calculated by the automatic algorithm, and the accuracy of the algorithm is verified by the BIM model, and the perpendicularity of the pier is studied in depth.

2 BASIC PRINCIPLE OF PIER VERTICALITY DETECTION

The hollow thin-walled rectangular pier has very strong geometry, so the problem of finding the perpendicularity of the pier in reality can be transformed into the problem of finding the center line of three-dimensional structure in geometry. The line is made up of numerous points. In mathematics, it is much easier to find the coordinates of the points than to find the three-dimensional curves directly. Therefore, the author uses the dimension reduction method to convert the problem of finding the three-dimensional center line of the pier into the problem of finding the plane center point, which is the characteristic point of the horizontal section of the pier. The pier is cut into several small rectangles according to the horizontal direction, and then the geometric principle is used to solve the centroid of the rectangle. Finally, the distance between the centroid of each section and the theoretical center point is obtained, that is, the deviation value of the pier.

The solutions to the problem of pier perpendicularity on the geometric level are given above. A point cloud is a collection of numerous points with three-dimensional coordinates. When processing point cloud, it can be roughly divided into three stages: data preparation, data analysis and result analysis. See Figure 1 for data processing flow chart.

3 POINT CLOUD DATA COLLECTION AND PREPROCESSING

3.1 Point cloud data collection method

The point cloud data needed by the Research Institute comes from a highway bridge under construction. The bridge pier is 66.53 m high. See Figure 2 for site scanning of piers. It can be seen from the figure that there must be deviation in the process of segmental pouring of piers. The traditional method of measuring the perpendicularity of piers cannot fully reflect the deviation of piers. 3D laser scanning technology can fully capture the surface characteristics of the structure, give the structure location parameters, and make up for the shortcomings of single point measurement of total station.

In order to ensure the accuracy of scanning, six stations and three groups of target ball points are set when the pier is scanned by a method such as x-330

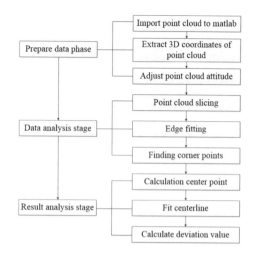

Figure 1. Data processing steps.

Figure 2. Field scan.

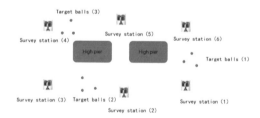

Figure 3. Scanning scheme.

scanner. As shown in Figure 3 Survey stations 1–6 are around the high pier and about 30m away from the high pier. Three sets of target balls are placed between six stations. The first group of target balls kept intervisibility with the test station 1.2. The second group of target balls kept intervisibility with the measuring station 1.2.3. The third group kept intervisibility with the measuring station 3.4.5.6. In this way, at least one set of target ball point cloud data can be collected for each station, which is convenient for point cloud splicing.

a. Point cloud raw data   b. Point cloud after processing

Figure 4. Point cloud processing.

3.2 Point cloud model point cloud data preprocessing

The data collected from field work contains a lot of noise that cannot be used for research, such as fence point cloud at the bottom of pier, tower crane point cloud, etc. The original data of point cloud is shown in Figure 4 (a). When analyzing the perpendicularity of the bridge pier, additional miscellaneous points will certainly have a certain impact on the algorithm, so it is necessary to reduce the noise of the point cloud. There are two common noise reduction algorithms, namely statistical filtering algorithm and radius filtering algorithm. Lu et al. (2019) made a comparative analysis on the noise reduction effect of the two methods. For the pier structure, the statistical filtering algorithm can better complete the noise reduction work, but due to the limitations of the algorithm itself, the point cloud of the auxiliary facilities cannot be removed, so it is necessary to import the point cloud into Geomagic to manually delete the noise, and the point cloud is as shown in Figure 4 (b).

4 POINT CLOUD DATA PROCESSING

4.1 Extract segment center point

The initial position of the point cloud in the coordinate system is not conducive to the calculation of the center point of the segment. Before extracting the center point of the segment, the attitude of the pier needs to be adjusted so that the vertical direction of the pier is parallel to the Z coordinate axis. Firstly, the point cloud is imported into Matlab, and the rotation formula is used to adjust the point cloud

$$\begin{bmatrix} x' \\ y' \\ z' \\ 1 \end{bmatrix} = \begin{bmatrix} 1 & 0 & 0 & 0 \\ 0 & \cos\theta & -\sin\theta & 0 \\ 0 & -\sin\theta & \cos\theta & 0 \\ 0 & 0 & 0 & 1 \end{bmatrix} * \begin{bmatrix} x \\ y \\ z \\ 1 \end{bmatrix} \quad (1)$$

$$\begin{bmatrix} x' \\ y' \\ z' \\ 1 \end{bmatrix} = \begin{bmatrix} \cos\theta & 0 & \sin\theta & 0 \\ 0 & 1 & 0 & 0 \\ -\sin\theta & 0 & \cos\theta & 0 \\ 0 & 0 & 0 & 1 \end{bmatrix} * \begin{bmatrix} x \\ y \\ z \\ 1 \end{bmatrix} \quad (2)$$

$$\begin{bmatrix} x' \\ y' \\ z' \\ 1 \end{bmatrix} = \begin{bmatrix} \cos\theta & -\sin\theta & 0 & 0 \\ \sin\theta & \cos\theta & 0 & 0 \\ 0 & 0 & 1 & 0 \\ 0 & 0 & 0 & 1 \end{bmatrix} * \begin{bmatrix} x \\ y \\ z \\ 1 \end{bmatrix} \quad (3)$$

Formula (1) (2) (3) is the rotation formula around the x, y, z coordinate axes, where x' y' z' is the coordinate value after rotation, x y z is the coordinate value before rotation, and θ is the angle to be rotated.

The point cloud is divided into 308 segments, each segment is 0.216 m high. The point cloud of pier section is shown in Figure 5 (a). The Z coordinate of segment point cloud is eliminated, and the 3D point cloud data is reduced to 2D. Extract the point cloud on any side of the rectangle, as shown in Figure 5 (b). You can choose the line function as the fitting curve, because the point cloud is approximately near a straight line, that is

$$p(x) = a_0 + a_1 x \quad (4)$$

In formula (4), $p(x)$ is the polynomial to be fitted, a_0 and a_1 are polynomial coefficients.

The point cloud data is taken into the normal equation of polynomial fitting:

$$\begin{bmatrix} m+1 & \sum_{i=0}^{m} x_i & \cdots & \sum_{i=0}^{m} x_i^n \\ \sum_{i=0}^{m} x_i & \sum_{i=0}^{m} x_i^2 & \cdots & \sum_{i=0}^{m} x_i^{n+1} \\ \vdots & \vdots & & \vdots \\ \sum_{i=0}^{m} x_i^n & \sum_{i=0}^{m} x_i^{n+1} & \cdots & \sum_{i=0}^{m} x_i^{2n} \end{bmatrix} \begin{bmatrix} a_0 \\ a_1 \\ \vdots \\ a_n \end{bmatrix} = \begin{bmatrix} \sum_{i=0}^{m} y_i \\ \sum_{i=0}^{m} x_i y_i \\ \vdots \\ \sum_{i=0}^{m} x_i^n y_i \end{bmatrix},$$

$$i = 1, 2, \ldots, n \quad (5)$$

In formula (5), M is the number of point clouds to be fitted, and N is the number of polynomial terms.

If we solve a_0 and a_1, we will get the fitting line equation of an edge. The linear equations of the other three sides are obtained in turn, and then the linear equations of the adjacent sides are combined to obtain a corner of the rectangle, as shown in Figure 5 (c). After obtaining four corner points, use the known two diagonal points to solve the diagonal equation, and then set up the diagonal equation to get the center point of this section, as shown in Figure 5 (d).

4.2 Analysis of perpendicularity deviation

The center point of the section represents the actual center position of this section of the pier. Figure 6 shows the plan of the cloud at the pier. The blue point represents the center point of the pier, and the red point represents the cloud at the pier. Figure 7 is the fitting diagram of the center line, the blue far point represents the center point of the pier section, and the red represents the fitting center line. It can be clearly seen from the figure that the deviation of the pier. The actual center line of the pier is an irregular curve. When the

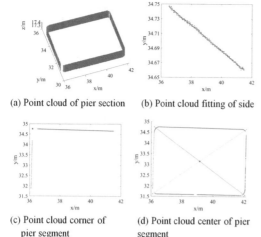

(a) Point cloud of pier section (b) Point cloud fitting of side

(c) Point cloud corner of pier segment (d) Point cloud center of pier segment

Figure 5. Data analysis stage diagram.

Table 1. Coordinates of pier center point

number	X/m	Y/m	number	X/m	Y/m
1	39.1503	33.1234	11	39.1538	33.1187
2	39.1519	33.1227	12	39.1536	33.1186
3	39.1522	33.1228	13	39.1548	33.1182
4	39.1523	33.1227	14	39.1542	33.1167
5	39.1527	33.1214	15	39.1543	33.1161
6	39.1548	33.1161	16	39.1543	33.1155
7	39.1529	33.1206	17	39.1545	33.1149
8	39.1527	33.1214	18	39.1546	33.1145
9	39.1528	33.1202	19	39.1548	33.1138
10	39.1539	33.1192	20	39.1552	33.1132

Due to the lack of theoretical center line as a reference, it is impossible to obtain the deviation value of each section of the pier. The author takes the average value of 20 points at the bottom of the pier as the theoretical center point, based on which the deviation analysis is carried out.

Table 1 shows the coordinates of the bottom 20 center points, the mean values in X direction and Y direction respectively.

$$\overline{X} = \frac{1}{n}\sum_{i=1}^{n} X_i = 39.15 \quad \overline{Y} = \frac{1}{n}\sum_{i=1}^{n} Y_i = 33.12$$

The point (39.15, 33.12) is the theoretical center line of the pier, and the equation of the center line of the pier is as follows:

$$\left. \begin{array}{l} X = 39.15 \\ Y = 33.12 \end{array} \right\} \quad (6)$$

The distance between a point and a line can be obtained by knowing the line equation. Because the straight line is parallel to the Z coordinate axis, in order to simplify the calculation process, the length of the straight line is set to 1 m, and the distance from the center point to the theoretical center line is obtained by using Helen formula

$$\begin{array}{l} p = (a+b+c) \div 2 \\ S = \sqrt{p(p-a)(p-b)(p-c)} \\ h = 2S \div c \end{array} \quad (7)$$

In formula (7), a, b and c are the three side lengths of the triangle, S is the area of the triangle, and h is the height of the triangle, that is, the deviation value.

In order to facilitate observation, the deviation Figure 8 of pier perpendicularity is obtained after the theoretical center line is set to zero.

Figure 8 (a) shows the distribution of the deviation values along the height. It can be seen from the figure that when the height of the pier is less than 10 m, the deviation value of the pier is within 1 cm, and then the deviation value expands with the increase of the height of the pier. When the height of pier reaches 30 m, the deviation of perpendicularity is unstable and fluctuates

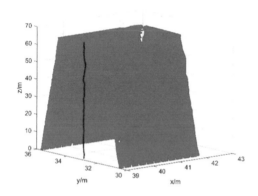

Figure 6. Point cloud section of bridge pier.

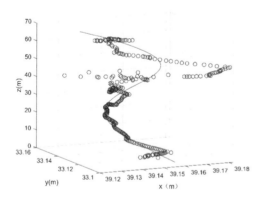

Figure 7. Centerline fit.

construction height of the pier does not reach 40m, the deviation value does not change significantly. When the construction height of the pier reaches 40 m, there will be a large deviation and the deviation value will change significantly, and then some correction will be made in the construction process.

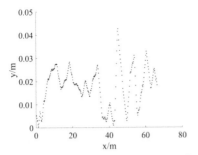
(a) Distribution diagram of deviation extension height

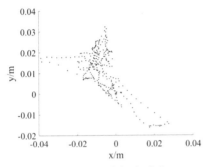
(b) Top view of pier deviation

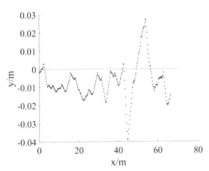
(c) Deviation in transverse direction

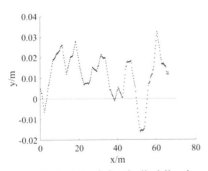
(d) Deviation in longitudinal direction

Figure 8. Deviation analysis chart.

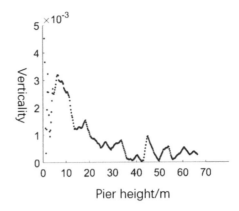

Figure 9. Verticality of pier.

of bridge piers in the transverse direction and along the bridge direction are within 3 cm. The inclination of the pier is not along the direction of the transverse bridge or along the bridge, but in both the longitudinal and transverse directions, and the deviation trend is relatively uniform.

Figure 8 (c) shows the deviation values of piers in the transverse direction. When the height of bridge pier is within 50 m, the deviation value of transverse bridge pier is small and the fluctuation of deviation value is small; when the height of bridge pier exceeds 45m, the fluctuation range of deviation value becomes larger.

Figure 8 (d) shows the deviation values of piers along the bridge. In the whole process of pier construction, the deviation value is unstable and fluctuates greatly.

(Note: the X-axis in Figure 8 (a) (c) (d) represents the pier height, and the Y-axis represents the deviation value. Figure 8 (b) X axis represents the deviation value in transverse direction, Y axis represents the deviation value in longitudinal direction, and red asterisk represents the theoretical center point).

According to the formula, the perpendicularity of 308 characteristic points of the pier is calculated, as shown in Figure 9.

It can be seen from Figure 9 that the perpendicularity of piers tends to decrease with the increase of pier height. The perpendicularity of the bottom of the pier is larger, because the height difference of the characteristic points at the bottom of the pier is small, and the perpendicularity is larger under the same deviation. Compared with Figure 8 (a), it can be seen that the place with the largest deviation is not necessarily the place with the largest perpendicularity.

4.3 Accuracy verification

The deviation values of the piers in the transverse and longitudinal directions are calculated by the above algorithm, and the maximum deviation values of the piers are calculated, and the variation diagram of the perpendicularity of the piers with the height of the piers is given. In order to verify the accuracy of the

obviously. When the construction height reaches 45 m, the maximum deviation occurs.

Figure 8 (b) is the top view of pier deviation. It can be seen from the figure that the deviation values

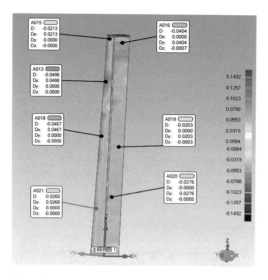

Figure 10. Comparison of 3D deviation.

above algorithm, the author analyzes the accuracy of the automatic algorithm by using the characteristics of BIM model, such as strong interaction and high modeling accuracy.

The BIM model of the pier is imported into the Geomagic point cloud processing software, and the point cloud and BIM model are aligned with the bottom of the pier as a reference. The deviation between point cloud and BIM model is calculated by the 3D analysis function of the software, as shown in Figure 10. The maximum deviation of the pier calculated by the algorithm is 0.046 m, and the maximum deviation obtained by the automatic analysis of Geomagic is 0.049 m, the error rate is about 6%, which meets the accuracy requirements.

5 CONCLUSION AND PROSPECT

3D laser scanning technology can obtain all geometric information of bridge pier structure. In this paper, the geometric characteristics of the rectangular thin-walled pier structure are analyzed, the perpendicularity of the high pier is analyzed, and the method to analyze the perpendicularity of the pier is put forward. According to the data relationship, the following conclusions are obtained:

1. Through 3D laser scanning to obtain point cloud data, the perpendicularity deviation of each height point of pier can be obtained. The deviation of perpendicularity increases with the increase of pier height, and the deviation is unstable finally.
2. The variation range of the deviation value of the pier is small in the direction of the transverse bridge (the long side of the rectangular pier) and large in the direction along the bridge (the short side of the rectangular pier).
3. The inclination of the pier is roughly along a diagonal direction of the rectangular pier.
4. The maximum horizontal deviation of the characteristic point of the pier is not necessarily the maximum perpendicularity of the pier, and the occurrence of the maximum horizontal deviation is random and unpredictable, so the accuracy of the result obtained by measuring the perpendicularity of the pier with the total station single point is insufficient.

After calculating the actual center line of the pier, using the actual center line data to reverse modify the modeling parameters of the BIM model of the pier, the BIM model corresponding to the actual state of the pier is obtained. Record the pier perpendicularity data under each working condition, and provide information support for the future bridge maintenance. The method of modifying BIM model with real 3D data is not only suitable for bridge piers. Bridge abutment, precast beam and other structures can be modified, but the corresponding algorithm needs to be developed to extract structural parameters automatically and efficiently, and more efforts need to be made in this respect.

ACKNOWLEDGMENTS

The related research and funding of this article are supported by Chongqing Smart City and Sustainable Development Academy (SCSDA). The authors would like to acknowledge the support of the SCSDA.

REFERENCES

Yang Guangqiang & He Fei 2014. Structural Design of 195 m Su per-high Pier of Hezhang Bridge. *China and Foreign Highway*, 34 (2): 152–156.

Liu Yajie, Ding Keliang & Guan Shixin 2018. Research on fast detection method of verticality of bridge high pier columns. *Journal of Shandong Jianzhu University*, 33 (2): 33–37.

Tie Huaimin & Jiang Xinyu 2017. Effects of Bridge Pier Construc tion Deviations on the Performance of High Pier Bridges. *Transportation Science and Engineering*, 33 (2): 31–36.

Ding Keliang, Liu Mingliang & Liu Yajie 2019. Non-contact rapid detection method and accuracy analysis of verticality of expressway high pier bridges. *Bulletin of Surveying and Mapping*, (6): 121–125.

Qiu Dongwei, Wang Laiyang, Wang Tong & Liu Yajie 2018. Application of Improved Least Squares Fitting in Bridge Pier Verticality Detection. *Bulletin of Surveying and Mapping*, (S1): 214–217.

Deng Xiaolong & Tian Shizhu 2018. Bridge Deformation Detection and Data Processing Based on 3D Laser Scanning. *Laser and Optoelectronics Progress*, 55 (7): 280–285.

Huang Chengliang 2012. Research on the Application of Three-dimensional Laser Scanning Technology to Bridge Pier Verticality Measurement. *Chinese Society of Surveying and Mapping*. 60–63.

Lu Dongdong & Zou Jingui 2019. Comparative study of noise re duction algorithms for 3D laser point clouds. *Bulletin of Surveying and Mapping*, (S2): 102–105.

Application of railway topology for the automated generation of geometric digital twins of railway masts

M.R.M.F. Ariyachandra & I. Brilakis
University of Cambridge, Cambridge, UK

ABSTRACT: The digitisation of existing railway geometry from point clouds datasets, referred to as "twinning", is a labourious task, which currently outweighs the perceived benefits of the resulting model. State-of-the-art methods have provided promising results, yet they cannot offer large-scale element detection required over kilometres without forfeiting precision and labour cost. The authors exploit the potential benefits of railway topology to automate the twinning process. The preliminary step involves automatically detecting masts as their positions are critical for the subsequent element detection. The method first removes vegetation and noise. Then it detects masts relative to the track centreline using the RANSAC algorithm and delivers final models in IFC format. The authors validated the method on an 18 km railway point cloud dataset and the results yielded an overall detection accuracy of 90.1% F1 score and reduced the manual twinning time by 98.6%. The proposed method lays the foundations to efficiently generate geometry-only digital twins of railway elements with no prior information.

1 INTRODUCTION

This paper discusses an automated twinning process to generate geometric digital twins of railway masts from airborne light detection and ranging (LiDAR) point cloud data. The authors define railways masts as trackside poles, either made of highly weathered metal or wood, which hold the overhead cables in place. The other pole-like objects consist of tree trunks/branches, signals, traffic sign and light (road) poles, and columns built-in rail infrastructure. The authors define the object detection as the combination of object localisation by determining the orientation and location of an object; clusterisation by segmenting point cloud dataset into sub-point clusters and finally classification by labelling the segmented point clusters such as rails, sleepers, masts. The challenge that the research addresses is how to efficiently generate geometric models [referred to in this study as a geometric digital twin (GDT)] of railway masts such that the perceived benefits of the GDT outweigh the investment made to create it. This is a significant challenge because of the potential value of GDTs in the construction, operation and maintenance of railways.

The UK has the fastest-growing railway network in Europe, with an increase in passenger numbers of 40% expected by 2040 (Office of Rail and Road 2020). In light of this increased demand, £48 billion funding has been recently approved for Network Rail including 16% for maintenance and 34.6% for the renewal of existing railway (Department of Transport 2017). The railway system in the UK is the oldest in the world and is a patchwork of overlapping designs, built at different times. Hence, the majority of existing and recently renovated railways dates back to 1830 and were originally constructed before the initiation of computer-aided design (CAD) in 1977. It is, therefore, challenging to make economic and efficient use of resources due to the absence of up-to-date digital representations of the railways to assist maintenance or renovation operations (NAO, 2014). Railway survey companies have already explored the potential benefits of laser scanned data to support these operations. Yet, the resulting laser survey data are often unstructured and do not contain any meaningful information of the documented railway. The digitisation of rail infrastructure mapping from point clouds generally referred to as the "twinning" process, is therefore introduced to utilise the need for high-level digital representation in a structured format. However, this process remains a daunting task, since it takes years rather than months before data collected reaches the database in a useable format (HM Government 2020). The authors argue that this establishes the need to create and maintain up-to-date digital twins (DT)s of rail infrastructure using quicker and more efficient approaches. An up-do-date GDT employs advanced data tools to provide back-and-forth connection between the twin and its physical asset while describing its geometry in real-time.

Leading software vendors like Autodesk, Bentley, Trimble, AVEVA and ClearEdge3D produce advanced semi-automated commercial solutions for DT generation. Yet, these software are tailored for generic or pre-defined geometries and are far from being fully automatic (Agapaki & Brilakis 2018a, 2018b; Wang et al. 2015). The authors' previous work (Ariyachandra & Brilakis 2019) reviewed the existing problems of the twinning of rail infrastructure in detail by investigating the entire workflow of GDT generation. The analysis summarised that the twinning of non-standardised geometries of railway elements requires extensive manual costs to extract and model sub-point clusters. This demands 95% of the total modelling time on extracting points corresponding to railway elements that are extremely thin and often overlapped with vegetation as well as customising shapes and fitting them to the sub-point-clusters which generally extend over kilometres. There is no single software that can offer a one-stop DT generation solution. Modellers have to shuttle intermediate results in different formats back and forth between different software packages during the modelling process, giving rise to the possibility of information loss. This explains why very few assets today have a usable DT. Hence, the authors contend that there is a pressing need to create less labour-intensive railway modelling techniques that can automate the twinning process with overall reduced costs and timescales.

In this paper, the authors address a core step in creating GDTs of rail infrastructure, i.e. the generation of GDTs of railway masts. Reliable mast detection is extremely useful for other railway element detection. It is the only vertical element which is in regular spatial offset on the trackbed. Hence, the detection of masts would provide the relative positional layout for the rest of the elements. Yet, automatic GDT generation of masts is often challenged by the presence of vegetation surrounding masts, the extremely thin shape of the sought object and the similar shape of tree trunks/branches. The following section analyses state-of-the-art methods proposed to streamline the twinning processes of masts and other pole-like objects.

2 BACKGROUND

The geometrical shape of the mast and other pole-like objects in railway point clouds (i.e. signal poles, traffic sign poles) are quite similar. Hence, this section analyses both masts and other pole-like object detection methods.

Only one method exists for detecting points belonging to masts (Pastucha 2016). The method first detects rail catenary objects into different classes and then localises the mast positions relative to previously detected catenary objects. However, this method uses geometrical distances from the trajectory of the scanner to set the thresholds and required manual user inputs to specify the geometric properties of the objects. Moreover, the initial vegetation and noise were manually removed. This method is impractical to use, as the automation achieved is small compared to the manual work needed.

Methods exist that can automatically remove the vegetation and noise of the point cloud. For example, a two-dimensional (2D) horizontal slicing based method can remove tree crowns and upper/lower structures (i.e. signal boards, traffic lights) attached to vertical pole-like objects (Huang & You 2015; Pu et al. 2011). Pu et al. (2011) used a percentile-based method to detect poles as pillars. Their results achieved 63% precision and 60% recall. Yet, this method requires the main pole to be an isolated pole-shaped object, without any structural attachments. In railroads, masts are always connected to cables and cantilevers. Hence, rather than being an isolated pole-shaped object, a mast is always a part of a structure. On railway point clouds, trees are often located close together. Consequently, tree crowns often overlap and can occlude masts. This is why Pu et al's (2011) method classified tree trunks as pole-like objects and did not remove them. Fukano and Masuda (2015) used patterns of the scan lines of mobile laser scanning (MLS) data as a basis to segment walls, roads and poles separately. However, if the poles were closely located, the scan lines belonging to the same group represented one pole or combined multiple poles. Hence, the precision was reduced to 76%, causing ambiguous results.

Existing semi-automated approaches can remove the vegetation and other noise to a certain extent. Yet, these methods do not provide a clear and concise approach to distinguish trees from other pole-like objects and are still dependent on the scanner profile information. In addition, the presence of vegetation, ground and facades that connect every object in a point cluster impose a huge computational load on these methods (Fukano & Masuda 2015; Yadav et al. 2015). These limitations highlight the need for automated filtering and segmentation of data at the initial stages of the process. For instance, El-halawany and Lichti (2013) used a segmentation method that employed 2D density-based calculations for the removal of the ground plane. Next, they applied vertical region growing to extract upright objects and then merged segments that belonged to the same object. The major problem left unresolved was that their ground removal method was sensitive to point densities and to the scanner trajectory. This method did not perform well when detecting poles surrounded by trees, distinguishing pedestrians from poles, segmenting incomplete poles and poles close to building facades.

Li et al. (2018) addressed these limitations using a three-step procedure to automatically decompose road furniture (including poles) into different components based on their spatial relations. This included ground plane removal relative to the scanner profile, and finally a slicing-based method, a random sample consensus (RANSAC) line fitting method and a 2D density-based method to extract vertical objects. However, the method required high-quality point clouds (35 points/m^2 to 350 points/m^2). Therefore, the performance of the algorithm was not promising for poor quality datasets. Likewise, this method recognised small booths supported by pillars as pole-like road furniture at both test sites. The detection algorithm was not enough to distinguish the difference.

The slicing-based segmentation steps segmented trees as poles when the trees were connected to the pole-like road furniture. The method did not categorise trees and road poles into separate groups.

Cabo et al. (2018) and Rodríguez-Cuenca et al. (2015) applied 3D voxelization to isolate poles from other noise data. They analysed the horizontal sections of the voxelized point clouds using 2D plane projection analysis. Yet, the method did not perform well when the poles were affected (a) by severe occlusions from large objects such as vehicles or large bins, parapets of bridges; (b) by the existence of other features such as pedestrians nearby; (c) when the poles were surrounded by bushes, or (d) when poles were too close to guardrails or walls. Li et al. (2019) investigated the continuity of surface roughness as a basis by which to differentiate poles from trees considering the various attachments of man-made poles (i.e. lamps, traffic lights, signboards etc.). The method could not detect poles with linear attachments (i.e. cables in masts). Also, trees that were occluded by other trees in front of them were wrongly detected as poles.

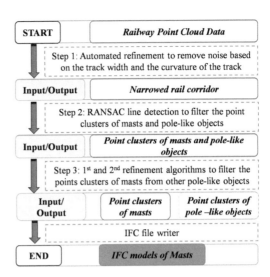

Figure 1. Workflow of the proposed method.

2.1 Knowledge gaps, objectives, and research questions

The review provided in the previous section demonstrates that existing pole detection methods have been tailored to MLS data (Arastounia 2015; Fukano & Masuda 2015; Li, Elberink & Vosselman 2018; Pastucha 2016; Zhu & Hyyppa 2014). Therefore, they cannot be directly applied to Airborne Laser Scanning (ALS) data. The ALS data is unorganised, meaning it does not contain any profile information and has arbitrary position and orientation. In addition, these methods are not robust in relation to occlusions and sparseness (El-halawany & Lichti 2013; Li et al. 2018; Pu et al. 2011). Railway point clouds are noisy and imperfect, suffering from both occlusions and sparseness. Masts are thin and hence often have few to no points representing them. Mast detection is a very difficult problem also due to the presence of vegetation and tree trunks. These factors render existing methods ineffective.

Despite the growing state-of-the-art, a fully automated railway mast twinning process is still in its infancy. This requires the development of an automated method to generate GDTs of masts from railway ALS data, as, in this case, no method in the literature meets all user requirements. To tackle this challenge, the authors propose an automated twinning method for masts in existing railways, aiming to meet the following objectives:

- *Objective 1:* Automatically remove the vegetation and other noise data surrounding railways without using any additional prior information such as neighbourhood structures, scanning geometry and intensity of input data.
- *Objective 2:* Automatically detect masts in the form of point clusters by differentiating masts from other pole-like objects in imperfect railway point clouds where occlusions and varying point density exist.

3 PROPOSED SOLUTION

3.1 Scope

The proposed method twins only the typical double-track railways because they make up 70% of the existing and under-construction railway network in the UK and Europe (Eurostat 2019). The proposed method exploits railway topology knowledge as guidance to directly extract point clusters corresponding to masts. Railways are not perfectly straight or flat and they usually contain varying horizontal and vertical elevations with curves and slopes. Nevertheless, railways are a linear asset type. Their geometric relations remain roughly unchanged often over very long distances. Close inspection of railway point clouds validates this effect, with repeating railway topological features such as: (1) the geometric relationships among masts, catenary and contact cables, and rails remain fairly unchanged along the railway corridor (Network Rail 2018); (2) the connections between masts and cables are placed in regular intervals (60 m intervals on average); (3) the main axis of the masts (Z-axis) is roughly perpendicular to the rail track direction (X-axis) [error tolerance is 11° (Network Rail 2018)]; and (4) masts are always positioned as pairs throughout the rail track. This study employs these four geometric features as railway topological relationships and uses these repeating geometric patterns as assumptions when developing the proposed method. The workflow of the proposed method is illustrated in Figure 1.

3.2 Step 1: Automated refinement to remove noise

The method initially uses principal component analysis (PCA) to find the principal axis of the point cloud and to align the railway point cloud such that the horizontal alignment of the rail track is positioned roughly parallel to the global X-axis. This enables easy exploitation of the point cloud features using various feature extraction algorithms because all features to be extracted in further steps lie in a global coordinated system. The Z-axis of the data is now parallel to the global Z-axis. Yet due to the horizontally curved alignments and vertical elevations of the rail track, the X and Y axes of the track continuously vary throughout the track. While PCA selects the most populated data axis parallel to the global X-axis, the track direction of the point cloud is not always parallel to the global X-axis. Thus, the centreline of the track does not reflect the true centreline of the rail track as it is occluded by many points especially those belonging to vegetation and the environment surrounding it. This restricts the usage of the centreline of the dataset to set a distance threshold to remove the noise. To address this challenge, the authors use an automated segmentation technique to align X and Y axes of datasets parallel to the global reference system. The following paragraph discusses each step in detail.

The proposed segmentation method first automatically crops the roughly aligned point cloud into near-straight pieces. Next, the method aligns these resulting pieces by computing PCA for each of these pieces and creates an axis-aligned bounding box around each in its principal direction. The result of this step yields a set of sub-bounding boxes (*SBBs*). The track direction of each *SBB* is now parallel to the global X-axis, and Y and Z axes are now parallel to global Y and Z axes. Prior to this step, the method requires an optimum *SBB* count, which: (a) provides near-straight pieces of the rail track; (b) removes the maximum number of vegetation and noise points, and (c) prevents the cropping of masts. The authors gauged the remaining number of masts as a percentage of the original number of masts with many *SBB* counts to decide on the optimum *SBB* count for each dataset. This finally gave 24, 30 and 17 as the optimum *SBB* for Dataset A, B and C respectively. In this paper, the authors have not illustrated the graphs representing calculations for these parameters due to limited space.

Following this step, the method gauges the minimum, the maximum, and the centre point ($q_{centreSBB}$) of each *SBB*. Note that $q_{centreSBB}$ is now aligned on the principal axes of the *SBB* and the width of the rail track (W_i) is now aligned to the Y-axis. Using $q_{centreSBB}$, the method determines a threshold distance (d_{SBB}) which is based on the W_i. W_i is used to set d_{SBB}, where d_{SBB} equals $W_i/2$. The method then uses d_{SBB} to remove the vegetation and other noise from the rail corridor data. The method computes d_{SBB} from $q_{centreSBB}$ on both sides along the Y direction (Figure 2) and removes the rest of the points of each *SBB*.

3.3 Step 2: Mast and other pole-like objects detection

Masts are now parallel to the global Z-axis, which is in-line with observation 3 mentioned in section 3.1.

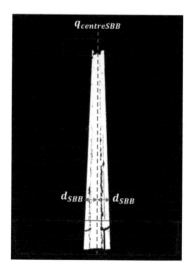

Figure 2. Resulting *SBB* after vegetation removal

Table 1. Performance matrices for RANSAC line detection.

Dataset	# of masts	TP	FP	FN	Pr	R	F1
A	212	134	24	78	84.8%	63.2%	72.4%
B	172	88	50	84	63.8%	51.2%	56.8%
C	188	69	76	119	47.6%	36.7%	41.4%
Avg.		291	150	281	66.3%	50.9%	57.6%

Hence, the method then detects masts as lines parallel to the global Z-axis using the RANSAC line detection algorithm since masts represent only long vertical elements remain after pre-processing and removal of vegetation. The proposed method allows for a deviation of $11°$ because the masts are not always perfectly parallel with the global Z-axis according to the railway design standards (Network Rail 2018). Prior to the RANSAC algorithm, the method uses the two-preprocessing steps outlined to reduce the computational load. (1) Removal of the ground plane to eliminate all ground points. This ensures that all points around masts are removed prior to further calculations and significantly reduces the points for faster computational performance. (2) Division of the remaining dataset into sub-boxes. This further increases the speed of RANSAC due to the small number of points considered each time.

The method then uses RANSAC for each *SBB*. The authors gauged the performance of the mast detection using performance metrics precision (*Pr*), recall (*R*) and F1 score (*F*1) as expressed below (Table 1):

- True Positive (*TP*): Masts were correctly detected as masts;
- False Negative (*FN*): Masts were not detected as masts and;
- False Positive (*FP*): Other pole-like objects were detected as masts.

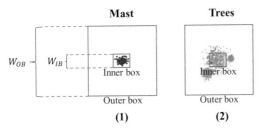

Figure 3. Ideal point distribution around masts (1) and other pole-like objects (2).

The detected vertical lines at this stage represent both masts and other remaining pole-like objects in railway point clouds. As a result, the detection accuracy was fairly satisfactory (57.6% F1 score), hence, further research is required to differentiate masts from other pole-like objects.

3.4 Step 3: Differentiate masts from other pole-like objects

The authors incorporated two refinement algorithms to differentiate masts from other pole-like objects based on the assumptions mentioned in section 3.1. At this stage, the surrounding of a mast has few or no points as all the ground points have been already removed at step 2, while other pole-like objects such as trees, bridge piers or walls usually contain few points that often belong to tree leaves, bridge columns and/or tree trunks. The authors used this observation to create the 1st refinement algorithm which consists of three steps to filter masts from other pole-like objects.

The 1st step creates an inner box (*IB*) around the detected lines, such that the point cluster of the detected line (a mast or other pole-like object) is tightly fitted to the *IB*. The *IB* only contains points belonging to a mast or other pole-like object. The 2nd step creates an outer box (*OB*) around the inner box such that this *OB* should only contain one mast and should not overlap with the other mast of the same pair. This box might contain any other points surrounding the pole which are usually caused by tree leaves, bushes, walls etc. The authors expect two different outcomes for masts and other pole-like objects as given in Figure 3.

(1) Since the ground plane is already removed, the area around the mast is sparse in terms of points. Almost no point should be detected in the immediate surroundings. This is the ideal point distribution around a mast. Therefore, the ratio of *IB* point count over the *OB* point count should ideally be 1. However, even after the removal of the ground plane, there might be points around a mast caused by other catenary system assets. This reduces the ratio to 0.9 or below.

(2) There are other pole-like objects located within the inner and outer boxes. Hence, even after the removal of the ground plane, there should be more points around other pole-like objects caused by other tree trunks and leaves. Therefore, the ratio of inner box point count over the outer box point count should be lower than outcome 1.

The outer box width (W_{OB}) value should not exceed the span between two masts of the same pair. In line with the railway design standards (Network Rail 2018), the authors hypothesized W_{OB} = 9.0 m and inner box width (W_{IB}) = 1.5 m. The authors confirmed these values using a point-based calculation method for different W_{OB} and W_{IB} values.

The 3rd step of the refinement algorithm is defining the threshold (R_D) which satisfies 0 < R_D < 1, to filter masts from tree trunks. The authors obtained the optimum R_D by computing F1 scores for different R_D values. According to the results obtained, the optimum R_D is 0.2 for all the datasets. Due to the limited space in this paper, the authors have not included the graphs representing calculations for R_D, W_{OB} and W_{IB}. Using R_D, the method filters masts from other pole-like objects in Dataset A. However, for Dataset B and C, this algorithm did not perform well (Table 2). When tree trunks, walls and rail bridges satisfy R_D this method recognises other pole-like objects as masts.

To remedy the resulting outcome, the authors used a 2nd refinement algorithm. This algorithm takes railway geometric observations into account and limits the region of search to a certain radius from the first pair of masts. Hence, the 2nd refinement algorithm starts from the left side of the track and repeats over the spans between masts on the right side of the track as explained below (Figure 4).

(1) The track direction is along the X-axis. Hence, the filtered point clusters of the previous refinement algorithm are first sorted (based on the X-coordinate) by automatically picking the leftmost line of the rail track which is the first in the row.

(2) The algorithm automatically picks a point (P_{L1}) on the first leftmost line (l_1) and searches for the next line, which should be located within the radius threshold (R_T). There are three possible outcomes:

- If l_1 represents a tree trunk or other pole-like object, presumably there should be more than one line adjacent to the l_1. But if l_1 is a mast, there should be only one other line (l_2) adjacent to l_1. Hence, if there are more than two lines positioned within R_T, the user needs to click the leftmost mast. This enters the coordinates of the l_1 into the refinement step.
- If there is one line (l_2) positioned within R_T, the algorithm assumes that these lines (l_1 and l_2) represent the first pair of masts.
- If there are none, this means only the leftmost line, l_1 is positioned within the R_T region.

(3) If l_1 and l_2 are already found, the algorithm automatically picks the next line (l_3) in the row and repeats Step 2 to find the next line (l_4). Now, four lines that belong to two pairs of masts are selected. If only one line is detected in Step 2, there will be two or three lines for further processing.

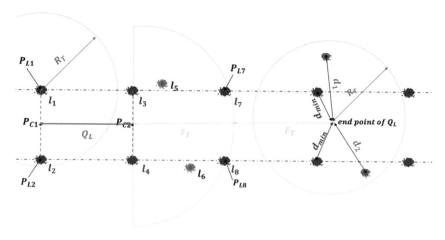

Figure 4. Second refinement algorithm to differentiate masts from other pole-like object

(4) The algorithm considers points (P_{L1} and P_{L2}) and (P_{L3} and P_{L4}) on (l_1 and l_2) and (l_3 and l_4). Using P_{L1}, P_{L2}, P_{L3} and P_{L4}, the refinement algorithm calculates midpoints (P_{C1} and P_{C2}) between two masts. If only one mast is detected by each pair of masts, rather than calculating the centre point of the two lines, the algorithm takes coordinates of the lines as P_{C1} and P_{C2}.

(5) The refinement algorithm then connects P_{C1} and P_{C2} to draw a line (Q_L). In a case of three lines, i.e., l_1 belongs to the first pair of masts and l_2 and l_3 are the two masts in the second pair, the starting point of Q_L is P_{L1} and the ending point is the centre point of l_2 and l_3.

(6) The algorithm defines an extend threshold (E_T), which is roughly equal to the regular intervals between two pairs of masts along the rail track (60 m on average).

(7) Using the coordinates of Q_L and E_T, the algorithm searches for the position of the next detected lines. All the lines which are positioned closer than E_T are discarded in this step because those lines represent other pole-like objects. The algorithm selects all lines that satisfy E_T. Instead of searching for lines in E_T along the X-axis, the proposed algorithm searches for a point appearing in an E_T radius region in relation to the second pair of masts, which is positioned on the right side.

(8) Supposing that the preceding step has discarded the next two lines (l_5 and l_6) in the row and selected subsequent lines (l_7 and l_8), the algorithm picks points (P_{L7} and P_{L8}) on l_7 and l_8. If one line has been detected, the algorithm picks a point on that line (i.e., if l_7 is detected, pick P_{L7}).

(9) The proposed algorithm extends Q_L by E_T along the rail track and obtains the coordinates of the endpoint of extended Q_L.

(10) Using Q_L, the algorithm searches for all possible lines located within the radius threshold (R_T) from the endpoint of extended Q_L.

Figure 5. A close up of the resulting IFC models of masts (viewed on Solibri Model Viewer).

(11) The refinement algorithm selects the two points that are most closely positioned to the endpoint of Q_L. These points belong to the next pair of masts. If there is only one, a point is picked on it.

(12) The method repeats steps 5–11 using the centre points of two most recently detected pairs of masts denoted as P_{C1} and P_{C2}.

This step gives the segmented point clusters of the masts, along with the position coordinates and heights of individual clusters. To deliver the final GDTs of segmented point clusters, the method uses implicit representation as the solid modelling approach which is based on the representation of 3D shapes using mathematical formulations. The method uses the shape definition of an I section and the resulting parameters of the 2nd refinement algorithm to define a mast (Figure 5).

4 EXPERIMENTS AND EVALUATION

An approximately 18 km-long portion of the rail track located between 's-Hertogenbosch and Nijmegen in the Netherlands served as the input of the proposed method. The size of the file was over 100 gigabytes, and hence, too large to process in terms of processor and memory capacity. The authors addressed this challenge by splitting the data file into three sub-point clouds, each roughly 6 km long, and termed as Dataset A, B, and C.

Table 2. Performance matrices for three data sets.

Sequence of steps	Dataset	Pr	R	F1
RANSAC	A	84.8%	63.2%	72.4%
	B	63.8%	51.2%	56.8%
	C	47.6%	36.7%	41.4%
	Average	66.3%	50.9%	57.6%
RANSAC with 1st refinement	A	94.7%	75.9%	84.3%
	B	97.1%	57.6%	72.3%
	C	70.5%	55.9%	62.3%
	Average	87.4%	63.8%	73.8%
RANSAC with 2nd refinement	A	96.5%	87.9%	92.0%
	B	97.4%	81.6%	88.8%
	C	87.6%	89.7%	88.6%
	Average	93.8%	86.6%	90.1%

The validation consisted of two parts. The first part was to experimentally define the optimal values of the key parameters (SBB_i, R_D, R_T, E_T) used in the proposed method. The second part was to assess the proposed method using performance metrics: precision, recall and F1 score. The authors followed the entire workflow of the mast twinning process as explained in (Ariyachandra & Brilakis 2019) to manually generate three ground truth (GT) datasets; each per one railway point cloud. The authors implemented the solution with the point cloud library (PCL) version 1.8.0 using C++ on Visual Studio 2017, on a laptop (Intel Core i7-8550U 1.8GHz CPU, 16 GB RAM, Samsung 256GB SSD).

Table 2 illustrates the results of mast detection. The authors used performance metrics as explained in section 3.3 to gauge the performance of the proposed method. The average segmentation accuracy for mast class was 90.1% F1 score (Table 2). The removal of vegetation took 12.5 sec/km to deliver a narrowed aligned rail corridor for each dataset. The RANSAC line segmentation with 1st refinement algorithm needed 20 sec/km while 2nd refinement required 3.3 sec/km. Finally, the generation of IFC models took 2 sec/km. Hence, the processing time of the proposed method was on average 37.8 sec/km hence, reduced the manual twinning time by 98.6%.

5 CONCLUSION

This paper presents a novel railway topological approach to develop a fully automated railway mast twinning process from airborne LIDAR data. Firstly, the authors explained why this was an unsolved problem by reviewing current industry applications and state-of-the-art methods that have been proposed to streamline the twinning process. The proposed method was tested on three railway point cloud datasets, with lengths over 18 km. The validation outcome showed that the method is quite reliable. Given the high performance of the method on real railway point clouds containing occlusions and sparseness, the authors contend that the method is reliable, scalable, and is independent of scanning technology. The method outperforms state-of-the-art methods and manual operations given the high segmentation accuracy and low run-time performance. The contributions of this research are as follows:

The proposed method:

(1) Can deal with complex, real railway topologies, such as varying rail track elevations and curved horizontal alignments of the rail track, meaning this method can detect masts despite the slope of the track.
(2) Can handle challenging scenarios such as occlusions, extreme vegetation around the track, and local variable densities of points. Although some inputs were very noisy (i.e. dataset B and C) due to the extreme vegetation surrounded the track, the method still achieved quite good performance in these datasets.
(3) Drastically reduces the computational cost by breaking down a lengthy railway point cloud into sub-bounding boxes. In this way, large-scale object detection can be significantly improved without sacrificing precision and manual cost.

However, the proposed method does not intend to be a cure-all. More railway data with different overhead electrification structures and single and quadruple tracks should be included and investigated in future studies. In short, this proposed method can significantly reduce the modelling cost and will accelerate the adoption of GDT for railway infrastructure mapping in existing railways. The future planned research will focus on overcoming of abovementioned limitations and addressing some of the assumptions, upgrading the algorithm to scale up to more complex railway configurations and detecting more rail asset components.

ACKNOWLEDGEMENTS

The authors express gratitude for Peter Apostle from Fugro NL Land B.V. who provided data for evaluation. The research leading to these results has received funding from the Cambridge Commonwealth, European & International Trust. Bentley Systems UK plc. partially sponsored this research under the grant agreement RG88682AH. The authors gratefully acknowledge the collaboration of all academic and industrial project partners. Any opinions, findings and conclusions or recommendations expressed in this material are those of the authors and do not necessarily reflect the views of the institutes mentioned above.

REFERENCES

Agapaki, E. and Brilakis, I. (2018a) 'State-of-practice on As-Is Modelling of Industrial Facilities', in Smith, I. and Domer, B. (eds) *Advanced Computing Strategies* for Engineering. EG-ICE 2018. Lecture Notes in Computer Science. Lausanne, Switzerland: Springer. doi: 10.1007/978-3-319-91635-4_6.

Agapaki, E. and Brilakis, I. (2018b) 'State-of-practice on As-Is Modelling of Industrial Facilities', in *European Group for Intelligent Computing in Engineering EG-ICE 2018*.

Arastounia, M. (2015) 'Automated recognition of railroad infrastructure in rural areas from LIDAR data', *Remote Sensing*, 7(11), pp. 14916–14938. doi: 10.3390/rs71114916.

Ariyachandra, M. R. M. F. and Brilakis, I. (2019) 'Understanding the challenge of digitally twinning the geometry of existing rail infrastructure', in *12th FARU International Research Conference (Faculty of Architecture Research Unit)*. Colombo, Sri Lanka: Faculty of Architecture Research Unit, Univeristy of Moratuwa, pp. 25–32. doi: 10.17863/CAM.47494.

Cabo, C. et al. (2018) 'An algorithm for automatic detection of pole-like street furniture objects from Mobile Laser Scanner point clouds', *ISPRS JOURNAL OF PHOTOGRAMMETRY AND REMOTE SENSING*, 87(January 2014), pp. 47–56. doi: 10.1016/j.isprsjprs.2013.10.008.

Department of Transport (2017) *New £48 billion funding for Britain's railways*. Available at: https://www.gov.uk/government/news/new-48-billion-funding-for-britains-railways (Accessed: 26 October 2020).

El-halawany, S. I. and Lichti, D. D. (2013) 'Detecting road poles from mobile terrestrial laser scanning data', 1603. doi: 10.1080/15481603.2013.866815.

Eurostat (2019) *Railway transport – length of lines, by number of tracks*. Available at: https://data.europa.eu/euodp/en/data/dataset/zibSXeaTsrM57MTXruQVvw (Accessed: 2 February 2018).

Fukano, K. and Masuda, H. (2015) 'Detection and Classificaton of Pole-Like Objects from Mobile Mapping Data', *ISPRS Annals of the Photogrammetry, Remote Sensing and Spatial Information Sciences*, II-3/W5, pp. 57–64. doi: 10.5194/isprsannals-II-3-W5-57-2015.

HM Government (2020) *Results of Competition?: SBRI?: Innovation in Automated Survey Processing for Railway Structure Gauging , Phase 1 Competition Code?: 1912 _ SBRI _ NR _ MMM _ GAUGINGP1*.

Huang, J. and You, S. (2015) 'Pole-like object detection and classification from urban point clouds', *Proceedings - IEEE International Conference on Robotics and Automation*, 2015-June(June), pp. 3032–3038. doi: 10.1109/ICRA.2015.7139615.

Li, F., Elberink, S. O. and Vosselman, G. (2018) 'Pole-Like Road Furniture Detection and Decomposition in Mobile Laser Scanning Data Based on Spatial Relations'. doi: 10.3390/rs10040531.

Li, Y. et al. (2019) 'Localization and Extraction of Road Poles in Urban Areas from Mobile Laser Scanning Data', *Remote Sensing*, 11(February), p. 401. doi: 10.3390/rs11040401.

NAO (2014) 'Lessons from major rail infrastructure programmes', *National Audit Office Report*, (HC267), p. 40. Available at: http://www.nao.org.uk/report/lessons-from-major-rail-infrastructure-programmes/.

Network Rail (2018) *Catalogue of Network Rail Standards*. Available at: https://www.networkrail.co.uk/wp-content/uploads/2018/04/Network-Rail-Standards-Catalogue.pdf (Accessed: 5 January 2019).

Office of Rail and Road (2020) *Office of Rail and Road (ORR)*. Available at: https://dataportal.orr.gov.uk/ (Accessed: 7 October 2020).

Pastucha, E. (2016) 'Catenary system detection, localization and classification using mobile scanning data', *Remote Sensing*, 8(10). doi: 10.3390/rs8100801.

Pu, S. et al. (2011) 'ISPRS Journal of Photogrammetry and Remote Sensing Recognizing basic structures from mobile laser scanning data for road inventory studies', *ISPRS Journal of Photogrammetry and Remote Sensing*, 66(6), pp. S28–S39. doi: 10.1016/j.isprsjprs.2011.08.006.

Rodríguez-Cuenca, B. et al. (2015) 'Automatic Detection and Classification of Pole-Like Objects in Urban Point Cloud Data Using an Anomaly Detection Algorithm', *Remote Sensing*, 7(September). doi: 10.3390/rs71012680.

Wang, C., Cho, Y. K. and Kim, C. (2015) 'Automatic BIM component extraction from point clouds of existing buildings for sustainability applications', *Automation in Construction*, 56, pp. 1–13. doi: 10.1016/j.autcon.2015.04.001.

Yadav, M. et al. (2015) 'Pole-Shaped Object Detection Using Mobile LIDAR Data in Rural Road Environments', *ISPRS Annals of the Photogrammetry, Remote Sensing and Spatial Information Sciences*, II-3/W5, pp. 11–16.

Zhu, L. and Hyyppa, J. (2014) 'The use of airborne and mobile laser scanning for modeling railway environments in 3D', *Remote Sensing*, 6(4), pp. 3075–3100. doi: 10.3390/rs6043075.

Project controlling of road construction sites as a comparison of as-built and as-planned models using convex hull algorithm

T. Walther, M. Mellenthin Filardo, N. Marihal & H.-J. Bargstädt
Bauhaus-Universität Weimar, Weimar, Germany

ABSTRACT: Continuous project control is crucial for the benefit of a smooth progress in construction projects. It implies the comparison of the target model with the actual situation of the building site. The aim of this work is to gain a comprehensive insight into project controlling via UAS photogrammetry and to give an example of implementation by a real case project. The method uses an approach that uses as-built point clouds and the as-planned model to estimate the degree of completion of road constructions work to using convex hull algorithm. The algorithm is a so-called cluster algorithm and is used in the data mining area. By enlarging this BIM model with the scandate and to 5D the costs are also monitored during the construction phase. Thus, an automated and practical project controlling will be developed. This implementation proves that a photogrammetric UAS controlling for degree of completion in road construction projects is feasible.

1 INTRODUCTION

1.1 Background

Project progress monitoring and controlling is one of the most important tasks in construction project management. Monitoring the actual status of the project can enable decision-makers to determine the deviations from the as-planned design and take corrective measures in case the project is behind schedule or over budget. Even though project control is very important, the architecture, engineering, construction, and operation (AECO) industry does not have integral and efficient monitoring systems, if compared to other industries. One reason is that traditional methods of collecting data on the progress of construction projects are still mainly manual, as they are usually carried out through on-site visual inspections. Computer technologies have great potential to improve construction management practices, e.g. through automation. Automating the construction progress helps to improve accuracy and reduce manual effort. Building Information Modeling (BIM) can help to automate this process (Mahami et al. 2019).

Rapid advances in BIM offer new opportunities to improve the efficiency and productivity of the construction process and increase the use of emerging technologies throughout the lifecycle of the project, not only in buildings but also in infrastructures. Using the BIM method adequately thorough visualization of the entire construction phase can be generated, thus redefining the scope of work, creating high-quality design schemes, supporting four-dimensional (4D) scheduling and five-dimensional (5D) cost estimation, and optimizing facility management and maintenance.

There are different approaches to compare the as-built and as-planned models, where different challenges are also associated with the aforementioned approaches and none of the available solutions achieve a robust and transferable performance (e.g., independent of building type or indoors/outdoors situation, within the available budget, adequate accuracy, etc.) (Kopsida et al. 2015), which indicates that further research on the topic is of vital importance.

1.2 Research problem statement

Current technologies and solutions for infrastructure projects cannot satisfy all project requirements and their participants, since infrastructure projects require engineering judgments during modeling if compared to building projects, due to a larger parameter range. These often-unsatisfied demands stimulate more research and development in the academic area, which can in turn be implemented in practical infrastructure projects. During the literature review, research papers involving the comparison of as-designed 5D BIM with e.g. an as-planned point cloud was limited, which indicates that this area is still a broad research field. The main motivation of this project is comparing point cloud imaging converted from drone-based photogrammetry of an as-built structure with the 5D BIM model of an as-planned model, to track the construction progress by analyzing the Percentage of Completion (POC) of the project.

Due to its extensive area, road construction projects offer an adequate case study for drone-based photogrammetry and thus, the entire proposed workflow.

1.3 Objectives of the study

A literature review gives us proof that the automation of progress monitoring will give us efficiency and quality in infrastructure projects. This methodology aims at allowing knowledge of the construction progress and getting an estimation of additional expenses and losses incurred, if there is any delay in the construction.

The proposed workflow uses the as-built point cloud data and the 3D model to evaluate the progress on the construction site. Collecting as-built data on a regular basis would allow project management to plan solutions more efficiently, therefore improving the project's schedule. It has the potential of providing detailed up-to-date progress reports and simultaneously generating as-built models for the project, if done using data acquisition technologies. It also promotes accurate and effective coordination between project team members, enabling the efficient implementation of the project. It allows quantitatively evaluating the health and safety dimensions of both static architecture geometry with further inclusion of 3D, 4D, and 5D data within building information models.

Throughout the creation and updating of quantity take-offs and cost forecasts, most of the working time engaged in cost control is accumulated, so any system or computerized system such as the proposed method that can simplify or optimize this process can produce a significant advantage and increase turnover times for bidding (Bradley et al. 2016).

2 LITERATURE REVIEW

2.1 BIM overview in the construction industry

Conventionally the on-site construction inspection is carried out visually and the quality of the progress data depends not only on the inspector's experience, but also on soft constraints like on-site distracting factor as well as time pressure. This conventional approach is error-prone, labor-intensive and time-consuming. Among recently proposed monitoring approaches, potential discrepancies between as-built and as-planned models are the main factor to facilitate decision making for corrective actions (Kopsida et al. 2015).

One approach is light detection and ranging (LIDAR) which is a digital photogrammetry technique, also known as laser imaging technology. In this approach, data is gathered in the form of point clouds which are shapes and dimensions of objects in real space converted and represented as a set of points in a digital 3D environment. There are, however, two main issues in this technique. First, geometric knowledge such as lines and surfaces cannot be extracted easily from object data in millions of points, due to manual processing. Second, a limited number of suitable scanners are available (Sepasgozaar et al. 2017). For object detection, the distance between the object and the laser sensor with very low laser return intensity is also an important factor (Kavulya et al. 2011).

The next approach is augmented reality (AR). In 1992, Koskela defined the direct relationship between BIM and AR as construction integrated by computer and automation in construction. Several prototypes and framework proposals are allowing BIM and AR integration using head-mounted devices, smartphones, tablets, and computers. Han and Golparvar-Fard (2015) have proposed a method for as-built progress monitoring and model visualization using BIM and daily construction images.

Calderon-Hernandez and Brioso (2018) have shown some investigation results on BIM and AR for future work. The main problem of AR systems is the correct synchronization of computer-generated and real-world data, which relies on the precision of monitoring the visual orientation and location of the user (Kopsida et al. 2015). Golparvar-Fard and Peña-Mora (2007) introduced a concept of semi-augmented reality, which includes color, texture, and structure for comparison. However, they need to be developed further in order to be applicable to a variety of construction materials.

Radio-frequency identification (RFID) is one of the approaches used for inspection purposes to gather on-site data, which enables the inspector to get information automatically by scanning the tag using a smartphone or tablet PC. Although this approach facilitates data acquisition compatible with available BIM-based software, it still requires installation and maintenance of RFID tags, accompanied by additional time and monetary investment (Kopsida et al. 2015).

Among the mentioned methods, photogrammetry is the most economical and efficient technique for obtaining a true 3D point cloud data. Laser-scanning is an expensive option requiring highly skilled operators as well as time-consuming data processing, usually resulting in lack of texture. Videogrammetry may also have problems and challenges, such as the high data volume and the similarly time-consuming processing needed on the film, as well as more noises in the final point cloud due to a reduction in image quality, if compared to still images (Mahami et al. 2019). Wang et al. proposed a method for the automatic removal of geometries from unorganized point clouds. The raw data collected undergoes data reduction, boundary detection, and categorization of building components, which results in building components being recognized as individual objects and their visualization as polygons (Wang et al. 2015).

2.2 BIM in infrastructure

Monitoring construction progress is a tedious time consuming and manual labor, especially regarding infrastructure projects spanning over large areas. While comparing as-planned designs and as-built progress, site managers and their teams face challenges regarding planning and allocating budgets, reporting to clients, and ultimately completing the project on schedule.

Although BIM implementation in the AECO industry is expanding, implementation of the BIM method in operation is still in its formative stage. In road construction, BIM is often reduced to 3D modeling by laser scanning to obtain a surroundings model and mitigate environmental impacts during the construction stage as well as quality control in the design stage and coordination among different stakeholders (Shou et al. 2015).

Puri and Turkan (2020) proposed a framework using 3D mobile LIDAR point clouds and 4D design to track bridge construction progress continuously. This method includes recording the as-built and as-designed data in the same coordinate system. The next step involves segmenting the point cloud and recognition of objects, in order to detect bridge elements. The difference between the expected and the actual work is quantified based on the object recognition tests.

Brilakis and Middleton (2018) used a slicing algorithm for generating bridge models from point clouds based on detaching the deck assembly from the pier assemblies, which then detect and segment pier caps using their normal surface, and girders with aligned boundary boxes and histograms of density. Finally, the system merges into individually named point clusters over segments.

Kivimäki and Heikkilä (2015) proposed a method in which 3D design surfaces are generated and stored on a central collaboration cloud-based platform. All the measuring devices used on site are integrated in the collaboration platform via internet connection. Using total stations, GNSS-based rovers (Global Navigation Satellite System) or other machine control systems, geometric measurements can typically be done on a construction site. All measurements are transferred to the proposed cloud-based platform immediately over the internet and automatically compared to designs for height differences and color-coding based on the required tolerances. The measurements can be viewed and approved on a map display using a regular web browser on PCs or tablets. This approach relies heavily on a comprehensive internet coverage of the site (Kivimäki & Heikkilä 2015).

When it comes to buildings, detailed geometry and component data can be said to be the most useful, providing the ability to perform clash detections, model co-ordination and generate linked costs and subsequent tasks. Whereas in infrastructure detailed geometry data is less important as the analysis it makes possible (e.g. clash detection) is of less benefit and is reliant on accurate data from other domains such as utilities. The most beneficial data on an infrastructure project comes from what can be termed non-graphical data such as cost information, material specifications, and component performance data (Bradley et al. 2016).

2.3 Data collection

It is established that photogrammetry is an effective and economical approach compared to other approaches detailed in the literature above. Photogrammetry is primarily concerned with making precise measurements of three-dimensional objects and terrain features from two-dimensional photographs (Aber et al. 2010). The literature in the following has been reviewed in two subsections: (i) terrestrial (with the camera handheld or on a tripod) photogrammetry and (ii) aerial (with the camera in the air) photogrammetry.

2.3.1 Terrestrial photogrammetry

Terrestrial photogrammetry is also known as close-range photogrammetry, which deals with object distance up to ca. 200 m (Aber et al. 2010). Being a remote sensing technique, physical touch is not involved in getting the image of the object (Bhatla et al. 2012). The benefit of this technique is the prevention of certain occlusions caused by large machines due to the fact that the procedure is no longer utterly automated and therefore the timing for the photographs can be chosen manually. But this approach is as labor-intensive as the visual inspection by a person, since the images are taken while walking around the construction site (Braun et al. 2015).

2.3.2 Aerial photogrammetry

Unmanned aerial systems (UAS) also known as drones are a promising approach for photogrammetry (Ham et al. 2016). By defining numerous positions over a construction site, a UAS can fly to those positions at each monitoring time step and can take images from nearly an identical position periodically (Braun et al. 2015). The goal of a UAS-driven monitoring technique is to (1) gather images or videos from informative views of the projects, (2) analyze those images and comparing with the BIM model regarding performance deviations, (3) monitor ongoing operations, (4) characterize the as-designed conditions, (5) rapidly visualizing the most updated state of work-in-progress (Ham et al. 2016).

3 RESEARCH METHODOLOGY

3.1 Preliminary design

3.1.1 Site information modeling (3D BIM)

1To begin with, the preliminary design of the target area, a 3D model is designed. For this purpose, the road components, terrain, and vertical alignment are to be considered. Data of the target area can be obtained in multiple ways. The location can be obtained by entering the coordinates in the software used for the preliminary design. The preliminary design is used in the research only for presentation purposes of the model, while the detailed technical information is added to the model in further steps.

3.1.2 Point cloud data and photogrammetry

The as-built point cloud data obtained via UAS photogrammetry captures not only target area images but also the surrounding objects such as buildings, trees,

Figure 1. Indication of the unwanted objects within the point cloud captured during the collection of point cloud data by using UAS, before cleaning and filtering.

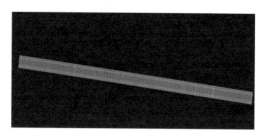

Figure 3. Representation of the 3D model of the as-planned road created for one stretch of the road above shown in Figure 2.

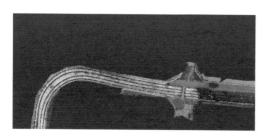

Figure 2. The point cloud data required for the implementation after cleaning and filtering of the unwanted data.

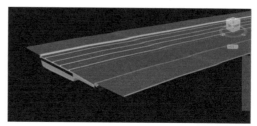

Figure 4. Cross section of the 3D model of the as-planned stretch of road.

vehicles, etc. as shown in Figure 1. These unwanted objects are not necessary for the proposed framework and can cause obstructions as well as misleading data for further steps. This unwanted data is to be eliminated by cropping the infill areas consisting of buildings, parks, and vehicles between the roads by using the limit box option is cropped as shown in Figure 2. This process is repeated manually, until all the unwanted data is eliminated from the point cloud. The proprietary .RCP file generated is then saved as an .E57 file for further processing. This format was chosen, as the .E57 file format is compact and provides a broad interoperability. The file was converted to this format for further registration and segmentation of point clouds with the as-planned model to select the area of interest.

Figure 5. Converting the as-planned .STL file created from the 3D model and converted into a point cloud by sampling the mesh.

3.2 Detailed designing of 3D BIM model

4After creating the as-planned road and its components in a preliminary design, the model is imported for technical detailing to 3D modeling software by selecting the area of interest and by defining accurate coordinates the same as in the software used for the preliminary design so that the object is placed in same geographic positions to avoid any offsets. Other objects such as terrain, planning roads, and existing roads can also be specified using the preliminary designed model.

5After exporting the files, a detailed design is created. The vertical profile of the road, the sub-assemblies, assembly, corridors, and alignments are added to the 3D model, as depicted in Figure 3 and Figure 4. After creating the corridor, the solids are extracted and saved as a new .DWG file. The .DWG

Figure 6. Close view of the as planned point cloud after converting the as-planned .STL file created from the 3D model and converted into a point cloud by sampling the mesh.

file is now converted to a triangulated mesh Stereo lithography (.STL) format to enable extracting points from the as-planned 3D model as shown in Figure 5 and Figure 6.

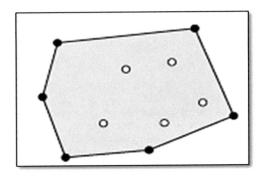

Figure 7. Convex hull representation for purpose of explanation

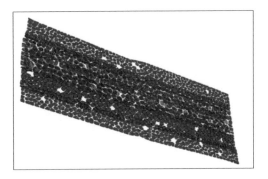

Figure 8. The obtained results of the surface from the as-built point cloud after running the convex hull algorithm.

3.3 Calculation of percentage of complete for progress tracking

6The progress tracking of the project can be achieved by automatic calculation using a concept of percentage of completion (POC). For this purpose, the K-means clustering and convex hull algorithms were implemented.

3.3.1 Clustering algorithm

Firstly, the point clouds have been processed using K-means algorithm. The clustering algorithm is an effective technique for data mining. Clustering is the process of splitting the data points into several groups, in such a way that data points in the same groups are more comparable to other data points in the same category than others. Simply put, the goal is to segregate and allocate groups with similar characteristics into clusters. The common approaches are as follows: Hierarchical clustering, partition clustering, and density clustering (Zhang et al. 2015).

K-mean clustering algorithm is a partitioning algorithm that grouped data into pre-defined number of clusters. It starts with random initialization of cluster centroids then assign data points to the closest (highly similar) centroids. The same operation is continuously repeating until the occurrence of termination criterion (either given number of iteration completed or clusters show no change after certain no of iteration) is met. It is centroid-based algorithm. Partitioning techniques have almost linear time complexity (Naeem 2018).

K means is the most common algorithms for clustering. It is widely used because of its simplicity, comprehensiveness and effectiveness. It is an iterative algorithm for clustering which seeks to find local maxima in each iteration (Theljani et al. 2013). It is a multi-dimensional, clustering algorithm (Zhang et al. 2015).

3.3.2 Convex hull algorithm

Subsequently, the convex hull algorithm was implemented to achieve the surface mapping of the point clouds. Convex hull is a widespread and frequently used concept in computational geometry.

8As definition, the convex hull of a discrete data set is the minimum area-convex polygon containing this

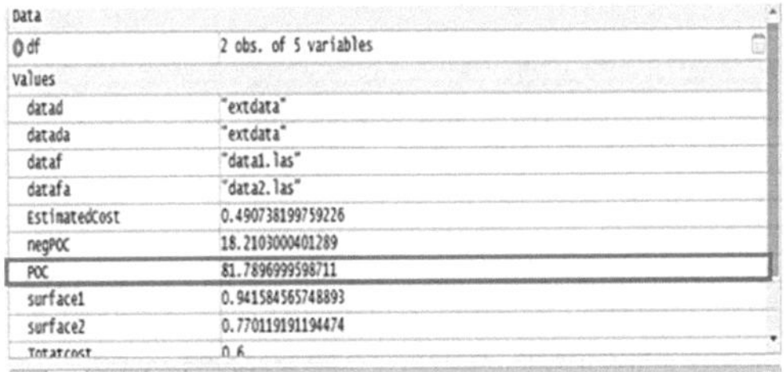

Figure 9. Representation of the results after running the algorithm, which indicates the percentage of completion along with the estimated cost respectively.

set, as depicted in Figure 7. It finds practical applications in various areas including pattern recognition, image processing, engineering, computer graphics, design automation and operation research (Theljani et al. 2013).

If we have a fixed set of points P consisting of n points in the plane, we now measure what is called the convex hull of P. The convex hull is instinctively what you achieve by driving a nail in each point of a plane and, wrapping a string around the nails as shown in Figure 7. This approach applies even to a polygon, or to any other number of lines, the hull of which is similar to its vertex point hull set. There are various uses of a convex hull, for example, obstruction avoidance, the awareness of concealed objects, and the study of forms (Kenwright 2014).

3.4 Comparison of as-planned and as-built models

10R is an open-source programming language, which was used to generate an algorithm for calculation of POC and estimated costs (EC). Some of the R packages help in processing the point cloud data. The K-means algorithm represented in section 3.3.1 was used for clustering the point cloud data of the as-planned and as-built models. Then, using the convex hull algorithm, a loop is formed around the clusters indicating the area of interest. After running the algorithms, the as-built and the as-planned areas were obtained. The area formed after running the algorithm was used to compare the as-built and the as-planned

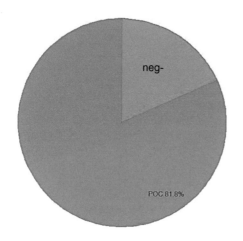

Figure 10. Results obtained after the comparison of the asbuilt and the as-planned point clouds represented in the form of a pie chart indicating the percentage of completion (POC).

objects. For a particular element, the surface area of the as-planned element is always greater than the as-built. Since the as-built element is under construction, its surface area is lesser. The difference in the surface area gives the difference between the two elements. Furthermore, the obtained area difference between as-planned and as-built point clouds was further utilized to obtain the POC and EC, which are calculated as follows:

$$POC = \frac{A_1}{A_2} \times 100 \qquad (1)$$

Where A1 = Area of as-built point cloud
A2 = Area of as-planned point cloud

$$EC = \frac{POC}{100} \times Budgeted cost \qquad (2)$$

If the value of the surface area for both as-built and as planned is the same, then the POC value is 100% which means that the element construction is completed.

4 RESULTS AND DISCUSSION

The proposed workflow was validated on a segment of a road construction project in Germany, that was selected as a case study. The date on which the UAS was flown to obtain the point cloud of the as-built model was taken as reference for the completion of the as-planned model. The models then were converted in .LAS files and the algorithm was run to obtain the differences on the selected date, subsequently giving the percentage of completion of the road for the scheduled date. It was noted that the element was 80% completed according to schedule, which was confirmed by the obtained POC of 80%. Also, an approximation of the cost estimate was done for the resulting POC, based on the cost required for the completion of the 100% construction work and accordingly the cost for 80% completion was approximated. The following are the results obtained from the research.

5 CONCLUSION AND OUTLOOK

The automated comparison of the as-planned and as-built point cloud models gives a percentage of completion of the construction works, therefore, validating that the proposed workflow can be adapted for automated progress tracking of 4D- and 5D-based construction with the help of BIM, resulting in a huge leap for progress monitoring in construction. Though the process provides accurate results, there are several aspects that affect the quality of the progress tracking such as density of the point clouds, the precision while eliminating the unwanted data captured by the UAS as well as the long-term automation of this step. Also, removal of certain temporary equipment can cause patches in the point cloud. The limitations of the proposed framework in road construction progress monitoring concentrate around the unavailability of data of road layers (sections) from the drone photogrammetry. Without information about the layers, the comparison of POC in as-planned and as-built models is not as robust as when applied to other geometries, such as bridge and building components.

The original K-means algorithm cannot be applicable to point cloud without density constraints. Especially, since the higher the value of K, the longer the algorithm spends on clustering operations. Highly complex and degenerate vertices arise in practice making it difficult to generate a reliable convex hull (Zhang et al. 2015).

Future works regarding this research include the aforementioned modeling of layer information as well as automating the elimination of unwanted data. The workflow implemented in this paper produces a general approach for generating a convex hull for the entire structure of the road without differentiation in the layers, hence giving us the percentage of completion for the entire road and not for every layer of the road. This workflow can be further developed to monitor the progress of each layer of the road during construction, provided that proper detailing of the as-built drone photogrammetry is available.

REFERENCES

Aber, J.; Marzolff, I. & Ries, J. 2010. Chapter 3 Photogrammetry. *Small-Format Aerial Photography*: 23–39.

Bhatla, A.; Choe, S. J.; Fierro, O. & Leite, F. 2012. Evaluation of Accuracy of As-Built 3D Modeling from Photos Taken by Handheld Digital Cameras. *Automation in Construction*: 116–127.

Bradley, A.; Li, H.; Lark, R. & Simon, D. 2016. BIM for Infrastructure: An Overall Review and Constructor Perspective. *Automation in Construction* 71: 139–152.

Braun, A.; Tuttas, S.; Borrmann, A. & Stilla, U. 2015. Auto-mated Progress Monitoring Based on

Photogrammetric Point Clouds and Precedence Relationship Graphs. *The 32nd International Symposium on Automation and Robotics in Construction and Mining* (ISARC 2015): 1–7.

Calderon-Hernandez, C. & Brioso, X. 2018. Lean, BIM and Augmented Reality Applied in the Design and Construction Phase: A Literature Review. *International Journal of Innovation, Management and Technology:* 60–63.

Golparvar-Fard, M. & Peña-Mora, F. 2007. Application of Visualization Techniques for Construction Progress Monitoring, In: *Computing in Civil Engineering*: 216–223.

Ham, Y.; Han, K.; Lin, J. & Golparvar-Fard, M. 2016. Visual Monitoring of Civil Infrastructure Systems via Camera-Equipped Unmanned Aerial Vehicles (UAVs): A Review of Related Works. *Visualization in Engineering* 4 (1):1.

Han, K. & Golparvar-Fard, M. 2015. Automated Monitoring of Operation-Level Construction Progress Using 4D BIM and Daily Site Photologs. *Construction Research Congress 2014:* 1033–1042.

Kavulya, G.; Jazizadeh,F.; & Becerik-Gerber, B. 2011. Effects of Color, Distance, and Incident Angle on Quality of 3D Point Clouds. 2011 *ASCE International Workshop on Computing in Civil Engineering*:169–177.

Kenwright, B. 2014. Convex Hulls Surface Mapping onto a Sphere, December 6: Article ID 40352480.

Kivimäki, T. & Heikkilä, R. 2015. Infra BIM Based Real-Time Quality Control of Infrastructure Construction Projects. *Proceedings of the 32nd International Symposium on Automation and Robotics in Construction and Mining:* 877–882.

Kopsida, M.; Brilakis, I. & Vela,P. 2015. Review of Automated Construction Progress Monitoring and Inspection Methods. *Proceedings of the 32nd CIB W78 Conference on Construction:* 421–431.

Lu, R.; Brilakis, I. & Middleton, C.R. 2018. Detection of Structural Components in Point Clouds of Existing RC Bridges. *Computer-Aided Civil and Infrastructure Engineering:*1–22.

Mahami, H.; Nasirzadeh, F.; Hosseininaveh, A. & Nahavandi, S. 2019. Automated Progress Controlling and Monitoring Using Daily Site Images and Building Information Modelling. *Buildings 9, 70*: 1–20.

Naeem, S. 2018. Study and Implementing K-Mean Clustering Algorithm on English Text and Techniques to Find the Optimal Value of K. *International Journal of Computer Applications Volume 182 – No. 31*: 7–14.

Puri, N. & Turkan, Y. 2020. Bridge Construction Progress Monitoring Using Lidar and 4D Design Models. *Automation in Construction:* 102961.

Sepasgozaar, S.; Shirowzhan, S. & Wang, C. 2017. A Scanner Technology Acceptance Model for Construction Projects. *Procedia Engineering*: 1237–1246.

Shou, W.; Wang, J.; Wang, X. & Chong, H. P. 2015. A Comparative Review of Building Information Modelling Implementation in Building and Infrastructure Industries. *Archives of Computational Methods in Engineering* 22 (2): 291–308.

Theljani, F.; Zidi, S.; Ksouri, M. & Laabidi, K. 2013. Convex Hull Based Clustering Algorithm. *International Journal of Artificial Intelligence 9(A12)*: 51–70.

Wang, C.; Cho, Y. K. & Kim, C. 2015. Automatic BIM Component Extraction from Point Clouds of Existing Buildings for Sustainability Applications. *Automation in Construction 56:* 1–13.

Zhang, K.; Bi, W.; Zhang, X.; Fu, X.; Zhou, K. & Zhu, L. 2015. A New K means Clustering Algorithm for Point Cloud. *International Journal of Hybrid Information Technology* 8 (9): 157–70.

City and building information modelling

A trend review on BIM applications for smart cities

A. Pal & S.H. Hsieh
Department of Civil Engineering, National Taiwan University, Taipei, Taiwan

ABSTRACT: To respond to the challenges like population growth, climate change, economic instability, cities are taking a data-driven, modern technology-oriented approach to provide better service, improved quality of life and a sustainable future. With the advancement of technologies, such as Building Information Modeling (BIM), applications of information integrated 3D models have started gaining popularity among the architecture, engineering, and construction (AEC) researchers. Considering the benefits of BIM and the requirements of smart cities, the amalgamation of BIM and smart city research was obvious. As the researches in both fields have become mature, a review of the BIM applications in the context of smart cities is much needed at this point. Ninety-two (92) papers from two major online databases were retrieved and analyzed through a systematic literature review approach. Results show that present researches are mostly focused on technology advancement. Six major research directions are observed. Sustainability, facility management, and construction management related researches have been discussed most frequently.

1 INTRODUCTION

Modern cities are facing various challenges like traffic congestion, high energy consumption, air pollution, excessive population pressure and so on. To enhance the quality of life in cities and to ensure a sustainable future, the cities are being equipped with various advanced information and communication technologies (ICT). These cities are known as Smart Cities. The latest technologies like the Internet of Things, Big Data, cloud computing, etc. have eased the process of monitoring and management of smart cities. On the other hand, Building Information Modelling (BIM), which is capable of storing geometric and semantic information of a building throughout its lifecycle, is a growing trend in modern research and possesses a lot of potential for catalyzing the smart city development. The integrated use of BIM in a smart city project, ranging from the household to the city level, is expected to bring better visualization, transparency, stakeholder engagement, information management, resource management in the domain of transportation, communication, service, infrastructure, and technology.

Although both smart city and BIM researches have gained a lot of attention among the built environment researchers and practitioners, the application of BIM for smart cities is not explored in its full potential yet (Al Sayed et al. 2015). As a lot of research related to technology advancement has taken place throughout the world, it is the right time to review those research in the context of smart city development. To bridge the gap this paper aims at reviewing existing literature till the end of 2019 to analyze the existing research focusing BIM application for smart cities. Section 2 of the paper describes the methodology adopted for a systematic literature search; Section 3 focuses on the importance of BIM for smart cities; Section 4 explains the research trends so far. Finally, Section 5 concludes the study.

2 METHODOLOGY

2.1 Literature search

In this paper, the application of BIM in its individual capacity or integrated form with other technologies was analyzed in the context of smart cities. The review was done in three steps. The first step was to analyze the literature's content type, year of publication and the participation of authors from various countries. In the second step, the literature was classified based on the KPIs of the smart cities, described in ISO 37122- 2019, to analyze the importance of BIM for smart cities. Finally, the contents were reviewed to highlight the current research trends. The selected papers were classified based on smart city applications or aspects. Top 6 applications that cover 90% of the literature were further analyzed for understanding the technological advancement concerning smart city development.

To conduct the systematic review, literature related to BIM and smart city were retrieved from two major online databases, namely Web of Science ™core collection and Scopus. The searched literature was consisting of journal articles, conference papers, and book chapters. A combination of keywords "BIM", "Building Information Modelling" and "Smart City/

Figure 1. Publication statistics.

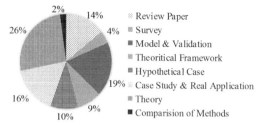

Figure 3. Research methodology analysis.

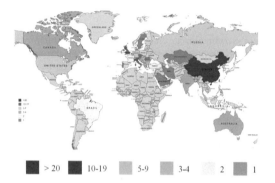

Figure 2. Publication distribution map globally.

Cities" with the operator "AND" was used for the search. For searching the topic of literature, title, abstract and keywords were included and the language of the literature was kept limited to English. As a result, 97 unique kinds of literature were retrieved. Five (5) of them were removed after content analysis because of less significance with the subject study. Finally, 92 literatures consisting of 31 articles, 60 conference papers, and a book chapter were selected for review. They were published in 27 different scientific journals and 37 conference proceedings.

2.2 Literature analysis

It is observed that literature related to BIM for smart cities started publishing in 2014. Although the publication rate was relatively low for the initial 3 years, it has started gearing up from 2017 onwards. It is noteworthy that 35 literatures (38%) from our literature pool has been published only in the last year 2019. Also, the citation rate for the literature is booming from 2018 onwards. This clearly shows the importance of this kind of research in recent times. Figure 1 shows the statistics.

The geographic distribution of the research papers shows that the authors are from 90 universities spreading across 29 countries in the world. However, China is taking a lead role with a contribution in 26 papers from 20 different universities in the country, followed by the UK and Italy, each with 13 and 10 papers respectively. Hong Kong, Germany, Russia, and the USA are the countries that have contributed more than 4 papers. Eleven (11) universities from the UK, 7 from Italy, 6 from Russia, 4 from USA and 3 from Hong Kong and Germany each have participated in BIM for smart city researches. Figure 2 shows the distribution of BIM for smart city research globally. In terms of citation, Italian researches were cited most frequently till 2019 followed by Korea, Hong Kong, Japan, and Australia's research papers.

Figure 3 shows the research methodology analysis for the selected literature. It is observed that almost 14% of literatures are review articles and another 26% are theoretical research in this domain. This result certainly reveals the need for more application-oriented future research.

3 BIM FOR SMART CITIES

The research direction of the BIM application for smart cities was primarily classified into three main groups: research focusing on technology advancement, research on citizen's and people's engagement for smart cities and research on smart governance and policy development. In addition to the trend analysis, an effort was made to link the researches with smart city performance indicators as suggested by the International organization for standardization (ISO) for sustainable development of communities.

3.1 Linking smart city performance indicators with BIM application research.

International Organization for Standardization (ISO) published two important standards (ISO 37122 and 37123) under ISO 37120:2018 (Indicators for city services and quality of life). These standards are expected to help cities to implement smart city policies, technologies, and practices for providing better quality life to citizens and achieving sustainability goals. ISO 37122 (Indicator for smart cities) first edition published in 2019 has been designed for assessing the performance of smart cities in terms of achieving sustainability goals during city development. This standard has put forward 20 key performance indicators (KPIs) and its measurement procedures. In this study, an effort was made to analyze the content of the literature and link them with one or more KPIs of smart cities. Figure 4 shows that BIM application explored in almost 85% of the literature is addressing only 6 smart city KPIs: report & record maintenance, urban planning, governance, energy, environment & climate change, and housing. Another 8 KPIs where

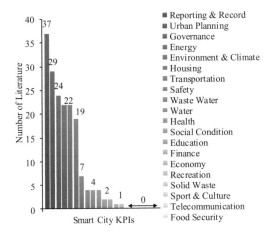

Figure 4. Literature addressing smart city KPIs.

BIM application is observed was only explored by 15% of the research papers and the rest 6 KPIs were not addressed by any of the literature studied.

Only a few research items addressed the urban infrastructure-related KPIs like transportation, water, and wastewater, health, solid waste, and telecommunication. On the contrary, KPIs associated with housing and buildings were explored mostly. Al Sayed et al. (2015) also proposed a theoretical framework to highlight the dependencies between smart homes/ smart cities with data-driven building information models using the novel Pearson correlation method. The use of the dependency network provided an empirical basis for predicting the performance of the cities. Ghosh (2018), a researcher from India, suggested the requirement of city-specific building codes for specifying the engineering and design requirements of city components. BIM was expected to take an important role in the formulation of those standards.

3.2 *Overview of the BIM and smart city research*

While the research trend was analyzed, it was observed that only 10% of the papers were focusing on citizens & people and government & policies, the rest 90% were focusing on technology advancement for smart city solutions.

Heaton and Parlikad (2019) from Cambridge University proposed a conceptual framework for aligning infrastructure assets of smart cities to citizen's requirement to ensure a citizen-centric development. During this study, they studied various standards related to BIM implementation and smart city frameworks published throughout the world. Parallel but isolated developments of BIM and smart city standards were observed. They emphasized the application of ICT technologies for capturing citizen's requirements, citizen's satisfaction and wellbeing within the smart city framework. They also mentioned that although the technology has been advanced significantly for smart city solutions, the technology for data analytics, data capture and integration is still limited. The research tried to link the functional output of infrastructure assets with the citizen's needs and satisfaction. Heidrich et al. (2017) also discussed the user's engagement concept while reviewing the adaptability of buildings combined with new technologies like BIM. Another way of engaging citizens and peoples in the smart city development process is by educating them. Hammi and Bouras (2018) proposed an interdisciplinary course in universities focusing on the integration of cybersecurity, blockchain, and BIM in the context of smart cities. Civil, architecture and computer science department's students and faculties were taken part in the course.

Cities are producing a large amount of data every day from various sources like people, machines, infrastructure, sensors, etc. Managing and systematically analyzing this information is another challenge for researchers and practitioners. BIM is playing a vital role in categorizing and managing such information. Web of building (WoB) proposed by Anjomshoaa (2014) is an information space to connect BIM resources and buildings located in a geographic location. Howell et al. (2016) explored a 3D game engine to support the decision-making process in the smart city environment by integrating the information from BIM, IoT enabled data sources and web services. This research promotes effective human-machine interaction for better city development. Correa (2015) feels that BIM alone is not capable enough to handle the big data generated from smart cities.

BIM integrating with other latest technologies has significantly helped in developing the solutions for current and future smart city applications. Figure 5 shows the BIM integration frequency with other technologies. GIS and IoT are the two most studied technologies within the selected set of literature. IFC, 3D point cloud, City GML, API, and big data are some of the technologies that have been studied with a considerable amount in the context of smart city development. The application of these BIM integrated solutions are capable of making the cities smarter (Tang et al. 2019). Some of the field applications of these latest technologies can be witnessed in the study by Travush et al. (2018). The current study identified 6 major and 4 minor smart city aspects or applications where BIM integrated technologies were embedded. Figure 6 shows these smart city applications with their occurrence. The research trends of the six (6) major smart city applications, which cover 90% of the technology advancement research are discussed in the next section one by one.

4 TECHNOLOGICAL ADVANCEMENT TRENDS FOR SMART CITY DEVELOPMENT

4.1 *Sustainability*

With no surprise, in 22 papers (around 25%) the BIM applications in smart city context were to address the sustainability issues of the cities. For achieving sustainability, efficient energy management was found to

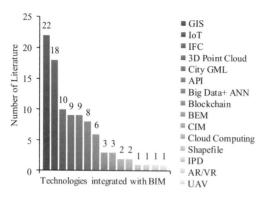

Figure 5. Frequency of BIM integration with other technologies in the context of a smart city.

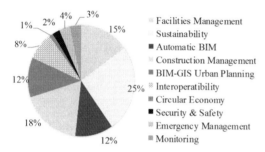

Figure 6. Smart city applications with occurrence.

4.2 Facilities management

Thirteen (13) papers (around 15%) studied facilities management (FM) solutions for smart cities, where BIM and other integrated technologies have been applied significantly. Smart cities are expected to possess a huge volume of assets. So, managing those assets effectively is a daunting task for facility managers. Six (6) out of 13 papers (46%) explored the asset management aspect. Technology-enabled platforms are expected to ease the process of facilities management (Bröchner et al. 2019). BIM-enabled operation & maintenance for projects was studied by Pelipenko and Gogina (2017), Lin et al. (2017b), Ding et al. (2018), and Min (2017).

BIM, GIS, IoT, and AR/VR applications are constantly increasing in recent years. BIM/GIS/IoT-based applications are leveraging the possibilities of real-time asset monitoring whereas AR/VR is increasing the user and system interaction. Two papers (15%) by Carneiro et al. (2018) and Baracho et al. (2018) explored these opportunities.

To manage that information for future FM, the role of big data analysis is very crucial. Five (5) papers (39%) studied big data analysis and information management aspects. For example, Ram et al. (2019), Garyaev & Garyaeva (2019), and Kuzina (2019) studied the big data infrastructure for smart cities and artificial intelligence for big data analysis.

be the core requirement. Buildings, the key component of the cities and a huge consumer of energy, were studied the most in 12 literatures out of 22 (55%). Building energy modeling (BEM) is the method for predicting the energy consumption of the building. Internet of things (IoT), wireless sensor networks (WSN) and building information modeling (BIM) together have made the process of energy management much systematic. Sava et al. (2018), Rinaldi et al. (2018) and Pasini et al. (2016). Corekci et al. (2019) provided a framework and Scheffer et al. (2018) explored the use of Open Data Model (ODM) for integrating the WSN data with BIM for real-time monitoring. Both Lee et al. (2016) and Giuda et al. (2016) modeled BIM-based energy management systems for improving the energy efficiency of the buildings. Lancaster et al. (2019) used smart meters to collect the energy consumption data and compared that with energy models. Ugliotti et al. (2016) used BIM API for energy analysis.

Two (2) research papers (9%), Rinaldi et al. (2016) and Habibi (2017), have considered occupant's behavior and occupant's comfort while developing energy management platforms. In 6 papers (27%), city-scale energy management was studied. For efficient energy management in city-scale, incorporation of Geographic Information System (GIS) was found mandatory Two papers (9%), Olawumi and Chan (2018, 2019), tried to identify the critical success factors for integrating sustainability practices and BIM.

4.3 Construction management

With 16 research papers (around 18%), the application of BIM for construction project management (CPM) in the context of smart city construction shows its importance among researchers. The application of BIM in connection with the latest information and communication technologies (ICT) has opened a new way of CPM research. Both Wu and Cai (2017) and Mo (2017) proposed an integrated application of BIM and ICT technologies for better project management during smart city constructions. BIM-enabled project management strategies were reviewed by Abdelhameed (2018), Lianguang (2017), Kisel (2017) and Mgbere et al. (2018). BIM together with Integrated project delivery (IPD) has shown a significant benefit in terms of construction project management. Su (2017), Hu and Zhang (2019) and Liu et al. (2019) used BIM in project management for real construction cases. Lin et al. (2017a) applied BIM for schedule management. Both Mo (2018) and Han (2018) highlighted that BIM models were found capable of accurate cost estimation and useful for cost control. Song et al. (2017), Ma and Ren (2017) studied BIM-GIS integration for quality management, process management, time and cost control, health safety and environment management, information management and coordination between different sectors. The review of blockchain application in construction by Li et al. (2019) highlighted that blockchain technology can bring enough trust in the construction management process by incorporating a smart contract. This blockchain technology can also play a vital role in smart city management processes.

4.4 Automatic reconstruction of BIM models

The automatic reconstruction of BIM models is recently gaining more importance among the researchers mainly in the context of smart cities. In 10 papers (around 12%) the topic was studied. Reconstruction of a comprehensive BIM model of an already built facility is very time consuming and labor-intensive. Technology has been developed for constructing an as-built BIM model or city information model (CIM) from 3D point cloud data for cities needing digital information models for further engineering. The application of such technology is also observed in the reconstruction of BIM models for management (e.g. conservation and maintenance) of ancient historic monuments structures, or buildings (Angelini et al. 2017, Chenaux et al. 2019). Use of various devices like augmented reality (AR) smartphone, light detection and ranging (LiDAR) scanner, unmanned aerial vehicle (UAV) or drone has been observed for collecting the 3D point cloud data. Xue et al. (2019a, 2019b) described the reconstruction of the BIM model from the 3D point cloud system. Chen et al. (2018), Tan et al. (2019) and Cui et al. (2019) tried to construct a BIM model from LiDAR scanner data and mobile laser scanning (MLS) data. Few researchers have taken the effort to explore the use of 3D GIS for capturing city information in a three-dimensional space.

4.5 BIM-GIS Integration for urban planning

BIM-GIS integration for urban planning is a very important direction from the smart city point of view. In the current review, 10 papers (around 12%) discussed the related topic. BIM models work on local coordinates whereas GIS can connect the city components with earth's coordinates. It is capable of providing an integrated interdependent city model. These integrated platforms can collect data to analyze the current demand of utility resources and to inform the future demand for better urban planning across all spatial scales. Cities can be designed as an integrated system of all utility resources like water, gas, electricity. These systems are capable of providing opportunities to the decision-makers to simulate various urban scenarios and their outcomes. Thus it will make the decision-making process easy and more effective. Six (6) out of ten (10) papers (60%) have discussed BIM-GIS applications. BIM-GIS integrated application was also observed in the smart city's intelligent transportation systems proposed by Mirboland and Smarsly (2019), and Bao and Wang (2019). A similar integrated platform for the electrical system was studied by Farooq et al. (2017) and Gilbert et al. (2018). BIM-GIS integration was also observed for the 3D pipe network in smart cities (Wang et al. 2019).

Three papers (30%) explored the feasibility of BIM-GIS integrated data mining, data analysis, and management. In a smart city system to manage the city infrastructure like transportation, electricity, water, healthcare, a huge amount of information/big data is being collected using city information models, IoT, sensors smartphones, etc. So, it has become necessary to analyze that information using data analytics or machine learning algorithms to infer some patterns of the data (Kiavarz et al. 2018; Mazroob Semnani et al. 2018).

While BIM-GIS related researches are gaining importance among the researchers for smart city development, it is high time to standardize its application. One paper (10%) by Janečka (2019) highlighted some ongoing international standardization activities focusing on 3D cadaster and BIM.

4.6 Interoperability

Despite we have advanced technology, smooth interoperability between various systems still needs our attention. Seven research papers (around 8%) have explored the interoperability issues. Digital models are at the center of the smart/digital city development. BIM generated data are exchanged using the mostly used data format, namely Industry foundation classes (IFC). However, GIS systems do not support IFC format directly without any format transformation. Shapefile is one of the popular data formats for GIS. Zhu et al. (2018) researched various methods of IFC to shapefile transformation. Shahrour et al. (2017) described interoperability as the key issue for digital smart city modeling. Wang (2015) took an effort to integrate the information collected from various sensors with 3D models using IFC, CityGML, and IndoorGML. Scheffer et al. (2018) proposed the use of the open data model (ODM) for the exchange of information among the stakeholders without any interoperability issue. Anjomshoaa (2014) and Howell et al. (2016) both proposed the use of web-based platforms to make an interoperable urban decision support system. For the internal analysis within a BIM tool, application programming interfaces (API) are more useful. Macros, plug-ins, scripts are the common usage of API. These APIs can help in reading and manipulating the properties of the BIM models (Ignatova et al. 2018).

5 CONCLUSION

Considering the growing trend of smart city and BIM research, this paper has made an effort to review the literature published until the year-end of 2019 through a systematic literature review method. Ninety-two (92) research papers have been analyzed to highlight the evolution of the research. Results show that China has published the maximum number of research papers whereas Italian research papers have been cited maximum time. Classification of research as per smart city

KPIs reveals that 85% of literature has addressed only 6 KPIs out of 20 KPIs proposed in ISO 37122: 2019. As 90% of the selected literature discusses the technology advancement aspect so the major emphasis was given to the same in this paper. Six (6) current application aspects of smart cities, where BIM and other integrated technology have been applied most, have been discussed in detail. Despite significant technological advancement, interoperability amongst various systems within the built environment domain remains as a major challenge.

REFERENCES

Abdelhameed, W. (2018). "BIM from conceptual model to construction." *2018 International Conference on Innovation and Intelligence for Informatics, Computing, and Technologies, 3ICT 2018*, IEEE, 18–20 November 2018, Sakhier, Bahrain, 1–4.

Al Sayed, K., Bew, M., Penn, A., Palmer, D., and Broyd, T. (2015). "Modelling dependency networks to inform data structures in BIM and smart cities." *SSS 2015 – 10th International Space Syntax Symposium*, 13–17 July 2015, Bloomsbury, United Kingdom, 1–15.

Angelini, M. G., Baiocchi, V., Costantino, D., and Garzia, F. (2017). "Scan to BIM for 3D reconstruction of the papal basilica of saint Francis in Assisi In Italy." *International Archives of the Photogrammetry, Remote Sensing and Spatial Information Sciences – ISPRS Archives*, 22–24 May 2017, Florence, Italy, 47–54.

Anjomshoaa, A. (2014). "Blending building information with smart city data." *CEUR Workshop Proceedings: 13th International Semantic Web Conference (ISWC 2014)*, 19 October 2014, Riva del Garda, Italy, 1–2.

Bao, L. X., and Wang, Y. Q. (2019). "Research on expressway network tolling platform technology based on GIS+BIM." *International Symposium for Intelligent Transportation and Smart City (ITASC) 2019 Proceedings*, Springer Singapore, 9–11 May 2019, Shanghai , China, 241–250.

Baracho, R. M. A., De Assis Cunha, I. B., and Pereira, M. L. (2018). "Information modeling and information retrieval for the Internet of Things (IoT) in buildings." *IMCIC 2018 – 9th International Multi-Conference on Complexity, Informatics and Cybernetics, Proceedings*, March 13–16; 2018, 13–16 March 2018, Orlando, Florida, USA, 85–91.

Bröchner, J., Haugen, T. I., and Lindkvist, C. (2019). "Shaping tomorrow's facilities management." *Facilities*, 37(7/8), 366–380.

Carneiro, J., Rossetti, R. J. F., Silva, D. C., and Oliveira, E. C. (2018). "BIM, GIS, IoT, and AR/VR Integration for Smart Maintenance and Management of Road Networks: A Review." *2018 IEEE International Smart Cities Conference, ISC2 2018*, 16–19 September 2018, Kansas City, MO, USA.

Chen, K., Lu, W., Xue, F., Tang, P., and Li, L. H. (2018). "Automatic building information model reconstruction in high-density urban areas: Augmenting multi-source data with architectural knowledge." *Automation in Construction*, Elsevier, 93(May), 22–34.

Corekci, M., Pinarer, O., Servigne, S., Smart, N., and Monitoring, H. (2019). "Designing a Novel Smart Home Monitoring Systems with the Integration of BIM and WSN." *CSC 2019?: 5th International Conference on Connected Smart Cities 2019*, 17–19 July 2019, Porto, Portugal.

Correa, F. R. (2015). "Is BIM big enough to take advantage of big data analytics?" *32nd International Symposium on Automation and Robotics in Construction and Mining (ISARC 2015): Connected to the Future, Proceedings*, 15–18 June 2015, Oulu, Finland.

Cui, Y., Li, Q., and Dong, Z. (2019). "Structural 3D reconstruction of indoor space for 5G signal simulation with mobile laser scanning point clouds." *Remote Sensing*, 11(19).

Ding, M., Yang, Q., Xing, J., and Xie, L. (2018). "Research on underground device operation and maintenance management system based on BIMserver." *Proceedings of the International Conference on Smart City and Intelligent Building (ICSCIB 2018)*, Springer Singapore, 15–16 September 2018, Hefei, China, 601–608.

Farooq, J., Sharma, P., and Sreerama Kumar, R. (2017). "Applications of building information modeling in electrical systems design." *Journal of Engineering Science and Technology Review*, 10(6), 119–128.

Garyaev, N., and Garyaeva, V. (2019). "Big data technology in construction." *E3S Web of Conferences: XXII International Scientific Conference "Construction the Formation of Living Environment" (FORM-2019)*, 18–21 April 2019, Tashkent, Uzbekistan.

Ghosh, S. (2018). "Smart homes: Architectural and engineering design imperatives for smart city building codes." *International Conference on Technologies for Smart City Energy Security and Power: Smart Solutions for Smart Cities, ICSESP 2018 – Proceedings*, IEEE, 28–30 March 2018, Bhubaneswar, India, 1–4.

Gilbert, T., Barr, S., James, P., Morley, J., and Ji, Q. (2018). "Software systems approach to multi-scale GIS-BIM utility infrastructure network integration and resource flow simulation." *ISPRS International Journal of Geo-Information*, 7(8).

Giuda, G. M. Di, Villa, V., Piantanida, P., Tagliabue, L. C., Rinaldi, S., Angelis, E. De, and Ciribini, A. L. C. (2016). "Progressive Energy Retrofit for the educational." *2016 IEEE International Smart Cities Conference (ISC2)*, 12–15 September 2016, Trento, Italy.

Habibi, S. (2017). "Micro-climatization and real-time digitalization effects on energy efficiency based on user behavior." *Building and Environment*, Elsevier Ltd, 114, 410–428.

Hammi, A., and Bouras, A. (2018). "Towards Safe-Bim Curricula Based on the Integration of Cybersecurity and Blockchains Features." *INTED2018 Proceedings*, 5–7. March 2018, Valencia, Spain, 2380–2388.

Han, L. (2018). "Research on Construction Engineering Cost Management Based on BIM Technology." *Proceedings – 2018 3rd International Conference on Smart City and Systems Engineering, ICSCSE*, IEEE, 29–30 December 2018, Xiamen, China, 214–217.

Heaton, J., and Parlikad, A. K. (2019). "A conceptual framework for the alignment of infrastructure assets to citizen requirements within a Smart Cities framework." *Cities*, Elsevier, 90(January), 32–41.

Heidrich, O., Kamara, J., Maltese, S., Re Cecconi, F., and Dejaco, M. C. (2017). "A critical review of the developments in building adaptability." *International Journal of Building Pathology and Adaptation*, 35(4), 284–303.

Howell, S., Hippolyte, J. L., Jayan, B., Reynolds, J., and Rezgui, Y. (2016). "Web-based 3D urban decision support through intelligent and interoperable services." *IEEE 2nd International Smart Cities Conference: Improving the*

Citizens Quality of Life, ISC2 2016 – Proceedings, IEEE, 12–15 September 2016, Trento, Italy, 1–4.

Hu, C., and Zhang, S. (2019). "Study on BIM technology application in the whole life cycle of the utility tunnel." *Proceedings – International Symposium for Intelligent Transportation and Smart City (ITASC) 2019*, Springer Singapore, 9–11 May 2019, Shanghai, China, 277–285.

Ignatova, E., Zotkin, S., and Zotkina, I. (2018). "The extraction and processing of BIM data." *IOP Conference Series: Materials Science and Engineering*, 365(6).

Janeèka, K. (2019). "Standardization supporting future smart cities- A case of BIM/GIS and 3D cadastre." *GeoScape*, 13(2), 106–113.

Kiavarz, H., Jadidi, M., Rajabifard, A., and Sohn, G. (2018). "BIM-GIS oriented intelligent knowledge discovery." *13th 3D GeoInfo Conference: International Archives of the Photogrammetry, Remote Sensing and Spatial Information Sciences*, 1–2 October 2018, Delft, The Netherlands, 79–82.

Kisel, T. (2017). "Features of the Organization of Construction with the Use of Technologies of Information Modeling." *MATEC Web of Conferences: International Science Conference SPbWOSCE-2016 "SMART City,"* 15–17 November 2016, St. Petersburg, Russia.

Kuzina, O. (2019). "Conception of the operational information model of smart city control system." *E3S Web of Conferences: XXII International Scientific Conference "Construction the Formation of Living Environment" (FORM-2019)*, 18–21 April 2019, Tashkent, Uzbekistan.

Lancaster, Z. S., Binder, R. B., Matsui, K., and Yang, P. (2019). "Developing a theory of an object-oriented city: Building energy for urban problems." *Energy Procedia: 10th International Conference on Applied Energy (ICAE2018)*, 22–25 August 2018, Hong Kong, China, 4210–4217.

Lee, D., Cha, G., and Park, S. (2016). "A study on data visualization of embedded sensors for building energy monitoring using BIM." *International Journal of Precision Engineering and Manufacturing*, 17(6), 807–814.

Li, J., Greenwood, D., and Kassem, M. (2019). "Blockchain in the built environment and construction industry: A systematic review, conceptual models and practical use cases." *Automation in Construction*, Elsevier, 102(January), 288–307.

Lianguang, M. (2017). "Study on Project Information Management Based on Building Information Modeling." *Proceedings – 2016 International Conference on Smart City and Systems Engineering, ICSCSE*, IEEE, 25–26 November 2016, Hunan, China, 238–240.

Lin, L., Li, J., Huang, M., Sun, Y., and Song, X. (2017a). "The Application and Exploration of the Foundation Construction Management in High-Rise Building Based on BIM Technology." *Proceedings – 2016 International Conference on Smart City and Systems Engineering, ICSCSE*, IEEE, 25–26 November 2016, Hunan, China, 304–307.

Lin, L., Song, X., Huang, M., Sun, Y., Li, J., Zhu, Y., and Yang, Z. (2017b). "The Creation and Exploration of Revit Family Based on BIM Technology." *Proceedings – 2016 International Conference on Smart City and Systems Engineering, ICSCSE*, IEEE, 25–26 November 2016, Hunan, China, 241–244.

Liu, B., Cai, T., Xiao, S., Fu, H., and Chu, W. (2019). "Research on application of BIM technology in municipal road construction." *Proceedings International Symposium for Intelligent Transportation and Smart City (ITASC) 2019*, Springer Singapore, 9–11 May 2019, Shanghai, China, 265–276.

Ma, Z., and Ren, Y. (2017). "Integrated Application of BIM and GIS: An Overview." *Procedia Engineering: Creative Construction Conference 2017, CCC 2017*, The Author(s), 19–22 June 2017, Primosten, Croatia, 1072–1079.

Mazroob Semnani, N., Kuper, P. V., Breunig, M., and Al-Doori, M. (2018). "Towards an intelligent platform for big 3d geospatial data management." *Proceedings of ISPRS TC IV Mid-term Symposium "3D Spatial Information Science – The Engine of Change"*, 1–5 October 2018, Delft, Netherlands, 133–140.

Mgbere, C., Knyshenko, V. A., and Bakirova, A. B. (2018). "Building information modeling. A management tool for Smart City." *2018 IEEE 13th International Scientific and Technical Conference on Computer Sciences and Information Technologies, CSIT 2018 – Proceedings*, IEEE, 11–14 September 2018, Lviv, Ukraine, 177–182.

Min, X. (2017). "Study on Interior Design Based on Building Information Modeling." *Proceedings – 2016 International Conference on Smart City and Systems Engineering, ICSCSE*, IEEE, 25–26 November 2016, Hunan, China, 76–78.

Mirboland, M., and Smarsly, K. (2019). "A semantic model of intelligent transportation systems." *CEUR Workshop Proceedings: 26th International Workshop on Intelligent Computing in Engineering*, 30 June- 3 July 2019, Leuven, Belgium, 1–10.

Mo, L. (2017). "The Application of IT in Engineering Management." *Proceedings – 2nd International Conference on Smart City and Systems Engineering, ICSCSE 2017*, 11–12 November 2017, Changsha, China, 93–95.

Olawumi, T. O., and Chan, D. W. M. (2018). "Identifying and prioritizing the benefits of integrating BIM and sustainability practices in construction projects: A Delphi survey of international experts." *Sustainable Cities and Society*, Elsevier, 40(February), 16–27.

Olawumi, T. O., and Chan, D. W. M. (2019). "Critical success factors for implementing building information modeling and sustainability practices in construction projects: A Delphi survey." *Sustainable Development*, 27(4), 587–602.

Pasini, D., Mastrolembo Ventura, S., Rinaldi, S., Bellagente, P., Flammini, A., and Ciribini, A. L. C. (2016). "Exploiting internet of things and building information modeling framework for management of cognitive buildings." *IEEE 2nd International Smart Cities Conference: Improving the Citizens Quality of Life, ISC2 2016 – Proceedings*, IEEE, 12–15 September 2016, Trento, Italy, 1–6.

Pelipenko, A., and Gogina, E. (2017). "Real application of BIM in the engineering system design for energy management." *IOP Conference Series: Energy Management of Municipal Transportation Facilities and Transport – EMMFT 2017*, 10–13 April 2017, Khabarovsk, Russian Federation.

Ram, J., Afridi, N. K., and Khan, K. A. (2019). "Adoption of Big Data analytics in construction: development of a conceptual model." *Built Environment Project and Asset Management*, 9(4), 564–579.

Rinaldi, S., Bittenbinder, F., Liu, C., Bellagente, P., Tagliabue, L. C., and Ciribini, A. L. C. (2016). "Bi-directional interactions between users and cognitive buildings by means of smartphone app." *IEEE 2nd International Smart Cities Conference: Improving the Citizens Quality of Life, ISC2 2016 – Proceedings*, 12–15 September 2016, Trento, Italy.

Rinaldi, S., Flammini, A., Pasetti, M., Tagliabue, L. C., Ciribini, A. C., and Zanoni, S. (2018). "Metrological issues in the integration of heterogeneous Iot devices for energy efficiency in cognitive buildings." *I2MTC 2018 – 2018*

IEEE International Instrumentation and Measurement Technology Conference: Discovering New Horizons in Instrumentation and Measurement, Proceedings, IEEE, 14–17 May 2018, Houston, TX, USA, 1–6.

Sava, G. N., Pluteanu, S., Tanasiev, V., Patrascu, R., and Necula, H. (2018). "Integration of BIM Solutions and IoT in Smart Houses." *Proceedings – 2018 IEEE International Conference on Environment and Electrical Engineering and 2018 IEEE Industrial and Commercial Power Systems Europe, EEEIC/I and CPS Europe*, IEEE, 12–15 June 2018, Palermo, Italy, 1–4.

Scheffer, M., Konig, M., Engelmann, T., Tagliabue, L. C., Ciribini, A. L. C., Rinaldi, S., and Pasetti, M. (2018). "Evaluation of Open Data Models for the Exchange of Sensor Data in Cognitive Building." *2018 Workshop on Metrology for Industry 4.0 and IoT, MetroInd 4.0 and IoT 2018 – Proceedings*, IEEE, 16–18 April 2018, Brescia, Italy, 151–156.

Shahrour, I., Alileche, L., and Alfurjani, A. (2017). "Smart cities: System and tools used for the digital modelling of physical urban systems." *2017 Sensors Networks Smart and Emerging Technologies, SENSET 2017*, 12–14 September 2017, Beirut, Lebanon, 1–4.

Song, Y., Wang, X., Tan, Y., Wu, P., Sutrisna, M., Cheng, J. C. P., and Hampson, K. (2017). "Trends and opportunities of BIM-GIS integration in the architecture, engineering and construction industry: A review from a spatio-temporal statistical perspective." *ISPRS International Journal of Geo-Information*, 6(12), 1–32.

Su, L. (2017). "Digitalization and application research of BIM-based power plants lifecycle information." *Smart Innovation, Systems and Technologies*.

Tan, T., Chen, K., Lu, W., and Xue, F. (2019). "Semantic enrichment for rooftop modeling using aerial LiDAR reflectance." *2019 IEEE International Conference on Signal Processing, Communications and Computing, ICSPCC*, 20–22 September 2019, Dalian, China.

Tang, S., Shelden, D. R., Eastman, C. M., Pishdad-Bozorgi, P., and Gao, X. (2019). "A review of building information modeling (BIM) and the internet of things (IoT) devices integration: Present status and future trends." *Automation in Construction*, Elsevier, 101(January), 127–139.

Travush, V. I., Belostosky, A. M., and Akimov, P. A. (2018). "Contemporary Digital Technologies in Construction Part 2: About Experimental & Field Studies, Material Sciences, Construction Operations, BIM and 'smart' City." *IOP Conference Series: Materials Science and Engineering: VII International Symposium Actual Problems of Computational Simulation in Civil Engineering*, 1–8 July 2018, Novosibirsk, Russian Federation.

Ugliotti, F. M., Dellosta, M., and Osello, A. (2016). "BIM-based Energy Analysis Using Edilclima EC770 Plug-in, Case Study Archimede Library EEB Project." *Procedia Engineering: World Multidisciplinary Civil Engineering-Architecture-Urban Planning Symposium 2016, WMCAUS 2016*, The Author(s), 13–17 June 2016, Prague, Czech Republic, 3–8.

Wang, H. (2015). "Sensing information modelling for smart city." *Proceedings – 2015 IEEE International Conference on Smart City/SocialCom/SustainCom together with DataCom 2015 and SC2 2015*, IEEE, 19–21 December 2015, Chengdu, China, 40–45.

Wang, S., Sun, Y., Sun, Y., Guan, Y., Feng, Z., Lu, H., Cai, W., and Long, L. (2019). "A hybrid framework for high-performance modeling of three-dimensional pipe networks." *ISPRS International Journal of Geo-Information*, 8(10).

Wu, Y., and Cai, J. (2017). "Research on New-Type Smart City in China Based on Smart Construction Theory." *International Conference on Construction and Real Estate Management 2017*, © ASCE, 10–12 November 2017, Guangzhou, China, 347–354.

Xue, F., Lu, W., Chen, K., and Webster, C. J. (2019a). "BIM reconstruction from 3D point clouds: A semantic registration approach based on multimodal optimization and architectural design knowledge." *Advanced Engineering Informatics*, Elsevier, 42(July), 100965.

Xue, F., Lu, W., Chen, K., and Zetkulic, A. (2019b). "From Semantic Segmentation to Semantic Registration: Derivative-Free Optimization-Based Approach for Automatic Generation of Semantically Rich As-Built Building Information Models from 3D Point Clouds." *Journal of Computing in Civil Engineering*, 33(4), 1–16.

Zhu, J., Wang, P., and Wang, X. (2018). "An Assessment of Paths for Transforming IFC to Shapefile for Integration of BIM and GIS." *International Conference on Geoinformatics*, IEEE, 28–30 June 2018, Kunming, China, 1–5.

Matching geometry standards for geospatial and product data

H. Tauscher
Dresden University of Applied Sciences, Dresden, Germany

ABSTRACT: Integration of data from the construction and geospatial domain relies on successful integration of the respective geometric components. In this paper we report on some issues encountered during conversion of building models in IFC format to city models in CityGML format. Our results reflect on the ISO geometry standards in the two domains and we present a framework to successively extend the analysis from single observations to the whole standards.

1 INTRODUCTION

1.1 Motivation

Integration of data from the construction and geospatial domain is increasingly required to carry out holistic analyses covering different scales of measurement. Clearly, geometric integration is an important part of this endeavour. In this paper, we report on geometric issues encountered during conversion from building models in IFC format to 3D city models in CityGML format. These issues date back to discrepancies in the underlying standards — subsets of the conceptual geometry models defined in ISO 10303-42 (2014) and ISO 19107 (2005) respectively.

Many concepts from those two standards can be correlated easily and straight forward, but during the development of our conversion procedures we discovered subtle differences. For example, both IFC and CityGML allow for groups of solids to be assigned to products or features as their representations. However, the specifications differ in the topological requirements they pose upon these groups. While CityGML only allows composite solids, such that the group is a valid solid again, IFC does not state such requirement.

1.2 Related work

BIM-GIS integration, and in particular the integration of building and city models, is a prevalent topic of recent research both in the geospatial and the building information modeling domains. The GeoBIM benchmark project aims to extensively evaluate the current state of the art and technology on both the building and city model side and for bi-directional conversion (Noardo et al. 2019). With synthetic test files covering a wide range of geometries the project puts particular emphasis on geometry.

Some researchers such as Arroyo Ohori et al. (2017) do exclusively focus on geometric aspects and challenges for BIM-GIS integration resulting from geometric and topological issues. These works carry out their analysis on the level of typical building model instances exported from CAD software and as such their observations pertain to issues introduced in the CAD authoring process or during processing and export in various software applications. The results are of high practical relevance but do not isolate issues resulting from incompatible standards.

To our knowledge there is no review of both ISO 10303-42 (2014) and ISO 19107 (2005) as implemented for building and city models. ISO 15926-3 (2009) specifies a geometric and topological model for the industrial automation of process plants and does indeed refer to both standards. However, even though its purpose is a very specific use case, it does not take any domain-specific standards into account, whereas we look at domain-specific subsets in IFC and CityGML. On the other end of the spectrum, McGlinn et al. do include various domain-specific standards even beyond IFC and CityGML, but they don't advance to the ISO level. Ding et al. (2020) carry out text mining on type names and descriptions.

1.3 Framework for schema correlation

In this paper, we mainly report the observations from the conversion project, but the ultimate goal is to extend this initial study towards a comprehensive analysis and documentation of spatio-semantic mismatches in IFC and CityGML and, beyond that, in the conceptual models of ISO 10303-42 (2014) and ISO 19107 (2005). To arrive there, we work in a bottom-up fashion on both domain sides:

Level 1 We start from geometries existing in common IFC4 files exported from CAD software. Hence, these are the geometries that conform to a specific model view definition (MVD), in this case reference view (RV). We may also cover design transfer view (DTV) and coordination view (CV) for IFC2x3.

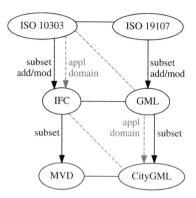

Figure 1. Relationships between MVDs, IFC, and ISO10303-42 and CityGML, GML, and ISO19107. The black continuous lines show relationship with regard to formal-structural aspects. The red dashed lines show the relationships with regard to the application domain.

On the CityGML side we start from those CityGML3 geometries we are actually creating, because they are required in given use case scenarios.
This corresponds to the two-fold approach taken in the conversion development where we come from both the requirements of the CityGML side and the IFC4 geometries encountered in typical building data to establish correspondence.
Level 2 From level 1 we extend the scope to look at the whole possible range of IFC and CityGML geometry definitions and extend the scope on the geospatial side to GML. CityGML as a specific application domain restricts the GML definitions.
Level 3 In final step we broaden the scope to the full ISO 10303-42 (2014) for the product model side and ISO 19107 (2005) for the geospatial side. It must be noted that at this stage, we not only look at supersets of the previous level schema definitions, but at specifications in a more intricate relationship. Both IFC and GML do indeed implement a subset of the broader conceptual geometry specifications, but at the same time add concepts or modify specifications.

It may seem more natural to look at IFC and ISO 10303-42 (2014) the same way as at CityGML and GML, because both embody a generalization of the application domain, but from the point of schema analysis, the structural aspect, hence the commonality of a subset/superset relationship between the product model and geospatial schemas being present (level 2) or not (level 3) is more relevant. Figure 1 shows these two ways to relate the standards on the various levels. The correlation along the application domain (red dashed lines in Figure 1 also corresponds to the level of implementation: GML is an XSD-implementation of ISO 19107, which contains the conceptual model, while ISO 10303-42 contains both a conceptual model and a specification in EXPRESS.

The following sections pertain to **Level 1** of this classification. Even more specifically, we are covering that flavour of the RV-MVD, which Revit exports for our sample projects and settings. In addition to these, there are geometric concepts that are used in synthetic test data sets which were created working "backward" from the CityGML requirements to the IFC inputs which would be needed to fulfil them. As such they don't correspond to existing MVDs but could establish a new geospatial MVD.

2 HOW TO EXPRESS SCHEMA CORRELATIONS

In this section, we describe how we approach the matching of the schemas, with the ultimate goal of integrating model instances. First of all, we can express the schema correlation through a simple mapping of corresponding types. Second, we can approach the mapping on a more sophisticated manner with Triple Graph Grammars.

2.1 Type correlations

Table 1, Table 2 and Table 3 show corresponding types that were extracted from our conversion process. The mappings in Table 1 cover the basic spatiosemantic concepts – solid spaces and semantic surfaces. These are commonly seen as two complementary spatiosemantic paradigms which pertain to the construction and geospatial domain respectively. Both concepts exist in IFC as well as CityGML, but on each side the main domain-specific paradigm has more consistently evolved and is mor mature.
The remaining mappings in Table 2 and Table 3 are divided into so-called "entry" and "exit" pairs. The "exit" pairs are the pairs, where a corresponding CityGML element is actually created during the conversion process. The "entry" pairs are only used to guide the conversion process. Two particular entry and exit pairs belong together in the way that the correlated types of one pair are subtypes or supertypes of the other pairs. The distinction will become clearer in the following section.
It is obvious that such a mapping is not sufficient to describe the correlation between schemas when we want to use the mapping for conversion, because it is not injective in any direction. For instance IfcPolyline has more than one correspondent on the CityGML side and likewise gml:Shell has more than one correspondent on the IFC side. Why is that? One reason is, that some mappings depend on the context of one of the two types which can't be captured with such simple mapping tables. Another reason is, that one of the models may be semantically richer than the other.
In the case of uni-directional transformation, the first reason applies to cases with varying mapping on the target side, while the second reason applies with varying mapping on the source side. In our example, the IfcPolyline type will be mapped to different types in CityGML, depending on its

Table 1. Correlations of spatio-semantic structure.

IFC types	CityGML types
IfcProduct	core:AbstractSpace
IfcRelSpaceBoundary	con:AbstractConstructionSurface

Table 2. Correlations of geometry from entry pairs.

IFC types	GML types
IfcConnectedFaceSet	gml:CompositeSurface
IfcConnectedFaceSet	gml:Shell
IfcCurveBoundedPlane	gml:MultiSurface
IfcFaceBasedSurfaceModel	gml:MultiSurface
IfcIndexedPolyCurve	gml:MultiSurface
IfcPolygonalFaceSet	gml:MultiSurface
IfcShapeRepresentation	gml:MultiCurve
IfcShellBasedSurfaceModel	gml:MultiSurface
IfcShellBasedSurfaceModel	gml:Solid
IfcSurfaceOfLinearExtrusion	gml:MultiSurface
IfcTriangulatedFaceSet	gml:MultiSurface

Table 3. Correlations of geometry from exit pairs.

IFC types	GML types
IfcCartesianPointList2D	gml:Polygon
IfcCartesianPointList3D	gml:Polygon
IfcFace	gml:Polygon
IfcIndexedPolygonalFace	gml:Polygon
IfcPolyline	gml:CompositeSurface
IfcPolyline	gml:LineString
IfcPolyline	gml:LinearRing
IfcPolyLoop	gml:LinearRing
IfcArbitraryClosedProfileDef	gml:Shell
IfcArbitraryProfileDefWithVoids	gml:Shell
IfcClosedShell	gml:Shell
IfcExtrudedAreaSolid	gml:Solid
IfcIndexedPolyCurve	gml:Shell
IfcManifoldSolidBrep	gml:Solid
IfcOpenShell	gml:CompositeSurface
IfcPolygonalFaceSet	gml:Shell
IfcRectangleProfileDef	gml:Shell
IfcShapeModel	gml:CompositeSolid
IfcShapeModel	gml:MultiCurve
IfcSolidModel	gml:Solid
IfcSurfaceOrFaceSurface	gml:MultiSurface

context. And simultaneously, the whole richness of surface representations on the IFC side collapses into gml:MultiSurface on the CityGML side.

Figures 2–4 show the mapped types in the context of the respective type graphs. Effectively, these are subgraphs of the IFC and CityGML type graph — and exactly the minimal sub type graphs that all instance graphs covered by our conversion process conform to. In other words, if our conversion (the geometry part of it) covers a particular IFC or CityGML instance graph, then this instance graph will conform to one of these three type graphs. These type graphs illustrate how we can think of mapping pieces of the type graphs instead of single types and attributes.

These graphs help to understand the m:n relationship between types, beyond the simple type mapping such as carryid out by Ding et al. (2020). Let us return to the 1:n and m:1 examples. It can be seen from the graph in Figure 1 that the IfcPolyline type appears in different roles: directly as member of the IfcShapeRepresentation.Items, as IfcSurfaceOfLinearExtrusion.Curve, or as IfcCurveBoundedPlane.OuterBoundary. These different roles (and thus different contexts in the graph) yield different mappings to CityGML types as can be seen in Figure 5 which will be explained in more detail in the next section. For the various IFC-counterparts of gml:Shell, the issue is different (Figure 6). Here the context on the CityGML side is identical, but the richer semantics on the IFC side yield the various mappings. The graph in Figure 4 reveals this. Even though the mapped IFC types share some semantic commonalities (that's why they can be mapped to the same CityGML type in the first place), they do not have a common supertype that is specific enough to warrant a combined mapping. What we see here, is one of the potential sources for schema mismatches: different classification and thus generalization/specialization of types on the source and target sides.

2.2 *Triple graph grammar*

The correlation between two typed graphs can much better be described with triple graph grammars (TGG), where a triple graph consists of two graphs of different type with their correspondence graph. A triple graph grammar specifies a whole set of consistent triples, hence all sensibly linked instances of two schemas, by production rules. The advantage of such a grammar is that it can serve as the base for conversion in both directions as well as synchronization and generation of the correspondence graph. The two main graphs are usually denoted as source and target graph, but these identifiers are absolutely random and only serve the purpose of clear distinction between the directions (forward and backward) in bi-directional conversion.

We have shown how this method can be applied for IFC-to-CityGML conversion and how operational forward conversion rules can be derived from the general TGG rules (Stouffs et al. 2018). We have implemented a web application to create and maintain rulesets (Tauscher 2019) and a conversion client to apply rule sets to given IFC input.

For the actual implementation of the conversion we have, however, directly specified the forward transformation rules and as such all the rules given as illustration for the mismatches are forward transformations, e.g. those shown in Figures 5 and 6. These rules can be read as instructions of what to look for in a triple graph (preconditions, light grey) and what to produce (postconditions, dark grey) during application of a rule (aka production).

It can be seen that these rules expose a fairly regular pattern (even though some rules are actually more complex and simplified for the illustrative purpose).

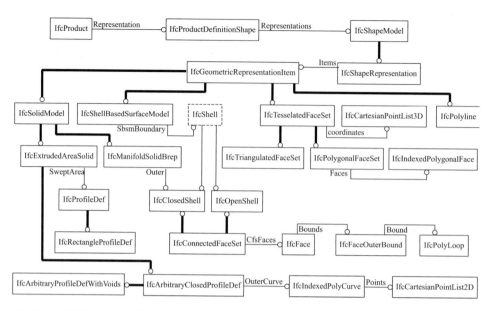

Figure 2. Covered IFC type graph for product geometry.

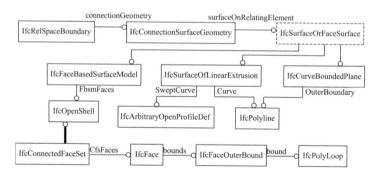

Figure 3. Covered IFC type graph for space boundary geometry.

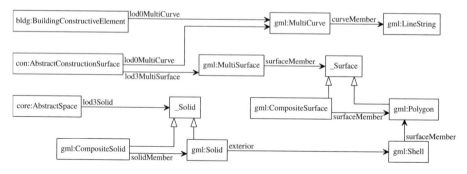

Figure 4. Covered CityGML type graph for both semantic surfaces and 3D spaces.

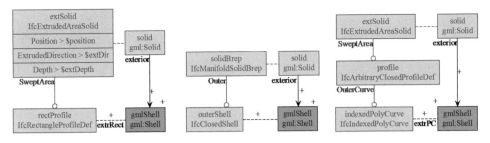

Figure 5. Different IFC graph templates mapping to the same CityGML type gml:Shell.

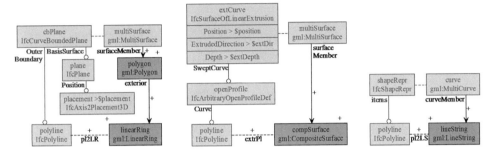

Figure 6. The same IFC type IfcPolyline mapping to different CityGML types depending on its context.

This pattern consists of an IFC-CityGML-pair on top of the rule which is supposed to already exist and an IFC-CityGML-pair at the bottom with an IFC node supposed to already exist and a CityGML node to be created. These are the entry and exit pairs mentioned earlier. It can be imagined how two such rules "fit together" for successive application when the concrete types of a pair in the instance graphs (the match) are subtypes of the types in both the exit pair of one rule and the entry pair of another rule which is applied thereafter. See Figure 7 for an example of such an application chain.

2.3 Layers of correlation

We correlate the concepts and thus design our rules on different levels.

The spatio-semantic rules define how geometry and topology are combined with semantics. There can be various configurations with regard to the sought dimensionality and spatio-semantic paradigm. The spatio-semantic paradigms (as described in Section 2.1) appear as different entry points to the geometry mapping on the IFC and CityGML side (Tauscher & Stouffs 2019). The two competing paradigms are reflected in Table 1 and the type graphs: Figures 2 and 3 pertain to the details of each paradigm on the IFC side (top-left-most types), whereas Figure 4 shows the paradigms for CityGML (three left-most types).

The main bulk of rules relevant for this paper is concerned with the actual geometry, which has its semantics defined in the ISO standards for geometry in product and geospatial models.

Some part of the correlation is currently hidden in so-called "converters", where part of the conversion process is programmatically described in an imperative or functional style instead of declarative as through the rules (Tauscher 2019).

Here we mainly focus on the first two layers, but still have to look into the converters to get a complete picture.

3 MISMATCHES

In this section, we describe some of the difficulties and issues found during conversion between the IFC and CityGML geometries.

3.1 Solid constraints

A preliminary study on the quality of IFC data and the impact of errors on conversion has identified validity of solid geometry as one of the main issues (Biljecki & Tauscher 2019), but has not considered where these constraints are defined and whether consistent geometry can be achieved at all. Here, we found that solid constraints do not match in the specifications of IFC and CityGML geometry standards.

Figure 7 shows a typical combination of three rules which would be applied successively during a forward transformation. The solid condition on the IFC side is defined for a IfcRepresentationItem, here occuring as IfcExtrudedAreaSolid. Hence, while the representation type (solid model) is specified for Ifc-ShapeModel, the actual condition is only on the level

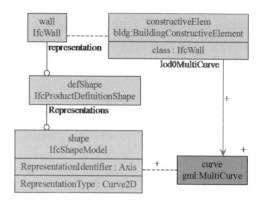

Figure 8. Wall axis representation.

3.3 Cardinality mismatch in aggregation relation

One quite common issue is that of varying cardinality for otherwise corresponding concepts on both sides. As an example, look at the objectified IFC relation IfcRelAggregates which seems to correspond to CityModelMember in CityGML. A closer look, however, reveals that cardinalities are not lining up at either side, thus it is not possible to relate these two types.

4 CONCLUSION

We have shown how triple graph grammars help to understand and represent the correlations between two different conceptual models such as the geometries in IFC and CityGML in particular or product and geospatial models in general. In addition, we have described some of the issues we observed in mapping IFC and CityGML with TGGs.

4.1 Types of mismatches

We can classify these mismatches, first of all whether they directly pertain to the geometry and topology or to the spatio-semantics, that is how geometries are charged with semantics, to which non-geometric concepts they are attached to and in which way. We can further classify mismatches with regard to the nature of the mismatch:

– missing concepts on either side (Results in incomplete mapping.)
– semantic granularity of concepts (How coarse/fine is the specialization?)
– granularity of instances (e.g. IfcSpace footprint versus lod0MultiCurve of semantic surfaces)
– matching structure with slightly deviating semantics

Even when only looking at the number of geometry types in IFC and CityGML, it becomes clear that there is an imbalance. In CityGML 2.0.0 (2012), p.247, an overview of the employed GML3 geometry classes is given — and these fit on one page. Much less types existing on the CityGML side to begin with, but in addition, we have reduced coverage due to our goal of a unidirectional conversion. We would need larger

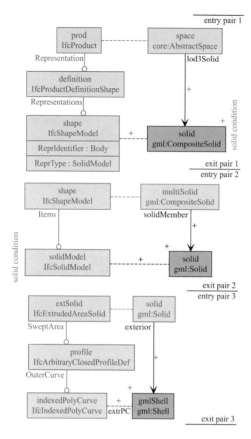

Figure 7. Three rules to map an extruded area solid.

of particular items of the representation. This is also backed by a formal rule which ensures that the items are of particular type. There is no constraint, however, that the whole set of items must be a solid again. On the CityGML side this is different, because the LOD3 geometry must mandatorily be a solid. Thus if we were to combine such a representation from multiple items, then it must be a gml:CompositeSolid. There is also gml:MultiSolid with relaxed solidity constraints, but this is not derived from the gml:Solid type.

3.2 Floorplan geometry

One of the floorplan representations suggested in Konde et al. (2018) requires axis representation of walls. Rooms are abstracted to points with a rough location and not geometry. This is not possible with CityGML neither for the walls nor for the rooms, the attribute shown in red in Figure 8 is not available.

As a workaround, a semantic surface could be introduced between the wall and the geometry, which sits in the center plane of the wall, but semantically this would be sloppy because the semantic surface would need to be attached as boundary to the wall, which would be wrong even if two surfaces with opposite orientation in the same plane would be used. A cleaner solution would use an ADE to add this element.

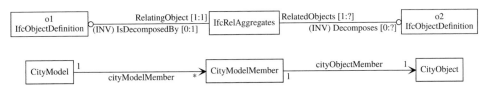

Figure 9. Mismatch in cardinalities: 1:1 on the left of the objectified relationship in IFC, but on the right in the respective candidate in CityGML. Note that in EXPRESS-G, names and cardinality is noted at the declaring end, while in UML (used for the CityGML graph presentation) multiplicities are notated on the declared side. We try to help the reader by placing corresponding naming and constraints either below or above the association line.

schema-coverage when attempting backward conversion, depending on the nature and scope of CityGML input.

4.2 Semantic concepts in the different standard layers

With regard to the different layers of standards as laid out in Figure 1 we can notice that a concept that does not map well on the lower level may well have a counter-part on a higher level.

For instance, concepts in an MVD (or in IFC) may relate to GML, but are not in CityGML. Same the other way around: Concepts in CityGML (or in GML) may relate to IFC, but are not in the MVD. In such cases one solution would be to define an MVD or ADE as needed to reintroduce the respective concepts.

4.3 Future work

In future work the relation between the different levels must be carved out more clearly and the correlations between the construction and geospatial domain be completed by working the way up from including correlations from converters on level 1 up to levels 2 and 3.

Another goal is the conversion of the rules to proper TGG rules, including rewriting of converters as proper rules down to the level of constraints. This will bring out the correspondences more clearly and eliminate skews due to the conversion direction. On that base, it can be formally analysed how schemas and subschemas are covered by a correlation or grammar.

REFERENCES

Arroyo Ohori K., Biljecki F., Diakité A., Krijnen T., Ledoux H., Stoter J. (2017): Towards an Integration of GIS and BIM Data: What Are the Geometric and Topological Issues? In: *Proceedings of the 12th 3D GeoInfo Conference*, University of Melbourne; ISPRS, Melbourne, Australia, 2017.

Biljecki F., Tauscher H. (2019): Quality of BIM–GIS Conversion. In: *Proceedings of the 14th 3D GeoInfo Conference 2019*, pp. 35–42, Singapore, 2019.

CityGML 2.0.0 (2012): *OGC City Geography Markup Language (CityGML) Encoding Standard* G. Gröger, T.H. Kolbe, C. Nagel, & K.-H. Häfele (Eds.). Open Geospatial Consortium (OGC).

Ding X., Yang J., Liu L., Huang W., Wu P. (2020): Integrating IFC and CityGML Model at Schema Level by Using Linguistic and Text Mining Techniques. *IEEE Access*, Vol. 8, pp. 56429–56440.

ISO 10303-42 (2014): *Industrial Automation Systems and Integration. Product Data Representation and Exchange. Part 42: Integrated Generic Resource: Geometric and Topological Representation.* International Organization for Standardization, Geneva, Switzerland.

ISO 15926-3 (2009): *Industrial Automation Systems and Integration. Integration of Life-Cycle Data for Process Plants Including Oil and Gas Production Facilities. Part 3: Reference Data for Geometry and Topology.* International Organization for Standardization, Geneva, Switzerland.

ISO 19107 (2005): *Geographic Information: Spatial Schema.* International Organization for Standardization, Geneva, Switzerland.

Konde A., Tauscher H., Biljecki F., Crawford J. (2018): Floor Plans in CityGML. In: *ISPRS Annals of Photogrammetry, Remote Sensing and Spatial Information Sciences: Proc. Of the 3D GeoInfo Conference*, pp. 25–32, Delft, Netherlands, 2018.

McGlinn K., Wagner A., Pauwels P., Bonsma P., Kelly P., O'Sullivan D. (2019): Interlinking Geospatial and Building Geometry with Existing and Developing Standards on the Web. *Automation in Construction*, Vol. 103, pp. 235–250.

Noardo F., Biljecki F., Agugiaro G., Ohori K.A., Ellul C., Harrie L., Stoter J. (2019): GeoBIM Benchmark 2019: Intermediate Results. In: *Proceedings of the 14th 3D GeoInfo Conference 2019*, pp. 47–52, Singapore, 2019.

Stouffs R., Tauscher H., Biljecki F. (2018): Achieving Complete and Near-Lossless Conversion from IFC to CityGML. *International Journal of Geo-Information (IJGI)*, Vol. 7, No. 9, p. 355.

Tauscher H., Stouffs R. (2019): Extracting Different Spatio-Semantic Structures from IFC Using a Triple Graph Grammar. In: Intelligent and Informed: 24th Annual Conference of the Association for Computer-Aided Architectural Design Research in Asia (CAADRIA 2019), pp. 605–614, Wellington, New Zealand, 2019.

Tauscher H. (2019): Creating and Maintaining IFC–CityGML Conversion Rules. In: *Proceedings of the 14th 3D GeoInfo Conference 2019*, pp. 115–122, Singapore, 2019.

City and building information modelling using IFC standard

V. Shutkin, N. Morozkin & V. Zolotov
Ivannikov Institute for System Programming of the Russian Academy of Sciences, Moscow, Russia

V. Semenov
Ivannikov Institute for System Programming of the Russian Academy of Sciences, Moscow, Russia
Moscow Institute of Physics and Technology, Dolgoprudny, Russia
Higher School of Economics, Moscow, Russia

ABSTRACT: The modern construction industry is moving towards the wide adoption of Building Information Modelling (BIM) technologies and their further evolution for City Information Modelling (CIM) problems. As these concepts are developed to solve multidisciplinary problems, open standards IFC and CityGML are becoming increasingly important. Despite numerous attempts to integrate data driven by these standards, software scalability and interoperability issues remain critical due to the complexity and heterogeneity of BIM and CIM models. In this paper an alternative approach to CIM is proposed and explored. The approach implies a hierarchical decomposition of a large CIM model into smaller models represented as linked IFC project datasets. The approach allows Levels of Detail (LOD) generation and visualization methods, provides high scalability for applications, and eliminates the need to harmonize the standards. This approach has been successfully validated by the development of a software application designed to visualize large-scale urban environments at different levels of detail.

1 INTRODUCTION

The modern construction industry is moving towards the wide adoption of Building Information Modelling (BIM) technologies and their further evolution and employment for City Information Modelling (CIM) problems.

The building information model is a digital representation of physical and functional characteristics of a facility. The model contains geometric information, time-related information (project schedule), resource and cost estimates, geographic information, and different quantities and properties of building elements. It simplifies the exchange of data between software applications and helps professionals to collaborate and make decisions throughout the whole life cycle of the facility: design, planning, construction, operation, maintenance and demolition (Eastman et al. 2008). Ultimately, BIM technologies decrease project cost and time requirements, increase productivity and quality (Azhar 2011). Due to the advantages, many governments require the use of BIM technologies for complex construction projects and large infrastructure programmes, and many companies have successfully adopted them in their businesses.

The generalization of BIM for city scale tasks such as urban planning and management is the city information modelling. City environments are very complex not only from an architectural, but also from a social and economic point of view. Therefore, city information models should be semantically rich, functionally comprehensive and accurate enough so that professionals can retrieve useful information from massive amount of data (Stojanovski 2013; Xu et al. 2014).

The city models obtained by laser scanning and photogrammetry technologies are widely used in Geographic Information Systems (GIS). The focus here is mostly on geometric representations of buildings and therefore such models do not fully meet the CIM requirements due to the lack of necessary semantic information. Luckily, meaningful and accurate information models exist in the BIM world.

Since both BIM and CIM concepts are developed to solve multidisciplinary problems by means of various software applications, open standards are especially important for the interoperable applications when exchanging and sharing common data.

The Industry Foundation Classes (IFC) developed by buildingSMART alliance is an open file format intended to represent and exchange architectural, building, and construction industry data. Many popular BIM tools can import and export IFC files, which makes the standard the best solution for BIM interoperability (Steel et al. 2012). The most prominent data model and exchange format for 3D city models and landscapes is the CityGML standard. It provides necessary semantic information and levels of detail (LODs) for geospatial and city objects (e.g. buildings, bridges, roads, rivers, vegetation and city furniture).

There have been many attempts to integrate IFC and CityGML for CIM purposes. The studies have focused on converting IFC building models to CityGML representations of a given LOD (Donkers et al. 2015; Floros et al. 2018), consolidating and cross-linking IFC and CityGML models (Vilgertshofer et al. 2017) and defining a median schema for conversion operations between IFC and CityGML models (El-Mekawy et al. 2011; Knoth et al. 2018). All these approaches have the same shortcoming: they either imply the conversion of models from one schema to another, which often leads to the loss of some data and their misinterpretation, or the necessity to work with two different schemas.

In this paper the possibilities of using the IFC standard for both building and city information modelling problems are explored. These possibilities are nontrivial for many reasons, the key of which is the need to manage city model complexity while using IFC files. Even for a single facility the size of the IFC file can reach up to several gigabytes, making it difficult to store, transmit, parse, validate and visualize. To address this issue, one could use file compression and geometry simplification (Hu et al. 2019). However, for CIM models, containing thousands or even millions of buildings and infrastructure objects, this opportunity is quite limited, and a more scalable approach is required.

The proposed approach is a hierarchical multi-level decomposition of a large CIM model into smaller models, i.e. the city is subdivided to districts, districts to blocks, blocks to buildings, buildings to building elements and so on. We assume that each model in the CIM hierarchy is driven by the IFC schema and obeys all of its formal and informal rules. Each model is an independent IFC project, which nevertheless may have cross-references to other projects using document references provided by the IFC schema. The cross-references enable to establish relationships between IFC projects and, thereby, allow to define the entire CIM model decomposition. In addition, cross-references can be used to link many other sources of information, including unstructured and semi-structured documents, which are often required for comprehensive information models of both buildings and cities.

It is worth mentioning that IFC files are based on the standard STEP Part 21 file format that allows to distribute a large dataset across multiple files. With this feature, recently introduced in the latest revision of the standard, IFC data can be stored in multiple files. However, such representation prevents independent operations with individual file data considering the need to maintain the integrity of the entire dataset as a single IFC project. The proposed decomposition approach is more flexible, since each model is a separate IFC project, regardless of data storage and representation.

In this paper, we argue that the proposed approach has the following advantages over the alternatives:

– there is no need to harmonize CityGML and IFC schemas and to resolve numerous interoperability issues for CIM and BIM tools;

– CIM decomposition hierarchies can be naturally combined with LOD (Levels of Detail) generation and visualization methods (Luebke 2001) and with Levels of Development (Latiffi et al. 2015). City, district, block and building levels of the hierarchy can represent levels of detail for the CIM model;

– proper scalability of CIM applications is reached due to the high fragmentation of the entire city model, and the ability to simultaneously load and process a limited number of child models stored in separate IFC files.

The presented approach has been validated by developing a software application and conducting a series of experiments with visualization of large-scale urban environments at different levels of detail.

This paper is structured as follows. Section 2 provides a brief comparison of the IFC and CityGML standards. The proposed CIM decomposition approach and the complementary IFC project linking mechanism are specified in Section 3. Section 4 clarifies how the model decomposition can be combined with levels of detail and development. The approach validation results are presented in Section 5. In Section 6 we conclude with the summary of the benefits of the proposed approach.

2 COMPARISON OF IFC AND CITYGML STANDARDS

Although IFC is primarily intended for BIM and CityGML — for CIM, they have much in common. Many researchers have focused on interoperability between IFC- and CityGML-compliant applications. They have identified common entities and proposed semantic mappings between IFC and CityGML schemas. Both schemas define data types to represent buildings and their elements, such as walls, doors, windows, roofs, etc.

Apart from that, CityGML focuses mainly on outdoor objects such as roads, bridges, tunnels, water bodies, vegetation, city furniture and relief (Xu et al. 2014), while IFC — on indoor environments. At the same time, significant extensions for IFC have been proposed and developed, following the BIM concept for infrastructure (Bradley et al. 2016) and including bridges (Yabuki et al. 2006), road structures (Lee & Kim 2011), and railways (Gao et al. 2016). Combined with support for geodetic Coordinate Reference Systems (CRS) (Markiè et al. 2018), this can render IFC applicable to CIM.

1An advantage of IFC over CityGML is the ability to represent very detailed and versatile information about buildings and infrastructure, including information about construction processes which is the basis for recent spatio-temporal (4D) planning and modelling technologies (Zolotov et al. 2014). At the same time, it should be recognized that in some cases, for example, in 3D cadastres, the use of CityGML may be preferable.

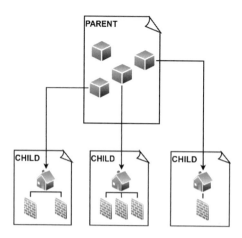

Figure 1. An example of two-level CIM decomposition.

3 LINKING AND COMPOSING IFC PROJECTS

As mentioned, the presented approach to CIM is based on the decomposition of parent models into child models, each of which is a separate IFC project dataset. A common mechanism for linking IFC projects is needed to represent parent-child relationships between models. For this purpose, it is proposed to employ the association structures already available in the IFC schema. These structures enable to associate any product, i.e. construction site, building or building element, with external documents, including IFC files.

Let us consider the proposed linking mechanism in more detail. IFC schema defines the relationship entity called *IfcRelAssociatesDocument*. This relationship allows to assign an external document reference (entity *IfcDocumentReference*) to an instance of type *IfcObjectDefinition*. These can be instances of the entities *IfcActor*, *IfcControl*, *IfcGroup*, *IfcProcess*, *IfcProduct* and *IfcResource*. Upon further consideration, we will focus on the product concept, which covers spatial structures defined by the *IfcFacility*, *IfcBuilding*, *IfcSite* entities and elements defined by the *IfcBuildingElement*, *IfcCivilElement*, *IfcFurnishingElement*, *IfcGeographicElement*, *IfcTransportElement* entities.

2The entity *IfcDocumentReference* has the following attributes: *Location* (an electronic location provided as an URI), *Identification* (a unique identifier of the referenced item within the external document in a form of a key, handle, UUID or GUID), optional human-readable *Name* and *Description*. Document metadata can also be captured through an auxiliary information defined by the entity *IfcDocumentInformation*.

The linking mechanism can be illustrated and clarified using a two-level decomposition of the CIM model presented in Figure 1. The root of the decomposition hierarchy is an IFC project corresponding to the entire city and containing simplistic LOD representations for all buildings of the city. The leaves of the decomposition hierarchy are IFC projects, each providing detailed information about a single building. To form the CIM decomposition hierarchy, so-called "parent-child" relationships between the root project and the leave projects must be specified.

For this purpose, buildings (*IfcBuilding* instances) of the parent and child projects can be associated with each other via adding a document reference (*IfcDocumentReference* instance) to the parent project. Its attribute *Location* must be initiated by the URL of the child project IFC file, its attribute *Identification* — by the globally unique identifier of the child project building (Figure 2). To interpret the document reference as a reference to the child project with higher LOD model, some additional assumptions and conventions are needed. For example, the availability of the child model can be declared using the attribute *Description* of the document. If it is complemented by an additional information (*IfcDocumentInformation* instance), the child model can also be declared using the attributes *Purpose* and *IntendedUse*.

Finally, an association relation (*IfcRelAssociatesDocument* instance) must be added to the parent project and initialized so that the building and the document reference are considered interrelated. The same linking mechanism applies not only to buildings, but to all concepts whose definitions are specializations of the *IfcObjectDefinition* entity.

The choice to put the association in parent project is typically the best option. Children are often self-sustained BIM projects; they don't have to be aware that they are parts of the CIM hierarchy. Processing of the hierarchies is typically implemented as traversal starting from the parent and descending to the children. However, if required by implementation, referencing parent from children can be established using the same mechanism.

Moreover, various specializations of the *IfcRelAssociates* entity can be used to link many other sources of information, including unstructured and semi-structured documents, classification items, approval information, etc. These capabilities are not prevented by the described linking mechanism but can also be valuable for BIM and CIM applications in need of comprehensive information.

The linking mechanism can help organize IFC product model libraries. Such libraries can be deployed and effectively used to model complex construction projects and large urban environments made up of typical products. For example, if the same building, furnishings, equipment or piping elements occur in BIM models repeatedly, they can be organized into the IFC product library and then reused by referencing from the IFC parent projects which represent these BIM models. Manufacturers can replenish the library with their own products, while BIM models can reference them. Figure 3 shows some of the mentioned ways of using the IFC project linking mechanism.

Each model of the CIM decomposition hierarchy must be a valid IFC project dataset that satisfies all the semantic rules and assumptions imposed by the IFC

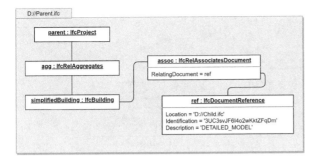
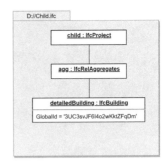

Figure 2. IFC object diagram illustrating the linking mechanism for parent and child projects.

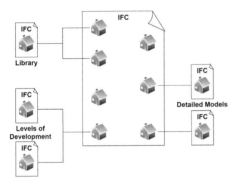

Figure 3. IFC project linking use cases.

standard. Each IFC project dataset can use own default units, coordinate systems, space dimensions, accuracies of geometric models and geospatial coordinate reference systems (CRS).

Therefore, a disappointing result can be obtained if IFC projects are linked directly. In such cases, the IFC project datasets must be pre-aligned so that the CIM decomposition models use the same reference data and its IFC projects remain relevant to the entire CIM model. This can be achieved, for example, by transforming the referencing data (e.g. sizes, coordinates, positions) of the child projects to the corresponding units, coordinate systems and accuracies of the parent project.

Sharing a common coordinate system and using absolute coordinates to position geometric objects in all IFC projects is the easiest, but the most burdensome way. If the underlying reference data has changed in the parent project, the referencing data of its child projects must be corrected and aligned with the parent project. A preferable approach is to use relative coordinate systems so that the objects of the child project can be modelled in local coordinate systems. These coordinate systems could be interpreted as implicitly related to the parent project and, thereby the child projects are positioned relative to the coordinate system of the parent project.

This approach has an obvious advantage if large-scale urban environments represented by typical facilities are modelled. In such cases, the same IFC files can be repeatedly referenced and reused to model similar facilities located on different sites, since their relative placements are specified in the parent project.

Let us recall that the relative placements in IFC datasets are specified using *IfcLocalPlacements* instances. They are assigned to sites (*IfcSite* instances), buildings, and building elements in such a way that building elements are set relative to buildings, buildings — relative to sites and sites — relative to a world coordinate system given by the *WorldCoordinateSystem* attribute of a geometric representation context (*IfcGeometricRepresentationContext* instance). The structure of typical IFC project is presented in Figure 4. The contexts are directly associated with a single project (*IfcProject* instance) through which the entire IFC datasets are usually traversed and processed. The proposed approach becomes feasible if the world coordinate systems of the child projects are interpreted as relative coordinate transformations applicable to particular products (*IfcProduct* instances) of the parent project, taking into account their local placements (properly assigned *IfcLocalPlacement* instances). It is also worth mentioning that, if needed, the mapping of projects coordinate system to geodetic CRS can be defined using *IfcMapConversion* entity.

The IFC project linking mechanism does not impose any principal restrictions on the CIM decompositions which can consist of more than two levels. As an example, the hierarchy presented in Figure 5 consists of four levels: the first level captures the entire city, the second corresponds to its districts, the third — to buildings, and the fourth — to building elements. In general, the number of levels can vary, models of the same level can belong to different product entities, and the decomposition hierarchy does not have to be balanced.

4 COMBINING MODEL DECOMPOSITIONS WITH LODS

CIM decomposition hierarchies can be combined with LOD (Level of Details and Level of Development) generation methods in such a way that higher models represent complex products with simplified specifications, and, conversely, lower models — simpler products with more detailed specifications. IFC standard does not impose restrictions on the detailing and

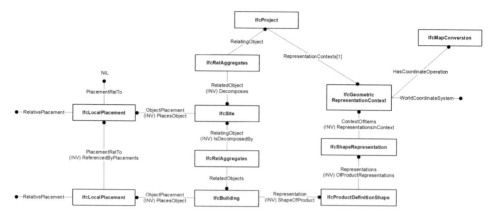

Figure 4. IFC schema diagram illustrating the entities intended for absolute and relative placements of products.

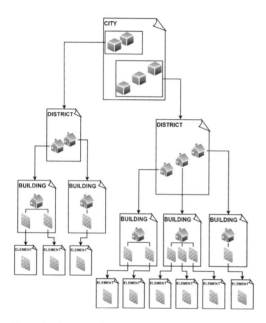

Figure 5. An example of four-level CIM decomposition.

therefore each decomposition model can be represented by a corresponding IFC file. As mentioned, information about referenced child projects and LOD models can be stored in parent projects and used by IFC-compliant applications to load and display the desired level of details and development for a modelled product.

Being represented as a single IFC project, consisting of billions of objects, the CIM model is unlikely to be effectively managed when it is processed in the main memory, stored in hard drives, exchanged between applications on the network or requested via IFC servers. Syntactic parsing, semantic validation, query execution, and visualization can require huge amounts of RAM, CPU, and GPU resources.

The main advantage of the proposed approach is the potential scalability of CIM applications, which can be achieved due to the high fragmentation and hierarchical decomposition of CIM models into relatively small models presented as separate IFC project datasets. When using the approach, there is no need to proceed with the entire CIM model. Instead, only its relevant models with limited dataset volumes should be loaded and processed simultaneously.

To illustrate the advantages let us consider an application intended for visualization of large urban environments. CIM models are suggested to be represented as two-level decompositions combined with an LOD technique such that each building has a simplistic tessellated shape representation in the parent project and a detailed geometric representation in the corresponding child project.

When observing the urban environment from remote locations (for example, to obtain bird's-eye views), only one parent IFC project with simplistic shape representations of buildings is required. Being loaded and rendered the representations enable to generate a series of images with the desired resolution and realism. When zooming to a particular building, IFC project linking mechanism is identifying a file containing its detailed data and the corresponding child project data is getting loaded additionally. There is no need to process the CIM model entirely to visualize the building against the urban background. For the use case under consideration, it is enough to process one parent project and one child project, which together contain a relatively small amount of data compared to the CIM model.

Thus, combining the decomposition approach with LODs helps manage the complexity of CIM models. In this regard, another problem arises of how to form CIM decompositions. For brevity, let us return to the two-level CIM decomposition, shown in Figure 1. It is assumed that the city model and building models are given by parent and child IFC files respectively. The child files can be generated by exporting BIM models to IFC files using popular CAD/CAM/CAE applications. The parent file can be obtained by converting CityGML representation of the city into IFC format and aligning the coordinate systems, placements and geometric models of both the parent and child projects with each other.

Because the exported child projects contain accurate geometric models of buildings, they can be used to obtain simplified geometric models needed for the parent project representation. The degree to which simplification of geometric models should be applied may depend on the complexity of the entire CIM model, e.g. the number of buildings, the total number of building elements, the total number of primitives in geometry models of building elements, etc. Obviously, proceeding with large CIM models is impossible without simplifications of the parent project, since it will be loaded and processed as a single dataset. From the opposite point of view, excessive simplification will lead to primitive shapes that will make the visual identification of buildings onerous. Thus, the complexity of the parent and child project models must be balanced to achieve both high application performance and acceptable quality of LOD models, in particular, the realism of visualised urban environments.

The topic of geometric and polygonal simplifications has been carefully studied by many researchers, and various methods have been proposed and successfully validated in computer graphics applications (Luebke 2001; Thakur et al. 2009). Most methods can be applied directly to the problems discussed if the geometric models are preliminarily aggregated into a single representation, which then undergoes simplification procedures. Special methods for urban scenes have also been developed (Salinas et al. 2015; Verdie et al. 2015).

It is worth emphasizing that not only geometric models should be processed, but also other IFC-driven data, so that LOD models remain significant for various AEC/FM disciplines. For example, applications of spatio-temporal (4D) modelling of construction projects, visual project planning and scheduling use nested representations of project tasks (so-called work breakdown structures) with assigned planned and actual time parameters (Semenov et al. 2013). The IFC schema defines all the data types needed to specify the nested project tasks. In particular, these are *IfcTask*, *IfcWorkCalendar*, entities and simple datatypes *IfcDate*, *IfcDateTime*, *IfcDuration,* etc. Some project management systems, such as Bentley Synchro, can export project plans and schedules to IFC files, so that the project tasks are related to building elements and the construction processes can be simulated in both spatial and temporal dimensions by designating the installations of building elements as planned, started or already finished. Child project data can be aggregated and simplified to become part of the desired spatio-temporal LOD model for the city parent project.

The easiest way is to bring the top work breakdown structures of child projects into the parent project and interrelate them with the corresponding simplified geometric models of buildings. Simulation of the entire city's evolution becomes possible due to the time parameters of the top work breakdown structures responsible for the construction of individual buildings throughout their makespans.

However, this method is not suitable for the relevant visualization of the intermediate phases of construction processes by storeys, spaces or in any other way. It can be overcome by presenting the construction processes for each building not only by one task and one related geometric model. Spatio-temporal clusters, each of which aggregates both a spatial model and its temporary behaviour, should be employed for such purposes. The method for automatic clustering of large pseudo-dynamic scenes has been also developed (Semenov et al. 2019). This allows to generate hierarchical dynamic levels of detail (HDLOD) and use them as simplified spatio-temporal models for those IFC project datasets which contain planning and scheduling information.

Finally, so far we have assumed that BIM models for urban buildings are available, and they may be subject to simplification procedures before being included in the city parent project. In practice, for many existing buildings, BIM models do not exist (Volk et al. 2014), however, 3D shape representations for buildings can be prepared using laser scanning and photogrammetry technologies. The decomposition approach allows some buildings and facilities to stay without more detailed or accurate models. In such cases, 3D shapes as geometric representations of buildings are placed directly in the parent project, and document references to child projects are not created. This use case is also shown in Figure 3.

Thus, the proposed decomposition approach is naturally combined with LODs, while covering various applications and use cases.

5 VALIDATION OF THE DECOMPOSITION APPROACH

To validate the proposed decomposition approach, a software application designed to visualize large-scale urban environments has been developed using C++ programming language and OpenGL API. A multi-window graphical interface of the application allows to visualize a CIM model associated with the urban environment in several 3D views from various cameras and with varying levels of detail. The user can explore the environment by using well-known 3D navigation techniques such as zooming, panning, rotating, as well as by selecting individual product elements and displaying their properties. This is achieved through the decomposition of the large CIM model into smaller models each of which is represented as an IFC project dataset and can be loaded and processed by the application as needed.

Once the user navigates close to the building (or its product element) with a link to the child IFC project with a more detailed model, that project gets loaded. When loading is finished, the detailed model replaces the original one. As the user moves away from the building (or its product element), the original

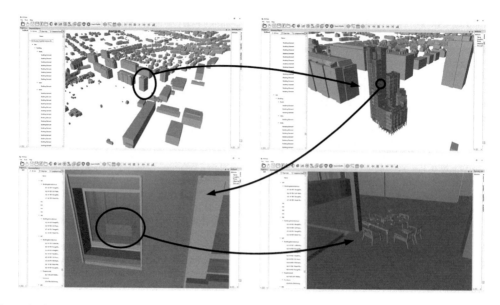

Figure 6. Screenshots of software application developed to visualize large-scale urban environments at different levels of detail.

model located in the parent IFC project substitutes the detailed model and the child IFC project dataset can be unloaded.

IFC files should be loaded and unloaded fast enough so that the user can smoothly navigate over the modelled environment without any interruptions. For this purpose, the application has been implemented in such a way that the files are loaded asynchronously in separate threads. In addition to commonly used STEP Part 21 file format, the application can save IFC project datasets in its own custom binary format, that corresponds to the internal C++ data model of the application. This enabled direct serialization and deserialization of the data and avoided CPU-consumable parsing and semantic validation.

Screenshots of the application are presented in Figure 6 to illustrate a scenario for navigating the Brussels CIM model. The model was prepared by converting a public city model available in the CityGML format (UrbIS Download) to an IFC file and using it as the top parent project of the CIM model. Several IFC files were prepared for individual urban buildings and were used as child projects. The parent and child projects were linked using the mechanism specified above. In turn, each building project was considered as a parent project for the furniture elements whose detailed geometric models were also preliminary converted to IFC files and linked with the corresponding building projects. Thereby, the CIM model was formed as a three-level decomposition hierarchy.

The developed application exhibited reasonable performance when navigating through the CIM model. At the same time, the model could be easily detailed and scaled by providing and linking more and more IFC files. This is a promising result for modelling large-scale urban environments.

6 CONCLUSIONS

Thus, the original approach to CIM problems has been presented in this paper. The approach implies a hierarchical multi-level decomposition of a large CIM model into smaller models, each of which is represented as an independent IFC project dataset. To form the decomposition hierarchy, the project datasets are linked with each other using the reference structures provided by the IFC standard schema.

Compared with other data integration methods, the proposed approach allows Levels of Detail and Levels of Development generation and visualization methods, provides high scalability for BIM and CIM applications, eliminates the need to harmonize the IFC and CityGML standards, simplifies the organization and reuse of product model libraries, and finally, can be extended with unstructured and semi-structured documents typically involved in comprehensive CIM models.

This approach has been successfully validated by the development of a software application designed to visualize large-scale urban environments at different levels of detail. Due to the decomposition of the city model and the proposed software optimizations the application exhibited high performance when navigating through the model. It proves the feasibility of the approach and makes it promising for the introduction to the AEC/FM industry.

REFERENCES

Azhar, S. 2011. Building Information Modeling (BIM): Trends, Benefits, Risks, and Challenges for the AEC Industry. *Leadership and Management in Engineering* 11(3): 241–252.

Bradley, A., Li, H., Lark, R., & Dunn, S. 2016. BIM for infrastructure: An overall review and constructor perspective. *Automation in Construction* 71(2): 139–152.

Donkers, S., Ledoux H., Zhao J. & Stoter, J. 2015. Automatic conversion of IFC datasets to geometrically and semantically correct CityGML LOD3 buildings. *Transactions in GIS* 20(4): 547–569.

Eastman, C., Teicholz, P., Sacks, R., & Liston, K. 2008. *BIM Handbook: A Guide to Building Information Modeling for Owners, Managers, Designers, Engineers, and Contractors*. John Wiley & Sons Inc., Hoboken, New Jersey, USA.

El-Mekawy, M., Östman, A., & Shahzad, K. 2011. Towards Interoperating CityGML and IFC Building Models: A Unified Model Based Approach. In Kolbe T., König G., Nagel C. (eds), *Advances in 3D Geo-Information Sciences. Lecture Notes in Geoinformation and Cartography*: 73–93. Springer, Berlin, Heidelberg.

Floros, G.S., Ellul, C. & Dimopoulou, E. 2018. Investigating Interoperability Capabilities between IFC and CityGML LOD 4 — Retaining Semantic Information. *The International Archives of the Photogrammetry, Remote Sensing and Spatial Information Sciences* XLII-4/W10: 33–40.

Gao, G., Liu, Y.-S., Wu, J.-X., Gu, M., Yang, X.-K., & Li, H.-L. 2016. IFC Railway: A Semantic and Geometric Modeling Approach for Railways based on IFC. Retrieved from http//cgcad.thss.tsinghua.edu.cn/liuyushen/vmain/pdf/GaoGe_ICCCBE2016_364.pdf.

Hu, Z.Z., Yuan, S., Benghi, C., Zhang, J.P., Zhang, X.Y., Li, D., & Kassem, M. 2019. Geometric Optimization of Building Information Models in MEP Projects: Algorithms and Techniques for Improving Transmission, Storage and Display. *Automation in Construction* 107: 102941.

Knoth, L., Scholz, J., Strobl, J., Mittlboeck, M., Vockner, B., Atzl, C., Rajabifard, A., & Atazadeh, B. 2018. Cross-Domain Building Models—A Step towards Interoperability. *International Journal of Geo-Information* 7(9): 363.

Latiffi, A.A., Brahim, J., Mohd, S., & Fathi, M.S. 2015. Building Information Modeling (BIM): Exploring Level of Development (LOD) in Construction Projects. *Applied Mechanics and Materials* 773–774: 933–937.

Lee, S.-H. & Kim, B.-G. 2011. IFC Extension for Road Structures and Digital Modeling. *Procedia Engineering* 14: 1037–1042.

Luebke D.P. 2001. A Developer's Survey of Polygonal Simplification Algorithms. *IEEE Computer Graphics and Applications* 21(3): 24–35.

Markič, Š., Donaubauer, A., & Borrmann, A. 2018. Enabling Geodetic Coordinate Reference Systems in Building Information Modeling for Infrastructure. *Proceedings of the 17th International Conference on Computing in Civil and Building Engineering*. Retrieved from https://publications.cms.bgu.tum.de/2018_Markic_ICCCBE.pdf.

Salinas, D., Lafarge, F., & Alliez, P. 2015. Structure-Aware Mesh Decimation. *Computer Graphics Forum* 34(6): 211–227.

Semenov, V., Anichkin, A., Morozov, S., Tarlapan, O., & Zolotov, V. 2013. Visual planning and scheduling of industrial projects with spatial factors. *Proceedings of the 20th ISPE International Conference on Concurrent Engineering*: 343–352.

Semenov, V., Shutkin, V., & Zolotov, V. 2019. Visualization of Complex Industrial Products and Processes Using Hierarchical Dynamic LODs. *Transdisciplinary Engineering for Complex Socio-technical Systems* 10: 655–664.

Steel, J., Drogemuller, R. & Toth, B. 2012. Model Interoperability in Building Information Modelling. *Software & Systems Modeling* 11(1): 99–109.

Stojanovski, T. 2013. City Information Modeling (CIM) and Urbanism: Blocks, Connections, Territories, People and Situations. *Simulation Series* 45: 86–93.

Thakur, A., Banerjee, A.G., & Gupta S.K. 2009. A survey of CAD model simplification techniques for physics-based simulation applications. *Computer-Aided Design* 41(2): 65–80.

UrbIS Download. http://urbisdownload.gis.irisnet.be/en.

Verdie, Y., Lafarge, F., & Alliez, P. 2015. LOD Generation for Urban Scenes. *ACM Transactions on Graphics* 34(3): 1–14.

Vilgertshofer, S., Amann, J., Willenborg, B., Borrmann, A., & Kolbe, T.H. 2017. Linking BIM and GIS Models in Infrastructure by Example of IFC and CityGML. *Computing in Civil Engineering 2017*: 133–140.

Volk, R., Stengel, J., & Schultmann, F. 2014. Building Information Models (BIM) for existing buildings – literature review and future needs. *Automation in Construction* 38: 109–127.

Xu, X., Ding, L., Luo, H., & Ma, L. 2014. From Building Information Modeling to City Information Modeling. *Journal of Information Technology in Construction* 19: 292–307.

Yabuki, N., Lebegue, E., Gual, J., Shitani, T., & Zhantao, L. 2006. International Collaboration for Developing the Bridge Product Model "IFC-Bridge". *Joint International Conference on Computing and Decision Making in Civil and Building Engineering June 14–16, 2006, Montréal, Canada*: 1927–1936.

Zolotov, V.A., Kazakov, K.A., & Semenov, V.A. 2014. Effective spatial reasoning in complex 4D modelling environments. In A. Mahdavi, B. Martens, & R. Scherer (eds), *eWork and eBusiness in Architecture, Engineering and Construction: ECPPM 2014*: 181–186. London: CRC Press.

Environmental, social and economic dimensions of sustainability

Evaluating the concept and value of smart buildings for the development of a smarter procurement strategy

J. Olsen
Client Advisor, Rambøll Denmark A/S, Copenhagen, Denmark

J. Karlshøj
Head of Section DTU BYG, Department of Civil engineering, Technical University of Denmark, Kgs. Lyngby, Denmark

ABSTRACT: The digital revolution has created new smart technologies like, smart phones, smart services, contactless payments, that have changed the way people live and interact with each other. These technologies have created a frictionless lifestyle for the modern people. These features and technologies are available because of 'big data' approaches and Internet of Things (IoT). Today the building industry is moving towards building designs with more complex and integrated systems. Drivers for this change are efficiency, convenience, cost savings, energy savings, enhanced communication etc. This research is focusing on a strategy for procurement of smart buildings with the aim of highlighting potentials and challenges with the collaboration of the stakeholders in the building industry especially with the focus of bringing the facility management in focus and defining critical requirements in the early design phase.

1 INTRODUCTION

1.1 The evolution of smart buildings

The design and expected performance of buildings have been changing throughout history. Newly constructed buildings often try to minimize environmental impacts through best practice design solutions by, for example, minimizing energy sources and appropriate material selection but also in order to score highly on different types of green building rates, including certificates. Sustainable buildings have often been characterised with the ability to comply with different metric systems such as LEED, BREEAM and DGNB, which cover environmental, social and economic dimensions of sustainability (Ghaffarianhoseini et al. 2017). In the building industry, these types of buildings are considered to be energy efficient. Great focus has been placed on trying to maximize the energy efficiency of buildings and, for some people; a smart building is synonymous with an energy efficient building (Jadhav 2016). In fact, the definition of what constitutes a smart building by BPIE is a building which is highly energy efficient (Groote el al. 2017).

A green building rating has, for many building owners, measured the success of development and design and has, in many cases, shown that environmentally friendly practices can be good investments that can generate higher incomes for building owners. The building industry has tried to push the boundaries of the sustainability agenda, which has been the mind-set shaping visionary buildings over the past two decades (Jadhav 2016).

Today, the building industry is emerging towards a new era of smart buildings, where 'best practices' are no longer enough, and design must apply 'next practice' design solutions. Alongside good environmental sustainability are high-performance buildings being designed to prioritise user experience and well-being with the goal of boosting work productivity and creating an attractive environment (McCraven 2017). These are developments on previous visions of the era of environmental performance that pushed for full operational efficiency, but this vision is now emerging and developing into one that looks to promote the experience of the building user. This new era that the building industry is entering may be described as an era with the central focus on user experience and functionality in the building.

There is, at the moment, no agreed shared definition of a smart building and this makes it difficult for stakeholders to grasp the extent of smart buildings.

2 A SMART BUILDING DEFINITION

2.1 Definitions of a smart building

This research shows that stakeholders in the building industry have different expectations of defined

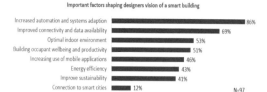

Figure 1. Research results from smart building definition survey.

goals and definitions for smart buildings. Many definitions often reflect the values of specific companies and organisations' visions and strategies. For the purpose of this research, a survey has been conducted to gather insights into what the building industry defines as a smart building. The results are shown in Figure 1.

The graph shows that designers focus significantly on automation and the implementation of Internet of Things (IoT) devices to improve indoor air quality, reduce energy consumption and to interact with building occupants. Some designers mention the connection to smart cities and sustainability. A similar research by Verdantix based on real estate inputs states that real estate is more focused on space utilisation than designers are, but both industries acknowledge the importance of improving occupant well-being and productivity (Yate 2017). The increased use of mobile applications is becoming a vision from both stakeholders, thus a vision of building users being more in control of their environment seems to be emerging.

2.2 Moving towards a smart building definition

The framework of a smart building is based on its ability to operate intelligently, adapting, learning, connecting, communicating and be sustainable. An important factor is adaptability which can be achieved with the use of IoT devices acting with the building systems combined with an external control signal.

More important than ever is to future proof buildings. The building systems need to be adaptable so they can accommodate, for example, new set points automatically based on data analytics which may even mean smart city integration. Trends show that the focus is on how the building occupants are experiencing the building. User-centred innovation is facilitated when humans, technology and business come together (Indo-German Smart Initiative 2017). The users should define what they want, thus what is desirable to have in the building. The facility management and designers must find feasible designs to accommodate this. However, the business economics need to be part of the equation to make sure that the design solutions are viable in terms of obtaining the expected return on investment for the building owner or facilities management organisation.

2.3 A new smart building definition

A definition for smart buildings characterising the value proposition and expectations from key stakeholders is necessary in the building industry. A lack of clarity of what constitutes a smart building will potentially create confusion rather than a direction, as companies and organisations start to embed their own values in the definitions.

Based on studies of scientific papers, information available from the market, industry talks and interviews with professionals from the AECO industry, a smart building definition is proposed as follows:

"A smart building is a building that incorporates data collection, analysis, asset control, and action to respond to defined goals for a particular outcome. It supports the potential of emerging, cutting-edge digital technologies to promote a user-centric, sustainable and continuously adaptable design for an optimised high performance with a responsive environment that is attractive and stimulating for the occupants."

The goal with this proposed definition is to initiate a discussion regarding a single standard definition of a smart building, sharing the expectations from all stakeholders in the building industry.

3 STRATEGY FOR SMART BUILDING PROCUREMENT

Collaboration between stakeholders in a building project is essential in order to develop a smart building with integrated systems that can adapt to the occupants' needs, but this is not easily solvable with the current process of designing buildings. The current process is not prioritised for communication and collaboration, in fact it is designed more for contractual agreements.

Smart buildings will only be able to deliver the expected outcomes and benefits that the market promise if the smart building is specified, procured and commissioned properly (King 2017).

Building owners have the potential of taking the lead in creating a valuable development of smart buildings and start requesting the integration of, for example, digital tools and ICT agreements with more focus on the operation phase allowing for better construction, smarter operation and better services for occupants. It will prove necessary to change the incentive structure in the building value chain in order to change the current short-term view of the operation and maintenance of buildings to one which focuses instead on the long-term. Most building owners stand alone and isolated as the only stakeholder with an obvious incentive to handle the operation in the design phase and one of the issues is that building owners have little opportunity to share the necessary costs with those who benefit from the improvements (Olesen et al. 2018).

The general purpose of a procurement process is to select the bidder best able to deliver the services required by a client within the target budget which

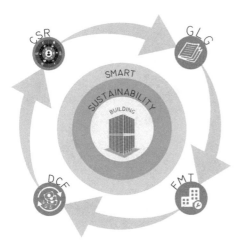

Figure 2. A strategy for a smart building procurement to incorporate sustainability and smart in buildings with key stakeholders in mind.

often refers to capitalisation of the investment for real estate (RICS Facilities Management Professional Group Board 2018). A successful procurement process must include clear descriptions of the objectives of the goals and requirements from the client, only then will all stakeholders have a clear understanding of the requirements.

A concept of how to incorporate sustainability and smart digital solutions in a building and how to include stakeholders collaboratively starts to form a circle starting with the needs of the clients, as illustrated in Figure 2.

3.1 Corporate Social Responsibility (CSR)

A clear vision with the concept of operations and expected outcomes from a building must be clearly stated by the client to create a smart building. In the planning process of procurement, it is important to clearly understand the objectives that reflect or support the wider strategy of the organisation (RICS Facilities Management Professional Group Board 2018). Corporate Social Responsibility (CSR) refers to a company's social responsibility towards, for example, the environment, CO2 and user comfort. This means that CSR is part of the company's business strategy with social initiatives meant to create social and economic value for customers, employees, and potential shareholders (Redlein & Zobl 2014). This includes focussing on initiatives that contribute to the sustainable development of society including the environment, comfort, social conditions, and climate. The CSR is defined by a company and initiatives supporting this should be part of the requirements that are expected for the building.

Research shows that employees today place far greater requests on a decent working environment and sustainability (Imperatori 2017). Furthermore, expectations are rising for smart technologies in the workplace (Clarke & D'Arjuzon 2018). For companies, it is essential to focus on the workplace to keep on attracting millennials and talents. An attractive workplace environment may ultimately mean whether a person will work for a business or not. The increasing interest in creating smart and sustainable buildings with a more people-centric focus rather than engineering, construction, and technology-centric focus, has increased the awareness of understanding the use of buildings in new ways and sustainable practices (Jadhav 2016).

3.2 Green Lease Guide (GLG)

The initiatives that are agreed in a dialogue with the client must be written into a lease contract. This means that a building must be able to explain through a lease agreement, what influence it has on sustainability and user comfort or how it supports the strategy of the organisation. This must be documented in order to answer which initiatives have been agreed within the lease. It is important to specify the requirements using clear language to make the evaluation easier when it comes to contract management. For this, a Green Lease Guide (GLG) can help. A GLG is a guide that gives instructions on how green initiatives are written into a lease. A GLG can consist of checklists with strategies for green and social initiatives that can be implemented in a building and these actions are often referred to and consistent with sustainability standards and certificates. An example is the GLG developed by Investas. This GLG includes some aspects that are expected of a smart building such as a comfortable, productive and healthy indoor environment, but also aspects such as reporting to tenants on core building environmental performance (Investas 2007).

3.3 Facility Management Technical company (FMT)

In order to provide a solution to the requirements that have been written into a lease contract, a Facilities Management Technical Company or organisation (FMT) must provide a service and make these requirements a reality in the building. Often, a company takes on a lease and it is the professionals from the FMT discipline who ensure that the building can live up to the lease contract. It is important that enough data is provided to allow bidders to understand the requirements from the client (RICS Facilities Management Professional Group Board 2018). The operation of a building is, in general, a process of descriptions of what the client is asking for and agreements by the supplier that services can be provided to support these requirements. Regarding requirements for reducing energy consumption in a building, it is worth mentioning Energy Service Companies (ESCO), which is a form of collaboration to ensure the reduction of energy consumption. ESCO partners guarantee that the agreed energy savings are real and can be documented and compensate for any lack of energy savings compared with the guaranteed energy savings. This has become a trend in the green building industry and has become useful for securing building performance and a step into sustainable and smart buildings (Jadhav 2016).

It is important that designers and contractors give the building owner and facility management the tools and documentation to implement smart strategies. Smart building specialists and facility managers need to be part of the early design stages of a building design to affect the strategic development of the smart building (King 2017). These stakeholders need to be able to affect the critical requirements for the operation of the building, thus a smart building strategy and a brief containing these requirements must be incorporated early in the planning process to ensure the successful design of a smart building.

The process does not end at the handover stage of the building as the legacy must continue to create value in the whole value chain. The specifications for the performance of a building changes over time, user requirements change, even the purpose of the building might change. Smart voting with IoT devices will be important in this context to measure, for example, atmospheric temperature and occupant patterns to ensure that the building stays attractive.

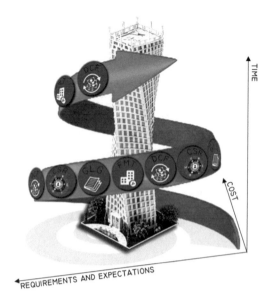

Figure 3. Smart building procurement strategy in a spiral containing the aspects of building life cycle and stakeholder's involvement.

3.4 Discounted Cash Flow (DCF)

Discounted Cash Flow (DCF) refers to a real estate agent talking about the capitalization of an investment and this must yield a return before the building is considered attractive. If the expected return cannot be achieved, then it is not considered a good investment. This goes back to CSR, as the DCF focuses on what interest ultimately comes to the real estate agent with the initiatives and technologies that are implemented in the building. However, suppliers also have a great influence on this. A solution might be viable with the right suppliers in the procurement process.

Smart buildings place greater requests on flexibility and adaptability, the individual user's experience in the building and transparency of the building's consumption. The building industry would, with the concept shown in Figure 2 and a possible extension of the Green Lease Guide or by introducing a new Smart Lease Guide, be able to deliver on the needs for smarter buildings. Just as ESCO partners guarantee that the agreed energy savings can be documented, IoT sensors and the right service platforms for facility management and occupants would also be able to document and visualize that the building's consumption and indoor air quality are guaranteed.

Moving on to the life-cycle phases of a building, it is also important to recognise the nature of changes in the requirements of the building through the design phase. Dr. Arto Kiviniemi suggests that the iterative nature of the design process and the usually large number of changes during the process increase the complexity of the problem (Kiviniemi & Fischer 2014). He argues that the process should be described as partly parallel activities, including requirement management through the whole process and in several stages.

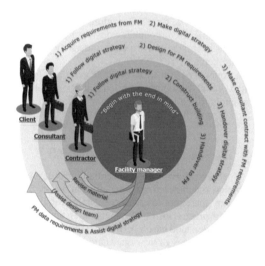

Figure 4. A concept of "begin with the end" to bring the facility manager in play.

Requirements and expectations in a project evolve during the design and construction phases. Taking this into account, the concept shown in Figure 2 evolves from a 2D graph to a 3D graph including time, expectations and costs. As the expectations for the building evolve, so do the costs. The concept is shown in Figure 3.

If the process is tied together in this way, then the strategy of the client's organisation expressed in the CSR will be integrated with the delivery requirements that the facilities management organisation must then deliver. With the building owner's capitalisation of the investment as part of the process, the integration and innovative initiatives proposed based on the CSR, vision and strategies within areas such as sustainability and digitalisation can be obtained to a greater extent. The concept seems simple, but it contributes to an expanded dimension of the challenges of procurement of smart and sustainable solutions in buildings.

The concept works for new buildings but also for retrofitting existing buildings which might be the smart buildings of tomorrow. In new buildings, the scope of the brief must be clear. The challenge is to avoid the temptation to over-specify with, for example, digital technologies, but also making sure that the buildings are not obsolete when handed over to the operation phase.

In existing buildings, the challenges are going to be integrating new and old technology. In this scenario, designers and facility managers are not starting out with the same blank canvas as with new buildings, since there are constraints with existing buildings that they need to work around.

To properly pursue the concept of procurement of smart solutions in buildings, the occupant or user experience of the building should be considered during the initial stages of any new construction or major retrofit project (Machinchich 2018).

4 ACHIEVING CLIENT GOALS

"Data created in the planning and design phase of a building must be utilized throughout the whole value chain so that it creates value both in the construction phase and in the subsequent operational phase. Only then will we get the full potential of digitisation and create value for the customers in the building and construction sector" Michael H. Nielsen, Director of the Confederation of Danish Industry (DI).

The built environment needs to understand the challenges which arise with the loss of knowledge and information during the handover of a project or when moving on to a new phase (Eastman et al. 2011). One of these challenges is that, during the design phase, designers fill models with a significant amount of information that is not relevant to the operation phase but is relevant to the design phase (Ridley 2018). Equally important is to understand the challenge of not specifying enough information about the operation in the design phase. Clements-Croome has identified the causes of failure in building design (Clements-Croome 2018)

- An inadequate brief and specification
- A genuine lack of data regarding certain aspects
- Poor communication between the client and the design team and between members of the design team themselves
- Inadequate analysis or synthesis of solutions
- Unpredictable quirks of human behaviour by users
- Poor selection of equipment, inadequate installation or facilities management

Considering the principle "begin with the end in mind" seems to be an important approach for building designs to achieve successful smart buildings. The concept is shown in Figure 4. It will be difficult, if not impossible, to reuse data from each phase with inadequate briefs and poor communication between the stakeholders involved (Clements-Croome 2018).

Generally, the problem is a lack of clear requirements regarding the implementation of FM data requirements throughout the early stages of a project and one of the biggest challenges is to define the data requirements (Yalcinkaya & Singh 2014). The role of the facility managers could potentially be revolutionised in this era of smart buildings.

The role of facility manager is crucial for the operation phase and it is anticipated that future facility managers will be included in the early design process to participate with the design teams and represent the interests of the building owner and future occupants (Imperatori 2017). If the traditional role of facility manager, of simply taking over the already completed construction continues, then there is little scope for facility managers to influence the efficiency of the building (Pašek & Sojková 2018).

If designers and building owners start thinking about operational data in the design from the outset, then it will be possible to make a connection to facility management and therefore to smart buildings. Doing this enables designers to design for the real reason for building design, which is to design for people. This will be a step towards the era of smart buildings with a user-centric focus (Clements-Croome 2018).

4.1 Description of services

In the Danish built environment, many companies use the document template "Description of Services – Building and Planning 2018" as an agreement between clients and designer to ensure services with the purpose of creating consistency in the delivery of projects. This document was developed by the Danish Association of Consulting Engineers (FRI) and the Danish Association of Architectural Firms (The Danish Association of Consulting Engineers and The Danish Association of Architectural Firms 2018). The document is not a mandatory requirement but can be used optionally in building projects as a requirement for different services. The document works as a template that follows the traditional phase model, to define which services to expect in different phases.

The Danish Association of Construction Clients (DACC) could potentially publish a similar document template, so that clients in the industry would start making greater requests of what should be delivered in buildings, for example, operations. Today, what clients set by way of requirements for the operation of the building in terms of digital delivery is limited, meaning often consultants do not have rules or specific guidelines specified by the client for the operation (Gullberg 2018).

One of the challenges for clients when requesting new smart solutions for the operation of their buildings is that many of the new digital solutions have not been sufficiently tested and the facility management organisations do not have the opportunity to approve solutions before the tender is delivered (Gullberg 2018). Below is section 1.1.2 of the new "Description of Services – Building and Planning 2018" that is relevant for the operation phase.

The brief must contain the client's specific requirements and wishes for the commissioning of the building, testing of technical installations and installations and operation (The Danish Association of Consulting Engineers and The Danish Association of Architectural Firms 2018).

This section states that the requirements of the operation of the building should be considered early in the design process and preferably in the brief. This can potentially increase the value creation for all stakeholders in the building industry and be the key to designing smarter buildings. In the future, revisions of the "Description of Services – Building and Planning" or a potential document developed by DACC could include more focus on smart solution implementation and strategies as part of the agreement.

5 CONCLUSION

A smart building is designed to deliver enhanced performance in areas of user-centric, sustainable, and continuously adaptable design as well as supporting organisational goals. The evolution of smart buildings indicates that buildings are becoming a space that is responsive to users in order to promote a stimulating and attractive environment. To do this, emerging cutting-edge technologies supporting data collection, analysis, asset control and activities which allow highly optimised performance of the buildings.

The findings in this research show that the smart building is designed and delivered around the client's visions and strategies. This process should incorporate FM and building users from the early beginning of the project to determine specific requirements.

Smart buildings increase the complexity of designing buildings with integrated systems, thus making a smarter procurement and 'next practice' design solutions even more important.

REFERENCES

Clarke, S. & R. D'Arjuzon 2018. Smart Building Technology Global Survey 2018: *Facilities Optimization Management Brands*. Verdantix.

Clements-Croome, D. 2018. The role of feedback in building design 1980–2018 and onwards. *Building Services Engineering Research and Technology* 40, 1–8.

Eastman, C. et al. 2011. BIM Handbook: A Guide to Building Information Modeling For Owners. Wiley.

Ghaffarianhoseini et al. 2017. Intelligent or smart cities and buildings: a critical exposition and a way forward. *Intelligent Buildings International* 10 (2), 122–129.

Groote el al. 2017. Smart buildings decoded. [Online]. Available: http://bpie.eu/wp-content/uploads/2017/06/PAPER-Smart-buildings-decoded_05.pdf. [03-09-2018].

Gullberg, D. 2018. Digital aflevering, byggeri – hvad tænker du, når du læser det? [Online]. Available: https://www.linkedin.com/pulse/digital-aflevering-byggeri-hvad-t%C3%A6nker-du-n%C3%A5r-l%C3%A6ser-det-gullberg/. [06-01-2018].

Imperatori, B. 2017. Engagement and Disengagement at Work: What's New, pp. 5–18. Springer.

Indo-German Smart Initiative 2017. Integrated urban development and co-production for indian cities. [Online]. Available: https://www.igsi.info/. [15-01-2019].

Investas 2007. Green lease guide for commercial office tenants (2 ed.). Investas.

Jadhav, N. 2016. Green and Smart Building Trends, Springer Singapore.

King, O. 2017. Smart Working – Smart Buildings and the Future of Work. UnWork.com Ltd.

Kiviniemi, A. and Fischer, M. 2014. Requirements Management Interface to Building Product Models. VTT Publications.

Machinchich, T. 2018. The holistic view of a commercial building needs to expand. [Online]. Available: https://www. navigantresearch.com/news-and-views/the-holistic-view-of-a-commercial-building-needs-to-expand. [02-01-2018].

McCraven, M. 2017. Keeping up with the evolution of buildings part 3 – buildings of the future. [Online]. Available: https://blog.schneider-electric.com/building-management/2017/07/12/keeping-evolution-buildings-part-3-buildings-future/. [26-12-2018].

Olesen, T., Larsen et al. 2018. Skyld eller ansvar? hvem er bannerfører for driftsøkonomien? [Online]. Available: http://www.dk-gbc.dk/nyheder/seneste-nyt/building-green-messen-i-koebenhavn-4-okt-1-nov-2018/. [02-01-2018].

Pašek, J and Sojková, V 2018. Facility management of smart buildings. International Review of Applied Sciences and Engineering 9, 181–187.

Redlein, A. & Zobl, M. 2014. Contribution of facility management to sustainability and corporate social responsibility. In ScieConf 2014.

RICS Facilities Management Professional Group Board 2018. Procurement of facility management. Royal Institution of Chartered Surveyors (RICS) and International Facility Management Association (IFMA).

Ridley, J. 2018. Digital twins are shaking the foundations of the property industry. [Online]. Available: https://www.linkedin.com/pulse/digital-twinsshaking-foundations-property-industry-joshua-ridley/. [06-01-2018].

The Danish Association of Consulting Engineers and The Danish Association of Architectural Firms 2018. Description of Services – Building and Planning 2018. The Danish Association of Consulting Engineers and The Danish Association of Architectural Firms.

Yalcinkaya, M. and Singh, V. 2014. Building information modeling (bim) for facilities management – literature review and future needs. Fukuda, S. et al. (Eds), IFIP International Federation for Information Processing, PLM 2014, IFIP AICT volume 442, 1–10.

Yate, I. 2017. Global Real Estate, Energy & Facilities Survey 2017: Budgets, Priorities & Tech Preferences. Verdantix.

From linear to circular: Circular Economy in the Danish construction industry

T.S. Rasmussen, R.J. Esclusa, E. Petrova & K.D. Bohnstedt
Aalborg University, Aalborg, Denmark

ABSTRACT: The construction industry relies predominantly on the take-make-dispose principle. That contributes significantly to resource depletion and creates a considerable negative environmental impact. Therefore, a paradigm shift is necessary, which helps achieve sustainable processes and minimises the negative contributions. Circular Economy is based on the principle that the value of products, materials and resources is maintained as long as possible by reusing and thereby minimising waste. However, Circular Economy is still to be adopted at a larger scale. This paper presents a methodology for assessment of the circularity level of AEC organisations and enables the discovery of specific actions which can lead the way to a circular industry. The method provides a measure of the current state of the company processes in accordance with five Circular Economy pillars. The results show that implementing Circular Economy is associated with motivating the companies to take action, strengthening collaboration, and implementing circular processes.

1 INTRODUCTION

Human activities are constantly increasing the pressure on the global climate and resources, and rapid resource depletion is now a reality. Currently, the available resources are being spent at a rate of 1.75 times what the Earth can regenerate (Global Footprint Network 2019). A trend which also includes negative contributions from the Architecture, Engineering and Construction (AEC) industry, which accounts for more than 35% of the total waste produced in Denmark and 40% of the raw material usage in the European Union (EU) (Danish Environmental Protection Agency 2019). Besides being a significant resource consumer, the AEC industry is also responsible for one third of the global CO_2 emissions and 40% of the total energy use in Europe (International Energy Agency 2013).

As is the case with many other industries, the AEC industry relies on the take-make-dispose principle. An unsustainable principle, as most developments are driven by the extraction and use of new resources and raw materials. Therefore, an alternative approach is required to achieve sustainable processes and minimise the negative contributions, i.e., one utilising new methods and technologies to design new industry standards that fulfil the world's current and future needs.

Circular Economy (CE) has been proposed as a solution to the goal mentioned above, as it offers a new way to acquire and use resources. Currently, most of the efforts within the industry are focused on managing raw materials and process outputs (Silva et al. 2019). That includes studying products (Mangialardo & Micelli 2018), managing supply chains to minimise wasted resources (Nasir et al. 2017) or revolutionising techniques and processes in the building life cycle (Finch et al. 2019, Finkbeiner et al. 2019). However, all these efforts are still to be adopted at a larger scale, which is what is needed to change the industry and create a better future (Nuñez-Cacho et al. 2018).

1.1 *Research objective*

In response to the outlined challenges, this study aims to create a methodology for the assessment of the levels of circularity of AEC companies. The purpose is to aid industry representatives in discovering and acknowledging the steps which can lead the way to a circular construction industry, i.e. steps which companies can tailor to and implement in their processes.

To be able to identify possible new CE initiatives, it is necessary to investigate and understand the current state of the industry, in terms of levels of circularity, collaboration efforts, as well as the constraints for CE implementation. Thus, we investigate the potential for CE implementation in the Danish construction industry and the steps needed to bridge the gap between the current system and a circular one. Ultimately, the focus is on the possibilities for the companies to improve their level of circularity and thereby contribute to a circular industry. Efforts, which are already being addressed actively, both on national and European levels (European Commission 2016).

CE is based on the principle that the value of products, materials and resources is maintained as long as possible by reusing and thereby minimising waste (European Commission 2014). As previously outlined, current research focuses mostly on the physical resources and materials involved in the construction process, including how improvement and recycling of materials in the industry can be facilitated, and how this can be done more sustainably and efficiently. However, resources such as time, money and labour are usually overlooked due to the importance of raw materials. Yet, they are also significant factors in the improvement of the sustainability and circularity of the industry. Furthermore, the use of CE principles for improving inter-organisational processes and minimisation of the waste of time, financial and labour resources is still unexplored in research. This paper, therefore, looks into resources and waste in all aspects of the AEC processes. This allows a new perspective on how to improve processes and help companies and the AEC industry implement the CE principles.

The remainder of the paper is organised as follows. Section 2 presents the adopted definition of CE in this paper and the related CE pillars. Section 3 describes the methodology adopted in this study. Section 4 presents the results from the qualitative data collection and analysis. Section 5 outlines the developed methodology for assessment of the level of circularity of AEC companies. Section 6 discusses potential recommendations for implementation of CE principles in the industry based on the performed assessment. Finally, Section 7 concludes the paper.

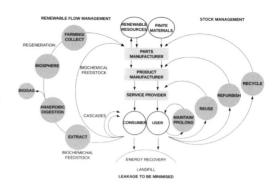

Figure 1. Circular Economy Cycles, after Ellen MacArthur Foundation (2013).

2 DEFINITION OF CIRCULAR ECONOMY

In this paper, CE has been defined following the principles set out by the Ellen MacArthur Foundation (2013) related to repair, reuse, recycling, etc., under the frame of sustainability. This CE definition is outlined in five pillars covering the following five areas.

1. Design Out Waste (Optimisation)
   This pillar is concerned with minimising the waste created during the development of projects. In this context, waste can be "anything", e.g. materials, time, money, resources, etc. Therefore, this pillar concerns the optimisation of the design and the related processes so that the waste can be minimised. This first pillar will be referred to as Optimisation throughout the remainder of the paper.
2. Build Resilience Through Diversity (Diversification)
   Resilience is understood as the ability for a company to recover from, or easily adapt to, changes. Resilience is achieved by having a broader range of options when faced with a problem. The ability to adapt will allow companies to survive and even thrive during changing circumstances.
3. Shift to Renewable Resources (Renewability)
   This pillar is about removing the dependency on fossil fuels and shifting to renewable energy sources. The pillar looks at both the energy input and output in terms of the use of material waste for the creation of renewable energy (factors in the context of the renewability of the company as a whole).
4. Think in Systems (Systems Thinking)
   To think in systems within a company or a project means to take a holistic approach, i.e., each activity is considered in connection with the surrounding ones, rather than as an independent activity, thereby creating links between the decisions made in one activity to the next. Fulfilling this pillar requires to consider the entire industry as a unit, and not each actor as an individual piece. This will create optimised processes and a better understanding of the different parts of the project, ultimately creating optimal project results.
5. Think in Cascades (Recycling)
   This pillar relies on the notion that the waste generated by a project of one company constitutes the resources needed by another. Therefore, to complete this pillar, the companies must have understanding of how they can recycle materials in their processes and how their waste can be recycled by them or others to optimise the resource use in the industry. This pillar aims for creating the highest value from every stage in the life cycle of a resource or material. The recycling of materials is the core of this pillar.

In this study, CE is considered to be of higher value, the tighter the circle of reuse is (see Figure 1). This means that the highest level of circularity is achieved when 1) a resource/product is maintained and repaired, followed by 2) its reuse and/or redistribution, then 3) the refurbishment or remanufacturing and finally the 4) recycling at the lowest level. For example, the best approach from the perspective of circularity is to repair a building before any other steps, the next level providing value would be to refurbish it, and finally, the recycling of the materials of the building is the lowest level of circularity, while still an improvement to the prospect of demolishing and disposing (see Figure 1).

Finally, the definition of circularity adopted in this paper also assumes that for something to be circular, it also has to be sustainable. Not all circular measures are sustainable, as sustainability has to consider all aspects of the acquisition and production of the materials, not just the reuse of existing materials. This is determined in accordance with the overall aim, which is to create a more sustainable industry through the implementation of CE. Therefore, there is no purpose in investigating measures, which are circular, but not sustainable.

These prerequisites apply to ensure the appropriate focus of the research and to evaluate only measures, which are deemed beneficial to the CE development of the industry. Therefore, the outlined CE definition is used to define a method that is both implementable and consistent.

3 METHODOLOGY

Measuring the circularity of companies is an assessment which can be used to motivate change and facilitate the CE transformation in and between companies. Using a reliable measurement of the circularity enables the understanding of possible measures, and the effects of these, that can be established for any company. It's thereby necessary to evaluate the current status of AEC companies before they can improve or start implementing CE measures. Thus, an assessment method is developed, which can also be used for self-evaluating and can create a basis for comparison within the industry.

To achieve the above-mentioned, this paper relies on a literature review and qualitative interviews to develop the methodology for the assessment of the circularity level. The literature review allows for establishing the gap in the research connecting CE and company processes and provides insights into the definitions, understandings and tools defined in CE research. The starting point for the study is the assumption that the AEC industry is not particularly circular, which is validated through the literature review.

Second, qualitative data is collected to develop further the understanding of the current situation in the AEC industry. This includes an understanding of the existing processes and issues AEC companies experience in moving toward circularity. The data is collected through semi-structured qualitative interviews, held with different AEC discipline representatives. The interviews are performed in three stages. The first stage is an interview conducted to create an understanding of each discipline, company and the industry in general. This round forms the basis for comparative analysis and categorisation of the research. The first interviews also aim to investigate further the hypotheses established from the literature review. The approaches *Coding, Content Analysis,* and *Meaning Condensation* are used in the analysis to apply a flexible approach which enables the capturing of essential understandings from the raw data.

The initial round of interviews is based on a set of identical questions for all companies. The obtained information is collected separately for each company to later determine trends within the separate disciplines and areas of the industry.

The second round of interviews focuses on digging deeper into the outlined issues in each company. These interviews are performed to point to specific solutions that can be implemented in a broader spectrum for similar companies, fitting into their goals and values. The interviews are unique for each company, based on the results of the first interview.

The third round of interviews discusses possible solutions and establishes a conversation with each company for the assessment of options and feasibility. The collected data is analysed and mapped to the CE pillars defined in Section 2. Finally, the results are used to establish an assessment mechanism for the level of circularity.

A method which is both unbiased and reliable is necessary to ensure that the process can be recreated and reused in other companies, and even other industries, with similar results. The creation of an unbiased method is crucial within this research, as it is based on qualitative data and has to ensure a valid and reliable measuring process which can be applied repeatedly. The data collection and analysis outlined in the following section has been developed to ensure exactly such a process.

4 DATA COLLECTION AND ANALYSIS

The data used in this paper is collected from a sample of nine companies representing the AEC industry in Denmark. This sample consists of two architectural companies, one engineering company, four general contractors and two clients. The companies have been chosen to achieve a realistic reflection of the AEC industry. The profiles of the companies are as follows.

Architect 1: A medium-size architectural company, a front runner in regards to sustainability and circular economy approaches. The company focuses on developing innovative solutions to enhance the implementation of sustainability in the industry.

Architect 2: A large conventional architectural company, focused on large scale projects, mainly within the public sector.

Engineer: A large company which works with engineering projects and client advising. The company maintains a focus on innovation in the industry and on developing new methods for efficient project development.

Contractor 1: A regional contractor company which only handles the management of projects and has no employed tradesmen. Additionally, the company focuses on large-scale projects.

Contractor 2: A construction company with a focus on the development of public projects while enforcing social sustainability into the work processes.

Contractor 3: A large contractor company with the ability to develop and innovate in project development. It is a significantly well-established company which mainly works with large public projects

Contractor 4: A contractor company working solely with prefabricated wooden constructions. The company focuses mainly on off-site construction, i.e., manufacturing building blocks in a factory which are assembled on-site.

Client 1: A housing union which develops and manages assets on behalf of the residents. The housing union has a focus on sustainability, but it is very limited by the tried and tested, as the company acts on behalf of clients who are not within the trade.

Client 2: A private developer with the ability to invest in project development as the company deems fit. The focus of the company remains on making developmental decisions which will benefit the company both short-term and long-term depending on the final ownership of the project.

The three rounds of interviews described in Section 3 were conducted with each of these companies, thereby obtaining different perspectives on the areas of concern in the research.

The questions for the first round of interviews are developed on the basis of hypotheses extracted from the literature review. These hypotheses are: 1) private companies can benefit from circular economy and can push the system (Ellen MacArthur Foundation 2013); 2) there are no 'one-size-fits-all' solutions (Ellen MacArthur Foundation 2013, Mont & Heiskanen 2015, Elia et al. 2017); 3) companies in construction have a linear approach (Adams et al. 2017, Pomponi & Moncaster 2017, Leising et al. 2018); 4) there is a surplus of individualistic company processes (Adams et al. 2017, Leising et al. 2018); companies are short-sighted (Lieder & Rashid 2016, Adams et al. 2017, Thelen et al. 2018); and, larger companies have a long-term approach so they innovate to stay relevant (Thelen et al. 2018).

The collected data is analysed with the purpose of establishing understanding and generalisation of findings based on the interviews. Coding allows for attaching keywords (codes) to the text (transcribed interviews) to identify statements or opinions. This makes it possible to prepare the information for categorisation (Kvale 2007). Content Analysis enables the quantification of the data. Through the grouping of codes, it is possible to count each code just once, regardless of the frequency of it being mentioned in the interview. This way, the data is reduced into a few main pillars. Thus, through categorising the codes into pillars, the long qualitative interviews can be summarised into a few tables and figures, which create the basis for comparisons and assumption testing. Finally, Meaning Condensation is used to analyse the data by condensing full sentences and long statements into brief ones. This approach helps the process, as it simplifies the search for the natural meaning and clarifies main themes within the raw data (Kvale 2007).

This analysis approach is taken throughout both the first and the second round of interviews, where understanding, issues, possibilities etc. are expressed in codes. Ultimately, the collected data is mapped to the five pillars of CE, as defined in Section 2: (1) Optimisation, (2) Diversification, (3) Renewability, (4) Systems Thinking and (5) Recycling.

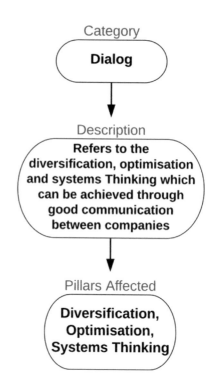

Figure 2. Example of the process of mapping interview findings to the CE pillars.

This categorisation is completed by including the coding results in a table and assessing which pillars are related to these codes, as one code can be related to more than one pillar. Codes are grouped and quantified under each pillar, giving way to a scaling system, which is the basis for company analyses. An example of this approach is provided in Figure 2.

The analysis gathers data points for each company in each of the five pillars. This is expressed through the collection of data points in a table where each code is aligned with the categories and results in a number of points in the relevant pillars. For instance, a code regarding the lack of feedback loops results in adding to the categories Linear, Processes and Dialog, which then add points to the pillars. As a result, the pillars offer an insight into the main problems experienced by the companies. The issues faced by each company can be compiled and used to develop tailored solutions and improvements.

5 A METHODOLOGY FOR ASSESSMENT OF THE CIRCULARITY LEVEL OF AEC COMPANIES

The developed method provides a measure of the current state of the company processes in the form of a scale measuring this state in accordance with the five CE pillars. Measuring the circularity of companies on the different pillars will create grouped measures which each company can use to focus on the problematic areas they experience. This means that the scales will allow companies to focus their efforts where they will be the most effective and create the highest impact. The ultimate objective of this approach is to offer alternative targeted recommendations towards improving the circularity of companies and thereby achieve a more circular industry. With the retrieved data, a few things should be noted for each company that is analysed. First, a clear image of the current situation within the company and within the overall industry should be created. Furthermore, the issues and obstacles experienced by the company should be identifiable through the ratings of the pillars. Finally, with consideration to the CE pillars, it should be clear what direction the company can take to improve the level of circularity.

The analysis underlines which CE pillar(s) should be the most beneficial for the company to focus development on. The results of the analysis are expressed visually with the scales for each of the nine companies in Figure 3.

In the assessment scale, the letters on the X axis refer to the first letter of each of the five CE pillars (see Section 2). Each number on the Y axis is defined in the following way: (5) Optimal: implies complete circularity, (4) Great: great practice but room for improvement, (3) Good: circularity in a developmental level, a lot to improve on, (2) Positive: shows intent or capability to implement circularity, (1) Positive Unawareness: potential to implement, but unaware of practices, (0) No answer, (-1) Negative Unawareness: practices with potential to stifle circularity, (-2) Negative: resistance to the implementation of circularity, (-3) Bad: practice of linear thinking at a developmental level, (-4) Damaging: practice of linear thinking and processes at a company root level, alternate solutions are not considered, and (-5) Abysmal: linear processes negatively influencing the whole company, with a disregard to outside practices and new process developments.

The scale illustrates the scores for each company in each of the five CE pillars. The assessment provides an overview of the company, which can serve as a basis for decision-making concerning CE implementation. Also, the scale provides a visual way to compare the company score to that of other companies, to the trends and averages within the different discipline groups of actors, and within the industry as a whole.

For example, it can be beneficial for a company to understand if they are significantly at a disadvantage within one pillar compared to other companies in the same discipline. This will increase the understanding of their competitive advantages and disadvantages.

Based on the synthesised information, a list of main issues within the industry is also established. These are issues and barriers related to CE implementation pointed out directly in statements by the interviewees and through the extracted information from the interviews. The barriers are related to the following main categories: 1) Motivation (governmental pressure); 2) Communication (collaboration amongst the different stakeholders in the industry), and 3) Industry Efforts

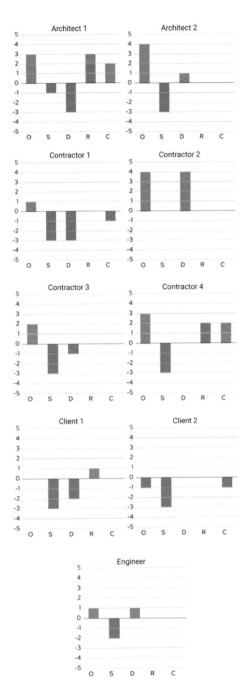

Figure 3. Results of the nine-company sample analysis.

(available tools which have not been implemented at a larger scale yet).

6 DISCUSSION

The collected data allows for identifying trends in the industry related to the implementation of CE principles. These include a generally low score in Systems

Thinking in the industry overall, which correlates with the expressions of low collaboration and fragmented project processes by the interviewees. Additionally, a high score in Optimisation can be seen in the group of the architectural companies, which correlates with their position in the value chain and the ability to influence the project development. Even though these trends are clear in the collected data, it is necessary to perform the study with a far more substantial test group in order to draw any deeper conclusions with regards to the overall and grouped trends in the industry. Additionally, these trends are an excerpt from the Danish construction industry. Results may vary, if the method is applied internationally or regionally.

6.1 *Recommendations*

Based on the developed understanding of the main issues and barriers related to CE implementation (Motivation, Communication and Industry Efforts), recommendations targeting potential solutions are created. The recommendations rely on two main criteria: Practicality and Low cost of implementation. The criteria selection is based on the collected data, which pointed to concerns in regards to large investments and uncertainties in the implementation processes.

Based on the understanding of the industry established through the literature review and the interviews, the Industry Efforts will only be implementable and efficient in improving the industry sustainability when the communication within the industry has been improved. Additionally, the Motivation needed to push the CE in AEC the industry must come, primarily, from the government, as it was stated by most participants in the initial assessment. Based on these observations, the recommendations which this paper focuses on and which have been presented to the interviewed companies, are within the Communication area.

6.2 *Communication*

The recommendations aim to improve the communication between the actors in the industry (in companies and even across industries) and thereby improve the industry as a whole. This area is further divided into four different Communication subcategories: Information Management, Collaboration, Standardisation and Strategy. These are created to establish main focus areas for the analysed companies. The Communication subcategories are all correlated with the pillars outlining the company issues. Through this correlation, any company can, through the understanding of their own issues, identify the Communication subcategories in which the best solutions for them can be found.

6.3 *Connection to CE pillars*

Each of the Communication subcategories is connected to one or more of the CE pillars defined in this paper. The connection between the recommendations and the issues will aid the companies in identifying appropriate steps which can be taken to improve the circularity of the company. The division into subcategories helps to identify a list of possible solutions for each company, which connect directly to the issues that have been identified. Thus, it is possible to not only apply solutions that work, but also to focus the resources of the companies most efficiently as the scale ranks the impact of the issues within each pillar.

The Communication subcategories are connected to the pillars as follows. Information Management is connected to, most importantly, Diversification, but also Optimisation, Systems Thinking and Recycling. That is because this category evaluates how information should be attained, kept and shared. Improving this category enables companies to collaborate better, thereby improving on the mentioned pillars.

Additionally, it increases the possibility for improvement of processes and interactions between companies. Possible solution areas within this subcategory are: (Internal) Information Sharing, Post-Mortem Evaluation and Use of Open Data Models.

Collaboration answers to issues within the pillars of Diversification and Systems Thinking. These pillars can be improved upon by the use of solutions within this category as here the value chain is considered as a whole. The solutions within this category improve the communication with other actors and thereby create collaborations, which resemble teams more than groups. Furthermore, they improve the relations in the industry and the possibility to work with new partners, thereby improving the Diversification. Examples of solution domains within this category are: Partnering, Co-location, Shared Digital Models.

The category Standardisation mainly answers to the pillar Optimisation but also touches on the pillar Recycling. The category improves the Optimisation pillar through the focus on increasing efficiency by standardising. Through standardisation, the work processes can be optimised, and the waste can be minimised as a result. The related recommendations can also improve the Recycling pillar through the reuse of processes. The potential solutions, which answer to this category are, among others: Timeframe model for the Involvement of Actors, Early Involvement of Contractors, Standardising the Reuse of Processes.

Finally, Strategy connects to the improvement of the pillar Diversification through communication, as it enables the company to branch out and become more resilient through different types of projects. In addition, it connects to the pillar Systems Thinking, as it focuses on how to include the client as a part of the value chain and thereby influence the entire system. The relevant recommendations and possible solutions hereby include: Client Workshops, Developing Proof of Feasibility and Life Cycle Analysis.

It is recommended that for a company to gain the optimal benefit from this analysis, appropriate solutions within the category which answers to the lowest scoring pillar for the company are identified and applied, as this will improve the company by the most significant margin. As seen in the analysis,

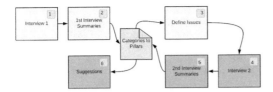

Figure 4. An example of the analytical process of extracting information from the collected data and transforming it into recommendations for the companies.

each of the Communication subcategories connects to multiple pillars and the solution areas overlap to a high degree. This also means that when improvements are made within one area, in order to better one pillar, the other pillars will also be affected. This underlines the importance of understanding which areas will benefit the most from being improved, e.g. which pillar has the lowest score, as improving it will give some improvements to the remaining pillars and thereby an improvement of the entire organisation.

Figure 4 demonstrates the general approach to identifying the potential solution domains and recommendations to each company. For instance, in case of a low score in the pillars Systems Thinking and Diversification, it is recommended that the companies look into Communication solutions related to Information Management, Collaboration and Strategy.

Based on the insight into the company, specific solutions which will have the biggest potential and the easiest implementation should be chosen as the initial focus areas. Steps such as working in Co-Location with other project participants and using Net Present Value Calculations to convince clients of the benefits of the circular model can highly benefit the companies and improve the lowest scoring pillars. For example, the solution Co-Location answers to the Communication subcategory Collaboration by improving the ease of collaboration. The Collaboration category, as mentioned, improves on the Systems Thinking and the Diversification pillars. Through the improvement of these, the company as a whole will improve their circularity levels, and push the possibilities and perspectives of the industry by proving that the Circular Model is not only possible but also beneficial.

Ultimately, the pillar Renewability did not have as much relevance as the other pillars in the improvement of the company process circularity levels, as this pillar refers to the energy sources of the companies and not the company processes themselves. Therefore, it is not discussed further in this analysis.

7 CONCLUSION

This paper provides a method for assessment of the level of circularity of AEC organisations. The method can be used to gain valuable insight into the internal company processes, aid CE implementation and improve the circularity level of the industry as a whole.

The development of a score for circularity within the five different CE pillars: Optimisation, Systems Thinking, Diversification, Renewability and Recycling provides an easily accessible view of the areas, in which AEC companies can benefit the most from improving. The developed method allows for both a comparative analysis between companies and profiling the industry as a whole. The proposed method can help provide deep insights into the circularity of the AEC industry processes and to how these can be improved.

In this study, the number of interviewed companies has been the main limiting factor, as a sample size of nine companies has been used to collect and analyse data of the current processes and test the method in practice. Larger sample size will enable concluding on trends and allow generalisations within the industry. When performed on a larger sample group, the analysis will also provide valuable insights into the trends of the different actors and disciplines in the AEC industry. These trends can outline general issues, and success factors in the industry and further underline where focus should be concentrated from both companies, but also governmental institutions, in order to improve the circularity of the industry as a whole.

REFERENCES

Adams, K., Osmani, M., Thorpe, T. & Thornback, J. 2017. Circular Economy in Construction: current awareness, challenges (enablers. *Waste and Resource Management* 170(1): 15–24.

Danish Environmental Protection Agency. 2019. *Affaldsstatistik*. https://www2.mst.dk/Udgiv/publikationer/2019/09/978-87-7038-109-3.pdf (visited on 11/04/2020).

Elia, V., Gnoni, M.G. & Tornese, F. 2017. Measuring Circular Economy Strategies Through Index Methods: a Critical Analysis. *Journal of Cleaner Production* 142(4): 2741–2751.

Ellen MacArthur Foundation. 2013. Towards The Circular Economy: Economic and Business rationale for an accelerated transition. Ellen Macarthur Foundation.

European Commission. 2014. *A Circular Economy: A Zero Waste Programme For Europe*. Brussels: European Commission.

European Commission. 2016. Next Steps For A Sustainable European Future. European Action For Sustainability. Strasbourg: European Commission.

Eurostat. 2019. *Construction and Demolition Waste*. https://rmis.jrc.ec.europa.eu/uploads/scoreboard2018/indicators/19._Construction_and_demolition_waste.pdf (visited on 11/04/2020).

Finch, G., Marriage, G., Pelosi, A. & Gjerde, M. 2019. Applications and Opportunities of Timber Space Frames in the Circular Economy. In *CIB World Building Congress 2019*. Hong Kong SAR, China.

Finkbeiner, J., Günter, K., Meissner, J., Merz, C., Stahl, F., Kruse, D. & Ernst, M. 2019. Woodscraper - Highrisc According to the Circular Economy. In *IOP Conference Series: Earth and Environmental Science* 225. Brussels, Belgium.

Global Footprint Network. 2019. EU Overshoot Day. Living beyond nature's limits. Brussels: WWF.

International Energy Agency. 2013. Transition to Sustainable Buildings: Strategies and Opportunities to 2050. International Energy Agency.

Kvale, S. 2007. *Doing Interviews*. London: Sage.

Leising, E., Quist, J. & Bocken, N. 2018. Circular Economy in the building sector: Three cases and a collaboration tool. *Journal of Cleaner Production* 176: 976–989.

Lieder, M. & Rashid, A. 2016. Towards Circular Economy implementation: a comprehensive review in context of manufacturing industry. *Journal of Cleaner Production* 115: 36–51.

Mangialardo, A. & Micelli, E. 2018. Rethinking the Construction Industry Under the Circular Economy: Principles and Case Studies. In Bisello A., Vettorato D., Laconte P. & Costa S. (eds) *Smart and Sustainable Planning for Cities and Regions. SSPCR 2017. Green Energy and Technology* (pp.333–344). Cham: Springer.

Mont, O. & Heiskanen, E. 2015. Breaking the Stalemate of Sustainable Consumption with Industrial Ecology and Circular Economy. In Reisch, L. & Thøgersen, J. (eds) *Handbook of Research on Sustainable Consumption* (pp. 33–48). Cheltenham: Edward Edgar.

Nasir, M.H.A, Genovese, A., Acquaye, A.A., Koh, S.C.L. & Yamoah, F. 2017. Comparing Linear and Circular supply chains: A case study from the construction industry. *International Journal of Production Economics* 183: 443–457.

Nuñez-Cacho, P., Górecki, J., Molina-Moreno, V. & Corpas-Iglesias, F.A. 2018. What Gets Measured, Gets Done: Development of a Circular Economy Measurement Scale for Building Industry. *Sustainability* 10(7): 2340.

Pomponi, F. & Moncaster, A. 2017. Circular economy for the built environment: A research framework. *Journal of Cleaner Production* 143: 710–718.

Silva, R.V., de Brito, J. & Dhir, R.K. 2019. Use of Recycled Aggregates Arising from Construction and Demolition Waste in New Construction Applications. *Journal of Cleaner Production* 236: 117629.

Thelen, D., van Acoleyen, M., Huurman, W., Thomaes, T., Brunschot, C., Edgerton, B. & Kubbinga, B. 2018. *Scaling the Circular Built Environment: pathways for businesses and government*. Geneva: World Business Council for Sustainable Development.

A BIM-based tool for the environmental and economic assessment of materials in a building within early design stages

Q. Han, N. Zhang & C.D. Van Oeveren
Eindhoven University of Technology, Eindhoven, The Netherlands

ABSTRACT: Sustainable decision-making is gaining momentum in building impacts assessment for improving the design, specifically from environmental and economic perspectives. Fast and accurate assessments can be made for the determination of building materials in earlier designs by integrating the environmental and economic assessment principles with building information modelling (BIM). Currently, there is a lack of assessment tools that can be applied for the assessment of environmental and economic impacts simultaneously within the early design stages of a building. This paper identifies the potentials of integrating BIM with life cycle assessment (LCA) and life cycle costing (LCC) by developing a tool that allows designers to optimize their designs based on impacts assessment in early design stages. The findings suggest that BIM has an interesting potential for the integration of LCA and LCC, and aids designers in making well-considered environmental and economic decisions related to the building materials in early design stages.

1 INTRODUCTION

In the past decades, the urge to become sustainable has been rising. The interest in the development of methods to understand and reduce environmental impacts of a building have been aroused in the Architecture, Engineering, and Construction (AEC) industry. One of these methods is the life cycle sustainability assessment (LCSA), which allows practitioners to organize complex environmental, economic, and social impact data in a structured form (Santos et al. 2019). The application of LCSA in the AEC-industry is still limited because it is time-consuming and requires subject matter experts to carry it out. LCSA studies are usually carried out after the completion of the building, which results in a full description of the true nature of the environmental and economic impacts of the building. A way of overcoming the descriptive nature of LCSA is the utilization of Building Information Modelling (BIM), specifically for determining the environmental and economic impacts (Anand & Amor 2017; Guinée 2016; Iacovidou et al. 2018; Nwodo et al. 2017; Soust-Verdaguer et al. 2017; Zuo et al. 2017). BIM can be used as an effective way of modeling and managing the project while enabling design changes to be made in a fast and effortless way. Combining LCA and LCC through the use of BIM-models by Industry Foundation Classes (IFC), the execution of LCA and LCC studies in early design stages is simplified, which enables designers to evaluate environmental impacts and economic variations more accurately (Bank et al. 2010).

Recent research shows the importance of LCA and LCC within a BIM-based environment to facilitate efficient decision-making (Ajayi 2015; Akbarnezhad et al. 2014; Akinade et al. 2015; Jalaei & Jrade 2015; Santos et al. 2019). This integration is crucial in the early design stages since most changes in the design occur there. Moreover, this integration allows for changes to be implemented within the model by assessing their environmental and economic impacts and ensure that all aspects meet the requirements of the different project participants. At this moment, there are limited tools on the market that help non-experts to assess their designs based on LCA and LCC principles (Najjar et al. 2017). These tools often provide knowledge, mainly from the environmental impact perspective. In general, these tools are not user-friendly and do not have any feedback-related features that are used as guidance for optimizing designs. In order to optimize the application of LCA and LCC in the early design stages, a decision support tool should be developed, which enables users to get a better understanding of the embodied effects of their decisions in design with regards to LCA and LCC.

This paper starts by providing insight into the subjects LCA, LCC, and BIM, which ultimately forms the literature research (Section 2). Section 3 proposes a technical approach that forms the base for the BIM-based environmental and economic impact assessment tool. This technical approach is explained by the usage of a system's architecture. Section 4 demonstrates the functionality of the tool through a case study of an apartment building in the early design stages. The conclusion of the paper is provided in Section 5, thereby leading to Section 6, which provides direction for future work.

2 LITERATURE REVIEW

The development and implementation of LCA and LCC within the AEC-industry have been significantly slower compared to other industries due to the sector's conservative nature and the reluctance of changes (Khasreen et al. 2009). However, because of the increasing awareness of the importance of environmental impacts in the AEC-industry, it can be stated that the significance of LCA and LCC has increased in the past decades. Therefore, a decision support tool is key to implement LCA and LCC in the AEC-industry.

This section provides a theoretical background of LCA and LCC studies (Section 2.1), and how the integration of LCA and LCC can be achieved through the usage of BIM (Section 2.2). Section 2.3 finishes with various challenges of integration between LCA and LCC by BIM.

2.1 Types of LCA and LCC studies

Following the LCA and LCC principles, there are four types of studies that can be conducted: (i) screening, (ii) simplified, (iii) complete, and (iv) streamlined. During the EeBGuide Project, conducted by Wittstock et al. (2012), it was identified that the use of the LCA studies in the AEC-industry could not be executed with the same level of detail as in other sectors because the application of an LCA-study within a building is time-consuming and complicated. The types of studies differ in terms of completeness of assessment, data representativeness, documentation, and communication of results.

In the early design stages, a screening study can be used for the initial assessment of the environmental impacts of buildings and products. However, LCC-study is highly dissuaded since, at these stages, not much data is available yet to perform a screening. Instead of conducting a screening study in the early stages, a simplified LCC-study is generally performed in more advanced stages when more data becomes available. Therefore, for the integration of LCA with LCC, a simplified study approach can be used. With regards to a complete study, a comprehensive assessment of the environmental impact and costs of the building or product can be performed, covering the entire lifecycle of the building or product. Hence, a complete study is more suitable for the final stages of a building project. The streamlined study can be described as an ad hoc version of a standard LCA and LCC approach in which the decision-maker selects the most suitable boundaries and categories for their environmental and economic study. In other words, the streamlined study can be performed by using both generic or specific LCA and LCC data and is considered explicitly suitable for early design stages.

By looking at the AEC-industry, the foundation of quality BIM-models is laid within the early design stages, where mainly the locations of elements are specified. Evolving with the design process, more elements will be specified in terms of materialization. Considering the many design changes in the early design stages, it is not always possible to perform any of the data-required studies. Because of the flexibility and comprehensiveness, a streamlined study approach is considered best practice for the performance of LCA and LCC studies in early design stages.

2.2 Integrating LCA and LCC through BIM

In the literature on BIM integration with LCA and LCC, three main approaches can be distinguished. The first approach is the utilization of a variety of tools to conduct the LCA and LCC analysis separately (Fauzi et al. 2019; Zuo et al. 2017). This approach is labor-intensive and requires more expertise in a variety of tools. Therefore, this approach is not recommended to make a well-considered analysis combining LCA and LCC.

The second approach, in which the integration of LCA and LCC within a BIM-based environment utilizes a data exchange standard, is a more streamlined approach. It includes the Bill of Quantity (BoQ) and other properties of BIM model elements, which are processed in an external tool or database (Antón & Díaz 2014; Bueno et al. 2018; Santos et al. 2019). This BoQ provides project-specific measured quantities of the items of a design, which can be measured in number, length, area, volume, weight, or time, for example.

By only using the BoQ from a BIM-model, BIM-based LCA and LCC tools only need to have the right classification of data in order to link the data with LCA and LCC databases. Thus, the import of the BIM-models within an LCA and LCC tool must be done through a corresponding classification. A drawback of this second integration approach is that BIM-LCA and LCC tools require more information than is generally comprised of BIM-models. This higher level of information requirement means that, in order to make a more accurate analysis, the BoQ should contain all the detailed information present in a design. A calculation can still be made based on generic data, and it is recommended to have a BoQ with a higher level of development (LOD) to make more accurate calculations. Besides, incorporating valuable data within a BIM-model means that the BoQ can only be exported from a single BIM software package. By using a single BIM software package, the interoperability between the use of different BIM-models from different software packages can be a challenge. This challenge can easily be solved by linking external databases that are corresponding to the BIM-model. Linking different components with each other through different platforms can help to manage and to calculate the environmental and economic impacts in a structured way.

A third approach suggests the inclusion of environmental and economic data within BIM, which could be used to represent an initial step towards assessment integration in BIM (Díaz & Antón 2014). Instead of using methodologies where BIM-models are used for their geometric and material data, efforts must be made to incorporate valuable data within a BIM-model (Antón & Díaz 2014; Santos et al. 2019). Santos et al. (2019) mention that if a BIM model includes data that is related to the environmental and economic impact of materials, an automatic streamlined analysis of the whole project should be possible. Nevertheless, for a complete study, it would be required to insert specific data of the project manually. Such data must be structured, consisting of the aspects of the system's boundary, types of LCA and LCC data, and the environmental and economic impact assessment method used. However, an IFC schema does not contain suitable properties to store the type of information required by each environmental impact category yet. According to Santos et al. (2019), a minimum of 137 IFC-properties is required for including LCA and LCC data, where at least 128 additional properties need to be created and included in the IFC schema. Therefore, to implement a third integration approach for all design stages, additional property types that contain the different LCA and LCC data need to be created.

2.3 Challenges of integration

The challenge of combining LCA and LCC lies within the linkage of underlying properties and principles. Some literature has addressed the challenge and attempted to harmonize the principles of LCA in combination with LCC (Hunkeler et al. 2008; Swarr et al. 2011). When considering LCA, decision-makers must also take the economic consequences of alternative products or designs into account. However, integrating economic analysis within LCA implies that LCA should not only consider the economic cost as another flow, but it also requires the additional variable of time to the economic modeling dimension, such as cashflows. It requires the ability to utilize variables that have no causal dependence upon inventory flows, and the ability to deal with probabilistic situations resulting in alternations in design based on the risks of costs. A way of overcoming this challenge is by providing an interrelated classification system. Through this classification system, which links the BIM data to LCA and LCC data simultaneously, environmental and economic calculations and considerations can be made. With regards to the early design stages, this means that this classification system should also take the standard category of materials into account.

Another challenge is how the data from the BIM-model is used to make calculations in the early design stages. Current approaches are mainly focusing on integrating LCA or LCC data within the parameters of a model within single BIM-software packages (Bueno et al. 2018; Röcket et al. 2019). Because in the early design stages, the LOD is low, only streamlined studies are recommended to be performed to make a quick and comprehensive environmental and economic assessment. Exporting quantities of an IFC-model allows for quick modeling and the possibility to use this data to make an ad-hoc, comprehensive environmental and economic assessment.

Another common issue when applying LCA and LCC in a design process is that in early design phases, the options for selecting solutions are numerous, and the available information on the products and materials within a design is scarce. LCA and LCC studies require detailed information to make accurate calculations. Existing BIM-LCA and LCC tools lack in terms of descriptive, useful feedback options for non-LCA and LCC experts to support their decision making. This way of working is considered inefficient, since modifying the design to achieve a high standard of environmental and economic performance criteria often results in backtracking (Santos et al. 2019; Schlueter & Thesseling 2009). Although most tools provide feedback in terms of reliable results, these tools do not provide any guidance to practitioners in how to resolve issues and improve their design (Nwodo et al. 2017). Comprising approaches to get better information about alternative options early in the design process and to get faster predictions and calculations of rough results are required. By enabling useful feedback, designers can make informed and faster decisions, which will positively affect the way of decision-making. Therefore, more extensive use of BIM models, especially data from BoQs, enables the possibility of generating a data flow between different databases, tools, and BIM models. This integration will support the simplification and implementation of LCA and LCC in the AEC-industry.

3 TECHNICAL APPROACH

The proposed system tool, the so-called Life Cycle Environmental Impact and Costing Assessment System (LEICAS)-tool, is a proof-of-concept that enables the user to assess the different interior- and exterior walls and floors of a building design in early design stages. The system consists of two processes. The internal processes deal with the impact assessment function, whereas the external processes are handling the user preferences. The data extracted from the BoQ of the IFC-model combined with the environmental and economic data from an external database is used to calculate and analyze the environmental and economic impacts of the design. This tool provides feedback and recommendation on the results of the environmental and economic impacts to support the user to make environmentally and economically considered decisions which can be used to optimize the design.

Figure 1 shows the system's architecture. It includes several modules which each deal with one function from input, analysis, output, to feedback For the linkage of the information flow, the programming language of Python is utilized. The information flow starts

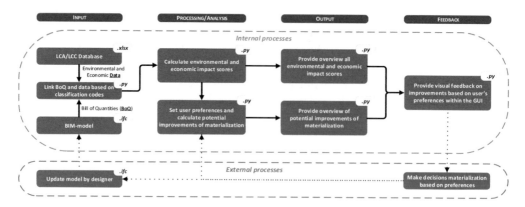

Figure 1. The system's architecture of the proposed tool.

from the input data from a BoQ of a BIM model and an external LCA-LCC database, proceeds to the potential output that supports the users to make informed decisions for the design. Every part of the tool is explained in this section. Section 3.1 provides insight into the required data input for the impact calculations and how the integration of this data is organized (Section 3.1). Within Section 3.2, the level of calculations is further explained, followed by the comparison of results and their corresponding feedback in Section 3.3.

3.1 Data input and integration

The data input comprises of two input sources, the BoQ from a BIM model, and a combination of an external database consisting of both environmental and economic data on building products or materials. For the environmental impact, the LCA data in the database consists of eleven indicators, based on Dutch regulations. These indicators form the bases of the environmental impact of certain building materials. For the linkage of environmental and economic data to the BoQ data, the database consists of an interrelated classification code. In order to make the environmental and economic calculations based on the BoQ, an IFC-model must consist of at least the classification codes, areas, and widths of building elements.

An interrelated classification system is needed that enables the reference of the building elements in a BIM-model as well as the environmental and economic data in the external database. This coding system enables the linkage between the BoQ of the IFC-model, LCA data, and the LCC data. If this coding system is incorrectly formulated in either the external database or the IFC-model, the calculations cannot be made. Figure 2 shows a specific part of the system regarding this link, which explains this integration of the BoQ with the environmental and economic data.

Within the Dutch AEC-industry, the NL/Sfb classification system is used as a standard for classifying building elements. The coding system of the NL/Sfb divides a building into products that are subsequently divided into different categories. The main structure of

Figure 2. Example of the integration of LCA and LCC data with the BoQ data by an interrelated classification system.

the coding system contains two groups of two digits each, in which the first group of digits denotes the location of the building element, such as walls and floors, and the second group of digits indicates the function of the building element. Based on these 4-digits, only *generic* calculations can be made. To specify the different materials in a building and by following the Dutch Milieu Prestatie Gebouwen (MPG)-method, the required environmental database utilizes additional digits for the material specification. This environmental database, the *Nationale Milieu Database* (NMD), provides materials information based on 7- or 9-digits. Starting with the 4-digits of the NL/Sfb classification system, the last few digits provide insight into what type of materials are used in that particular building element. The different environmental and economic impact calculations can be made and are interlinked to each other based on the NL/Sfb classification codes.

3.2 Level of calculation: Generic vs. specific

The *generic* level of calculation is based on all the unspecified, generic building elements in the IFC-model. The LCA and LCC data that is used to calculate the environmental and economic impact is based on the mean values of all available building materials per building element in the database. This data is extracted from the database based on the 4-digits NL/Sfb coding. By using the square meters related to the unspecified, generic building elements from the IFC-model and linking them to the mean values of the same building elements from the impacts database, the generic environmental and economic impact can be calculated. For example, all the materials related to the exterior load-bearing wall type starting with the classification code "28.01" are used to determine the mean values of the environmental and economic impact. After the mean values are determined, the impacts will be combined with the square meters related to the exterior load-bearing wall type starting with the classification code "28.01" of the IFC-model. This combination creates generic results based on the different building elements in a design.

Because the design process evolves, decisions on which material type of load-bearing walls are used can be made. The classification of different material types means that a more *specific* level of calculations can be made. Each specified building element with associated materials is calculated separately and is merged with the other specified building elements into a total specific impact score. As follow up of the previous example, the data from both the external database and the IFC data related to the specified building elements with materials starting with classification code "28.01" will be calculated separately and are summed up into a total score for the environmental and economic impact of the specific calculations. In other words, only the materials in the BoQ consisting of the 7- or 9-digits classification are used for the specific calculations.

3.2.1 LCA calculations

Following the Dutch regulations, when performing calculations for an LCA-study, the Dutch *Milieu Prestatie Gebouwen* (MPG) should be used. The result of an MPG calculation is an environmental profile of the building consisting of eleven environmental impacts expressed in their equivalence units. The environmental impacts of each product component for the eleven impact categories are calculated according to the specific calculation rules. The sum of the results per product component per impact category is then added up per impact category, resulting in an environmental profile of a specific product.

In order to get a comprehension of what the environmental impacts of building materials are, following the MPG methodology, LCA studies use a method called monetization. The monetization method refers to the expression of different environmental impacts in terms of shadow costs, which are identified as the theoretical valuation of costs that are acquired to reverse the damage caused by using the different materials on the environment. The environmental impacts are expressed in equivalence units and are multiplied with shadow cost values per environmental impact. By combining all individual shadow costs into one value, a weighted score is merged into one final indicator.

3.2.2 LCC calculations

A standard method used to assess the LCC valuation is the *net present value* method (NPV). The NPV can be described as the sum of the discounted future cash flows related to costs and profits. An essential variable of the NPV-method is the discount factor, which indicates the time value of money that is used to convert cash flows occurring over time. The NPV-method consists of the annual net cash flow, which is based on the revenues, expenses, and capital costs of a building for over one year. The total annual net cash flow is the combined annual net cash flow over the entire life cycle of a building.

In general, considering the early design stages of a building, it must be stated that only rough estimations can be made to the costs after the completion of the building. The costs relating to the use, maintenance, and recycling of building components corresponding to the building's life cycle will become more evident during the development of a design. This implies that applying LCC in the early design stages is difficult. The only component of the calculation that is fixed and can be calculated in early design stages are the acquisition costs of the building in the first year of a building's life cycle. If early design stages are considered within a streamlined LCC analysis, then a BIM object should only be linked to acquisition cost information. The acquisition costs are defined as the costs of materials and their construction costs.

3.3 Compare impacts & provide recommendations

As mentioned earlier, one of the weaknesses in current BIM-based LCA and LCC tools is that these tools struggle to implement user feedback functions. In order to make informed environmental and economic decisions, feedback should be provided based on input data related to the designer's preference. By enabling designers to provide information to the system about the preferred material use or the importance of impact, the system can make calculations and provide feedback fit to the designer's requests. Depending on the *generic* and *specific* data present in the BIM-model, the system calculates the environmental and economic impact of the building elements and provide insight on which materials are the most optimal to choose with regards to the unspecified building elements containing a 4-digits classification code.

An essential feature of the LEICAS-tool is the ability to provide feedback based on the preferences of a designer on its design, such as the preference of having an environmentally considered design. By using these preferences, the tool can provide the most optimal solution for the materialization of the unspecified building elements. This optimal solution is provided based on three different outcomes: (i) the lowest environmental impact; (ii) the lowest economic impact; and (iii) the best solution that fits the preference of the designer. The calculations for the optimal solutions are based on the available materials and building system choice. Besides, on a scale from one to ten, the designer can decide on how environmentally or economically considerate the design should be.

4 CASE STUDY

A demonstration of the potential of the LEICAS-tool is provided by assessing a schematic design of a 4-story apartment building design. The main structure of the apartment building consists of four exterior load-bearing walls, four floors, and various types of interior walls. The building consists of a central core following standard grids and has a total gross floor area of 1080 square meters. The BIM-design model was modeled in BIM-software package Revit 2019. Figure 3 provides the user interface of the proposed LEICAS-tool and shows the results of the case study. This section provides information on the content of the tool by elaborating on the results of the case study. The presentation and differences in results are explained in Section 4.1. Section 4.2 elaborates on the comparison of different impacts and provides insight on how to use the results as feedback for informed environmental and economic decision-making.

4.1 Differences in results

The results of the calculations are presented in twofold. On the left sidebar, the results are presented in a comprehensive overview which shows the overall results in a numerical display of the current situation. Furthermore, the left sidebar consists of a feature that checks for the acceptance or rejection of the required environmental impact score based on the Dutch policies.

On the right, bar charts are plotted to represent the environmental impact (green bars) and the economic impact (blue bars) of materials per building element location. The content of the bar charts changes depending on the different preferences of the designer. When the calculation option is set to *generic*, the bar charts only present the results per material based on the overall square meters of the particular building elements. When the user chooses the *specific* calculation option, the bar charts will show a distinction in the results of both specific results as well as generic results.

4.2 Decision-making based on results and feedback

For the comparison of the influence of the environmental and economic impacts, the bar charts provide information regarding the difference between the current situation and the potential new situation. If no materials have been specified in the BIM-model, only generic calculations can be made. Based on the generic, 4-digits building elements in a BIM-model, the tool makes calculations for the specification of these building elements. The presentation of these results is provided in the form of percentages and grand totals. The percentages are based on the increase or decrease of the total impact scores of the materials compared to the current total situation, consisting of the total impact scores of both generic building elements and, if present, specific building elements with materials. Within each of the bar charts, the bar on the far left always presents the current situation of the calculations. The remaining bar charts present the different materials and their environmental and economic impact on the design.

Table 1. Optimization of unspecified building elements by specification of materials using the NL/Sfb classification scheme.

	Int. walls	Floors	Ext. walls
Lowest env. impact	22.03.016	23.01.004	28.01.028
Lowest eco. impact	22.03.001	23.01.003	28.01.029
Best comb. situation	22.03.005	23.01.004	28.01.034

To give an example, Figure 4 provides a comparison of the environmental and economic impacts of the interior walls. In this figure, only five materials are displayed and categorized by their classification codes. The bar charts show that selecting a material based on a specific impact means that the other impacts are neglected. For example, when choosing interior wall type "*22.03.016*", the most environmentally considered material is selected. However, this wall type shows the highest economic impact, meaning that this decision is expensive.

Since the bar charts do not provide concise feedback on how to optimize the design, the tool provides a table that gives an overview of the best solutions for materializations. As mentioned in Section 3.3, the tool provides results on the lowest environmental and the lowest economic impact and the combined environmental and economic impact based on the designer's preferences. For the case-study, an equal weight of environmental and economic preference was chosen. In Table 1, this preference provides the designer with a balanced recommendation of material choice. Within this table, the designer can see in an unambiguous way which materials are the most optimal to choose with its NL/Sfb classification code for certain building elements in the design.

5 CONCLUSION AND DISCUSSION

This paper shows prospects for integrating LCA and LCC with BIM in the early stages of a design by giving insight on how to integrate the principles of LCA and LCC with a BIM-based environment and how to optimize the design based on comprehensive feedback generation. Integrating LCA and LCC with BIM enables for automated environmental and economic impact assessment of embodied impacts of building elements during early design stages.

Decision-making in the early design stages can be improved by using the proposed tool since it provides information about the environmental and economic impacts are of a particular design. The proposed tool can import IFC-models, combine them with an external LCA and LCC database, and displays the results in a user-friendly interface. Through this interface, users can gain insight into the results regarding the

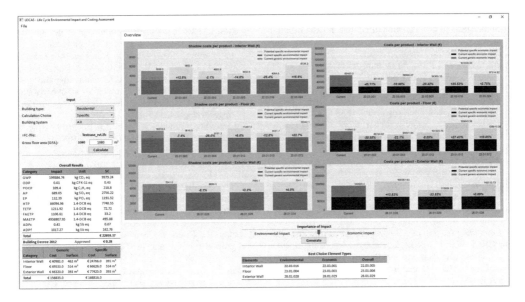

Figure 3. The results of the case study, propositioned in the LEICAS-tool.

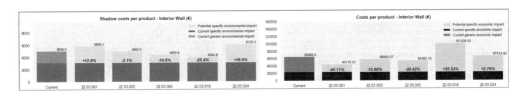

Figure 4. Bar charts representing the current situation of the design and improved or deteriorated situations of materialization.

environmental and economic impacts of their designs and decide on how to optimize and improve certain building elements in their design based on their materialization.

Multiple types of research have been conducted in the integration of LCA and LCC through BIM.

This research focuses on the usage of the streamlined study approach for the environmental and economic impact assessment in the early design stages based on the generic and specific data of a BoQ of an IFC-model. The proposed tool is an initial step towards the integration of LCA and LCC through BIM and the implementation of such an integrated impact assessment tool in the AEC industry.

Within this research, two main pillars of LCSA were considered: LCA and LCC. Since LCA and LCC form the most researched content of LCSA, the integration of LCA and LCC within this research forms an initial step to implement LCSA within the AEC-industry. However, to complete the LCSA approach, the social LCA pillar should have been considered as well. By implementing social LCA into the impact assessment tool, a more considered design can be developed. Although social LCA is challenging to research, the first step of implementing social LCA in the AEC-industry is by looking for a way to integrate BIM with social LCA, which could use a similar approach to this research.

For the extension of feedback, a future version of the tool integrates a feature that performs an export of the results and feedback in a BIM collaboration format, which allows different project participants to collaborate on issues in a project. By having such an export, consisting of data to improve the design based on the users' preference, both from environmental as well as an economic perspective, other parties can update the design directly.

REFERENCES

Ajayi, S.O. 2015. Life cycle environmental performance of material specification: A BIM-enhanced comparative assessment. *International Journal of Sustainable Building Technology and Urban Development* 6(1): 14–24.

Akbarnezhad, A., Ong, K. & Chandra, L. 2014. Economic and environmental assessment of deconstruction strategies using building information modelling. *Automation in Construction* 37: 131–144.

Akinade, O., Oyedele, L., Bilal, M., Ajayi, S., Owolabi, H., Alaka, H. & Bello, S. 2015. Waste minimisation through deconstruction: A BIM-based Deconstructability Assessment Score (BIM-DAS). *Resources, Conservation and Recycling* 105(Part A): 167–176.

Anand, C. & Amor, B. 2017. Recent developments, future challenges and new research directions in LCA of buildings: A critical review. *Renewable and Sustainable Energy Reviews* 67: 408–416.

Antón, L. & Díaz, J. 2014. Integration of life cycle assessment in a BIM environment. *Procedia Engineering* 85: 26–32.

Bank, L., McCarthy, M., Thompson, B. & Menassa, C. 2010. Integrating BIM with system dynamics as a decision-making framework for sustainable building design and operation. *Proceedings of the First International Conference on Sustainable Urbanization.* Hong Kong: ICSU.

Bueno, C., Pereira, L. & Fabricio, M. 2018. Life cycle assessment and environmental-based choices at the early design stages: an application using building information modelling. *Architectural Engineering and Design Management* 14(5): 332–346.

Díaz, J. & Antón, L. 2014. Sustainable Construction Approach through Integration of LCA and BIM Tools. *Computing in Civil and Building Engineering*: 283–290.

Fauzi, R., Lavoie, P., Sorelli, L., Heidari, M. & Amor, B. 2019. Exploring the Current Challenges and Opportunities of Life Cycle Sustainability Assessment. *Sustainability* 11(3): 1–17.

Guinée, J. 2016. Life Cycle Sustainability Assessment: What Is It and What Are Its Challenges? In R. Clift, & A. Druckman, *Taking Stock of Industrial Ecology*: 45–68. Guildford: Springer.

Hunkeler, D., Lichtenvort, K. & Rebitzer, G. 2008. *Environmental Life Cycle Costing*. SETAC.

Iacovidou, E., Purnell, P. & Lim, M. 2018. The use of smart technologies in enabling construction components reuse: a viable method or a problem creating solution? *Journal of Environmental Management* 216: 214–223.

Jalaei, F., & Jrade, A. 2015. Integrating building information modeling (BIM) and LEED system at the conceptual design stage of sustainable buildings. *Sustainable Cities and Society* 18: 95–107.

Khasreen, M., Banfill, P., & Menzies, G. 2009. Life-Cycle Assessment and the Environmental Impact of Buildings: A Review. *Sustainability 2009* 1: 674–701.

Langdon, D. 2007. Life cycle costing (LCC) as a contribution to sustainable construction - Guidance on the use of the LCC Methodology and its application in public procurement. *Davis Langdon Management Consulting*.

Najjar, M., Figueiredo, K., Palumbo, M. & Haddad, A. 2017. Integration of BIM and LCA: Evaluating the environmental impacts of building materials at an early stage of designing a typical office building. *Journal of Building Engineering* 14: 115–126.

Nwodo, M., Anumba, C. & Asadi, S. 2017. BIM-Based Life Cycle Assessment and Costing of Buildings: Current Trends and Opportunities. *ASCE International Workshop on Computing in Civil Engineering*.

Röck, M., Passer, A., Ramon, D. & Allacker, K. 2019. The coupling of BIM and LCA - challenges identified through case study implementation. *IALCCE 2018 - The Sixth International Symposium on Life-Cycle Engineering:* 841–846. London: Taylor & Francis Group.

Santos, R., Costa, A., Silvestre, J. & Pyl, L. 2019. Integration of LCA and LCC analysis within a BIM-based environment. *Automation in Construction* 103: 127–149.

Schlueter, A., & Thesseling, F. 2009. Building information model based energy/exergy performance assessment in early design stages. *Automation in Construction* 18(2): 153–163.

Soust-Verdaguer, B., Llatas, C. & García-Martínez, A. 2017. Critical review of BIM-based LCA method to buildings. *Energy and Buildings* 136: 110–120.

Swarr, T., Hunkeler, D., Klöpffer, W., Pesonen, H.-L., Ciroth, A., Brent, A. & Pagan, R. 2011. Environmental life-cycle costing: a code of practice. *International Journal of Life Cycle Assessment* 16(5): 389–391.

Wittstock, B. G., Lenz, K., Saunders, T., Anderson, J., Carter, C., Gyetvai, Z. & Sjostrom, C. 2012. *EeBGuide Guidance Document - Part B: Buildings*. EC.

Zuo, J., Pullen, S., Rameezdeen, R., Bennetts, H., Wang, Y., Mao, G. & Duan, H. 2017. Green building evaluation from a life-cycle perspective in Australia: a critical review. *Renewable and Sustainable Energy Reviews* 70: 358–368.

Housing energy-efficient renovation adoption and diffusion: A conceptual model for household decision-making process

H. Du, Q. Han & B. de Vries
Eindhoven University of Technology, Eindhoven, Netherland

ABSTRACT: Residential buildings have great potential for energy saving and reducing greenhouse gas emissions by energy-efficient renovation (EER). EER technologies are proven to be economically viable. However, the housing EER diffusion process falls behind the expectations. A literature review is conducted to find what the models used to describe the EER adoption decision-making process are and what the influences are. Behaviour theory and social network theory are introduced as complementary information for the review. Findings of the review suggest that there are differences among the research findings and there are limitations when applying behaviour theory in EER field. A conceptual model is presented based on the review. The model includes influences such as household background, experiences, ability, subjective norm, opportunities and information and physical resources. It contains the usual influences from behaviour theories and emphasizes the influences proven have significant impact.

1 INTRODUCTION

The building sector accounts for about 40% of total energy consumption around the world (Hong et al. 2015), and more than 80% of the energy consumption happens during the operation stage. Hence, there is a great potential for energy saving and reducing greenhouse gas emissions by energy-efficient renovation (EER) (Moglia et al. 2017). It is well-recognized that energy efficiency in the building sector is the key to energy system transformation (Torgal et al. 2013). Numerous literature proved that many household EER technologies are economically feasible. However, the diffusion of EER is slow, and the penetration falls behind the expectations (Byrka et al. 2016). According to IEA, the typical energy renovation rate in the building sector is 1–2%, and building energy intensity improvements is less than 15% (IEA 2019).

The improvement of building energy efficiency is dependent on households' decisions, and various factors influence this decision process. Researchers have been trying to understand, describe the decision-making process, and determine crucial influences of the process (Ebrahimigharehbaghi et al. 2019; Wilson et al. 2015, 2018). These decision-making models and identified important influences are tightly connected with local situations. Both similarities and differences exist among these models and findings. There are two ways of developing new EER decision-making models. First is referring to previously published articles; however, referring to only one or several existing literatures may be partial and result in unreliable or false study results. The second way is using existing behaviour theories or social network theories. Several studies use agent-based modelling to simulate the EER diffusion process, and behaviour theories are used to build the individual decision-making process (Hesselink & Chappin 2019). However, these behaviour theories or innovation diffusion theories could not be very suitable for the decision-making process of EER as it is not verified by empirical researches or proved there are limitations (Priest et al. 2015).

The overall objective of this study is to understand the EER decision-making process of households and understand the slow uptake of EER. A conceptual model of the decision-making process is presented to provide a framework for future EER decision-making process studies.

2 METHODOLOGY

In order to understand the major influences of the EER decision-making process, a systematic analysis of the EER decision-making process and adoption influences is conducted, and a conceptual model of the decision-making process is presented. First, models and identified influences from literature are described; the research findings will be compared. Behaviour theories and social network theory will also be presented to incorporate more information about possible influences of EER adoption behaviour. Afterward, a conceptual model for the EER decision-making process is put forward. Overall, this research provides insights into the EER decision-making process and finds opportunities for future research. The analysis of this research consists of three steps, and it is described as follows.

2.1 Overview of housing EER adoption

General exploration of EER adoption is conducted before the systematic review to gain an integrated understanding of EER, which includes its definition, development barriers, policies or incentives, and plans or targets for different countries.

Before further discussion, the definition of energy-efficient renovation for this study should be given. The meaning for EER adoption for this research is the adoption of energy-efficient products or renewable energy technology and retrofits for improving housing insulation. The definition means that housing-related measures such as P.V. panel installation, adopting energy-efficient heating or boilers, choosing energy-saving household appliances and improving the insulation of roof, wall or floor are included in the framework of EER in this study. While using an electric vehicle or renewable energy power generation in a power station is not included in the research since these measures are not both housing-related and household measures, even though they are also energy-efficient technologies.

As described previously, the diffusion of housing EER is not processing as expected. Numerous barriers hindered the adoption and diffusion of housing EER are identified, and there are four kinds of barriers: structural barriers, economic barriers, behavioural barriers and social-behavioural barriers (Hesselink & Chappin 2019). Among these obstacles, low returns, uncertain quality assurance and information barriers are frequently brought out as main barriers (Allcott & Greenstone 2012; Weber & Wolff 2018).

Another common topic in this field is the policies encouraging, promoting the adoption of housing EER, and aiming at break previously mentioned barriers. These policies include finical incentives, tariffs, information or education, and legislative measures. These measures were studied for their effectiveness and future improvement (Blumstein 2010). The government always plays an important role in these measures, and in recent years, the significance of local governments has been emphasized. Their role should be increased since they have a closer relationship with households, which are the primary decision-maker of housing EER (Ebrahimigharehbaghi et al. 2019).

2.2 Search terms and filters for article collecting

Keywords were selected for the literature search after the general review of housing EER studies. This search was conducted using the advanced search option of the Scopus scientific library. The following search queries were used:

- ("energy efficient" AND "household OR housing" AND "renovation OR retrofit OR renovate" AND "decision-making process");
- ("energy efficient" AND "household OR housing" AND "renovation OR retrofit OR renovate" AND "adoption");
- ("energy efficient" AND "household OR housing" AND "renovation OR retrofit OR renovate" AND "social network or diffusion").

Other limitations introduced in the queries include that the language of the literature is in English, and the publishing year should be later than 2010 (including). This query results also only includes peer-review journals and conference papers. Scopus scientific library provided 55 results with queries defined above (59 search results with 4 duplicates). A check was made to avoid missing relevant papers after the query. After the check, the titles and the abstracts of the papers were read to exclude undesired literature which focuses on renovation options, renovation impacts, or renovation decision for housing cooperatives and housing associations. This reading procedure leads to a list of 2 papers.

2.3 Paper evaluation and analysis

The 2 papers were evaluated and analysed using the following review procedure. This review starts with the introduction of the models, what kind of model is presented and what is the feature of the model. An example of a frequently used model is described in this part of the review. The second step is introducing and evaluating the influences of the EER decision-making process. Research results of significant influences are compared to find similarities and differences. Last, the theories used in the paper collection will be summarized. Some well-recognised theories not included in the collection are also introduced to have a relatively thorough picture of decision-making and behaviour theories.

3 EER DECISION-MAKING MODELS AND INFLUENCES

3.1 Household EER adoption decision-making models

Among the 20 selected papers, 8 of them present the decision-making models. These models demonstrate the household EER decision-making process from different aspects such as transaction cost barriers, social network, etc., for various study purposes. These models are divided into three kinds.

The first kind of model uses different phases to describe the decision-making process. Five studies present this kind of model. These phases described in the literature can be categorized as before the decision and after the decision. Phases before the decision include considering or thinking about the renovation and planning for the renovation; phases after the decision are implementing and experiencing. It is worth mentioning that some studies introduced a "phase 0" to represent not interested or considering EER renovation (Pettifor et al. 2015; Wilson et al. 2018). While others use terms such as "get interested" or "consideration" to describe if a household will enter

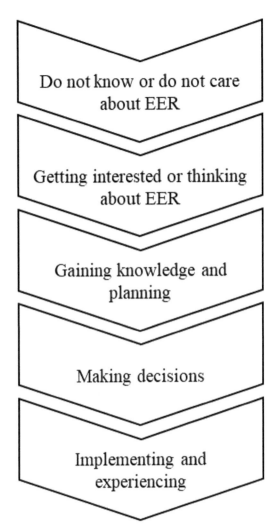

Figure 1. EER adoption decision-making model with phases.

the thinking and planning phase (Broers et al. 2019; Ebrahimigharehbaghi et al. 2020). In particular, one of the models focus on the barriers of the adoption and view all the influences/barriers as the transaction costs for each phase. Studies use different words for these phases, but the meaning is similar. The summarized EER adoption decision-making model with different phases is shown in Figure 1. The last phase in Figure 1 is a combination of implementing and experiencing, which usually are two separate phases in models from reviewed literature.

The second kind of model introduces the decision-making process from a more general angle, and the households in the model only have two statuses, adopters, and non-adopters. These models tend to focus on studying the different influences for the decision while households' specific behaviours for each phase are overlooked. There are two of this kind of adoption models. The first is developed upon expert interviews using fuzzy cognitive mapping. The respective strength of different influences is determined by standardizing them into effect coefficients (Seebauer et al. 2019). The other one is a more typical decision-making model. It illustrates influences based on interviews. These influences are classified and cover different aspects, including attitudes, habits, personal capability, and contextual factors. The first two aspects describe households' personal sphere while the last two are contextual spheres.

There is only one model, including contextual spheres, among the literature selection. This model not only simulates the households' decision-making process but also demonstrates the dynamic diffusion process of household EER. It aims at studying the effectiveness of social network interventions. The decision-making part of the model is similar to regular general decision-making models. It includes both personal benefits and social benefits in the model. However, households' attitudes may alter, and adoption behaviour may change during the diffusion process (Bale et al. 2013).

3.2 EER adoption influences

All the 20 studies analysed EER adoption influences to some extent. Some present a model to give a more structured demonstration, use the model to analyse interventions, or simulate possible future EER adoption diffusion, as shown in Section 3.1. Others give closer attention to determining the key influences and studying the impact of households' particular preferences.

The most common and widely investigated influences are socio-demographic factors, dwelling or housing factors, attitudes towards the environment, and information access. Socio-demographic factors usually include homeowners' age, household size, income, occupation, and education. The research results about the significance of these socio-demographic factors have both similarities and differences. For example, nearly all researches agree that the younger generation has a greater tendency to adopt EER compare with the older generation (Abreu et al. 2019; Mortensen et al. 2016). However, the research results do not agree on influences such as education and household income. Some studies suggest that education is a major influence (Schleich et al. 2019), while some research results show that the impact of education is not significant (Ebrahimigharehbaghi et al. 2020; Mortensen et al. 2016). As for household income, findings of this factor indicate that households with medium to high incomes are more likely to invest EER in general (Trotta 2018). However, when the household's income reaches certain level, the interests for EER adoption drops when the annual income increases (Mortensen et al. 2016). It may be due to that the households with considerable high income do not benefit much from EER adoption as profitability is another crucial influence (Achtnicht & Madlener 2014). Besides these findings, other household-related influences such as marriage status (Pelenur & Cruickshank 2012; Trotta 2018), number of children (Mortensen et al. 2016; Schleich et al. 2019), household size and retirees (Wilson et al. 2018) are also proven to be a crucial influence in some studies.

Housing-related factors contain housing physical situation, building type, housing location, etc., which also play a major role in the decision-making process. For instance, semi/detached house owners are more likely to conduct EER compared with their counterparts live in flats. What is more, housing comfort, such as thermal comfort, is an important influence that motivates EER adoption (Broers et al. 2019; Long et al. 2015). Another evidence is that it is reported that households are more likely to adopt EER when there are opportunities, such as the heating system needs replacement (Achtnicht & Madlener 2014). In other words, EER normally stems from necessity needs (Abreu et al. 2019). A study in the U.K. finds that property size and property age have no significant influence on the adoption (Wilson et al. 2018). Research findings of housing aesthetics are worthy of attention. It is considered as a barrier in studies both in the UK (Galvin & Sunikka-Blank 2014) and in the Netherlands (Broers et al. 2019) while it is viewed as a motivator for EER adoption in a study in Portugal (Abreu et al. 2019).

Householders' attitudes towards EER or environment are widely recognized as vital for the adoption (Abreu et al. 2019; Trotta 2018). A positive attitude towards the environment and EER could be believing in saving the environment is the right thing; protecting the environment is important or wanting energy use to be greener etc. People with responsible energy attitudes are more confident about overcoming information barriers, achieving energy-saving with housing energy renovation (Long et al. 2015), and more likely to adopt EER (Trotta 2018). However, research also suggests that environmental awareness alone may not lead to household EER adoption behaviour (Pelenur 2018).

Information access is another crucial influence as it is presented that imperfect information is an important barrier (Achtnicht & Madlener 2014; März 2018), and access to reliable information is viewed as one of the major transaction cost (Ebrahimigharehbaghi et al. 2020).

Influences such as previous experience with EER, perceived risks in future energy-saving and risk aversion are also reported as having impacts on EER adoption behaviour. A study suggests that households' attitudes could change after EER adoption, and they will be more willing to accept deeper renovation (Abreu et al. 2019; Long et al. 2015). While risk-averse and loss-averse consumers are less likely to adopt EER (Qiu et al. 2014; Schleich et al. 2019).

These differences among research findings described above suggest that significant influences for people live in different communities or from different backgrounds could vary to a large extent. The differences also imply that when researchers are trying to identify major influences, they should have a list as thorough as possible before the survey or interview.

3.3 Behaviour theories and social network theories

Two of the selected articles adopted existing theories. One incorporates the diffusion of innovations theory, and the other one used social network theory.

The key concepts of diffusion of innovations theory are utilized in an applied project for energy-efficient retrofit (Priest et al. 2015) in the paper selection. This theory was introduced by Rogers (1962). It believes that innovation diffusion is often a social process, as well as a technical matter, and there are four main elements in the innovation diffusion process: innovation, communication channels, time, and the social system.

Another utilized theory is social network theory. It is commonly used to investigate the influences of human interaction (Du et al. 2016), which could be both direct and indirect. According to complex system theory, the social network is tightly connected with individual behaviours (Bale et al. 2013). The importance of social networks is well recognized as network interventions could accelerate behaviour change and be utilized for forming energy policy (Bale et al. 2013) even though its role in the spread of EER information is a relatively new area for research. There are four basic types of networks: regular lattice, small-word network, scale-free network, and random network. These networks could describe the real-world situation, and they are sometimes utilized for describing the social network. However, it is suggested that these networks not always reflect social networks (Hamill & Gilbert 2008). Social network influences for EER diffusion sometimes are described with other terms such as peer effect and social learning (Bollinger & Gillingham 2012; Lee & Son 2019).

Besides models described in Section 3.2, several behaviour theories and social network theories are usually used when trying to simulate EER adoption and diffusion with computer models for prediction and policy evaluation.

The most commonly applied is the Theory of Planned Behaviour by Ajzen (1991). It is an extension of the Theory of Reasoned Action (Ajzen & Fishbein 1980); hence there are some similarities between two theories. Core constructs for the theory of planned behaviour include attitudes toward behaviour, subjective norm, and perceived behaviour control. In contrast, the theory of reasoned action only contains the first and the second constructs. Attitudes towards behaviour refer to the extent that a person favors or agrees with certain behavior. Social norm is the influence or pressure people received from others to conduct the behavior. Perceived behavior control is peoples' idea about the ease or difficulty of performing a behavior, which is a reflection of both past experience and anticipated barriers (Ajzen 1991). This theory is applied for developing decision-making in agent-based models of EERs such as heating system (Maya

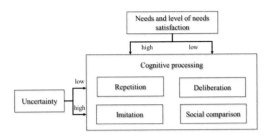

Figure 2. Basics Jager's conceptual meta-model of behaviour.

et al. 2011, 2013) and solar photovoltaic (Rai & Robinson 2015), etc. The reason why this theory is so popular could be its flexibility and elegance, and it is relatively easy to apply (Hesselink & Chappin 2019).

Another theory is a conceptual model presented by Jager (2000) for simulating consumer behaviour. The basic content of this model is shown in Figure 2. It is concluded that there are four cognitive processing styles determined by the level of need satisfaction and the uncertainty of the behavior. The four cognitive processing are repetition, deliberation, imitation, and social comparison. Different behaviour theories could apply to these cognitive processes. Such as classical conditioning theory and operant conditioning theory could apply to repetition; decision and choice theory, attitude and perceived control part of the theory of planned behaviour apply to deliberation; social learning theory and theory of normative conduct could be used to describe the imitation cognitive process while social comparison theory, relative deprivation theory and social norm part from the theory of planned behaviour apply to social comparison.

As mentioned previously, these theories provide some ideas about what could influence the EER adoption decision-making process; however, they originate from other social behaviour fields and consider behaviours from a general view. Hence, sometimes these theories could not fit in the context of EER adoption to some extent, and struggles could appear during the application of these theories in the EER field.

4 CONCEPTUAL MODEL FOR EER DECISION-MAKING PROCESS

A conceptual model, as shown in Figure 3, for the EER decision-making process is developed to give a thorough outline of EER adoption influences. The presentation of the model aims to tackle the problems met during the application of behaviour theories and give a structured description of the EER adoption decision-making process. Figure 3 shows the main influences and indicates whether the influence is external or internal though the position of the arrow below the influences. It also suggests household interactions through the social network.

The conceptual model contains two levels, one is households (micro-level) and the other is macro-level

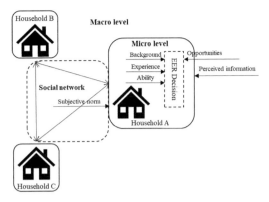

Figure 3. Conceptual model for EER decision-making process.

which consist of social network and general society. Influences are categorized into 3 kinds, micro-level influences, social influences and opportunities. The micro-level includes households' background, experiences with EER and ability, which are all internal influences. Backgrounds refers to households' socio-demographic features such as homeowners' age, education, household type and size, etc. Household attitudes towards EER are also included in the category. The background influences describe the basics of households. Experience with EER consists of both whether the household had EER adoption experience before and if any household members work in EER related field. Ability refers to households' financial ability and physical ability. It is worth noting that physical ability contains two parts, which are household members' health, fitness strength and possessed tools or circumstances. Health situations could link to the requirement of housing, such as reducing the mould and damp and influence the EER adoption decision (Long et al. 2015). Tools or circumstances mainly refer to the physical situation of the housing, such as property size, age etc. as these factors could determine if it is convenient or possible to adopt some energy-efficient measures.

The second kind is social influences. To be more specific, it consists of subjective norms and information and physical resources. Subjective norm is the perceived social pressure to perform particular behaviour or not, in this case, is the adoption of EER, from important others. This influence mainly comes from house households' social network, as shown in Figure 3. This impact is mutual as one households' adoption decision is impacted by other households and could influence others' decisions. Information and physical resource refer to the EER information availability from mass media and accessibility of EER adoption physical resource, such as if there are any contractor provide EER service in the local region.

The opportunity influences could be both internal and external, as shown in Figure 3. It could be households' internal changes or key moments in life such as inheritance, new relationship, new child or second child, etc. (Abreu et al. 2019). External opportunities could be governments' new incentive policy or new housing regulations.

The model covers the frequently investigated influences described in Section 3.1 and 3.2, including socio-demographic influences, household attitude toward EER or general environment, property-related influences, and information. Risk preference and loss aversion are not incorporated in the model as they are connected with influences such as housing physic situations and opportunities such as moving plans (Qiu et al. 2014), which is incorporated in the model. The conceptual model emphasizes the information and opportunity influences compare with the theory of planned behaviour as these influences could create the major barrier or motivator, and their importance is not obvious in the theory of planned behaviour. A dynamic decision-making process could be realized with the introduction of the social network. Hence it is possible to incorporate the interaction among households when applying the conceptual model for computer simulation compared with diffusion theory and theory of planned behaviour.

5 CONCLUSION

In this paper, a review of the EER adoption decision-making process is conducted. Models for the EER adoption decision-making process are introduced and described. Most models in the article selection describe this process using various phases and investigate the influences for each phase. Several general models and social network models are also utilized for describing the decision-making process. Influences for the EER adoption decision-making process are identified and categorized into four kinds, socio-demographic influences, housing-related factors, attitudes towards the environment, and information access. Besides, there are less investigated influences, such as previous experiences, risk preference, and loss aversion. Afterward, theories commonly used for developing the EER adoption decision-making process are introduced. Review findings suggest that the impact of influences varies, and a specific influence could have opposite effects in different contexts.

What is more, general behaviour theories could be used for developing EER decision-making models. However, limits exist during the application. A conceptual model is developed based on the literature review. The conceptual model captures possible important influences of the decision-making process and demonstrate these influences in a structured way. The conceptual model includes background, experiences, ability, subjective norm, opportunities and information, and physical resources influences. The model also indicates whether the influences are internal or external.

Pathways for future studies are model applications. More specific EER adoption decision-making models will be developed for the chosen context to validate the model.

REFERENCES

Abreu, M. I., De Oliveira, R. A. F. & Lopes, J. 2019. Energy-related housing renovations from an everyday life perspective: Learning from Portuguese homeowners, *WIT Transactions on Ecology and the Environment*, 237, pp. 63–74. doi: 10.2495/ESUS190061.

Achtnicht, M. & Madlener, R. 2014. Factors influencing German house owners' preferences on energy retrofits, *Energy Policy*, 68, pp. 254–263. doi: 10.1016/j.enpol.2014.01.006.

Ajzen, I. 1991. The Theory of Planned Behavior, Organizational Behavior and Human Decision Processes. doi: 10.15288/jsad.2011.72.322.

Ajzen, I. & Fishbein, M. 1980. Understanding attitudes and predicting social behavior. Englewood cliffs: Prentice-Hall.

Allcott, H. & Greenstone, M. 2012. Is There an Energy Efficiency Gap?, *Journal of Economic Perspectives*, 26(1), pp. 3–28. doi: 10.1680/ener.2008.161.4.145.

Bale, C. S. E. et al. 2013. Harnessing social networks for promoting adoption of energy technologies in the domestic sector, *Energy Policy. Elsevier*, 63, pp. 833–844. doi: 10.1016/j.enpol.2013.09.033.

Blumstein, C. 2010. Program evaluation and incentives for administrators of energy-efficiency programs: Can evaluation solve the principal/agent problem?, *Energy Policy. Elsevier*, 38(10), pp. 6232–6239. doi: 10.1016/J.ENPOL.2010.06.011.

Bollinger, B. & Gillingham, K. 2012. Peer effects in the diffusion of solar photovoltaic panels, *Marketing Science*, 31(6), pp. 900–912. doi: 10.1287/mksc.1120.0727.

Broers, W. M. H. et al. 2019. Decided or divided? An empirical analysis of the decision-making process of Dutch homeowners for energy renovation measures, *Energy Research & Social Science*, 58(September), p. 101284. doi: 10.1016/j.erss.2019.101284.

Byrka, K. et al. 2016. Difficulty is critical: The importance of social factors in modeling diffusion of green products and practices, *Renewable and Sustainable Energy Reviews*, 62, pp. 723–735. doi: 10.1016/j.rser.2016.04.063.

Du, F. et al. 2016. 'Modelling the impact of social network on energy savings, *Applied Energy. The Authors*, 178, pp. 56–65. doi: 10.1016/j.apenergy.2016.06.014.

Ebrahimigharehbaghi, S. et al. 2019. Unravelling Dutch homeowners' behaviour towards energy efficiency renovations: What drives and hinders their decision-making?, *Energy Policy. Elsevier Ltd*, 129(June 2018), pp. 546–561. doi: 10.1016/j.enpol.2019.02.046.

Ebrahimigharehbaghi, S. et al. 2020. Transaction costs as a barrier in the renovation decision-making process: A study of homeowners in the Netherlands, *Energy and Buildings*, 215. doi: 10.1016/j.enbuild.2020.109849.

Galvin, R. & Sunikka-Blank, M. 2014. The UK homeowner-retrofitter as an innovator in a socio-technical system, *Energy Policy*, 74(C), pp. 655–662. doi: 10.1016/j.enpol.2014.08.013.

Hamill, L. & Gilbert, N. 2008. A simple but more realistic agent-based model of a social network, Structure. Available at: http://epubs.surrey.ac.uk/1600/

Hesselink, L. X. W. & Chappin, E. J. L. 2019. Adoption of energy efficient technologies by households – Barriers, policies and agent-based modelling studies, *Renewable* and *Sustainable Energy Reviews*. Elsevier Ltd, 99(March 2018), pp. 29–41. doi: 10.1016/j.rser.2018.09.031.

Hong, J. et al. 2015. Uncertainty analysis for measuring greenhouse gas emissions in the building construction phase: A case study in China, *Journal of Cleaner Production*. Elsevier Ltd, 129, pp. 183–195. doi: 10.1016/j.jclepro.2016.04.085.

IEA 2019. Tracking Buildings. Available at: https://www.iea.org/reports/tracking-buildings

Jager, W. 2000. Modelling consumer behaviour, Inhoud ISSN 0033-3115. Available at: http://www.tijdschriftdepsycholoog.nl/assets/sites/6/DePsycholoog_nr10-2001.pdf#page=26

Lee, S. & Son, Y. J. 2019. Extended decision field theory with social-learning for long-term decision-making processes in social networks, *Information Sciences. Elsevier Inc.*, 512, pp. 1293–1307. doi: 10.1016/j.ins.2019.10.025.

Long, T. B. et al. 2015. The impact of domestic energy efficiency retrofit schemes on householder attitudes and behaviours, *Journal of Environmental Planning and Management*, 58(10), pp. 1853–1876. doi: 10.1080/09640568.2014.965299.

März, S. 2018. Beyond economics—understanding the decision-making of German small private landlords in terms of energy efficiency investment, *Energy Efficiency*, 11(7), pp. 1721–1743. doi: 10.1007/s12053-017-9567-7.

Maya, B., Kl, C. A. & Hertwich, E. G. 2011. Exploring policy options for a transition to sustainable heating system diffusion using an agent-based simulation, 39, pp. 2722–2729. doi: 10.1016/j.enpol.2011.02.041.

Maya, B., Klöckner, C. A. & Hertwich, E. G. 2013. Adoption and diffusion of heating systems in Norway?: Coupling agent-based modeling with empirical research, *Environmental Innovation and Societal Transitions*. Elsevier B.V., 8, pp. 42–61. doi: 10.1016/j.eist.2013.06.001.

Moglia, M., Cook, S. & McGregor, J. 2017. A review of Agent-Based Modelling of technology diffusion with special reference to residential energy efficiency, *Sustainable Cities and Society*. Elsevier B.V., 31, pp. 173–182. doi: 10.1016/j.scs.2017.03.006.

Mortensen, A., Heiselberg, P. & Knudstrup, M. 2016. Identification of key parameters determining Danish homeowners' willingness and motivation for energy renovations, *International Journal of Sustainable Built Environment*. The Gulf Organisation for Research and Development, 5(2), pp. 246–268. doi: 10.1016/j.ijsbe.2016.09.002.

Pelenur, M. 2018. Household energy use: a study investigating viewpoints towards energy efficiency technologies and behaviour, *Energy Efficiency*, 11(7), pp. 1825–1846. doi: 10.1007/s12053-018-9624-x.

Pelenur, M. J. & Cruickshank, H. J. 2012. Closing the Energy Efficiency Gap: A study linking demographics with barriers to adopting energy efficiency measures in the home, Energy, 47(1), pp. 348–357. doi: 10.1016/j.energy.2012.09.058.

Pettifor, H., Wilson, C. & Chryssochoidis, G. 2015. The appeal of the green deal: Empirical evidence for the influence of energy efficiency policy on renovating homeowners, *Energy Policy*, 79, pp. 161–176. doi: 10.1016/j.enpol.2015.01.015.

Priest, S. H. et al. 2015. Rethinking Diffusion Theory in an Applied Context: Role of Environmental Values in Adoption of Home Energy Conservation, *Applied Environmental Education and Communication*, 14(4), pp. 213–222. doi: 10.1080/1533015X.2015.1096224.

Qiu, Y., Colson, G. & Grebitus, C. 2014. Risk preferences and purchase of energy-efficient technologies in the residential sector, *Ecological Economics*, 107, pp. 216–229. doi: 10.1016/j.ecolecon.2014.09.002.

Rai, V. & Robinson, S. A. 2015. Agent-based modeling of energy technology adoption: Empirical integration of social, behavioral, economic, and environmental factors, *Environmental Modelling and Software*. Elsevier Ltd, 70, pp. 163–177. doi: 10.1016/j.envsoft.2015.04.014.

Rogers, E. M. 1962. Diffusion Of Innovations. Simon and Schuster.

Schleich, J. et al. 2019. A large-scale test of the effects of time discounting, risk aversion, loss aversion, and present bias on household adoption of energy-efficient technologies, *Energy Economics*, 80, pp. 377–393. doi: 10.1016/j.eneco.2018.12.018.

Seebauer, S., Kulmer, V. & Fruhmann, C. 2019. Promoting adoption while avoiding rebound: Integrating disciplinary perspectives on market diffusion and carbon impacts of electric cars and building renovations in Austria, *Energy, Sustainability and Society*, 9(1). doi: 10.1186/s13705-019-0212-5.

Torgal, F. P. et al. 2013. Nearly Zero Energy Building Refurbishment: A Multidisciplinary Approach. London: Springer.

Trotta, G. 2018. Factors affecting energy-saving behaviours and energy efficiency investments in British households, *Energy Policy*, 114, pp. 529–539. doi: 10.1016/j.enpol.2017.12.042.

Trotta, G. 2018. The determinants of energy efficient retrofit investments in the English residential sector, *Energy Policy*, 120(May), pp. 175–182. doi: 10.1016/j.enpol.2018.05.024.

Weber, I. & Wolff, A. 2018. Energy efficiency retrofits in the residential sector – analysing tenants' cost burden in a German field study, *Energy Policy*, 122(January), pp. 680–688. doi: 10.1016/j.enpol.2018.08.007.

Wilson, C., Crane, L. & Chryssochoidis, G. 2015. Why do homeowners renovate energy efficiently? Contrasting perspectives and implications for policy, *Energy Research and Social Science*. Elsevier Ltd, 7, pp. 12–22. doi: 10.1016/j.erss.2015.03.002.

Wilson, C., Pettifor, H. & Chryssochoidis, G. 2018. Quantitative modelling of why and how homeowners decide to renovate energy efficiently, *Applied Energy*. Elsevier, 212, pp. 1333–1344. Available at: https://www.sciencedirect.com/science/article/pii/S0306261917317002 (Accessed: 30 January 2019).

Digital technologies as a catalyst to elevating IPD+BIM synergy in sustainable renovation of heritage buildings

B.F. Brahmi & S. Sassi-Boudemagh
AVMF Laboratory University of Constantine3 –Salah BOUBNIDER-, Constantine, Algeria

I. Kitouni
Department of Basic Computing and its Applications, University of Constantine2 –Abdelhamid MEHRI-, Constantine, Algeria

A. Kamari
Department of Engineering, Aarhus University, Aarhus, Denmark

ABSTRACT: The implementations of Integrated Project Delivery (IPD) and Building Information Modeling (BIM) for sustainable renovation of heritage buildings are considered to be very efficient to achieve a balance between sustainable design and the historical preservation, as well as to enhance productivity/efficiency in processes, catalyzed by digital technologies. The aim of this paper is to explore, identify, and discuss the relevant digital technologies as a catalyst to elevating IPD+BIM synergy in sustainable renovation of heritage buildings. The research adopts a mixed methodology through application of an analytical framework (with a set of defined variables consisting of 46 criteria, classified into 15 categories, and grouped into five thematic strands - people, process, policy, technology, and product) besides triangulation approaches for data collection and validity of the research work. We investigate on how the digital technologies open up new possibilities and where they add or are expected to add value.

1 INTRODUCTION

Sustainable renovation of heritage buildings is a dynamic intervention that takes place in complex contexts involving interactions of multi-disciplinary fields (Fouseki & Cassar 2014). As such, the need to facilitate the access of the different actors and manage more and more complex information requires the development and application of new innovative solutions and holistic decision-making frameworks (Kamari et al. 2018a,b, 2020) to enhance the existing practices. That will help professionals develop and decide on the most appropriate renovation solutions (Kamari et al. 2019a,b), concerning the requirements for reaching sustainability together with preserving the heritage buildings' characteristics, features, and values (Tomšič et al. 2017).

Information Technology (IT) is widely discussed within the emergence of large, ambitious, and complex projects in the architecture, engineering, construction, and operations (AECO) industry, due to the new requirements of sustainability that need efficient information exchange between the project's participants and stakeholders on a regular basis through the whole project lifecycle (Oesterreich & Teuteberg 2016). Nowadays, all industries are becoming increasingly reliant on IT to uncover previously unexplored value potential. Like a wide range of industrial sectors, the Fourth Industrial Revolution, "Industry 4.0" (Lasi et al. 2014) is transforming the AECO sector. The digitalization and automation of the construction, also referred as Construction 4.0, has changed the supply chains management and products (Dallasega et al. 2018), through the adoption of innovative and disruptive technologies including building information modeling (BIM); cloud computing; big data analytics; internet of things; virtual/augmented/mixed reality; as well as autonomous robots. Industry 4.0 is seen as a major enabler of productivity improvement, as well as more sophisticated and integrated design and construction (Oesterreich & Teuteberg 2016). It allows the holistic adoption and implementation of green and/or sustainable business models and promotes a circular economy (CE) performance (Ramakrishna et al. 2020).

In the framework of 'Industry 4.0' in the AECO (Architecture, Engineering, Construction, Operation) sector, BIM concept has become a topic of great interest in the heritage renovation projects (Kamari et al. 2017) within the developed digital technology and methods, notably 3D laser scanning and photogrammetry. The BIM technology generates a new evolution of integrated and efficient information management for the conservation process due to its capability to store interrelated semantic information on favoring the dissemination of the intangible values of a building during its lifecycle (Angelini et al. 2017). Nevertheless, little progress has been made to address the use of BIM technologies for managing the whole intervention design and the renovation processes, such as the generation and evaluation of various design alternatives (Pocobelli et al. 2018).

Numerous studies proposed Integrated Project Delivery (IPD) as the best project management method to leverage BIM functionalities (Rowlinson 2017). Like BIM, IPD has emerged to improve the quality of the construction projects, increase their performance, and eliminate weaknesses of current project delivery systems (Rowlinson 2017). The synergy between BIM and IPD can remove collaboration barriers, and enables the project team to provide more pragmatic/effective solutions to complex project issues and enhance sustainability performance (Fischer et al. 2017). In the previous research work by the authors of the current paper based on an in-depth investigation of the potential advantages of the synergy between BIM and IPD for sustainable renovation of heritages (Brahmi et al. 2020), the novel methodology was considered to achieve the balance between sustainable design and historic preservation as well as to enhance productivity/efficiency, via preparing better collaborative and integrating processes, as a key for the successful delivery of building project (more details will be described further in section 2.2).

Despite our study of the BIM and IPD synergy for sustainable heritage renovation in (Brahmi et al. 2020), it is observed that the potential and limits of the digital technologies in this area are not studied adequately, and there is a significant unexploited existing and growing potential for the use of digital technologies yet to be explored for these projects. In this view, the aim of this paper is to explore, identify, and discuss the relevant digital technologies as a catalyst to elevating IPD+BIM synergy in the sustainable renovation of heritage buildings. The paper is organized in Section 2, which summarizes the concept of BIM and IPD, and their use in heritage renovation via reviewing the relevant literature in these areas. Section 3 presents an overview of the used research methodology on use of a mixed methodology through application of an analytical framework besides triangulation approaches for a comprehensive and systematic exploration of the research field. Section 4 investigates the potential of further digitalization through the application of IPD+BIM, where it has a potential to enhance the renovation context to attain collaboration and performance outcomes. Section 5 discusses the findings on what and in which direction the digital systems can be further expanded. Finally, section 6 presents a brief conclusion and sets out recommendations for further research work.

2 BACKGROUND

2.1 *IPD and BIM*

IPD and BIM are two emerging and innovative approaches that reveal strategic changes and have the potential to stimulate an integrated intervention design and construction process for achieving project purposes in a collaborative environment (Rowlinson 2017). Succar (2009) considers BIM as integration of product and process modeling, besides a disparate set of technologies and processes, which generate a systematic approach for managing critical information within a unique and shared platform, forming a reliable basis for decisions throughout the building life cycle (Succar 2009). In recent years, more emerging technologies have been integrated into BIM for different utilization. Among the others, Zhao (2017) identified: mobile and cloud computing, laser scan, augmented reality, safety rule and code checking, semantic web technology, and automated generation as the hot topics of BIM recent researches.

IPD is an alternative delivery method based on a relational contract, where the project's goals are outlined in a collaborative environment within multiple participants (AIA & AIA California council 2007). The jointly shared risks and rewards among the stakeholders allow developing a sense of commitment, increase communication and creativity, and reduce litigation costs and efficiency (Ashcraft 2012). The highly cited Simple Framework of Fischer et al. (2017) positions the product as a starting point in the IPD framework. The authors combine four key elements: integrated organization, process integration, integrated information, and finally integrated system to create a high-performing building through virtual design and construction (VDC).

2.2 *IPD and BIM synergy for sustainable renovation of heritage buildings*

Megahed (2015) recommends BIM as support for IPD in heritages renovation to allow model-based collaboration between people, systems, and business structures and practice. Likewise, according to the previous research work by the authors of the current paper in (Brahmi et al. 2020), the application of novel and innovative combined methodologies of BIM and IPD were evaluated as to be very efficient in achieving a balance between sustainable design and the historical preservation values for heritage renovation projects, as well as to enhance productivity/efficiency in actual renovation processes.

In this regard, the authors (Brahmi et al. 2020) develop primarily an analytical framework consisting of a set of defined variables grouped into five thematic strands. The framework strives to encompass the multifaceted perspectives of the IPD and BIM synergies and facilitates the complex understanding of the sustainable renovation design process, given its highly complex value profile and many heterogeneous stakeholders.

By applying the analytical framework to existing heritage renovation projects, the shared collaborative practices across the projects and the degree to which the teams are able to implement the IPD and BIM tools and processes were determined. IPD processes drive collaboration, produce better solutions, and ensure on-time/budget project delivery. At the same time, BIM's tool is an enabler of IPD that foster collaboration and causes the multiple participants to provide complex analyses and sustainable design simulations at an early stage through interoperable platforms and software. It can be concluded that the IPD+BIM synergy provides sustainability achievement and better value to the project due to better collaboration and integration processes.

3 METHODOLOGY

As it was described in section 1, the aim of this paper is to explore, identify, and discuss the relevant digital technologies as a catalyst to elevating IPD+BIM synergy in the sustainable renovation of heritage buildings. In doing so, the research adopts a mixed methodology through the application of the analytical framework (Brahmi et al. 2020) as was described in section 2.2 besides triangulation approaches from (Denzin 1978; Love 2002). Triangulation is used for data collection (refers to the review of existing literature and analysis of case studies), with a set of defined variables consisting of 46 criteria, classified into 15 categories, and grouped into five thematic strands (people, process, policy, technology, and product), and to increase the validity of the research work.

The authors use Likert scale of five points from 1 (not important at all) to 5 (extremely important) to evaluate the importance of the digital technologies use related to the 15 categories of IPD and BIM synergy for sustainable renovation of heritage buildings. Similarly, the lack of digital technologies and the potential for further digitalization are evaluated using Likert scale from 1 (not relevant) to 5 (extreme lack of technology). Then, the authors demonstrate through radar charts the relative graphical positioning of the comparative assessment of the digitalization importance and then discuss the lack/need for improvement of digital systems concerning the five strands of people, process, policy, technology, and product. The authors identify and investigate how the digital systems open up new possibilities and where digital technologies add or are expected to add value.

4 RESULTS

The research begins by exploring the importance and lack of the application of digital technologies in an IPD+BIM framework in the sustainable renovation of heritages. Table 1 presents evaluation of the digitalization importance in this context, while Table 2 presents their lack. The evaluation is done based on the reviewed literature, previously studied cases in (Brahmi et al. 2020), including the renovation of: Renwick Gallery of the Smithsonian Art Museum (Washington, USA/ February 2012-July 2015); Centre Block of the Parliament Hill National Historic Site (Ottawa, Canada/ 2018- in progress); and Wayne Aspinall Federal Building (Colorado, USA/ June 2010-February 2013). We build on our many years of experience in the field of renovation of heritages as well.

Table 1. Importance Likert scale of the digital technologies used for elevating IPD+BIM synergy in the sustainable renovation of heritages.

Strands	Categories	Not important at all	Slightly important	Moderately Important	Very important	Extremely important
People	Team organization				X	
	Team selection & Capabilities		X			
	Team behaviors & Social dimensions	X				
Process	Project planning				X	
	Quality assurance & Commissioning					X
	Lean system				X	
Policy	Contract	X				
	Regulations					X
	Guidelines			X		
Technology	Software					X
	Hardware				X	
	Network				X	
Product	Nonstructured output				X	
	Structured output: Physical components					X
	Structured output: Virtual components					X

In order to address the findings systematically, and by following the developed analytical framework, here, the discussion is categorized using the five strands (people, process, product, policy, and technology). The following subsections elaborate on each strand.

The Radar charts are displayed below, showing each strand and its three categories. The scores around the triangles are listed counter-clockwise in descending. Two different line styles are used to represent the assessment related to each scale. The blue line represents the digital technologies application importance, and the red line represents the state of the current use or lack.

4.1 People

As demonstrated in Figure 1, the scores given to the three categories of Team Organization, Team Selection & Capabilities, Team Behaviors & Social Dimensions are low regarding both the importance and lack of digitalization, which represent the less important need for further digitalization on the people strand. Here, the use of digital technologies is limited to the involvement of different stakeholders in the renovation process.

Simeone et al. (2014) investigated the potential impact of BIM adoption in heritage renovations in order to enhance the collaboration among specialists and knowledge management. The authors conclude that similar to new construction projects, the BIM

Table 2. Likert scale of the digital technologies' lack for elevating IPD+BIM synergy in sustainable renovation of heritages.

Strands	Categories	Not important at all	Slightly important	Moderately important	Very important	Extremely important
People cline2-7	Team organization		X			
	Team selection & Capabilities	X				
	Team behaviors & Social dimensions		X			
Process	Project planning			X		
	Quality assurance & Commissioning					X
	Lean system				X	
Policy	Contract	X				
	Regulations				X	
	Guidelines		X			
Technology	Software					X
	Hardware			X		
	Network		X			
Product	Non structured output		X			
	Structured output: Physical components					X
	Structured output: Virtual components				X	

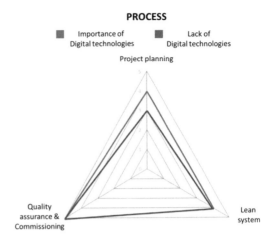

Figure 2. The relative graphical positioning of the comparison assessment importance/lack of digital technologies used for heritage renovation on the Process strand.

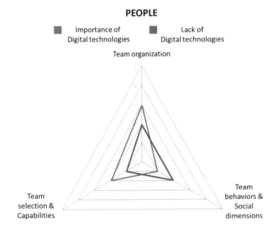

Figure 1. The relative graphical positioning of the comparison assessment importance/lack of digital technologies use of for heritage renovation on the People strand.

models ensuring the availability, accessibility, consistency, and coordination of all the knowledge related to a historical artifact and shared by the different actors involved in the investigation/conservation process; which support the decisions on developing the relevant interventions (Simeone et al. 2014).

As an example, in the Renwick Gallery's renovation, the BIM models are used to integrate the MEP engineering directly within architectural team, as well as to communicate with the building owners and facility managers. All stakeholders could see what was proposed through a full virtual-construction model, which streamlined decision-making and approvals, and allowed challenging spatial and historic preservation constraints to be addressed from the very beginning of the project.

4.2 *Process*

As illustrated in Figure 2, the scores given to the three categories are very high regarding both the importance and lack of digitalization, especially for quality assurance and commissioning. There is significant potential for further digitalization in these categories.

Among others, decision support is a crucial topic in heritage renovation. Gigliarelli et al. (2017) developed a holistic and multi-scalar methodology for energy intervention at a historic center and buildings of a town in southern Italy. The authors proposed a decision support systems (DSS) using a Multicriteria Analysis (MCA) tool (the Analytical Hierarchy Process) based on four key criteria: compatibility with restoration principles, economic affordability, energy efficiency, and environmental sustainability. The proposal was used to evaluate and select the best retrofit solution for a historical pilot building among various alternatives, involving experts of the research team and stakeholders. However, the authors highlight a number of limitations still present in the interoperability between software.

In terms of environmental sustainability, the team members of the three study cases used a set of sophisticated technologies, including Design Software, Energy Simulation Software, and lighting simulation software, to make assumptions required for early parametric building energy simulations. Contrariwise, Wong and Zhou (2015) claim that there are limited research effort for managing environmental performance at the renovation projects, along with the lack of 'cradle-to-grave' comprehensive BIM-based environmental sustainability simulation tool, as well as insufficient consideration given to the current cloud

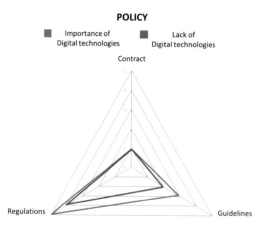

Figure 3. The relative graphical positioning of the comparison assessment importance/lack of digital technologies use of for heritage renovation on the Policy strand.

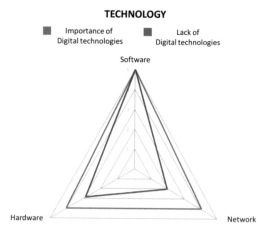

Figure 4. The relative graphical positioning of the comparison assessment importance/lack of digital technologies use of for heritage renovation on the Policy strand.

computing technology and 'Big Data' management within the green BIM tool.

In the same context, it is evident that the Lean approach is still limited to renovation projects. The analyzed case studies revealed that Lean construction principals and techniques had been incorporated into the projects in different degrees and a limited manner, even though it has been widely discussed within the Industry 4.0 context, as the digital technologies may further empower lean manufacturing applications to maximize value and decrease wastes (Ramadan & Salah 2019).

4.3 Policy

The analyses outcome in Figure 3 shows that there is a vital need for further digitalization in the regulation category; this requisite for tools like Rule-based Code Checking (Pocobelli et al. 2018) that provides coordination and standardization of policies and controls incorporating the environmental/energy performance and historic preservation codes, as well as automate the Leadership in Energy and Environmental Design (LEED) process for green building certification. With a relational data model that has a unifying control set, it will be possible to collect data and look at it without having to collect it again, to rationalize controls and reduce redundant efforts to comply with multiple regulations, so that the same information can be applied to multiple assessments and audits.

4.4 Technology

According to Figure 4, the scores given to the three categories are very high regarding the importance of digital tools in this strand; however, the software category has the most need for further digitalization.

Recent studies proposed a methodology for linking together Heritage-BIM with different digital technologies and simulations like the use of building performance simulation (BPS) and the computational design (Gigliarelli et al., 2017). However, (as mentioned in the process strand), they reveal the lack of open source platforms for BIM in heritage and the limitation of interoperability between different software environments as either gbXML files or IFC file (Cheng 2015; Gigliarelli et al. 2017). The energy modeling processes utilized for the Aspinall Federal Building are an example of how BIM and building analysis software data can be appropriately viewed and exported in a limited and controlled manner to help the process of designing a net-zero energy building. However, the additional analysis requires tools to calculate the upgrades' thermal performance of existing building components, and there are currently no workflows directly from a BIM to these tools.

In the same context, Kassem et al. (2015) discuss that the benefits for BIM and FM have yet to be established, especially for existing buildings. They argue the lack of open systems and standardized data libraries that can be utilized as a bridge between BIM and computer-aided facilities management (CAFM) technologies.

4.5 Product

As demonstrated in Figure 5, there is a large need for advance digital technologies to enhance the high-performance outcomes and energy efficiency of the renovation projects. Concerning the existing condition models, Murphy et al. (2013) identify the new approach of utilizing parametric objects "Historic Building Information Modeling" (HBIM) as a solution to the geometric primitives issue of a heritage building in the fact of variety and complexity of its

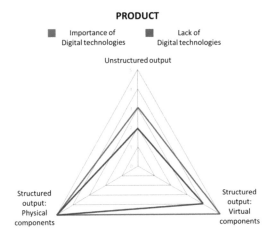

Figure 5. The relative graphical positioning of the comparison assessment importance/lack of digital technologies used for heritage renovation on the Product strand.

smart objects which are not representative for current typical BIM software libraries (Logothetis et al. 2015). HBIM system is a plug-in for BIM involves a reverse engineering process for modeling historic structures to represent heterogeneous and original existing morphologies (Murphy et al. 2013).

Nevertheless, many researchers have reported the lack of BIM implementation in heritage buildings due to challenges of high modeling/conversion effort from captured building data into semantic BIM objects (Pocobelli et al. 2018), where the unavailability of automated processes and the restrictively of using BIM in the case of a specific architectural style that is not present in parametric smart objects libraries and need to be manually modified (López et al. 2018). In addition, there are no open-source platforms that exist for HBIM (Cheng 2015).

In the Centre Block rehabilitation (Canada) case study, the team project in the beginning had a challenge with the process for verifying the model created from point cloud data involved creating multiple sectional views along with elements in Revit and measuring the deviations that appeared to be the greatest between the point cloud and the model element. This method was time-consuming, and it limited the verification of the model to specific section locations. Therefore, a verification system was developed that resulted in significantly enhanced communication and collaboration efforts amongst team members; the system increased the speed and workflow of the translation of data into building components as well as assisted in determining the integrity and accuracy of the model through visual quality control checks. It is worth reporting that, recent researches in this regard indicate the need for more studies to identify the limitations of this approach as new libraries for historical styles are developed. In addition, there is a need to develop and adjust of simulation software to accurately represent the conditions of heritage buildings and allow accurate environmental simulations (Khalil 2017).

5 DISCUSSION

According to the radar charts presented in the previous section, the importance and lack of digitalization vary in the five strands of people, process, policy, technology, and product. The analysis primarily shows both the high importance and lack Industry 4.0 concepts in facilitating the integrated processes and assuring the quality of the final product of renovation heritage projects concerning BIM application, especially in measurement and verification, post-occupancy performance, and modeling of the existing heritage building conditions.

The findings of the case studies and the literature review reveal that the integration of other emerging technologies within BIM enables to gain *in automation and further data manipulation* at different phases of a project's life cycle. In addition to the *knowledge sharing* opportunities (people) through the introduction of technologies that more efficiently support information sharing, the *interoperability* between BIM applications and energy simulation tools (technology) improve the *visualization* and *virtual simulation* of the renovation practices, as well as the operation of the renovated building (process and product). That can lead to more effective decision-making with the *standardization* of design practices (policy) to facilitate these processes on exploring and selecting among a large number of renovation alternatives and approaches available in the mark, and thus leading to cost savings, time-saving, and improving quality and sustainability.

The overall results reveal that there is a crucial need for further digital technologies to elevating the IPD+BIM synergy in heritage renovation, particularly in order to cope with multiple criteria and deal with the projects complexity and values concerning the 3D documentation of the heritage building, the simulation of efficient environmental performance analyses and sustainability enhancement, as well as assuring the post occupancy performance. Besides, it seems also to be vital, a need to create an approach that combines different methodologies, techniques and software to open up new possibilities to attain sustainability and high-performance outcomes. Such need can mainly categorize through the following three thematic areas:

Development of a conceptual framework of a *Cloud-BIM-Based Decision Support System*, that includes multiple criteria decision making to allow faster complex analyses, commissioning and make the appropriate decisions, through advancing interoperability between design team applications.

Development of *Rule-based Code Checking* to implement design verification and validation comparing BIM models against current codes and regulations translated into parametric rules.

Conception of *intelligent algorithms* capable of automatically converting point clouds into parametric objects.

6 CONCLUSION

The Industry 4.0 is perceived as a major enabler of sophisticated and integrated design and construction to promote sustainability and productivity. For sustainable renovation of heritage buildings, there is a significant unexploited existing and growing potential for the use of digital technologies yet to be explored. This paper presented an in-depth study of the current state of art of the existing digital technologies as a catalyst to elevating IPD+BIM synergy in the sustainable renovation of heritage buildings. The research adopted a mixed methodology through the application of an analytical framework (with a set of defined variables consisting of 46 criteria, classified into 15 categories, and grouped into five thematic strands - people, process, policy, technology, and product) besides triangulation approaches for data collection and validity of the research work.

The major findings confirm that moving towards Industry 4.0 promotes collaboration and sustainability in heritage renovations, similar in new construction projects. Likewise, the findings of the review literature show the existing gap in the current digital technologies substantially concerns the use of BIM for heritage renovation projects through the whole lifecycle, due to the interoperability issues and, more importantly, lack of open source platforms. That requires further research and development to extend beyond semantic object properties to include more facilities management, business intelligence, lean construction principles, green policies, and whole lifecycle costing data. It is observed that digital technologies such as decision support systems are still at a formative stage, as prototypes and applications are being developed for mainstream use in heritage renovation. On the other hand, the analysis of studied cases in this research study reveals that many developments have occurred during the last years in existing practices and practical ways for the adoption of new technologies to digitize and automate the renovation process to enable the different BIM functionalities. Meanwhile, the understanding of the high-performance outcomes generated from other digital technologies investment is appeared to be fundamental, as well as overcoming barriers. In this regard, future efforts should focus on examining more case studies that implement technological innovations, their successful processes/actions, and challenges.

This contribution is relevant to researchers and technology or software developers, who can use the resultant to advance better the digital technology used for the sustainable renovation of heritage buildings. In doing so, the advancement of digitalization can be used as a basis for the industry 4.0 adoptions in the new/existing building (or the manufacturing industry in general) for benchmarking the effects of digital technologies in the sustainable renovation of heritages. However, to enable the different tasks, it essentially needs to include the standard deliverable information requirements for heritage renovation at three levels: data modeling, data exchange, and process modeling. As such, it is required to further develop standard Level of Development (LOD) and Level of Information (LOI) for heritage metric survey specifications and model production.

REFERENCES

AIA National, AIA California council. (2007). Integrated Project Delivery: A Guide. Version 1, California. Washington, DC, 57 pp.

Angelini, M. G., Baiocchi, V., Costantino, D., & Garzia, F. (2017). Scan to BIM for 3D Reconstruction of the Papal Basilica of Saint Francis in Assisi in Italy. *The International Archives of the Photogrammetry, Remote Sensing and Spatial Information Sciences*, XLII- 5/W1, 47–54.

Brahmi, B.F., Boudemagh, S.S., Kitouni, I. and Kamari, A. (2020). IPD and BIM-focused Methodology to Renovation of Heritage Buildings. *Construction Management and Economics* (submitted for review).

Chang, R. (2019, January) The Renwick Gallery of the Smithsonian American Art Museum, Washington, D.C. High Performing Buildings. Available from: http://www.hpbmagazine.orgCase-StudiesThe-Renwick-Gallery-of-the-Smithsonian-American-Art-Museum-Washington-DC/.

Cheng, H.M, Yang W.B, & Yen, Y.N. (2015). BIM applied in historical building documentation and refurbishing. *The International Archives of the Photogrammetry, Remote Sensing and Spatial Information Sciences*, XL-5/W7, 85–90.

Chow, L., and Fai, S. (2017). Developing Verification Systems for Building Information Models of Heritage Buildings with Heterogeneous Datasets. *The International Archives of the Photogrammetry, Remote Sensing and Spatial Information Sciences*, Volume XLII-2/W5, 2017 26th International CIPA Symposium 2017, 28 August–01 September 2017, Ottawa, Canada.

Dallasega, P., Rauch, E., and Linder, C. (2018). Industry 4.0 as an enabler of proximity for construction supply chains: A systematic literature review. *Computers in Industry*, 99, 205–225.

Denzin, N.K. (1978). Sociological methods: A sourcebook. McGraw-Hill, New York, NY.

Eastman C. M., Teicholz P., Sacks R. and Liston K (2008). BIM handbook: A guide to building information modeling for owners, managers, architects, engineers, contractors, and fabricators. Wiley, Hoboken, N.J.

Fischer, M., Ashcraft, H., Reed, D. and Khanzode, A., (2017). Integrating Project Delivery. Hoboken: NJ: Wiley & Sons.

Fouseki K. and Cassar M. (2014). Editorial: Energy Efficiency in Heritage Buildings-Future Challenges and Research Needs, *The Historic Environment: Policy & Practice*, 5(2), 95–100.

Gigrliarelli, E., Calcerano, F., & Cessari. L. (2017). Heritage BIM, Numerical Simulation and Decision Support Systems: an Integrated Approche for Historical Buildings Retrofits. *Energy Procedia*, 13 (3), 135–144.

Kamari, A., Corrao, R., Petersen, S. & Kirkegaard, P.H. (2017). Sustainable Retrofitting Framework: Introducing 3 levels of Integrated Design Process Implementation and Evaluation. In PLEA 2017 (Passive Low Energy Architecture) conference, Edinburgh, UK.

Kamari, A., Jensen, S., Christensen, M. L., Petersen, S., and Kirkegaard, P. H., (2018a). A Hybrid Decision Support System (DSS) for Generation of Holistic Renovation Scenarios—Case of Energy Consumption, Investment Cost, and Thermal Indoor Comfort. *Sustainability*. 10 (4), 1255.

Kamari, A., Laustsen, C., Petersen, S., and Kirkegaard, P. H., (2018b). A BIM-based decision support system for the evaluation of holistic renovation scenarios. *Journal of Information Technology in Construction*, 23 (1), 354–380.

Kamari, A., Jensen, S.R., Corrao, R., & Kirkegaard, P.H. (2019a). A Holistic Multi-methodology for Sustainable Renovation. International Journal of Strategic Property Management, 23, 50–64.

Kamari, A., Schultz, C., and Kirkegaard, P. H., (2019b). Constraint-based renovation design support through the renovation domain model. *Automation in Construction*. 104, 265–280.

Kamari, A., Kirkegaard, P.H., and Schultz, C. (2020). PARADIS: A process integrating tool for rapid generation and evaluation of holistic renovation scenarios. Journal of *Building Engineering*. DOI:10.1016/j.jobe.2020.101944

Kassem, M., Dawood, N., Serginson, M and Lockley, S. (2015) BIM in facilities management applications: a case study of a large university complex. *Built Environment Project and Asset Management*, 5 (3), 261–277.

Khalil, H.B. (2017). HBIM and environmental simulation: Possibilities and challenge. New York, NY: Routledge. *Heritage building information modelling*. 190–202.

Lasi, H., Fettke, P., Kemper, H., Feld, T., and Hoffmann, M. (2014). Industry 4.0. Business & Information Systems Engineering, 6, 239–242.

Logothetis, S., Delinasiou, A., & Stylianidis, E. (2015). Building Information Modelling for Cultural Heritage: A Review. ISPRS Annals of the Photogrammetry, Remote Sensing and Spatial Information Sciences, II-5/W3.

López, F. J., Lerones, P. M., Llamas, J., Gómez-García-Bermejo, J., & Zalama, E. (2018). A review of heritage building information modeling (H-BIM). Multimodal Technologies and Interaction, 2(2), 21.

Love, P. E. D., Holt, G. D., and Li, H., (2002). Triangulation in construction management research. Engineering, Construction and Architectural Management, 9 (4), 294–303.

Megahed, M. A., 2015. Towards a Theoretical Framework for HBIM Approach in Historic Preservation and Management. *International Journal of Architectural Research*, 9 (3), 130–147.

Murphy, M., McGovern, E & Pavia, S. (2013). Historic Building Information Modelling – Adding intelligence to laser and image based surveys of European classical architecture. *ISPRS Journal of Photogrammetry and Remote Sensing*, 76, 89–102.

Oesterreich, T. D., & Teuteberg, F. (2016). Understanding the implications of digitisation and automation in the context of Industry 4.0: A triangulation approach and elements of a research agenda for the construction industry. *Computers in industry*, 83, 121–139.

Pocobelli, D. P., Boehm, J., Bryan, P., Still, J., & Grau-Bové, J. (2018). BIM for heritage science: a review. *Heritage Science*, 6(1), 30.

Ramakrishna, S., Ngowi, A. and Awuzie, B.O. (2020), "Guest editorial", Built Environment Project and Asset Management, 10 (4), 485–489.

Ramadan, M., & Salah, B. (2019). Smart Lean Manufacturing in the Context of Industry 4.0: A Case Study. *International Journal of Industrial and Manufacturing Engineering*, 13, 174–181.

Rowlinson, S., (2017). Building information modelling, integrated project delivery and all that. *Construction Innovation*, 17 (1), 45–49.

Simeone, D., Cursi, S., Toldo, I., & Carrara, G. (2014). BIM And Knowledge Management for Building Heritage. Proceedings of ACADIA 14, Los Angeles.

Succar, B. (2009). Building information modelling framework: a research and delivery foundation for industry stakeholders. *Automation Construction*, 18 (3), 357–375.

Tommasi, C., Achille, C., & Fassi, F. (2016). From Point Cloud to Bim: A Modelling Challenge in the Cultural Heritage field. *The International Archives of the Photogrammetry, Remote Sensing and Spatial Information Sciences*, XLI-B5.

Tomšič, M., Mirtiè, M., Šijanec Zavrl, M., & Rakušèek, A. (2017). Energy renovation of cultural heritage buildings "by the book". *Procedia Environmental Sciences*, 38, 212–219.

U.S. General Services Administration, Public Buildings Service. (2012). "GSA BIM Guide for Energy Performance." GSA BIM Guide Series, 05.http://www.gsa.gov/graphics/pbs/GSA_BIM_Guide_Series_05_Version_2.pdf

Zhao, X. (2017). A scientometric review of global BIM research: Analysis and visualization. *Automation in Construction*, 80, 37–47.

BIM education and training

Analysis of digital education in construction management degree programs in Germany and development of a training model for BIM teaching

M. Pieper & S. Seiß
Chair of Construction Engineering and Management, Bauhaus University Weimar, Weimar, Germany

A. Shamreeva
5D Institut GmbH, Stuttgart, Germany
Chair of Intelligent Technical Design, Bauhaus University Weimar, Weimar, Germany

ABSTRACT: The Building Information Modeling (BIM) methodology plays a key role in the digital transformation of the architecture, engineering, and construction (AEC) industry. The development of digital technologies, especially software programs for 3D design, engineering and construction, model-based quantity determination, calculation and scheduling or application of artificial intelligence in AEC, supports BIM use cases that were previously not feasible. Nevertheless, a human being must be able to master the digital technologies to implement the BIM methodology and achieve BIM goals. Well-trained specialists with BIM qualifications are in demand on the AEC job market today. The aim of this article is to analyze construction management degree programs in Germany regarding their anchoring of BIM in the curriculum. Based on the analysis and the derived requirements, a concept for a training model for the targeted BIM teaching of construction managers at higher education is proposed.

1 INTRODUCTION

1.1 Problem statement

Despite a successful completion and execution of research projects, the advancement of national BIM standards and the development of BIM tools, the current shortage of professionals trained in BIM remains a barrier to the universal adaptation of this digital methodology in Germany. The AEC industry demands people with new skills. Some studies maintain changes in the role of AEC professionals (Macdonald 2013), others define BIM job types like BIM project manager, director, BIM manager, BIM coordinator, BIM designer, senior architect, BIM mechanical, electrical and plumbing (MEP) coordinator, and BIM technician (Uhm et al. 2017). Although pioneering universities and colleges are replacing their drawing classes with courses that educate architects and engineers in BIM, there is a lack of well-trained construction professionals in the field (Nejat et al. 2012). The focus of this paper is therefore on understanding what BIM content is taught in construction management (CM) and how this education can be developed and improved.

1.2 BIM methodology

The BIM methodology describes the way in which buildings and civil engineering facilities are digitally designed, constructed, and operated. Encompassing both technology and process of design, construction, and facility management, BIM has a strong influence on the AEC industry and has been adopted to improve the construction process. Thereby BIM technology offers numerous benefits to the construction industry, e.g., improvement of information sharing and delivery, time and cost savings, improved quality, productivity gains, transparency in decision making, increased sustainability, reducing field construction problems and the number of requests for information, etc. (Borrmann et al. 2018; Eastman et al. 2018; Lee & Hollar 2013).

On the one hand, one of the most common definitions of the term BIM is modeling technology and bunding of processes to produce and analyze building models that contain all appropriate information created or gathered about that building (Eastman et al. 2018). On the other hand, BIM does not simply mean a technology for producing building models, but a collaborative process. The most important value to achieve using BIM is the support of effective communication and cooperation among project participants throughout the entire life cycle of a building.

1.3 BIM implementation in construction management

Both the technology and the business processes of the BIM methodology enable the productive use of BIM in the field of CM. A construction manager is located between design or engineer organization, owner organization, building company, and subcontractor companies in the conceptual diagram representing an AEC project team (Eastman et al. 2018).

The main tasks of a construction manager might contain quantity determination, scheduling, task management, cost estimation, cost planning, quality management, billing, etc. (Bargstädt & Steinmetzger 2016). Using the model-based approach of the BIM methodology, the information about cost and scheduling is linked with building model objects. That allows, e.g., model-based analysis of cost estimation, scheduling, or collision detection. A construction manager in the BIM process must be able to edit model data, perform analyses required for the project, and communicate with other project participants using the BIM collaboration format (BCF) to follow the openBIM approach. If the model data is created based on the employer's information requirements (EIR) and model objects have consistent attributes, the task to link the model with other information and to make model-based analysis might be simple. This shows that, on the one hand, the use of BIM extends the scope of activity of a construction manager, but on the other hand, it also offers the advantages of using modeling standards.

As far as the construction industry in Germany is concerned, it has the following special features that can affect BIM implementation in CM: the circumstantially structured legal regulations of the construction industry (e.g. Official Scale of Fees for Services by Architects and Engineers (HOAI 2013)), a variety of BIM tools and lack of interoperability, the absence of consistent BIM standards, the fragmented, modular structure of the building construction sector as well as the diversity of small construction companies.

2 RESEARCH FRAMEWORK

2.1 BIM education International

Studies show the lack of qualified employees as one of the main obstacles worldwide to the implementation of BIM in practice (Uhm et al. 2017). To improve this, numerous training programs for users in the field have been developed, which mostly resemble software training from a specific provider. This focus on technology, which means primarily on programs and models, is one major issue in the discussion about BIM. People and processes play a secondary role.

The same problem is also found in different studies on BIM education (Macdonald 2013; Nejat et al. 2012). In most cases, the description and usage of software is the first mentioned issue. However, users and their collaboration and management, as well as workflow processes including software, and standards and guidelines for modeling and data exchange, are important. Therefore, BIM education should be based on the three pillars of people, process, and technology (Forsythe et al. 2013).

Although BIM training is essential to promoting the construction industry and there are a few engineering schools that have been teaching BIM since 2000 (Barison & Santos 2012), integration of BIM education in construction management degree programs has its challenges. The main obstacles are a lack of consensus on what the curriculum should focus on (Nejat et al. 2012), software interoperability (Barison & Santos 2012), the choice of BIM technologies and the sharing of object-oriented models through the digital application (Adamu & Thorpe 2016), the lack of reference materials for teaching (Sabongi 2009), a shortage of well-trained personnel (Macdonald 2013) as well as academic differences.

In order to advance BIM teaching at universities, some studies propose integrated models in construction teaching and learning, e.g. virtual construction models (Forsythe et al. 2013), an integrated project delivery concept in the course curricula (Nejat et al. 2012), cooperation between industry and academic world (Lockley 2011) or recommendations for course design by experienced construction companies (Mutai & Guidera 2010).

However, most studies on BIM training agree that no single module can cover a wide range of BIM tasks. Some researchers propose BIM teaching in one or several courses (Hietanen & Drogemuller 2008) or integrating BIM with traditional or mainstream courses (Lee & Hollar 2013). Barison and Santos (2010) highlighted, e.g., following categories of BIM courses: introductory, intermediary, and advanced. They described (2012) the integration of an advanced BIM course with a CM course or with an interdisciplinary-collaborative design studio course. In their study (2011), they recommended identifying the responsibilities, functions, and competencies of BIM specialists before planning a BIM curriculum.

For CM students, the following main topics can be summarized for BIM-related courses: interdisciplinary team collaboration, 3D visualization and modeling for detailed conditions, clash detection modeling, 4D schedules, 5D model-based cost estimation and control, spatial trade coordination, planning logistics and materials and the site logistics (Barison & Santos 2012; Lee & Hollar 2013; Nejat et al. 2012).

2.2 Status of BIM education in Germany

In Germany, training guidelines have already been created to standardize BIM teaching in practice and to impart not only software skills but also roles and responsibilities in the BIM process (Borrmann et al. 2018). On the other hand, the further need for improvements in the training of BIM skills is demonstrated by the federal government's demand to implement BIM in academic education (Brokbals et al. 2017). Institutes of higher education are obliged to impart BIM-relevant knowledge and skills during the education of future professionals in AEC project management programs (Forsythe et al. 2013).

Relevant publications on this topic are from Brokbals et al. (2017) and Khorrami et al. (2015), whereby the former deals exclusively with BIM courses in Germany and the latter also considers BIM training in different countries and compares it to the German higher education. The studies investigated both civil engineering and architecture programs. Both publications analyzed the existing range of BIM courses and defined necessary contents and requirements from practice and research. They summarized the need for action for BIM teaching, but they did not develop a model lecture. Deubel et al. (2018) relied on these investigations and developed an interdisciplinary project course for the application of BIM for civil engineers and architects.

However, they do not describe how they defined the competences and learning goals accordingly for the participating architects and engineers, so construction managers were no target group, nor do they describe a didactic approach from a pedagogical point of view.

Besides that, the dynamics of the development of the BIM methodology over the last years justifies a new investigation of BIM modules and their training guidelines in CM degree programs in German higher education under distinguished aspects.

3 MATERIALS AND METHODS

3.1 Research method

The research methodology of this paper can be divided into the following steps:

1) *Data collection*
   (BIM courses at German higher education)
2) *Data analysis*
   (Text mining and cluster analysis with R)
3) *Comparison*
   (Analysis of the BIM education for construction managers at Bauhaus University)
4) *Didactic concept*
5) *Model lecture*
   (Design of a model lecture for the targeted BIM teaching of construction managers)

For the first step, the research area to be investigated was delimited before the data collection. Since the competences of the different professional fields are diverse (Uhm et al. 2017), the authors concentrate exclusively on civil engineering and construction management programs. The German higher education system is divided into universities of applied science and regular universities. Regular universities focus on theoretical knowledge and research. In contrast, universities of applied science concentrate on practice-oriented courses.

The collected data is based on the module manuals of the corresponding study programs, which were available online. It should be noted that the data quality depends on the actuality and correctness of the available module manuals, which is assumed. Lecture descriptions can be misinterpreted, and the lecturer can put them into practice in different ways. Nevertheless, these descriptions can be used as a first basis for an interpretation of the status of BIM in higher education.

Table 1. Data collection.

1. Type of institution	University/University of applied science
2. Study program	Civil engineering/Construction management
3. Degree	Bachelor/Master of Science/Engineering
4. Title and type of module	Compulsory/elective
5. Learning goals	Copied from the module manuals
6. Module description	

Table 2. Research objectives.

1. Is the teaching of BIM focused on …	2. What are the main learning contents and objectives of BIM teaching in …

a) master or bachelor programs?
b) universities or universities of applied science?
c) compulsory or elective modules?
d) informatics in construction or construction management?

3. Are the course contents and competencies taught according to the AEC disciplines?

BIM-related courses were searched and collected from the available module manuals, using "BIM," "Building Information Model," and "Building Information Modeling" as keywords (Table 1).

Based on this research, the developed database is analyzed in a second step, using text mining and cluster analysis. The following, underlying research objectives (Table 2) were defined:

3.2 Collected data

A total of 139 BIM modules at 52 institutions were registered. Based on a total of 60 institutions examined, it can be said that BIM is already integrated into German higher education studies. The modules can be classified according to the institution of higher education, the type of module, and the desired degree (Table 3). The higher number of modules at universities of applied science depends on the twice as high share of these universities in the study. Therefore, a ratio is also given.

The table illustrates that universities of applied science, unlike universities, make BIM-based lectures mandatory in their curriculum. The data also shows that BIM lectures at universities of applied science are more likely to be part of the bachelor's program. Universities consider BIM education to be a specialization within the master's program, which is based on elective modules.

It is difficult to generalize what BIM related content is taught in Germany. Figure 1 shows a word cloud of all 139 modules related to BIM that have been analyzed based on the module descriptions and goals.

As mentioned above, most BIM modules deal with technologies such as CAD, modeling, application, data, software, and tools. However, process- and

Table 3. Collected data.

Criteria	Universities Total	%	Universities of applied science Total	%	Total
Institutes	17	28%	35	58%	60
Modules	32	23%	107	77%	139
Compulsory	11	14%	66	86%	77
Bachelor	6		44		50
Master	5		22		27
Elective	21	34%	41	66%	62
Bachelor	5		15		20
Master	16		26		42

Ratio compulsory to elective modules in bachelor programs
1,2   2,9
Ratio compulsory to elective modules in master programs
0,3   0,8

Figure 1. Word cloud of all BIM modules.

person-related vocabulary is also revealed in the outer word cloud area.

4 RESULTS AND DISSCUSIONS

4.1 Analysis and comparison of BIM courses

At first, the differences between bachelor's and master's programs as well as elective and compulsory modules are analyzed. A comparison between the surveyed lectures was carried out by text mining and clustering the frequency of the words of the lecture contents and objectives from the module manuals. It was found that the bachelor modules mostly focus on the basics of planning, project management, and the preparation of service specifications, calculations, and scheduling. These topics are related to the fundamentals of digital information exchange and digital information processing using BIM.

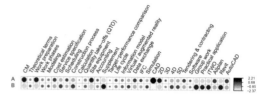

Figure 2. Chi-square correlation of analyzed lecture content.

Interesting for bachelor modules is also the frequent use of the word CAD, which is not so common in master modules. Bachelor modules therefore focus more on constructive tasks. In comparison, master modules concentrate on the application of digital tools in planning and their data exchange as well as the management of construction projects over the entire life cycle.

The study shows that group work or software-related work is equally represented in elective and compulsory modules. It has also been shown that project- or software-related work is increasingly used in master's programs, in 2 out of 3 modules compared to the bachelor's program with 1 out of 2 modules. This supports the thesis that master modules are designed to be more application oriented.

The further analysis of the modules was based on 34 keywords. These are established on the text mining of the module contents and objectives and reflect the teaching content of CM from the tender to the project handover. The relation was analyzed by a chi-square test of independence (Figure 2). The relationship of the keywords of the 139 modules was analyzed according to bachelor (A) and master programs (B). For a given cell, the size of the circle is proportional to the amount of cell contribution. Positive residuals are shown in black and negative residuals in grey.

It becomes clear that CAD, 2D, work preparation as well as construction site equipment are of high relevance for bachelor modules. Master modules are more concerned with the service specifications, the construction process as a whole, and supplement management. How the content is taught and whether the content is conveyed using BIM as a methodology cannot be assessed.

With reference to the BIM information level, the master modules aim to link time and cost (4D/5D). There is a trend that specializations such as augmented and virtual reality are becoming increasingly important in master's programs.

Less relevant in bachelor programs are topics like teamwork, practical relevance, and software applications. On the other hand, these points are important in the master modules, which also speaks for a larger application focus in these modules.

In addition, a correlation analysis was performed with the 34 keywords. This has shown that the following learning contents are strongly correlated and taught together: allocation of construction services, service phases, service specifications, scheduling,

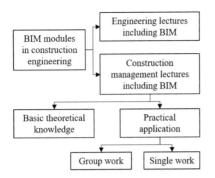

Figure 3. Dendrogram of module types.

work preparation, construction process, calculation, quantity takeoff, site setup, accounting, supplemental management, and controlling. This corresponds to the classical fields of construction management and work preparation.

The term data exchange is only mentioned in connection with IFC, simulation, CAD, 2D or 3D. In total, only 32 modules deal with data exchange. In the field of construction management, zero modules deal with that topic. A connection between CM topics and data exchange can therefore not be established, although this is very relevant for the different fields of activity. However, the use of the term IFC regarding data exchange suggests that an open BIM approach is taught. The correlation of the words "life cycle" and "IFC" also supports this assumption. Another interesting connection exists between the programs iTWO and Revit. It seems that in cases where the iTWO program is used, Revit is used as the modeling software.

The examined lectures can be categorized into informatics, engineering, or architectural lectures, which the authors have summarized as "engineering lectures including BIM" (Figure 3). On the other hand, modules with a strong focus on BIM-based CM are of great interest. These are divided into practical and theoretical lectures. This manual separation allows a better examination of the teaching contents and methods.

The study resulted in 29 lectures on BIM-based CM and 110 lectures on other areas of civil engineering. This fact illustrates the dominance of civil engineering compared to the field of CM in the study programs investigated.

The focus of the engineering lectures in the bachelor's program is on basic engineering, construction and design of components, and general informatics. The lectures on CM deal with project management in the construction industry, costs and times, service specifications and process modelling. Similarly, the master's programs focus on the digitization of project management in the construction industry, with the engineers focusing on information exchange and coordination and the construction managers focusing on work processes and methods. After the survey, the professors responsible for the 29 modules were asked about the content and didactic design of their lectures.

The most important results of the investigation, which are relevant for the conception of the model lecture, can be summarized as follows:

– Construction informatics lectures dominate the BIM teaching in Germany
– Modeling (3D) is often the focus of teaching although this step is more relevant for the architects and not the engineers
– There is a missing focus on data exchange
– Only one construction management lecture teaches explicit the usage of the German BIM-specification DIN SPEC 91400, a component-oriented classification and description system for BIM and IFC data exchange.

4.2 BIM education at Bauhaus University

The third stage of the research methodology described in this chapter is a BIM course that was conducted in the summer semester 2020 at the Bauhaus University in the master's program "Management [Construction Real Estate Infrastructure]".

The lecture was divided into two parts. The first part focused on industrial science including work design and business organization. The second part focuses on the use of BIM, starting with the tendering process and ending with the start of construction. For this purpose, a BIM-based architectural design as well as a prefabrication design, a quantity deduction, and a cost estimation with the tool DBD-BIM according to the national standard "DIN SPEC 91400" was performed in Revit. Afterwards, the precast model and the architecture model including the service specifications based on the BIM container according to the national standard "DIN 91350" are exchanged. The next step in RIB iTWO was a model check, quantity calculation, cost estimation, and scheduling. Working with an architectural model and a precast model serves to compare variants and to think through the assembly method from a planning, process, and calculation perspective.

During the course, students were trained in seminars in theory and practice of digital construction processes. Core of the seminars was the practical processing of a project work, which was given to the students at the beginning of the lecture and follows the above-mentioned process steps. This gives the students the opportunity to implement the theoretical approaches and the demonstrated software functions directly in their individual projects during and after the seminar. The instructions for handling the software were prepared in written and graphic form and made available to the students for individual processing. The strengths and problems identified during the semester can be summarized as follows:

Strengths:

– Based on the literature research, the designed module corresponds to the state of the art in BIM
– End-to-end process for a delimited part of the CM from modelling to scheduling

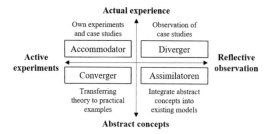

Figure 4. Learning types and examples of training sequences (Gerlach & Squarr 2015).

- Application of various software solutions that are relevant in practice (including data exchange)
- Group work for interdisciplinary exchange

Weaknesses:
General problems:

- Heterogeneity of students in the master's program: students come from different institutions or study programs that do not focus on digitization or CM (different starting points)
- Complicated process description: BPMN (business process model and notation) diagram necessary for students to understand better and faster which process steps build on each other (to avoid openly formulated tasks)

Content-related:

- No consideration of the entire construction phase (for time reasons), e.g. no design of the site equipment (only calculated)

Covid-19 related problems:

- No use of computer labs possible
  - Complications in software installation on students' devices
  - Very high supervision effort from a distance

The research results lead to the conclusion that it is necessary to revise the concept of the lecture and improve its didactic design.

4.3 Requirements and didactic concept

Before the content of the lecture can be designed, some basic conditions must be defined. The students of a lecture form a heterogeneous group with different levels of knowledge and learning types, which must be addressed equally. The following considerations are based on the concept of four learning types (Figure 4).

Some people find it easiest to understand new learning content by means of concrete examples, demonstrations, or experiments. Others find it easiest to understand new facts with the help of abstract models, diagrams, or theories. Thus, the different types of learners are addressed by differently structured learning sequences. Speed and level of learning must therefore be adaptable to the individual group in the process of the lecture. The learning process

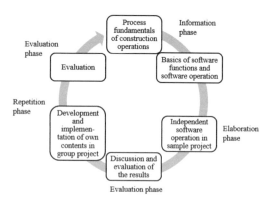

Figure 5. Learning phases in the module process.

should be based on a high proportion of exercises and repetitions. For the handling of software, it is especially important to teach system correlations and basic principles instead of detailed knowledge (Gerlach & Squarr 2015). The concept of the model lecture is based on the idea to address the 4 learning types most effectively.

According to Gerlach and Squarr (2015), learning units consist of different phases (Figure 5). The information phase provides the basic theoretical knowledge. In the following elaboration phase, tasks are solved with the acquired knowledge. In the evaluation phase, the solution methods are discussed and possible need for repetition is determined. Finally, these repetitions serve to consolidate the learned knowledge.

According to these learning phases, the concept of the module can be divided into three parts.

(1) At the beginning of the semester, students must think about what type of learner they are and what construction and BIM knowledge they have or lack. In this way, students can prepare themselves for the requirements of the module in a similar way to a preparation course. In addition, a short introduction or review of essential contents of construction management and BIM is given.

(2) Each seminar is introduced with a theoretical explanation of the topic, followed by an example task. Afterwards, the task is partially solved by the teacher and then reproduced individually by the students. Finally, individual group project work enables the students to apply what they have learned in practice. Videos of case studies and written instructions for software work are provided for the individual group work. Based on the good experiences, a growing FAQ, an exchange platform for students, regular consultations, and interim presentations will complement the lecture.

(3) Presentation of group work during and at the end of the course for feedback and to increase the exchange of knowledge between the students.

4.4 Structure of the model lecture

All three parts are processed according to the activities in the following BPMN-scheme (Figure 6). The

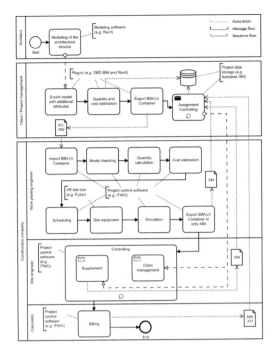

Figure 6. BPMN-scheme according to the CM activities.

horizontal lanes describe the role of the students according to those responsible in practice. Within the lanes, the different tasks of the students (process steps) as well as the software applications ([...]) and the data formats (visualized as text sheets) exchanged between them are defined.

The course is part of a master's program, so basic knowledge of CM and BIM is a prerequisite for participation. The process diagram focuses on the data exchange during the construction phase. For this purpose, common German exchange standards are used, e.g. BIM-LV container according to DIN SPEC 91350 as well as the GAEB DA XML file standards: X82 (service specifications), X84 (submission of tender), X31 (quantity determination and measurement), X89 (billing). A BIM-LV container is a subset of an IFC, GAEB, and link file. The link file links the IFC objects to the service specifications, which is described in a predefined XML schema (GAEB DA XML).

In the first lecture the students' knowledge of BIM-based design and data management is repeated. Accordingly, students will learn how to change attributes and geometries of a parameterized, example model, which is worked on in group work. The main goal of the course is to demonstrate a digital handling of the process steps of CM. The students are not taught in modeling or design, as this task is traditionally performed by the architect, which defines the starting point of a project.

Students take on the role of the project manager (PM) and use the tool DBD-BIM (plug-in for Revit) for model enrichment, quantity takeoff, and estimation. Then they learn how to export the model according to the BIM-LV container. The PM is also responsible for storing and sharing data objects, e.g., through Autodesk 360 Docs, a project documentation, and collaboration platform (visualized as database). It is used to exchange the group's project data.

Before the change into the responsibility of the building contractor, a short presentation is given by the students. Afterwards, the lecture deals with work preparation. For this purpose, the students have to learn how to use a BIM-LV container, check the model, perform a quantity measurement, a cost estimate, schedule, and site equipment planning, as well as the standardized exchange of the tender. All these activities are carried out in a project management software, e.g. RIB iTWO, except for planning and simulation of the construction site equipment. This activity is represented by a VR tool, e.g. Fuzor. Fuzor enables students to visualize and simulate 4D and 5D models. Furthermore, it provides a catalog of construction equipment objects, which can be used to visualize the construction process in detail.

On the following track, the site engineer takes over the project and must deal with the control activities. These are divided into supplementary management and claim management. The handling of claims and supplements is taught to students in theory. For practical supplement management, students receive group-related changes and deficiencies from the lecturer. Due to the changes, the students must create a supplement and exchange it with the PM. In addition, the lecturer initiates complaints in accordance with the project management activity and activates the management process of this for the students. The students must process and document the claim as well as the supplements in the CM program. Finally, the last service billing ends the process of the module. In this step, the amounts and final costs are submitted to the project management or the client.

5 CONCLUSION AND FUTURE RESEARCH

In this overview, an attempt was made to examine content in the BIM training of civil engineering and construction management programs. A basic scheme for a BIM-enabled construction management lecture was designed. The paper is the first with a concrete focus on a standardized BIM teaching in the relevant process steps of construction management. The presented BPMN model is kept general and can be applied to all courses in this area. An international adaptation of the scheme for the development of BIM lectures in CM is therefore conceivable. For this purpose, national standards such as exchange formats, basic principles of quantity calculation, and contractual conditions must be observed and adapted.

Data exchange and cooperation is a central element of the BIM methodology, which is to be taught to students in addition to pure software use. The listed software represents only one possible solution, the teachers can choose to use other tools. However, due to the difficulties of installing some programs on students' devices, cloud databases and web-based work platforms should be used.

The developed lecture will be integrated and verified in the curriculum at the Bauhaus University in the coming semester to generate further improvements in cooperation with the students. The integration of further tools and the use of gamification to increase student motivation are conceivable. Due to the current Covid-19 pandemic, which also affects education, good digital teaching is becoming increasingly important. This is especially true for this topic and should be seen as an opportunity, as it illustrates very well the digital communication of project participants using the BIM methodology.

Demand for such education is also reflected in the development of independent study programs such as construction information technology, which combines business informatics and construction management, thus addressing the industry's requirements for specialists in this field. In addition, it remains important to train civil engineers and construction managers as generalists who need basic knowledge to design the BIM methodology in their field. The lecture presented in this paper can contribute to this.

REFERENCES

Adamu, A.Z., Thorpe, T. 2016. How universities are teaching BIM: a review and case study from the UK. *Journal of Information Technology in Construction – ISSN 1874-4753, ITcon Vol.* 21: 119–139.

Bargstädt, H.-J., Steinmetzger, R. 2016. *Grundlagen des Baubetriebs.* Weimar: Schriften der Professur Baubetrieb und Bauverfahren Nr. 65.

Barison, M.B., Santos, E.T. 2010. Review and analysis of current strategies for planning a BIM curriculum. *Proceedings of the CIB W78 2010 27th International Conference*: 1–10.

Barison, M.B., Santos, E.T. 2012. BIM teaching: current international trends. *Gestão e Tecnologia de Projetos* 6: 67–80.

Borrmann, A., König, M., Koch, C., Beetz, J. 2018. *Building Information Modeling. Technology Foundations and Industry Practice.* Cham: Springer Nature.

Brokbals, S., Cadež, I. 2017. BIM in der Hochschullehre. Entwicklung – Status Quo – Handlungsbedarf. *Bautechnik* 94(12): 851–856.

Deubel, M., Zelling, I., von Both, P. & Hagsheno, S. 2018. Interdisziplinäre Projektbearbeitung mit BIM in der Hochschullehre am Beispiel des Karlsruher Instituts für Technologie (KIT). *Tagungsband zum 29. BBB-Assistententreffen – Fachkongress der wissenschaftlichen Mitarbeiter der Bereiche Bauwirtschaft |Baubetrieb| Bauverfahrenstechnik.*

Eastman, C., Sacks, R., Lee, G., Teicholz, P. 2018. *BIM Handbook – A Guide to Building Information Modeling for Owners, Designers, Engineers, Contractors, and Facility Managers.* Third Edition, Canada: Wiley John + Sons.

Forsythe, P., Jupp, J. & Sawhney, A. 2013. Building Information Modelling in Tertiary Construction Project Management Education: A Programme-wide Implementation Strategy. *Journal for Education in the Built Environment* 8(1): 16–34.

Gerlach, S., Squarr, I. 2015. *Methodenhandbuch für Softwareschulungen.* Heidelberg: Springer Vieweg.

Hietanen, J., Drogemuller, R. 2008. Approaches to a university level BIM education. *IABSE Conference, Helsinki.*

Khorrami, N. 2015. Implementierung der Methode BIM in der Aus- und Weiterbildung in Deutschland. *Tagungsband zum 26. BBB-Assistententreffen – Fachkongress der wissenschaftlichen Mitarbeiter der Bereiche Bauwirtschaft | Baubetrieb | Bauverfahrenstechnik.*

Lee, N., Hollar, D.A. 2013. Probing BIM Education in Construction Engineering and Management Programs Using Industry Perceptions. *49th ASC Annual International Conference Proceedings, Associated Schools of Construction.*

Lockley, S. 2011. *BIM and Education.* BIM Report, London: Riba Enterprises LTD.

Macdonald, J.A. 2013. A framework for collaborative BIM education across the AEC disciplines. *37th Annual Conference of the Australasian Universities Building Educators Association (AUBEA), The University of New South Wales, Australia:* 223–230.

Mutai, A., Guidera, S. 2010. Building Information Modeling in Construction: Current Practices and their Implications for Construction Engineering Education. *Conference Proceedings of 2010, ASEE Annual Conf.*

Nejat, A., Darwish, M.M., Ghebrab, T. 2012. AC 2012-5582: BIM teaching strategy for construction engineering students. *American Society for Engineering Education.*

Sabongi, F.J. 2009. The integration of BIM in the undergraduate curriculum: an analysis of undergraduate courses. *Proceedings of the 45th ASC Annual Conference, Florida.*

Uhm, M., Lee, G. & Jeon, B. 2017. An analysis of BIM jobs and competencies based on the use of terms in the industry. *Automation in construction* 81: 67–98.

Experiences from large scale VDC-education in Norway

E. Hjelseth
Norwegian University of Science and Technology (NTNU), Norway

M. Fischer
Stanford University, USA

ABSTRACT: This study investigates how a new approach for large scare education can solve the need for increased performance in design and construction projects. A large-scale course in Virtual Design and Construction (VDC) with 200+ participants in Norway is used as a case for assessing this type of learning environment. The study is done by collecting assessment of student's feedback and student reports from real projects. The results are analyzed through the lens of the "Community of Practice" (CoP) theory. This theory explains why the learning outcomes in knowledge, skills, and attitude can be relatively higher than with traditional classroom courses with 20 participants. The impact of organizing the learning environment as a CoP can contribute to fill the competency gap in the construction industry in a very efficient way. A good course is not only about content, but to create confidence to change current practice and performance.

1 INTRODUCTION

This study shares course participants' experiences in implementing large-scale Virtual Design and Construction (VDC) education comprising of over 200 participants from the Architecture, Engineering, Construction/Facility Management (AEC/FM) industry in a single course. The course is offered to fill a competency gap in the AEC/FM industry to attain high performance in design and production of the built environment. It is well documented that the AEC/FM industry needs to improve project performance. Statistics of the AEC/FM industry show that there is a slight decrease in productivity (effectiveness) in most countries (Norway: SSB (VDC-NTNU 2019), USA: reference to national statistics) whereas other industries show continuous improvement. The gap in productivity as compared to the other industries is partly due to the lack of competency and collaboration. The challenge is to make improvements of course delivery to increase practitioner competency.

This study explores experiences from a large-scale (200+ participants) VDC course in Norway by comparing it to traditional classroom (20 students) courses, addressing the followings research questions:

- What are the experiences with large-scale VDC courses?
- What is the impact of a large-scale course for creating a community of practice?

This study uses a large-scale VDC education program in Norway. Likewise, VDC-educations with 100+ participants have just been started in Peru and Switzerland and are considered in other countries. This study focuses on this type or practice-related course. Experiences should be transferable from and to similar learning environments, independent of professional content. The impact or scaling up and still obtaining high learning outcomes can contribute to fill practitioners' gap in competency. The outcome of this study should be of interest for the entire AEC/FM-industry.

2 CASE DESCRIPTION AND METHOD

The Norwegian University of Science and Technology University (NTNU) and the Stanford Center for Professional Development have in collaboration developed a 9-month VDC certification course worth 15 ECTS (credits). The first course had 202 students and started with a 5-day workshop in August 2019, followed monthly reports and a 2-day mid-term workshop in January 2020. The final 2-day workshop is in May 2020 online due to the coronavirus pandemic. This study is based on the students' experiences until March 2020 including all 6 periodical (monthly) reports. The majority of the course is completed; as of the writing of this paper, the final report and last workshop are yet to be completed.

The students (course participants) are practitioners from the AEC/FM industry. The reason for calling the practitioners "students" is because the VDC course is a university course at the Master's level with pre-defined learning outcomes approved by the NTNU educational committee. The students who pass the reports do also achieve a VDC Certificate of Achievement from the Stanford Center for Professional Development. Number of students answering, N = 202.

The student distribution was:

- Owners/clients: 7%
- Architects: 2%
- Consulting engineers: 48%
- Contractors: 43%

With N = 202, 40% are from the civil engineering sector (road, rail, infrastructure) and 60% from the buildings sector. The 6 software providers were classified according to affiliation of their software. The students can therefore learn from each other across domains, N = 178, 24 did not answer.

Work experience,

- 0–1 year: 4%
- 1–4 years: 30%
- 5–10 years: 32%
- more than 10 years: 34%

The majority of students have a moderate to high level of work experience. Sharing their experience is therefore a significant source for learning and reflection.

The gender distribution, N=202, was 46 women (23%) and 156 (77%) men.

The VDC course is focused on use of VDC for real projects. When the course participants were asked about their current project and whether they are able to apply VDC in practice, the response was: "Do you have a project to implement (try our) VDC?". N = 178, 24 did not answer.

Own project:

- I have a project I will lead: 17%
- I will participate in a project: 42%
- Expect to lead a project: 11%
- Expect to be on a project: 22%
- I'm not working on real projects: 6%

As part of the course, the participants develop 6 monthly (periodic) reports and one final report. These reports are based on practical experiences with implementing VDC in their own projects. Each student is part of a group of about 20 students that is guided by one mentor, who gives individual feedback on the reports. The level and progress in the reports is assessed based on 19 pre-defined assessment criteria. Each mentor also meets at least once per month with the students sub-groups. The 10 mentors in this course are practitioners with significant experience leading the implementation of VDC on their own projects and in their companies. They are part of the "mentor team", with regular meetings for sharing experiences and insights.

3 SHORT INTRODUCTION TO VDC

This section briefly introduces VDC (Virtual Design and Construction). VDC was conceptualized at CIFE (Center for Integrated Facility Engineering) at Stanford in 2002 to combine emerging information management methods like BIM (Building Information

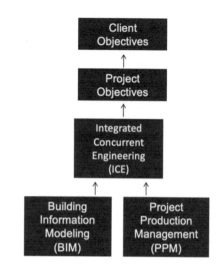

Figure 1. VDC framework.

Modeling) with process management and collaboration methods in pursuit of project excellence.

Every project team wants to create a high-performing building for its client(s) and users. Note that "building" is understood very broadly here to mean any built structure, from roads and other infrastructure assets to industrial plants to commercial and residential buildings, etc. This means that clients and users have to define what high-performance means to them. In other words, they have to establish how the building should perform during the use phase, which includes use, operations, and sustainability objectives. In the VDC framework (see Figure 1), these objectives are called "client objectives". They are the raison d'être for the building and can only be measured while the building is in use. To show that the building project was successful at the handover of the building to the client, the project team has to define measurable project objectives. These objectives should support the client objectives. Since they can be measured when the building is turned over to the client, they become the main measure of success for the project team. Project objectives address the buildability of a building and typically fall into the commonly used four categories of project objectives: safety, quality, schedule, and cost (with cost understood broadly, i.e., in terms of money, of course, but also environmental impact or cost, etc.).

One of the main tasks, if not the main task, of the project leadership team is to design the project organization and its working methods to make it as likely as possible that the project objectives and eventually the client objectives are reached. This includes, of course, the selection of key project participants, i.e., disciplines, how they'll collaboration, the workflows they'll follow, and how they'll create and use information, since information is the lifeblood of any organization. Note that the project and client objectives must motivate the selection of the key project participants. Hence, in the VDC framework, the collaboration method called ICE (Integrated Concurrent

Engineering) is at the heart of the framework, since it's the collaboration of the key disciplines and the decisions they make that put the project on the road to success or disappointment. When the key disciplines work in an integrated and concurrent way to engineer good solutions, it is much more likely that the project objectives will be met than when they work in a fragmented, sequential way. Such highly effective project teams must be able to rely on information they can trust and workflows or processes that consider the use of digital tools and the advantages of ICE sessions.

We find that many leading professionals and many project teams avail themselves of some of these methods, e.g., some lean process management methods are used, or BIM is deployed for some phases by some disciplines, but we see very few teams that leverage all the VDC methods fully to maximize the chances for project success. Enabling more and more professionals and project teams to leverage VDC holistically, not just in bits and pieces, all the time is the main motivation behind the large-scale VDC programs.

4 RESEARCH METHODS

This study is based on multiple methods. Assessment of the workshops are based on a net-based survey at the end of each day in the workshops. The progress in learning is based on the analysis of students' results of their submitted reports. The learning outcome is based on thematic analysis of these reports.

This was based on exploring a randomized selection of 25 reports from the latest (fifth of the six) periodical (monthly) reports. In addition, we include a selected number or "single sentence statements" reported in the check-in with their mentors on the last reporting. A short reflection regarding net-based / blended learning is included in the discussion section.

Community of Practice" (CoP) (Wenger 1998, 1991) is used as a theoretical lens to analyze and reflect over the findings in the discussion section. The authors have the role as course managers, and their background is inherently part of reflections made in this paper. The role of the authors on the course can of course influence the assessment of the VDC-course. The scope in this study is therefore not to prove that this single VDC-course was a success, but to provide a deeper understanding of the course delivery i.e. how this type of large-scale practitioner related courses impacts the AEC/FM industry. The outcome of this study should therefore be relevant for all competency programs in the industry.

5 RESULTS

5.1 Learning as part of a community

The results show that the students are very satisfied with this type of course. This includes both the way the course is organized and its professional content. Learning by working on practical implementations of VDC under the guidance of the mentors contributes to align theory and practice.

The large course offers a significant advantage over a small course because it doesn't only enable the students to learn and report by applying principles in their own projects, but they do so as part of a community. Such an experience is difficult to obtain in traditional courses.

The students are positive about being part of a multi-disciplinary course with workshops. Working together and sharing experiences in presentations (selected reports) both workshops with all other students, in workshops with their mentor team of 20, and in the regularly monthly check-in with parts or their mentor team. The student share domain, practice and community. The structure of the learning environment has contributed to establish a VDC-community of practice.

5.2 Experiences for the introductory week workshop

Figure 2 indicates that the students are pleased with the VDC course. The net-based survey was done by all two hundred students using Mentimeter (Mentimeter 2019) in the introductory week workshop from 19.–23. August 2019. There is one bar for each course day.

An interesting finding (also related to CoP) is that the students are more pleased with their groups' efforts than with their own effort. Another finding, related to "Today's objectives were clearly decided", is that it takes time to grasp the holistic view of VDC as it is a systemic framework. The students' maturity in understanding is therefore increasing during the entire workshop; evident from the lowest starting point (Monday) and highest final score (Friday).

The use of metrics on the themes shown in Figure 2 illustrate that development of qualitative targets can be measured. The use of free digital tools like Mentimeter (mentimeter.com) and others like Kahoot (kahoot.com/kahoot.it) demonstrate that this is applicable to implement in real projects.

5.3 Quantified VDC statements

As part of preparing the students' final report, all students were asked to write one-sentence statements about the impacts they have observed in their own projects. The list below presents a selection of statements where the impact was quantified:

- 40% reduced design time using the VDC methodology.
- With VDC, we cut the design time by 30%, reduced the risk, and improved the interdisciplinary collaboration.
- VDC ensured that the construction project, including 30% required additional tasks, was completed in time for the customer to move in 2 months earlier than originally planned.

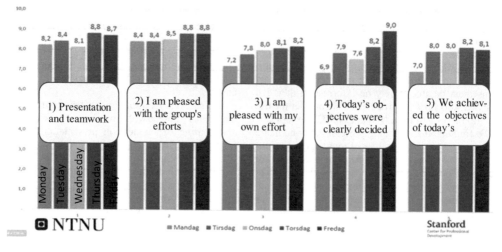

Figure 2. Student assessments of the Introduction week workshop at the VDC-course.

– Active use of VDC in collaboration with external contractors associated with construction shows potentially a 5% reduction in design time and a 20% reduction in production time on site time."

5.4 Qualitative VDC statements

Not all effects of VDC can be expressed with quantitative values. The students have expressed the importance to focus on the process and "soft factors". Examples of these types of statements are presented below:

– VDC has contributed to increased interdisciplinary understanding, greater ownership to the project, in addition to better management through continuously measurements.
– Carefully considered measures based on VDC measurements allow for flawless delivery and satisfied customer.
– Implementing ICE and PPM, in addition to having well-defined customer and project goals (we have implemented BIM in the past), has made us better equipped to meet customer wishes and needs.
– The VDC framework provides a logical and structured collaboration platform that keeps goals, value, and follow-up close together with collaboration. It is a collaborative platform that embraces all project types.
– The VDC framework breaks down professional barriers and promotes good interdisciplinary discussions. This results in better scorecard, increased quality, and more inspired employees.
– VDC provides increased understanding and knowledge of the project's critical points in terms of quality, cost, and execution, while ensuring goal achievement.
– VDC provides a framework for managing a successful project delivery.

– VDC has helped me formulate and develop goal parameters that have increased the quality of meetings and decision-making processes.

5.5 Reflections from students

The following are longer reflections from the students' 5th (of 6) monthly reports. Comparative studies have not been part of the course, but we have one relatively reliable example where a student working as a project manager in a large construction company reports:

> "Current project has 40% shorter design time compared to the last similar project. We saved working hours for the project manager, with a value of approx. EUR/USD 35,000. In addition, he can do other income-generating works. Our aim is to reduce construction time by over 10% and the current status is a 4% reduction."

Another comment focuses on the time for advance preparation versus time spent on the ongoing project:

> "I find that we have much higher benefits than expected, at a slightly lower cost than expected. I spend a lot more time on assignments with VDC methodology than in traditional implementation (though probably part of my time can be attributed to the study itself), but overall, we have gains".

However, the ongoing VDC course does not cover every aspect; the following reflection draws attention to a missing part:

> "The importance of human factors in the implementation plan has been underestimated. The assumption that the participants in the project have expertise about VDC and the use of BIM, ICE, and PPM is greatly overestimated. A

way to measure maturity in the VDC group should be introduced. Here, I do not have clear measurements for myself, and how to do this."

Metrics is one of the five fundamental VDC elements (see Figure 1). Metrics is one of the aspects of implementing VDC that students report to have most impact in the project performance. All projects have an increased number of metrics compared to traditional projects. The number of metrics reported vary from 4 to 14 per project. The number of metrics has shown to be increasing during the course period.

Visualization is another aspect that is highlighted as important for project management and coordination. Visualization enables a joint understanding in a short time and is therefore an important part of good commination about the status and progress of the project.

Client objectives: Increased focus on the client, both as including the client more actively in the design and production process. It is also important to have Client Objectives as a defined objective, not only as part of the project objectives, as in traditional projects.

Multi-disciplinary: VDC contributes to a more organized ways of working with multiple disciplines due to structured coordination and decision making through ICE sessions in combination with breakout sessions for dedicated tasks.

Reduction of latency: Improved processes for developing solutions and making decisions in ICE session are frequently highlighted as a benefit of VDC.

Motivation: VDC is about people collaborating, where optimal motivation is critical. Awareness of motivation, and variation of motivation during the project is included as a metric in the reports. Even if implementing VDC can be a challenge, there are no reports indicating that one wants to go back to the old way of working. Some students reported in latest report reduced resistance to change, which may be connected to the time required to gain a matured holistic understanding of VDC.

6 DISCUSSION

It is challenging to explain the learning outcome of the VDC course just by the presented course content itself. There "must" be something more – and, in our understanding, this must be related to the establishment of a "self-enforcing learning environment" because the students "make each other good". This impact is connected to how students express and share their competency with each other.

The scale of the course implies increased probability for "good reports". When students share and explain their experiences this contributes to learning for all. It is a unique aspect when interpreting the results and understanding why the results improve as the course progress. In traditional pedagogy, the ratio of teacher/student is an indicator of educational quality. Both NTNU and Stanford have experiences with traditional courses where one teacher has 20 students

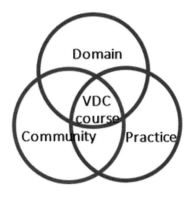

Figure 3. Community of Practice (Lave & Wenger 1991), adapted to VDC by the authors.

with relatively positive feedbacks from the students. In our learning environment can therefore the students be regarded as "peer teachers"

Awareness of the new learning environment that large scale education creates is essential for understanding the impact of this course, even though the course is about individual competency and certification. This discussion explores social learning versus individual learning. "Community of Practice" (CoP) is therefore used as a theoretical lens for understanding the relatively high performance in learning objectives documented in the student's project reports.

The normal size of courses is around 20 students, so the lecturer can follow up with each student individually if needed and adapt the level of professional content to the students (in addition to recruit students with a similar level and background). In the large-scale VDC course, where there are more than 200 students with very diverse backgrounds (refer to the case description above), adaptation to each student is not applicable. Scaling up to 200 should therefore be typically regarded as a high risk for the course providers and also the students.

CoP is defined by Wenger (1998) as groups of people who share concern ("domain") or a passion for something, which makes them learn how to do it better ("practice") as they interact regularly ("community"). These relations are illustrated in Figure 3.

Studies by Wenger et al. (2002) confirm that CoP influence change in practice, which requires a change in practitioners' behavior.

The social and cultural context within which CoPs operate is likely to influence impact. CoP is also a mechanism to share tacit knowledge, spark innovation, and create a steep learning curve. These effects are observed in the large-scale VDC course. It is important to be aware that the students are actually implementing VDC (in varying degrees) in their own projects.

Practitioners gain confidence by the fact that 200 other practitioners are implementing the same methods in their projects. The impact of community of practice is also supported by Bui (2020) in studies within implementation of digitalization. Being part of a community creates a force to try out ways of working and co-working that not would be applied into practice if you were doing this alone as the first pioneer.

This study demonstrates the importance of including a holistic approach to establish a learning environment. It will therefore be interesting to share experiences with other educational programs.

7 CONCLUSION

Large-scale VDC education is offering the participants a different type of learning environment than traditional courses. Large-scale VDC education has not only the capacity to fulfill the learning objectives, but also to create an environment for change and continuous improvement in project performance.

The extraordinary results achieved in the large-scale course are explained by the establishment of a "Communities of Practice" where the students and the mentors collaborate, co-learn, and connect in a self-reinforcing way. The impact of large-scale educational programs can contribute to continuous learning and improvements in each student's own projects. The total impact can result in significant improvements for the AEC/FM industry.

Another impact is the relatively short time it takes for the students to reach a mature level which allows them to not only earn to the VDC certificate, but also gain confidence to change the performance of their team and project deliverables.

REFERENCES

Bui, N. (2020). Implementation of Digital Engineering Methods in Infrastructure Construction, Phd. thesis in progress, dissertation scheduled 2020, University of Agder, Faculty of Social Sciences, Norway.

Lave, J & Wenger, E. (1991). Learning in Doing: Social, Cognitive and Computational Perspectives, Cambridge University Press, ISBN: 9780521423748.

Mentimeter (2019). Everything you need for interactive presentations., https://www.mentimeter.com/.

VDC-NTNU (2019). VDC-Certificate Program Norge (NTNU – Stanford Center for Professional Development) https://www.ntnu.edu/studies/courses/BA6280#tab=omEmnet https://www.ntnu.no/videre/gen/-/courses/nv18109

Wenger, E. (1988). Communities of Practice: Learning, Meaning, and Identity. New York: Cambridge University Press.

Wenger, E., McDermott, E.A. & Snyder, W. (2002). Cultivating Communities of Practice: A Guide to Managing Knowledge, Harvard Business Press, ISBN 9781422131107.

The efficacy of virtual-reality based safety training in construction: Perceptions versus observation

M. Poshdar, Y. Zhu, N. Ghodrati, H. Alqudah & J. Tookey
Auckland University of Technology, New Zealand

V. A. Gonzáles
University of Auckland, New Zealand

S. Talebi
University of Huddersfield, UK

ABSTRACT: The construction industry is widely regarded as a high-risk environment that is also known for its high fatalities and injury rates. The lack of appropriate health and safety training is a leading cause of accidents on construction sites. In recent years, several industries have experimented and gradually employed Virtual Reality (VR) technology to enhance safety training. Unlike traditional health and safety training that relies on a one-way instructional method of delivering information, VR provides experienced-based learning by placing the user inside a drill. This paper reports on a preliminary study that sought to assess the size of the efficacy of VR-based training within the construction industry context. By utilizing a mixed-method case-control research methodology, the study compared achievements made through the traditional methods of health and safety training such as the use of PowerPoint presentations against VR-based training. The achievements were evaluated based on the immediate perception of the trainees compared to the actual observations afterwards. The study focused on two particular aspects of the training outcomes: (1) impact on the knowledge retention by the trainees to identifying risks on construction sites, and (2) influence on their future risk-taking behavior. Eight residential construction sites in Auckland, New Zealand were selected as the indicative clusters of construction activities in the region. Correspondingly, a simple random method was applied to select one of the sites and sample ten trainees from the nominated site. Accordingly, a control group of five professionals received approximately 25 minutes of traditional health and safety training. An intervention group of another five professionals received the same length of health and safety training while using VR technology. After each training session, the groups were interviewed through the use of face to face semi-structured strategy twice. In both rounds of the interviews, the respondents were quantitatively assessed by rating their responses on a 5-point Likert scale. The first round was conducted immediately after the training session. It graded the perception of the respondents about their level of engagement and satisfaction, besides the impact of the session on their future safety behavior. In order to examine the genuine achievements of the training sessions, the second round of interviews was instituted the day after the training session. It assessed the participants' recall of the training contents and their risk-taking behavior in a hypothetical scenario. Despite the strong belief from trainees about the effectiveness of the VR technology in improving their health and safety learnings, no significant difference was observed in their risk-taking behavior and risk identification ability. Furthermore, some cases of nausea and dizziness were observed when using the VR headsets. The results can be generalized only if the future studies with bigger sample size show a similar outcome.

1 INTRODUCTION

The construction industry is generally considered as a high-risk domain with high accident numbers and injury rates. The industry acknowledges the lack of effective health and safety training as a leading cause of accidents (Toole 2002). Therefore, novel possibilities are opening up to bring in emerging pedagogies into the training sessions.

Safety training traditionally would include one-way information delivery in the form of lectures, toolbox talks, handouts, audio-visual materials, computer-based instructions (Gao et al. 2019, Blanchard & Simmering 2014). In recent years, several industries have experimented Virtual Reality (VR) technology to enhance training, e.g. see (Cha et al. 2012, Chittaro & Buttussi 2015, Nedel et al. 2016, Norris et al. 2019). VR is the use of computer technology to create a simulated environment (Sacks et al. 2013). Unlike traditional health and safety training that relies on a one-way transfer of information, VR places the user inside a drill (Cha et al. 2012, Norris et al. 2019, Zhao & Lucas 2015).

However, the educational effectiveness of VR training has remained a matter of question in the construction sector (Gao et al. 2019, Sacks et al. 2013, Choi et al. 2017). This paper reports on a preliminary study that sought to assess the size of the efficacy of VR-based training systems within the construction industry.

2 TYPOLOGY OF VR SYSTEMS

The VR systems can be categorized based on the levels of immersion offered to their users. Immersion refers to a computer-generated display that allows the user to have a sense of presence in an environment other than the one they are actually in, and to interact with that environment (Schroeder 1996). Accordingly, three types of VR systems can be recognized:

- Non-immersive VR, where the user interacts with the contents on display through some input devices, similar to a traditional video game. In this type of interaction, the users can keep control over the physical surrounding while being aware of their external environment (Mujber et al. 2004).
- Semi-immersive VR systems allow users to experience three-dimensional virtual environments while remaining connected to the real-world surrounding, similar to a typical flight simulator. The Cave Automatic Virtual Environment (CAVE) that uses projectors to create three-dimensional images is a typical setup of the semi-immersive VR experiments (Norris et al. 2019).
- Fully immersive VR presents an artificial environment that replaces users' real-world surroundings. The system typically uses a head-mounted display that relays a three-dimensional visual environment, where the user can control the objects. The virtual scenes get updated as the user moves. It removes the perception or awareness of the real environment (Mujber et al. 2004).

3 THE USE OF VR FOR SAFETY TRAINING

Over the past two decades, several studies have experimented the use of VR for safety training purposes. The following sections summarize the prominent characteristics reported so far:

3.1 Offering a safe working environment

The hazardous nature of construction sites makes onsite training difficult and prevents the experience of failure (Sacks et al. 2013). A lack of experience with adverse outcomes from risky behavior often leads to unrealistic optimism (Weinstein 1984). The VR systems enable simulating the real-world (Guo et al. 2012, Leder et al. 2019), where the trainees can effectively rehearse tasks in a safe environment (Zhao & Lucas 2015). Therefore, VR technology allows training through the experience of failure without suffering possibly life-changing consequences (Sacks et al. 2013, Felicia 2011). The trainees can gain firsthand knowledge of circumstances in this environment that is not easily achievable in the traditional training methods such as a classroom setting (Norris et al. 2019). A simulated environment also allows the learner to observe cause and effect (Bandura 2001), which has proven to be useful in developing hazard recognition and intervention abilities (Zhao & Lucas 2015, Bosché et al. 2015, Dawood et al. 2014).

To sum up, the outcomes of training in the safe work environment provided by a VR experience can be two-folds (Norris et al. 2019):

1. The learners gain a better understanding of the job tasks and the associated risks,
2. They will often pay a higher respect to the safety procedures post-training.

3.2 Engagement and satisfaction

Evidence shows the importance of engagement in the success of safety programmes. Higher user engagement can lead to better learning outcomes (Greuter & Tepe 2013, Smith & Ericson 2009). Overall dissatisfaction with the engagement of the traditional training materials was reported by Wilkins (2011) in a survey of 105 construction workers who had taken a 10-hour safety training course. Burke et al. (2006) reviewed 95 studies involving over 20,000 participants, where they confirmed the enhancement of the training outcomes by using VR-based systems instead of conventional passive methods. VR platforms enable interaction between trainees and the training material that can leverage the engagement (Chao et al. 2017, Felicia 2011, Osberg 1995, Bhagat et al. 2016, Chittaro & Sioni 2015). Most of the users who attended a study by Pena and Ragan (2017) to contextualize construction accident reports noted the engagement level as a distinct point between the VR experience and the traditional training methods. The level of engagement, however, is suggested to correlate with the degree of immersion provided by the system (Warburton 2009, Steuer 1992).

Another stream of studies has argued the fact that the workforce in the construction sector tends to have lower educational attainment on average compared to other industries (Loosemore & Andonakis 2007). It highlights the importance of developing text-free materials for the training purpose (Gao et al. 2019). For example, Zeitlin (1994) recorded the inadequacy of written instructions in ensuring safe-conducts. The VR-based training removes the constraining requirements to language proficiency and literacy to understand the contents (Lin et al. 2012). Therefore, they are expected to increase the level of engagement.

3.3 Familiarity with the situation and confidence

Zhao and Lucas (2015) attributed a significant percentage of construction incidents to worker unfamiliarity with hazardous situations. Hazard perception is a multi-component cognitive skill that can improve with gaining experience over the years (Deery 1999, Knight 2012, Norris et al. 2019). As discussed, VR systems give the trainees an opportunity to expedite practicing hazard identification in specific scenarios (Cha et al. 2012, Norris et al. 2019). The users involve a hazardous situation with multiple senses in real-time that can boost the training outcomes (Zhao & Lucas 2015). Given that, a dramatic increase in the self-efficacy of participants was reported by Buttussi and Chittaro (2017). They interpreted these results as higher confidence in the participants to deal with the real situations.

VR technology not only enables simulating routine operations, but it also enables experimenting low-probability/high-consequence events, emergency responses, abnormal operations, upset scenarios, and critical integrity procedures all in a safe and controlled environment (EON Reality UK 2018). This ability is unique, with no alternative training method (Gao et al. 2019). It allows the trainees to develop a default correct response in a completely safe environment (Norris et al. 2019).

In a nutshell, the use of VR-based training systems can enable organizations to cultivate a workforce whose experience in hazardous situations is both robust and varied (Norris et al. 2019).

3.4 Repeatability

The hands-on confined construction training requires tremendous preparation and setup, which prevents learning through repetition (Sacks et al. 2013). In contrast, VR training systems enable endless replication until the users achieve mastery of the task (Norris et al. 2019).

3.5 Disadvantages

The studies have also reported some drawbacks for VR-based training. Some leading examples are as follows:

- The learning curve for using the software packages and for troubleshooting can present a usability issue (Norris et al. 2019);
- The challenges associated with an ageing workforce were highlighted by Sacks et al. (2013). The VR safety training can become struggling for the older trainees, who are unaccustomed to computerized learning systems, which may result in reduced training outcomes;
- Norris et al. (2019) also raised concerns about the inflexibility of the VR scenarios. The Programmes that create the scenarios can turn to be inflexible, particularly for off-the-shelf VR modules. Developing company-specific or location-specific VR training Programmes can be costly;
- Last but not least, LaViola Jr (2000) documented cybersickness, a symptom similar to motion sickness besides vertigo, ataxia, disorientation, headache, eye strain and nausea as some side effects that could appear in the users of a VR system.

4 THE EFFECTIVENESS OF TRAINING: PERCEPTION VS OBSERVATIONS

Despite all the characteristics reported about a VR training session, the measurement of its effectiveness is still a controversial topic in construction. "Effectiveness can be defined as the extent to which a training approach yields desired outcomes" (Ho & Dzeng 2010). Gao et al. (2019) suggested that the desired outcome of safety training can be crystallized in the following three aspects:

1. Increasing knowledge acquisition,
2. Decreasing the unsafe behavior, and
3. Reducing in the injury rate.

A variety of approaches have been adopted by different studies to measure the effectiveness of the training. Among them, the following three approaches stand out:

4.1 Qualitative assessments based on narrative interpretation

Pre- and post-training questionnaires have been a standard data collection method. The data might be self- or supervisor-reported (Probst 2004).

Guo et al. (2012) used post-training self-reports to make a comparison between the level of engagement in a VR-based training session and a traditional session. They draw the results based on a qualitative interpretation of the reports. The same approach was used by Jeelani et al. (2017), Chao et al. (2017), Azhar (2017), Pena and Ragan (2017). All concluded clear advantages for the VR technology.

4.2 Quantitative assessment of the learning outcomes

Sacks et al. (2013) made a quantitative assessment of the worker's learning and recall in identifying construction safety risks to compare the effectiveness of VR training with conventional methods. The assessment confirmed the effectiveness of the VR technology. Later on, Leder et al. (2019) conducted three quantitative assessments to measure the effectiveness of VR-based trainings in the following aspects:

1. The improvement in the participants' risk perception by assessing their judgments of the probability of accidents and accident severity,
2. The level of their knowledge acquisition by assessing their ability to recall, and
3. The improvement in the decision making quality by asking the participants to recognize a safe machine from an unsafe one.

They concluded that the effectiveness of safety training only depends on the quality of the displayed materials rather than the technology used for training purposes.

4.3 Quantitative assessments based on the field observations and historical data

Shamsudin et al. (2018) tested the technicality and usability aspects of VR-training for hazard recognition by using quantitative assessments of the ability of trainees in hazard identification post-training. Their study showed a success rate between 20% and 80% based on the characteristics of the trainees. Nevertheless, as noted by Leder et al. (2019), the studies that have addressed the effects of different safety training approaches on actual decisions are scarce.

In summary, a review of the existing studies shows that most of them have used the narrative interpretations that is able of measuring the perception of the trainees about the extents of the effectiveness of the training method. Very few studies have taken the other two methods that can observe and measure the genuine contribution of the VR technology in the safety behavior alteration of the trainees.

5 RESEARCH METHOD

This study used a mixed-method case-control research methodology to compare the achievements made through the traditional methods of health and safety training, such as the use of PowerPoint presentations against VR-based training. The perception of the trainees was evaluated based on self-reports immediately after the session and compared to their actual ability to recall the safety aspects and risk-taking attitude one day after the session. Accordingly, the study focused on two aspects of the training outcomes: (1) Knowledge retention, and (2) behavior alteration.

5.1 Sampling method

Eight residential construction sites in Auckland, New Zealand were selected as the indicative clusters of construction activities in the region. Correspondingly, a simple random method was applied to select one of the sites and pick the group members from the nominated site. The sampled company had 18 employees, including 15 full-time and three part-timers. The study involved ten of them. The demographics of the sampled group is presented in Table 1.

The participants were divided into two groups of five, namely the control group and the intervention group.

5.2 Data collection process

The control group received approximately 25 minutes of traditional health and safety training using a PowerPoint presentation. The intervention group

Table 1. Demographic of the groups.

	Control group	Intervention group
Number of respondents	5	5
Gender	All male	All male
Average age	35 years	33 years
Average work experience	2.5 years	2.0 years
Average education level	level 6*	level 6*

* based on NZ qualifications framework.

received the same length of health and safety training using immersive VR technology.

The large VR devices such as CAVE are generally costly, and they can be discouraging for construction companies. Therefore, this study preferred HTC VIVE (Viveport 2020) that is relatively cheap equipment (approximately 1000 NZD per set). It provides 2160×1200 resolution with a refresh rate of 90 HZ. Therefore the users can experience a fully immersive first-person experience. The hardware set included one headset, two controllers and two base stations. The VR training software was developed by OM company, and it was based on the SteamVR platform (Viveport 2020). The platform enabled trainees to see a 360-degree simulated version of the site environment by using the headset. The trainees could use the controller to interact with the simulated environment. The controllers could vibrate to give a higher realistic experience to the trainees. Besides, the trainee could hear detailed safety knowledge from the earphones, which allowed them to learn while 'playing the game'.

5.3 Training contents

The training materials were selected based on the guidelines from WorkSafe New Zealand (WORK-SAFE 2020). The materials contained the most common causes of accidents in the New Zealand construction industry, which ensure that the safety training had a high correlation with real-life situations.

Therefore, the following aspects were covered in the training session. General site safety, including:

– Safety knowledge about heavy vehicles on site (Figure 1);
– Safety knowledge about electricity on site (Figure 2);
– Safety knowledge about working at height (Figure 3);
– Safety knowledge about fire safety and fire evacuation (Figure 4);
– Safety knowledge about scaffolding (Figure 5);
– The use of tools and equipment (Figure 6), including:

  – The correct use of the table saw;
  – The correct use of safety harness and belt.

• And the identification of safety violations.

Figure 1. Screenshot of the training scene about heavy vehicles on site.

Figure 2. Screenshot of the training scene about the electricity safety.

Figure 3. Screenshot of the training scene about working at height.

Figure 4. Screenshot of the training scene about fire safety.

Figure 5. Screenshot of the training scene about scaffolding and safety.

Figure 6. Screenshot of the training scene about the use of tools and equipment.

5.4 Measuring indicators

After each training session, the groups were interviewed through the use of face to face semi-structured strategy twice. The first round of the interviews was conducted immediately after the training session that graded the perception of the respondents about the impact of the session on their recall ability of the training contents. This round involved the following three questions:

Q1. To what extent did you find the safety training engaging?
Q2. To what extent did you find the training session satisfactory?
Q3. How much impact does safety training have on your attitudes to safety?

While the first two questions evaluated the general perception of the trainees about the effectiveness of the session on their knowledge acquisition, the third question assessed their perception about the impact of the session on their future safety behavior. In answer to all these three questions, the interviewees rated their responses on a 5-point Likert scale.

The second round of interviews was instituted one day after the training session that examined the actual ability of knowledge retention and behavior alteration of the participants. The following eight questions were used to quantitatively assess the extent of their knowledge retention about the risk factors:

Q1. What is the right extinguisher for different types of building materials?
Q2. What is the correct procedure in the case of fire?
Q3. What are common causes of vehicle-related accidents on a construction site?
Q4. What is the correct procedure for using the safety harness and belt?
Q5. What are the safety guidelines for using construction tools?
Q6. What are the safety guidelines for scaffolding?

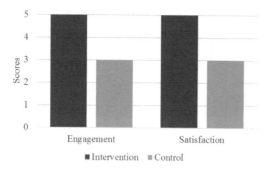

Figure 7. Comparison between the perceptions of the trainees in regards to their engagement and satisfaction.

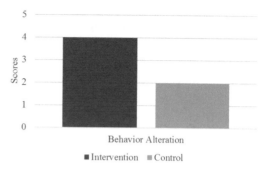

Figure 8. Comparison between the perceptions of the trainees about the impact of the session on their risk-taking behavior.

Q7. What are the safety regulations for using a table saw?

Q8. What are the hazards of working at height?

All factors had been discussed in the training session.

A scenario-based question was also used to assess the risk-taking behavior of the interviewees as follows quantitatively:

Q9. Assume the project manager reaches you to carry out a construction activity without protection. You can get an additional NZD 200 as a bonus, while the probability of an accident will only be 1 %. Indicate the extent of your willingness to undertake the task on a scale of 5.

6 RESULTS AND DISCUSSIONS

The results of the study are discussed in two separate parts: perceptions and real observations.

6.1 *The perception of the trainees*

The following results are drawn based on the interpretation of self-reports post-training.

A comparison between the perception of the control group and the intervention group about their level of engagement and satisfaction is presented in Figure 7. As can be seen, the VR-based training session received a full score 5 of 5 in both aspects. The statistical mode of the scores given to the traditional experience was 3 for both questions. It presents a score of 40% lower than those given by the control group.

One of the interviewees in the intervention group stated that "I have participated in various training in New Zealand, such as the company safety training, site induction, safety training from Site Safe. Although I could admit the great efforts of the organizers, I would usually lose my attention within 10 minutes. The VR safety training gave me a feeling of playing computer games, and it makes the day go by much faster." These impressions are consistent with those of Sacks et al. (2013), who reflected on the importance of the sense of presence in VR-based training. In the meantime, one member of the intervention group reported dizziness that had gave the trainee some hard times.

As Figure 8 shows, a significant difference was also observed between the perception of the control group and the intervention group about the effectiveness of the training session on altering their behavior. The intervention group ranked the effectiveness of the session 40% higher than the control group.

This difference can be linked to the fact that VR technology creates a simulated environment, which can give the trainees a firsthand experience of the hazards in the real environment. During the experiment, most trainees responded to the accidents intensely and actively. A member of the intervention group stated that "I do not have any experience of the accident. This is my first time to 'experience' an accident. I am a bold person, so sometimes I stand on the timber frame to install trusses when the safety net is not ready. Now I realized that as my work experience grows, my work efficiency is getting higher, but I am no longer pay attention to safety issues. I believe I will pay more attention to safety issues in the failure."

This observation conforms to those of Weinstein (1984) that suggested the lack of experiencing accidents could lead to gradual ignorance of the seriousness of the consequences and cause people to be too optimistic about themselves. Rosenstock et al. (1988) further offered that the experience does not need to be first hand, even seeing someone else experiencing an inevitable outcome can result in the same outcome.

6.2 *Assessing the true effects*

The study results indicated no significant difference in the safety knowledge retention between the control group and the intervention group (Figure 9). Both groups answered close to 90% of the questions correctly. These results are similar to those of Leder et al. (2019). They tested the trainees' knowledge retention from VR safety training twice. Even though their first tests showed a slight advantage for the VR-trained group, the second tests did not show any significant advantage.

In the meantime, as Chittaro and Sioni (2015) suggest the inter-personal differences in intellectual,

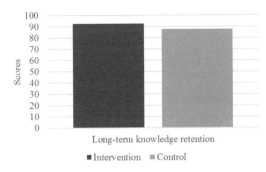

Figure 9. Comparison between the long-term knowledge re-tention observed among the trainees.

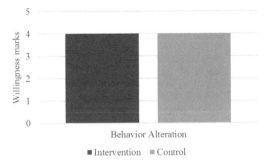

Figure 10. The observed behavior alteration in the two studied groups.

physical, mental, emotional and social aspects may affect the outcome of this assessment. It is difficult to find two groups of trainees with the same level of intelligence and memory.

No significant difference in behaviour alteration was also observed between the two groups (Fig. 10). Both groups marked their willingness to carry out a construction activity without protection as four on a 5-point scale subject to a good financial incentive. Risk perception is known as a precondition for people to deal with risk, so it plays a vital role in their decision making. According to Weinstein (1984), the risk perception mainly includes two aspects: assess the likelihood of an accident (probability perception) and assess the severity of the consequence for an accident (risk-severity perception). The higher the likelihood of occurrence and the higher the severity of the consequences, people will pay more attention to the risks. They are also more likely to take some action to deal with these risks. Future investigations may consider surveying a scenario with a higher probability of occurrence. The feedback from the interviewees, however, showed other factors such as the personal belief could have been contributed to such a weak response too. One of the interviewees stated that "risks exist with everything in the world, such as driving a car or stock investment. Even if you do a lot of safety measures, you still cannot guarantee 100% safety. I do not believe there is an absolute safe job in the world, so if the accident probability is only 1%, I can accept it. However, I will think about the worst scenario, then make the final decision." This diversity in the determinants of the risk-taking behaviour is also discussed in other studies such as in (Figner & Weber 2011, Hanoch et al. 2006, Schoemaker 1993).

7 CONCLUSION

Despite the firm belief of the trainees about the effectiveness of the VR technology in improving their health and safety performance, no substantial evidence was found in their knowledge retention and risk-taking behavior. However, the factors affecting risk-taking behaviour are complicated, that include risk perception, situational factors, nature of the risk and personality.

8 LIMITATIONS AND FUTURE RESEARCH

In this research, the experiment scale was limited by resources and funds. Only ten respondents were involved in the experiment. A limited experiment sample may result in a deviation in the experimental results. Besides, the experiment adopted a general construction safety training software, which may have been too simple for some respondents with higher working experience.

The results can be generalized only if the future studies with bigger sample size show a similar outcome.

REFERENCES

Azhar, S. 2017. Role of visualization technologies in safety planning and management at construction jobsites. *Procedia engineering*, 171, 215–226.

Bandura, A. 2001. Social cognitive theory: An agentic perspective. *Annual review of psychology*, 52. 1–26.

Bhagat, K. K., Liou, W.-K. & Chang, C.-Y. 2016. A cost-effective interactive 3D virtual reality system applied to military live firing training. *Virtual Reality*, 20. 127–140.

Blanchard, P. & Simmering, M. 2014. Training delivery methods. *Encyclopedia of Business*.

Bosché, F., Ahmed, M., Turkan, Y., Haas, C. T. & Haas, R. 2015. The value of integrating Scan-to-BIM and Scan-vs-BIM techniques for construction monitoring using laser scanning and BIM: The case of cylindrical MEP components. *Automation in Construction*, 49. 201–213.

Burke, M. J., Sarpy, S. A., Smith-Crowe, K., Chan-Serafin, S., Salvador, R. O. & Islam, G. 2006. Relative effectiveness of worker safety and health training methods. *American journal of public health*, 96. 315–324.

Buttussi, F. & Chittaro, L. 2017. Effects of different types of virtual reality display on presence and learning in a safety training scenario. *IEEE transactions on visualization and computer graphics*, 24. 1063–1076.

Cha, M., Han, S., Lee, J. & Choi, B. 2012. A virtual reality based fire training simulator integrated with fire dynamics data. *Fire Safety Journal*, 50. 12–24.

Chao, C. J., Wu, S. Y., Yau, Y. J., Feng, W. Y. & Tseng, F. Y. 2017. Effects of three-dimensional virtual reality and traditional training methods on mental workload and training performance. *Human Factors and Ergonomics in Manufacturing & Service Industries*, 27. 187–196.

Chittaro, L. & Buttussi, F. 2015. Assessing knowledge retention of an immersive serious game vs. a traditional education method in aviation safety. *IEEE transactions on visualization and computer graphics,* 21. 529–538.

Chittaro, L. & Sioni, R. 2015. Serious games for emergency preparedness: Evaluation of an interactive vs. a non-interactive simulation of a terror attack. *Computers in Human Behavior,* 50. 508–519.

Choi, B., Hwang, S. & Lee, S. 2017. What drives construction workers' acceptance of wearable technologies in the workplace?: Indoor localization and wearable health devices for occupational safety and health. *Automation in Construction,* 84. 31–41.

Dawood, N., Miller, G., Patacas, J. M. D. L. & Kassem, M. 2014. Construction health and safety training: the utilization of 4D enabled serious games. *Journal of Information Technology in Construction,* 19. 326–335.

Deery, H. A. 1999. Hazard and risk perception among young novice drivers. *Journal of safety research,* 30. 225–236.

EON REALITY UK 2018. How Can VR and AR Address Challenges in Health and Safety Training? Manchester, UK.

Felicia, P. 2011. *Handbook of research on improving learning and motivation through educational games: Multidisciplinary approaches: Multidisciplinary approaches,* iGi Global.

Figner, B. & Weber, E. U. 2011. Who takes risks when and why? Determinants of risk taking. *Current Directions in Psychological Science,* 20. 211–216.

Gao, Y., Gonzalez, V. A. & Yiu, T. W. 2019. The effectiveness of traditional tools and computer-aided technologies for health and safety training in the construction sector: A systematic review. *Computers & Education,* 138. 101–115.

Greuter, S. & Tepe, S. Engaging students in OH&S hazard identification through a game. DiGRA Conference, 2013.

Guo, S., Hu, X. & Wang, X. On time granularity and event granularity in simulation service composition (WIP). 2012. Society for Computer Simulation International.

Hanoch, Y., Johnson, J. G. & Wilke, A. 2006. Domain specificity in experimental measures and participant recruitment: An application to risk-taking behavior. *Psychological Science,* 17. 300–304.

Ho, C.-L. & Dzeng, R.-J. 2010. Construction safety training via e-Learning: Learning effectiveness and user satisfaction. *Computers & Education,* 55. 858–867.

Jeelani, I., Han, K. & Albert, A. 2017. Development of immersive personalized training environment for construction workers. *Computing in Civil Engineering 2017.*

Knight, F. H. 2012. *Risk, uncertainty and profit,* Courier Corporation.

Laviola JR, J. J. 2000. A discussion of cybersickness in virtual environments. *ACM Sigchi Bulletin,* 32. 47–56.

Leder, J., Horlitz, T., Puschmann, P., Wittstock, V. & Schütz, A. 2019. Comparing immersive virtual reality and powerpoint as methods for delivering safety training: Impacts on risk perception, learning, and decision making. *Safety science,* 111. 271–286.

Lin, K., Migliaccio, G., Azari, R., Lee, C. & De La Llata, J. 2012. Developing 3D safety training materials on fall related hazards for limited English proficiency (LEP) and low literacy (LL) construction workers. *Computing in civil engineering (2012).*

Loosemore, M. & Andonakis, N. 2007. Barriers to implementing OHS reforms–The experiences of small subcontractors in the Australian Construction Industry. *International Journal of Project Management,* 25. 579–588.

Mujber, T. S., Szecsi, T. & Hashmi, M. S. 2004. Virtual reality applications in manufacturing process simulation. *Journal of materials processing technology,* 155. 1834–1838.

Nedel, L., De Souza, V. C., Menin, A., Sebben, L., OLIVEIRA, J., Faria, F. & Maciel, A. 2016. Using immersive virtual reality to reduce work accidents in developing countries. *IEEE computer graphics and applications,* 36. 36–46.

Norris, M. W., Spicer, K. & Byrd, T. 2019. Virtual Reality: The New Pathway for Effective Safety Training. *Professional Safety,* 64. 36–39.

Osberg, K. 1995. Virtual reality and education: where imagination and experience meet. *VR in the Schools,* 1. 1–3.

Pena, A. M. & Ragan, E. D. Contextualizing construction accident reports in virtual environments for safety education. 2017 IEEE Virtual Reality (VR), 2017. IEEE, 389–390.

Probst, T. M. 2004. Safety and insecurity: exploring the moderating effect of organizational safety climate. *Journal of occupational health psychology,* 9. 3.

Rosenstock, I. M., Strecher, V. J. & Becker, M. H. 1988. Social learning theory and the health belief model. *Health education quarterly,* 15. 175–183.

Sacks, R., Perlman, A. & Barak, R. 2013. Construction safety training using immersive virtual reality. *Construction Management and Economics,* 31. 1005–1017.

Schoemaker, P. J. 1993. Strategic decisions in organizations: rational and behavioural views. *Journal of management studies,* 30. 107–129.

Schroeder, R. 1996. *Possible worlds: the social dynamic of virtual reality technology,* Westview Press Boulder, CO.

Shamsudin, N. M., Mahmood, N. H. N., Rahim, A. R. A., Mohamad, S. F. & Masrom, M. 2018. Virtual reality training approach for occupational safety and health: a pilot study. *Advanced Science Letters,* 24. 2447–2450.

Smith, S. & Ericson, E. 2009. Using immersive game-based virtual reality to teach fire-safety skills to children. *Virtual reality,* 13. 87–99.

Steuer, J. 1992. Defining virtual reality: Dimensions determining telepresence. *Journal of communication,* 42. 73–93.

VIVEPORT. 2020. *Vive* [Online]. Available: https://www.vive.com/nz/ [Accessed 10.Apr.2020].

Warburton, S. 2009. Second Life in higher education: Assessing the potential for and the barriers to deploying virtual worlds in learning and teaching. *British journal of educational technology,* 40. 414–426.

Weinstein, N. D. 1984. Why it won't happen to me: perceptions of risk factors and susceptibility. *Health psychology,* 3. 431.

Wilkins, J. R. 2011. Construction workers' perceptions of health and safety training programmes. *Construction Management and Economics,* 29. 1017–1026.

WORKSAFE. 2020. *WORKSAFE, Mahi Haumaru Aotearoa* [Online]. Available: https://worksafe.govt.nz/ [Accessed 10.Apr.2020].

Zeitlin, L. R. 1994. Failure to follow safety instructions: Faulty communication or risky decisions? *Human Factors,* 36. 172–181.

Zhao, D. & Lucas, J. 2015. Virtual reality simulation for construction safety promotion. *International journal of injury control and safety promotion,* 22. 57–67.

Applying activity theory to get increased understanding of collaboration within the VDC framework

E. Hjelseth
Norwegian University of Science and Technology, Trondheim, Norway

S.F. Sujan
Norwegian University of Science and Technology, Trondheim, Norway
University of Liverpool, Liverpool, UK

ABSTRACT: Improving process performance in projects is challenging; systemic change requires methods that combine prioritised elements to optimise continuous improvement. Projects that have implemented the Virtual Design and Construction (VDC) framework document improvements in process and product performance. These projects have made multiple changes in their way of working and collaborating. However, there is a difference in observing "what" is done in practice, to explain "why" changes brought by e.g. VDC have these effects. This paper uses Activity Theory to understand and demystify VDC. Theory-based reflections and experiences from 200 students in a Norwegian VDC-course are used as reference for critical reflections. The results of these reflections show that there are connections between multiple activities that contribute to systemic improvement. The impact of this type of understanding makes implementation of the VDC framework clearer. Combining theory and practice gives potential to get better results with less effort.

1 BACKGROUND

'The best way to explain it is to do it'
(Carroll 1865)

This paper intends to demystify "why" Virtual Design and construction (VDC) is working. So far, it has been utilised in some AEC industries such as in Norway, however, there is limited understanding of "why" it improves practice. In this paper, VDC is explored and analysed by use of Cultural Historical Activity Theory (CHAT) as a theoretical framework. The impact of this increased understanding of "Why" can contribute to improve current practice in the Architects, Engineers, Contractors and Facility Management (AEC/FM) industry by practical implementation of VDC in real projects.

According to Fischer (2019, 2020) and Hjelseth (2020), practical experiences from the use of Virtual Design and construction (VDC) in real AEC-FM projects show positive impact in performance.

Majority of previous studies focus on it's definition and it's benefits, presenting as relatively substantial improvements in large projects branded as "VDC" (or Collaboration, Joint, Integrated or similar) projects. In these projects, all five elements in the VDC framework (shown in Figure 3) are included with significant effort put on all of them (indicating their interdependence and importance). Even though, this is acceptable in larger projects, for smaller projects with limited resources, it is critical to give priority to the most important elements in the right way. Furthermore, VDC has shown significant improvement in project delivery and it is unclear as to the way that it does this. Therefore, adopting activity theory to structure the understanding of the influence of VDC on BIM and project management is timely.

Activity Theory (AT) has been used widely to get a deeper understanding; map out and understand the way that changes in technology affect people and processes at multiple levels e.g. the individual, project, organizational or industrial level. It's wide use makes it a suitable theoretical lens to structure discussion with.

The contextual origin comes from the recent widescale introduction of VDC in the Norwegian industry. There are currently over 200 practitioners participating in a VDC course conducted by NTNU. The course participants come from several different companies and projects across the country, with the predominant aim of uniting the competency of the information management process.

Therefore, the aim of this paper is as to understand how VDC works at the project level by utilizing activity theory. Furthermore, to use this understanding to discuss the gaps and influences of VDC in the organisational and industrial levels.

2 THEORETICAL BACKGROUND

2.1 *Activity theory*

Activity Theory (AT) was developed in psychology, now being applied in information systems

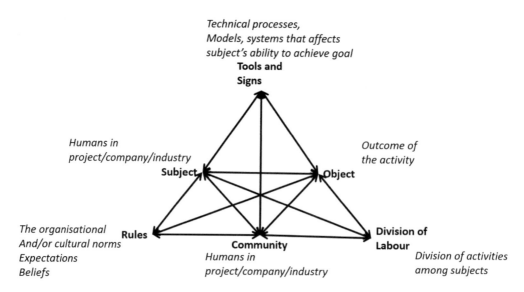

Figure 1. Framework for use of Activity Theory (Engeström 1987).

(Kaptelinin & Nardi 2006), study of innovations (Miettinen 1999) and now within design collaboration (Hartmann & Bresnen 2011). Figure 1 shows the definitions of the activity structure.

The AEC industry has been undergoing transformation in the way it delivers solutions. The transformation is encompassed by digitalisation where practices surrounding information exchange using Building Information Management (BIM) models are commonly used. A few studies have also studied the use of BIM tools e.g. (Nørkjaer Gade et al. 2018). It is particularly useful to utilize activity theory when a holistic view of interacting phenomena considering social and technical aspects are considered. In essence, the theory can be applied to understand how and why people utilize tools/methods.

The use of AT in this paper is to conceptualize and understand humans, their actions within the social, cultural and technical contexts in utilizing technology and VDC in construction projects (Willis et al. 2007).

Miettinen and Paavola (2014) find that BIM implementation implies learning by experimenting. Further, they say that the process the practitioners and users use, play a key role. BIM development and implementation are an open-ended expansive process.

This means that the BIM based project process is variant as the process is defined in project depending on the context of a project and the participants' capabilities; VDC sets part of this process. It is suggested that designers tend to ensure the progression of the design development and not the use of BIM-tools (Nørkjaer Gade et al. 2018); improvisations caused inconsistencies in BIM models. This is asymptomatic of the lack of coordination and cooperation between teams as a result of lacking ability to communicate efficiently.

In similar research surrounding BIM implementation, Akintola et al. (2019) define the outcome as 'deliver building and construction project to client's satisfaction'. The outcome is defined in a similar way in this paper as 'high performance building that optimally meets client needs'. In this context, it does this by the definition of client and project objectives.

AT is a framework that connects multiple aspects into a holistic solution. AT can (as Figure 2 illustrates) be applied to understand utilization of BIM. The interactions are mapped and discussed within the activity structure such as done by Akintola et al. (2019).

2.2 *Virtual Design and Construction (VDC)*

VDC is the use of integrated multi-disciplinary performance models of design-construction projects to support explicit and public business objectives (Kunz & Fischer 2012; Kunz & Fischer 2020). VDC does this by focusing on the definition of a combination of digital based methods at the interface of design, site and project management that streamlines coordination between project teams. In simple terms, VDC connects the creation and management of building information (BIM) to process and project management (PPM), and decision-making base on Integrated Concurrent Engineering (ICE).

Below is an overview of the way of acting (working, Figure 3); The ICE-session is the starting position (*"Who level"*) for well-organised collaboration and problem solving. The next levels (*"What level"*) is BIM for joint understanding of the project, and PPM for how the work is scheduled and managed. This supports the *"Why level"* to fulfil Project (PO) and Client Objectives (CO). This includes both process and product (what is going to be built) Metrics to enable fact-based continuously improvements of Figure 3 is a modified version based on a presentation by Fischer (2020).

"Current project has 40% shorter design time compared to last similar. We saved working

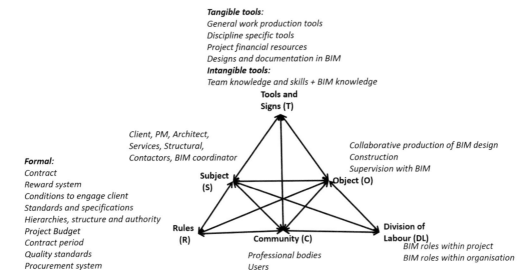

Figure 2. Interactions mapped into activity structure in AT (Akintola et al 2019).

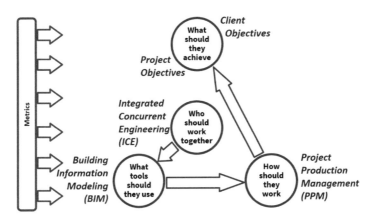

Figure 3. Overview of the VDC framework (Fischer 2020, modified by authors).

hours for project manager, value approx. EUR/USD 35,000…we can do other income-generating works. Our aim is to reduce build time by over 10% and current status is 4% reduction."

(VDC Course participant)

Experiences from projects and postgraduate education students implementing VDC in their projects document following values (Hjelseth 2020):

− VDC framework combines ICE and PPM to define, measure and achieve objectives
− Enables multidisciplinary process
− Streamlining processes of decision making and information use/development with the aim to create measurable outcomes driven by objective targets
− Flexibility and agility in visual planning and collaborative decision making
− Reduction in rework and increase in productivity on site
− More revenue for the client from long term strategic decision making by utilising the project team's collaborative potential
− 5% reduction in production cost
− Schedule efficiency increase
− Reduced risk for significant delays in time, cost and quality (improved control).

One of the characteristic elements in VDC is it's use of ICE-sessions, a highly agile form of interactive meetings. The way meetings are organised depends on the needs of the task. The multidisciplinary participation and the solution/decision focused approach in

ICE sessions has significant improvements in latency. VDC shows it's strength by solving reciprocal tasks, however, there are gaps in understanding:

> 'The importance of human factors in the implementation plan has been underestimated. The assumption that the participants in the project have expertise about VDC and the use of BIM, ICE and PPM is greatly overestimated'
> (VDC course participant)

The VDC facilitator manages discussion to understand the dependency of tasks. Once these are achieved, reciprocal tasks need the most coordination and therefore the larger meetings. Whereas pooled and sequential tasks require smaller meetings. Therefore, ICE is a way to manage interaction between people to make decisions collaboratively by understanding the nature of the decisions. In an ICE room (BigRoom), it is required to have tangible tools, e.g. planning tools, BIM and measurements to enhance their capability to make decisions collaboratively. Break-out sessions; working (not meeting) in smaller groups is used where this solves the task in a better way than ICE-sessions.

Project Production Management (PPM) is the planning part of VDC, that brings together the information about schedule, cost and task dependency. PPM is actively updated depending on the decisions made at ICE and the current state of the project.

BIM in the context of VDC refers to managing the building information that is created by merging multiple models that are developed individually in each design team. The decisions of the requirements and how the design is developed is done in ICE. The novelty is in the ability to make decisions based on interdependencies. The combination of BIM and PPM forms a better foundation for presenting and assessing multiple alternative solutions. The quality of these solutions are based on their capacity to fulfil project- and client-objectives.

It is critical to note that VDC is an intervention at the project level, therefore affects the individual and depends on their knowledge of the VDC framework. This implies that VDC does not require changes on the organisational and industrial levels to be implemented, although there will be supportive factors. However, wide implementation of VDC in projects has the capacity to evolve organisations and industry.

2.3 Metrics as the applied side of theory

Metrics (qualitative or quantitative) are the foundation for research and fact-based improvements. Metrics are one of the five VDC elements that professionals struggle most to get implemented in projects (Hjelseth 2020). A study by Belsvik et al. (2019) gives an overview of metrics in both real projects and literature. The motivation for metrics is to enable continuous improvements. This has applied references to the way or thinking by the hermeneutical circle, the Deming's circle by its focus on Plan-Do-Check-Act (Deming 1989; Moen & Norman 2006). The core contribution by introducing theory is to sharpen the scope of metrics and link other factors that have their own metrics. By this approach, one could develop a foundation for a holistic understanding and more dedicated efforts.

2.4 Research gap

There is a lot of research in organisational science. A characteristic of case/real studies is that they are based on a theory, or a theoretical foundation. This works/acts as "theoretical lens". The research included by this approach explains "Why", not only a description of "What and How". The pragmatic nature of case studies within AEC/FM related research mostly focuses on "What and How" descriptions. It is of course interesting to learn from others' success, but there is a challenge to recreate these experiences in other contexts such as in their own construction projects. Best practice-based research is one type of research approach, with it's own pros and cons. However, to explain why, one needs a theoretical foundation, where the quality of the research is based on the validity and reliability of the scope of the scientific research, not the results of the project. There is also relatively limited research on projects that are not regarded as a success. The results reported in research do not correspond to the economical results (or project/client objectives) in the real world. This can be illustrated by the fact that there are limited studies published about projects intending to implement VDC (in high or middle degree) who has failed. The impact of this unbalanced research landscape is that a lot of learning outcomes is missed out. To bridge this gap in this paper, the authors aim to share understanding of how VDC works by applying activity theory.

3 RESEARCH APPROACH

The research approach used in this paper can be referred to as 'problematisation' (Sandberg & Alvesson 2011) where the authors question the possible reasons for the success of VDC in practice. The approach aims to draw reasoning based on existing conceptual integrations from activity theory to make a contribution. The limited literature on VDC makes it a clear gap, and the highly common use of activity theory in many applications makes it suitable choice to apply and engage with other researchers.

In addition to a review of literature on activity theory and BIM based approaches, the methods used in the paper involve critical reflection from the authors accompanied by studies of student's periodical reports from the ongoing VDC-course in Norway. The development of the applied activity structure come from both the views of the VDC course participants and the authors. The authors are also involved in the delivery of the course allowing active discussions, a form of critical reflection. Critical reflection is a reasoning process to make meaning of an experience. Critical reflection is a "meaning-making process" that helps researchers

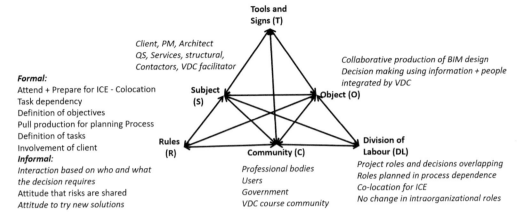

Figure 4. VDC elements integrated in the AT framework.

set goals, use what they have learned in the past to inform future action and consider the real-life implications of their thinking. It is the link between thinking and doing, and at its best, it can be transformative (Dewey 1923; Rodgers 2002; Schön 1983).

4 EXPLORING VDC AND AT

4.1 What are the VDC elements in the AT framework?

Figure 4 shows how the interactions of the VDC elements within the wider activity structure by defining methods and making connections within them i.e. PPM and BIM are defined and used in ICE to make communication and decisions more efficient. Therefore, reducing the intangible assumptions regarding expectations and understanding from varying skills and knowledge. The physical tools have an influence in this perception as people behave more interactively when they have physical objects and information to refer to when communicating. Therefore, VDC shows its systemic nature from having multiple types of methods interconnected. However, this requires significant changes in competency, definitions and meaning.

4.2 Holistic view of activity structure

In this section, the activity structure is looked at holistically to explain what VDC does. Figure 5 shows an overview of activity structure applied to VDC, where BIM is used as example. It should be noted that the following can be deduced by developing this activity structure and comparing it with BIM:

– VDC defines and aligns the communication of people by definition of methods

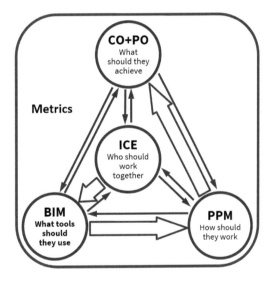

Figure 5. Multiple comments, direct and indirect, be-tween all VDC elements.

– VDC aligns perception of intangible and informal aspects as people are collocated when making interdependent/reciprocal decisions
– VDC is mainly a change at the project level, there is minimal effect at the organisational level
– VDC facilitator is a new role as compared to BIM, differences in the role are important to consider
– VDC requires people to be competent with the definitions of methods, objects within the framework
– Integration of information at ICE meetings is critical so that people can share their perception of the data when co-located

– VDC follows the principle of the Plan-Do-Check-Act (Moen & Norman 2006) and provides tangible methods to do this.

4.3 Interactions in the activity structure

In this section, the activity structure is dissected to describe some of the interactions in the activity structure. The connection between the aspects is illustrated in Figure 4.

Tools <-> Object: *combination of tools to achieve objectives*

Combination of tools is mandatory; this is how VDC creates value by connecting decision making to actual actions in PPM and BIM that relate to individual teams' work and output. They are interdependent, the tangible objects that unite people's project vision and their ability to communicate. Without streamlined methods that form the objects that can facilitate discussion, expectations are tacit and not communicated enough. The effect is an intricate combination of tools to achieve objectives by influencing human behaviour and decision making

Tools <-> Subject <-> Rules: *Know the assumptions*

People need to gain competency in the definition of VDC methods which allow an integrated holistic vision of how objectives are defined, planned for and met. In other words, people need to know and learn the rules of VDC by inductively utilizing it in a project. Some VDC participants explain their experiences of learning. A common opinion is that VDC gives them more knowledge of the intangible aspects in project delivery e.g. the importance of trust and respect. This supports the plausible hypothesis that intangible expectations and rules are made more tangible in VDC projects. E.g. using decision making as an example, the dependencies are not assumed individually but collectively.

Rules <-> Object: *Streamlined Decisions*

VDC uses classification of tasks, and links it to the way people make decisions; decision-process dependency. Task dependency in project management allows interaction to be done based on the requirements of a task therefore making tacit expectations into non-tacit rules.

Subject <-> Tools <-> Division of Labour: *Process Based Interactions*

ICE + co-location make roles more flexible as some decisions are interdependent. Division of labour is not static all the time; how people communicate and make decisions depends on the dependency of tasks ICE + Colocation makes this easier to discover

Tools <-> Rules <-> Object: *Process Definitions as Assumptions*

Streamlined process enables more detailed language leading to in depth communication of project vision. E.g. VDC specific language such as TAKT, line of balance etc. Rules are made more tangible about the process. This requires people to learn the same definitions of process which allows more efficient communication about the process and execution plan.

5 DISCUSSION

This paper raises the awareness in use of theory in scientific research to explore and explain "Why" a specific way for working and collaborating gives practical results. This deep understanding is required for dedicated improvements and adaptations to various contexts/projects.

VDC is currently an intervention at the project level. However, project collaboration has dependencies from the organisational and industrial levels. It is therefore necessary to understand the influence of VDC at these levels.

This paper has shown that VDC aligns methods and people's competency of them to create a systemic outcome that enables development of metrics. At the industrial level, the current BIM methods are highly variant between projects, VDC can be the driver to align project methods competency. VDC is increasingly being implemented in Norway, US, Switzerland and other parts of the globe. The driver is in its ability to show tangible and quantitative improvements in project productivity.

At the organisational level, internally, many firms have their own methods to integrate their operations e.g. use of continuous improvement and lean. However, this is not connected to VDC (in projects). The potential of making this connection relies on defining methods and objects in a unified manner between them. The impact would be systemic as internal organisational productivity can be connected to their productivity in the project (VDC) lens. At the moment, the intraorganizational interventions of lean are disconnected from the VDC approaches at the project level. Philosophically, they are similar and therefore it is possible to hypothesize that connections between them are possible, however, it depends on common definitions and assumptions that systems are designed with; it is driven by people's motivation and ability to share common understanding and perspectives.

Practical use of the VDC-framework in real projects represents an intricate combination of tools to achieve objectives by influencing human behaviour and decision making. The complexity in collaboration in real projects is demonstrated by Sujan (2020) where human psychological and cultural aspects interact with other factors to represent behaviour in reality. Deep understanding is needed via thorough studies to create insight for systemic changes. However, on an applied level in real projects, one needs a simplified framework enabling a focus on controllable factors; keeping in mind that there are uncontrollable factors that affect them.

AT represents an analytical approach which explores the connection between multiple elements illustrated in Figures 1 and 4. The challenge for project managers, and project participants, is how this can be transformed into practical action in real projects. In this respect, does the VDC-framework contribute with an applicable systematic and standardised approach that motivates for continuously improvements? VDC is a standardised framework including five elements: ICE, BIM, PPM, Project-/Client Objectives and metrics.

VDC is illustrated as a hierarchical model, whereas AT is illustrated as a network model. Figure 5 intends to exemplify that in real projects, VDC acts, or is understood as a dynamic model. By giving priority to only controllable factors, the connections between elements are simplified into a form that is easier to implement. The holistic/systemic approach enables the practitioner to start where they are and look at the most relevant element in connection to their position. In Figure 5, the large arrow represents the connections in VDC that are communicated inductively with course participants. The small arrows represent interdependency between elements when viewed as a network like in activity theory, this understanding comes from the application of principles in practice as it requires practitioners to take deductive lessons in their inductive action.

Experiences from VDC-courses show that it is crucial for course participants to apply the principles in real projects. This understanding emerges through proactive discussion with students and reflections, and needs to be supported by theoretical studies (pedagogic strategy is out of the scope of this paper). This learning by doing approach can be illustrated by the quote from Dodo in Alice in Wonderland, *"The best way to explain it is to do it."* (Carroll 1865).

The emergence of this reflection shows that the connections between elements of VDC are interdependent. In which, course participants go through an active inductive and deductive process to increase their own competency. This is strengthened by reflection from applying principles in practice. The VDC course in its current form focuses on the inductive aspects of competency giving the participants the methodological tools to base its application in practice. This represents the need to apply a combination of theories to represent reality as described by Sowa (2006) as a 'knowledge soup'.

6 SUMMARY

6.1 *Outcome of this study*

The VDC framework can be regarded as an applied and modified version of AT. This implies that there is an established theoretical foundation for understanding why VDC is working in practice. Theoretically, VDC integrates methods to produce/structure information (BIM), manage tasks (PPM), make decisions (ICE) for an aligned purpose (client/project objectives); by having this combination, there is dramatically higher impact, reiterating the systemic nature of technology use in the AEC-FM industry.

This foundation is both relevant for further research where researchers can go into depth and explore and explain details (why, how, where, when, how much), in a way that enables transfer into other contexts (projects/industries). It is possible that this can also contribute to more studies on projects that are not profiled as a success. This will speed up the learning process by acting as a bridge between practice and research. It is important to note that whether it is called VDC, Lean, Scrum, the streamlining of methods and competency has a similar conceptual and philosophical background; this means that by using the same theories to centralise discussion, it is possible to learn from one another despite differences in naming and/or approaches in delivery. This paper contributes by exemplifying activity theory for this application, suggesting that researchers studying approaches to streamline methods can utilise it. This would enable easier transfer and comparison between different approaches to draw and improve them.

There is a lack of use of theories on studies on VDC and its effect on the existing practice. In this paper, it has been attempted to explain the novelty of VDC and what makes it successful. VDC has shown that process maturity in digitalisation needs to increase in the sector. VDC provides a systematic project planning and control method, however this needs to be reproduced on multiple projects, once this level of competency is developed, continuous improvement principles can be applied.

6.2 *Directions for further research*

In this paper the analysis and discussions are plausible hypothesis that need further research. The following are these hypotheses:

- VDC uses methods and tangible tools to make tacit knowledge transferrable by focusing on non-tacit objects
- VDC streamline of methods allows for development of metrics to assess the efficiency of projects
- VDC definition of methods needs alignment with intra-organisational integration strategies
- VDC has influence at the industrial level by streamlining competency in BIM and PM oriented processes
- Traditional "Best practice" research is of course useful for crating motivation for changes in the AEC/FM-industry. On the other hand, uniting approaches in organisations and projects will also add value although it is currently a gap in research and in practice it is causing some confusion in definitions and system design
- The lack of theory applied to VDC as a practical tool. It is not isolated; activity theory can be used to understand this complexity and explain why VDC has been successful
- High quality social science based research can give precise research within these interactions

– Increased integrated research approach. One cannot do research on VDC alone – it is in a complex system, integrated and dependent on Lean, IPD, BIM etc.

There is need to both "Demystify" VDC and integrate understanding of it with other existing approaches; activity theory can help bridge understanding and develop VDC further. For example, bridging interventions in projects to ones inside organisations. Furthermore, VDC is used in some dedicated projects in the AEC/FM industry needs to become an applied framework for utilizing it's significant contribution; VDC's way of digital technology/BIM in combination with good process for collaboration is an established solution for process- and production management to fulfil both project and client objectives.

REFERENCES

Akintola, A., Venkatachalam, S. & Root, D. 2019. Understanding BIM's impact on professional work practices using activity theory. *Construction Management and Economics.*, pp. 1–21.

Belsvik, M.R., Lædre, O. & Hjelseth, E. 2019. Metrics in VDC Projects *In: Proc. 27th Annual Conference of the International Group for Lean Construction (IGLC)* [Online]. M.Sc student, NTNU – Norwegian University of Science and Technology, Trondheim, Norway, + 47 982 07 858, matildeb@stud.ntnu.no, pp. 1129–1140. Available from: http://iglc.net/Papers/Details/1749/pdf.

Carroll, L. 1865. Alice in Wonderland, Chapter 3, A Caucus-Race and a Long Tale. Available from: https://bookriot.com/2017/11/24/alice-in-wonderland-quotes/.

Deming, W.E. 1989. Out of the Crisis. Quality, Productivity and Competitive Position. *Massachusetts Institute of Technology, Cambridge, MA.* 81, p.82.

Dewey, J. 1923. *Democracy and education: An introduction to the philosophy of education.* Macmillan.

Engeström, Y. 1987. *Learning by expanding. An activity-theoretical approach to developmental research* [Online]. Available from: http://lchc.ucsd.edu/mca/Paper/Engestrom/expanding/toc.htm.

Fischer, M. 2019. VDC course presentation.

Hartmann, A. & Bresnen, M. 2011. The emergence of partnering in construction practice: an activity theory perspective. *Engineering Project Organization Journal.* 1(1), pp. 41–52.

Hjelseth, E. 2020. *VDC Course – Student Reports.*

Kaptelinin, V. & Nardi, B.A. 2006. *Acting with technology: Activity theory and interaction design.* MIT press.

Kunz, J. & Fischer, M. 2012. Virtual design and construction: themes, case studies and implementation suggestions. *Center for Integrated Facility Engineering, Stanford University.*

Kunz, J. & Fischer, M. 2020. Virtual design and construction. *Construction Management and Economics.*, pp. 1–9.

Miettinen, R. 1999. The riddle of things: Activity theory and actor-network theory as approaches to studying innovations. *Mind, Culture, and Activity.* 6(3), pp. 170–195.

Miettinen, R. & Paavola, S. 2014. Beyond the BIM utopia: Approaches to the development and implementation of building information modeling. *Automation in Construction.* 43, pp. 84–91.

Moen, R. & Norman, C. 2006. Evolution of the PDCA cycle.

Nørkjaer Gade, P., Nørkjaer Gade, A., Otrel-Cass, K. and Svidt, K. 2018. A holistic analysis of a BIM-mediated building design process using activity theory. *Construction Management and Economics.*, pp. 1–15.

Rodgers, C. 2002. Defining reflection: Another look at John Dewey and reflective thinking. *Teachers college record.* 104(4), pp. 842–866.

Sandberg, J. and Alvesson, M. 2011. Ways of constructing research questions: gap-spotting or problematization? *Organization.* 18(1), pp. 23–44.

Schön, D.A. 1983. *The reflective practitioner: How professionals think in action.* Routledge.

Sowa, J.F. 2006. The challenge of knowledge soup. *Research trends in science, technology and mathematics education.*, pp. 55–90.

Sujan, S.F. 2020. *A holistic and systemic model of collaboration in the AEC industry.* University of Liverpool.

Willis, K., Daly, J., Kealy, M., Small, R., Koutroulis, G., Green, J., Gibbs, L. & Thomas, S. 2007. The essential role of social theory in qualitative public health research. *Australian and New Zealand journal of public health.* 31(5), pp. 438–443.

A conceptual method to compare projects by combining assessment of controllable and non-controllable factors

S.F. Sujan
Norwegian University of Science and Technology, Trondheim, Norway
University of Liverpool, Liverpool, UK

E. Hjelseth
Norwegian University of Science and Technology, Trondheim, Norway

ABSTRACT: Systemic methods to assess and compare project processes considering their context are missing in the Architectural Engineering Construction (AEC) sector. Project assessments are highly dependent on the factors that define their context, making comparisons challenging. Inspired by interventions from macroeconomics that use empirical factors to simplify contextual complexity, project comparison can be integrated into assessments. This paper discusses a conceptual method to combine statistical methods of project context and numerical measurements of project process in Virtual Design Construction (VDC) projects. The method proposed suggests ways to break the bottleneck of assessing projects independently using streamlined big data sources from real projects. The study raises awareness to integrating numerical (process) and statistical (context) data. By this combination, a concept for intelligent performance metrics is discussed.

1 BACKGROUND

To make significant improvements in the AEC industry, methods/metrics to compare projects are missing. Comparisons are challenging due to contextual differences. The systemic nature of the industry means that multidimensional datasets are needed to make comparisons between projects.

The paper contributes by shifting focus on digitalisation and methods to integrate the large amounts of data created on projects for performance measurement. Currently the data is not connected between projects as neither people nor process are fully connected. Developing these connections can be a first step in the direction of applying machine learning to connect project process data and contextual indicators. This paper raises the awareness for identifying and specifying (filtering) relevant data from real projects for data-driven decision support within and beyond the project level using machine learning applications.

The transition from an 'one project' data to continuous improvement between a number of projects (industrial view) is needed to enable industry 4.0; the perspective and use of data needs to be long-term.

The construction management domain is fragmented like the industry (Smyth et al. 2019) as knowledge is contextually dependent (Vandenbroeck et al. 2014). Researchers understand the contextual factors that make projects different from one another (Sujan 2020), however, the methods to generate real data and make sense of it in a reproducible manner is still a gap.

In industry, practitioners create and develop data to manage and execute projects with the primary goal of tracking project progress and planning. Virtual Design and Construction (VDC) is a way to streamline process by uniting practitioner competency/communication. For example, VDC uses classifications of process such as pooled, reciprocal and sequential where each have their own characteristics. This presents an opportunity for researchers. How can project data be developed with a long-term research perspective? In this paper, a conceptual combination of methods is proposed:

- Real project process tracking data generated in projects
- Statistical indicators of a project's context, reliability assessed with qualitative methods, iteratively developed by researchers.

Without the second part, the first part is subjective to the particular project. With this combination it is then arguable that it is possible to adjust the process dataset to a baseline using contextual indicators (empirical factors) and therefore, enable comparison between projects at the industrial level.

In this paper, the possibility of this rich integration of data is discussed, taking inspiration from developments in macroeconomics who have also used empirical factors to add context to economic models.

The primary aim of this paper is to present a long-term vision where continuous improvement can be utilized at the industrial level with machine learning. In doing so, the barrier is structured data about project processes; it needs to be aligned and developed using a research perspective.

2 LITERATURE REVIEW

2.1 Collaboration and digitalisation

Integrated Design and Delivery Solutions developed an abstract representation of AEC projects involving process, people and technology (PPT) (Owen et al. 2013). However, IDDS is industry oriented and the acknowledgement and representation of context was not directly discussed. (Sujan 2020) finds that the way people behave in projects depends on their expectations with regards to initiatives put across in terms of PPT; this means that factors forming the context are important in comparing projects. Figure 1 shows an expanded representation of IDDS to show the dependency of PPT on the context of the project.

Holistic Model of Collaboration in the AEC industry (HMC-AEC) suggests that collaboration is complex with number of interdependent factors that are filtered through psychological and social factors (Sujan 2020; Sujan et al. 2019;. It shows that the complex connections between the factors make it challenging to study factors independently.

Furthermore, collaboration in a project depends on people's expectations and partly, how these expectations are managed influences their attitudes. In general, the reality of project collaboration is as a result of interaction between collaboration factors (e.g. contracts, client, industrial, intraorganizational) that form the context of a project. This makes it challenging to compare collaboration in multiple projects as the typical researcher feels that projects are unique and therefore cannot be compared without in-depth case studies. In this paper, a method using metrics is suggested as a driving force to connect contextual project elements to project process productivity. In the long term, it is with the aim to compare projects by dealing with contextual elements just as done in other complex domains such as macroeconomics, further explained in section 2.4.

Research in a domain that is studying service providers suggests there is a short-term mentality.

2.2 Virtual Design and Construction (VDC)

VDC is the use of integrated multi-disciplinary performance models of design-construction projects to support explicit and public business objectives (Kunz & Fischer 2012). VDC does this by focusing on the definition of a combination of digital based methods at the interface of design, site and project management that streamline coordination between project teams. In simple terms, VDC connects the creation and management of building information (BIM) to process- and project management (PPM), and decision-making base on integrated concurrent engineering (ICE).

Part of the framework depends on understanding the interdependencies that tasks have, pooled sequential and reciprocal, as initially identified by Thompson (1967). Depending on these interdependencies, project managers decide how to progress in the project.

In light of the wide adoption of VDC in Norway, this aligns the language and competency regarding digitalisation methods and decision making.

2.3 The value in integrating information

The focus on information owes to the generalization that all projects create information for its delivery which can be used for research purposes. However, this information needs to be streamlined to make comparison possible; the role of VDC in this process.

Moody and Walsh (1999) developed seven laws of information and value. The sixth law is the most relevant to this paper; the value of information increases when combined with other information (Moody & Walsh 1999) where benefits of integration can be achieved by standardizing a relatively small percentage of data.

In the AEC setting standardizing information is predominantly about building parts/objects (i.e. BuildingSmart object standards) not standardised process information. This needs to be extended to process and people from IDDS which are part of the contextual aspects of the project to attain a holistic view.

2.4 Solow-swan model from macroeconomics

Solow-swan model attempts to explain long-run economic growth by looking at capital accumulation, labor/productivity growth and increases in

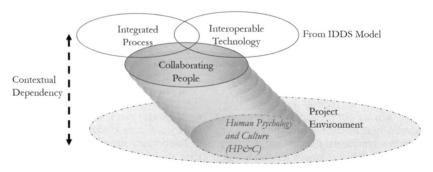

Figure 1. Contextual dependency of People, Process and Tools (Sujan 2020).

technological progress (Acemoglu 2009). The model formed a starting point for various extensions because of its simpler mathematical characteristics. The development of this model involved adding extensions to simplify complexity from complex sociotechnical factors showing that trends can be used to develop empirical indicators.

As part of the use of the model there are two types of variables; endogenous and exogenous variables. The exogenous variables are ones that are external to the model that are imposed on the model. The endogenous variables are within the model.

An example of an exogenous variable is the 'solow residual' which is an empirical factor. From the Solow-Swan model, there was need to adjust the labor according to the context of the industry i.e. they differ greatly between developed and developing countries. The differences come from very complex socio-economic factors making it impractical to go into more specific detail. To avoid loss of generalization capability, the solow residual was developed taking into account some socio-economic indicators such as unemployment.

3 METHODOLOGY

'the need for multi-methods and perspectives to attempt to secure an in-depth understanding of phenomena in question'

(Denzin & Lincoln 1994)

The aims of this paper are to discuss a concept method of utilizing metrics and indicators of project context.

The CIFE horseshoe framework was developed to guide researchers process in turning methodology into methods. It includes an iterative process between:

1. *Observing a problem*
2. *Intuition*
3. *Theoretical Point of Departure*
4. Research Methods
5. Research Questions
6. Research Tasks
7. Validation/Results
8. Claimed Contributions
9. Predicted impact.

The purpose of this paper is to focus on the first 3 parts of the CIFE Horseshoe (in italics) as scholars call for 'proper research design must be performed before data is collected' (Toole 2006).

The methods utilized to meet this methodological purpose were critical reflection and discussions with practitioners from the VDC course in Norway. Critical reflection is an overall process of learning from experience, with the aim of improving professional practice (Fook 2011). Discussions with practitioners were developed informally by 'bouncing ideas' back and forth from them. The purpose was to understand the point of view of industry and forecast challenges and barriers of developing project data for research purpose.

In this paper, a point of departure is suggested as the conceptual method which will be used to contact practitioners and gauge interest. The following steps in the research process are detailed in section.

4 A CONCEPTUAL METHOD

4.1 Overview

Figure 2 shows an overview of the metric. **VDC process data (1)** surrounds the concepts for classifications

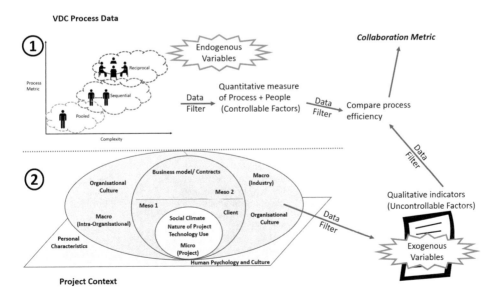

Figure 2. Method for data-driven analysis of project process and context.

that VDC depends on; reciprocal, sequential and pooled. Using these classifications, project data can be developed to track 'who did what and when'. Currently project data shows tasks and time, therefore, the foundation for this **dataset** is already existing, however, methods vary even though VDC is being used. Connecting 'Who' to a higher level of classification of 'what' produces the value – connecting people and process enables connections of transactional project elements e.g. contracts and business models.

This dataset represents the endogenous variables of the model; people, task type, based on dependency, and time. It provides a valuable holistic view of the project comprising of its complexity in real time, which is highly dependent on the **Project Context (2)**; without considering project context, VDC process data (1) will not be grounded, comparisons made irrelevant. Currently metrics are used mainly to show how much the project has improved, however, this is contextually dependent.

Construction projects are highly unique and complex environments, however, the reflection made with macroeconomics shows that it is possible to make simplifications of complex environments by developing indicators and mathematically applying them to the endogenous variables.

The **project context (2)** depends on a number of factors, of which some are suggested in Table 1. The factors come from a doctoral thesis of Sujan 2020 where the HMC-AEC model was developed. From this model it is inevitable that these factors are highly subjective and complex, however, there are some factors that are more important than others.

Research in the CM domain has found it challenging to justify comparisons between the subjective elements as methodology and subjectivity appear as bottlenecks. However, in other domains such as macroeconomics, it has shown to be overcome by the use of trend finding with indicators.

A significant part of the model involves the client and business model/contracts as they represent the transactional elements. However, in terms of context, other important indicators are ones that affect the nature of work that people have to do on the project i.e. comparing a prefabricated building to a museum needs to take into account design creativity.

Table 2 shows the basic three task interdependencies that VDC utilises and its common traits. From a research point of view this is a generalization of process. In the conceptual method suggested in this paper, these assumptions form the central part of the real project data about process. By which, contextual aspects act like attachments to initially explain and allow for development of indicators that can one day be utilized to generalize projects and allow wider comparison between them. However, researchers and academics cannot highly justify what influences the efficiency of these types of tasks.

or example, this makes it possible to see if a specific type of business model can influence how reciprocal tasks are executed and planned for. I.e. a design bid build project means contractors come on later than design and build projects. This means that changes to design can occur once a contractor is appointed as it is justifiable that conZtractors have more knowledge of the possibilities of buildability with respect to their own resources. Therefore, it is expected that more reciprocal tasks will present itself when contractor joins the project. Keeping in mind that this is an oversimplification just to show the concept, combined with other types of contextual factors as shown in Table 1 a holistic (non-reductionist) approach to research connecting people to process can be done with a numerical basis.

Table 1. Factors, trends and data types.

Factor	Examples of Multi-Project Trends	Data Type
Business Models/ Contracts	Time of involvement	Structured
	Shared risk/liability	Mixed
	Extrinsic Motivation	Structured
Personal Characteristics	Personality of key stakeholders	Mixed
	Intrinsic Motivation	Mixed
Client	Trust in Leadership	Mixed
	Knowledge/Requirements	Structured
	Hierarchy in client organisation	Structured
Organisation Culture	Hierarchy in organisations	Mixed
	Size of organisations	Structured
	Profession oriented	Mixed
Social Climate	Commitment	Mixed
	Trust	Unstructured
	Relationships	Unstructured
Nature of Project	Design Creativity	Mixed
	Buildability	Structured
	Size of Project	Structured
Technology Use	Information sharing maturity	Structured

Table 2. Process classifications used in VDC.

		Reciprocal	Sequential	Pooled
Task Traits	Interdependency	High (>2 teams)	Medium (2 Teams)	Low (1 team)
	Resource efficiency or productivity	Low	Medium	High
	Creativity	High	Medium	Low
	Dependence on context = complexity	High	Medium	Low

5 DISCUSSION

5.1 Inspiration from macroeconomics

The development of macroeconomic models to compare the economies across the world have had to find ways to simplify the complexity e.g. the Solow-Swan model. To apply a similar approach in the construction management domain, this means that the endogenic variables are ones that are controllable and measurable in projects, whereas, the exogenic variables are socio-technical indicators of context.

One key difference is that with the economics example, the data was available to the researchers, consistently developed from a long period time i.e. statistics from various countries. In the AEC industry data is developed differently between projects with a short-term perspective in mind, only on the micro-level.

Therefore, the focus should be on developing streamlined data from projects. A large part of project delivery is in developing and managing data, however, to enhance its use at a macro level. There needs to be more streamlined way of managing data. This paper attempts to raise the problem and provides a next step utilizing VDC as an enabler for this change.

Reflections on the principles used in developing the Solow-Swan model are used to show that complexity can be simplified into indicators to streamline and compare construction projects despite their large differences. But first, streamlined data is needed.

To streamline data based on the VDC categorization of processes, the classifications and definitions need to be further developed to ensure ambiguity is avoided.

5.2 The methods-data bottleneck

A holistic view in method design with quantifiable data is necessary to make assessments possible – researchers and practitioners have to collaborate in the long term to enable streamlined creation of data.

Projects create data in a similar manner with the predominant aim of project execution. However, there are aspects that are left out as practitioners do not have the same intentions as researchers. They are motivated to improve the single project or their own firms' execution of the project.

For researchers, the bottleneck is evident by the lack of streamlined data available to researchers. Every project appears unique, although the VDC framework's intervention in uniting project-level competency makes this possible in the near future as its adoption becomes the norm.

The bottleneck of not having the right blend of systemic data can be linked to methods. Symptomatically, many researchers study a few projects at high level of detail, reductionistic disconnection of complex concepts, which makes it lose practicality, or the best of which combine critical reflection with qualitative methods.

Whereas practitioners only keep record of some aspects that appear necessary to them e.g. number of change orders but with missing connections to the other characteristics of the project's people and process.

Interventions to collaborate in the long term with practice are required to make significant progress in the sector. Projects collect and manage data – why is this data not structured and developed for research purpose at the industrial level?

The method suggested is to show that this type of integration is required to be able to compare projects accounting for their contextual features. Can researchers compare productivity in projects the way economies are compared in macroeconomics?

5.3 Concept model +VDC process definition

The conceptual method described in this paper is based on the hypotheses developed by the authors as a result knowledge of VDC and the complexity of collaborative practice. By combining process information: who is doing what and when, with project context information: what is affecting people's decisions, using empirical factors, it is hypothesized that a holistic quantifiable view is possible like in macroeconomics. However, this is challenging, without a long-term perspective to research and practice collaboration, it is easy to fall into the trap of reductionistic assumptions which isolate key factors with variously structured datasets. It is the author's view that isolating factors without clear data and reasoning should be avoided, however, new factors adopted as part of one united research method with one universal dataset.

5.3.1 Barriers, enablers, opportunities, drivers for change

The key driver for improvements in metrics is in improving the evidence base at the industrial level to make decisions strategically about industrial policy – needs to be done with respect to modelling of multiple project data (Owen et al. 2013). By having data to compare projects by making predictions with contextual indicators, decision makers can make more justifiable decisions without relying solely on experience and psychological judgement.

The enablers of improvements in metrics in the context of Norway is the VDC unification of process, providing a pool of projects for implementation with people who have knowledge of VDC. Enabling an opportunity to streamline the data. Furthermore, the collaborative implementation of VDC with academic bodies in Norway (e.g. NTNU which allows a more research oriented perspective and development) means that researchers can work closely with industry.

The Barriers of developing streamlined metrics come from the short term nature of construction project relationships and firms' mentalities. There is clear resistance to innovative action that does not provide short term benefits as for example seen with other aspects implemented e.g. BIM. Another possible barrier is the hidden risks from doing an explorative study of project context to find trends that can be used to develop empirical factors. It is challenging to predict all the future stages of this research as it depends on the results in the previous stage; there is need to have some characteristics of design thinking in the research approach. A more technical dilemma that needs clarification is that VDC process classifications can be ambiguous. The classification needs to be expanded on with more detail which can lead to development of a prototype of the metric.

The opportunities are many, of which, the most relevant are the possibility of making comparison between projects by accounting for contextual factors. This can then be used to create large datasets for machine learning and more complex research explorations which can improve the accuracy of metrics and trend finding. The opportunity to make coherent information flow and knowledge reuse between projects is an opportunity to transfer lessons and knowledge between complex projects. Streamlining data between projects using contextual factor indicators can also make comparisons at the industrial level possible therefore, bringing opportunity to enhance policy development.

5.4 Industry perspective and future possibilities

The nature of the construction industry is transient and complex, and as a result, practitioners have short term goals limited to their own firm as part of their own projects. Confined to developments within their organisations, there are very innovative uses of technology inside organisations, however, this is not the case at the project inter-organisational level; VDC fills this large gap.

For example, there are initiatives such as lean applied in firms, however, not extended to other firms' teams at the project level or in the other players of the supply chain. This shows that there needs to be a focus on the project level, however, this is challenging because it then depends on a number of stakeholders working together for a short time; showing that clients are important and need to be motivated to allow long term research activity using the predominant aim of continuous improvement.

VDC influence on practice is in streamlining competency and methods at the project level. By collaborating with practitioners and showing conceptual methods like in this paper, from an industrial point of view, continuous improvement between projects can be enabled. However, this starts with streamlining data with research motivations feeding into comparison between projects using machine learning, big data.

6 SUMMARY

The clashing perspectives of industry and research means that the vision of the research perspective had to be communicated. The authors envision uniting real-time data from projects with the development of deduction-based indicators of context.

From the industry point of view, this addition means defining and classifying data in an optimal way for research which would require some dedication.

VDC is an enabler for streamlined process, so what is limiting developing this data also with a research purpose?

The VDC framework uses the categorization of tasks by interdependency, the authors suggest that this can be a central dataset that shows the reality of 'who is doing what and when'. By understanding the context of this data, the end goal is to develop indicators of context. Taking inspiration from macroeconomics, these indicators (empirical factors) can then be used to adjust the project data to allow comparisons between projects.

7 FUTURE WORK

The focus should be on data, whilst large datasets exist, they lack the structure needed for research. By structuring the data produced in projects, it can be used

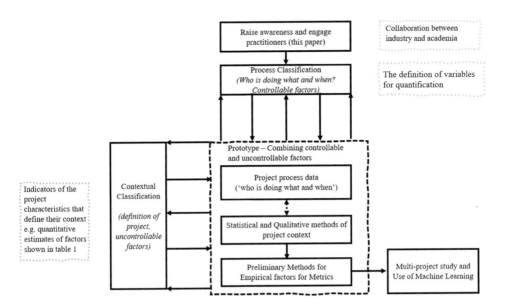

Figure 3. Further research.

for industrial level (multi-project) research. The purpose of this paper has been to raise awareness and seek collaboration with practitioners in industry.

By developing datasets that are created in the project, contextual empirical indicators can be utilised to bridge the gap between controllable and uncontrollable factors like done in macroeconomics.

The following are the next steps (first three presented in Figure 3):

1. Share vision (as in this paper) with industry to gauge interest
2. Develop classifications of tasks to avoid ambiguity keeping in mind the VDC methods should be the baseline
3. Prototype
   a. the use of these classifications on a real project and iteratively design them
   b. start the development of statistical studies (using qualitative reliability) to measure/understand the context
4. Review the conceptual method and present findings aiming to seek wider adoption
5. Expand to more VDC projects
6. Shift focus of automating methods using machine learning
7. Revisit empirical factors design
8. Develop a metric for practitioners using empirical factors.

Collaboration between researchers and practitioners can enable development of intelligent performance metrics that consider the uncontrollable factors in projects; by combining structured real multi-project data and deductive studies of context from the same projects. Although this is an ambitious target, reflection with macroeconomics show that it is possible to use indicators of context to simplify the uncontrollable factors that make project comparison complex.

REFERENCES

Acemoglu, D. 2009. The Solow growth model. *Introduction to modern economic growth.*, pp. 26–76.

Denzin, N.K. & Lincoln, Y.S. 1994. *Introduction: entering the field of qualitative research.* Thousand Oaks, CA: Sage.

Fook, J. 2011. Developing Critical Reflection as a Research Method BT – Creative Spaces for Qualitative Researching: Living Research *In*: J. Higgs, A. Titchen, D. Horsfall and D. Bridges, eds. Rotterdam: SensePublishers, pp. 55–64. Available from: https://doi.org/10.1007/978-94-6091-761-5_6.

Kunz, J. & Fischer, M. 2012. Virtual design and construction: themes, case studies and implementation suggestions. *Center for Integrated Facility Engineering, Stanford University.*

Moody, D.L. and Walsh, P. 1999. Measuring the Value Of Information-An Asset Valuation Approach. *In: ECIS.*, pp. 496–512.

Owen, R.L., Amor, R., Dickinson, J., Prins, M. and Kiviniemi, A. 2013. *Research roadmap report: Integrated Design and Delivery Solutions (IDDS).* Rotterdam: International Council for Research and Innovation in Building and Construction.

Smyth, H., Razmdoost, K. & Mills, G.R.W. 2019. Service innovation through linking design, construction and asset management. *Built Environment Project and Asset Management.* 9(1), pp. 80–86.

Sujan, S.F. 2020. *A holistic and systemic model of collaboration in the AEC industry.* University of Liverpool.

Sujan, S.F., Kiviniemi, A., Jones, S., Mannis, A., Wheatcroft, J., Meda, P. & Hjelseth, E. 2019. A Multi-Level Model of Collaboration – Lessons Learnt from Social Scientific Interpretive Research *In: CIBW78 Conference.* Newcastle: CIB.

Thompson, J.D. 1967. *Organizations in action: social science bases of administrative theory.* [Online]. McGraw-Hill. Available from: https://liverpool.idm.oclc.org/login?url= https://search.ebscohost.com/login.aspx?direct=true&db=cat00003a&AN=lvp.b1043139&site=eds-live&scope=site.

Toole, T.M. 2006. A primer on social science research methods in construction *In*: *2 nd Specialty Conference on Leadership and Management in Construction.*, p. 300.

Vandenbroeck, P., Dechenne, R., Becher, K., Eyssen, M. and Van den Heede, K. 2014. Recommendations for the organization of mental health services for children and adolescents in Belgium: use of the soft systems methodology. *Health policy.* 114(2), pp. 263–268.

The practice of VDC framework as a performance measurement system for projects

S.B.S. Ahmad & E. Hjelseth
Norwegian University of Sciences and Technology, Trondheim, Norway

ABSTRACT: The use of Virtual Design and Construction (VDC) methods, tools and practices coupled with powerful business analytics can provide a solution to the construction industry's long-standing performance measurement problems. The study opts for a pragmatic approach to investigate the perception of the VDC practitioners to use VDC framework as a performance measurement system for projects through an online survey. It further explores the hidden relationships in the perceptions of the VDC practitioners through the application of data analytic tools. Results indicate that the VDC framework has the potential to be practiced as a performance measurement system for the construction projects. The study is unique in the approach and is targeted at the VDC practitioners and academics aiming to digitalize the performance measurement in construction industry.

1 INTRODUCTION

The application of analytics to manage the performance of the business is referred to as Business Performance Analytics (BPA). It is defined as the management and control of the organizations strategic dynamics and performance by the application of internal and external data analytic methods (Silvi et al. 2010). BPA allows the decision makers and analysts to directly interact with data to gain insights, draw conclusions, make better decisions, and improve business performance through improved strategy and implementation.

One of the basic prerequisites for the BPA is the Performance Measurement System (PMS) with the ability to consistently generate, structure, and mine the relevant digital data for performance analytics. However, construction industry has long been criticized for its performance issues and limited measurement practices (Egan 1998; Latham 1994), and business analytics are hardly used in construction research. According to Neely et al. (1995) PMS is 'a set of metrics used to quantify both efficiency and effectiveness of actions'.

It is difficult for the managers to comprehend if they have achieved their intended objectives or they will be able to do so (Neely et al. 1996). Even with the incredible rates of advancements in the computing power and storage capacity in the recent decades. The analysis of data is still inconsistent and messy in the performance measurement domain of the construction projects.

To extract the relevant and valuable performance information from the flood of digital data, several construction companies are increasingly investing in the BPA. However, these investments often get cornered up in collection, cleansing and storing of different forms of data (Zhang et al. 2015). This compromises the focus on data understanding and the possible value it can deliver to the support and management of organizational performance (Economist Intelligence Unit 2013). Furthermore, the poor and limited interaction between the databases and PMS adds to the difficulty of extracting the full potential of the BPA.

Since its conception in 2001, the Virtual Design and Construction (VDC) has been attracting increased attention from the construction academics and practitioners (Alarcón et al. 2013; Kam et al. 2013; Knotten & Svalestuen 2014). The VDC methods provide virtual or computer-based performance models, and is defined by Kam and Fischer (2004) as 'The use of multidisciplinary performance models of design and construction projects, including product (i.e. facilities), work process and organization of the design construction operation team in order to support business objectives'.

Performance measurement is at the vortex of the VDC framework that binds the multidisciplinary project teams together with technology, techniques, tools, and social practices to achieve the performance objectives. As VDC performance models are virtual or computer based, this study is an attempt to explore the perceptions of the VDC practitioners of using VDC framework as a PMS. Because it is only after the appropriate PMS has been adopted that BPA can be effectively applied and tested. Thus, this study aims to answer the following research questions:

1. What are the perceptions of the VDC practitioners about using the VDC framework as a PMS for the building and construction projects?
2. What are the hidden relationships in the perceptions of the VDC practitioners?

2 THE VDC FRAMEWORK

2.1 *VDC tools, methods and techniques*

The VDC framework counters the challenges of performance measurement through a set of tools, methods and techniques which typically include 1) Integrated Concurrent Engineering (ICE), 2) Building Information Modelling (BIM), 3) The Last Planner System (LPS), 4) Model Maturity Index (MMI), 5) Lean Construction and 6) Performance Metrics.

In order to reduce the decision latency and achieve concurrency in design activities for 'faster, better, cheaper' design, the Jet Propulsion Laboratory (JPL) at NASA developed the ICE practice (Mark 2002). According to Chachere et al. (2009) ICE is a rapid combination of expert designers, advanced modeling, visualization and analytical tools, a set of consistent social processes, and specialized design facility to create preliminary designs of complex systems.

BIM is the virtual product of the information required for facilities design, construction, and operations, from conception to the all the life-cycle stages (Eastman et al. 2011). LPS is a pull planning tool that has reported to be successful in increasing the accuracy of the design plan by several authors e.g. Ballard and Koskela (1998), Hamzeh et al. (2009), Hattab and Hamzeh (2013), Wes et al. (2013). With targeted attempts to measure the design, Ballard (1999) introduced Percentage Plan Completed (PPC) for measuring the status of completion of the planned tasks in LPS. PPC provided the managers with better control and documentation over unfinished tasks. According to Ballard (1999) PPC addressed both design and construction phases of the project.

As BIM is central to VDC realization, and virtual BIM models are realized through sophisticated information flows and much more information in form of attributes connected to objects than traditional paper drawings it causes challenges for planning and control of design process (Hooper 2015). To counter the issue of design progress measurement the VDC framework advocates for measuring the Model Maturity Index (MMI) together with design Percentage Plan Completed (PPC). MMI reflects the design readiness of a zone area in the project for different design disciplines (Eray et al. 2018).

The application and adaption of the concepts and principles of the Toyota Production System to construction is referred to as Lean Construction (LC) (Sacks et al. 2010). LC focuses on increasing the customer value, reducing the waste and continuous improvement (Lichtig 2006). For the detailed guide to the application of VDC to the lean project delivery process Khanzode et al. (2006) published a guide. The basic concept in the guide is to deliver a construction project as a lean production system.

2.2 *Performance measurement with VDC*

Performance measurement is a well-discussed topic in construction literature (Beatham et al. 2004; Egan 1998; Horta et al. 2009; Kim & Huynh 2008; Skibniewski & Ghosh 2009; Yeung et al. 2012). However, the fundamentals of performance measurement in VDC framework are based upon the performance metric methodology. According to Fischer et al. (2017) the project objectives can be achieved through metrics.

The VDC framework facilitates the performance measurement through 1) Output metrics and 2) the controllable factors. The VDC output matrices serve as performance targets and are motivated with questions like what to build, who should build it, how and when to build it (Fischer & Kunz 2004). Whereas, the controllable factors are the factors that the VDC practitioners use to control and satisfy the output metrics.

The output metrics represent the objectives at different hierarchical levels of the project such as client metrics, project metrics, and production metrics. Nevertheless, the challenge is to define meaningful relevant metrics that can relate to project goals and objectives (Fischer et al. 2017), maintaining the alignment between the different hierarchical metrics.

To solve this challenge the VDC framework provides the specifications to develop the client metric from the client objectives, which is the reflection of the project team (explicit target) on what they would typically be judged against by the client (Kunz & Fischer 2020). The client metrics is the representation of the client stakeholders such as project users, operators and executive objectives in measurable values. Based on the client metric the project team develops the project objectives and metric that could satisfy the client objective metric. The project metric forms the basis for the production objectives that could meet the project and ultimately client objectives. Thus, an alignment is maintained with the client metric through project definition and delivery.

The objective metrics (client, project and production) serve as a yard stick to evaluate the overall performance. Whereas the daily or weekly milestones and objectives are followed up by the Product Organization and Processes (POP) models to measure the performance on the process level. The VDC framework specifies the guidelines to establish multidisciplinary Product Organization and Process (POP) models (Kunz & Fischer 2020). The VDC models are virtual and provide more interactive and flexible experience to the project teams.

The virtual POP models are based on the three controllable aspects under management control: the design of the Product (P) to be built, the design of the organization (O) that does the design and construction, and the design and the design-construction Process (P) that the organization follows (Kunz & Fischer 2020).

However, in a PMS once a measure is defined it goes through four stages of data creation, data collection, data analysis and information distribution (Bourne et al. 2000; Marr & Neely 2002; Nudurupati et al. 2007). This strengthens the view that more systemized and standardized the data creation and collection are the more effective the data analysis (performance analytics) would be.

In the VDC framework the data is created by the multidisciplinary project teams working collaboratively over time. These multidisciplinary teams apply the POP models to specify, create, and check the project activities. Furthermore, these teams also adjust the POP models incrementally over the project lifetime and keep it aligned with the project and client metrics (Kunz & Fischer 2020).

The VDC tools, methods and techniques such as ICE, BIM, LPS, MMI and LC are the controllable factors and satisfy the production metrics. The production metrics include design and construction controllable factors or metrics to control and evaluate the design and production. The VDC framework does recommend the guidelines to establish meaningful production metrics and controllable factors. But it is the project team that establishes the production metrics and chooses the controllable factors to satisfy the production metrics.

Khanzode et al. (2006) recommended the use of PDCA (Plan, Do, Check, Act) metric for the production activities based on the lean construction philosophy. Knotten and Svalestuen (2014) suggest measuring the MMI with the PPC for the design activities for design production control. Belsvik et al. (2019) reviewed the literature of design management to list 14 design phase production metrics that can be used to measure design production in case MMI is not measured in the project. These 14 design phase metrics are further classified in basic and supplementary metrics groups to measure the design production control.

3 METHODOLOGY

The study uses two different methods to answer the research questions 1) investigate the perception of the VDC practitioners about the VDC framework's ability to be used as a PMS through an online survey, 2) apply the data analytics tools to explore the hidden relationships in the perceptions of the VDC practitioners.

An online questionnaire was sent to the participants of the Stanford-NTNU VDC certification program in Oslo 2019-20. 210 construction practitioners were involved in the program and the survey was available for participation from November 22, 2019 to December 20, 2019. Valid responses were received from 95 participants.

The sample group was selected due to their active assignments of applying the VDC framework on their respective projects. Rather than providing the 5-point Likert Scale of strongly agree to strongly disagree the respondents were provided with a numerical point scale from 0-100. The numerical scales provide more relevant responses and safeguard against forced and obvious choices with the Likert scale. After the survey responses were received the responses were clustered in 0-20 strongly disagree, 20-40 disagree, 40-60 neutral, 60-80 agree, and 80-100 strongly agree.

As the respondents are affiliated by the VDC certification program, there could be the potential of biasing the results towards being more exaggerated in favor of the VDC's potential. However, the biasness could go either way in favor or against because the pool consists of seasoned professionals trying to implement a different way of working and change is not always appreciated by all. It was deemed necessary to use this pool of respondents to 1) get perceptions from those trying to implement VDC on their projects 2) serve as a comparison baseline for the future surveys in the similar program.

In order to explore the hidden relationships among the survey participants perceptions data analytics tool Power BI was applied on the survey data. The Power BI interactive visualizations helped in deeper understanding of data. However, the limitations of applying the data analytics to the data makes it difficult to present the interactive visual analytics in a research paper format. This also limits the readers and editor's interaction with data, and a final picture hardly represents the full story upon which the researcher banks its conclusions and contributes to the body of knowledge.

4 RESULTS AND DISSCUSSION

The Figure 1 presents the respondents categorization in terms of project type. Almost 51% of the respondents of the online survey reported to be working on infrastructure projects and 42% on the public building projects. Where a mere 6% of the respondents were associated with the commercial building and theater/museum projects.

Similarly, Figure 2 indicates the job designations of the survey respondents. With 32% consultants were the most responsive group, followed by the 30% design

• Infrastructure • Public Building • Commercial Buildings • Theater/Museum

Figure 1. Respondent classification by project type.

• Consultant • Design Manager • Project Manager • Construction Manager
• Top Management

Figure 2. Respondent classification by job designation.

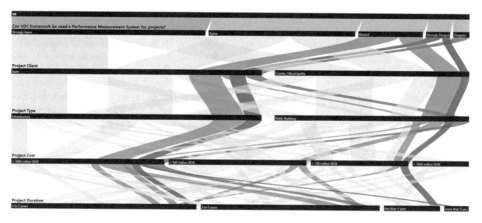

Figure 3. VDC practitioners perceptions based on project client and type.

managers, and 27% project managers. The top managers and construction managers both represent 5% in the data sample. Furthermore, the data sample represents that the 92% of the survey respondents are working on the state and municipality projects. As the data is marginal for other types of clients, such as private clients, the data analysis and discussions are only focusing on the state and municipality projects.

4.1 *Practitioners perceptions*

The construction practitioners responded with a score of 70.4 in reflecting the VDC frameworks ability to be used as a PMS on their projects to achieve high performance, which means that on average there is an agreement that VDC framework can be used as a PMS. However, this result does not reflect the whole story about the score of 70.4 i.e. capturing the perceptions of the construction practitioners based on their business sectors, affiliations, project types, client types, roles and responsibilities. Furthermore, it can be a very tedious task to find the hidden relationships contributing to this score manually. Whereas with the help of data analytic tools these relationships are easier and faster to explore.

4.2 *Hidden relationships in perceptions*

The Figure 3 presents the response with the project classification. Although there is wide agreement that VDC framework can be used as a PMS for projects, 27% neutrality and disagreement could be traced to public buildings and 18% to the infrastructure projects. As the number of respondents associated with the infrastructure projects is higher than that of public building projects. The weighted neutrality and disagreement amongst the construction practitioners working on the public building projects is 33%.

Thus, in terms of percentage the lower confidence is found amongst the construction practitioners working on the public building projects. But the data sample indicates that the 58% of the municipality projects are public buildings and 42% are infrastructure projects. Where none of the survey respondents on the infrastructure project with municipality reported responded with neutral or disagreeing scores. This further means that with municipality project skepticism can only be traced to public building projects. A much deeper analysis reveals that strong disagreement is only reported on small municipality projects i.e. projects with a total duration of less than a year and cost of under 100 million NOK.

A similar exercise for the state projects reveals that the 66% of state sponsored projects are infrastructure and 34% public building in the data sample. Both state sponsored infrastructure and public projects report 26% of skepticism with the VDC frameworks ability to be used as a PMS for the projects. However, the strong disagreement can only be traced to the state sponsored infrastructure projects with total cost between 500–1000 million NOK.

Furthermore, Figure 1 also indicates that strong disagreement is only found amongst the participants working on the projects on extreme sides of the spectrum i.e. big projects with longer duration (more than 5 years), and the projects with very short duration (less than an a year). The participants with medium duration projects (3–5 years) did not reflect strong disagreement. In terms of cost the participants with projects of total cost of over 1000 million NOK (43%) and under 100 million NOK (18%) reported the highest proportion of low scores. This indicates that the trend follows in terms of project duration and costs, and that the highest confidence is found amongst the project participants where the total project costs are between 500–100 million NOK and duration is between 3–5 years.

Based on the results presented in the Figure 3, the hidden relationships in the VDC practitioner's perceptions are:

1. The VDC practitioners associated with the municipality sponsored projects for public buildings tend to have lower confidence in VDC frameworks ability to be practiced as a PMS;

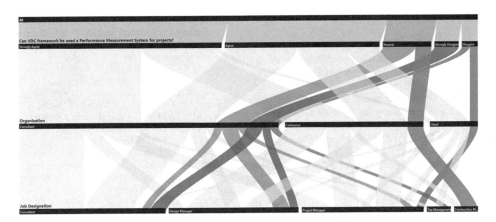

Figure 4. VDC Practitioners Perceptions based on organization type and job designation.

2. The VDC practitioners working on state infrastructure projects have higher skepticism in using VDC framework as PMS;
3. The total duration and cost of the project have an influence on the perception of the VDC practitioner to practice it as a PMS for the project.

It might be that the small projects with shorter durations and low cost do not allow the VDC practitioners with the time to resources to fully practice VDC. The skepticism reflected from the practitioners involved with the projects of over 5 years duration is interesting. One of the reasons that might explain this relationship is the nature of these projects.

The projects with over 5 years duration might be less defined on front-end and inherent the change of scope. Where one government can initiate a project with a certain scope and next government can totally change the scope of the project. Given that the VDC framework is practiced with locked performance targets with early involvement of the stakeholders, the change of stakeholders might be difficult to deal with and may require re-modelling of the existing POP-models in the projects.

The basis of defining the stakeholders objectives and POP models with the VDC framework are the questions of what to build, who should build how, and when to build it (Fischer & Kunz 2004). This might be the other reason that the participants from these projects have reflected lower confidence. Because these metric defining questions might just not be enough for these projects.

This also motivates future research into the metric formulation method of VDC framework for such projects, and an inquiry to explore how VDC metrices and POP models respond to the change of scope in the projects.

The Figure 4 presents the perceptions of the construction practitioners using VDC framework for PMS on their projects in relation to the organization type and job designations. Amongst the construction practitioners, 20% of the consultants represent the neutrality and disagreement in VDC frameworks ability to be used as a PMS for the projects. Whereas, 16% of disagreement is found amongst the clients associated with the job designation of top management. Only 18% of the survey participants working for as construction contractors reported to be neutral, and this neutrality is only found amongst the survey participants working as construction managers.

An interesting trend to note in the Figure 4 is that 43% of the design managers working for the consultants reflected low confidence, whereas the design managers working for the contractors reflected very high scores and all of them were of agreement that VDC framework can be used as a PMS for their projects. While there is no clear explanation of this contrasting variation amongst the design managers. A reason might be the scope of design management, where the consultant design manager might be responsible for one or few design disciplines, the contractor design manager is responsible for all the multidisciplinary design of the project and station themselves close to the production. The contractor design manager is also responsible for the multidisciplinary design control, clash detection and BIM coordination providing him/her more influence over the project design.

It can be concluded with the data analysis that the most positive group is the contractor design managers and the most negative group is the consultants design managers about the VDC frameworks ability to be used as PMS for projects. The element of future research in this context is the contract type and practice of VDC framework. Because the consultants often are contracted with hourly rate, especially for the bigger state projects, where the motivation for time reduction might be compromised. Whereas the contractors are often contracted with lump-sum or turn keys projects where time and cost reduction are often rewarded. Furthermore, 83% of the consultants were positive about using VDC framework as a PMS for their projects, 11% reported strong disagreement. This 11% is associated with the big state infrastructure projects costing more than 1000 million NOK.

5 CONCLUSION

The construction practitioners actively implementing the VDC framework were engaged through an online survey to find out their perceptions about using VDC as a PMS on their projects. The survey reflected that 78% of the VDC implementers believe that the VDC framework provides them an appropriate PMS for projects, where 14% of the VDC practitioners were neutral and 8% reflected disagreement. Data analytic tools were applied to find the hidden relationships in the perceptions amongst the respondents. The notable trends that emerged with the data analytics were:

1. The total cost and duration of the project has an influence on the perception of the VDC practitioner to use it as a PMS for the project. The VDC practitioners associated with the projects ranging from 3-5 years were found to be most supportive to use VDC as PMS;
2. Some skepticism is found among the VDC practitioners associated with public building projects sponsored by the municipality and the infrastructure projects sponsored by the state;
3. The contractors design managers are the most positive group of people about the VDC's ability as a PMS, whereas the consultants design manager reflected the highest disagreement to use VDC as PMS.

The overwhelming confidence in the performance measurement ability of the VDC framework makes it the best candidate for applying project level business performance analytics. The consensus in its ability amongst the different construction businesses (clients, consultants and contractors) also makes its suitable for cross-organizational and industry level performance benchmarking and analytics. Moreover, increased implementation of VDC as a PMS has the potential to support systematic and fact-based performance improvement of the construction industry.

REFERENCES

Alarcón, L., Mandujano, R., Maria, G. & Mourgues, C. 2013. Analysis of the implementation of VDC from a lean perspective: Literature review. *21th Annual Conference of the International Group for Lean Construction*. Fortaleza, Brazil.

Ballard, G. 1999. Improving work flow reliability. *International group for lean construction*. Berkley, CA, USA.

Ballard, G. & Koskela, L. 1998. On the agenda of Design Management research. *6th Annual Conference of the International Group for Lean Construction*. Guarujá, Brazil.

Beatham, S., Anumba, C., Thorpe, T. & Hedges, I. 2004. KPIs: a critical appraisal of their use in construction. *Benchmarking: an international journal*, 11, 93–117.

Belsvik, M. R., Lædre, O. & Hjelseth, E. 2019. Metrics in VDC Projects. *27th Annual Conference of the International. Group for Lean Construction (IGLC)*. Dublin, Ireland.

Bourne, M., Mills, J., Wilcox, M., Neely, A. & Platts, K. 2000. Designing, implementing and updating performance measurement systems. *International journal of operations production management*.

Chachere, J., Kunz, J. & Levitt, R. 2009. The role of reduced latency in integrated concurrent engineering. Center for Integrated Facilities Engineering (CIFE): Stanford, USA.

Eastman, C. M., Eastman, C., Teicholz, P., Sacks, R. & Liston, K. 2011. *BIM handbook: A guide to building information modeling for owners, managers, designers, engineers and contractors*, John Wiley & Sons.

Economist Intelligence Unit 2013. The Data Directive: Focus on the CFO. London: Economist Intelligence Unit.

Egan, J. 1998. *Rethinking construction*, Department of Environment, Transport and the Region.

Eray, E., Haas, C., Rayside, D. & Golparvar-Fard, M. 2018. A conceptual framework for tracking design completeness of Track Line discipline in MRT projects. *ISARC. Proceedings of the International Symposium on Automation and Robotics in Construction*. IAARC Publications.

Fischer, M., Ashcraft, H. W., Reed, D. & Khanzode, A. 2017. *Integrating project delivery*, Wiley Online Library.

Fischer, M. & Kunz, J. 2004. The scope and role of information technology in construction. Proceedings-Japan Society of Civil Engineers, DOTOKU GAKKAI, 1–32.

Hamzeh, F. R., Ballard, G. & Tommelein, I. D. 2009. Is the Last Planner applicable to design? a case study. *17th Annual Conference of the International Group for Lean Construction*. Taipei, Taiwan.

Hattab, M. A. & Hamzeh, F. 2013. Information Flow comparison between Traditional and BIM-based Projects in the Design Phase. *21th Annual Conference of the International Group for Lean Construction*. Fortaleza, Brazil.

Hooper, M. 2015. Automated model progression scheduling using level of development. *Construction Innovation*, 15, 428–448.

Horta, I. M., Camanho, A. S. & Da Costa, J. M. 2009. Performance assessment of construction companies integrating key performance indicators and data envelopment analysis. *Journal of Construction engineering Management*, 136, 581–594.

Kam, C. & Fischer, M. 2004. Capitalizing on early project decision-making opportunities to improve facility design, construction, and life-cycle performance—POP, PM4D, and decision dashboard approaches. *Automation in construction*, 13, 53–65.

Kam, C., Senaratna, D., Mckinney, B., Xiao, Y. & Song, M. 2013. The VDC scorecard: Formulation and validation. Center for Integrated Facility Engineering (CIFE): Stanford University.

Khanzode, A., Fischer, M., Reed, D. & Ballard, G. 2006. A guide to applying the principles of virtual design & construction (VDC) to the lean project delivery process. Center of Integrated Facilities Engineering (CIFE): Stanford University.

Kim, S.-Y. & Huynh, T.-A. 2008. Improving project management performance of large contractors using benchmarking approach. *International Journal of Project Management*, 26, 758–769.

Knotten, V. & Svalestuen, F. 2014. Implementing VDC i Veidekke – Using simple metrics to improve design management process. *22th Annual Conference of the International Group for Lean Construction*. Oslo, Norway.

Kunz, J. & Fischer, M. 2020. Virtual design and construction. *Construction Management and Economics*, 1–9.

Latham, M. 1994. Constructing the team: final report of the government/industry review of procurement and contractual arrangements in the UK construction industry. Hmso, London.

Lichtig, W. A. 2006. The integrated agreement for lean project delivery. *Construction Law,* 26, 25.

Mark, G. 2002. Extreme collaboration. *Communications of the ACM,* 45, 89–93.

Marr, B. & Neely, A. 2002. Balanced scorecard software report, a business review publication from Cranfield school of management with contributions by Gartner. Inc. Connecticut, USA.

Neely, A., Gregory, M. & Platts, K. 1995. Performance measurement system design: a literature review and research agenda. *International journal of operations production management,* 15, 80–116.

Neely, A., Mills, J., Platts, K., Gregory, M. & Richards, H. 1996. Performance measurement system design: should process based approaches be adopted? *International journal of production economics,* 46, 423–431.

Nudurupati, S., Arshad, T. & Turner, T. 2007. Performance measurement in the construction industry: An action case investigating manufacturing methodologies. *Computers in Industry,* 58, 667–676.

Sacks, R., Koskela, L., Dave, B. A. & Owen, R. 2010. Interaction of lean and building information modeling in construction. *Journal of construction engineering management,* 136, 968–980.

Silvi, R., Moeller, K. & Schlaefke, M. 2010. Performance management analytics-the next extension in managerial accounting.

Skibniewski, M. J. & Ghosh, S. 2009. Determination of key performance indicators with enterprise resource planning systems in engineering construction firms. *Journal of construction engineering management,* 135, 965–978.

Wes, J. G. B., Formozo, C. T. & Tzotzopoulos, P. 2013. Design Process Planning and control: Last Planner System adaptation *21th Annual Conference of the International Group for Lean Construction.* Fortaleza, Brazil.

Yeung, J. F., Chan, A. P., Chan, D. W., Chiang, Y. & Yang, H. 2012. Developing a benchmarking model for construction projects in Hong Kong. *Journal of construction engineering management,* 139, 705–716.

Zhang, J., Yang, X. & Appelbaum, D. 2015. Toward effective Big Data analysis in continuous auditing. *Accounting Horizons,* 29, 469–476.

Exploring the degree of automated process metrics in construction management

K. Rashasingham & E. Hjelseth
Norwegian University of Science and Technology, Trondheim, Norway

ABSTRACT: Process metrics are commonly used in many industries to monitor and track different business processes, but in the construction industry there is currently no unified method. However, there is a trend towards establishing both qualitative and quantitative metrics, mostly on product performance. Identifying process metrics are crucial for all project teams that want to track their performance, provide documentation on progress and to detect areas of improvement. The essence is therefore to collect data and to provide information about the progress toward reaching client and project objectives. This paper aims to explore the possibilities to automate different process metrics that is already measured by students, in their respective projects and organizations, in a large-scale Virtual Design and Construction (VDC) course. A methodology to assess the *degree of Automated Process Metrics* (APM) is developed.

1 INTRODUCTION

The increasing pressure on engineers and managers to track and document progress during a project demands a set of process metrics to be automated. With emerging technologies within Internet of Things (IoT) and tremendous amount of new software applications one should imagine that metrics and measurement systems should be fully automated and easily applied. However, that is not the case, yet.

Industry 4.0 aims to utilize information and communication technologies (ICT) together with production methods to create many opportunities for growth offered by digitalization, interconnectedness and emerging technologies. According to a study, there is the combination of several technologies that is vital enabler to make the transition from the third industrial revolution to Industry 4.0, e.g. cyber-physical systems (CPS) (Nowotarski & Paslawski 2017). The spirit of such combination of technologies is to innovate the current ways of doing business, i.e. work smarter.

New systems and technologies give the construction industry the possibility to connect the whole industry, from asset-owners, designers, suppliers, to the products themselves and the final end-users. As of today, the use of ICT and integrated software in project management is limited and, seen from the point-of-view of project management. The use of stand-alone software is widely common.

The method of Virtual Design and Construction (VDC) is used to achieve client and project objectives such as scope, schedule, budget and quality by high use of virtual methods i.e. product visualization tools for 3D modelling to visualize products, and process modeling tools for 4D modelling to get a better understanding of how the product will be developed (Khanzode et al. 2006).

Research on process metrics is been going on for several years, on mapping process metrics and establishing in-depth knowledge of the impact of these on performance of the organization or project, to determine the areas that are doing well and to map the areas of improvements (Butler et al. 1997; Kagioglou et al. 2001). To the impact it is interesting to observe that this domain has been limited on exploring the possibilities to automate these and is still very manual oriented.

A study by Belsvik et al. (2019) present an overview of metric based on literature review of VDC related paper and on a real case-study of a VDC project. 13 metrics were presented, and 6 were recommended as basic metrics in all projects. However, metrics as a manual task is the default methodology.

This paper explores the possibilities to automate process metrics and emphasize construction management to embrace Industry 4.0 and to innovate the current way of doing business. In this respect can automated metrics be a support for continuous improvement of performance.

2 CONCEPTUAL RESEARCH

Two different methods have been utilized in this paper. The methods are:

1. Short literature review in use of metrics
2. Review of metrics used currently in large scale VDC course in Norway

Firstly, a brief literature review was performed where the search was based on both worldwide web and among books. The aim of conducting the brief literature review was to get an overview on latest trends and discussed topics in three different industries:

1. The construction industry
2. The manufacturing industry
3. The information technology (IT) industry

There is much literature on empirical studies. However, this is not the case regarding conceptual research. That being said, according to Kohtari conceptual research is to "develop new concepts or to reinterpret existing ones" by exploring "abstract idea(s) or theory (Kothari 2004).

In this paper the conceptual method is based on describing and explaining real-world examples to its simplest form. The benefits of doing a conceptual research is that the research "depend heavily on real-world description, thereby serving as a check on the external validity of our research findings" (Meredith 1993).

Secondly, the assessment done in this research is combining real-world examples from different industries with the aim of mapping potential applicability to the construction industry. Real-world examples from the construction industry are collected from student enrolled in further education course in Virtual Design and Construction run by the Norwegian University of Science and Technology in collaboration with Stanford university (VDC-NTNU 2020). There are 200 students enrolled in the course. They represent companies from whole supply chain, asset owners, consultants, contractors, software developers and suppliers. The course started in August 2019 and is scheduled to be finished in May 2020. During this period, all students submit monthly, except for in December, and report to document their use of VDC in projects. Students are grouped into smaller groups of 20 where they have their own mentor. All the mentors are required to have the VDC certificate from Stanford University and several years of experience from VDC-projects. The students must submit a monthly report based on a PowerPoint template on their work and progress on VDC that mentors evaluate and score according to a scoring template. The students can do some modifications to the PowerPoint-slides but are urged to follow the pre-set template. There is a section in this template dedicated to 'metrics'. This study is based on report number five (report #5).

A thoroughly systematic review of 10 student reports was done. The systematic review was performed to only focus on the part where 'Metric' is elaborated in each report. A pre-set PowerPoint template was given to the students. The part about metrics was given a dedicated slide. This allowed the systematic review to be convenient and quickly done.

2.1 Exploring the IDEF0 framework

From the manufacturing industry, a concept such as Integration DEFinition (IDEF) is used to modelling business processes. The IDEF0 is used to model

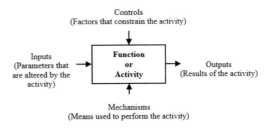

Figure 1. IDEF0 (Kim & Jang 2002).

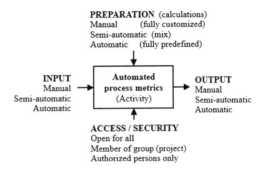

Figure 2. The Automated Process Metrics.

the need of information to "minimize errors and the unplanned evolution of activities" (Kim & Jang 2002). The five elements of IDEF0, shown in Figure 1 are inputs, controls, mechanisms, function or activity and output.

The box in the center represents the activity or function. "Inside the box is the breakdown of that activity into smaller activities", which can be understood as a process. A process can be understood as a set of activities or steps that needs to be performed in order to achieve a desired output.

Figure 2 (APM) illustrates how APM is adapted to the concepts in IDEF0 framework to specify the input, preparation and output and map them in 'degree of automation' and is divided into three levels:

1. Manual: meaning that there is high human involvement in either input or output, e.g. manual typing and registration or manual processing of collected data
2. Semi-automated; there is some human involvement but is somewhat supported by digital tools and software
3. Automated; there is almost no human interference meaning that the digital tools and software are on shared platform and exchanging data and information. The human intervention is only needed to start or end the processing, e.g. upload a file or convert a file or simple clicks on buttons.

Access (or security) to metrics is embedded in APM. Metrics can contain industry secrets, and therefore not presented to everyone. Digital solutions can offer these

Table 1. The methodology of 'degree of automation of processes.

Process (activity)	Degree of automation			% (I+P+O)/3
	Input	Preparation	Output	
Today				
Task or activity	20%	40%	40%	33%
Planned implementation				
Task or activity	40%	80%	90%	70%

Table 2. Degree of automation in metrics related to coordinating meetings/ICE-sessions.

Process (activity)	Degree of automation			% I+P+O /3
	Input	Preparation	Output	
Attendance in meetings	20	90	55	55%
Number of disciplines attending the meetings	25	50	55	43%
Invitation to meetings sent to all participants	65	80	80	75%
Meeting summary sent to all	50	55	65	57%
Attendance of key individuals in meeting	15	50	55	50%
Evaluation of meetings – plus / delta *) See more in section 4.1	20	40	40	33%
Map the use of BIM during ICE meetings	50	80	90	73%
Allocate correct roles to correct persons before/during meetings	15	55	50	40%
Number of new cases vs planned cases	30	45	50	42%
Number of planned cases that are closed during the meeting	30	45	55	43%
Time to close a case	30	45	55	43%
Print and scan documents for meetings	45	40	10	32%

types of services in a more dynamic way than paper passed systems. The ISO 19650 standard is an international standard for managing information over the whole life cycle of a built asset using building information modelling (BIM) (Kerr 2020). User of automated solutions will enable the implementation of ISO 19650 series of standards.

2.2 The degree of automation of processes (APM)

A traffic light rating system, color-scale, accordingly to scores is established:

1. Red (here: dark grey), 0–29%; manual
2. Yellow (here: grey), 30–59%; semi-automated
3. Green (here: light grey), 60–100%; fully automated

The 'degree of automation' will be indicated below each attribute, input, preparation and output, and sets the base for the conceptual discussion and analysis.

Table 1. The methodology of 'degree of automation of processes showcases the methodology based on an example, e.g. task or activity. The breakdown of each element in section 'Today' is based on the student reports and the 'traditional' way of doing things, while the section 'Planned implementations' is suggested by the authors.

3 RESULTS

This section presents the findings from the literature review and from the systematic review on the different student reports. The literature review indicates that other industries are a step ahead on automation and digitalization than the construction industry. Based on the systematic review there is a high number of metrics which measure processes. The published material presents only a selected part of these.

The literature review shows that the health care industry in collaboration with the ICT industry is finding novel ways of doing patient consultations and monitoring surgical performance. Firstly, researches from the health care industry using robots and camera in combination have proposed a way of monitoring performance. The study on robot-assisted radical prostatectomy had a setup with a "dVlogger", similar to a black box, and a camera to monitor the process to train the Machine Learning Algorithms to predict the outcome (Sears 1994). Today, machine learning is more applicable, but old principles are still valid.

Another study explores the benefits of Cloud of Things (CoT) and how it can contribute towards better remote health care. The study explores remote patient monitoring system that is supported by mainly two technologies: IoT and CoT. Within IoT the authors have included technologies such as RFID, real-time localization, sensor networks and short-range wireless communication, while the CoT consists of IT-related resources such as network, storage to enable virtualized services. The authors conclude that combination of such technology can disrupt the industry and be an enabler for novel ways of conducting health care (Shah et al. 2016).

The findings indicate a high number of metrics; however, these vary in the degree of automation. Applying the methodology of degree of automation of processes, the findings are graded in Table 2. Degree of automation in metrics related to coordinating meetings/ICE-sessions and Table 3. Integrated Concurrent Engineering (ICE) is a very structured session or meeting for coordination – and decision-making. The assessment of risk metrics gives an indication on the focus being pinpointed on mapping the risk towards keeping scheduled progress.

Table 3. Degree of automation in metrics related to risk assessment.

Process (activity)	Degree of automation - Input	Preparation	Output	% I+P+O /3
Order material so it's on construction site to the right time	10	30	50	30%
Documentation of finalized work on construction site *) See more in section 4.1.4	15	45	55	38%
Keep track on what is happening on the construction site *) See more in section 4.1.3	15	15	45	25%
Handing over drawings to the workers so they can do their job *) See more in section 4.1.2	15	55	20	30%
Number of postponed cases	10	45	55	37%
Map needs for more resources	25	45	55	42%
Delays in project	50	40	55	48%
Monthly coordination on project plan / progress	45	50	65	53%

There was a long list of other tasks and process metrics which were observed in student reports. The list below presents possible task and process to be automated:

All drawings handed to relevant parties; Number of cancelled meetings; Number of drawings produced; Allocate software licenses to those who need it; Estimate savings; Up to date information on how to perform a job; Numbers of IFC files that are shared with the relevant parties; Base and information needed for decision; Issues resolved in software; Efficient communication; sender and receiver understands each other; Execute planned meetings; Cost of exceeding the schedule; MMI (Model Maturity Index) gives the project members an indicate how far the in the design phase the project has reached; Efficiency during meeting; 5D BIM; Always connect to right VPN; Time needed to prepare ourselves before meetings; Work on digital platforms/different software to reduce mails being sent back and forth with big files; Paying consultants after approving their work; Control and checking consultants drawings; 4D BIM.; Changes and implementation in BIM-model; Clash detection in models; Up to date information on when to perform a job; avoid double booking of a meeting room; Order material so it's on construction site to the right time.

As the list above demonstrates, the list indicates that automation will be relevant to reduce time and effort to get the metrics.

4 CONCEPTUAL ANALYSIS AND DISCUSSION

The conceptual analysis and discussion are built up in mainly two parts:

1. The first part aims to elaborate on the assessment of four real-world examples from the student deliveries.
2. Secondly, we discuss whether this is applicable to the construction industry.

This discussion elaborates on the findings from previous section.

At first sight, the results show that there is a huge variation regarding the level of automation of processes. With today's emerging presence of technology, one could be surprised that it seems the construction industry to be so little automatized but there are other organizational and cultural mindset that comes into play.

Using the results from the students and learning from the other industries and researchers, the imagination is the only constraint limiting this conceptual research.

Based on literature and somewhat the student reports, there is no clear indication that processes are being automated. In this context, can the implementation of the proposed APM improve the current situation?

4.1 Real-world examples indicate low degree of automation, but there is a suggested 'planned implementation'

4.1.1 Evaluation of meetings – plus/delta

Table 4. Exploring the degree of automation for evaluation of meetings.

Process (activity)	Degree of automation - Input	Preparation	Output	% I+P+O /3
Today				
Evaluation of meetings – plus / delta *	20	40	40	33%
Planned implementation				
Evaluation of meetings – plus / delta **	75	60	65	67%

*) Today:
Input: Everyone shares their thought orally, one person writes every opinion on the whiteboard.
Preparation: Written down on notes.
Output: Sent on e-mail.
**) Planned implementations
Input: Use online live poll, e.g. Mentimeter, https://www.mentimeter.com Preparation: Upload it on intranet / project hotel.
Output: Presented on a digital dashboard with suggestions on improvements.

4.1.2 Implementing BIM-kiosk will increase the degree of automation

Table 5. Exploring the degree of automation when handing over drawings to the workers.

Process (activity)	Degree of automation			%
	In-put	Prepa-ration	Out-put	I+P+O /3
Today				
Handing over drawings to the workers so they can do their job *	25	25	35	28%
Planned implementation				
Handing over drawings to the workers so they can do their job **	75	85	95	85%

*) Today:
Input: A .pdf drawing.
Preparation: Printed copy of latest drawing.
Output: In-person and physical gathering and handover.
**) Planned implementation:
Input: A BIM-container/kiosk placed on construction site, where the BIM is synched during night, that all workers have access to. One could also imagine a robot that is placed on construction site with same rules of synchronization.
Preparation: BIM-model is synched overnight so that the workers have the latest drawings. Open BIM in given software, e.g. Solibri Model Viewer, explore the model.
Output: Workers get back to the BIM-kiosk and indicate that the work is done. Work accordingly to the model.

4.1.3 Project Information Model (PIM)

Table 6. Exploring the degree of automation to keep track on what is happening on the construction site.

Process (activity)	Degree of automation			%
	In-put	Prepa-ration	Out-put	I+P+O /3
Today				
Keep track on what is happening on the construction site	25	55	55	45%
Planned implementation				
Keep track on what is happening on the construction site **	85	75	75	78%

*) Today:
Input: Daily/weekly meetings or inspections.
Preparation: Taking notes and updating project hotel.
Output: Presented on a stand-alone software, e.g. MS Project or similar Gantt-chart / schedule-oriented tools.
**) Planned implementation:
Input: A robot scanning the construction site or camera mounted on several places.
Preparation: Uploading new constructed elements that are detected during scanning.
Output: Daily updated Project Information Model (PIM) accordingly to progress on construction site.

Table 7. Exploring the degree of automation when documenting finalized work on construction site.

Process (activity)	Degree of automation			%
	In-put	Prepa-ration	Out-put	I+P+O /3
Today				
Documentation of finalized work on construction site site*	25	45	55	42%
Planned implementation				
Documentation of finalized work on construction site site**	85	65	80	77%

*) Today:
Input: Daily/weekly meetings or inspections.
Preparation: Taking notes and updating project hotel.
Output: Presented on a stand-alone software, e.g. MS Project or similar Gantt-chart / schedule-oriented tools.
**) Planned implementation:
Input: A robot that continuously scans the construction site and updates the needed BIM-models.
Process: Engineers operate a digital solution to create, and update needed Facility Management and Operation (FMO) documents.
Output: As-built Asset Information Model (AIM) presented on the project hotel or intranet.

4.1.4 Asset Information Model (AIM)

Solution increased APM van also be connected, where on solution built on previous one.

4.2 Fully applicable to ongoing projects

Observing 200 students going through the VDC course and documenting below average level on automated process metrics is a sad read. The importance to dig deeper in the framework of degree of automation is not only to improve the different processes but also to improve the industry itself.

Going into the future there is no doubt that technology such as ICT and CPSs will play a bigger part, and the need to map todays 'business as usual' and explore the thoughts of 'tomorrow and future' is a needed duty. The suggested method of APM will help project engineers and staff members to map their processes as of today and elaborate on how to automate these.

The suggestions given in 'planned implementation' witnesses that there are some processes that are easily automated, at least taking it a step further and reducing the many hours of manual typing, and some suggestion requires more planning and investments. Such as a BIM-kiosk is not easily assessed, it requires better logistical planning on the construction site depending on progress and it is only relevant during the construction phase. Further, it requires a purchase investment to be done, educate relevant workers to utilize software and an infrastructure of power and internet.

Anyone can apply Mentimeter to their meetings as of tomorrow. Mentimeter is a web-based software that allows the host or meeting facilitator to create pulls, interactive presentation, and survey to be done during meetings, so that it "removes that process of handing out and collecting coting devices, thereby saving valuable time for teaching and learning…" (Little 2016). Again, the interest of automation process metrics is to release many work hours that goes into very repetitive manual tasks, and to achieve a higher degree of automation.

In the Norwegian Construction Industry, VDC has emerged and played a huge role to adapt new mindset and culture. As metric is a core element of VDC, more people in the industry are aware of establishing measurement system to map current state and detect areas of improvements and to set goals achieving higher standards.

The idea of also having updated BIM-model during construction, Project Information Modelling (PIM), and for facility management and operation, Asset Information Modelling (AIM) is to always have the correct information at any given time through of the lifecycle product e.g. a building. This means that documenting finished work is crucial and doing it on handwritten paper or word documents saved only in one laptop is not desired. The desire is to use BIM dedicated software and project management software instead so it's easily accessible for all parties of a project. Going into the future, both contractual models based on Integrated Project Delivery (IPD) and partnership will be trending in the construction industry and will be a driver for higher degree of digital inclusion, meaning all parties collaborating and performing more work using technology.

5 CONCLUSION

This study has developed a framework for identifying the degree of automated process metrics (APM). The presented framework could help construction management to automatize their processes. Some of the presented and discussed findings in previous sections are easily automated and suggestions for 'planned implementation' can be easily applied as these suggestions are based partly on existing tools.

Lastly, APM shall be a strategic metric on the management level to raise the awareness of continuous investment in automation and digital tools. In terms of companies being more aware of its current state, these kinds of assessment will impact, not only giving competitive advantages but open up for new business models where the service is to provide insight in current state of doing business and areas of improvements, with focus on automating different processes. Improvements can by this be a systematic and systemic process. By APM assessment within projects the different parties can set the base to attract future customers, attract future employees by selling the high degree of automation, and improve organization reputation, in addition to the benefit of improved process performance in each project.

6 RECOMMENDATION FOR FURTHER RESEARCH AND IMPLEMENTATION

This research presents a methodology to assess the degree of automated process metrics. The need to explore process metrics is crucial to further develop conceptual ideas using the APM. The research recommends:

1. Revisit today's ways of registering performance metrics with the aim of finding APM that suits a companies' cost-benefit analysis.
2. Further investigate the possibilities to develop integrated applications that allow different software to exchange data and information.

This type of research should be multidisciplinary and should be combined with real case implementation. Therefore, the authors do welcome collaboration with software and construction industry partners and researcher.

REFERENCES

Belsvik, M., Lædre, O. & Hjelseth, E. 2019. *Metrics in VDC Projects*.

Butler, A., Letza, S.R. & Neale, B. 1997. Linking the balanced scorecard to strategy. *Long Range Planning*. 30(2), pp.153–242.

Howard Kerr 2020. *A Passport to Global Opportunities and Transformative Collaboration* [Online]. Available from: https://www.bsigroup.com/en-HK/iso-19650-building-information-modelling/.

Kagioglou, M., Cooper, R. & Aouad, G. 2001. Performance management in construction: a conceptual framework. *Construction Management and Economics*. 19(1), pp. 85–95.

Khanzode, A., Fischer, M., Reed, D. & Ballard, G. 2006. A guide to applying the principles of virtual design & construction (VDC) to the lean project delivery process. *CIFE, Stanford University, Palo Alto, CA*.

Kim, S.-H. & Jang, K.-J. 2002. Designing performance analysis and IDEF0 for enterprise modelling in BPR. *International Journal of Production Economics*. 76(2), pp.121–133.

Kothari, C.R. 2004. *Research Methodology: Methods and Techniques* [Online]. New Age International (P) Limited. Available from: https://books.google.no/books?id=hZ9wSHysQDYC.

Little, C. 2016. Technological Review: Mentimeter Smartphone Student Response System. *Compass, Journal of Learning and Teaching*. 9(13), pp.64–66.

Meredith, J. 1993. Theory building through conceptual methods. *International Journal of Operations & Production Management*.

Nowotarski, P. & Paslawski, J. 2017. Industry 4.0 Concept Introduction into Construction SMEs. *IOP Conference Series: Materials Science and Engineering*. 245, p.52043.

Sears, A.L. 1994. Automated metrics for user interface design and evaluation. *International Journal of Bio-Medical Computing*. 34(1), pp.149–157.

Shah, T., Yavari, A., Mitra, K., Saguna, S., Jayaraman, P.P., Rabhi, F. & Ranjan, R. 2016. Remote health care cyber-physical system: quality of service (QoS) challenges and opportunities. *IET Cyber-Physical Systems: Theory & Applications*. **1**(1), pp.40–48.

VDC-NTNU 2020. BA6280 – VDC-Certificate Programme Norway NTNU – Stanford. Available from: https://www.ntnu.no/studier/emner/BA6280#tab=omEmnet.

Organizational, perceptual and technological issues of BIM adoption

DigiPLACE: Towards a reference architecture framework for digital platforms in the EU construction sector

A. David & A. Zarli
ECTP, Brussels, Belgium

C. Mirarchi
Politecnico di Milano, Milan, Italy

N. Naville
CSTB, Marne-la-Vallée, France

L. Perissich
Federcostruzioni, Roma, Italy

ABSTRACT: With the increase of digitalisation and the spread of digital tools and applications, the Construction sector faces a need for digital platforms and hubs that should allow the development and generalisation of common digital services and data for all stakeholders of the value chain, in all phases of Construction processes. These digital platforms should provide a fluent communication of semantic data models and instances between computer-based systems, a seamless integration of 3rd-party services and ensure a secured data management, based on agreed standards. The H2020 DigiPLACE project aims at devising a European-level consensus on a Reference Architecture Framework for Construction digital platforms and a roadmap for future rapid deployment and use. This paper introduces its preliminary results, namely an inventory and evaluation of existing digital platforms used in the Construction sector and selected other industrial sectors, the identification of success factors and key expected characteristics of such digital platforms functionalities.

1 INTRODUCTION AND CONTEXT

Similarly to other industrial sectors, Construction is now seeing its own "digital Revolution", having previously benefited from only modest productivity improvements (McKinsey & Company 2017). Digital tools and Building Information Modelling (BIM) are being progressively adopted by different parts of the built environment value chain as strategic tools to deliver cost savings, increased productivity and operations efficiency, improved infrastructure quality and better environmental performance. Indeed, BIM and object-oriented representation of AEC (Architecture Engineering and Construction) assets and processes have started to offer the perspective of seamless communication of semantic data models between computer-based systems used from the design stage to the operation of the facilities, and up to the deconstruction phase. Still, the fragmentation of the sector and the diversity of actors and practices has hampered the development and use of shared collaborative environments (the so-called CDEs: Common Data Environments) and associated standards (ISO 2018a, 2018b). This led all stakeholders to call for a European-wide – and even world-wide – consensus on the definition, structure, accessibility and usability of such environments, thus moving from a project perspective to an inter-company and sector chain perspective (Rezgui et al. 2010).

Enabling and generalizing the digitalization of the construction sector is now about building collaborative digital frameworks providing vendor-neutral data exchange, object-orientated organisation of information, CDE-based sets of services, and the assignment of responsibilities for data and information management. This is a pillar for collaborative information sharing standard practices for the use in domestic or international markets. DigiPLACE[1] is a European project aiming at creating a roadmap for a Digital Platform for Construction in Europe: it relies on a key pillar which is the ongoing definition of a Reference Architecture Framework (RAF) for such a platform, based on an EU-wide consensus involving a large community of stakeholders within the construction sector as well as from nearby engineering and software fields. The objective is to identify and respond to all construction stakeholders' needs in terms of digital tools for dematerialised processes, leveraging on available

---

[1] Website of the project: https://www.digiplaceproject.eu

appropriate software applications and services, with a special focus on SMEs and micro-businesses.

A primary objective is to provide Data Access, which refers to the set of rules that should ensure that data sourced and accessed easily 'for all'. This raises a certain number of questions and expectations:

– How could the accessibility, availability and use of data be increased (at least public data, open products data as well as non-sensitive data), thus 'fighting' market and technological barriers? And how to ensure security (i.e. no loss and no inappropriate change of data), validity and quality of data?
– What are the future data-driven solutions and services that will capture the value of data? What are the related business models (including data ownership, sharing agreement, cost of using/monetising data, etc.)? How to define a 'fair access to data', e.g. ensuring fair competition among service providers who need to access this data, as well as to federate their own data with data from other services to generate increased added value?
– How could existing initiatives on interoperability and standardisation be supportive in further data access and construction stakeholders engagement? How to ensure the quality and usability of data, as mentioned above?

An obvious topic of interest is that those DPs, in order to develop a thriving data economy in constructions, must provide data-driven mechanisms and solutions to ease access to data and provide software interoperability to boost technological diffusion, to create new services/business models throughout the whole construction value chain, and potentially generate new markets or extend/generalise the current ones. DPs also raise tricky questions as:

– How the framework and role of DPs should be developed so as to allow a fair access to data and fair competition at level of new markets for applications and services?
– What are the criteria defining a common 'denominator' for all DPs? Including the type and quality of digital services available through these DPs?
– What are the rules to be established for the governance of DPs, including rules for data access and exchange, and universality of data not being locked-in with respect to (potentially proprietary) platforms?
– How these DPs can help stimulating the creation of new data-driven services – in particular based on big data/data lakes being accessible through technological innovations (AI, Blockchain, advanced simulations, etc.)? How to generate insights from data, leading to informed actions?
– How can such DPs allow relevant information for assets being created, stored and updated in an ethical and safe manner that stays relevant for all value-chain actors including policy makers, stakeholders and users alike for the whole lifecycle, giving them also the possibility for future flexibility and adaptability?

To answer all these questions, DigiPLACE has primarily performed a global description of the context of the digitalisation in the construction sector, in comparison with other industrial sectors, through different inventories, analyses and interactions with stakeholders, as described in the section 2 below. This paper provides with the preliminary results of the project, offering an overview of the digitalisation of the European construction sector through a global picture of the level of implementation of digital tools and services, exhibiting a typology of the various existing platforms and tools, and a detailed analysis of selected digital platforms to describe what functionality they provide and how.

From this vision of the actual market configuration, we can already highlight the key aspects of digital platforms to be taken into account and identify success factors, setting the basis for the future development of the DigiPLACE RAF, with agreed options in terms of services offered, standards, interoperability, usability, etc.

2 OVERALL METHODOLOGY

In order to provide such a global state-of-the-art of the construction sector, as well as of selected other industrial sectors, in terms of digitalisation, the approach followed in these initial activities was organised around the 3 following main tasks:

1) Assessing the level of implementation of digital technologies and hubs/platforms in the European construction sector and other industrial sectors:

  – *Identification of 300+ digital platforms and tools used in the construction sector.*
  – *1st characterisation and clustering of around 200 of them into 15 platforms categories.*
  – *Through the diffusion of an online survey targeting end-users and interviews with key stakeholders: description of end-users' digital practices, needs, expectations, etc. across Europe within the construction sector and other industrial sectors, and description of the level of digitalisation of the whole value chain in several EU countries.*

2) Realising a technical and comparative analysis of existing digital platforms in the construction sector and other industrial sectors:

  – *Creation of a detailed list of evaluation criteria (KPIs) for digital platforms and tools.*
  – *Technical analysis of representative digital platforms and tools with those evaluation criteria.*
  – *Identification of key features proposed or that should be proposed in a European digital platform.*
  – *Identification of good practices or examples in other industrial sectors (agriculture, healthcare, automotive, aerospace).*

This task aimed in particular at answering questions such as: what functionality do platforms enable (i.e. what features do they propose)? How are these features made available and implemented? How do the platforms work? Are they adaptable to the different kinds of users of the construction sector? How difficult and costly is it to use the platforms?

3) Identifying success factors and realising an impact analysis, for the identification of potential points of knowledge transfer towards the construction sector:
   – Identification and description of 20 success factors.
   – Assessment of how these success factors are currently addressed or taken into account in the existing digital platforms or tools, both in the construction sector and other industrial sectors (aerospace, healthcare, automotive).
   – Realisation of a first level analysis of international markets (with a particular focus on the USA, China and Australia), to describe the global context of digital platforms outside Europe and to compare it with the European market.
   – Listing of R&D orientations for each success factor to be considered for the future development of DigiPLACE RAF and the related implementation Roadmap.

As a result, this work should provide a set of inputs for the future development of the RAF and implementation Roadmap:

– Levels of implementation of digital technologies in the construction sector
– Technical description and comparative analysis of existing platforms in different industrial sectors
– Identification of stakeholders
– 1st challenges and key aspects or interrogations to consider
– 1st use cases and limitations
– Success factors & possible ways of implementation
– Possible points of knowledge transfer
– Suggestion of further research orientations

3 DIGITALISATION IN THE CONSTRUCTION SECTOR AND EVALUATION OF DIGITAL PLATFORMS

3.1 Level of implementation of digital technologies and digital industrial hubs/platforms in the EU construction sector and other industrial sectors

DigiPLACE partners have listed 300+ existing digital platforms and tools. Most of them are specific to the construction sector, others are general and could be used in any other sector. A first analysis of around 200 of them has been performed, thanks to a set of evaluation criteria, including a 1st clustering of platforms in which each category reflects a use case corresponding to one or several phases of a construction project. These categories and their occurrences in studied platforms are reflected in Figure 1. It is worth specifying that this first analysis included a wide area of possible interpretations of the terms platform and digital tool according to an exploratory research methodology.

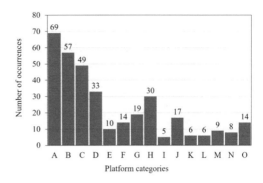

Legend (platforms categories)	
A	Collaborative platform - Generic (all phases)
B	Collaborative platform - Design phase
C	Collaborative platform - Construction phase
D	Collaborative platform - Operation & maintenance phase
E	Collaborative platform - Demolition & renovation phase
F	Object catalogues
G	Product catalogues
H	Data and analytics platform
I	Materials re-use (circular economy)
J	Inventory & supply chain management
K	IoT Platforms
L	Marketplace for materials
M	Marketplace for digital services
N	Risk management
O	Other

Figure 1. Occurrence of platforms categories in the platforms studied.

A certain number of observations can be made about existing digital platforms:

– Even though most existing platforms come from American developers, an important number originates from Europe. In particular from Finland, France, Germany, the Netherlands and the UK.
– Most of them enter in the global category of "collaborative platforms" (i.e. platforms used by different team members in order to exchange data and documents, and that provide a better communication and a more effective project management), whether it is for the design, construction or operation & maintenance phase, or for a generic use. This last sub-category that is mainly represented.
– Several categories seem to be much less developed: collaborative platforms for the demolition & renovation phase, platforms for materials re-use, IoT platforms, marketplaces for materials or digital services and risk management platforms.
– A large majority of these platforms are private, not open-source, hosted on the cloud, used collaboratively by several construction subsectors of activity and have and international reach.

Then, an online survey targeting practitioners (i.e. professionals of the construction sector) has been disseminated. It provided a global picture of what digital technologies and tools they use, as reflected

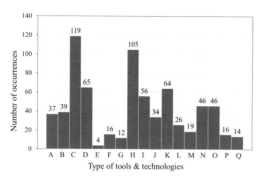

Legend (types of tools and technologies)	
A	3D printing
B	3D scanning
C	BIM
D	BIM objects digital library
E	Blockchain
F	BMS - Building Management System
G	City GML
H	Digital 3D drawing/modelling (for design per discipline)
I	Digital tool for bill of quantities/estimation
J	Digital tool for planning / 4D simulations (based on BIM)
K	Digital tool for design coordination
L	Digital tools to capture data on site
M	Digital tool for facility management and user feedback (based on BIM)
N	Document Management Systems
O	GIS
P	Thermal scanning
Q	Other

Figure 2. Occurrence of advanced digital tools & technologies used by practitioners.

in Figure 2, for which purpose, with whom, etc., together with a reflection of their related needs and expectations.

It has especially been observed that[2]:

– Nearly 80% of subjects use "advanced" digital technologies and tools. The main ones are BIM (in particular Autodesk Revit) and digital 3D drawing/modelling, whereas blockchain or City GML are poorly represented.
– 70% of the subjects use digital platforms. The generic platform Dropbox comes far ahead of construction platforms: BIM 360, RIB Byggeweb, etc.
– Many of those who do not use advanced digital tools do not see any added value for their business. The implementation cost, both in terms of money and time, as well as the technical complexity are strong barriers for them.

---

[2] Though the methodology and templates remain valid, these results must be considered with caution. A significant part of the 185 answers came from Danish architects (who are especially digitalized) and more data is necessary to be fully exhaustive.

– Digitalised businesses raised a number of advantages (project efficiency, better management, reduction of cost, easier tendering and communication with clients, standardisation of procedures, etc.) and obstacles (lack of common standards/formats which hinders data exchanges partners, high cost of software, lack of knowledge in the construction community which hinders the frontrunners, etc.).

Finally, consultations with some stakeholders allowed to obtain their views on the digitisation of the construction sector in their countries, to exchange on good practices and to hear about potential examples or needs in terms of national policies. Based on these discussions only, several observations could be generalised at a European scale, e.g.:

– The level of digitisation in the construction sector is still very low.
– Major gaps are observed between large companies (good knowledge of digital tools & technologies) and SMEs (very little knowledge, except for some frontrunners). Also, important differences exist between the actors of the supply chain, unrelated to their size. The most digitalized one seem to be the consultancy firms, followed by architectural studios, planners and engineers and, finally, general contractors and specialized contractors.
– Digital technologies bring real economic benefits only when considering their impact on the whole value chain. Mostly large companies can see this impact, since they intervene on the whole – or most of the – value chain.
– Digital technologies make tendering easier, as they enable to fulfil the more and more frequent requirement of clients to digitalise their projects.
– Digitising the construction sector is not a technology problem, but rather a business case issue: technologies are available or can be addressed/developed by many stakeholders. The market on the other hand is not ready yet and a convincing business case is necessary for SMEs to support the adoption of digital solutions at a large scale.
– Public and private initiatives for the digitisation of industrial sectors – i.e. supporting companies in adopting digital solutions – seem to emerge in all countries, both at national and regional scale (e.g. with the development of Industry 4.0 in many countries). Private companies and associations also provide services and training.

3.2 *Comparative analysis of existing platforms in the construction sector and in other industrial sectors*

This analysis aimed at highlighting key aspects of digital platforms that will have to be taken into account during the future development of DigiPLACE RAF, as well as the different implementation options existing on the market in terms of services offered, standards, interoperability, services provided, formats, lifecycle, costs, usability, etc.

To perform a meta-analysis of platforms, a set of evaluation criteria has first been created. They have been defined according to the objectives of DigiPLACE and to the necessity of assessing their technical characteristics in terms of architecture, functionality, services provided, standards, adapted levels of details, etc. The criteria have been divided into 4 categories:

– Use case & services: features proposed by the platform and how they are implemented (40+ items)
– System functioning: description of how the platform works in terms of data management, architecture, etc. (30+ items)
– Usability: how the platform is adaptable to different kinds of users and needs (10+ items)
– Economic factors: cost of the platform and packages available (8 items).

A dozen of representative construction platforms and tools[3] have then been analysed using the previous criteria. It was not intended to provide a fully exhaustive listing of all features and configurations existing on the market, but a representative set of elements and use cases that will have to be taken into account when developing the DigiPLACE RAF.

These analyses have shown that "collaborative platforms" tend to offer a vast array of use cases and services, under many different forms. If not directly integrated into the platform or tool, it is often still possible to access services through the use of APIs. Some use cases are less frequently provided, such as blockchain, building permits, integration of object catalogues, etc. Also, it confirmed the use of many formats, extensions, linked services, standards, etc. across software.

The general practices and some main platforms of 4 other industrial sectors (agriculture, healthcare, aerospace and automotive) have also been studied with these same criteria, in order to compare them with the construction sector and to highlight some main issues:

– Some services and features are not properly answered in the construction sector and could take example from developments in other sectors. E.g. for machine learning, Product Lifecycle Management (PLM) or requirements for documentation.
– Two major issues for the development of a large-scale digital platform have been described. They are common to all the sectors analysed: the lack of standardisation and the low interoperability of software and services. The further development of the IFC standard and the use of semantic web technologies could be an answer. But other approaches in other sectors could be considered, such as the protocol-based standards in healthcare.
– The aerospace sector started its digital transition a couple of decades ago. Thus, its maturity represents an interesting source of examples.
– The concept of horizontal platform, especially present in the healthcare sector, seems particularly interesting. It aims at enabling many people to access the same data and work from it for multiple purposes, with many deriving services. Such platform is characterized, among others, by the use of APIs to allow connection with other tools and platforms, a large level of flexibility, unified repositories, and the guaranty of data security and property.
– The notion of integrations & development platforms could be favoured. It focuses on allowing the integration with other platforms and applications – the main functionalities being to acquire, normalize, store and expose data – but providing as well an environment and tools to develop applications inside the platform.

Thanks to this work and the accompanying discussions, a preliminary agreement on the definition of a digital platform – at least in the context of DigiPLACE project – emerged. A digital platform would be considered as a knowledge sharing system based on the connection with other systems devoted to collect and manage data, information and knowledge, and able to provide services and allow the connection of other tools and services via APIs.

For further details about all the analyses presented in this section, the reader is invited to refer to deliverables D3.1 (ECTP et al. 2019a, 2019b) and D3.2 (ECTP et al. 2019a, 2019b) of DigiPLACE project.

4 SUCCESS FACTORS AND RESEARCH ORIENTATIONS

The consortium identified 20 success factors, presented in Table 1, i.e. focus points that the DigiPLACE RAF should consider in order to fulfil the objectives of the DigiPLACE project.

These success factors have then been analysed, in order to assess how they are currently implemented or taken into account in some representative existing digital platforms and tools in several industrial sectors (construction, healthcare, automotive, aerospace).

Most of the success factors are actually addressed by the above-mentioned industries – yet through different manners – since they represent widely shared concerns such as the need for standardised formats and practices, the increasing amount and complexity of data to deal with, the need to comply with new regulations, etc. Inspiration could be taken from other sectors for specific features, e.g. regarding the protection of private data in the healthcare sector.

---

[3] Revit, BIM360 (Autodesk), KROQI (CSTB), usBIM.platform, Primus (ACCA Software), CUSP (Cardiff Univ.), Built Environment Platform (IBM), BIMserver.org (TNO), 3D Experience (Dassault Systèmes), Archicad, BIMcloud (Graphisoft), WorksManager (Trimble). Also studied – though not with the evaluation criteria: E-Construction (Government of Estonia)

Table 1. Success factors identified.

TECHNICAL ASPECTS
1 Interoperability and sustainability
2 Collaboration enabler
3 Single entry point
4 Capacity to connect several platforms both at regional and national levels
5 Integration of both public and private data
6 Easier circulation of / access to services and products
7 Maintenance of data
8 Maintenance and update of the services
9 Adequate backup of data
10 Be customizable
11 Be scalable and dynamic (provide an environment able to integrate new/existing tools)
12 Efficient and fast data management and data queries

DEMAND/REGULATORY ASPECTS
13 Capacity to check compliance with regulations & certifications
14 Capacity to answer the demand/needs of every kind of stakeholder
15 Relying on the national level, by interconnecting with national platforms

ECONOMIC ASPECTS
16 Identification of clear funding mechanisms/systems (analysis of the economic sustainability of the platform)
17 Identification of business cases for all stakeholders
18 Increase of the competitiveness for all the value chain

SECURITY ASPECTS
19 Information and data security
20 GDPR compliance

However, some major differences in the comprehension of the purpose and characteristics of platforms have been highlighted. The need for interconnected platforms, with an important role for public authorities at regional and/or national level, seems more adapted to fragmented sectors such as construction, where markets and regulations have mostly remained national. On the opposite, interconnected national platforms are not really considered in sectors like aerospace and automotive, where the market is dominated by fewer and bigger international service providers or companies having set up their own proprietary solutions.

Further recommendations or research and development orientations for the development of the RAF and its associated Roadmap have been proposed. They should be investigated in order to meet the success factors and the global objectives of DigiPLACE. They consist in different propositions and alternatives of implementation to be considered in the RAF for each success factor. This means that focussing on a single success factor, two different ideas might be introduced that sometimes reveal being in opposition or incompatible. Also, some objectives or suggestions might appear only partially applicable (e.g. only to part of the stakeholders) depending on the choice that will be made later during the project on the structure, characteristics and scope of the RAF.

For a complete description of success factors and of the recommendations and further R&D orientations, the reader is invited to refer to the deliverable D3.3 (ECTP et al. 2020) of DigiPLACE project.

5 TOWARDS A REFERENCE ARCHITECTURE FRAMEWORK FOR EUROPEANS CONSTRUCTION PLATFORMS

As already mentioned, the previous analyses do not intend to directly provide the answer to the question of what DigiPLACE output should be. However, it seems clear that it should not be a publicly funded free-access alternative to services that the market is already providing in a satisfying way. Nor will it be a mere European-wide extension of public platforms existing at regional/national level. Even though it could propose the generalisation of identified best practices, none of these existing platforms encompasses all the aspects DigiPLACE is intended to cover. Most importantly, they don't address the very objective of better European integration of the construction sector and markets. Instead, the analyses allowed to identify all the questions to be answered by European stakeholders in order to build a consensus on what this European RAF should be.

Beyond the analysis of existing platforms, the upcoming identification of persisting barriers to the digitalisation of the construction sector, will help clarifying what is missing in the current situation and market. Some of these shortcomings were already cited in this paper: lack of interoperability and communication between platforms, barriers created by heterogenous regulations, low availability of data hindering use of artificial intelligence tools, etc. This identification of barriers and mitigation options will be crucial to complete the definition of DigiPLACE use cases and of a RAF adapted to answer these use cases.

This RAF is likely to be composed of a mix of different types and degrees of integration:

– The creation of a proper European web platform, publicly funded, that could be inspired from existing national public platforms, and would give access to similar services at a European scale. As already mentioned, there is an ongoing reflexion among stakeholders on whether DigiPLACE platform should be a collaborative project platform (i.e. directly providing collaborative services) or not. In this reflexion, the leading principles are to ensure the vendor neutrality, to look for complementarity with what the market can provide, and to foster fruitful competition rather than hindering it.
– The connection of existing platforms by providing some common services that could federate them by either easing access to these platforms, or allowing

them to easily deploy at European scale. An example of this would be a common repository of national construction regulations, provided in a harmonised way, and through APIs, in order to allow integration into existing platforms
- The definition of common guidelines that would apply both to private and public platforms: standards to be used, guidelines to favour interoperability and communication, etc.

To synthetize the core activities of the project, the key use cases to be answered by DigiPLACE RAF will be derived from: i) the analysis of existing platforms and success factors for knowledge transfer from other sectors and ii) the identification of persisting barriers to digitalization, and mitigation options, iii) keeping in mind that several types and degrees of European integration will have to be mixed, depending on the issues addressed.

Even though the identification of key use cases is yet to be performed, the work described in this paper already allows to outline several themes that will be included in the discussions, of which we can cite a few examples:

- Data sharing and analytics: sharing of private data in order to improve common knowledge and apply artificial intelligence tools is a central aspect of digital platforms in other sectors (e.g. agriculture, healthcare, aerospace). The construction sector appears to be lagging behind. Numerous uses can be imagined, such as the sharing of environmental or cost performance data to help improve designs.
- Environmental and product data: one of the objectives will be to favour an efficient adoption of the new frameworks (e.g. LEVELS, Construction Product Regulation, building passport), and provide tools to help exploit their full potential, related to the previous point, one can think of improved data sharing to progressively implement a comprehensive system of environmental impact accountability for the construction sector. Another topic could be the mutualisation of efforts aiming at providing object libraries to model new or existing buildings.
- Interoperability, standards, connectivity: even though the need is clearly identified, the challenge will be to clarify the potential added value of DigiPLACE RAF to promote the use of standards, and support the emergence of a common language for a digitalised construction sector across Europe, in as a complement to regulations and standardisation bodies.
- Tools for SMEs: helping SMEs across Europe benefit from digitalization is one DigiPLACE key objective. Some platform developed at national levels such as INNOVance, KROQI, BIMrel, Drumbeat, etc. represents examples of public initiatives with a special focus on SMEs. It will be discussed whether to have a similar approach at European level or not.
- Supply chain management: the move towards more streamlined and digitalized supply chains is tightly connected to the Industry 4.0 perspective. Emerging services allow to improve the link between design and manufacturing. These evolutions could be a significant driver for increased European integration of the sector, and the potential role of DigiPLACE is to be assessed.

6 CONCLUSION

This paper presents the results of an extensive analysis on existing digital platforms in the construction sectors and in other four industrial sectors that was developed in the context of the DigiPLACE project. The analysis identified a high number of solutions available in the market highlighting the interest on this topic on the industry side. Other studies devoted to understand the characteristics of collaborative environments can be found in the scientific literature. For example, (Alreshidi et al. 2016) proposed a set of requirements for the development of a GovernBIM platform. (Wong et al. 2014) proposed a literature review of cloud-based BIM technologies presenting the main characteristics and functions of the analysed ones. (Shafiq et al. 2013) analysed existing model collaboration systems proposing a classification and list of features including user's perspectives.

On the policy side, the Final Report of (Working Group 2 – Digital Industrial Platforms 2017), within the Digitizing European Industry Initiative of the European Commission (EC), aimed at supporting the creation of next-generation digital platforms by defining possible ones, reflecting on how building platforms should be approached at European level and considering how existing and planned EU-wide, national, and/or regional platform development activities could contribute.

The work proposed in this paper extends the existing research by integrating an extensive analysis of existing solutions in the construction sector with the study of four other industrial sectors, i.e. healthcare, automotive, aerospace and agriculture. This allowed the identification of a set of twenty success factors that arose from common features identified both in the cross-sector analysis and in the discussion of results with key stakeholders in the market.

In parallel, a first level analysis of international markets enabled to describe the global context of digital platforms outside Europe and to compare it with the European market. It highlighted a worldwide and multisectoral push for digitalisation, considered as an indispensable way of remaining competitive in the fast-changing markets, and a worldwide development of BIM solutions, often under the impulsions of public authorities.

Some interesting initiatives at different levels have been spotted in the countries focused in this study (the USA, Australia and China). Platforms are considered as a beneficial alternative to the traditional use of software or software-as-a-service. However, platforms in the sense of DigiPLACE still do not fully draw attention internationally. The wide interconnection of public or private platforms, with the access to their services, is not a direct incentive for these initiatives.

Of course, the results proposed in this paper only have for objective to create a basis for the development of the RAF, as a main output of the DigiPLACE project. Nevertheless, they constitute a useful source for all those involved in the study of digital platforms in the Construction sector. The success factors and crucial features identified in the proposed analysis can serve as a common basis for future analysis on this area of research.

The natural prosecution of this work is the understanding of the barriers and needs of the end-users. Hence, the next steps of the project will help describing further some of the issues and solutions encountered from the end-user (functional) and implementation (technical) points of view. Future activities will feed from these inputs, make further analyses about their potentials, compare them, and ultimately provide final answers through the development of the DigiPLACE RAF and associated Roadmap.

In this context, the RAF proposed by the DigiPLACE project might be among the first comprehensive construction digital platforms. Further research is recommended, but it seems that its quick implementation could provide a strategical advantage to the European market in front of other big players. It would also allow the construction sector to contribute, with other sectors, to the creation of a common European Industrial IoT, Data and AI Ecosystem as planned by the EC (Zwegers 2019).

Such strategy will be reinforced and developed through the DigiPLACE Roadmap that will evaluate how to develop and integrate the proposed solutions in the different European and national contexts.

ACKNOWLEDGEMENT

The authors would like to thank the European Commission for supporting the DigiPLACE project, which has received funding from European Union's H2020 research and innovation programme under Grant Agreement N. 856943.

For their contributions to the initial work presented here, the authors also thank all the other Partners and Linked Third Parties of the DigiPLACE consortium, in addition to those represented in this paper (ACE, ANCE, BAM, BBRI, BMVI, BuildingSMART, CECE, CEREMA, CPE, CU, EBC, EFCA, FIEC, INDRA, LIST, MEEM, MIT, NTNU, Tecnalia, TNO, UL, VTT), as well as the members of DigiPLACE Advisory Board for their valuable support.

REFERENCES

Alreshidi, E., Mourshed, M. & Rezgui, Y. (2016) 'Cloud-Based BIM Governance Platform Requirements and Specifications: Software Engineering Approach Using BPMN and UML', *Journal of Computing in Civil Engineering*, 30(4). doi: 10.1061/(ASCE)CP.1943-5487.0000539.

ECTP et al. (2019a) *D3.1 – Level of implementation of digital technologies and digital industrial hubs/platforms in the European construction sector and in other industrial sectors.*

ECTP et al. (2019b) *D3.2 – Comparative analysis of existing platforms in the construction sector and in other sectors.*

ECTP et al. (2020) *D3.3 – Impact analysis and success factors for the identification of possible points of knowledge transfer.*

ISO (2018a) *ISO 19650-1:2018 – Organization of information about construction works – Information management using building information modelling – Part 1: Concepts and Principles.*

ISO (2018b) *ISO 19650-2:2018 – Organization of information about construction – Information management using building information modelling – Part 2: Delivery phase of the assets.*

McKinsey & Company (2017) *Reinventing construction: a reoute to higher productivity.*

Rezgui, Y., Hopfe, C. J. & Vorakulpipat, C. (2010) 'Generations of knowledge management in the architecture, engineering and construction industry: An evolutionary perspective', *Advanced Engineering Informatics*. Elsevier Ltd, 24(2), pp. 219–228. doi: 10.1016/j.aei.2009.12.001.

Shafiq, M. T., Matthews, J. and Lockley, S. R. (2013) 'A study of BIM collaboration requirements and available features in existing model collaboration systems', *Journal of Information Technology in Construction*, 18, pp. 148–161.

Wong, J. et al. (2014) 'A review of cloud-based BIM technology in the construction sector', *Journal of Information Technology in Construction*, 19(September), pp. 281–291.

Working Group 2 – Digital Industrial Platforms (2017) *Digitising European Industry*. Available at: https://ec.europa.eu/futurium/en/system/files/ged/dei_wg2_final_report.pdf.

Zwegers, A. (2019) 'Towards a Common European Industrial IoT and Data Ecosystem?'. Available at: www.internationaldataspaces.org/wp-content/uploads/dlm_uploads/2019/07/20190625-1500-Common-European-Industrial-IoT-by-Arian-Zwegers.pdf.

ND
"We need better software" – the users' perception of BIM

A. Rekve & E. Hjelseth
Norwegian University of Science and Technology (NTNU), Trondheim, Norway

ABSTRACT: This paper explores how the users' perception of BIM impacts the implementation of BIM. Semi-structured in-depth interviews were conducted with BIM users in the Real Estate department in a Norwegian municipality. The goal is to uncover in-depth understanding of BIM and related practices of BIM stakeholders. Our findings show that the BIM-users' perception of BIM has a significant impact on their activities. BIM is seen as an add-on to traditional roles and responsibilities, where you must use BIM-software to be a BIM user. This study recommends, based on deep studies in innovation research, a shift to invest in people rather than software to increase the speed of BIM adoption.

1 INTRODUCTION

1.1 Why the perception of BIM is important?

Your perception of BIM affects your actions in BIM. If you believe the world to be flat, you would not venture towards the horizon in fear of falling off the edge of the world. Your belief about the world affects what actions you make. This study focuses on discovering the users' perceptions of BIM – and on explaining the logic behind them. In doing so, we increase our understanding of why perceptions are hard to change. If we want to change how people perceive something, it is not enough to explain the new perception and expect people to "convert". We need to address the realities the users face and relate the new perceptions to what these realities are.

If you believe there to be a causal relationship between performing work for a client and getting paid by a client, then you are likely to perform the work. If you believe "green" credentials are of value, you are likely to pursue them. If you believe BIM to be unfruitful, not your responsibility, or too difficult, or a type of software only architects and engineers are using during design – you are unlikely to pursue it and explore new possibilities. We have therefore included a large number of references to increase the awareness and to substantiate a deeper understanding regarding this "tacit" causality. This study tries to find the "blind spots" in the Johari window (Luft & Ingham 1955).

On an industry level, the semantic meaning of BIM has not been very clear. Even if the abbreviation "BIM" is expressed as Building Information Modelling where Model*ing* is used to indicate the process perspective, this is not confirmed based on actions related to use of BIM (Hjelseth 2017a).

Through experience and in literature, we observe that BIM adoption in the AEC/FM is slow process (Gu & London 2010; Migilinskas et al. 2013; Ullah et al 2019). It is "[46] years since C. Eastman theorized what would later become known in the world as BIM" (Daniotti et al. 2019). Some of the barriers to BIM adoption are: interoperability and technology issues (Matarneh et al. 2019b); missing information requirements (Matarneh et al. 2019a; Munir et al. 2020); complexity of construction processes (Sujan et al. 2018). We feel there is a gap in literature when investigating the barriers to slow BIM uptake. BIM is a tool for digitalization, which means that *people* will use it. How do they see the world? *What are the perspectives of the BIM-users themselves?*

1.2 Research question

RQ: What do BIM-users themselves view as the relevant issues as to why BIM-adoption is slow?

1.3 Impact of the perception of BIM

From a policy standpoint in Norway, BIM is a tool which will facilitate the digitalization of the built environment (Ministry of Local Government and Modernisation 2011). This government white paper explains and justifies Building Information Modelling (BIM) by some of the main issues it is sought to solve: The productivity issue and the environmental issue. The paper states that compared to other sectors, the productivity growth of the Architect, Engineering and Construction (AEC) is lagging behind the other sectors (Ministry of Local Government and Modernisation 2011), though the issue of productivity is likely more nuanced because of the complexities of measuring it (Sezer & Bröchner 2014). The potential for increased productivity coupled with the size of the sector means that even moderate productivity increases will lead to large value gains, increasing the tax base. And being the largest real estate purchaser and operator in the country, higher productivity would lead to more economical and higher-quality buildings. The environmental aspect is the building and construction sector accounting for 36% of global energy use and 39% of energy related CO_2 emissions (IEA 2019). Being the largest emitter of greenhouse gasses of all sectors, the UN sustainability goal puts the onus on the built environment to reduce overall CO_2 emissions (IEA 2019). And BIM is a central tool in for efforts to support green buildings (Lu et al. 2017).

In the sub-units of government, such as government owned businesses and local and provincial administrations, we find the people who are expected to execute and operationalize government policies. These people, such as the BIM-users in the Norwegian municipality, are therefore in a very good position to influence developments. They are in a way both *bound to* and *enabled by* the relevant government BIM policies. And as an organization, they have a sizable procurement power to exercise in the marketplace. And exercising procurement power is one of the main strategies in Norway for promulgating BIM in the AEC/FM sector (Ministry of Local Government and Modernisation 2011).

2 CASE DESCRIPTION AND METHOD

2.1 *Research question type*

This study uses gap analysis to explore additional aspects as to why BIM-adoption is slow by focusing on the BIM-users' perspectives. The focus on the people in the process and their views is an overlooked area. This makes our research question a neglect spotting mode (Sandberg & Alvesson 2011). Our BIM-users have a broad definition. In short, we define them as users within the municipality whom *are* and *could use* information in BIM for their work processes, were BIM available to them. This way we capture the perspectives of people who *could be* BIM-users, but for some reason are not now. What are their reasons – in their own words?

2.2 *Methodology*

We are interested in learning about how a collection of people see and experience the world they operate in. Questions such as *what* and *how* are well suited for qualitative methods (Silverman 2010) and where we are after rich descriptions of complicated processes (Creswell 2007). We used a case study approach since this gave us the opportunity to perform "[...] an empirical inquiry that investigates a contemporary phenomenon within its real-life context [...]" (Yin 2009).

As a case we chose Bærum municipality in Norway. Neighboring the capital Oslo it is 5th largest municipality by population as well as the richest and most educated. The Real Estate (RE) department has about 230 employees and manages 565.000 m² real estate, 2100 housing unit and several commercial buildings.

The RE department is responsible for the whole life cycle of their built environment. They define, procure and follow the delivery of new assets, mostly through Engineering, Procurement and Construction (EPC) types of contracts, as well as operations and facilities management. They cover all aspects of the life cycle.

It should be highlighted that Nordic (and Swiss) local governments have a high degree of autonomy from central government. Especially compared to local governments in the UK and Ireland (Ladner et al. 2016). This means that the ambitious municipal strategies covering topics such as real estate; climate; innovation and digitalization are a product of the municipality itself. Which means ownership (and accountability) of the issues lies with the municipality itself. Our argument is that findings from our purposive sampling where we have "[sought] out groups, settings and individuals where ... the processes being studied are most likely to occur" (Silverman 2010), are not overly influenced by politics, and should be transferable to other organizational settings where the same phenomena happen.

The interviewees were chosen based on Lead User theory, which states (von Hippel 1986):

1. "lead users face needs that will be general in a marketplace – but face them months or years before the bulk of the marketplace encounters them *and*
2. lead users are positioned to benefit significantly by obtaining a solution to those needs"

We want to talk to people who are doing the BIM activities; have hands on knowledge of the relevant issues; and who are incentivized to utilize BIM to begin with. The value of the involvement of the lead user is closely related to the benefits of user involvement in increasing user information satisfaction and system usage in information system development (Bano & Zowghi 2015; Baroudi et al. 1986; Bokhari 2005; Robey & Farrow 1982). In short, we want to learn from the holders of the relevant information, and we want our findings to be relevant.

Yin argues case study research should employ several sources for evidence (Yin 2009). We have performed 6 in depth semi-structured interviews with key BIM-users who were also unit heads or assisting unit heads in the technical services department; planning department; or real estate department and also a project manager for a leading development project. The interviews were between 60–120 minutes and were taped and transcribed. Our primary goal with the interviews was to enable the BIM-user to control the conversation as much as possible. We were interested in how and what topics they brought up. Too much guidance or control from our side would likely be counterproductive. In addition, we had three shorter informal interviews with three key stakeholders in the organization with insight and influence on digitalization in the municipality. These interviews happened over lunch or coffee in the municipal office building and were not planned, but rather a result of talks around the "water-cooler" that progressed further. Notes were taken by the researcher after these meetings. In addition to this, extensive reading on relevant national, municipal and industry strategy and policy documents helped inform on the context in which our interviews objects were operating.

Ideally, we would have liked to interview all department heads as well as several more unit heads. We tried to get interviews with four additional BIM-users which would have taken from 9 to 13 interviewees, and closer to the expected number of interviewees that should provide saturation in the data set (Guest et al. 2006). During the process of transcription and familiarization with the data (which happened concurrently with the interviewing) we used Microsoft Excel to develop initial codes, which where iteratively coded and developed into themes in a thematic analysis (Saldaña 2013). This process happened concurrently with setting up interviews, which meant that we saw the data quickly saturating around the same codes and set of sub-themes. We therefore felt at this time that the number of interviews required was determined by the research question, and that the those we had were enough (Silverman 2010).

3 RESULTS

Using thematic analysis, we have identified four main themes, which represent the issues for slow BIM adoption according to BIM-users in Bærum municipality:

– Organizational
– Technological
– Perceptional
– Systemic

3.1 *Organizational*

The organizational theme was the most dominant theme in terms of having the most sub-themes. It encompasses the sub-themes in which the municipality, the organization itself, can affect. The most prevalent sub-theme was matters related to *information management*. That is, where the problem of slow BIM adoption could be solved by better information management by the organization. Specifically, the issues were that:

1. The steps in the value chain do not know each other's information requirements
2. The municipality had not set any information requirements

Both issues affect the potential for relevant information flow from one step in the value chain to the next. If there were no codified information requirements set, then the preceding step in the value chain would be unlikely to provide the next step with information that would help in their work, since they have not been required to do so. But even if their natural disposition was to share information with the next step, it is hard for them to know *what* to share. From the interviews it seemed this was a result of the increased specialization happening within each step in the value chain. Information is contextual, which the following anecdote from one of the interviewees illustrated:

"When the municipality updates the municipal maps, the air photos are sent to low-cost countries for processing. When the maps come back, almost all the villas have swimming pools in them, which does not make sense in temperate Norway. Trampolines in back yards are interpreted as swimming pools by data processors coming from tropical regions (India). They don't have an understanding a of how Norwegians live. Context is important. If you specialize in one thing, it is difficult for you to know what information is relevant for the next person."

Missing processes were instances where the municipality did not have a process for organizational learning that captured experience data which was relevant to BIM. This could be processes that captured experience data from operations/FM, which would inform the BIM parts of future tendering processes. For example, the municipality could have poor experiences with a certain type of HVAC system, but the process for capturing this experience and utilizing it in EPC contract tendering processes was not a streamlined process. Another example of missing processes was related to considerations around information security which complicated the process of establishing the necessary ICT infrastructure for Revit network servers. Certain types of information held by the municipality are under strict information security protocols, which means that there was no available space for Revit network servers in existing infrastructure. This led to long delays whilst the infrastructure issues were remedied.

Missing technology refers to reflections around the amounts of data generated during municipal construction projects. For example, municipal laptops were not sufficiently powerful to use Solibri or other 3D software tools. But also, reflection around how the municipality is supposed to scale their ICT infrastructure in order to cope with a doubling in the amount of data every 4 years.

Path dependency was a sub-theme that materialized during the iterations between coding and identifying themes. It will be discussed more extensively later. As a sub-theme it is not exclusive to a single main theme, rather it has occurred as a sub-theme to three of the four main themes. For *organizational* development, the path dependency sub-theme captures sources of "inertia", or resistance to change in the organization. Interviewees spoke about municipal workers being "set in their ways". This was exemplified by comments such as "well, in 10 years I will retire" and "how can I keep up [with the developments]". Or feelings such as "BIM in the organization requires BIM specialists".

Perception of roles as a sub-theme also occurred under more than one main theme. But with regards to the organizational main theme, the reflection was that the organizational units in the municipality had

Table 1. Table listing the identified codes, sub-themes and main themes.

Main themes	Sub-themes	Codes
Organizational development	Information management	Steps in value chain do not know each other's information requirements
		Municipality has not set information requirements
	Missing processes	Municipality has limited system for learning
		BIM is difficult because of security concerns
	Missing technology	BIM requires updated technology
	Path dependency	Municipal workers are "set in their ways"
		Management positions are locked into a certain set of issues and tasks
	Perception of roles	Municipal Real Estate department is production organization; all others are operation organizations
Perceptional development	Perception of BIM	BIM is software which will solve the complexities
		BIM is for large projects
		What is BIM?
	Perception of roles	What are our roles?
		BIM is for young people
Technological development	Perception of BIM	BIM is software which will solve the complexities
	Path dependency	Existing software competencies are a qualification in itself
Systemic development	Path Dependency	Value chain is complex
		Municipal work tasks are determined by law
		AEC/FM workers are "set in their ways"

different roles. Specifically, the unit concerned with developing and project managing the development of new projects could be characterized as a production organization. This was opposed to many of the other units and department which were more aptly characterized as organizations concerned with the operations of the various assets. The view was that the incentives of these organizational roles were not aligned and led to sub-optimal outcomes. Especially since a cost saving during production would often lead to a cost increase in operations and vice versa.

3.2 Technological

Path dependency in this sense was that the direction and development of new or existing software was dependent of previous and other types of software. The existing software exist in an ecosystem of other software and is a result of sometimes decades of continuous development, where some functionality is limited by choices made long time ago. It is also path dependent since people had invested a lot of time and energy in learning to use the existing systems. So, learning something entirely new becomes costly.

3.3 Perceptional

The *perception of BIM* and the *perception of the roles* were two closely linked sub-themes and perhaps the most powerful in explaining slow BIM uptake in the organization. The most frequently repeating code in the thematic analysis was the perception or understanding that BIM was 1) in one way, shape or form a primarily a software; and 2) that this software would solve the observed complexities of the work task in question. Other related codes were *what are our roles; BIM is for young people; BIM is for large projects* and honest reflections around *what is BIM?* We find this finding to be good at explaining our research question and will discuss this in detail in the following.

3.4 Systemic

A few issues point to a systemic *path dependency*. By this we mean that the environment in which the BIM-users are operating in, has a certain set of conditions and requisites that dictate how they operate. The most common theme was that *municipal work tasks are determined by law and regulations*. An illustrative example is the building and planning processes, which are codified and based on laws and regulations the municipality must follow. So, when a project has started, the "machinery has started", and behaviors are locked in. The two other related themes were that *the value chain is complex* which was the reflection that the value chain was disjointed with many interfaces, which meant that information handovers were difficult. Also, *the AEC/FM sector is "set in its ways"*. This referred to the sector being a large economic sector with discipline traditions, habits and norms that have settled over several decades, if not centuries.

Viewing the AEC/FM sector as a system is helpful because it highlights the qualities of the sector itself. BIM-users perceive a system upholding the status quo, acting as a resistance to change: Successful implementation of BIM is difficult, because it requires change. But change is difficult because of the inherent complexity in the system. In addition, existing behaviors are institutionalized by "hard" factors such as laws and regulations, "soft" factors such as traditions, habits and norms.

4 DISCUSSION

Integrated design and delivery solutions (IDDS) focus on the integration of collaboration between people, integrated processes and interoperable technology (Owen et al. 2013). The themes identified in this study have similar relationships to each other when it comes to the adoption of BIM in the municipality, the efforts are being made difficult though factors explained by the need for additional organizational, perceptional, technological and systemic development, as illustrated in Figure 1.

Of our four main themes, the organizational theme stood out as having the most sub-themes. It is tempting to try and derive meaning from this. For example, we could speculate that the number of different sub-themes meant that in the perspectives of the BIM-users, organizational factors were the premier reason for slow BIM adoption. This would fit well with previous research showing inadequate organizational processes and strategies as being a reason for slow BIM-uptake (Won et al. 2013). Although it is true that relevant processes for ensuring BIM implementation is lacking, it would be a mischaracterization to say Bærum municipality is a laggard. It is an organization with politically approved missions statements and strategies to be in the forefront of the digital and environmental shift (Bærum kommune 2020). There is a general political and organizational *will* to make changes to the organization to help on digitalization and BIM-adoption in the municipality. Seeing that the BIM-users we interviewed had senior positions within their respective units, we think this will to change was more a reflection of that fact.

But the contrast of the stated *will* to make organizational changes and how the BIM-users explained the causal relationships as to *why* their unit was not "fully BIM" was interesting. This was captured by the *BIM is software which will solve the complexities* coding in our thematic analysis. The general perception was that BIM adoption in their respective units was held back by lack of suitable software, but once suitable software is in place, the necessary organizational changes will be implemented. This meant that BIM-users see BIM as 1) software, 2) something that is externally developed. Hence the title of this paper. The implications of this is that BIM adoption is slow because the people best situated to institute the changes view the solution as something that will be supplied externally. But at the same time, complexities of the municipal tasks remain unknown to the software developers, because "the municipality is not very good at setting requirements to [the software providers], and getting them to develop [what we need]", as one of the BIM-users pointed out.

In other words, the necessary organizational and technological changes are not being made, because stakeholders' perception of what BIM is diverges too

Figure 1. BIM processes consisting of People, Process and Technology.

much. The heart of the problem is an unclear understanding of *what is BIM?* And where the prevalent misconceptions are that it is either a *software* or something that mostly is relevant for *larger projects*. And because of these views, the *roles* and responsibilities of stakeholders with respect to BIM remains equally unclear. The result often is the view that BIM is for the younger generation – exemplified by comments such as "in 10 years, I'll retire" and "how can I keep up [with the developments]?".

These quotes are descriptive of the underlying personal reality for many stakeholders in the AEC/FM industry. Years of working in the sector has led to the establishment of a set of problem-solving abilities that happen automatically and unconsciously. The stakeholders rely on their "System 1" for decision making (Kahneman 2011). Successful BIM adoption necessitates extensive use of the more taxing "system 2", where complicated considerations relating to technological, organizational and perceptional development need to be considered. With so much invested in the status quo, it is understandable that stakeholders protect their "cognitive sunk cost", and consciously or unconsciously "hold out" for the remaining part of their career.

Consider the time BIM has been around (Daniotti et al. 2019), the increased emphasis on BIM in education (Hjelseth 2017b), and during which there has been a continuous inflow of the younger generation and outflow of the older generation. Then perhaps we are nearing the inflection point where "the younger generation" has taken over, and we are able to implement the changed behaviors necessary for successful BIM adoption? Subconsciously we might visualize this inflection point as the tipping point on a scale, thereby deducing that once one side of the scale consists of 51% "younger generation" we have favorable conditions for establishing new behaviors. But research in experimental sociology show that this inflection point is 25%, not 51%. That is, when 25% of the people in the *system*, the AEC/FM sector in this case, exhibit a new behavior, then the remaining 75% will follow suit (Centola 2018; Centola et al. 2018).

Systemic development was the final theme in our analysis. In contrast to the "BIM is for young people" code discussed in the previous paragraph, the "set in their ways" code was more a statement of fact than a perception or opinion held by the BIM-user. Together with "the value chain is complex" and

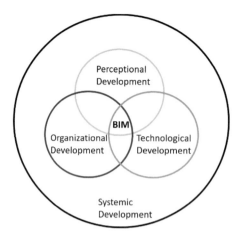

Figure 2. BIM processes consisting of People, Process and Technology within a system, such as the AEC/FM or the built environment.

"municipal works tasks are determined by law", these codes describe qualities of the system in which the BIM-users are operating. The common theme was a sort of path dependency, where the opportunities for making decisions and action were determined, or maybe limited, by events prior. This is path dependence in the general sense, where we take it to mean that history and previous decisions guide what options are available here and now (Liebowitz & Margolis 1995). For example, BIM could not be implemented because people were set in their ways, educated within a certain discipline and would not do new things; or BIM could not be implemented because the value chain is too complex; and BIM could not be implemented because municipal tasks were determined by law.

These are the *systemic developments* that need to be amended in the eyes of BIM-users. The theme has some similarities with the fourth component, *structure*, of Harold Leavitt's Diamond model, of which IDDS builds on (Leavitt & Bahrami 1988; Owen et al. 2013). But the municipality is one organization in a large system, and the "structure" component is not sufficient in explaining the interdependencies between the People, Process and Technology components in the larger system (the AEC/FM) to which the municipality is a part. In this sense, concepts belonging to innovation sciences offer alternative models.

Austrian economist Joseph Schumpeter defined innovation as the "new combinations of existing resources" (Fagerberg 2006). Municipal BIM-users' experiences that people were "set in their ways" and that the "value chain was too complex". These are examples of what Schumpeter called inertia, or "resistance to new ways" (Fagerberg 2006). In the case of this paper, the "inertia" is present in within the AEC/FM sector, which is an example of a "sectoral innovation system". This concept is that the role of innovation and transformation varies depending on economic sectors. Established processes in a sector is the product of the norms, rules, institutions, regulations, laws, expectations, policies that are inherent to it (Malerba 2006).

By applying the four themes identified from our interviews with the imperatives from IDDS and the concept of the sectoral innovation system we arrive at the following conceptual model to explain our results presented in Figure 2.

5 SUMMARY COMMENTS

The answer to the RQ: "What do BIM-users themselves view as the relevant issues as to why BIM-adoption is slow?" can simply be answered by the following three statements:

– "We need better software"
– "Software companies will soon develop the appropriate BIM software that will solve our problems"
– "BIM is not my business"

This study shows that slow BIM implementation is related to the stakeholders' understanding of what BIM is. This understanding is directly connected to whether they assess BIM as something: 1) relevant for themselves, or 2) not relevant for themselves.

Even if the stakeholders are aware of the benefits of BIM, they do not connect this to their own practice. It is not their practice – "their world", and "best practice" examples are therefore of limited value. The dominating understanding (view) of BIM is technology related, and by this mostly relevant for users (operators) of BIM software.

This study has explored, from mature disciplines such as innovation studies, the importance between principles (theories) and practice to enable development of a contextual understanding.

Digitalization in general, and BIM in special, is an immature domain in the AEC/FM-industry. By viewing BIM as an organizational innovation, we can depart from well researched concepts and frameworks from other research fields. The shift of perspective can increase the speed of BIM adoption by motivating the users to redesign the work environment, rather that only work with design of the built environment.

6 FURTHER RESEARCH

Our results indicate that the AEC/FM *system* itself is a resistance to change. Increased BIM adoption necessitates changes to the system. Schumpeter (Schumpeter 1976) meant that innovation was "new combinations of existing resources", and that the activity of innovating was the role of the *entrepreneur*. The entrepreneur had to overcome the inertia, this resistance to the new ways (Fagerberg 2006).

We know the AEC/FM market is large, complex, specialized and heavily regulated. But new business models, such as Airbnb and Uber (and many others) are great at forcing the relevant stakeholders to address legal, regulative and organizational issues that might stand in the way of change. With big data, IoT, smart buildings and cities and BIM touted as being answers to the vast environmental issues the sector faces – what *creative destruction* (Schumpeter 1976) can be inflicted on the AEC/FM sector to facilitate for change?

One place to start is by addressing what BIM really is. Firstly, focusing on the people in the process. Then the organizational development in order to address the informational foundation for all following processes. Further research will therefore explore the combination of innovation and organizational science for understanding of principles, and the use of ISO 19650 Information management series of standards as framework for practical actions in organizations and in the AEC/FM industry.

REFERENCES

Bærum kommune. 2020. "Stategiske Dokumenter." https://www.baerum.kommune.no/om-barum-kommune/organisasjon/styringsdokumenter/strategiske-dokumenter/ (April 12, 2020).

Bano, Muneera, & Didar Zowghi. 2015. "A Systematic Review on the Relationship between User Involvement and System Success." *Information and Software Technology*.

Baroudi, Jack J., Margrethe H. Olson, and Blake Ives. 1986. "An Empirical Study of the Impact of User Involvement on System Usage and Information Satisfaction." *Communications of the ACM*.

Bokhari, Rahat H. 2005. "The Relationship between System Usage and User Satisfaction: A Meta-Analysis." *Journal of Enterprise Information Management*.

Centola, Damon. 2018. "How Behavior Spreads." *How Behavior Spreads*.

Centola, Damon, Joshua Becker, Devon Brackbill, & Andrea Baronchelli. 2018. "Experimental Evidence for Tipping Points in Social Convention." *Science* 360(6393): 1116–19.

Creswell, J. W. 2007. *Qualitative Inquiry & Research Design*. Thousand Oaks, California: Sage publications Inc.

Daniotti, Bruno et al. 2017b. "Building Information Modeling (BIM) in Higher Education Based on Pedagogical Concepts and Standardised Methods." *International Journal of 3-D Information Modeling* 6(1): 35–50.

Daniotti, Bruno et al. 2019. *BIM-Based Collaborative Building Process Management*. Springer Tracts in Civil Engineering. http://www.springer.com/series/15088.

Daniotti, Bruno et al. *Springer Tracts in Civil Engineering BIM-Based Collaborative Building Process Management*. http://www.springer.com/series/15088.

Fagerberg, Jan. 2006. "The Oxford Handbook of Innovation." In *The Oxford Handbook of Innovation*, eds. Jan Fagerberg, David C. Mowery, and Richard N. Nelson. Oxford University Press, 1–26.

Gu, Ning, & Kerry London. 2010. "Understanding and Facilitating BIM Adoption in the AEC Industry." *Automation in Construction* 19(8): 988–99. http://dx.doi.org/10.1016/j.autcon.2010.09.002.

Guest, Greg, Arwen Bunce, & Laura Johnson. 2006. "How Many Interviews Are Enough?: An Experiment with Data Saturation and Variability." *Field Methods* 18(1): 59–82.

Hjelseth, Eilif. 2017a. "BIM Understanding and Activities." In *WIT Transactions on the Built Environment*, WITPress, 3–14.

IEA. 2019. *2019 Global Status Report for Buildings and Construction: Towards a Zero-Emission, Efficient and Resilient Buildings and Construction Sector*.

Kahneman, Daniel. 2011. *Thinking, Fast and Slow*. 1st ed. New York: Farrar, Straus and Giroux.

Ladner, Andreas, Nicolas Keuffer, &Harald Baldersheim. 2016. *Self-Rule Index for Local Authorities (Release 1.0)*. Luxembourg. https://ec.europa.eu/regional_policy/sources/docgener/studies/pdf/self_rule_index_en.pdf.

Leavitt, Harold J., and Homa. Bahrami. 1988. *Managerial Psychology?: Managing Behavior in Organizations*. University of Chicago Press.

Liebowitz, S J, and Stephen E Margolis. 1995. "Path Dependence, Lock-in, and History." *Journal of Law, Economics, & Organization* 11(1): 205–26. http://www.jstor.org/stable/765077.

Lu, Yujie, Zhilei Wu, Ruidong Chang, & Yongkui Li. 2017. "Building Information Modeling (BIM) for Green Buildings: A Critical Review and Future Directions." *Automation in Construction* 83: 134–48.

Luft, Joseph, & Harrington Ingham. 1955. "The Johari Window, a Graphic Model of Interpersonal Awareness." In *Proceedings of the Western Training Laboratory in Group Development*, Los Angeles: UCLA.

Malerba, Franco. 2006. "The Oxford Handbook of Innovation." In *The Oxford Handbook of Innovation*, eds. Jan Fagerberg, David C Mowery, and Richard N. Nelson. Oxford University Press, 380–406. https://www.oxfordhandbooks.com/view/10.1093/oxfordhb/9780199286805.001.0001/oxfordhb-9780199286805.

Matarneh, Sandra T. et al. 2019a. "BIM for FM: Developing Information Requirements to Support Facilities Management Systems." *Facilities*.

Matarneh, Sandra T. et al. 2019b. "Building Information Modeling for Facilities Management: A Literature Review and Future Research Directions." *Journal of Building Engineering* 24.

Migilinskas, Darius, Vladimir Popov, Virgaudas Juocevicius, & Leonas Ustinovichius. 2013. "The Benefits, Obstacles and Problems of Practical Bim Implementation." In *Procedia Engineering*, Elsevier Ltd, 767–74.

Ministry of Local Government and Modernisation. 2011. *Meld. St. 28: Gode Bygg for Eit Betre Samfunn: Ein Framtidsretta Bygningspolitikk*. Oslo.

Munir, Mustapha, Arto Kiviniemi, Stephen Jones, & Stephen Finnegan. 2020. "BIM-Based Operational Information Requirements for Asset Owners." *Architectural Engineering and Design Management*.

Owen, Robert et al. 2013. Publication 370 *Integrated Design & Delivery Solutions*. http://eprints.qut.edu.au/71070/1/pub_370.pdf.

Robey, Daniel, & Dana Farrow. 1982. "User Involvement In Information System Development – a Conflict Model and Empirical Test." *Management Science*.

Saldaña, Johnny. 2013. *The Coding Manual for Qualitative Researchers*. www.sagepublications.com.

Sandberg, Jörgen, & Mats Alvesson. 2011. "Ways of Constructing Research Questions: Gap-Spotting or Problematization?" *Organization* 18(1): 23–44.

Schumpeter, Joseph Alois. 1976. *Capitalism, Socialism and Democracy*. Routledge.

Sezer, Ahmet Anil, & Jan Bröchner. 2014. "The Construction Productivity Debate and the Measurement of Service Qualities." *Construction Management and Economics* 32(6): 565–74.

Silverman, D. 2010. *Doing Qualitative Research (3 Ed.)*. London: Sage publications Ltd.

Sujan, S. F. et al. 2018. "Holistic Methodology to Understand the Complexity of Collaboration in the Construction Process." In *EWork and EBusiness in Architecture, Engineering and Construction – Proceedings of the 12th European Conference on Product and Process Modelling, ECPPM 2018*, CRC Press/Balkema, 127–34.

Ullah, Kaleem, Irene Lill, & Emlyn Witt. 2019. "An Overview of BIM Adoption in the Construction Industry: Benefits and Barriers." In, 297–303. https://www.emerald.com/insight/content/doi/10.1108/S2516-285320190000002052/full/html.

Won, Jongsung, Ghang Lee, Carrie Dossick, & John Messner. 2013. "Where to Focus for Successful Adoption of Building Information Modeling within Organization." *Journal of Construction Engineering and Management* 139(11): 1–10.

Yin, R. K. 2009. *Case Study Research: Design and Methods*. Thousand Oaks, California: Sage Inc.

Development needs on the way to information-efficient BIM-based supply chain management of prefabricated engineer-to-order structures

P. Lahdenperä & M. Kiviniemi
VTT Technical Research Centre of Finland Ltd, Tampere, Finland

R. Lavikka
VTT Technical Research Centre of Finland Ltd, Espoo, Finland

A. Peltokorpi
Department of Civil Engineering, Aalto University, Espoo, Finland

ABSTRACT: Effective information management impacts the overall success of the construction project and is particularly vital in projects utilizing engineer-to-order (ETO) prefabrication, which is especially vulnerable to missing information. Therefore, this study focuses on ETO prefabricated structural steel and concrete components and how the related supply chain management could be improved by more comprehensive and effective application of the Building Information Modelling (BIM) environment for timely exchange of information between the parties involved throughout the delivery process. Current application of BIM falls short of its potential. Therefore, the study also identifies 25 key areas of development under five management categories drawing, primarily, on extensive dialogue with numerous experts from the industry. As these key development areas are strongly interrelated, their mutual interrelations and reasonable order of implementation are also discussed.

1 INTRODUCTION

Major construction projects are highly demanding undertakings due to their complexity. A vast range of expertise is involved and a huge number of interdependent tasks have to be carried out by numerous different actors. Orchestrating this effort requires sophisticated project management, which is both information-intensive and information-dependent. The performance of information management therefore impacts the overall success of the project (Back et al. 2001).

Information management refers to the collection, storage, processing and communicating of various types of project information with, as presented by Chen and Lu (2019), the ternary requirement of information adequacy, quality and accessibility. According to the authors, effective information management: minimizes manual work in the collection and processing of information, enables easy identification of needed information, presents the information visually, and makes the information available through a range of mobile terminals. New information emerges continuously whenever a task is completed or a new task is issued, which means that processing has to be made in a timely and continuous manner. A management practice that incorporates and makes full use of the above-listed features for the benefit of the project is information-efficient in the terms of this study.

Effective information management is particularly vital in engineer-to-order environments, where the level of uncertainty tends to be high; the adoption of prefabricated components requires intense exchange of information to integrate the production of components, logistics operations and site assembly (Bataglin et al. 2019) – especially when there are two or more prefabrication locations. Effective information management minimizes fragmentation between different parties, optimizes the flow of manufactured components at the stocking and dispatching stages, and enables the erection and positioning of massive components to be planned virtually (Ghasemi Poor Sabet & Chong 2020). It is especially needed in prefabrication-based projects, where flexibility is highly limited once the units are being transferred to site. Accordingly, Abanda et al. (2017) also consider information management benefits to be more meaningful in the case of prefabrication than traditional on-site construction.

Bataglin et al. (2019), Ghasemi Poor Sabet & Chong (2020) and Abanda et al. (2017) all refer specifically to the use of Building Information Modelling (BIM) instead of just effective information management. BIM has often been defined in reference to information related to the physical and functional characteristics of a facility (e.g. National Institute of Building Sciences 2007), but, arguably, this does not fully capture the comprehensiveness of potential

BIM applications. Instead, Hatem et al. (2018) broadly define BIM as an integrated and comprehensive system including whatever is related to a construction project and its stages; it represents a unified database for all project data through which project documents and in-formation are available to all stakeholders. Correspondingly, efficient exchange of information using a standardized machine-readable format has always been part of the related BIM vision (National Institute of Building Sciences 2007).

This view of comprehensiveness is supported by reported applications (Farnsworth et al. 2015) and, especially, bibliometric analyses of BIM-related R&D (Lemaire et al. 2019; Li et al. 2017; Santos et al. 2017) as an indication of future practice. The analyses reveal, for instance, that the early application of BIM in design and planning has extended to more recent attempts to make BIM reflect the real-life situation, as emphasized especially by Chen et al. (2015). Efficient management of a building process requires that the parties of the process have constant access to all relevant, up-to-date information on the current state of the project, both its plans and its realization.

On these grounds, this study focuses on engineer-to-order (ETO) products and how the related supply chain could be improved by more comprehensive and effective application of BIM for timely exchange of information between parties throughout the process. The study is conducted in the context of prefabricated structural steel and concrete elements, as the use of prefabricated concrete has an important role in Finland both today and historically (Huuhka et al. 2015). Finland has often been said to be one of the forerunners in the development and application of BIM (Aksenova et al. 2019; Halttula et al. 2015; Lejeune et al. 2016), which makes this study relevant to a wider audience.

In our study, we give special emphasis to inter-organizational information exchange, as it is generally known that efficiency gains are impeded by excessive fragmentation in the industry together with disparate project management processes and non-standardized information (Fulford & Standing 2014). BIM has accordingly been adopted primarily for productivity improvement by individual firms, but has not yet led to systemic change in the business environment (Aksenova et al. 2019).

The paper proceeds as follows. The next section briefly presents the study methods, after which current practice and its deficiencies in information-efficiency are examined. Then, the results of our study – the identified key areas of development – are presented before a more general discussion and conclusions.

2 STUDY METHODS

The supply chain process involves numerous actors that have come together to provide a technically challenging one-off project while operating in a multi-project environment under numerous constraints. Due to this challenging research context, we adopted an interpretive research approach to help understand and describe the social processes and complex factors at work (Merriam 2009; Schwandt 1994).

Firstly, eight practitioners were interviewed one at a time. The group of informants consisted of experts of different functions in two structural ETO companies and a design company. Of the ETO companies, a steel frame fabricator and a pre-cast concrete element supplier were involved equally. General contractor and information technology provider perspectives were also incorporated in the study through group interviews. Semi-structured interviews were designed to reveal the current practice and its deficiencies with respect to inter-organizational information exchange, and ideas for its improvement. Snowball sampling was applied in the selection of the interviewees (Biernacki & Waldorf 1981).

Secondly, after the material from the first stage had been analysed and tentative development needs determined, a number of participatory action workshops were organized to process the identified deficiencies and initial solution ideas into more specified concepts. The mini-workshops enabled not only validation of the initial results from stage one, but also development of a common understanding of eligible solutions. This could be achieved due to the comprehensive involvement of experts representing various functions in the participating companies. As regards the interviews and workshops, two or three researchers attended all of the events to increase the validity of data collection. Ideas and possibilities for development were fed to the process as appropriate (see the literature survey above).

A full list of interviews and workshops as well as the number and roles/titles of practitioners on each occasion is presented in Lavikka et al. (2020), who also report the used study methods in more detail.

3 CURRENT PRACTICES

3.1 *Contract management*

Fabricators become involved in building projects along with call-for-tender packages, which may include a structural model of the targeted building. In many cases, however, this only includes drawings of different component types and loadings and, in some cases, a bill of quantities. Therefore, the fabricator models the components for the tender whenever needed, but is unlikely to share the emerging model with other relevant parties. Moreover, if the fabricator wins the work, including detailed design, they usually start modelling from scratch again due to, for instance, changes agreed on in the contract negotiations. This practice impedes the comparison of the quantities of the realized structure to those of the tender. Design changes also take place during the subsequent process, but even then the modelling does not support the joint examination of quantities, which have to be dealt with separately.

As regards the scope of contracts, it is important to note that variations always exist in both the detailed design and the actual works on the construction site. The fabricator may be in charge of detailed design, or they may only fabricate the components based on the client's design. The latter practice may be rare internationally, but in Finland it is relatively common (although non-exclusive) with respect to precast concrete components. The same duality is true for the installation on site, as installation works can be either included in or left out of the ETO delivery contract. Organizational variation, again, means a need for more comprehensive standardization, as a bigger share of information exchange may potentially flow through inter-organizational interfaces.

3.2 General project management

Production planning in fabrication plants is actualized on two main levels. So-called rough planning begins immediately after the contract is completed based on the types and quantities of components in each installation section (block) of the entire project, and the planned order and timing of installation of the blocks. Rough planning also determines the necessary bulk of work input and the needed parts and materials to be ordered. The detailed design of each component in a production lot should therefore be completed 6 to 8 weeks before the manufacture of a given lot begins. However, too often this does not take place, and the fabricator has to inquire after critical information. At this stage information on individual components types is transferred into the Enterprise Resource Planning (ERP) system. Key components with long lead times nevertheless need to be purchased sufficiently in advance of the detailed component design.

Detailed planning is carried out 1 to 2 weeks prior to the start of manufacturing of a lot, by which time a more precise time of delivery should also be known. The planning is an ongoing/iterative process, which is complicated by the fact that the supplier is likely to serve a few or many different construction projects or customers simultaneously.

Although BIM is occasionally used also for scheduling, traditional project management software is much more commonly used, and the fabricator typically only receives a printed works schedule from the contractor. The start of installation at the construction site serves as the starting point for planning, with the preceding activities scheduled backwards from the installation. Correspondingly, the master schedule and installation schedule are separate documents that are coordinated, but not integrated or even less systematically updated.

Production planning is carried out per installation block. Definition of installation blocks is relatively straightforward initially. However, the late involvement of an installation contractor can potentially derail the plan from the perspective of the fabricator if that party considers it reasonable to install the components in a totally different order. This can often happen in cases where the installation is not the responsibility of the fabricator. It is also hard to anticipate changes, as no determination of the level of planning exists and parties tend to expect any plan that is circulated to be in a final state.

3.3 Management of detailed design

In detailed design, different disciplines work concurrently. However, if they do not work in the same BIM environment, they cannot be fully aware of the status of each other's design. On the other hand, a shared BIM environment is not desirable to companies who want to benefit from their own smart design algorithms that they have developed themselves, as a joint environment would mean giving others part of their competitive advantage (considering the competition for subsequent projects). A common environment might also allow designers to access each other's output, which is a potential risk. Joint consideration of separate discipline-specific models is possible using a separate application, but it requires extra effort and results in only a stand-alone temporal model. In other words, it does not offer an interactive real-time practice for concurrent design.

Other design-related challenges include divergent status classifications and new revisions that appear from time to time. In such cases, designers tend to avoid tentative versions by other designers due to uncertainty, as parts of the design might change in the near future. In addition, new versions typically provide no information on the underlining reasons for changes made, which is confusing and laborious. Furthermore, the initiator of a change is not traceable. The challenge of revisions is further intensified by the fact that they can occur even after the agreed due dates, which, on the other hand, is not surprising as designers do not receive status updates from the fabricator regarding the start of fabrication.

In addition, drawing and model coding practices and data submission procedures differ between fabricators, causing further disorder in the case of one-off projects where both the product and the organization building it are project-specific.

3.4 Fabrication and transport management

Needed data for production management and material purchase are transferred from BIM to ERP systems by means of data exchange files, but no transfer occurs in the opposite direction. For instance, information on the selection of component parts and materials made by the fabricator remains in the fabricator's possession and is not transferred to other parties. Consequently, this important data is often not made available for later facility management of the completed building and, even if it is supplied to the owner, it is not included in the BIM. In cases where the client or client's designer specifies parts and materials, those parts and materials are naturally used and the client holds the related information.

Transportation management is also conducted using the ERP system, making it separate from BIM. Calls by the installation manager to transport a set of elements to the construction site are currently made separately from each of these systems.

3.5 Installation management

Schedule inaccuracies are the biggest problem regarding installation. Preceding works on the site can often be delayed, but contractors typically do not inform fabricators of the delays. It can be difficult for the fabricator to get an exact estimate of the delay even on request, and related information is dealt with in site coordination meetings only. The contractor may not be willing to risk further delays by reporting the precise situation to the fabricator. However, at the same time, the fabricator might be struggling to deliver on time (in original schedule) due to uneven workload while serving numerous projects, and any flexibility in the delivery schedule would be welcome.

Synchronization is needed not only to meet fabricators' efficiency objectives, but also due to the limited space for storage on site and the need to avoid the extra work of moving components. Therefore, in addition to exact timing, also the precise order of installation should be known to enable the installation directly from trucks without intermediate storage. Although this practice is followed occasionally (esp. hollow-core slabs), it is not the prevailing practice. Commonly, the only determined day of installation (and, thus, due date for completion of prefabrication) for all components within an installation block is the start of installation of the block.

In addition, the installation order should be commonly known and agreed by all parties, as it is the designer who prepares the installation manual for the project. In the current practice, the installation manual is a kind of visualized narrative separate from BIM. The same concerns the handling of deviations. Tools for non-conformance reporting are in use, but they have not been integrated with BIM. Moreover, related information is not connected to status data, which further confuses the information value of the initially general status data.

4 IDENTIFIED KEY AREAS FOR DEVELOPMENT

4.1 General overview

This study identified numerous key areas that need to be developed for a better-performing supply chain of engineer-to-order construction components. These key development areas are presented as separate items in Figure 1 under five indicative management categories of the overall delivery process. *Contract management* covers the whole process, but the emphasis is on the early process. *General project management* steps in properly after the parties have entered into a contract. *Detailed design (management)* has to be completed before fabrication, although changes tend to appear subsequently. *Fabrication and transport management* is linked with the look-ahead planning of the project and, finally, *installation management* is linked with the contractor's short term or weekly planning. These management categories and the key development areas are presented below accordingly.

4.2 Contract management

As regards *contract management* (A), a BIM-based call-for-tender practice would foster BIM-based tendering and, eventually, contracting. This practice would also enable a quick start in delivery and lay a foundation for a subsequent BIM-based contract management process. Automatically updated quantity take-offs would improve the everyday work and pave the way to fluent final settlement of accounts when both contracting parties recognize the validity of the take-offs as should be the case in solid BIM-based practice. Any issues arising in due course of the process are to be taken into account, since a process rarely proceeds without any deviations (change orders, compensations, etc.). More specifically, the development tasks are as follows:

A1. BIM-based invitation to tender. Inclusion of BIM in a call-for-tender package is needed to facilitate proposers' quantity surveying in tendering and to foster subsequent BIM-based management. This requires the definition of content requirements for the pertinent BIM data as a Model View Definition (MVD) and related checking protocols to ensure the model meets the set requirements. Continuity of element identification throughout the process is essential.

A2. BIM-based contracting. Traditional document-based contracting hinders active use of BIM in the process, but BIM-based contracting might change this course. The practice requires determination of the requirements for such a model. Furthermore, the contract conditions need to be determined as regards BIM-based design and information management to ensure common understanding of reasonable project practices.

A3. Shared up-to-date quantity take-offs. BIM-based quantity take-offs (QTOs) that are common to contracting parties and automatically updated to correspond to the prevailing design solution would improve contract management and final cost prediction as post-contract design changes frequently occur in the delivery of an ETO system. Underlying causes of changes should also be traceable, which links this key development area to issue management (A4).

A4. Integrated issue management system. An integrated issue management system is needed for raising, dealing and settling events and deviations during the process, i.e. missing source information, change orders, deficiency in quality, delays, etc. (utilizing the Building Collaboration Format, BCF). The system is intended to serve inter-organizational project management for the most part, although reclamations

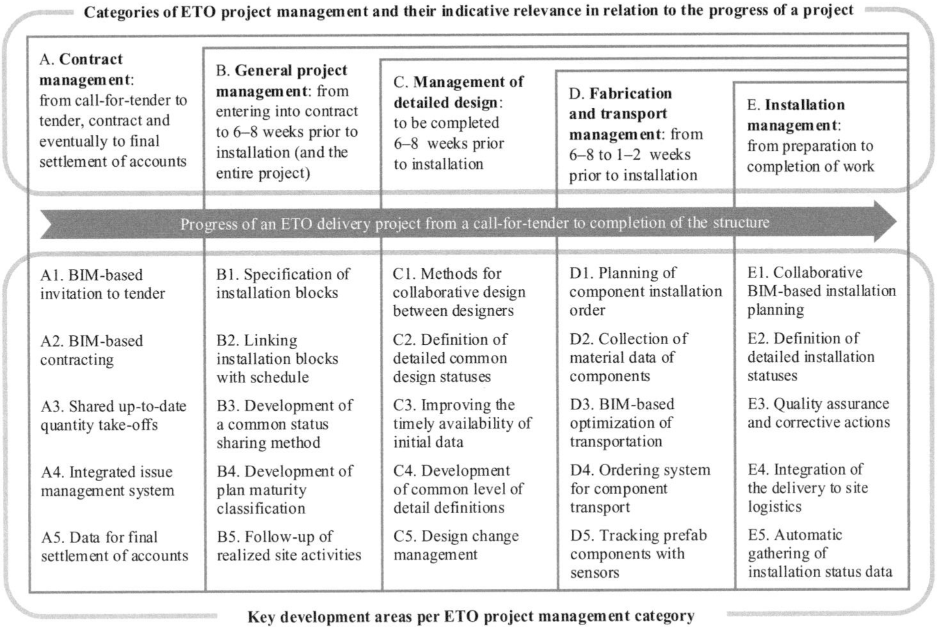

A. Contract management: from call-for-tender to tender, contract and eventually to final settlement of accounts	B. General project management: from entering into contract to 6–8 weeks prior to installation (and the entire project)	C. Management of detailed design: to be completed 6–8 weeks prior to installation	D. Fabrication and transport management: from 6–8 to 1–2 weeks prior to installation	E. Installation management: from preparation to completion of work
Progress of an ETO delivery project from a call-for-tender to completion of the structure					
A1. BIM-based invitation to tender	B1. Specification of installation blocks	C1. Methods for collaborative design between designers	D1. Planning of component installation order	E1. Collaborative BIM-based installation planning
A2. BIM-based contracting	B2. Linking installation blocks with schedule	C2. Definition of detailed common design statuses	D2. Collection of material data of components	E2. Definition of detailed installation statuses
A3. Shared up-to-date quantity take-offs	B3. Development of a common status sharing method	C3. Improving the timely availability of initial data	D3. BIM-based optimization of transportation	E3. Quality assurance and corrective actions
A4. Integrated issue management system	B4. Development of plan maturity classification	C4. Development of common level of detail definitions	D4. Ordering system for component transport	E4. Integration of the delivery to site logistics
A5. Data for final settlement of accounts	B5. Follow-up of realized site activities	C5. Design change management	D5. Tracking prefab components with sensors	E5. Automatic gathering of installation status data

Key development areas per ETO project management category

Figure 1. Identified development needs in ETO supply chain information management.

and cost issues are bilateral between the contracting parties.

A5. Data for the final settlement of accounts. Due to scattered information, the final settlement of accounts between contracting parties is a laborious effort in a typical project. Integration of a system, which in the case of changes and deviations updates QTO lists automatically (A3), with the related issue management (A4) would ease the effort significantly. Such a system would enable filtering of relevant, up-to-date information for final settlement of accounts.

Development should begin with key areas A1 and A2. The development of the related tendering and contracting practice is relatively straightforward and also likely to yield an immediate advantage. The results also form a starting point for the other listed activities that tend to become considerably more demanding towards the end of the itemized list.

4.3 General project management

From the overall or *general project management* perspective (B), it is essential to focus on the specification and planning of installation blocks, which form a breakdown structure for the element assembly. The blocks are to be agreed between the parties and incorporated into the schedule and, hence, this structuring guides the synchronization of tasks in general. In other words, changes in order and timing occur and, therefore, the management system should be flexible in this respect. On the other hand, the schedule is only a plan, and actual realization may deviate from it. Thus, it is important to keep related information apart and develop status tracking and sharing methods so that the up-to-date situation picture is continuously available to all major players of the project. The development tasks are the following:

B1. Specification of installation blocks. Specification of installation blocks (i.e. production segments) is needed for early scheduling of the fabrication and installation of components. If the specifications of the parties vary, they have to be matched and made known to others. BIM offers means for related coding and exchange of information with Enterprise Resource Planning (ERP) systems. The practice should be independent of any specific tool or software.

B2. Linking installation blocks with schedule. Transfer of schedule information between the contractor and the supplier/fabricator (ERP) had to be based on interoperability of the systems. This keeps the supplier aware of the prevailing schedule and enables the contractor to evaluate the feasibility of any change that occurs in the course of the process. From the point of view of the situation picture, it is of critical importance to keep apart the plan and actualization.

B3. Development of a common status sharing method. To efficiently share element-specific up-to-date status data in a multi-actor project organization, interoperability between the information systems of those actors has to be enabled. In other words, the systems have to interpret any information similarly; thus, a generic, cross-platform ontology for presenting information is needed (i.e. machine-to-machine readable practice, M2M).

B4. Development of plan maturity classification. Installation blocks and their mutual order and timing of installation are drafted early in production planning, but changes take place, and the information becomes fixed only later. Thus, it should be ensured that preliminary information is not used as final. Development and use of a classification of maturity (cf. level of detail/development in design) might be needed for production planning to minimize misunderstanding.

B5. Follow-up of realized site activities. The supplier should have access to up-to-date status information on activities realized on the site to estimate whether the forthcoming component installations can be realized as planned, or if there is a reason to expect some sort of deviation. For that reason, dependencies between various tasks and preconditions for installations also have to be determined. This practice enables the parties to improve their look-ahead planning.

Much can be done to improve the schedule management of projects. Scheduling at the level of installation blocks is, however, of the first priority. Therefore, B1 and B2 are the measures to start with in order to get a project well on track initially, although some adjustments still need to be coordinated traditionally. The other key development activities should then follow.

4.4 Management of detailed design

Where the management focus switches from general project management to management of the specific phases of delivery, *management of detailed designing* (C) emerges as one of the key development areas. As the design work is carried out by separate parties, smooth combination of different models and design phases is necessary in order to coordinate them and ensure compatibility. Collaborative design also requires the definition and use of detailed design statuses and levels of detail that are common to all parties, thus helping them to adjust their activities to the progress of the entire design process. Moreover, change management practice is also important and interplays with the more general issue management system as part of contract management. Thus, development efforts should be targeted as follows:

C1. Methods for collaborative design between designers. Systems for combining the models of different designers should be developed to enable continuous compatibility review (when editing) and to keep parties informed about the evolution of each other's solutions. Currently, viewing other designers' models requires the use of a common modelling environment (which may not be appropriate) or a separate tool (allowing an ad hoc static view of a version only).

C2. Definition of detailed common design statuses. Detailed design statuses that are common to all project parties need to be defined to enable communication of the degree of readiness and trustworthiness of related information. Current status definitions (if used) tend to vary and be company-specific so that 'completed' is the only common status. In some cases, a status could also be registered automatically based on the procedures completed by the designer in BIM.

C3. Improving the timely availability of initial data. Due to mutual information needs and dependencies between design disciplines, the coordination of the inter-organizational design process needs improvement. The use of up-to-date status definitions and automatic notices and reminders would foster the timely availability of initial information and a more efficient process with less rework. Information on the actual situation should naturally be tracked separately from plans.

C4. Development of common level of detail definitions. Level of detail/development (LOD) is an established concept, but its application is often deficient. LOD definitions should be based on the purpose of use of results from certain design stages and determine the pieces of information that have to be included, e.g., in initial data or complete design. The information should be available in BIM where software algorithms could verify its comprehensiveness.

C5. Design change management. Design changes are common in construction projects, and a related management system is at the core of the targeted issue management system (A4). Standardization of existing administrative and decision-making procedures offers a starting point for development. The practice and its applications should also be able to ensure that the most recent versions of certain model views (drawings) are in use automatically.

In the concurrent design process, the improvement of interactive design practice, where designers have a real-time understanding of each other's basic work, is key. C1 and C2 therefore call for the earliest attention. Other development activities should then be planned to make full use of the BIM-based practice and to avoid misunderstandings.

4.5 Fabrication and transport management

Another specific area of development is *fabrication and transport management* (D). The installation order of elements serves as a basis for temporal planning of fabrication and transport of each element. However, the optimization of fabrication (batches) and transportation (shipments) is a likely reason to depart from installation order alone. Tools need to be developed for these purposes, especially if different suppliers are involved in the same project. The same is true for the practice of calling elements to site and dealing with delivery related information. The state of elements in terms of properties and location should also be made available to the parties involved to improve their situation awareness. Accordingly, the itemized list of key development areas is as follows:

D1. Planning of component installation order. Determination of installation blocks (B1) is a more general scheduling issue, while installation order within a block is needed for mid-term synchronization of works. Exact sequences or, at least, daily output should be determined to allow the fabricator to time and optimize truck loads. This requested practice is already applied in some cases, but even then it is not based on the use of BIM as we suggest here.

D2. Collection of material data of components. Consideration of quality in the construction and subsequent facility management phases presumes improved traceability of parts and materials used and built into the building. Much too often this information remains solely in the fabricator's possession, although it should be incorporated into BIM components to offer access to the information as needed later on (e.g., mix proportion and sampling results of precast concrete).

D3. BIM-based optimization of transportation. Major projects require a large number of prefabricated components, which are large and heavy. Thus, from the cost and ecological perspectives, transportation is a key challenge. This is even more so in large projects where components are purchased from different suppliers. In such a case, a BIM-based optimization tool to allocate assignments to suppliers and components to shipments would be of considerable benefit.

D4. Ordering system for component transport. Calls by the installation or site manager to transport a set of elements to the construction site are currently typically made by e-mail and without any specific protocol. A BIM-based tool should be developed to enable a more efficient practice and to make all events visible to interested parties by allowing related calls and supply receipts to make corresponding status changes in the BIM. Deviations may need special attention.

D5. Tracking prefab components with sensors. Sensor-based tracking of prefabricated components is an essential part of the sustaining/updating of a real-time situation picture. It can reveal information on the location of an element and, in some cases, its properties, condition and circumstances. A continuous flow of information on the location of a shipment during transportation also allows site managers to anticipate the time of arrival to the site.

Determination of an installation order within a block and allowing the transportation to be optimized and organized accordingly is a useful step to take immediately. This practice means prioritizing D1, D3 and D4 over the other developments. Sensor-based tracking of components (D5) and the accumulation of material information for future reference (D2) serve as further development activities.

4.6 Installation management

Finally, *installation management* (E) practices also need to be developed. Various plans and standards govern installation work and their integration into BIM would improve information transparency. The work also consists of multiple tasks and phases, and communication and organization of daily activities would benefit from a more detailed conception of those tasks and related statuses. In addition to this, automatic tracking and recording of status information is also needed to keep the status information up to date throughout the installations. Naturally, this information has to be in a machine-to-machine readable general format, which, again, connects the challenge to the broader development of information systems. An itemized list of tasks that cover this development is as follows:

E1. Collaborative BIM-based installation planning. Short-term installation planning incorporated into the BIM environment would offer an integrated management solution for collaborative work planning on the (order of) installation of individual elements. The development of such a BIM application should be based on clear operational principles (including acceptance procedures) and should connect the installation works to other related tasks to be carried out on the site.

E2. Definition of detailed installation statuses. Broad status categorization is not sufficient in the case of prefabricated elements. Quality assurance and realization of subsequent works require more detailed information on the assembly, e.g., whether a component is only put in place or end-fixed (torque tightened). Concrete joints require precast inspections, and hardening of the concrete (with moulds in place) is also an installation status. Pertinent practices need to be developed.

E3. Quality assurance and corrective actions. Even with cautious work and handling, quality defects in elements delivered to the construction site still occur. A procedure must therefore be in place to ensure that defects are repaired efficiently. Another reason for this needed development is to prevent the repetition of similar mistakes. The key development area is strongly connected to that of more general BIM-based issue management and also to possible use of quality assurance technologies.

E4. Integration of the delivery to site logistics. Integration of the component ordering system into the project's production model is needed to rationalize the planning of site activities. This interconnection enables automatic reservation of site resources, such as cranes, unloading areas and storage locations, as a part of delivery management. Improved awareness of the real-time situation on site facilitates optimization of the contents of shipments.

E5. Automatic gathering of installation status data. Automatic tracking of detailed installation statuses is needed to ensure the collection of status data from the work stages and to make use of this data. This can be done by tracing the location and status of workers, elements and equipment using reality capture technologies such as laser scanning, 3D imaging, in-situ sensing, electronic tagging and GPS with wireless web technologies.

Conformance of the installation with regulations and safe practice is always the first priority. Accordingly, development activity E2 should be given first priority. BIM-based planning and quality assurance logically come next, as they are relatively easy to implement compared to gathering status information, which requires the introduction of new tracking technology on site, i.e. including additional components and/or vehicles.

5 DISCUSSION

This study examined how to improve the efficiency of the supply chain of engineer-to-order structures through improved access to relevant, up-to-date information. Extended utilization of BIM is offered as a solution, and 25 identified key development areas under five management categories are presented.

These key development areas are strongly interrelated as the sought-after comprehensive situation picture of the parties involved and data interoperability between different systems are the two critical principles guiding the development. The goal of an industry-wide solution also gave reason to focus on mainstream practices and ignore company-specific or other separate applications, despite their possible sophistication. Such systems exist, but not as industry-wide solutions.

In addition to the targeted solution, also underlining motives were examined. Mutual interrelations of the development tasks are also discussed to support efficient implementation of the findings. The study thus adds new knowledge and provides implications for practice.

Regarding the limitations of this study, firstly, it was conducted in the context of the supply of prefabricated structural elements, and only a limited number of companies were involved in the study. Secondly, the study was carried out in Finland, and the results may not be generalizable to other markets. Moreover, this paper presents only conceptual ideas and is not able to stretch to the determination of technical solutions.

On the other hand, both a precast concrete element supplier and a steel component supplier participated in the study. Both companies are also leading operators in Finland and they operate in various ways with respect to scope of delivery and organizational role. Their market areas also extend to other Nordic and European countries. These factors support the validity of the results. A global information technology provider was also involved in the study, bringing technical and international expertise and enabling us to focus on state-of-the-art technology solutions, although the latter are not discussed in this paper.

Our examination of current BIM and ERP applications and related process deficiencies and the possibilities offered by utilizing state-of-the-art technology was iterative by nature due to the numerous interviews and workshops and the involvement of several experts from different positions. This should also support the validity of the results in terms of triangulation: both by data triangulation (multiple sources of expert views) and investigator triangulation (multiple researchers). Therefore, it is expected that the study serves well the development of management practices in its targeted primary application area.

In addition, the study result may also facilitate similar development in other areas and markets, even if their current levels of development or application are not necessarily the same as those presented here. In such cases, the view should be on the overall framework or classification of relevant development areas. This can be used as a set of criteria for evaluating the situation-specific state in order to determine the steps needed in each case. This is a logical approach, as, in essence, also the present study is an interim evaluation of the current state in Finland – what has been introduced in practice so far and what needs to be done in future – rather than an investigation of novel visionary technologies. This view of serving other markets is also supported by what has been said above regarding the development situation in Finland regarding the application of BIM.

6 CONCLUSIONS

This study concludes that management of the ETO supply chain and its incorporation into general construction project management should be developed in accordance with extended use of BIM and the identified key development areas presented here. Our results present, firstly, conceptual or principal solutions for fundamental aspects of information-efficient ETO management practice. Secondly, the results offer a comprehensive framework for planning, positioning, phasing and prioritizing ETO supply chain management development efforts while also maintaining the bigger picture in the case of a long-lasting stepwise development process.

Considering the comprehensiveness of the targeted solution, it would take time to implement the results in full, although experiments with more limited application (parties, scope) could be carried out more quickly. In construction, where temporary organizations implement one-off projects, the benefits of enhanced information management can be obtained only if most companies are able to exchange information in a machine-readable format. This requires the use of common status definitions, data models, etc., which, in turn, require numerous preceding experiments, positive experiences, industry-wide commitment and relevant standards.

Therefore, it is reasonable to expect that the time needed for such a comprehensive industry-wide change would be no less than a decade or so. Moreover, full implementation of the study results would also require the introduction of novel tracking technology in demanding site conditions, which is an additional challenge compared to the introduction of new software features – not least as it also involves a cultural change in work practices. Nevertheless, the change is one well worth making, as it would pave the way to a more efficient and reliable engineer-to-order process.

REFERENCES

Abanda, F.H., Tah, J.H.M. & Cheung, F.K.T. 2017. BIM in off-site manufacturing for buildings. *Journal of Building Engineering* 14: 89–102.

Aksenova, G., Kiviniemi, A., Kocaturk, T. & Lejeune, A. 2019. From Finnish AEC knowledge ecosystem to business ecosystem: lessons learned from the national

deployment of BIM. *Construction Management and Economics* 37(6): 317–335.
Back, W.E., Moreau, K.A. & Jones, J.A. 2001. Information management strategies for project management. *Project Management Journal* 32(1):10–19.
Bataglin, F.S., Viana, D.D., Formoso, C.T. & Bulhões, I.R. 2019. Model for planning and controlling the delivery and assembly of engineer-to-order prefabricated building systems: exploring synergies between Lean and BIM. *Canadian Journal of Civil Engineering* 47(2): 165–177.
Biernacki, P. & Waldorf, D. 1981. Snowball sampling: Problems and techniques of chain referral sampling. *Sociological Methods & Research* 10(2): 141–163.
Chen, K. & Lu, W. 2019. Bridging BIM and building (BBB) for information management in construction: The underlying mechanism and implementation. *Engineering, Construction and Architectural Management* 26(7): 1518–1532.
Chen, K., Lu, W., Peng, Y., Rowlinson, S. & Huang, G.Q. 2015. Bridging BIM and building: From a literature review to an integrated conceptual framework. *International Journal of Project Management* 33(6): 1405–1416.
Farnsworth, C.B., Beveridge, S., Miller, K.R. & Christofferson, J.P. 2015. Application, advantages, and methods associated with using BIM in commercial construction. International *Journal of Construction Education and Research* 11(3): 218–236.
Fulford, R. & Standing, C. 2014. Construction industry productivity and the potential for collaborative practice. *International Journal of Project Management* 32(2): 315–326.
Ghasemi Poor Sabet, P. & Chong, H.Y. 2020. Interactions between building information modelling and off-site manufacturing for productivity improvement. *International Journal of Managing Projects in Business* 13(2): 233–255.
Halttula, H., Haapasalo, H. & Herva, M. 2015. Barriers to achieving the benefits of BIM. *International Journal of 3-D Information Modeling* 4(4): 16–33.
Hatem, W.A., Abd, A.M. & Abbas, N.N. 2018. Motivation factors for adopting Building Information Modeling (BIM) in Iraq. *Engineering Technology & Applied Science Research* 8(2): 2668–2672.
Huuhka, S., Kaasalainen, T., Hakanen, J.H. & Lahdensivu, J. 2015. Reusing concrete panels from buildings for building: Potential in Finnish 1970s mass housing. *Resources, Conservation and Recycling* 101: 105–121.
Lavikka, R., Lahdenperä, P., Kiviniemi, M. & Peltokorpi, A. 2020. Digital situation picture in construction – Case of prefabricated structural elements. *Proc. of 18th Intern. Conf. on Computing in Civil and Building Engineering, (changed/ planned to be hold in virtual format) 18–20 August 2020*. Cham: Springer. (in press)
Lejeune, A., Nach, H., Rizkallah, G. & Aksenova, G. 2016. Finnish BIM pioneers: Like hackers architects in a community studio. *ISPIM Innovation Forum, Boston, 13–16 March 2016*.
Lemaire, C., Montréal, E., Rivest, C.L., Boton, C., Danjou, C., Braesch, C. & Nyffenegger, F. 2019. Analyzing BIM topics and clusters through ten years of scientific publications. *Journal of Information Technology in Construction (ITcon)* 24: 273–298.
Li, X., Wu, P., Shen, G.Q., Wang, X. & Teng, Y. 2017. Mapping the knowledge domains of Building Information Modeling (BIM): A bibliometric approach. *Automation in Construction* 84: 195–206.
Merriam, S.B. 2009. *Qualitative research: A guide to design and implementation*. San Francisco, CA: Jossey-Bass.
National Institute of Building Sciences. 2007. *National Building Information Modeling standard. Version 1 – Part 1: Overview, Principles, and Methodologies*. Washington, DC.
Santos, R., Costa, A.A. & Grilo, A. 2017. Bibliometric analysis and review of Building Information Modelling literature published between 2005 and 2015. *Automation in Construction* 80: 118–136.
Schwandt, T.A. 1994. Constructivist, interpretivist approaches to human inquiry. In Denzin, N.K. & Lincoln, Y.S. (eds.) *Handbook of Qualitative Research*: 118–136. Thousand Oaks, CA: Sage Publications.

# Analysis of the influencing factors for the practical application of BIM in combination with AI in Germany

A. Shamreeva
5D Institut GmbH, Stuttgart, Germany

A. Doroschkin
Institute of Construction Management, University Stuttgart, Stuttgart, Germany

ABSTRACT: By using Building Information Modeling (BIM) methodology in the architecture, engineering and construction (AEC) sector, construction participants create digital building models and produce large amounts of information during the entire life cycle of a building. This generated digital data may also be used for further evaluation, analysis and simulation of machine intelligence. The integration of AI methods into the construction process is feasible in various application areas, such as building design, industrial prefabrication, planning on the construction site and project management. The aim of this study was to identify the factors that influence the practical application of BIM in combination with AI in the German construction industry. In the course of this research, the most important aspects of the country-specific AI initiation relating to BIM data were gathered and evaluated, and, based on this analysis, the potential application-oriented goals of BIM and AI methods can be carried for AEC.

1 INTRODUCTION

1.1 Digitalization in the AEC sector

Climate change, sustainability, digitalization, demographic change, urbanization, and growth of infrastructure are all factors influencing the current development of the AEC industry. However, digitalization has the potential to change the construction sector significantly in the near future.

Using digital technologies in construction allows a new perspective of the construction process. The digital transformation is fundamentally changing the way companies create value. Rapid technological progress enables the digital networking of items – equipment, tools, or manufactured products. This digital transition of the construction industry introduces numerous improvements to traditional construction processes and also elevates the quality level of the end product of the construction process.

However, the development of digital technologies in AEC progresses very slowly in comparison with other sectors or industries, e.g., machine construction. Yet the construction industry is certainly aware of the importance of digital technologies, but their practical implementation is still a major problem.

Analyzing the causes of the hinderance for AEC, Schober and Hoff (2016) emphasized four key areas for the digitalization:

- *digital data* (electronic collection and analysis)
- *digital access* (mobile access to the internet and internal networks)
- *automation* (use of new technologies to create autonomous, self-organizing systems)
- *connectivity* (connection and synchronization of hitherto separate activities).

Therefore, they determined the following links in the construction value chain for its stakeholders:

- logistics
- procurement
- production and construction
- marketing and sales
- after-sales and end-customer marketing.

According to their survey, most construction stakeholders have proposed that production, construction and logistics would mostly benefit from automation and digital data.

At the same time, a number of authors emphasize building information modeling (hereinafter BIM) as the core of the digital technologies.

1.2 BIM methodology

The main objective of BIM is the permanent use of an information-loaded digital building model for the entire life cycle of a building – from its initial conception and planning to its demolition – whereby the information in the model is used as a database for the management of every stage of the product lifecycle.

BIM methodology made the building parties rethink the planning process: "first build virtual, then real" (BMVI 2015) where the main decisions about a

Table 1. Applications of the lifecycle BIM in the AEC value chain.

DESIGN AND ENGINEERING
Parallel and robust design and engineering

Parametric modeling and object libraries	Constructability and clash analysis	Coordination of design disciplines	Integrated design construction process

CONSTRUCTION
Real-time data sharing, integration, and coordination across stakeholders

Construction planning and scheduling	Efficient, information-rich tenders	Coordination of sub-contractors and suppliers	Continuous system integration across parties

OPERATIONS
BIM-enhanced operations and maintenance

Storage, maintenance, and use of building information

building's construction must be carried out during the earlier planning phases, similar to the conventional planning process. Nevertheless, BIM maintains the traditional tasks associated with all stages of planning. For example, using model data, architects and engineers can visualize the load bearing capacity of building materials over time. The visualization of construction subsections supports the accuracy of estimation for subcontractors. Hence, this visualization capability coupled with the opportunity of model-based quantity determination, ensures the success of the planning process and reduces the amount of time on a project.

So, BIM methodology can have various use cases, but only a few are usually utilized on a project. However, in view of the advantages of the BIM methodology, Gerbert et al. 2016 proposed the use of BIM for the entire life cycle of a building. The applications of lifecycle BIM along the AEC value chain are summarized in Table 1.

Much digital data is produced during these BIM implementation cases. Primarily, the development of a digital building model needs parametric modeling of objects with alphanumeric information, which come from object libraries. In infrastructure projects, such as road construction or road reconstruction, e.g., field data from laser scans is always available, which would be integrated into the project. In addition, the generated model data supports prefabrication and additive manufacturing, will be used in facility and asset management, and maintain reconstruction planning in renovation cases. Inputs to the BIM model provide the basis for the function of autonomous equipment. Moreover, partial models, e.g. models of subsections in building construction, help to coordinate the design disciplines, and to carry out clash detection and analysis.

The focus of this paper is on digital data, which continuously increases in the course of a BIM project. The collection and record of this information provides opportunity for the analysis of the current operation of any past BIM project. In this regard, the data can be applied for the improvement and optimization of work in the AEC sector. This paper describes the methods of artificial intelligence (AI) for processing with digital AEC data.

1.3 *AI and its sub-disciplines*

Many specialists from different economic sectors have recognized AI as an innovation driver. Thanks to recent technological improvements, database management and data sciences have significantly changed and enabled the further development of AI-based systems. Over a period of 20 years, the volume of researches in the AI field has grown dramatically. Having many active research topics and practical applications, AI has gained in importance for national competitiveness. For example, with its AI development plan, China has set a goal to become a world leader in AI theory, technology, and application (Elsevier 2018). In the USA and Europe, AI has also become a key topic within innovation and research policies.

Covering the areas of search and optimization, fuzzy systems, natural language processing and knowledge representation, computer vision, machine learning and probabilistic reasoning, planning and decision making, and neural networks, AI does not seem to have a universal definition (Elsevier 2018). However, general differentiation can be made between weak AI, i.e., machines that can simulate thinking to accomplish a specific task, and strong AI, i.e., intelligent thinking machines (Russell & Norvig 2016). The article examines exclusively the perspectives of weak AI systems. Specifically, some sub-disciplines, such as machine learning, open up many possibilities for the AEC industry.

Machine learning (ML) presents a technique that enables computer systems to improve themselves with experience and data. Using ML methods, it is possible to create AI systems that can operate in complicated real-world environments and have the ability to acquire their own knowledge by extracting patterns from raw data (Goodfellow et al. 2016). Some of the common machine learning techniques are regression, classification, clustering, anomaly detection, market basket analysis, etc. When applying the ML methods, the chosen algorithms define a form of learning model for computers. For this purpose, a specialist collects and prepares relevant data for system training. The choice of data representation has an enormous impact on the performance of ML algorithms (Goodfellow et al. 2016). After feature engineering, i.e., elaboration of training data representation, the training of a learning model with sample data takes place. As the next step, the results of the training are compared with the sample information in order to apply the trained model to the new data. It is important to note that if the

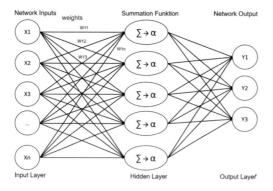

Figure 1. Structure of a deep neural network.

algorithms receive too much erroneous muster data, the system cannot return a correct response. Similarly, if the sample data is not representative, the output of the system should at least be considered uncertain.

According to the information from Kaggle, a leading platform for hosting public data and ML competitions, data scientists and ML experts often use (approx. 40% of the time), among other ML methods, artificial neural networks (Döbel et al. 2018). Motivated by models of the human brain and nerve cells, the artificial neuron offers a local memory and carries out localized information processing operations (Akash 2015). The neurons are usually organized in neural networks (NN), that are arranged in layers of which there are three types: input layer (for the reception of input signal and its transfer to the neurons in another layer), output layer (for the reception of a signal from input or hidden layer), and hidden layer (for a possible connection between layers) (see Figure 1).

A quantity of layers defines an architecture of a NN. A NN, which possesses one or more hidden layers, is called a multilayer network. Deep learning (DL), a sub-discipline of ML, belongs to multilayered (deep) NN. However, intensive development of DL methods has become possible only in recent years, because of the critical size of NN that requires high-performance hardware and software infrastructure. To achieve a better result, DL must process a large volume of data. Thus, increased computing capacity (attributed to cloud computing) and growth of digital data volumes are the mandatary factors for the creation of efficient deep NN.

On the contrary, one of the main problems of ML and its subarea DL is that it needs to learn from a large volume of past data. Using any ML methods, there is no possibility to connect seemingly disparate threads to solve unique problems without past experiences. Tackling novel situations puts a fundamental limit on the human tasks that can be automated by using AI-based systems. Another issue for AI applications is that many of the factors of variation influence every single piece of raw data. Many of these parameters can be identified only using nearly a human-level understanding of the information (Goodfellow et al. 2016).

Computer vision, e.g., which is one of the most active research areas for DL applications, is an enormously complicated task for machines, but the vision is effortless for humans.

In regard to AEC, ML can maintain the automatization of the production process, industrial robotics, building operations, and quality management. The processing of data allows the monitoring of equipment, design analysis and execution planning, and incident prognosis. However, despite the fact that many experts advocate AI-based systems and support their utility in the AEC sector, not many studies have identified the practical application of AI methods in the construction process. Some studies have articulated the most widely used AI methods in AEC and the most commonly addressed topic using AI concepts, but most of them are theoretical, not practical (Darko et al. 2020). The majority of studies have adopted some of the AI use cases in AEC in theoretical projects, giving a possibility for future use (Bloch & Sacks 2018; Cheng et al. 2020; Karan & Asadi 2019; Sun et al. 2020). Furthermore, the potential of robotics and its advantages in the automation of the construction process are often shown in a laboratory environment, which can be substantially different in practice or in the field (Grigoryan & Semenova 2020). Some of the studies and researches suggest specific applications of artificial NN e.g. self-organizing maps (Kosyakov et al. 2018; Popova 2013). Many papers describe DL methods such as convolutional NN for site safety: real-time detection of personal protective equipment (Nath et al. 2020) or inspection of construction material components (Tay et al. 2019). Therefore, the practical integration of ML methods in any construction company and the impact of using AI-based systems on the current working process are still vague and unclearly identified. If ML is an effective tool to support AEC and change the way of planning, construction, and operating processes, then its practical implementation in the construction sector should be studied in a more detailed manner.

This study aims to provide an insight into the obstacles blocking the way to AI-based systems with BIM data, and to evaluate the interference factors for the AI implementation in AEC. By investigating theoretical AI methods in planning and construction phases, the influencing factors for the practical application of BIM in combination with AI-based systems will be analyzed and summarized.

2 RESEARCH FRAMEWORK

2.1 Digitalization in AEC in Germany

This review circumscribes the German construction industry and reproduces its specific cultural features. According to the study "Transformation of the construction industry" (Baumanns et al. 2016), Germany is in first place in Europe for construction volume. Nonetheless, the productivity growth of the AEC sector is nearly twice lower than the overall productivity of the German industry in the period from 2000 to 2011, as reported by the Hamburg Institute of International Economics (ger. HWWI).

One of the main government programs "Industry 4.0" that enfolds all areas of the digital transformation in Germany aims to interlink production with the latest information and communication technologies and create intelligent connections between human beings, machines, and manufacturing processes. Relating to the AEC industry, this digital transition is called "Construction 4.0" or "Building site 4.0".

Based on one study by the Association of German Chambers of Commerce and Industry (ger. DIHK), 93% of companies indicate that digitalization will influence every one of their business processes. Regarding the AEC industry, the digitization status has grown above the average in nearly half of the interviewed companies (Liecke et al. 2017). Construction stakeholders concentrate principally on the digitization of their business processes, such as planning, construction, and logistics. Among digital opportunities, which have the most considerable influence on the business of the AEC sector, are analytics and big data in operation, production, marketing and logistics, additive manufacturing, robotics in production, human-machine collaboration, augmented reality in production, predictive maintenance, cloud-based logistics solution, etc. (Schober & Hoff 2016). At the same time, BIM as a central digital project platform is of great importance in the construction industry.

2.2 *Implementation of BIM methodology*

One of the most significant milestones in the implementation of the BIM method in Germany was the multiphase plan "Digital Design and Construction", initiated by the Reform Commission of Germany's Federal Ministry of Transport and Digital Infrastructure (BMVI 2015). This Road Map determines a common understanding of the BIM methodology, makes the use of BIM compulsory for public infrastructure projects as of 2020, and lays down the requirements for all project participants. These rules have similarities with the BIM strategy in the UK, the Netherlands, Denmark, Finland, and Norway. Thus, they obtain the international standards for managing information over the whole life cycle of a built asset using BIM (ISO 19650) and for standardization of the exchange data formats (ISO 16739). ISO 19650 describes the forms of project information deliverables and considers three areas of the information management: specification of the informational requirements by the asset owner or operator, planning of the information deliverables by the BIM coordinator, and delivery of the information by planning and construction companies. However, there are some issues for information management with BIM that ISO standards do not completely cover. Neither does this norm define how BIM tools can be integrated into the fundamental traditional processes of the German AEC sector, nor does it define the process for the exchange of information between all participants, and how the information can be exchanged and shared in a BIM process. Furthermore, the standards for common data environment and cloud-based working are not settled for the current BIM applications. The role of a BIM manager and the procedure of effective BIM implementation are not clearly defined. Therefore, this standard demonstrates the requirements of the asset owner and operator but does not suggest any particular BIM implementation approach for planning offices and construction companies.

With its model-based approach, BIM methodology is, of course, supported by industry professionals. Using the BIM method in planning and construction processes, stakeholders gain an important competitive advantage. According to a study by the Fraunhofer IAO institute, however, only 22% of participants from the AEC sector always or often use BIM as an object-oriented building model (3D), whereas, 41% have never used BIM for their processes. 5D BIM methodology, which incorporates time and costs as additional planning dimensions, is only used by a very small percentage of companies (Braun et al. 2015). Analyzing the current implementation status separate for planning offices and construction companies, constructors see more benefits in 5D BIM in comparison with planners.

However, as reality shows, the implementation of model-based processes is going very slowly in the German construction industry. One of the reasons slowing down the integration of BIM is the circumstantial structured legal regulations of the construction industry, e.g. Official Scale of Fees for Services by Architects and Engineers (HOAI 2013). These regulations split the planning process into nine service phases, which is much more complicated than in other countries. On the one hand, at the interfaces between the phases, one project participant can carry out his/her services according to the contract and refer the planning data to the next participant, resulting in possible partial loss of the project data during the construction progress. On the other hand, change management on the model data during the BIM life cycle poses a great challenge with lots of interfaces for data deliverables. Furthermore, the data quality of an information deliverable does not always correspond to the data requirements of other project stakeholders. The data exchange still depends on the software application because the standardized data exchange format (IFC) is, in reality, insufficient for some current working processes. Hence, data delivery does not work without losses between various software platforms. While not being able to work on the planning data from planning offices, many construction companies design, for instance, their own BIM model only for the purpose of their calculation process. Consequently, some of the main obstacles to collaboration in BIM in Germany are a lack of interoperability, hindering preconditions, and an absence of BIM consistent standards.

By analyzing the current situation for this research, it was noticed that planners and contractors do not share, as a rule, their experience with each other, and often develop the same tools or applications (e.g., a plugin for CAD software for the semi-automatic planning of the formwork in structural design) simultaneously. As a result, AEC stakeholders make the same mistakes during BIM implementation and have a long way to go before they reach their goals. The other obstacle to the integration of the BIM methodology into a company's own work process is false implementation strategy; the companies develop their BIM road maps in an incorrect sequence: firstly, they think about technologies, and secondly, about BIM goals and BIM guidelines for their processes. The right workflow, instead, primarily involves strategic planning first, followed by tactical measures. Finally, the technology is selected during the operative planning (Baldwin 2018).

Summarizing the problems faced by specialists involved in the development, implementation and practical application of BIM technologies, there was an emphasis on the minimal experience of construction stakeholders, a lack of the organization while working with BIM, and a shortage of education at the universities for future BIM tasks. The integrated digital process needs more time to be entirely established in the German AEC sector.

2.3 *AI research in Germany*

According to the Outline for a German Strategy for Artificial Intelligence, AI is identified as the key to digitalization (Harhoff et al. 2018). In the global scientific world, Germany remains the fifth largest producer of AI research (Elsevier 2018). Thanks to the German Research Center for Artificial Intelligence (ger. DIHK) and Fraunhofer Society for the Promotion of Applied Research, federal-state initiative "Cyber Valley", and numerous globally active industrial companies, the country has favorable conditions with regards to industrial fields of AI application.

In 2019, the Federal Ministry for Economic Affairs and Energy promoted a project called "AI meets BIM" in the research area of AI applications in AEC. The research agenda for the development and establishment of AI technologies included structure planning, planning of construction progress, automation of construction site, and operating phase. At the same time, the Cluster of Excellence on Integrative Computational Design and Construction for Architecture was established at the University of Stuttgart. The Cluster aims to make use of digital technologies in order to rethink design, fabrication, and construction processes. Among its main research units are computer science and robotics. At the Karlsruhe Institute of Technology, an AI research program has been started that incorporates two important issues: AI for interoperability and AI for practical application. Under the summary term of Smart Design and Construction are merged the following tasks for AI methods: qualification using object recognition, identification of pattern, automation of planning process, allocation of the construction services, cost planning, and time scheduling, supply chain, project control, quality control, identification of risks, etc.

In regard to applied research by construction companies, RIB Software SE, a German company providing construction and BIM software, for instance, is intensively searching for intelligence solutions for the simulation of construction site logistics with the company-specific data applying ML methods. Based on a company database, the programmable algorithms of the software shall support the execution planning by evaluating possible construction progresses.

While AI-based systems depend on data quality and volume, the management of Big Data, the organization of cloud services and digital ecosystems are important issues for the development of ML applications. Some German companies and organizations, e.g., Daimler, Audi, BMW, Industrial Data Space, SAP, have already initiated digital ecosystems for cooperative use by partners, clients, providers, and developers. The implementation of digital platforms is also supported by the government. According to the research of the German digital association, 37% of industrial stakeholders appreciate the potentials of digital platforms (Bitkom 2020). The results of the study about applied digital methods have shown that more than 50% of companies harness Big Data, but only 11% make use of AI-based systems (Bitkom 2018).

Furthermore, the implementation of AI methods requires a large volume of data. In the AEC industry, digital data can be delivered from a BIM process, or ML algorithms create data for a further BIM application. Nevertheless, the training process of an AI-based system needs often more raw qualified data than many construction companies are able to provide. Thus, one of the researches of Fraunhofer Society, dedicated to ML with few data, can influence the future implementation of AI methods.

3 MATERIALS AND METHODS

3.1 *Applications of AI in AEC*

The theoretical and practical fundamentals for the article are based on the numerous expert discussions with specialists from the software industry and construction sector (10 companies with more than 300 employees and more than 20 years of experience), digital and construction fairs, thematic publications and papers. Considering the research on this topic, the possible applications of AI along the AEC value chain are summarized in Table 2 (Müller & Shamreeva 2019).

The core of a BIM process and AI methods is digital data. BIM methodology provides for AEC this digital construction data that can be used in AI-based systems.

3.2 *Influencing factors for AI applications*

Based on the experience of BIM implementation at planning offices and construction companies, the

Table 2. AI applications along the AEC value chain.

DESIGN AND ENGINEERING

Modeling	Structure planning	Execution planning
Semi-automated modeling, intelligent object libraries	Design of unique structure, simulation of load bearing behavior	BIM model on site, comparison of actual state, adaptation as-built model

Survey of existing construction	Design planning	4D & 5D planning
Semi-automated extraction of geometric data from point cloud	Intelligent clash analysis based on past, experience	Forecast of time and costs, on site simulation optimization of bid/cost estimation

INDUSTRIAL PREFABRICATION

Manufacturing production	Robotics
Transfer of on-site processes to the prefabrication plant	Entire automation of the production process, robots on construction site

Logistics	Construction systems
Just-in-time production, delivery of materials and prefabricated components	Façade, door, window systems, apartment units, story units etc.

PLANNING ON SITE

Task management	Technical documentation
Intelligent assistance for prioritizing and recording the tasks personalized task distribution for a better productivity	Semi-automated recording of defects, mapping defect place to the model, update and approval of the data

Execution	Procurement
Update of the BIM model, semi-automated labeling, remarks & comments	Forecast for material delivery, tracking and just-in-time delivery of materials & equipment

Quality control	Safety control
Call up of plan section from BIM model during site inspection, updating & retracing of defects	Tracking & reporting of safety incidents on site, on-line guide on safety behavior & warnings for employers

Construction progress	Billing/Invoicing
Digital representation of construction site, delivery of execution plans & scheduling for each worker	Semi-automated reporting of performance & invoice quantities, updating of target/actual model

Table 2. (Continued.)

PROJECT MANAGEMENT

Tender/Competition	Database
Analysis of own projects and competitors, intelligent selection for the project team	Objects, image & text recognition, automated classification & storage of data

Contract management	Facility management
Checklist for compliance with contracts, updating of data records for all project participants with regard to contract conditions, tracking of payment processing	Monitoring of the construction & building services engineering, plant engineering, inventory

influencing factors for the practical integration of digital methods in the German construction industry, such as AI application, can be summarized as follows:

– *data availability* (the construction process involves a large number of people with diverse work steps, that can influence data deliverables)
– *data quality* (a number of unpredictable events, such as natural disasters and man-made mistakes can reduce data quality)
– *data sharing & data protection* (many companies have generally very strict regulations and data protection rules, thus inhibiting the sharing of experiences and working together on the common database for AI applications)
– *competitiveness* (many companies try to gain a supposed competitive advantage by developing and applying their own individual technologies or applications)
– *standards for conventional processes* (cost estimations and quantity surveys are carried out by planners and constructions in different ways. The current standards are not sufficient for consistent processes)
– *standards for BIM modeling* (standards for the modeling process such as objects, classification types and model element attributes are needed for uniform planning and construction phases)
– *standards for BIM management* (information deliverables must be defined)
– *detailed documentation* (each completed project must be digitally documented in detail and in a uniform format so that it can be used in the further process of building operation)
– *structure of the building construction sector* (presence of only a few main contractors, but lots of planning offices and construction companies)
– *diversity of construction companies* (different activities in the building construction value chain: work

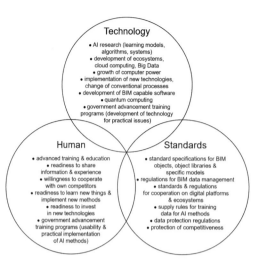

Figure 2. Influencing factors and their correlation for the practical application of BIM in combination with AI.

preparation, execution, operation, and renovation (Baumanns et al. 2016)
- *size of enterprises* (more than 70% of business volume in AEC is made by companies with less than 100 employees) (Baumanns et al. 2016)
- *AEC stakeholders* (fear of changes or readiness to learn new things)
- *level of prefabrication* (transfer of construction site processes to prefabrication)
- *level of digitization* (level of BIM usage)
- *legal effects* (regulations about data protection, digital privacy, digital platforms and ecosystems)
- *confirmability, explicability, transparency* of processing with AI systems (Döbel et al. 2018).

4 RESULTS AND DISCUSSIONS

4.1 *Analysis of the influencing factors*

Based on the listing above, it is possible to identify the main influencing areas for the implementation of AI-based systems with BIM data in AEC. Figure 2 illustrates the correlation of the influencing factors and demonstrates the necessity of considering all the parameters in the practical integration of BIM in combination with AI.

The result of the analysis of the impact factors shows that it is imperative that scientific research, government, and AEC companies cooperate to take steps for the success of the practical implementation of AI.

4.2 *Solution routes*

While evaluating the parameters for the successful integration of AI methods, it is necessary to shed light on issues that have been already handled or are in progress. Thus, the influence of positive factors must increase, and the impact of negative parameters must be minimized.

First, standards for BIM modeling and semi-automated processing of learning models in ML have a high potential to reduce the development efforts of AI applications. DIN BIM Cloud, the first German library of object attributes, for instance, implements pre-set primary standards for a modeling process according to German regulations. Second, a strong AI ecosystem can provide strong networks between science and AEC actors. A common language of the AI ecosystem will better connect them. Providing learning systems and platforms for AI was the main topic of Digital Summit in Germany in 2019. Powerful hardware manufacturers and cloud providers are encouraged to create a safe digital environment. Third, data pools and more advanced methods of anonymizing data must be developed for mobilizing data for AI. An alternative approach to AI that can operate with little data will also promote rapid AI implementation (Harhoff et al. 2018). Furthermore, the German government will simulate data sharing between AEC participants with governance models. Thanks to these models, the participants can exchange data and information for scientific and economic aims without thinking about data protection and being able to control the process. This can help to overcome the problem of diversity of construction companies. However, AEC participants must implement the concepts according to BIM application in combination with AI themselves. Moreover, new services in the conventional scope of works can emerge from applying AI methods. Clearly structured legal framework for working with Big Data, rights of use, and regulations for data protection will encourage construction stakeholders to act together with their competitors. For instance, an adaptation of copyright will enable companies to apply Text and Data Mining. Finally, changes in the current educational programs must be taken; study courses for different BIM roles shall include the methods of data processing such as ML and its sub-discipline NN.

5 CONCLUSION

This paper is intended to capture possible influencing factors for the practical application of BIM and AI methods in the AEC sector. The possible application-oriented goals of BIM in combination with AI for the German construction industry can be shown in the context of the application possibilities of AI along the AEC value-added chain, the analysis of the effect parameters and the perspectives for the further development of AI methods:

- improvement of project planning, site supervision, quality and risk management
- optimization of project supply chain
- increase in efficiency
- cost reduction.

To achieve these objectives, planning the establishment of future-oriented digital infrastructures is required, as well as inspiring transfer projects in cooperation between fundamental research and practical application.

REFERENCES

Akash, M.S. 2015. Artificial intelligence & neural networks. *International Journal of Soft Computing and Artificial Intelligence* Issue 2 (Volume 3).

Baldwin, M. 2018. *BIM Manager: Practical guidance for BIM project management*. Berlin: Beuth Verlag GmbH; German Institute for Standardization (ger. DIN), Mensch und Maschine Schweiz AG.

Baumanns, T., Freber, P.S., Schober, K.S., Kirchner, F. 2016. *Construction industry in transition – trends and potential up to 2020*. Munich: Roland Berger GmbH & UniCredit Bank AG.

Bitkom 2020. *Digital platforms*. Berlin: Federal Association for Information Technology, Telecommunications and New Media (ger. Bitkom).

Bitkom 2018. *Bitkom Research about current applied digital technologies*. Berlin: Federal Association for Information Technology, Telecommunications and New Media (ger. Bitkom).

Bloch, T. & Sacks, R. 2018. Comparing machine learning and rule-based inferencing for semantic enrichment of BIM models. *Automation in Construction* (Volume 91): 256–272.

BMVI 2015. *Road map for Digital Design and Construction (Introduction of modern, IT-based processes and technologies for design, construction and operation of assets in the built environment)*. Berlin: Federal Ministry of Transport and Digital Infrastructure (ger. BMVI).

Braun, S., Rieck, A., Köhler-Hammer, C. 2015. *Results of the BIM study for planners and contractors "Digital planning and production methods"*. Stuttgart: Fraunhofer Institute for Industrial Engineering and Organization (ger. IAO).

Cheng, J.C.P., Chen, W., Chen, K., Wang, Q. 2020. Data-driven predictive maintenance planning framework for MEP components based on BIM and IoT using machine learning algorithms. *Automation in Construction* 103087 (Volume 112).

Darko, A., Chan, A.P.C., Adabre, M.A., Edwards, D.J., Hosseini, M.R., Ameyaw, E.E. 2020. Artificial intelligence in the AEC industry: Scientometric analysis and visualization of research activities. *Automation in Construction* 103081 (Volume 112).

Döbel, I., Leis, M., Vogelsang, M.M., Neustroev, D., Petzka, H., Riemer, A., Rüping, S., Voss, A., Wegele, M., Welz, J. 2018. *Machine learning: an analysis of competences, research and application*. Munich: Fraunhofer Society for the Promotion of Applied Research (ger. Fraunhofer-Gesellschaft).

ELSEVIER 2018. *Artificial Intelligence: How knowledge is created, transferred, and used. Trends in China, Europe, and the United States*. Elsevier B.V.

Gerbert, P., Castagnino, S., Rothballer, C., Renz, A. & Filitz, R. 2016. *Digital in Engineering and Construction (The Transformative Power of Building Information Modeling)*. Munich: The Boston Consulting Group.

Goodfellow, I., Bengio, Y., Courville, A. 2016. *Deep Learning (Adaptive Computation and Machine Learning series)*. Cambridge: The MIT Press.

Grigoryan, E.A. & Semenova, M.D. 2020. Automation of the construction process by using a hinged robot with interchangeable nozzles. *Materials Today: Proceedings* in press.

Harhoff, D., Heumann, S., Jentzsch, N., Lorenz, P. 2018. *Outline for a German Strategy for Artificial Intelligence*. Berlin: Stiftung Neue Verantwortung e. V.

HOAI 2013. *Official Scale of Fees for Services by Architects and Engineers (in Germany)*. Duesseldorf: Werner Verlag.

Karan, E. & Asadi, S. 2019. Intelligent designer: A computational approach to automating design of windows in buildings. *Automation in Construction* (Volume 102): 160–169.

Kosyakov, E.D., Talipova, L.V., Romanovich, M.A., Roshkovanova, A.I., Simankina, T.L., Braila, N.V. 2018. Strategies for redevelopment of gray belt objects on the basis of neural networks. *Construction of Unique Buildings and Structures* 7(70): 31–42.

Liecke, M., van Renssen, L., Sobania, K. 2017. *Increasing challenges meet with growing optimism (Company barometer on digitization)*. Berlin & Brussels: Association of German Chambers of Commerce and Industry (ger. DIHK).

Müller, W., Shamreeva, A. 2019. Can civil engineers be replaced by artificial intelligence? Detlef Heck (ed.), *5. Congress for construction operations, construction industry, construction management (ger. 5. BBB Kongress); Proc. intern. symp., Graz, 19 September 2019*. Graz: Technical University of Graz.

Nath, N.D., Behzadan, A.H., Paal, S.G. 2020. Deep learning for site safety: Real-time detection of personal protective equipment. *Automation in Construction* 103085 (Volume 112).

Popova, O.N. 2013. Application of self-organizing maps (SOM) for the analysis of the housing stock in its complex reproduction. *Vestnik of the Irkutsk State Technical University* No. 10: 171–177.

Russell, S.J. & Norvig, P. 2016. *Artificial Intelligence. A modern Approach, 3rd ed.* New Jersey: Pearson Education, Inc.

Schober, K.S. & Hoff, P. 2016. *Digitization in the construction industry (Building Europe's road to "Construction 4.0")*. Munich: Roland Berger GmbH.

Sun, Y., Lei, K., Cao, L., Zhong, B., Wei, Y., Li, J., Yang, Z. 2020. Text visualization for construction document information management. *Automation in Construction* 103048 (Volume 111).

Tay, J., Shi, P., He, Y., Nath, T. 2019. Application of Computer Vision in the Construction Industry. *Proceedings of the 4th World Engineers Summit 2019, Singapore, 28–29 Aug 2019* Paper No. 046.

A systematic review of project management information systems for heavy civil construction projects

W. Chen & M. Leon
Scott Sutherland School of Architecture and Built Environment, Robert Gordon University, Aberdeen, UK

P. Benton
Cephas Project Management Ltd, Montrose, Angus, UK

ABSTRACT: Project success depends heavily on effective collaboration among organizations and individuals and efficient Information and Communication Technology (ICT). Among these tools, Project Management Information Systems (PMIS) are complex, distributed, multi-functional systems for decision-making support when planning, organizing, and controlling projects. However, these tools create new problems such as information overload, increased pressure and communication chaos caused by the complex and volatile nature of projects; they also neglect the particular needs of Heavy Civil Construction (HCC) section, a significant subdomain of the AEC industry where the challenges are much higher than typical projects. Through a systematic review, this paper outlines the landscape of PMIS and it presents the benefits and adoption barriers. The research highlights the lack of holistic applications and proposes new and innovative ad-hoc solutions for efficient management of construction and infrastructure projects to be adopted by the developers, via standardization of PMIS for HCC. The research informs the industry stakeholders, software vendors and policymakers of the industry gaps and the potentials for the development of collaborative solutions for PMIS standardization, similarly to the Building Information Modelling (BIM) technologies.

1 INTRODUCTION

1.1 Nature of construction information management

Project success heavily depends on effective collaborations within and among organizations, which is a major challenge for the Architecture, Engineering and Construction (AEC) industry due to its diverse, complex and volatile nature (Sears 2015). On top of that, every construction project is unique and demands a one-off management structure that has to be set up in a very short time (Sears 2015; Tezel et al. 2018; Wilkinson 2005). These characteristics may explain the observation that the construction industry being characterized as "inefficient", "wasteful" and "high-risk/low-profit" (Tezel et al. 2018). As the AEC industry is highly information-dependent, "late, inaccurate, inadequate or inconsistent information" and "poor communication practices" are serious delimiters to project performance (Dainty et al. 2007; Wilkinson 2005).

1.2 Project management information tools – benefits and limits

Traditional ways of information management have been revolutionized by the utilization of Information and Communication Technologies (ICT). Nowadays, ICT based Information System tools (IS) are widely implemented; the benefits are evident: information service is faster, easier, more accurate and reliable, and it provides support for project visualization, planning and decision-making (Papadonikolaki et al. 2018).

However, although the integration between ICT and construction management has been ongoing since the 90s, the outputting management tools/services are far from being panaceas – in fact, while solving problems, they may also create new ones such as information overload, increased pressure, communication chaos and diminished human-touch among others (Dainty et al. 2007; Tezel et al. 2018).

1.3 Particularity of Heavy Civil Construction (HCC) projects

This paper argues that the current development of ICT PM tools has neglected the particular needs of Heavy Civil Construction (HCC) section – a major subdomain of the construction industry. Compared to the projects in other construction areas, HCC projects deal with very different challenges and uncertainties in technical, social and environmental aspects (Moghayedi & Windapo 2019). Highway projects, for example, include engineering particularity in design and construction, like dependency on availability and suitability of natural material; mass earthmoving and haulage machinery; the impact from climate and weather conditions; latent underground conditions. These distinct characteristics lead to inexact and semi-empirical approaches during the design and construction process, thus undermining projects progression and outcomes (Moghayedi & Windapo 2019; Walsh et al. 2011).

1.4 Aim and objectives

This paper aims to provide an assessment of industrial acceptance and effectiveness of ICT PM tools in HCC area in comparison with other subdomains (typically building projects) with the following objectives:

- Investigate the acceptance of ICT PM tools in HCC area and comparing it with other areas.
- Examine of market volatility, implementation barriers and functions/features of PMIS tools.
- Identify of gaps specifically related to the particular nature of HCC projects.
- Summarise the results and proposal of potential solutions to the identified inadequacy and gaps.

2 LITERATURE REVIEW

2.1 Nomenclature

When referring to ICT systems, there are numerous synonyms and abbreviations regarding project management tools; these include terms from Total Information Transfer Management (Tam 1999), to Web-based project management systems (Nitithamyong & Skibniewski 2004), to concurrent engineering (CE) environment (Wilkinson 2005), with more recent terminology mentioning Project Management Software or Information Systems (Arnold & Javernick-Will 2012; Lee & Yu 2012) and Construction information management framework (Lee et al. 2018). One reason for such diversity in nomenclature is the ICT tools have been evolved to deal with problems far beyond their original purpose such as document management. It also reflects the divergence in goals, strengths and priorities between the developers and the researchers that investigate these areas. In this paper, the term Project Management Information System (PMIS) is used to refer to ICT tools as discussed above.

2.2 Development of ICT PMIS in construction

ICT had revolutionized the industry in three ways: faster information processing, easier information access and supporting better decision-making and control (Fryer et al. 2004). ICT integration specifically is categorized based on technology progression, business models and accomplished changes in IS. Regarding technology evolution, there are overall three eras; the first era being stand-alone tools for particular works (e.g. CAD); the second era being the development of communication technologies such as e-mailing and document management system (mid-90s); the third era being the pursuit of a cohesive overall system through integrating different applications (from early 2000) (Froese 2010). Scott et al. (2003), Nitithamyong and Pollaphat (2004) further distilled PMIS into three business models: Project Collaboration Network (PCN), Project Information Portal (PIP) and Project Procurement Exchange (PPE). Concerning the changes in IS, Laudon and Laudon (2013) summarized these into the prevalence of mobile platforms like smartphones, the growth of online software as service and "cloud computing". These technologies led to a substantial evolution in how we consider construction sector, with BIM being one of the most recent and paradigm shifting technologies.

2.3 PMIS within BIM and web-based solutions

Overall, there has been a strong correlation between PMIS usage and project performance, with increased usage leading to better performance (Pellerin et al. 2013). Furthermore, the advantage of using commercial PMIS products over in-house development has also been recognized (Nitithamyong & Pollaphat 2004).

As the internet reaches increased maturity, the advantages of Web-based PM-ASP are multiple and include, platform/application-independence, location-independence, synchronous/asynchronous communication, real-time collaboration and cost saving on time and resources, compared to traditional approaches (David et al. 2003; Lee & Yu 2012; Nitithamyong & Skibniewski 2004). The common denominator though, when considering PMIS tools, is that the information quality is essential to IS success and acceptance (DeLone & McLean 2003; Lee & Yu 2012).

However, when it comes to connecting PMIS and BIM, there is little direct connection. BIM is praised as a revolutionary force to tackle problems in productivity, information flow and collaboration, and is widely supported by government mandates in the UK and Europe. Nonetheless, Papadonikolaki et al. (2018) conducted a systematic analysis against project stages defined in the RIBA plan, and found that the BIM software ecosystem is fragmented across work stages. The investigation on IFC format compliance also revealed their inadequacy in terms of interoperability. Lee et al. (2018) also pointed out that BIM is effectively used for 3D models but not sufficiently linked to information management. The effort of information consolidation is lacking, especially within the construction phase and there is no straightforward process to effectively link structural information, non-structural information and BIM-based information databases.

3 METHODOLOGY

3.1 Research rationale

This paper applies a systematic review of the PMIS ecosystem currently available for the construction industry. The research involves an investigation of the acceptance of PMIS tools within HCC projects. Following that, the paper examines different aspects that impact PMIS uptake, from market volatility to barriers and functional features. It also identifies specific gaps in the market related to solutions for HCC projects.

3.2 Data collection

Nowadays, hundreds of PMIS products can be found online with keywords "construction collaboration", "Information system", "Project management", and other similar keyword combinations. However, there has not been a recognized intelligible categorization within the industry; for that reason, this research reviews the key identified software within the industry and professional resources and any previous research on the topic since 2000. Based on this premise, this paper located 7 review-based PMIS ranking lists from construction information websites, as illustrated in Table 1.

Following this generic search, a more detailed review of previous research indicates that many software are now disappeared, acquired or are under a different name; from 29 Project Collaboration Network (PCN) services listed by Nitithamyong & Skibniewski (2004) only 24% are still available. Meanwhile, another investigation, which follows Wilkinson's "established construction collaboration products" in the UK (2005), finds that 50% of PMIS products are still in service, 40% are under different names and only 10% discontinued. The first result indicates the evolving industrial needs and market volatility in past 20 years; the second result, on the other hand, shows the endurance of top products.

Table 1. ICT PM software ranking lists in 2019.

Websites	Key words/filters	No. of products
Software advices	heavy-highway contractor/cloud-based	127
Capterra.com	Construction management software/web-based (Exclude): commercial, residential	140
Finance on line	Top 10 construction management software	10
Project-management.com	Top 5 Construction Project Management Software in 2019	5
PAT Research	Top 19 construction management software	19
Technology Advice	Construction Project Management Software Products (Exclude) small business	16
G2.com	Construction Management Software (Exclude) small, middle business	50

3.3 Barriers to implementation and data analysis

Having this technology widely available does not translate into application; quite often, a number of obstacles occur that hinder the industry from implementing these solutions widely. Based on a survey by the UK Department of Trade and Industry (Wilkinson 2005), the obstacles of technology implementation were identified as setup and running costs, lack of skills and resources, lack of knowledge, staff reluctance and IT difficulties with systems integration. Nitithamyoung and Skibniewski (2004) highlighted problems of cost, technology limits and "non-technical" barriers such as resistance to change, collaborative maturity and density of communication channels. Further research indicated that the most important factor is management buy-in and cultural issues were among chief barriers (Ruikar 2005). Additional issues include ICT collaboration due to industry fragmentation, limitation in bandwidth and cost, and information overload (Dainty et al. 2007). Oesterreich and Teuteberg (2019) clustered the barriers into four dimensions: structure, people, technology and task. They find that social barriers relating to "people" and "structure" dimensions are the most important barriers such as people's resistance to change, compared to which, the pure technical issues are less critical. This conclusion echoes to studies of other researchers (Dainty et al. 2007; Eadie et al. 2014; Nitithamyong & Pollaphat 2004; Ruikar 2005;).

Based on Oesterreich and Teuteberg's (2019) categorization, an extensive review of the PMIS rankings from Table 1 isolated five most popular and shared software, that are above a critical acceptance threshold within the industry (Nitithamyong & Skibniewski 2004), which are afterwards analysed based on their features and applicability, hence, showcasing the functional fragmentation in mainstream market solutions.

4 DATA ANALYSIS

4.1 ICT integration barriers

Expanding upon the work of Oesterreich and Teuteberg (2019), the research so far indicates that these are typically distilled down to: lack of a standards and interoperability; issues with usability (i.e. operability and user-friendliness), time required for adoption of new technologies, setup/running costs and unclear Return on Investment (ROI) additional knowledge and skills prerequisites, poor quality of information, legal and contractual uncertainty, social and cultural barriers (for example, resistance to change), feasibility issues. Table 2 summarizes these results, with the findings indicating a logical merging of barrier concepts as follows;

P4 "lack of training" and T4 "time-consuming adoption" are in fact referring to the similar barriers, therefore can be merged into one and

Table 2. Matrix of implementation barriers.

Implementation barriers		DTI (Department of Trade and Industry) (2004)	Nititham young and Skibniewski (2004)	Ruikar (2005)	Dainty, Moore and Murray (2007)	Eadie, odeyinka et al (2014)	Oestereish and Teuteberg (2019)
External Structural dimension							
S1	Legal and contractual uncertainty		X	X		X	X
S2	Lack of demand					X	X
S3	Lack of awareness about functionality			X		X	X
S4	lack of government incentive and regulation						X
S5	Lack of necessity					X	X
S6	Non-widespread use			X		X	X
Internal People dimension							
P1	Resistance to change	X	X	X		X	X
P2	Lack of expertise	X			X	X	X
P3	Lack of skilled personal	X				X	X
P4	Lack of training				X	X	X
P5	Collaborative maturity		X				X
P6	Lack of management support					X	X
P7	Increased pressure and information overload		X	X	X		
Technology dimension							
T1	lack of standard and interoperability	X	X	X	X	X	X
T2	Insufficient infrastructure	X	X	X	X		X
T3	complexity of software				X		X
T4	Time-consuming adoption	X					X
T5	Lack of applicability and practicability						X
T6	Availability of Software						X
T7	Poor quality of model information		X				
T8	system reliability		X				
T9	system security			X			
T10	Management complexity		X				
Task dimension							
TA1	High investment costs	X		X	X	X	X
TA2	lack of proven benefits		X			X	X
TA3	lack of investment ca2ital						X

described as "time and resources for training and setup".

P2 "lack of expertise", P3 "lack of skilled people" and T10 "software management complexity" can be combined into one barrier as "required additional knowledge and skills".

P5 "Collaborative maturity", "increased pressure and information overload" and "resistance to change" rather belong to "soft" and "distant" barriers, about which the ICT developers cannot offer direct solutions as they should be sought from social and cultural perspectives.

T7 "poor quality of model information" had the lowest importance index of all barriers, (RII as 0.476, when the average is 0.730, according to (Oesterreich & Totenberg 2019)).

T3 "complexity of software" can be phrased as "Usability (i.e. operability and user-friendliness)".

TA1, TA2 and TA3 are interrelated and can be expressed as: "High setting-up and running cost and unclear Return on Investment (ROI).

4.2 Functions fragmentation in market solutions

Based on the industry identified software lists in Table 1, the five most popular and shared software that are above a critical acceptance threshold within the industry (Nitithamyong & Skibniewski 2004) were reviewed according to their functional structures as outlined from e-manuals and demo versions, based on which a PMIS function "matrix" was compiled in Table 3. Again, a wide diversity appears in ways of organizing functions, due to the divergent solution-seeking routes. These function structures are analysed according to activities level for regrouping them under function "features".

Table 3 indicates that many developers attempt to offer an all-solution package, covering information management, communication and coordination, plan, monitoring and control, estimation, financial analysis, among others. However, these efforts are lacking unified criteria. As a result, it is rather difficult to make a comparison. Meanwhile, new integration options

Table 3. Matrix of ICT PMIS software function/features.

Features		IBM 360	Oracle Aconex	Procore	Asite	Cmic
Document management	document control	x	x	x	x	x
	document distribution	x	x	x	x	x
	drawing management			x	x	x
	specification			x	x	
	supplier document		x			
Communication/workflow management	RFI management	x		x	x	x
	Transmittals/submittals	x	x	x		x
	mail management		x	x	x	
	meeting management			x		
Design collaboration	Design collaboration	x		x		
	Design change (constructability)	x	x	x	x	x
3D model visualization/management		x	x		x	
Construction collaboration	clash detection	x	x	x		
	trade coordination	x	x	x		
Quality management (field)	check list/issues	x	x	x	x	x
	punch list	x		x	x	x
Project management tool	reporting and dashboards	x	x	x	x	x
financial/cost management	budget control	x	x	x		x
	cost management			x		x
	invoicing			x		x
	Change orders	x		x		x
	Accounting			x		
Safety management	health and safety	x		x	x	
Schedule	Scheduling			x		x
Human resource				x		x
Supply chain management			x	x	x	
Bidding management			x	x		x
Asset management	asset management				x	
	inventory				x	x
	maintenance				x	
	material					x
Facility management					x	
IS system management				x		
Enterprise planning						x
Sitedatacapturing				x	x	

keep emerging with technology development such as 4D visualization, portable mobile platforms and cloud-computing. Nevertheless, the breaking-down of function structures in Table 3 provides a representative mapping, which is to be used as a basis for effective further investigations.

4.3 HCC projects requirements

The challenges in HCC project management have a distinctly different nature from the other sub-domains. These particularities require tailored ICT integration options. HCC projects are characterized by particular needs, with a focal point being material/earthmoving management. Axelsson et al. (2018) identified seven "wastes" in highway construction in terms of Lean Construction: Overproduction, Waiting, Transportation, Over-processing, Unnecessary motion and Defects. The first three "wastes" are mostly generated in earthmoving activities. Material managing consume the largest portion of time and resources in a road construction project. Unlike building projects, the raw materials are mostly self-produced/transported, hence, resulting in planning and controlling internal issues (Ji & Borrmann 2014).

Asset Management is also central within HCC, even though it appears to be side-lined in ICT PM tools, with only 2 of 5 identified PMIS tools having this feature (Table 3). Asset management refers to different scopes between HCC projects and building projects, with HCC projects normally needing to manage a large number of earthmoving equipment over many scattered locations.

Another distinct characteristic of HCC projects is the high demand for data capturing before and after the activities. Although automatic survey equipment has been widely used, the link between captured data and information system appeared to be missing (Chen et al. 2012; Perkinson et al. 2010). Figure 1 summarizes the functional requirements of HCC projects.

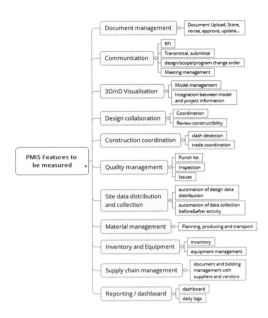

Figure 1. ICT PM requirements for HCC projects.

Based on these findings, this research proposes the integration of GIS and Laser scanning as one particular need of HCC projects, to cope with the constant demand of mass-data collection and design data setting-out. Further PMIS support is also required regarding material and transportation management since avoiding "wastes" in earthmoving activities has always been a challenge in HCC projects. Current BIM technology is insufficient to handle information interoperability for the horizontal infrastructure projects (Kenley et al. 2016).

5 DISCUSSION

5.1 Research impact

The research focus is twofold: first, to inform developers and contractors concerning the current PMIS landscape and ICT solutions, and the lack of support to HCC projects. Secondly, to impact software service providers and software vendors to bridge the gap between PMIS systems and HCC particularities.

Although numerous ICT PMIS products/services have been developed since the early 2000s, the industrial adoption is still at a low level. Compared to BIM, which has been elaborately standardized and widely promoted, PMIS remains as a vague managerial concept; though literature is numerous, the definition and scope discussed are incongruous.

5.2 Adoption barriers

Adoption barriers seem to be varied, from external structural dimensions (legal aspects, incentives and necessities), to internal people dimensions (resistance to change, lack of expertise and trained personnel), to a technological dimension (systems reliability, software availability, lack of technical infrastructure), up to tasks aspects (investment costs, return of investment). These are further summarized into: lack of a standards and interoperability, issues with usability, time required for adoption of new technologies, setup/running costs and unclear Return on Investment, additional knowledge and skills prerequisites, poor quality of information, legal and contractual uncertainty, social and cultural barriers; feasibility issues. This paper finds some earlier barriers, especially those linked to hardware limitations, are disappeared or diminished, many major barriers appear to non-technical, such as people's resistance to change.

6 CONCLUSIONS & FUTURE RESEARCH

Although the findings answered the objectives of this paper, it is acknowledged that this research could benefit with further quantitative investigations, exploring the industry practices worldwide, taking into account variables such as different types of projects, region, the role of organization, project size and procurement type.

Compared to manufacturing and information industries, PMIS in the AEC industry appears to be more like a vague managerial term; the understanding of its scope and functionality is incongruous. A reason for this as identified within this paper concerns the functions fragmentation in market solutions. It has also been recognized the lack of an internationally acknowledged standard for PMIS's scope. A collaborative effort at international level would not only provide more coherent industry solutions, but also would solve issues related to interoperability and communication, improving shared understanding among construction professionals while handling complex projects, like the HCC.

REFERENCES

Arnold, P. & Javernick-Will, A. 2013. Projectwide Access: Key to Effective Implementation of Construction Project Management Software Systems. *Journal of Construction Engineering & Management, 139(5), pp. 510–518.*

Axelsson, J., Froberg, J. & Eriksson, P. 2018. Towards a System-of-Systems for Improved Road Construction Efficiency Using Lean and Industry 4.0. *2018 13th Annual Conference on System of Systems Engineering (SoSE). 19–22 June 2018. IEEE. pp. 576–582*

Chen, G., Liu, Y., Chen, J. & Wu, J. 2012. Data-Driven 4D Visualization for Simulating Highway Construction Processes. *2012 ACM/IEEE/SCS 26th Workshop on Principles of Advanced and Distributed Simulation. 15-19 July 2012. IEEE. pp. 196–198*

Dainty, A., Moore, D. & Murray, M. 2007. *Communication in Construction: Theory and Practice.* Abingdon, Oxon: Routledge.

DeLone, W.H. & McLean, E.R. 2003. The DeLone and McLean Model of Information Systems Success: A Ten-Year Update. *Journal of Management Information Systems, 19(4), pp. 9–30.*

Eadie, R., Odeyinka, H., Browne, M., Mckeown, C. & Yohanis, M. 2014. Building Information Modelling Adoption: An Analysis of the Barriers to Implementation. *Journal of Engineering and Architecture, 2, pp. 77–101.*

Froese, T.M. 2010. The impact of emerging information technology on project management for construction. *Automation in Construction, 19(5), pp. 531–538*

JI, Y., Borrmann, A. & Wimmer, J. 2014. Coupling Microscopic Simulation and Macroscopic Optimization to Improve Earthwork Construction Processes. *EG-ICE 2011, European Group for Intelligent Computing in Engineering.*

Kenley, R., Harfield, T. & Behnam, A. 2016. BIM Interoperability Limitations: Australian and Malaysian Rail Projects. *MATEC Web of Conferences, 66, pp. 00102.*

Lee, D., Park, J. & Song, S. 2018. BIM-Based Construction Information Management Framework for Site Information Management. *Advances in Civil Engineering, 2018 Article ID: 5249548.* Hindawi.

Lee, S. & Yu, J., 2012. Success model of project management information system in construction. *Automation in Construction, 25, pp. 82*

Moghayedi, A. & Windapo, A. 2019. Key uncertainty events impacting on the completion time of highway construction projects. *Frontiers of Engineering Management, 6(2), pp. 275–298*

Nitithamyong, P. & Skibniewski, M.J. 2004. Web-based construction project management systems: how to make them successful? *Automation in Construction, 13(4), pp. 491–506*

Oesterreich, T.D. & Teuteberg, F. 2019. Behind the scenes: Understanding the socio-technical barriers to BIM adoption through the theoretical lens of information systems research. *Technological Forecasting and Social Change, 146, pp. 413–431*

Papadonikolaki, E., Leon, M. & Mahamadu, A. 2018. BIM solutions for construction lifecycle: A myth or a tangible future? *eWork and eBusiness in Architecture, Engineering and Construction: Proceedings of the 12th European Conference on Product and Process Modelling (ECPPM 2018). 09/03.* Copenhagen: CRC Press. pp. 321–328.

Perkinson, C.L., Bayraktar, M.E. & Ahmad, I. 2010. The use of computing technology in highway construction as a total jobsite management tool. *Automation in Construction, 19(7), pp. 884–897.*

Ruikar K. 2005. End-user perspectives on use of project extranets in construction organisations. *Engineering, Construction and Architectural Management, 12(3), pp. 222–235.*

Scott, D., Cheong, M. & Li, H. 2003. Web-based Construction Information Management System. *Construction Economics and Building, 3(1), pp. 2910; 43–52.*

Sears, S.K. 2015. *Construction project management: a practical guide to field construction management.* 4th ed. Hoboken, New Jersey: Wiley.

Tam, C., 1999. Use of the Internet to enhance construction communication: Total Information Transfer System. *International Journal of Project Management, 17(2), pp. 107–111*

Tezel, A., Koskela, L. & Aziz, Z. 2018. Current condition and future directions for lean construction in highways projects: A small and medium-sized enterprises (SMEs) perspective. *International Journal of Project Management, 36(2), pp. 267–286*

Walsh, I.D. et al. 2011. ICE Manual of Highway Design & Management – Section 4. Highway Design Principles and Practice: 31.1 Background. *ICE Manual of Highway Design and Management – 31.1 Background.* ICE Publishing.

Wilkinson, P. 2005. *Construction Collaboration Technologies: An Extranet Evolution.* 1st ed. Independence: Routledge.

Multi-stakeholder involvement in construction and challenges of BIM implementation

Z. Yazıcıoğlu
Istanbul Medipol University, Turkey

ABSTRACT: Project development is a complex process where many stakeholders work together. Employer and main contractor are the base stakeholders, whereas designer, engineer, sub-contractors, suppliers, supervisors and consultants are other stakeholders. BIM is a tool that should be considered by every stakeholder with the opportunities it offers. The main goal of this paper is to explore the problems associated with the adoption of BIM in multi-stakeholder projects. The paper is a conceptual study, summarizing the author's practical experience with design offices and construction firms working with BIM. In the transition period to BIM, three of the challenges are examined: the compatibility of supplier companies with BIM, the need for two-dimensional drawings, and contractual issues related to BIM. The paper reviews the literature on BIM usage and reviews the challenges in the transition stage to BIM. Finally, a number of suggestions for the future about accelerating BIM use is provided.

1 INTRODUCTION AND PURPOSE

Project development is a complex process in which many stakeholders work together. Stakeholders from many different fields, such as design teams, sub-contractors, suppliers, manufacturers and consultants, especially employers and main contractors work together. The complexity of the construction process and the involvement of many stakeholders lead to time and cost overruns. Failure to comply with the work schedule and inefficient use of resources in construction processes indicate that it is necessary to accelerate production and increase efficiency. The emergence of the Building Information Modeling (abbreviated as BIM) approach and computer software developed in this context are an important technological breakthrough in this era. The aim of BIM is create coordinated architectural, structural, mechanical and electrical projects. BIM is an approach that should be considered by every stakeholder with the opportunities it offers, such as minimizing construction errors, reducing construction time, and estimating the final construction cost.

The implementation of BIM is a process spreading to years. The main purpose of this article is to investigate the issues related to the adoption of BIM in multi-stakeholder projects. This is a conceptual study that summarizes the observations and practical experiences in companies working with BIM. There are various researches about BIM related implementation problems from the design phase of the project to the implementation phase in the field.

During the transition to BIM, three of these issues are discussed:

1. BIM compliance of supplier firms;
2. The need for two-dimensional drawings;
3. BIM related contractual issues.

A construction project using BIM is considered as a case study and the relationship between design-bid-build processes is examined by considering these problems. Finally, there are a number of suggestions for speeding up the use of BIM for the future.

2 WHAT IS BUILDING INFORMATION MODELING (BIM)?

BIM can be defined as a joint project production by creating a three-dimensional digital model of the building using a computer program, defining all the information required for the building's construction by working on the same model by all users, with all the administrative information related to the building's design and demolition.

According to the BIM definition of RIBA (Royal Institute of British Architects), BIM is the digital representation of the physical and functional features of a facility. It provides an environment where reliable and shareable information is created for the decisions to be taken during the life cycle from conceptual design to destruction.

According to The National Building Specification (NBS), which was part of RIBA until 2018 making accepted definitions about architecture and building, BIM is a way of working, the information management system that a team works with. It is a value formed as a result of the collaboration of people, processes and technology.

BIM also supports the Integrated Project Delivery (IPD) concept, a new project delivery approach to integrate people, systems, business structures and applications into a common process to reduce waste and optimize efficiency at all stages of the project life cycle (Glick & Guggemos 2009).

3 PROJECT PRODUCTION AND BIM AS A SAVER

Project production occurs in a process that is quite complex and has many stakeholders. It can be said that the project is unique due to the differentiation of the place where the building will be located. Since the production of the project has complex processes and a lot of time and effort is spent on it, accelerating productivity is required. One of the technological steps taken in this sense is the development of computer software for the Building Information Modeling approach, abbreviated as BIM. The use of BIM enables architectural, structural, mechanical, electrical and infrastructure projects to be drawn in a coordinated manner. It provides realistic environmental controls such as solar gain, energy sustainability, acoustic values and lighting requirements. It is a virtual environment that can be considered as a digital version of mock-up.

Two-dimensional drawings can be shared and coordinated between all disciplines, but it is observed that accurate results cannot be obtained in two-dimensional environments. When all disciplines share and overlap their two-dimensional projects, each discipline faces the risk of transferring missing information from its project to another project or overlooking revisions. For example, coordination problems, such as missing of transferring the columns from the static project to the architectural project, placing a water tank in the mechanical project in the areas where electrical equipment is placed in the electrical project, have been observed in various projects. These problems are tried to be overcome by coordination meetings. In addition, the problems experienced in the three-dimension are more likely to be missed and are noticed only in the construction phase in very large projects. An example of this has been observed in an airport terminal building, the construction work of which started in the Middle East in 2009. The fact that its project design was made in BIM environment, it was not necessary to prevent coordination problems. The determination of the height of the door without checking the location and height of the beam, the door hitting the beam in the open position could only be noticed during the construction phase. The process continued in the form of cessation, demolition, formation of new productions and new orders suitable for the situation encountered in construction. Additionally, resubmitting the projects after revisions and repetition of approval process are experienced. As a result, loss of time and money occurs due to coordination problems.

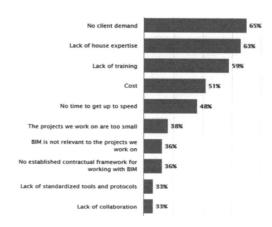

Figure 1. Factors affecting the use of BIM according to employees in the construction Sector in the united kingdom (Sönnichsen 2019).

The BIM approach should be taken into consideration by every stakeholder with the opportunities it offers, such as compliance with the work schedule and reducing the construction time, determining the cost from the beginning and complying with the determined cost, and minimizing the errors during the construction phase.

Although the history of the use of BIM is not very old, the studies of developed countries in this regard are pioneers for the whole world. Continental European countries such as England, Denmark and Norway, in addition, the United States have worked on making the project production in BIM environment in accordance with the legislation and made it mandatory under certain conditions. In recent years, the United Arab Emirates in the Middle East has made it mandatory to use BIM in buildings with certain conditions through legislative studies and legal regulations.

4 DIFFICULTIES IN TRANSITION TO BIM IMPLEMENTATION

BIM is a new technology for the entire construction industry. The proliferation of a new technology will take time to become practically understandable and usable by all stakeholders. A literature search was conducted on the difficulties experienced in the transition to BIM implementation. Information about this is given in Figure 1.

According to the Figure 1, the Employer's request regarding BIM is considered as the main restriction encountered in the use of BIM. The employer's failure to request the use of BIM does not encourage design teams and the contractor, primarily all other sub-teams. Due to the insufficient use of BIM in the construction industry and the current drawing methods being more practical, the transition to BIM cannot be achieved.

As a result of interviews and observations with construction companies, suppliers' failure to work in compliance with BIM, the need of two-dimensional drawings and contractual problems are identified as problems related to BIM.

4.1 BIM compliance of supplier firms

Materials and equipment used in the building are provided by the supplier firms. Depending on the nature of the business, the party whom the suppliers are responsible for is the employer, the main contractor or the design teams. The employer can guide the contractor to do the work on the selection of the supplier or the contractor can determine the supplier to buy the materials and equipment and submit it to the employer's approval. The material details and information of the supplier firm must meet the demands of published tender documents. Regardless of the situation, information of the materials provided by the supplier is used in application projects. If the details of the supplier are not used in the application project, problems occur on the field during the construction.

The fact that suppliers are not working in compliance with BIM is an issue that needs to be emphasized on behalf of the construction industry. The service provided by the suppliers has a key importance for the construction works. BIM, a system that allows different stakeholders to work on the same project without physically coming together, aims to increase efficiency in the construction industry. Construction work is a whole from design to application, from supply to maintenance. Therefore, encouraging suppliers to use BIM is as important as encouraging design teams and contractors.

In a situation where design teams or contractors use BIM but suppliers do not use BIM, it creates an additional workload for the party working on the BIM model. Design teams or contractors have to draw the details that the supplier did not draw in the BIM environment. In order to use BIM effectively, all kinds of materials, products and systems used in the project must be drawn by the manufacturers in the BIM environment and all suppliers working on the project must be able to use BIM and work on the cloud.

Another problem arises when suppliers do not work in accordance with BIM, is misunderstanding of the projects. It has been observed that if the design team works in the BIM environment, design is understandable only when it is examined in three-dimension environment. The fact that the first design is made in the BIM environment affects the level of intelligibility of the project. The designed structure, building element or building material can only be understood by analyzing the three-dimensional model. It may not be possible to express a structure or building element designed in three-dimensional environment with two-dimensional drawings. Difficulties in this way of handling a project produced in a BIM environment in two dimensions reveals the necessity of all stakeholders involved in the project to work in harmony with BIM.

As it is seen in Figure 1, the biggest factor in the failure of BIM to become widespread is that the employer does not demand, causing the contractor and subcontractor to move away from BIM due to the lack of obligation. If suppliers have to use BIM by the request of the employer or main contractor, the use of BIM can achieve its goal.

4.2 Need for two dimensional drawing

Two-dimensional drawing is the oldest known method of expression. With Computer Aided Design (CAD), drawing projects using computer software is an alternative method to manual drawing, but standard drawing techniques and expressions continued to be used in the same way. The tool used with CAD is computer software instead of pencil and paper. This form of expression can be understood by all stake-holders who do not have technical knowledge such as employer, administration, approval authorities, but are associated with construction works. It is observed that the digital files of the projects drawn in two dimensions are easily used by the stakeholders. It is not possible for a project produced in BIM environment to be viewed digitally by non-technical stakeholders. However, the model created with the BIM approach must be delivered digitally and con-trolled in this file. As a result, in order to approve the project by the local administrations, public authority that will approve must be equipped with BIM knowledge. If this condition is not met, the project produced in BIM environment should be expressed in two dimensions.

Considering the relationship between the employer and the contractor, it can be said that the use of BIM has a structure that does not match the Traditional Project Delivery system (Figure 2). One of the observed situations is that, the design team prepares a project in the BIM environment for the employer and the employer publishes Request for Proposal containing Project Drawings which consist of details and specifications. The contractor is determined with the Design-Bid-Build method. The problem is that the design team submits the project to the employer without the required BIM level. The project, drawn in an unimproved BIM environment for the application, makes it obligatory for the contractor to deal with all the BIM drawings from the beginning after winning the tender. After the tender is finished, the design team's work for design mostly ends, and if the contractor negotiates for the design control after the tender, they may have duties such as answering the contractor's questions and checking and approving the drawings drawn by the contractor. Therefore, the project production of the design teams in the BIM environment ends with the tender. The design teams and the contractor tend to form the model for their own purposes, this approach lies behind the unsuccessful BIM initiatives (Dogancay 2014).

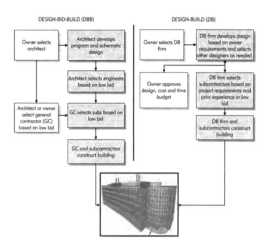

Figure 2. Design-Bid-Build and design-build work flow diagram (Eastman et al. 2008).

In the projects carried out with the Traditional Pro-ject Delivery System, the second issue observed re-garding BIM is that the design team publishes the project in two dimensions when the project will be published as a tender document, even if it is required to do the project in the BIM environment. The BIM file is not shared at the tender stage for various reasons such as the fact that the project file produced in the BIM environment is not wanted to be shared with all the tender participants, the design team can continue to work on the BIM model until the tender is concluded, and the BIM file cannot be examined by all participants due to the lack of information about the use of BIM.

Regardless of the Project Delivery System, the employer orders the preparation of shop drawings and the contractor is obliged to do this work. Shop drawings are more descriptive and informative than architectural drawings. Dimensions, details, material properties and junction details are indicated in the drawing for application (siliconec.co.nz 2020). Shop drawings are directly prepared for construction on site, so that the materials produced by the manufacturing companies for the construction can be produced correctly and sent to the field. The expectations of the Shop Drawing, which the employer set as a contract condition, actually questions the employer's trust in BIM projects. It can be said that if BIM can be used effectively, there is no need for shop drawings. Apart from architects, engineers and managers, if the suppliers involved in construction projects could use BIM efficiently, the shop drawing item can be re-moved from the contracts and may not be an additional obligation for the contractor. All drawings be-come a natural result of the project produced in BIM environment. As discussed in Section 4.1, since suppliers use two-dimensional environments, two-dimensional drawings will definitely be needed and no matter how effectively the design team and the contractor use BIM.

The conversion of the model created in the BIM environment into a two-dimensional drawing has been explained above. The detail quality of the drawing obtained by the two-dimensional conversion method is directly related to the detailing quality of the BIM model. The BIM model, which has not been sufficiently detailed in the BIM environment, has not been handled in an integrated manner by all design teams (architects and engineering disciplines), causes line confusion when converted into a two-dimensional drawing environment, and is observed as the overlapping of multiple layers and many lines in two-dimensional drawing.

In addition to all these issues, to reflect all the details of the products of all disciplines and supplier companies on the BIM model is not always possible technologically. Even if all the disciplines and all the suppliers involved in the project are competent to use BIM effectively, it is not always possible to transfer all the details to the project that is working on a single model in the cloud. Therefore, the details have to be produced and published in two-dimensional environments. If BIM model is not handled in sufficient detail level and integrated by all disciplines, the details are needed to drown in two-dimensional drawing software. The detail produced in low level BIM environment consists of complex lines, uncertainties about the materials, and incompatible representation. It indicates that there is a lack of detailed information in the BIM model. A collaborative BIM process requires all processes to interact with each other. A large amount of computer memory and processing power are needed to achieve this operation (Ireland 2009).

4.3 *Contractual issues regarding BIM*

One of the contractual issues under discussion with BIM implementations is about the ownership and management of the model. The fact that when a common accessible model is created by the stakeholders and available to them, it is also possible to block the access to the BIM model by any stakeholder involved in the BIM model. This situation questions the management of the BIM model. There has been a case brought to the British Technology and Construction Court about the prevention of access to the BIM model. Trant Engineering was commissioned by the Ministry of Defense for the Mid-Atlantic Energy Project worthing $72 million. Mott MacDonald has been appointed as design consultant and BIM coordinator. Mott McDonald's duties include controlling access to the shared data platform. Due to a payment dispute, Mott MacDonald suspended his service and blocked Trant's access to the shared data platform. The court found that blocking Trant's access was wrong, because a prevention of access to data platform will exceed the limit of damage that Mott MacDonald could afford (Wilkinson & Heywood 2018).

This case and similar cases make us ask questions about the management and the accessibility of the model. There should be an authority in projects where BIM technology is used, and to conclude that this information should be included in the contract. Azhar et al. (2012) commented on the access and ownership of the BIM model as follows:

The process-related risks include legal, contractual and organizational risks. The first risk is the lack of determination of ownership of the BIM data and the need to protect it through copyright laws and other legal channels. For example, if the owner is paying for the design, then the owner may feel entitled to own it, but if team members are providing proprietary information for use on the project, their proprietary information needs to be protected as well.
Thus, there is no simple answer to the question of data ownership; it requires a unique response for every project depending on the participants' needs.

It is seen that the terms of the contract have been re-evaluated in accordance with the use of BIM and studies have been carried out in this field. BIM Protocol is one of these studies. The BIM Protocol is an additional legal agreement included in construction contracts. The protocol covers issues such as licensing models and categorizing the permitted information for different stakeholders. Instead of specifying a specific use for each model, "Permitted Purpose" is defined to organize the use of the models (Croneri 2016).

BIM Manager is needed to regulate legal responsibilities and relationships among stakeholders. Model Manager can be determined as responsible person for the parties' model access, security, communication, archiving and BIM systems coordination. Legal issues regarding BIM Manager are about; the assigning authority, how to replace him/her, and who will cover the costs associated with this role. The general view on this issue is that the BIM Manager should be responsible to the employer, but it is also discussed that in Design-Build contracts he/she should report to the main contractor.

It is recommended that the parties make an appropriate choice according to the procurement methods. The authority and responsibilities of the Model Manager and the roles of other parties and stakeholders in the BIM process should be specified in the protocol. The relationship between the Model Manager and an Architect/Project Manager and other team members should be defined, since the Model Manager can be changed during the project. To avoid conflict, it is recommended that the Party, which the Model Manager reports to, has the responsibility about providing binding instructions on BIM related issues (Udom 2012).

According to Thompson et al. (2007), another contractual issue is about controlling the data entry into the model and the responsibility about the mistakes. Updating and correcting BIM model data is a great risk. Complex compensation claims made by BIM users, limited warranties, and designer's disclaimer offers are important negotiation issues that need to be resolved before using BIM technology. In addition, more time needs to be spent for BIM data entry, which is a new cost in the design and project management process. Although these new costs can be compensated by efficiency and time gains, it should be noted that there is still an additional cost. Therefore, before using BIM technology, usage risks should be defined and BIM implementation cost should be determined.

5 EVALUATION OF BIM IMPLEMENTATION IN A MULTI-STAKEHOLDER CONSTRUCTION WORK

The problems experienced in BIM implementation in international construction works will be conveyed based on the observations made in the companies performing international construction works.

5.1 *Observed case*

belongs to an internationally tendered office building. The employer in country A works with a design team in country C, for a project that is to be built in country B. The contractor firm with international business experience from country D has won the tender, with contract type Design-Bid-Build project delivery system.

5.2 *Documents of the project*

The design teams prepare the project drawings and documents for the employer in the BIM environment. Correspondingly two-dimensional drawings converted from the BIM model are published as the tender document. After the tender is concluded, all the drawings of the project are shared with the winner contractor, including the BIM model. It has been determined that 1/5, 1/10 scale details of the project are not transferred to the BIM environment.

5.3 *Project control*

The responsibility of the design teams in this work is to check the final application projects, BIM file, and shop drawings that the contractor obliged to draw. The Employer Representative defined as 'Engineer' in the contract is selected from the design team.

5.4 *Project approval*

The contractor completes the project drawings in BIM environment and gets approval. In addition, the contractor completes the shop drawings with contractual conditions and gets approval. Construction works begin after the approvals are completed.

5.5 *Subcontractors and manufacturers*

Contractor in country D works with subcontractors and manufacturers in accordance with the conditions specified in the contract and technical specifications. The location of these companies may vary according to the contract. Contractor works with suppliers and manufacturers from different countries for each material. For example, for stone works, agreements are made with the quarries in countries E and F. The anchorage system of the stones to be applied with the mechanical assembly system is carried out by the company in the country G.

5.6 *Shop drawings*

Generally, the details of the manufacturer companies are used in detail solution in the application projects and no shop drawings are made in addition to the application project. However, in the case examined, shop drawings are an additional project package that is submitted to the employer's approval for the start of construction works. Unlike the application projects, it does not only include adapting the manufacturer's details to the details, but the solutions of the manufacturer are shown in the entire project. Some shop drawing titles determined by the employer as a contract condition in the case study are: Reinforced concrete structural system, stone cladding works, plasterboard wall works, ceramic works, wood works and so on. In addition to the shop drawing package for each subject, the properties of the material, the quantities and cost must be provided and submitted by the main contractor and approved by the employer.

5.7 *Working in the BIM environment*

The level of BIM file prepared by design teams is quite insufficient. It is observed that the contractor company should continue to work on the BIM file after winning the tender. A process that contradicts the concept of design-bid-build project delivery system is occurred. Under construction, the contractor must check and ensure the coordination of all drawings, make the BIM file work, produce and have the shop drawings approved. While doing all these of that, it is not authorized to make any design changes.

5.8 *Drawing responsibilities*

The subcontractor and manufacturer firms which the contractor works with are also effective in fulfilling the responsibilities of the contractor company in drawing projects. The manufacturing company can make the shop drawings, and the main contractor can establish a team for this work or have it made by an external team. Some of the shop drawings listed above are done by the contractor and some of them are made by external teams. The shop drawings are drawn exactly in accordance with the explanations given in the technical specifications and with the de-tailed solutions of the manufacturer. Manufacturing companies and teams have prepared shop drawings without accordance with BIM. Instead shop drawings were prepared using two-dimensional drawings obtained from the BIM file. As mentioned earlier, due to the insufficient level of the BIM file, the two-dimensional drawings produced from it, has also poor quality.

5.9 *General evaluation*

The aim of the BIM approach is to obtain a model that each discipline works together on, from the design stage to the construction stage. The employer's design teams work in the BIM environment, but it is observed that a common model is not synchronized by all disciplines. The BIM model, which is submitted as a tender file, needs to be handled again by the contractor firm. There are problems and uncertainties that need to be resolved. The contractor is not alone in producing drawings, he works with external teams and manufacturing companies. Manufacturers and external teams that cannot work with BIM have to work with two-dimensional drawings derived from the BIM file and the quality of these drawings is very low. Additional time is spent by the contractor to improve these drawings and make them workable.

6 RESULTS AND DISCUSSION

In this article, the relationship between BIM and the design, tender, procurement and construction processes was examined in a case study where these difficulties were experienced, considering practical applications and the problems encountered.

Considering the project delivery systems, the parties vary that is responsible for the employer. In Design-Build system the contractor is the only party that is responsible for the employer. If the contractor firm stipulates the use of BIM in contracts with other teams it will work with, all stakeholders will have to work in the BIM environment and the BIM system can achieve its goal. When the compatibility of project delivery systems with BIM is questioned, it is concluded that the Integrated Project Delivery System (IPD) may be more compatible for BIM. IPD is a project delivery system where employers, design teams, managers, contractors and suppliers are all partners of the project. It is suitable for synchronized work on a common model. In the project delivery system, where design and construction parties such as Design-Bid-Build are clearly separated, all teams use the model for their own purposes and cause the BIM initiative to fail.

It can be said that the use of BIM by stakeholders will become widespread with the demands of employers. If local governments require the use of BIM in public projects, it can be a starting point for some countries. Depending on this, if the BIM model contains all targeted data, all drawings required for production become a natural output of the project. Time can be saved without having to set up a separate team for the preparation of shop drawings and new approval processes.

Making different levels of models in BIM environment requires different levels of computer hard-ware and software. It may be sufficient to use BIM software for the BIM model that will meet the minimum expectation, but as the expectations from the BIM application increase, the software and hardware features of the computers should be increased. For example, if there is an expectation such as clash detection with BIM application, an additional software should be used. This additional software will in-crease the storage capacity requirements of the computer and affect the processing speed (Sahil 2016). Similarly, adapting the details to the BIM model and working on the common model will require enhancement of the computer's hardware such as processor power, memory and graphics card above the standard.

REFERENCES

Azhar, S. 2012. Building Information Modeling (BIM): Now and Beyond. *Australasian Journal of Construction Economics and Building,* 12 (4) 15–28.

Croneri, 2016. Integrating BIM into Contracts. Retrieved from https://app.croneri.co.uk/feature-articles/integrating-bim-contracts.

Doğgançay, H. 2014. BIM Modelinin Sahibi Kimdir? Telif Hakkı Kime Aittir? http://halukdogancay.blogspot.com/2014/12/bim-modelinin-sahibi-kimdir-telif-hakk.html

Eastman, C., Teicholz, P., Sacks, R. & Liston, K. 2008. *BIM Handbook: A Guide to Building Information Modeling for Owners, Managers, Designers, Engineers, and Contractors.* New Jersey: John Wiley & Sons.

Glick, S. & Guggemos, A.A. 2009. IPD and BIM: Benefits and Opportunities for Regulatory Agencies. *Proc., 45th Associated Schools of Construction National Conference, Gainesville, FL.*

Ireland, B. 2009. Barriers to BIM. https://www.ecmweb.com/content/barriers-bim

Sahil, A. Q. 2016. Adoption of Building Information Modeling in Developing Countries: A Phenomenological Perspective.

Siliconec.co.nz, 2020. Architecture Shop Drawing. Retrieved from http://www.siliconec.co.nz/architectural-engineering/architecture-shop-drawing.html

Sönnichsen, N. 2019. Main barriers to implementation of building information modelling (BIM) according to construction industry professionals in the United Kingdom (UK) in 2019. https://www.statista.com/statistics/1019362/construction-industry-main-barriers-to-bim-usage-uk/

Thompson, D.B. & Miner, R.G. 2007. Building Information Modeling – BIM: Contractual Risks are Changing with Technology, http://www.aepronet.org/ge/no35.html

Udom, K. 2012. BIM: Mapping Out the Legal Issues https://www.thenbs.com/knowledge/bim-mapping-out-the-legal-issues.

Wilkinson, D. & Heywood, M. 2018. Which project party is the legal owner of BIM data? www.constructionweekonline.com/article-48038-which-project-party-is-the-legal-owner-of-bim-data.

Impacts of BIM implementation on construction management processes in Turkey

Y. Beslioglu & İ. Akyaz
İstanbul, Turkey

ABSTRACT: In the last decade, construction industry has witnessed emergence of a new tool that claimed itself to drive efficiencies, reduce costs and add long-term value to the development and management of built assets. Building Information Management (BIM) is an architectural, engineering, planning, control and facility management tool that contains all design and quantitative data necessary throughout the life cycle of a project. As a project planning and control tool, BIM occupies the field of traditional construction management systems by replacing planning tools based on two dimensional displaying methods. This research has been conducted to find out effects of BIM implementation on construction management processes by focusing on time, budget and contract management performances, from Turkish construction industry participants' perspective and it contributes to literature by depicting the current state of experiences and observations of Turkish construction industry practitioners on BIM implementation.

1 INTRODUCTION

BIM is a reflection in construction industry of a technological renaissance that shapes the world in early 21st century. It represents the transformation of construction design, planning, construction, control and management processes using technological tools and digital transformation.

In literature, there are various definitions on Building Information Modelling (BIM) that attempt to define it in its exact boundaries though its ever-evolving character prevents a universally accepted definition (RICS 2015). USA National BIM Standards Committee defines BIM as "a digital representation of physical and functional characteristics of a facility creating a shared knowledge resource for information about it and forming a reliable basis for decisions during its life cycle, from earliest conception to demolition" (NIBS 20) A practical definition focusing on outcomes defines BIM "as a set of systems that create digital three dimensional models, intended to provide more accessible and versatile design and cost data to identify efficiencies and improvements throughout the life of a built facility" (Kings College Centre of Construction Law 2016)

Information data lies at the heart of BIM system. Improving processes to get a more efficient construction system through a dynamic and data focused approach is what BIM aims to bring construction industry. Efficiency gains may be obtained through various ways in different stages of construction; from saving time through automated functions to saving money through more cost effective decision systems stemming from data availability. Construction industry participants' view reflects these gains in efficiency; Hardin cites a survey conducted to search the factors influencing the use of BIM in US in 2007 and 2012 and indicates the change in factors from owner requirements to interoperability, functionality and clearly defined deliverables (Hardin 2015). Eastman emphasises the effect of BIM usage on construction budget and stresses that construction projects that use BIM system has a high probability of lowering construction budget than a project that use traditional methods (Eastman et al. 2011) A research conducted to observe benefits of BIM using an analytical approach shows that the most successful outcome of BIM usage has been "budget control" while "time management" has been the second most successful outcome (Bryde et al. 2013) Positive effects on number of change orders and changes in project programmes have also been noted as effects of increased efficiency of BIM utilization (Barlish & Sullivan 2012). The aforementioned findings reveal that construction management process is coming forward as a major impact zone of BIM usage.

Besides operations such as estimating, logistics and safety, scheduling and project control operations contain the core of construction management activities. Time management, budget management and contract management activities constitute three main pillars of these scheduling and project control systems. An evaluation of time, budget and contract management performance parameters shall thus substantiate effects of a construction management system. From that perspective, BIM effects on construction management process can be identified by evaluation of these criteria. Besides its other benefits, BIM Handbook lists benefits of BIM in construction process as synchronizing design and construction planning, providing ability to implement lean construction and synchronizing procurement phases which all lead to time and budget management improvements. In terms of time management BIM implementation has the advantage of integration of contractor and subcontractor schedules as well as "model simulation" studies or "sequencing animations." Improved visual models, effective team communication means and inter-linked project planning tools provide BIM systems high skills for improving time management performances (Hardin 2015).

Budget control is another important aspect of construction management. Taking literature on BIM into account, Eastman defines BIM as an enabler of constructability in terms of reduced costs and project duration (Eastman 2008) Jrade and Lessard indicate that considerable cost reductions can be obtained using BIM by early detection of design clashes (Jrade & Lessard 2015) Hardin describes contributions of BIM in terms of budget control in two phases, first is using BIM derived data as a source of information from a design model as a base for budget planning and second as a dynamic model from field that uses daily updated data for cash and budget control (Hardin 2015).

As stated above, third main pillar of construction management system is contract management. As an inter-linked project management tool that enables users to distribute files and documents, saves these documents on an inter-linked platform BIM constitutes a convenient contract management system that enables parties to establish contractual expectations. In terms of disputes arising from delays, reducing design changes by early detection of design clashes, improvements in project control and scheduling systems and organized data sharing plans incorporated into contracts by BIM Project Execution Plans, using BIM is expected to reduce number of disputes (El Hawary & Nassar 2015) On the other hand BIM usage affects legal liability and contractual duties since "the aspirations of BIM users to improve collaboration and efficiency need to be framed within appropriate and insurable legal commitments" (Kings College Centre of Construction Law 2016).

Starting from early 2000s, Turkish construction industry has rapidly grown relying on private construction investments and international construction activities mainly in Middle East and Asia. Locally, construction industry benefited from a construction based growth model that relied on lowering interest rates and on international field, Turkish construction firms enjoyed benefits of high oil prices that resulted in increased construction spending by Middle East and Asian oil producer countries. Since 2013, decreasing local demand in residential construction projects and reduced construction spending in Middle East and Asia due to low oil prices resulted in a shrink in construction investments which has been largely compensated with large scale infrastructure spending often under PPP models (TCA 2020). In accordance with this construction activity Turkish construction companies are amongst largest in their sector globally. ENR 2018 records show that 44 Turkish contractors are listed in Top International Contractors list (ENR 2019). As global players of construction industry and contractors of large scale construction projects Turkish contractors adopt contemporaneous construction management methods although in a delayed period. Turkish contractors often used BIM as a design tool but did not have the habit of systematically using BIM in their projects (Akkoyunlu 2015; Aladag et al. 2016). Similar to researches cited by Hardin above, Turkish construction industry participants indicate customer demand as the main factor in adopting BIM. However, "need for collaboration, coordination, communication and interoperability between stakeholders" is indicated as second most significant factor to adopt BIM, which reveals the fact that Turkish construction industry is in a similar path with its Western counterparts (Aladag et. al. 2016). Another issue with the Turkish case is that BIM is implemented to construction projects lacking the directive and supportive guides that are available in US or UK, whether that lack of guidance caused any contractual problems, is yet a research topic to be identified.

In this context, this research has been conducted to find out effects of BIM implementation on construction management processes by focusing on time, budget and contract management performances, from Turkish construction industry participants' perspective. Although it is known from literature that BIM is being used by construction industry, its practical implications have not yet been identified. Past research shows that it has rather been used as a design tool though its implications from a construction management perspective have been lacking (Akkoyunlu 2015) One of the other problems that the research focuses on is the construction disputes. Obtaining data on effects of BIM usage on construction disputes and industry participants' view has been one of the main objectives of this research, as relationship between construction disputes and BIM usage is an area yet to be identified.

As data collection methodology, a survey has been conducted among Turkish contractors that carry out large scale local and international construction projects, architectural design companies and consultants. Survey covers twenty-two questions about effects of BIM on project budget and time management, BIM and contracts, construction disputes and comparison of construction management experiences using BIM with construction management experiences using traditional management tools. At the end of the survey, fifty answers were obtained from participants and results have been thoroughly discussed. Data obtained from interviewees is not claimed to be entirely representative of the industry as a whole though since it is conducted among participants that take part in large scale local and international projects, survey results are deemed to be indicative of industry practice.

2 RESEARCH

First 5 questions of the questionnaire were prepared towards understanding participant profile. 58 percent of the participants had more than 20 years of industry experience while 30 percent had between 11–20 years of experience and 12 percent had less than 10 years of experience. 40 percent of participants are holding BIM Director, BIM Engineer, BIM Coordinator positions in their companies, 34 percent stated that they are holding positions in technical office and the other 26 percent are planning engineers, member of board of directors or engage in other director positions.

As regards participant company profiles, 48 percent are contractors, 18 percent are project management and consultancy companies, 26 percent are design offices and others are investment companies.

58 percent of participant companies are in business for more than 20 years, 30 percent are in business more than 10 years and 12 percent are in business for less than 10 years.

90 percent of participants stated that they use BIM systematically in their projects while 10 percent do not systematically use BIM in their projects. Of those who use BIM, 100 percent find BIM useful in their projects. 94 percent of participants support the idea that using BIM helps gaining time in coordination between parties while 6 percent do not support that using BIM helps gaining time in coordinating with other parties.

92 percent of participants support the idea that BIM reduces the rate of errors in coordinating other parties. Thus, a very high majority of participants support that using BIM helps better coordination among parties.

Another question was about quantity measures, 88 percent of participants think that quantities are measured more accurately using BIM than traditional methods.

Participants were asked whether 3D models were integrated with schedules when they are using BIM and 66 percent of participants stressed that they use 3D models integrated with schedules. 34 percent, on the other hand, did not.

On the cost front, participants were asked to evaluate whether project costs were reduced using BIM, taking into account of their recent projects in which BIM is used. 24 percent of participants stated that their project costs were reduced 5 percent or less 20 percent stated that their project costs were reduced 10 percent or less, 18 percent stated that their project costs were reduced 15 percent or less, 12 percent stated that they had cost reductions about 20 percent or less and 26 percent stated that there have not been any cost reductions in project costs.

Participants were asked about their opinions on whether using BIM should be held obligatory in public projects in Turkey. 76 percent supported the idea

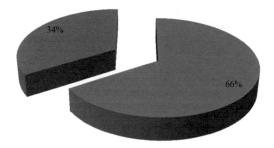

Figure 1. 66 percent of participants use schedule integrated 3D models.

that BIM should be obligatory in public projects, 14 percent stated that using BIM should not be an obligation, 10 percent stated that the decision should be left to contractors and public authority but it should not be an obligation. Some participants who supported the idea that using BIM should be obligatory in public projects stated that ownership requirements would encourage construction industry to use BIM and would elevate BIM consciousness among industry participants. It is also stated that unlike international contractors, local contractors who work with public bodies do not have sufficient experience in using BIM, whereas if they are held obliged to work with BIM they would get closer to international industry standards and from a public benefit perspective, reducing costs and time using BIM and preventing design clashes, public works will have an upper standard than they currently have. Participants who expressed that using BIM should be optional stated that using BIM could be helpful though lack of qualified personnel with BIM knowledge and lack of control mechanisms on public bodies' side during facility management phase would be major drawbacks.

Another question was about the apprehension of BIM concept among industry participants. 88 percent of participants stressed that they think the other parties do not have same apprehension of BIM during project execution. This means that architects, consultants and contractors have different approaches and expectations from BIM.

Participants were asked to express the benefits obtained from BIM on a 1 to 10 scale where 1 expressed the minimum and 10 expressed the maximum benefits. Participants noted relevant benefits as such;

– Almost 80 percent of participants noted 8 and over for comprehensibility of planning using BIM.
– 92 percent of participants noted 8 and over for usefulness of BIM in detecting design clashes.
– 60 percent of participants noted BIM as 8 or more for constructability analysis of projects.
– Only 32 percent of participants noted 8 or more for benefits of BIM for Health&Safety issues.
– 64 percent of participants noted 8 or more for benefits obtained from BIM in matters relating to budget and time issues.

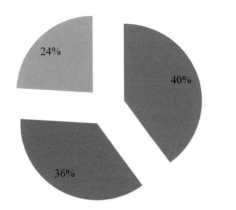

Figure 2. Distribution of answers on BIM usage in disputes

– 86 percent of participants noted over 8 or more for benefits regarding the ability of BIM to provide possibility to work on same platform for different parties.

Participants were also asked about the difficulties for adapting BIM in projects. According to answers, participants find following factors in order as the most important difficulties in adapting BIM;

– Lack of qualified personnel for using BIM
– Lack of educational resources for personnel
– Opinion of other parties that 2D is sufficient
– Time and human resources required to create and update model.
– Requirement to change organizational structure of companies
– High degree of detail needed for a healthy analysis in design phase.

Participants were also asked questions about contract management using BIM. 69% of participants stated that their contracts have express clauses on BIM related matters. In the projects where BIM is used 40 percent of participants stated that there have been disputes in their project. Of those 40 percent who experienced disputes, 40 percent stated that BIM helped them in solving dispute, 36 percent stated that BIM did not help solve the dispute and 24 percent stated that they do not have an opinion.

Participants were asked to compare projects in which BIM is used with projects in which BIM is not used in regard to number of change orders and claims. 42 percent of participants stated that there has been a decrease in number of change orders and claims, 58 percent, on the other hand, stated that there has not been any meaningful change in change order and claim numbers. Also, in regard to delays, 56 percent stated that they feel more confident to reach schedule targets when using BIM. In response to a question about copyrights, 50 percent of participants stated that their contract has clauses on ownership of copyrights of designs conducted using BIM. 80 percent stated that simple contractual requirements are not sufficient in themselves to adopt BIM to projects.

3 CONCLUSION

Survey results revealed the fact that many industry participants are aware of developments regarding BIM and are actively using BIM in their projects. Results also reveal that BIM contributes performance targets with respect to budget and time however contractual issues and contract management using BIM, still needs to be improved.

Most of the participants stated that they find BIM useful and of the benefits they obtained, "Detecting design clashes" is on the top. Participants found BIM benefits least on health and safety issues. However, as 3D models and planning/control stages are more integrated, health and safety benefits can increase.

It is seen that lack of experienced personnel and educational resources are seen as the most important obstacles in adopting BIM. These obstacles may be seen as temporary obstacles as universities are increasingly including BIM related topics in their courses and growing use of BIM results in increase in numbers of experienced personnel.

In regard to contractual matters, it is interesting to see that 31% of participants use BIM in projects where there is no clause in contract about BIM. Half of participants stated that they have no copyright clauses in their contract for design conducted using BIM. These answers point out the fact that there is still much way to go in relation to legal and contractual matters about BIM in Turkey. Large number of participants stating that claim numbers have not declined with BIM indicates that contract management using BIM still needs to be improved. Besides design and planning stages, using BIM in project control and documentation stages would improve the practice and lessen disputes. Participants mostly had the idea that public bodies should pioneer the process and using BIM in public projects should be obligatory. Given the obstacles in adopting BIM to all public projects, it would be helpful to adopt BIM to public projects in a stepped approach; starting from large scale projects to smaller ones, an adaptation strategy should rapidly be determined. It is suggested by researchers that under the leadership of public authorities and in collaboration with industry participants, a roadmap to an obligatory BIM utilization in Turkish public projects should be prepared, closely supervised and applied with care.

REFERENCES

Akkoyunlu T. 2015. *Kentsel Donusum 5 icin BIM Uygulama Planı Onerisi*. PhD Thesis. Istanbul Technical University Graduate School of Science Engineering and Technology.

Aladag H., Demirdögen G., Isık Z. 2016. Building Information Modeling (BIM) Use in Turkish Construction Industry. *Procedia Engineering* 161 (2016) 174–179.

Barlish, K., Sullivan, K. 2012. How to Measure the Benefits of BIM, A Case Study Approach *Automation in Construction* 24 149–159.

Bryde D. Broquetas M. Volm J.M 2013. "The project benefits of Building Information Modeling (BIM)" *International*

Journal of Project Management, Volume 31, Issue 7: 971–980.

Centre of Construction Law and Dispute Resolution, King's College 2016. *Enabling BIM Through Procurement and Contracts* London; available at https://www.kcl.ac.uk/law/research/centres/construction/assets/bim-research-report-1-jul-2016.pdf.

Eastman, C. M., Teicholz, P., Sacks, R., & Liston, K. 2011. *BIM Handbook: A Guide to Building Information Modeling for Owners, Managers, Architects, Engineers, Contractors and Fabricators*. NJ, USA; John Wiley and Sons.

El Hawary, A., & A. Nassar. 2015. "The Effect of Building Information Modelling (BIM) on Construction Claims." Int. J. Sci. Technol. Res. 5 (12): 25–33.

Engineering News Record 2019. *ENR's 2019 Top 250 International Contractors* 2019. Available at; https://www.enr.com/toplists/2019-Top-250-International-Contractors-1.

International BIM Working Group of the Royal Institution of Chartered Surveyors 2015. *International BIM Implementation Guide 1st ed*. London; Royal Institution of Chartered Surveyors (RICS).

Turkish Contractors Association 2020. *Insaat Sektoru Analizi*. available at; https://www.tmb.org.tr/arastirma_yayinlar/tmb_bulten_nisan2020.pdf.

US National BIM Standards Committee (NBIMS) 2014. *National BIM Standard:Version 2* Washington; National Institute of Building Sciences (NIBS).

Building information modeling warnings towards a deadline

L.V. Damhus & P.N. Gade
Architectural Technology and Construction Management, Aalborg, Denmark

R. Qian
Building Services, WSP, Auckland, New Zealand

ABSTRACT: In large projects, it is common to see BIM warnings accumulate into the thousands. These warnings can be design and reworks related errors. The objectives of this research are to test the hypotheses that most warnings occur up towards a deadline and identify whether the number of warnings increases according to model size. An empirical data collection was used on 33,106 warnings over 32 weeks, then analyzed with descriptive statistics to test the hypothetico – deductive method. The results found that towards a deadline there is an average increase of 9% in warnings from four weeks before the deadline. Although when the model size becomes a part of the equation there is a 4% decrease in warnings per megabyte file size in the same time period. This study finds that more warnings happen close to deadlines, not because of more mistakes, but because of an increase in the work being done.

1 INTRODUCTION

In recent years, the architecture, engineering and construction industry has adapted and utilized a wide range of technology to its benefit (Ghaffarianhoseini et al. 2017). However, this has also caused an ineffective utilization of the technology (Lopez et al. 2010) due to an unhealthy overdependence on computer-aided automation. This can lead to errors being committed whereby design consultancy firms overlook pragmatic considerations (Lopez et al. 2010) and ignore warning messages issued by the BIM software. In large complex projects, it is common to see the BIM warnings accumulate into the thousands (Lee et al. 2015). These warnings are the BIM software's ability to identify design-related errors and reworks by allowing for automatic detection of errors related to model elements (Lee et al. 2012; Love et al. 2011). BIM software issues real-time warnings messages to inform users of potential problems that can harm the key components of design information, such as the integrity of the model, the design intent, and the reliability of documentation (Davis 2011). BIM software allows the user to dismiss the warning messages when they are issued. If dismissed, the warnings are stored by the BIM software until the user revisits them at their convenience. When users wish to make corrective actions, the software allows for quick retrieval of previously dismissed warnings (Lee et al. 2015).

BIM software requires a high level of computing and excessive warnings in the thousands are known to significantly reduce the speed of model processing and increase file size (IMAGINIT 2013; Lee et al. 2015).

Therefore, users are required to diligently manage the accumulating warnings in order to prevent this useful feature from causing inefficiency during the design and modelling processes (Lee et al. 2015).

This leads to the question of why the warnings messages are not solved as they appear, when there are significant disadvantages to warnings accumulating. With reference to the legal profession, it has been found that patent attorneys' work quality is systematically lower for patent fillings close to deadlines (Balasubramanian et al. 2016).

Similar research on a systematic review on time pressure in software engineering (Kuutila et al. 2020) compares the findings to the Yerkes-Dodson law. This law states that performance increases with physiological or mental arousal (stress) but only up to a point. When the stress level becomes too high, performance decreases (Figure 1) (Green et al. 1908). In the systematic review, it is shown that quality decreases in most of the compressed scheduled projects, despite their effort increasing (Kuutila et al. 2020). These findings may also be applicable in areas of the architecture, engineering and construction industry that use BIM software.

2 RESEARCH OBJECTIVES AND METHODS

This study has two objectives: (1) to test a hypothesis that most BIM warnings occur up towards a deadline (the Yerkes-Dodson law); (2) to test a hypothesis that the number of BIM warnings increase in accordance with the BIM models file size.

In order to achieve the intended objectives:

1. Collect the deadlines from the timeline.
2. Identify deadlines that will fit the criteria.
3. Recreate the older versions of the weekly model reports to collect the BIM warnings.
4. Test the hypothesis that warnings occur up towards a deadline.
5. Collect the file sizes from the weekly model reports.
6. Test the hypothesis that the number of BIM warnings increases according to the BIM model file size.

To understand the context of BIM warnings the explanation from (Lee et al. 2015) is used to differentiate warning messages from warnings. In the present study, "warning messages" refer to the messages that are universally pre-defined by a BIM provider (e.g., Autodesk for Revit) while "warnings" refer to the individual warnings that are specific to each model. This means that when there is an issue with a model element, BIM issues a generic warning message that is specifically linked to that element. Accordingly, warnings are sorted by each warning message in the warning dialog box or warning reports, and each message likely pertains to multiple warnings for different model elements of BIM (Figure 2).

In the research, a scientific perspective has been used with the aim of creating an empirical data collection based on records and measurements within a project. Therefore, the quantitative method will be used to provide an opportunity to generalize the results, to read contexts and trends to achieve the research objectives. In order to test the hypotheses, the hypothetico deductive method is used. The hypothetico-deductive method cannot definitively confirm or deny whether the hypotheses hold, but merely make them probable, because there are likely to be multiple causes (Nola & Sankey 2014).

The data is collected through a weekly model report that is made on the project each week. To collect the data for previous deadlines, the weekly model report is recreated through a version history. To determine how much data to obtain, three samples of 10 weeks of data was used to find the optimal number of weeks of data to collect. The findings of the samples showed that there was often more than one deadline in 10 weeks. As such, data spanning four weeks was subsequently collected as there was only one deadline in this timeframe. Furthermore, service models were kept out as BIM software has a way to calculate in those areas, and if not used it will cause warnings. At times, the weekly model report has been inconsistent, which in some cases has led to three-week intervals between weekly reports. In this case, other deadlines were chosen. Data for eight different deadlines was collected. No further data could be collected due to time restraints. However, eight samples of data were deemed sufficient to discuss the research objectives. To keep the project anonymous, deadlines names are removed and naming conventions for the data samples were made to keep track of the data sets (see Figure 3).

"Project" shows which building model the data was collected in. "Model" shows where the data was collected, being an architectural (ARC) and structural (STR) model. Towards a deadline a "level of development" (LOD) for each element, that had to be followed. In the collected data the LODs are 200 (2), a mix of 200 and 300 (23) and 300 (3) (BIMForum 2019). The "Unique number" is made to separate some of the data sets through the stages or deadlines, because there could be more deadlines within the same LODs and models.

Descriptive statistics are used to analyze the collected data. Descriptive statistics is a method that

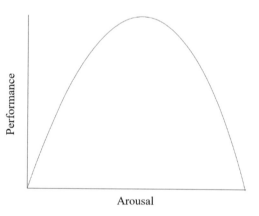

Figure 1. The York-Dodson law model.

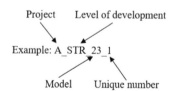

Figure 3. Naming convention.

Warning Messages	Model Elements Having Problems
Highlighted elements are joined but do not intersect.	Walls : Basic Wall : (22.1) Indervægge (lejlighedskel) (215mm) : id 593159 Walls : Basic Wall : (21.1) Sandwichelement 500 mm : id 917027
Highlighted elements are joined but do not intersect.	Walls : Basic Wall : (22.1) Indervægge (lejlighedskel) (215mm) : id 593289 Walls : Basic Wall : (21.1) Boksmodul - 578mm : id 917044
Conditions for wall embedding are no longer satisfied.	Walls : Basic Wall : (21.1) Sandwichelement 500 mm : id 287820 Walls : Curtain Wall : Curtain Wall : id 724961
Conditions for wall embedding are no longer satisfied.	Walls : Basic Wall : (21.1) Sandwichelement 500 mm : id 588581 Walls : Curtain Wall : Curtain Wall : id 724578

Figure 2. BIM warning report showing warning messages and corresponding model elements.

describes and summarizes the collected data. It only describes the collected data and does not generalize based on this data. Descriptive statistics results are presented in tabular form or by other graphical presentation (Shi & McLarty 2009). The graphical presentation will be presented with percentages to show all the statistics in the same graph. The highest number of warnings will be presented as 100% and the rest of the data with their accordingly percentages, in their respective weeks. To make an average of the graphs, the percentages each week will be used, so the percentages can be used as a common factor and therefore give a clear view of the statistics.

Table 1. Project information.

Project	Level*	Area*	Views*
A	−13 m	26340 m²	2582
B	0 m	13467 m²	1988
C	−33 m	31897 m²	2490

Note: *Level is the number of meters the train platform is below ground. Area* is the amount of square meters there are in the architectural model and Views* is the amount of views there is in the architectural model. This information was taken at the latest report at the time.

3 DESCRIPTION OF CASE PROJECT

For the purpose of this research, a total of 33106 warnings were collected from BIM models over time in a project valued at over a billion USD dollars and made up of several different buildings. The buildings are all metro stations. Each building is modelled in Revit by Autodesk within the last decade. Further information on each building is shown in Table 1.

4 RESULTS

The data is collected from an architectural and a structural model. The data is collected four weeks before each of the deadlines and then put into the same line chart to see if there is a correlation. The line charts are made from the data collection and presented with descriptive statistics to describe and summarize the data.

The data in Figure 4 and Figure 5 is highly consistent. Regarding when the data sets reached the highest number of warnings over the period of four weeks, only three out of the eight models were below the 90% mark (Figure 4), meaning that five out of the eight models had less than 10% change in warnings.

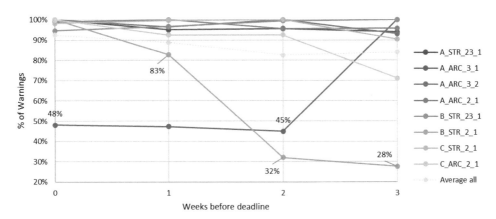

Figure 4. Percent warnings over four weeks 20–100%.

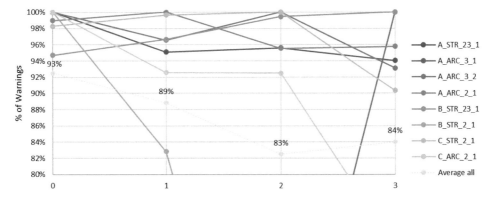

Figure 5. Percent warnings over four weeks 80–100%.

Table 2.

Data set	A_STR_23_1			A_ARC_3_1			A_ARC_3_2			A_ARC_2_1		
	H	A	L	H	A	L	H	A	L	H	A	L
Warnings	1393	1340	1310	2216	1331	994	960	935	894	2156	2104	2060
File size	477	430	412	518	490	460	432	420	392	380	350	308
W/MB		3.1			2.7			2.2			6.0	

Data set	B_STR_23_1			B_ARC_2_1			C_STR_2_1			C_ARC_2_1		
	H	A	L	H	A	L	H	A	L	H	A	L
Warnings	359	351	340	1509	915	417	289	282	262	1145	1020	814
File size	57	54	53	77	59	44	244	219	196	672	672	672
W/MB		6.5			15.5			1.3			1.5	

Note: H* The highest amount, A* The average amount and L* The lowest amount *W/MB Warning per megabyte.

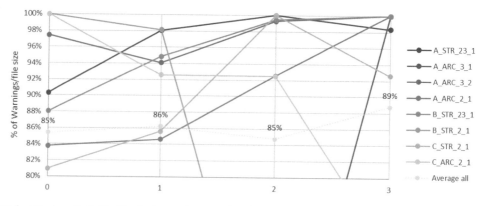

Figure 6. Warnings divided by file size in megabyte over 4 weeks 40–100%.

When testing research objective (1), the data shows that only A_ARC_3_1 and B_STR_23_1 had a lower number of warnings in the week where the deadline was due compared to three-weeks before the deadline (Figure 4) else the rest reached a higher number of warnings. Looking closer at the data in Figure 5, the average of all that data shows that three weeks before the deadline the number of warnings are at 84%; two weeks before the deadline, they decrease to 83%; one week before the deadline, they increase to 89%; and finally, in the week of the deadline, they increase to 93%. This is a total increase of 9% (Figure 5). The two data sets A_ARC_3_1 and B_STR_2_1 are outliers as the number of warnings increase and decrease more than 22% (Figure 4). A_ARC_3_1 had 2216 warnings three weeks before the deadline, which then decreases by 52% to when the deadline was due (Table 2, Figure 4). B_STR_2_1 had 417 warnings three weeks before the deadline, which then increased by 72% to when the deadline was due (Table 2, Figure 4).

The data in Figure 6 and Figure 7 is highly consistent. Regarding when the data sets reached the highest number of warnings divided by the file size in megabyte over the period of four weeks, only three of the eight models were below the 80% mark (Figure 6), meaning that five of the eight models had less than 20% change in the number of warnings per megabyte over the four weeks. When testing research objective (2) the data shows that only two of the data sets fit the objective. B_STR_2_1 and C_ARC_2_1 are the only sets where warnings increase more than file size (Figures'6, 7). Looking closer at the data in Figure 7, the average of all that data shows that three weeks before the deadline the number of warnings per megabyte are at 89%; two weeks before the deadline, they decrease to 85%; one week before the deadline, they increase to 86%; and finally, in the week of the deadline, they decrease to 85%. This is a total decrease of 4% (Figure 7).

The 2 data sets A_ARC_3_1 and B_STR_2_1 are again outliers as the number of warnings per megabyte increases and decreases more than 22% (Figure 6). A_ARC_3_1 had 4,8 warnings per megabyte three weeks before the deadline, which then decreases by 57% to 2,1 warning per megabyte when the deadline is due (Table 1, Figure 6). B_STR_2_1 had 9,5 warnings

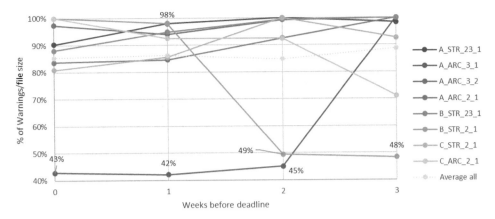

Figure 7. Warnings divided by file size in megabyte over 4 weeks 80–100%.

per megabyte three weeks before the deadline, which then increases by 52% to 19,6 warnings per megabyte to when the deadline is due (Table 2, Figure 6).

5 DISCUSSION

The results show that there are more warnings happening up towards a deadline. Figure 5 shows an increase of 9% over the four weeks up till the deadline, as the hypothesis in research objective (1) suggested. It cannot be definitively confirmed that the hypothesis hold; merely it is probable, as the hypothetico-deductive method describes (Nola & Sankey 2014).

The opposite of research objective (1) was found in research objective (2). Figure 7 shows there is a reduction of 4% in warnings per megabyte file size. This almost complies with the hypothesis in research objective (2), where the hypothesis was to test whether warnings would increase in accordance with the model size. This hypothesis can be confirmed if a 4% tolerance is allowed for. However, it cannot be definitively confirmed, as the hypothetico-deductive method describes (Nola & Sankey 2014).

Now the hypothesis of research objective (2) casts doubt on research objective (1). This opens up to more interpretations, as looking at the number of warnings relating to file size, the modellers are in fact receiving or ignoring 4% less warnings than three weeks before the deadline. As the York-Dodson law shows, up to a certain point people perform better (Figure 1). This is also seen in Figure 7 and Table 2, where the file size of the model gets bigger within the same timeframe, perhaps because the modellers are close to the deadline and are getting aroused (stressed) to the optimum amount. Therefore, their performance increases so they produce (perform) more in the model in terms of megabytes, but also warnings. However, looking at it in comparison to the number of warnings relating to file size, the modellers are in fact receiving 4% less warnings. These results could also be the impact of good management, especially when B_STR_2_1 and C_ARC_2_1 are excluded from consideration, as they contradict the general trend by having more warnings per file size towards the deadline. This may indicate bad management.

Looking at the York-Dodson law, the modellers do not become over aroused when coming up to a deadline. Rather, they reach the peak of their performance. This is similar to what was found in the systematic review on time pressure in software engineering (Kuutila et al. 2020), where the effort increased during compressed schedules. In this research, more modelling is being done (more effort is taking place), whilst the quality simultaneously increases (less warnings) which did not happen in Kuutila, Mäntylä, Farooq and Claes research reviews.

In this project it is important to mention that the deadlines used had two-three more stages to go before the models were going to be built. Within each stage, a BIM manager would go through the model report to see if the model fit the criteria of the BIM execution plan. In the BIM execution plan, it states that only 500 warnings are allowed in the weekly report as well as when it goes through a stage gate. Only two of the eight models had less than 500 warnings (Table 2). This could be because there are more stages and there is an agreement between the modeller and the BIM manager, where the modeller was allowed more warnings for a certain deadline. Another factor could be that some warnings are being ignored because they are not relevant. In research by Lee et al. (2015) on BIM warnings, it was found that around 28–39% (depending on the design stage) of the warnings had little to no impact (Lee et al. 2015).

The data collected was not intendent to be generalized as descriptive statistics were used to display the data sets. It is also important to mention all particulars of the project to create the most transparency and best comparability to similar projects, which was not done. The intent was to create awareness of the subject (which may only be applicable to this project), and have a way to track it and make suggestions to similar further research, where recommendation hopefully will be found.

The two data samples that increased in the number of warnings per megabyte, B_STR_2_1 and C_ARC_2_1, (Figure 6) are in the first design stage. In the first design stage, warnings are more likely to occur because more design options are being explored (Lee et al. 2015). However, C_ARC_2_1 doesn't change in file size, only in the number of warnings (Table 2). In order to rectify this, in further research it would be more suitable to look at model elements instead of file size. Another reason to support this recommendation is that using file size as a measurement may not give accurate results as the model template would be included in this measurement with no allowance for the template size. In particular, the template size can vary depending on the model discipline (e.g. architectural versus structural) and/or the template already containing warnings. Therefore, in further research it is preferable to use model elements as the measurement and categorize warnings in terms of importance, as was done in the research undertaken by Lee et al. (2015).

A practical benefit to be obtained from this research is that it can assist with tracking warnings and model size to see correlations between them, up towards a deadline. As suggested above, tracking model elements instead of file size gets more precise results. Additionally, if warnings are put into categories of importance, it makes it possible to see which warnings are relevant. These results would make it possible to go in depth and analyze if some of the warnings are "solved" on the sheets but not in the model. This may need to be fixed later in the project if the warnings need to be reduced to a certain number before the model can be delivered.

This research also connects the York Dodson Law with BIM modelling towards a deadline. BIM warnings per megabyte (or model elements) can be used as an indicator of performance and time can be an indicator of arousal. For example, the closer a person gets to a deadline, the more aroused they could be. However, this is hard to accurately measure and is subject to other variables, such as personal stresses or team environment. In essence, if the performance indicator under the York Dodson Law is low up towards a deadline, this could indicate lack of or over-arousal of the modeller.

6 CONCLUSION

The research presented within this paper was to test the hypotheses that most warnings occur up towards a deadline, which was the first research objective. This was confirmed, with an increase of 9% in warnings over the four weeks towards the deadline. The second research objective was to test whether the number of warnings correlated with the model size. It was found that the warnings per megabyte decreased by 4% over the four weeks of data. The confirmation of research objective one (that more warnings happen close to deadlines) is not necessarily because more warnings are being received or ignored than usual. It is simply because more modelling is being produced as the hypothesis in research objective two found. Practical benefits the research can provide is the guidance on tracking warnings towards a deadline and the connection with the York Dodson law that can be used as a performance and stress indicator.

REFERENCES

Balasubramanian, N., Lee, J. J., & Sivadasan, J. (2016). Deadlines, Work Flows, Task Sorting, and Work Quality. *Academy of Management Proceedings, 2016*, 14346.

BIMForum. (2019). *Level of Development (LOD) Specification Part 1 & Commentary For Buidling Information Models and Data*. (April).

Davis, P. (2011). *Introducing Autodesk Revit Architecture 2012*. John Wiley & Sons.

Ghaffarianhoseini, A., Tookey, J., Ghaffarianhoseini, A., Naismith, N., Azhar, S., Efimova, O., Raahemifar, K. (2017). Building Information Modelling (BIM) uptake: Clear benefits, understanding its implementation, risks and challenges. *Renewable and Sustainable Energy Reviews, 75*(September), 1046–1053.

Green, C. D., Yerkes, R. M., Dodson, J. D. (1908). Classics in the History of Psychology An internet resource developed by the relation of strength of stimulus to rapidity of habit-formation. *Journal of Comparative Neurology and Psychology, 18*, 459–482.

IMAGINIT. (2013). *Revit Modeling Best Practices*. (November), 1–4.

Kuutila, M., Mäntylä, M., Farooq, U., Claes, M. (2020). Time pressure in software engineering: A systematic review. *Information and Software Technology, 121*, 106257.

Lee, G., Park, H. K., Won, J. (2012). D 3 City project – Economic impact of BIM-assisted design validation. *Automation in Construction, 22*, 577–586.

Lee, H. W., Oh, H., Kim, Y., Choi, K. (2015). Quantitative analysis of warnings in building information modeling (BIM). *Automation in Construction, 51*(C), 23–31.

Lopez, R., Love, P. E. D., Davis, P., Edwards, D. j. (2010). Design Error Classification, Causation, and Prevention.pdf. *Journal of Performance of Constructed Facilities*, 10.

Love, P. E. D., Edwards, D. J., Han, S., Goh, Y. M. (2011). Design error reduction: Toward the effective utilization of building information modeling. *Research in Engineering Design, 22*(3), 173–187.

Nola, R., Sankey, H. (2014). 7. The hypothetico-deductive method. In *Theories of Scientific Method?: An Introduction*. Routledge.

Shi, R., McLarty, J. W. (2009). Descriptive statistics. *Annals of Allergy, Asthma and Immunology, 103*(4 SUPPL.).

The role of trust in the adoption of BIM systems

P.N. Gade
Architectural Technology and Construction Management, University College of Northern Denmark, Aalborg SV, Denmark

J. de Godoy
Department of Energy Technology, Aalborg University, Aalborg Ø, Denmark

K. Otrel-Cass
Institute of Education Research and Teacher Education, Universität Graz, Graz, Austria

ABSTRACT: The adoption of Building Information Modelling Systems (BIM-systems) has not been as straightforward as hoped, and questions must be raised why this is the case. A vital characteristic of the adoption is the "outsourcing" of manually processed information to digital systems and how users perceive data processing. For successful adoption, users need to trust that the digital systems process information according to their expectations. If expectations are not met, users lose their trust like the way when trust is lost between people. This article explores the relationship between trust in BIM systems more widely and specifically what this means for the current adoption. Trust in technology is explored qualitatively through interviews with various BIM-system users. We put forward characteristics to do with trust and digital technology that impact the adoption of BIM-systems.

1 INTRODUCTION

1.1 Trust, a critical component in technology use

It has been widely reported how digital technologies have penetrated many industrial sectors successfully, but in the construction industry, the incorporation of digital tools that can support the management process has not yet been fully explored (Fulford & Standing 2014). This is highly relevant since digital technologies may assist in creating smarter, more sustainable, energy-efficient buildings.

Trust is critical to creating a network of organizations with shared values that are prepared to invest in IT development and standardization of information and processes (Fulford & Standing 2014). Without a more rigorous understanding of how technologies like Building Information Modelling (BIM) can provide support, people may not fully trust the technology and resort to their old ways. Thus, research that addresses trust relations between humans and digital technologies in those environments is needed.

In general, having trust in technology has been described as a critical component for the construction industry's management process (Chalker & Loosemore 2016; Fulford & Standing 2014; Gad & Shane 2014). However, the topic of trust specific to digital technologies is underexplored. To understand the implications of what trust entails, it is essential to start with an understanding that trust in digital technologies comes with ethical implications. For instance, when algorithms are used (Mittelstadt et al. 2016; Tsamados et al. 2020), the technology used may make decisions that come with unintended consequences for individuals, groups, or society.

Unfamiliarity with the local environment of the projects, lack of complete specification, lack of uniformity of materials and work, and general unpredictability create uncertainties (Dubois & Gadde 2002), which reliable digital systems may be able to reduce while building trust.

However, this requires collaborative efforts from different people, teams, and organizations to work together to fulfill a shared ambition to create and run a building. For instance, it has been argued that trust among project participants, clients, contractors, and designers could lead to higher performance because it would lead to improved communication, knowledge exchange, and dispute handling (Gad & Shane 2014).

The kind of trust needed in such projects operates very similar to the way how interpersonal trust works, where the trustee delivers what has been promised to the trustor, so events that could be a source of low trust, distrust, or even a loss of trust and could make collaboration unfeasible, compromising the execution or quality of the project do not occur (Hoffman et al. 2013).

What typically challenges the collaborative work in the building industry is that people have to work with incomplete information from the beginning to the end (Tsamados et al. 2020). For instance, clients do not start with a complete specification of what building they want; designers do not fully detail the design they conceptualize, and builders do not fully describe how they anticipate creating the building. This means that the building process is typically messy and chaotic (Bertelsen 2003).

Incomplete information in the building's design may require radical changes during the process, and this means that clients need to trust their designers (Gero 1998). By introducing digital technologies to the decision-making process in building construction management, trust needs to be placed into digital systems (e.g., into the information they provide or how these systems process information).

Another challenge in construction projects is that contracts cannot completely specify the division of labor and possible risks associated with services (Chalker & Loosemore 2016). Insufficiently detailed blueprints, specifications, and/or contracts in creating buildings mean that it is not possible to provide complete sets of information a priori, increasing uncertainty. This includes the client's uncertainty and needs, which materials should be used, and how the site may constrain assembly (Zuboff 2019, p. 334).

Due to these complexities in the construction industry, digital automation technologies like Building Information Modelling (BIM) systems aim to reduce uncertainty by raising information sharing transparency between the different stakeholders. Thus, a portion of the necessary trust is deferred to the technology. However, the question remains how trustworthy digital technology can be?

1.2 *What is trust?*

Trust describes the human belief in reciprocity from people, organizations, or artifacts (Hardin 2002; McKnight & Chervany Norman 2001; Walker et al. 2010) and implies some degree of risk-taking, vulnerability or dependency because a person has to rely on someone or something (Ostrom & Walker 2003).

However, trust is experienced differently since it is subjective, and this makes it complex to understand and measure, especially if we want to distinguish whether the trust has been lost or is just low (Marozzi 2015). If events that generate distrust are not remediated, trust may be lost altogether. Thus, trust is not necessarily a "long-standing trait" but rather a phenomenon that may be generated and even manipulated in the short-term (Cappelen & Dahlberg 2018).

When it comes to digital technologies, not only should technology be "reliable safe and secure, the user must believe it is reliable" (Castelfranchi & Falcone 2010). Humans trust technological artifacts when they believe that the "designer has done a good job to avoid bad outcomes" (Coeckelbergh 2011). Similarly, designers trust that "users will make good use of it," thus using the "artifact for morally justified purposes" (Coeckelbergh 2011).

Since trust is primarily an attitude and depends on the trustee's motivation, it would seem that to trust technology; it is necessary to know who created it, e.g., the algorithms, the technological artifact. However, this is not necessarily so since humans can trust technologies under certain conditions. Although it is a highly sophisticated process, understanding such conditions can allow people and digital technologies to complement each other (Matzner 2019).

1.3 *The trustworthiness of digital technologies*

When individuals, groups, or communities start using new technologies, people usually make changes to already existing practices (Winner 2014). People tend to accept changes if there is a benefit associated with it. If we want to assess the benefit of new technologies in order to understand how trustworthy these technologies are, we need to take a closer look at their impact on people's lives and practices (Sandler 2014).

However, such an assessment is difficult when several uncertainties exist (for example, how decisions were made), making it challenging to identify possible risks reliably. The advantage of risk assessment is that it allows users to look for ways that minimize or reduce potential harm, which in turn can raise the trust placed in each technology. Unlike the trust that develops between people, trust in digital or automated technology involves factors that concern their specific functional limitations such as reliability, validity, utility, robustness, and false-alarm rate (Hoffman et al. 2013).

However, understanding why the new technology is used and deemed trustworthy is a lot more complicated than evaluating its functionality. Once people start using technology for specific purposes, the technology also develops idea-power in particular when it reshapes how people interact with each other and the world and "'redefine' what work means in that setting" (Winner 2014).

In our assessment of BIM systems' trustworthiness, we want to go beyond the technology's functionality and focus on extrinsic factors that shape different people's assessment of what digital technologies may afford to them, including protection, responsibility and accountability, fairness, autonomy, and transparency.

Protection – concerns the possible negative impacts of a given technology on human welfare and the nonhuman environment (Sandler 2014). The protection of users, non-users, and the environment from any harm include the protection of the data collected and processed as part of the interaction with the system, as well as the physical protection of users' and non-users, including that their rights and interests are not violated (to the extent possible). For the system to be able to ensure protection, the system itself should be reliable, robust, and predictable.

Responsibility and accountability – The adoption of new digital technology is often argued to strengthen individuals' liberties and often supports the arguments for the introduction of new technologies. However, one of the major concerns associated with digital technologies is how technology shapes our ways of acting and who should be made responsible when things go wrong (Jonas 2014). Jonas argues that ignorance of digital systems' complexity is not necessarily an acceptable response and that responsibility and accountability can only be assumed through knowledge and foresight.

Fairness – Benefits, and burdens are not necessarily equally distributed when digital technologies are utilized, and while they empower some people, they can achieve the opposite for others. In times of increased algorithmic sophistication, data justice can become a real problem when massive quantities of social and behavioral data are collected that may support bias (Hoffmann 2019).

This can be the case, for example, when training data that has been used produces outputs that disadvantage specific individuals or groups. The increasing use of big data that ought to support fair policy decision-making, when official statistics are collected that theoretically, should support through random measures (Gorur 2020). However, with low-quality data or out of contexts (Mittelstadt et al. 2016), the impact and level of fairness are not yet known.

Autonomy – Since technology gives (some) people power, it shapes our experiences (Sandler 2014). Autonomy implies that users can make decisions and that decisions are not being made against their will. Autonomy also implies that the system respects other parties'/persons' rights and freedoms.

Transparency – Transparency is one of the critical characteristics that achieve trust in digital technologies. To have transparent software, information must be accessible and comprehensible to guide user's decision-making processes. However, information about the functionality of algorithms is often not accessible (Mittelstadt et al. 2016), which may be due to software companies' competitive nature (Mittelstadt et al. 2016). Even when the information is accessible, it is often hard to comprehend for non-expert users.

Next, we will detail how we used the above five concerns to assess how different BIM community members described their trust in this technology.

2 METHODOLOGY

In this paper, we intend to problematize technologization (Teräs et al. 2020); that is, the idea that technology is neutral and there to resolve societal problems. We adopt a critical stance to reflect on people's choices when they are using BIM technologies to understand how this is related to their levels of trust in this technology. We evaluated the following research question: How do the algorithmic structures of BIM systems challenge trust relationships between users and digital technologies?

We interviewed six Danish engineering and architecture professionals who had a minimum of five years' work experience with BIM systems. The interviews were semi-structured and were conducted following Barriball and White's (1994) recommendations to ensure validity and reliability.

The interview questions were developed from the theoretical underpinnings presented in the introduction regarding trust. Each interview lasted around two hours and was audio recorded. The audio recording was transcribed, and then the transcribed interviews were coded and categorized based on the theoretical trust categories, protection, responsibility and accountability, fairness, autonomy, and transparency.

3 RESULTS

The five practitioners' interviews resulted in around 10 hours of material that were categorized according to the theoretical factors presented in section 1.3. The results of the interview are presented below.

Protection

The interviewees mainly focused on managing project risks, which included ensuring proper use of systems through understanding the system's limitations. In general, the interviewees did not consider BIM systems to be a source of data insecurity. Instead, one person expressed the new possibilities that came with implementing new systems to increase surveillance and control over their employees. The interviewee said that when they developed their systems to provide automation, they monitored the users.

"We can go in and see who uses it, what functions they use, and how many times they use it. It enables us to bang them on the head, now that we've made this system... Why don't you (their employees) use it when I can see that you spend oceans of time manually dealing with the information... It is a reasonably hard line." Carl.

This constitutes an example of a company that develops their systems, which also utilize the possibility to conduct employee surveillance in new ways not possible before.

Responsibility and accountability

The interviewees explained that the use of digital systems required extensive collaborative efforts in planning the responsibility of managing information. What muddled the responsibilities was that the information that was shared among project participants would be understood differently. When the information is passed around between the different stakeholders, it was difficult over time to identify who had the responsibility of, e.g., taking care of errors, even with IT-contracts in place, which should manage such responsibilities.

"We are often second in line when dealing with information," Mads.

Trust in the information that was shared in this process was based on who had responsibility for ensuring that the information was reliable at the time. An interviewee explained, however, that the reliability of the information that had been received at various stages of the building process was often a reason for disagreements. There would often be diverging opinions on the reliability of the information that was generated.

Ideas would often be vague or be interpreted as subjective, putting in question what degree of reliable information was needed to be trustworthy at a given point in time in the design process. An interviewee explained that some of the distrust was generated because of a lack of tracking over who had the responsibility of information reliability. If tracing the responsibilities were too opaque, design ideas that were generated would often be rejected.

Another aspect of responsibility was how much control users should have over the BIM systems. A general strategy when companies developed their own systems was, according to Mads, to restrict user's autonomy. If users were able to change everything, it would be difficult to identify where misuse of the systems would occur.

Fairness

The interviewees did not problematize fairness in relation to the BIM systems in general. However, a couple of interviewees noted issues related to how the systems allowed for user inputs that would skew fair competition. An interviewee explained that, e.g., using BIM systems that allow for exempting specific cost assessments of the projects and not highlighting exemptions in the documentation and result created an unequal basis on how the costs were assessed. So, when a system sets up rules to assess, e.g., economic aspects and would be too unspecific, the competition would often "optimize" how they would feed the system information to show the lowest price to get the contract.

Transparency

The most significant aspect that the respondents highlighted that was shaping their levels of trust in digital technologies is concerning transparency. A consensus was that in order to have trust in digital systems, it was important to understand how the system processes information. The interviewees expressed that they would arrange a "trust assessment" of the systems conducted by select experts.

"We have professionals in the organization to validate the code part and ensure that we have the output as we want it is set up the correct way." Carl

The interviewees indicated that while systems provided transparency (for example, allowing users to see the code the systems were programmed in), it was not necessarily possible that the users would be able to use this insight and understand how the system related to, e.g., local building codes or company standards. Process transparency is communicated using, e.g., a visual programming language that sets new requirements to the users. An interviewee gave an example of difficulties they had experienced applying the visual programming system named Dynamo.

"Dynamo is still a specialist tool that, for many, is very complicated." Geir

The approach to how to attain an understanding of the processes varied. In some cases where processes were black-boxed but highly critical, selected experts in the companies scrutinized the systems as much as possible. This is typically done by "playing around" with various inputs to foresee possible outputs of the system.

"We were not able to see the calculations behind the system...by trial and error we found that with area inputs above 100 square meters it (the system) multiplied the amount of domestic water by two" Lasse

On the other hand, when they obtained a sufficient understanding of how the system processes the information, they trusted the system. However, not understanding the processing due to the system's opaqueness can potentially cause problems.

"You could easily over-interpret the results that come out of it... Do not hang your hat on something that is a spit out of a program if it does not make sense with your professionalism. It's something we've been discussing, but it's not something we've experienced has been a problem" Carl

One interviewee argued for digitizing less. He highlighted the danger of processing information that people would not understand and that this could be a source of risk in the project.

"I did not think there is more that should be automated, maybe a little less, because it's as if you're driving a little over your head because you do not participate in it" Peter.

As a solution, he called for better feedback of the systems to ensure that users' understanding of the processes was kept high. Multiple interviewees explained that they often missed information about how the systems calculated, e.g., a design's sustainability score. In many cases, they were only informed about a score, but not how far the score was either increased or decreased with adjustments to the building design.

"What you get out of the results is not the exact truth; it is a simulation. You must be able to interpret this with your professionalism." Carl

It was also noticed that trust would be related between the companies developing a system and the system itself. However, an interviewee noted that they also take responsibility for the processing. While an interviewee explained that this was a way to obtain trust, it was not always enough for them to obtain enough trust in using the systems.

Autonomy

The interviewees generally agreed that they needed to have autonomy because they knew that the systems needed to be adjusted to their business and project context because many aspects of automation were categorized as "soft" and therefore required interpretation.

The interviewees argued that to use the systems successfully, they needed to adapt it to their environment. This meant sometimes changing specific parameters or even a larger part of the systems information processing. In general, the interviewees combined buying mature systems and developing their own. Interviewees from larger companies tended to develop their systems with the following argument:

"We spend much time developing our own code because each company requires different interpretations of processes" Mads

The ability to adjust and interpret their processes had a high priority. Nevertheless, if systems were externally developed, the processes were under strict quality assurance, and especially with system updates, they required to scrutinize the processes to adjust them to their context.

"When we get updates from bought systems, I need to go and adjust the processes, which often is very problematic for us." Mads

The quotes from Mads are related to the companies' need to assess the systems and their updates but based on the assessments; they need to re-adapt the processes to fit the company's and project's environment. A concrete example given was the safety margins on how the systems calculated, e.g., piped mentions.

4 DISCUSSION

In the preceding sections, we suggested that an analysis of trust may provide a useful framework for studying diverse practices with digital technologies in the building and construction industry. To unpack trust as an attitude, we focused on autonomy, fairness, transparency, responsibility and accountability, and autonomy. We analyzed experts' perspectives following interviews where they reflected on their practices with digital technologies (BIM systems) and their perspectives on whether such tools can advance the sector's cooperation and development.

We focus on trust because experts in the construction environment must deal with several uncertainties they have to manage. The interviewees expressed different approaches to deal with uncertainties and regarded the ability to scrutinize the systems' digital processing in different ways. Assessing the level of trustworthiness of digital technologies is central to understanding how the system acts.

Sharing information is expected from the users of the BIM system. Thus the willingness to do so is crucial even to adopt the BIM system (Guo et al. 2019). However, users may be concerned with disclosing sensitive information, the transparency over resources, and the cost and impact on the contract of the services (Guo et al. 2019).

Being able to maintain autonomy was reported as a critical factor. The user's ability to adapt the systems to suit their context maintains some degree of autonomy but requires that experts must identify selected algorithms they trust. In practice, incorporating contextual and external factors linked to cognitive actions to algorithms is limiting because humans are embodied in the project and cannot be reduced as information processing systems, as machines are (Matzner 2019). Thus, such powerless interactions can be a source of conflict.

If users' autonomy is restricted, it will limit if and how they apply and use those systems in their practices. Also, as highlighted by Davis and Harty (2012), the idea of "helping" the users enacts "proper" processing the developers idealized. Should we view the experts we talked to as "irrational" users who engage with a system capable of processing information "rationally," or is this a "transhumanist" dilemma where "faulty" humans should be ultimately replaced by technology (Matzner 2019)?

Similarly to the example, Matzner (2019) gave on the smart CCTV camera, thinking of such appliances as a much more objective device is only partly right. Ethical implications of those algorithms that are mediating social processes can range from epistemic concerns (like inconclusive or misguided evidence) to normative (unfair outcomes, e.g., low quality of data) ones (Mittelstadt et al. 2016).

The interviewees noted a potential issue of fairness involved in the digital systems based on the user's ability to conduct too much user input. That meant that the basis upon how the output of the systems was evaluated was potentially unequal. An interviewee expressed that he had experienced malicious use of these functionalities to "low ball" offers in a tendering process using a BIM-based system for cost estimation.

However, to the contrary, this could also indicate that the appendices on how the malevolent user did not provide enough documentation that would highlight such behavior. On the contrary, Reychav et al. (2017) argued that systems like BIM improve fairness because it calls for user involvement and further details the design at earlier stages. With this increased involvement and a more detailed design, it can reflect fairness towards different stakeholders.

Little has been explored regarding the protection and BIM systems. In the few articles treating this topic, Davis and Harty (2012) warned that the digital systems' persuasiveness would potentially increase the demand for imposing control over the users of systems. This persuasiveness could lead to violating individual protection of data ownership. Since 2012, this warning has not been responded to with much attention in the digital technology domain in the construction industry, but as seen in the interviews, our respondents' worry can somewhat be confirmed to continue to exist.

The problem relies on the fact that management control in the construction industry is a conventional method used to improve the system by aligning strategies between organizations and employees (Nieminen & Lehtonen 2008). Nieminen and Lehtonen (2008) explore the types of cybernetic control used, mentioning four ways: over the process, measuring the process, comparing measurements to performance standards, intervention over the process.

According to one of our interviewees (Carl), the interest in controlling their employees is to identify struggles the users are having with the platform. However, we could not identify any measures or standard measurement methods used to evaluate user-platform interactions in BIM systems. Thus, this could lead to misuse of evidence or analysis of a user's performance leading to distrust of events and people.

This is echoed in research that states that BIM is being used for "dark" purposes, as Davis and Harty (2012) also noted in an effort to enact control over users through the imposition of standardized practices through surveillance and monitoring. Ultimately, it is also an indicator of a lack of trust in their users, indicating a tendency to use surveillance to enforce control and standardized practices. The more processes are digitized, the more information we can get about these processes, and it is here that the persuasiveness ascends to monitor and extend control. To what degree the user's consent was sought was not evident here, but it is a highly relevant topic for future research.

5 CONCLUSION

This article looked at aspects that shape the trustworthiness of utilizing digital building construction technologies (i.e., BIM technologies). The intention was to go beyond paying attention to the functionality of digital artifacts and the processes they shape and instead look at how such technologies are responsible for the psychological, social, and political conditions of production (Winner 2014). We put forward characteristics to do with trust and technology that impact the adoption of digital technology.

BIM functionalities require relationships of trust between humans and the digital systems used. However, BIM includes systems that could create power imbalances, and there is some evidence suggesting how they can be used for surveillance practices. Security and ethical challenges could be overcome if protocols and standards of practices are addressed with transparency, giving autonomy to the users to decide when to share data in a way that does not compromise the construction management development and user's willingness to cooperate.

We have explored some of the conditions in which individuals trust or distrust digital technologies from a user's experts' point of view. Understanding the ideal conditions, like fairness, data protection, and autonomy of the BIM interface, is essential to design more trustable systems. A great danger would be that we did not consider these aspects of digital technology when we either develop or adopt it into our practices.

While we may not foresee all potential issues of trust that could harm the relationship between the technology we want to assist people and their companies, we risk harming such a relationship if we are not careful

with the result of such technology being rejected. However, as Hoffman (2013) notes, mistrust with technology can be very difficult to restore as with people. We need to give people the proper setting to trust great BIM systems with good intentions.

We are aware that further studies need to be done to understand, for example, how non-experts interact with BIM systems, and how to compare the levels of trust of those individuals with expert users, as well as how they perceive data processing, privacy, and security to understand how to develop BIM systems that address the trust dimensions between people and digital technologies.

REFERENCES

Barriball, K.L. & While, a. (1994), "Collecting data using a semi-structured interview: a discussion paper.", *Journal of Advanced Nursing*, Vol. 19 No. 2, pp. 328–335.

Bertelsen, S. (2003), "Complexity – Construction in a new perspective", *International Group for Lean Construction*, pp. 1–12.

Cappelen, C. & Dahlberg, S. (2018), "The Law of Jante and generalized trust", *Acta Sociologica (United Kingdom)*, Vol. 61 No. 4, pp. 419–440.

Castelfranchi, C. & Falcone, R. (2010), *Trust Theory: A Socio-Cognitive and Computational Model*, Trust Theory: A Socio-Cognitive and Computational Model, available at: https://doi.org/10.1002/9780470519851.

Chalker, M. & Loosemore, M. (2016), "Trust and productivity in Australian construction projects: A subcontractor perspective", *Engineering, Construction and Architectural Management*, Vol. 23 No. 2, pp. 192–210.

Coeckelbergh, M. (2011), "Humans, animals, and robots: A phenomenological approach to human-robot relations", *International Journal of Social Robotics*, Vol. 3 No. 2, pp. 197–204.

Davies, R. & Harty, C. (2012), "Control, surveillance and the 'dark side' of BIM", *ARCOM*, pp. 23–32.

Dubois, A. & Gadde, L.-E. (2002), "The construction industry as a loosely coupled system: implications for productivity and innovation", *Construction Management and Economics*, Vol. 20, pp. 621–631.

Fulford, R. & Standing, C. (2014), "Construction industry productivity and the potential for collaborative practice", *International Journal of Project Management*, Elsevier Ltd, Vol. 32 No. 2, pp. 315–326.

Gad, G.M. & Shane, J.S. (2014), "Trust in the Construction Industry: A Literature Review", No. October 2015, pp. 2136–2145.

Gero, J. (1998), "Conceptual designing as a sequence of situated acts", *Artificial Intelligence in Structural Engineering*, pp. 165–177.

Gorur, R. (2020), "Afterword: embracing numbers?", *International Studies in Sociology of Education*, Routledge, Vol. 29 No. 1–2, pp. 187–197.

Guo, H., Yu, R. & Fang, Y. (2019), "Analysis of negative impacts of BIM-enabled information transparency on contractors' interests", *Automation in Construction*, Elsevier, Vol. 103 No. July 2018, pp. 67–79.

Hardin, R. (2002), *Trust and Trustworthiness*, Russel Sage Foundation.

Hoffman, R.R., Johnson, M., Bradshaw, J.M. & Underbrink, A. (2013), "Trust in Automation", *IEEE Intelligent Systems*, No. aprIL, pp. 61–67.

Hoffmann, A.L. (2019), "Where fairness fails: data, algorithms, and the limits of antidiscrimination discourse", *Information Communication and Society*, Vol. 22 No. 7, available at:https://doi.org/10.1080/1369118X.2019.1573912.

Jonas, H. (2014), "Technology and Responsibility: Reflections on the New Tasks of Ethics", *Ethics and Emerging Technologies*, available at:https://doi.org/10.1057/9781137349088_3.

Marozzi, M. (2015), "Measuring Trust in European Public Institutions", *Social Indicators Research*, Springer Netherlands, Vol. 123 No. 3, pp. 879–895.

Matzner, T. (2019), "The Human Is Dead – Long Live the Algorithm! Human-Algorithmic Ensembles and Liberal Subjectivity", *Theory, Culture and Society*, Vol. 36 No. 2, pp. 123–144.

McKnight, D.H. & Chervany Norman, L. (2001), "Trust and distrust definitions: One bite at a time", *Lecture Notes in Artificial Intelligence (Subseries of Lecture Notes in Computer Science)*, Vol. 2246 No. January, pp. 27–54.

Mittelstadt, B.D., Allo, P., Taddeo, M., Wachter, S. & Floridi, L. (2016), "The ethics of algorithms: Mapping the debate", *Big Data and Society*, Vol. 3 No. 2, pp. 1–21.

Nieminen, A. & Lehtonen, M. (2008), "Organisational control in programme teams: An empirical study in change programme context", *International Journal of Project Management*, Vol. 26 No. 1, pp. 63–72.

Ostrom, E. & Walker, J. (2003), *Trust and Reciprocity: Interdisciplinary Lessons from Experimental Research, Trust and Reciprocity: Interdisciplinary Lessons from Experimental Research*, Vol. 9781610444347, available at:https://doi.org/10.1016/j.jebo.2003.07.005.

Reychav, I., Maskil Leitan, R. & McHaney, R. (2017), "Sociocultural sustainability in green building information modeling", *Clean Technologies and Environmental Policy*, Springer Berlin Heidelberg, Vol. 19 No. 9, pp. 2245–2254.

Sandler, R.L. (2014), *Ethics and Emerging Technologies*, Palgrave Macmillan UK, London.

Teräs, M., Teräs, H., Arinto, P., Brunton, J., Daryono, D. & Subramaniam, T. (2020), "COVID-19 and the push to online learning: Reflections from 5 countries.", *Digital Culture & Education*.

Tsamados, A., Aggarwal, N., Cowls, J., Morley, J., Roberts, H., Taddeo, M. & Floridi, L. (2020), "The Ethics of Algorithms: Key Problems and Solutions", *SSRN Electronic Journal*, pp. 1–32.

Walker, G., Devine-Wright, P., Hunter, S., High, H. & Evans, B. (2010), "Trust and community: Exploring the meanings, contexts and dynamics of community renewable energy", *Energy Policy*, Elsevier, Vol. 38 No. 6, pp. 2655–2663.

Winner, L. (2014), "Technologies as Forms of Life", *Ethics and Emerging Technologies*, available at:https://doi.org/10.1057/9781137349088_4.

Zuboff, S. (2019), *The Age of Surveillance Capitalism: The Fight for a Human Future at the New Frontier of Power*, Profile Books Ltd, Vol. 129, London, available at:https://doi.org/10.1093/sf/soz037.

Author index

Abualdenien, J. 76
Adounvo, J.D. 3
Ahmad, S.B.S. 495
Akyaz, İ. 558
Alqudah, H. 471
Alstad, T. 274
Althoff, K.-D. 268
Amor, R. 229
Arishin, S. 51
Ariyachandra, M.R.M.F. 373
Arora, H. 268

Barbero, A. 35
Bargstädt, H.-J. 381
Bargstädt, H.-J. 200
Belsky, M.E. 67
Benton, P. 544
Berger, C. 106, 171
Berthold, H. 313
Beslioglu, Y. 558
Bochukova, V. 106
Bohnstedt, K.D. 423
Boje, C. 299
Borrmann, A. 12, 59, 76
Bourahla, N. 261
Bourahla, Y. 261
Boutros, M. 3
Brahmi, B.F. 446
Breitfuss, D. 132
Breitfuß, D. 125
Brilakis, I. 373
Brötzmann, J. 313
Bui, N. 367
Burukhina, O.S. 190
Bushinskaya, A.V. 190

Cao, J. 147
Carbonari, A. 291
Carpinteri, C. 3
Châteauvieux-Hellwig, C. 76
Chen, F. 361
Chen, W. 544
Corneli, A. 291
Cui, Q. 184

Damhus, L.V. 563
David, A. 511
de Godoy, J. 569
de Vries, B. 439
Del Giudice, M. 35, 98
Dettori, M. 98
Dikbas, H.A. 194

Dimyadi, J. 229
Dong, Y. 367
Doroschkin, A. 536
Dsoul, A. 3
Du, H. 439
Durmus, C. 194

Eisenstadt, V. 268
Ellinger, A. 343
Engeland, B.D. 351
Erduran, E. 361
Erfani, A. 184
Esclusa, R.J. 423
Esser, S. 12, 59

Fauth, J. 42
Fischer, M. 465
Fürstenberg, D. 177

Gade, P.N. 563, 569
Galishnikova, V. 141
Gautier, P.E. 3
Gerhard, D. 305
Ghodrati, N. 471
Gonzáles, V.A. 471
Gulichsen, T. 177
Guo, J. 361
Guo, T. 351, 367

Hall, D.M. 147
Hamdan, A. 245
Han, D. 351, 357, 361, 367
Han, Q. 431, 439
Hempel, E.E. 361
Hjelmbrekke, H. 351
Hjelseth, E. 177, 274, 465, 479, 487, 495, 502, 519
Hoffmann, A. 83, 91, 211, 216, 223
Hosamo, H. 351, 357, 361, 367
Hsieh, S.H. 391
Hu, K. 351
Huhnt, W. 141
Huxoll, J. 305
Huyeng, T.-J. 83, 91, 119, 313

Jaud, Š. 12
Jaskula, K. 153
Johansen, K.W. 237

Kamari, A. 163, 446
Karlshøj, J. 417

Karoui, S. 3
Kastner, W. 125
Katranuschkov, P. 283
Kitouni, I. 446
Kiviniemi, M. 527
König, M. 305
Kosse, S. 305
Kovacic, I. 119
Kubicki, S. 299
Kukkonen, V. 320

Lædre, O. 177
Lahdenperä, P. 527
Langenhan, C. 268
Lassen, A.K. 357
Lavikka, R. 527
Legatiuk, D. 334
Leon, M. 544
Li, B. 163, 229
Lihai, L. 3
Lin, F. 253
Liu, J. 253
Lorenzen, S.R. 313
Lossev, K. 334

Magnano, S. 98
Mahdavi, A. 106, 113, 171
Marihal, N. 381
Mellenthin Filardo, M. 381
Mirarchi, C. 511
Mirtschin, J. 12
Morozkin, N. 406
Morozov, S. 21
Muhič, S. 12

Naticchia, B. 291
Naville, N. 511
Nielsen, R.O. 237
Nisbet, N. 29

Olsen, J. 417
Osello, A. 35, 98
Otrel-Cass, K. 569

Pal, A. 391
Pauwels, P. 328
Peltokorpi, A. 527
Perissich, L. 511
Petrova, E. 423
Petzold, F. 153, 268
Philibert, J.G. 3
Pieper, M. 457

Polter, M. 283
Poshdar, M. 471
Preindl, T. 125

Qian, R. 563

Rashasingham, K. 502
Rasmussen, T.S. 423
Rekve, A. 519
Rezgui, Y. 299
Rolfsen, C.N. 351, 357, 361, 367
Rüppel, U. 83, 91, 211, 216, 223, 313

Sassi-Boudemagh, S. 446
Sazonov, S. 21
Scherer, R.J. 245, 253, 283, 343
Schneider, J. 313
Schultz, C. 163, 229, 237
Seiß, S. 200, 457
Semenov, G. 51
Semenov, V. 21, 51, 406

Shamreeva, A. 457, 536
Shi, M. 83, 91, 211, 216, 223
Shutkin, V. 406
Šibenik, G. 119, 125, 132
Smarsly, K. 334
Sprenger, W. 83, 91, 119
Srećković, M. 125, 132
Sujan, S.F. 479, 487

Tafraout, S. 261
Talebi, S. 471
Tauscher, H. 399
Teizer, J. 237
Thiele, C.-D. 83, 91, 119, 313
Timashev, S.A. 190
Tookey, J. 471

Ugliotti, F.M. 35

Vaccarini, M. 291
Vala, P. 343
van Gool, S. 328

Van Oeveren, C.D. 431
Vogt, O. 305
Volkov, A. 334

Wagner, A. 83, 91
Walther, T. 343, 381
Wikström, L. 12
Wolf, M. 305
Wolosiuk, D. 113
Wörner, C. 343

Yang, D. 328
Yazıcıoğlu, Z. 551
Yifan, L. 3
Ying, C. 351, 357, 361, 367

Zahedi, A. 153
Zarli, A. 299, 511
Zhang, N. 431
Zhou, Y. 367
Zhu, Y. 471
Zolotov, V. 406